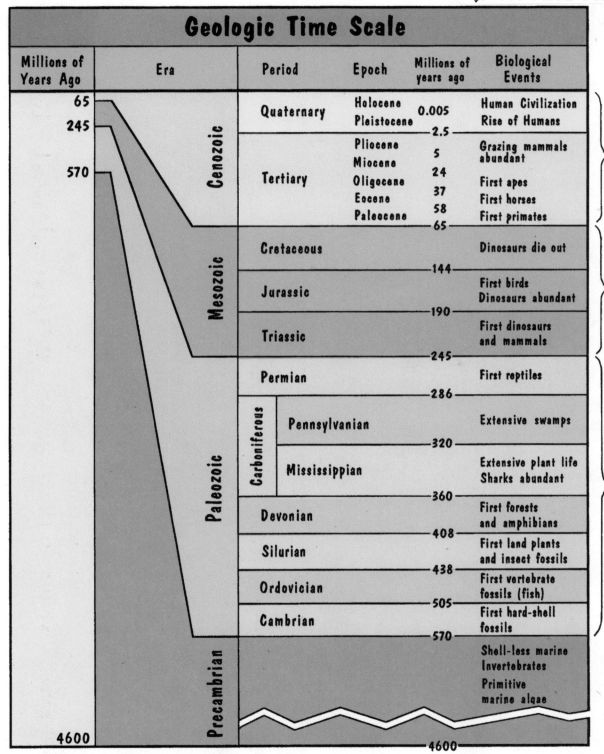

Geologic Time Scale

Millions of Years Ago	Era	Period	Epoch	Millions of years ago	Biological Events
65	Cenozoic	Quaternary	Holocene	0.005	Human Civilization
245			Pleistocene		Rise of Humans
570				2.5	
		Tertiary	Pliocene	5	Grazing mammals abundant
			Miocene	24	
			Oligocene	37	First apes
			Eocene	58	First horses
			Paleocene		First primates
				65	
	Mesozoic	Cretaceous			Dinosaurs die out
				144	
		Jurassic			First birds Dinosaurs abundant
				190	
		Triassic			First dinosaurs and mammals
				245	
	Paleozoic	Permian			First reptiles
				286	
		Carboniferous — Pennsylvanian			Extensive swamps
				320	
		Carboniferous — Mississippian			Extensive plant life Sharks abundant
				360	
		Devonian			First forests and amphibians
				408	
		Silurian			First land plants and insect fossils
				438	
		Ordovician			First vertebrate fossils (fish)
				505	
		Cambrian			First hard-shell fossils
				570	
4600	Precambrian				Shell-less marine Invertebrates Primitive marine algae
				4600	

Age of Mammals

Age of Reptiles

The Rise of Complex Life Forms

Earth formed about 4600 million (4.6 billion) years ago.

CONCEPTUAL
Physical Science

PAUL G. HEWITT
City College of San Francisco

JOHN SUCHOCKI
Leeward Community College

LESLIE A. HEWITT

HarperCollins*CollegePublishers*

Executive Editor: Doug Humphrey
Developmental Editor: Louise Howe
Project Editor: Carol Zombo
Design Administrator: Jess Schaal
Text and Cover Design: Kay Fulton
Cover Photo: Robert Ewell
Photo Researcher: Kelly Mountain
Production Administrator: Randee Wire
Compositor: Publication Services
Printer and Binder: R.R. Donnelley & Sons Company
Cover Printer: R.R. Donnelley & Sons Company

Cover Description: The cover photograph of Bryce Canyon powerfully demonstrates physical, chemical, and geological activity. The back cliffs are lit by the reflection of sunlight off a snowbank, the top of which is just visible over the first ridge in the foreground. The red color of the sandstone is due to the presence of iron oxide, which results from the reaction of iron with atmospheric oxygen and water.

Periodic Table Insert: John and Tracy Suchocki

Conceptual Physical Science
Copyright © 1994 by Paul G. Hewitt, John Suchocki, Leslie A. Hewitt

Library of Congress Cataloging-in-Publication Data

Hewitt, Paul G.
    Conceptual physical science / Paul G. Hewitt, John Suchocki,
  Leslie Hewitt.
        p.   cm.
    Includes index.
    ISBN 0-673-46379-6
    1. Physical sciences.   I. Suchocki, John.   II. Hewitt, Leslie.
  III. Title.
  Q158.5.H48   1993
  500.2—dc20   93-4561
                                                  CIP

    95 96 9 8 7 6 5 4

CONTENTS IN BRIEF

CONTENTS IN DETAIL

CONCEPTUAL PHYSICAL SCIENCE PHOTO ALBUM

• •

Conceptual Physical Science is a very personalized book, a family undertaking reflected in many photographs throughout the book. Charlie Spiegel, our mentor to whom the book is dedicated, opens the book facing page 1. Little Sarah Stafford who sits on his lap with the chickie is Charlie's great granddaughter. (In four editions of *Conceptual Physics,* a similar girl-with-chickie opening photo was of little Jenny Jones, of Pattaya, Thailand—daughter of friend BJ. Jenny is a bit older now, and poses with her dog in Figure 6.13, page 149.)

All eight part openers feature children who are friends and family. Part 1, page 15, is tiny friend Andrea Wu (who shows up again in Figure 3.19 on page 71); Part 2 on page 139 is nephew Terrence Jones; Part 3 on page 199 is Malchan Summerhill; Part 4 on page 247 is brother Dave Hewitt's granddaughter, Lillian Allen; Part 5, page 337, is author Paul's grandson (and Leslie's and John's nephew) Manuel Hewitt; Part 6 on page 397 is John and Tracy's son, Ian Suchocki (pronounced Suhock'-ee, with a silent c); Part 7 on page 531 is Manuel Hewitt again; and Part 8, page 675, is little Jasmine Crowe, daughter of University of Hawaii astronomer Richard Crowe.

Author Paul is shown with his buddy Pablo Robinson who is sandwiched between beds of nails in Figure 5.3 on page 114. Pablo is the author of the lab manual that accompanies the 7th Edition of Hewitt's *Conceptual Physics.* Many of those labs appear in the *Conceptual Physical Science* lab manual. Brother David Hewitt (no, not a twin) and his wife Barbara work the water pump shown in Figure 5.27 on page 126. Lifelong friend Paul Ryan dips his finger in molten lead on page 194.

Author John walks barefoot across red-hot coals in Figure 7.2, page 162 (his first time—an assignment from senior author uncle Paul!). Wife Tracy and son Ian grace Figure 2.26 on page 52. Ian also appears before birth in the sonogram of Figure 11.14, page 255. Figure 11.43 on page 271 is part of the wedding party of John and Tracy, including left to right, John's brother-in-law Butch Orr, sister Cathy Orr, bride and groom, sister Joan Lucas, parents John Suchocki and Marjorie Hewitt Suchocki (a theologian at Claremont College), and Tracy's parents Sharon and David Hopwood, friends Kellie and Mark Werkmeister, and Uncle Paul. Niece Alexandra Lucas is the model in Figure 17.5, page 407.

Author Leslie, at her dad's insistence, appears in three figures. The first is Leslie at the age of 16 in Figure 14.1, page 339 (the black and white photo that has been in the last five editions of *Conceptual Physics*). The second is more recent, Figure 25.4 on page 626, and the third is with her dad in Figure 22.39 on page 559. Leslie's mom fearlessly holds her hand above expanding steam in Figure 7.9, page 166, and Leslie's brother Paul pumps air into his bicycle in Figure 8.4 on page 184. Leslie's younger brother James (father of Manuel) is shown at the age of 10 in Figure 13.40 on page 324. James was tragically killed in 1988 at the age of 24 in a car accident in Salida, Colorado. On a brighter note, Leslie's fiancé Bob Abrams, who helped considerably with Part 7, is shown in Figure 23.28, page 587.

Author Paul's dear friends include former teaching assistants at CCSF. The first is Tenny Lim, now a design engineer at JPL in Pasadena, who is seen pulling back her arrow in Figure 3.13 on page 66. The second is Helen Yan, now an engineer at Lockheed, who did the hand lettering that adorns this book. Helen re-poses with the same black-hole box she posed with more than ten years ago, for the black and white photo that appeared in three editions of *Conceptual Physics.* She is now seen in color in Figure 7.13 on page 169. More recent CCSF teaching assistants are shown in the air track photo of Figure 3.10 on page 64. They are, left to right, Ralph Diaz and Glenda Gin, with colleagues Annette Rappleyea, who assisted with test bank materials, and Will Maynez, who designed and built the air track. His assistants are to his right; Ray Choi and Kumiko Furukawa.

Many of the photographs in this book were taken by Meidor Hu, talented art student at the University of Hawaii in Hilo, who stands electrified in Figure 9.10, page 207. Meidor is shown again in Figure 11.22 on page 261. Of the several photos Meidor took are one of her brother Herbert in Figure 10.9, page 231, and her sister Mei Tuck in Figure 11.10 on page 253.

Author Paul's best friend, Tim Gardner, is shown among the telescopes atop Moana Kea in Hawaii in the opening photo to Chapter 28, page 700. And lastly in Figure 28.26 on page 718 is little Melissa, daughter of friends Dennis and Tai McNelis.

The inclusion of these people who are so dear to the authors makes *Conceptual Physical Science* all the more our labor of love.

TO THE STUDENT

Physical Science is about the rules of the physical world--physics, chemistry, geology, and astronomy. Just as you can't enjoy a ball game, computer game, or party game until you know its rules, so it is with nature. Nature's rules are beautifully elegant and can be neatly described mathematically. That's why many physical sciences texts are treated as applied mathematics. But too much emphasis on computation misses something essential--*comprehension*--a gut feeling for the concepts. This book is *conceptual*, focusing on concepts in down-to-earth English rather than in mathematical language. You'll see the mathematical structure in frequent equations, but you'll find them *guides to thinking* rather than recipes for computation.

We enjoy physical science, and you will too--because you'll understand it. Just as a person who knows the rules of botany best appreciates plants, and a person who knows the intricacies of music best appreciates music, you'll better appreciate the physical world about you when you learn its rules.

Enjoy your physical science!

PAUL G. HEWITT

John Suchocki

Leslie A. Hewitt

TO THE INSTRUCTOR

Conceptual Physical Science is a melding of physics, chemistry, earth science, and astronomy, in a manner that captivates student interest. This book seeks to build a conceptual base in physics and chemistry, which is then applied to earth science and astronomy. For the nonscience student, it is a base from which to view nature more perceptively—to see that a surprisingly few relationships make up its rules. For the science student, it is this and a springboard to involvement in other sciences such as biology and health-related fields.

We begin the study of physical science with physics–the most basic of the sciences because it reaches up to chemistry, which in turn reaches to the earth sciences, and ultimately up to the life sciences. Physics is about the laws of motion, energy, electricity, heat, sound, and light. Physics is treated conceptually, which means its focus is on qualitative comprehension rather than on mathematical expression. Physics in this text is not treated as applied mathematics. We minimize the deterring mathematical language and mathematical problems that are roadblocks to many students. Although a flip though the pages will show that the equations are there, they are presented as guides to thinking rather than recipes for algebraic manipulation. Their derivations are addressed in the footnotes. The treatment of physics leads up to the realm of the atom—a bridge to chemistry.

Chemistry deals with the structure and behavior of matter. The conceptual treatment in this text emphasizes visual models. For example, we recognize that most physical science courses do not have the time to adequately delve into a quantum mechanical approach to chemistry. So we treat electron configurations via the easy-to-visualize shell model and chemical bonding in terms of overlapping atomic shells and Coulomb's law. Throughout our treatment we relate chemistry to the students' familiar world—the fluoride in their toothpaste, the Teflon on their frying pans, and how to combat the hotness of chili sauce. Students go beyond learning that everything is made of chemicals, including the food they eat—for many environmental aspects of chemistry are also addressed.

Earth science encompasses the sciences of geology and meteorology. Geology is the study of our home planet. This includes the formation of rocks and minerals, the study of the internal dynamics that have and continue to influence earth's surface, and the study of the water that nourishes it. The treatment is both conceptual and visual, focusing on the central concepts of geology with emphasis on processes. The theme of both geology and meteorology is that geological and atmospheric changes are ongoing—and that the present is the key to the past.

Applying physics, chemistry, and geology to other massive bodies in the universe brings us to astronomy. Astronomy deals with what happens "out there," where space and time differ from the students' everyday notions. Both special and general relativity make up most of concluding Chapter 29. So our tour of physical science begins with the physics of atoms, then proceeds to the chemistry of molecules, then to the geology of aggregates of molecules, and finally to the aggregates of matter in the cosmos—astronomy.

PEDAGOGY

At the ends of all chapters are Review Questions, Exercises, and in many chapters, Problems and Home Projects. All of the important ideas from each chapter are framed in relatively easy-to-answer Review Questions, grouped by chapter sections. They are, as the name implies, a review of chapter material. Their purpose is simply to provide a structured way to review the chapter. They are not meant to challenge the student's intellect, for in the vast majority of cases, the answers can be simply looked up. The Exercises, on the other hand, play a different role. Some of these are designed to prompt the application of physical science to everyday situations, while others are more sophisticated and call for considerable critical thinking.

The Problems are mainly simple computations that aid in learning concepts. Their numbers are relatively small, to decrease the possibility that they become the end-all to a student's experience with *Conceptual Physical Science*. The challenge to your students should be in the conceptual reasoning and critical thinking that are called for mainly in the Exercises. Although building confidence in math is a worthy goal, it is not what this book is about.

Units of measurement are not emphasized in this text. When used, they are almost exclusively expressed in SI (exceptions include such units as calories, grams per centimeter cubed, and light-years). Mathematical derivations are avoided in the main body of the text and appear in footnotes or in the appendixes.

ANCILLARY MATERIALS

More than enough material is included for a one-year course, which allows for a variety of course designs to fit your taste. These are suggested in the *Instructor's Manual,* which you'll find differs from most instructor's manuals. It contains many lecture ideas and topics not treated in the textbook, as well as teaching tips and suggested step-by-step lectures and demonstrations.

The *Instructor's Manual* also includes, among other important items, full-page (8 1/2″ × 11″) answers to Exercises and Problems. There are also chalkboard techniques and clip art that can enliven your classroom presentations.

The *Test Bank* is available both in booklet form and on Test Master software disks for IBM PC and Macintosh.

There is a lab manual for *Conceptual Physical Science,* written by the authors. In addition to interesting laboratory experiments, it includes a range of activities similar to the home projects in the textbook. These guide students to experience phenomena before they quantify the same phenomena in a follow-up laboratory experiment.

Be sure to consider the student booklet, *Practicing Physical Science,* which guides your students to a sometimes computational way of developing concepts. You can make this practice book available to your students via your bookstore (it's affordable!). It is a most unusual workbook, with a "user-friendly" tone that makes wide use of analogies and intriguing situations.

Last, but not least, videos of classroom demonstrations in physics and chemistry are also available through your HarperCollins sales representative.

Go to it! Your conceptual physical science course really can be the most interesting, informative, and worthwhile science course available to your students.

FEATURES

The Swinging Wonder

Momentum conservation is nicely demonstrated with the swinging wonder, the novel device shown in Figure 3.19. When a single ball is raised and allowed to swing into the array of other identical balls, a single ball from the other side pops out. When two balls are similarly raised and released, presto—two balls on the other side pop out. The number of balls incident on the array is always the same as the number of balls that emerge. We can see that *momentum before = momentum after*. That is, $mv = mv$, or $2mv = 2mv$, or $3mv = 3mv$, and so on. The intriguing question arises: When a single ball is raised, released, and makes impact, why cannot two balls emerge with half the speed? Or if two balls make impact, why cannot one ball emerge with half the speed? If either of these cases occurred, the momentum before would still be equal to the momentum after: $mv = 2m(\frac{1}{2}v)$; or $2mv = m(2v)$. Intriguingly, this never happens. Nor can it happen.

Why? Because something besides momentum must be conserved in this interaction—energy. Since the collisions are quite elastic, with very little energy transforming to heat and sound, to a good approximation the kinetic energy before equals the kinetic energy after. That is, $\frac{1}{2}mv_{above}^2 = \frac{1}{2}mv_{after}^2$. Consider dropping two balls with one emerging at twice the speed. Then will $\frac{1}{2}2mv^2 = \frac{1}{2}m(2v)^2$? The answer is no! If this case were to occur, there would be more energy after the collision than before (we'll leave it to you to figure out how much more). Give this some thought and you'll see there is a reason why, for identical balls, the number of balls that make impact will always equal the number of balls that emerge.

In any collision, elastic or inelastic, momentum before collision equals momentum after collision. In the special case of a perfectly elastic collision, where no energy is transformed to other forms, kinetic energy before collision equals kinetic energy after collision.

Why is this device called the swinging wonder? Because the unequal-balls situation and its impossibility has left many people wondering—and wondering—and wondering. But you know the reason why the number of incident and emerging balls must be the same. It's nice to know some physics!

Figure 3.19 The Swinging Wonder. The number of balls that are raised, released, and that make impact with the array of identical balls always equals the number of balls that emerge. Two conservation principles explain why.

As we draw back the arrow in a bow, we do work in bending the bow; we give the arrow and bow potential energy. When released, this potential energy is transferred to the arrow as kinetic energy. The arrow in turn transfers this energy to its target, perhaps a rigid wooden fence post. The slight distance the arrow penetrates multiplied by the average force of impact doesn't quite match the kinetic energy of the arrow. The energy score doesn't balance. But if we investigate further, we'll find that both the arrow and fence post are a bit warmer. By how much? By the energy difference. Energy changes from one form to another. It transforms without net loss or net gain.

The study of various forms of energy and their transformations from one form into another has led to one of the greatest generalizations in physics—the law of **conservation of energy:**

Energy cannot be created or destroyed; it may be tra[...] form into another, but the total amount of energy ne[...]

When we consider any system in its entirety, whether i[...] ing pendulum or as complex as an exploding galaxy, there is [...] change: energy. It may change form or it may simply be tran[...]

SPECIAL INTEREST BOXES
Special interest boxes in each chapter review important, timely, environmental and technological topics that relate directly to the physical sciences.

Figure 2.13 The ratio of weight F to mass m is the same for all objects in the same locality; hence, their accelerations are the same in the absence of air drag.

$$\frac{F}{m} = g \qquad \frac{2F}{2m} = g$$

We know that a falling object accelerates toward the earth because of the gravitational force of attraction between the object and the earth. We call this force the *weight* of the object.* When this is the only force—that is, when friction such as air resistance is negligible—we say that the object is in a state of **free fall**.

The attractive force between a more massive object and the earth is greater than that of a less massive object. The double brick in Figure 2.13, for example, has twice the attraction as the single brick. Why then, as Aristotle supposed, doesn't a double brick fall twice as fast? The answer is that the acceleration of an object depends not only on the force—in this case, the weight—but on the object's resistance to motion, its inertia. Whereas a force produces an acceleration, inertia is a *resistance* to acceleration. So twice the force exerted on twice the inertia produces the same acceleration as half the force exerted on half the inertia. Both accelerate equally. The acceleration due to gravity is symbolized by g. We use the symbol g, rather than a, to denote that acceleration is due to gravity alone.

The ratio of weight to mass for freely falling objects equals a constant—g. This is similar to the constant ratio of circumference to diameter for circles, which equals the constant π. The ratio of weight to mass is the same for both heavy and light objects, just as the ratio of circumference to diameter is the same for both large and small circles (Figure 2.14).

We now understand that the acceleration of free fall is independent of an object's mass. A boulder 100 times more massive than a pebble falls at the same acceleration as the pebble because although the force on the boulder (its weight) is 100 times greater than the force (or weight) on the pebble, its resistance to a change in motion (mass) is 100 times that of the pebble. The greater force offsets the equally greater mass.

$$\frac{F}{m} = g \qquad \frac{F}{m} = g$$

$$\frac{C}{D} = \pi \qquad \frac{C}{D} = \pi$$

Figure 2.14 The ratio of weight F to mass m is the same for the large rock and the small feather; similarly the ratio of circumference C to diameter D is the same for the large and the small circle.

Q UESTION

In a vacuum, a coin and a feather fall equally, side by side. Would it be correct to say that in a vacuum equal forces of gravity act on both the coin and the feather?

When Acceleration Is Less Than g—Nonfree Fall

Objects falling in a vacuum are one thing, but what of the practical cases of objects falling in air? Although a feather and a coin will fall equally fast in a vacuum, they fall quite differently in the presence of air. How do Newton's laws apply to objects falling in air? The answer is that Newton's laws apply for *all* objects,

A NSWER

No, no, no, a thousand times no! These objects accelerate equally not because the forces of gravity on them are equal (they aren't!), but because the ratios of their weights to masses are equal. Although air resistance is not present in a vacuum, gravity is (you'd know this if you stuck your hand into a vacuum chamber and a Mack truck rolled over it). If you answered yes to this question, let this be an alarm to be more careful when you think physics!

*Weight and mass are directly proportional to each other, and the constant of proportionality is g. We see that weight $= mg$, so $9.8 \text{ N} = (1\text{kg})(9.8 \text{ m/s}^2)$.

QUESTION AND ANSWER BOXES
Question and answer boxes throughout the text encourage critical thinking and reinforce chapter concepts.

MATH FOOTNOTES
Designed to introduce key mathematical concepts without interrupting the conceptual rhythm of the text, **math footnotes** present important formulas.

SUMMARY OF TERMS

Electromagnetic wave An energy-carrying wave emitted by vibrating electrons that is composed of oscillating electric and magnetic fields that regenerate one another.

Electromagnetic spectrum The range of electromagnetic waves extending in frequency from radio waves to gamma rays.

Transparent The term applied to materials through which light can pass in straight lines.

Additive primary colors The three colors—red, blue, and green—that when added in certain proportions will produce any color in the spectrum.

Subtractive primary colors The three colors of absorbing pigment—magenta, yellow, and cyan—that when mixed in certain proportions will reflect any color in the spectrum.

Complementary colors Any two colors that when added produce white light.

Diffraction The bending of light around an obstacle or through a narrow slit in such a way that fringes of light and dark or colored bands are produced.

Interference The superposition of waves producing regions of reinforcement and regions of cancellation. Constructive interference refers to regions of reinforcement; destructive interference refers to regions of cancellation. The interference of selected wavelengths of light produces colors known as *interference colors*.

Polarization The alignment of the electric vectors that make up electromagnetic radiation. Such waves of aligned vibrations are said to be *polarized*.

REVIEW QUESTIONS

The Electromagnetic Spectrum

1. Does light make up a relatively large part or a relatively small part of the electromagnetic spectrum?

2. What is the principal difference between a *radio wave* and *light*?

3. What is the principal difference between *light* and an *X ray*?

4. How much of the measured electromagnetic spectrum does light occupy?

5. What color do the lowest visible frequencies appear?

6. How does the frequency of light compare to the frequency of the vibrating electron that produces it?

7. How does the wavelength of light compare to its frequency?

Transparent and Opaque Materials

8. One tuning fork can force another to vibrate. How is this similar to light?

9. In what region of the electromagnetic spectrum is the resonant frequency of electrons in glass?

10. What is the fate of the energy in ultraviolet light that is incident upon glass?

11. What is the fate of the energy in visible light that is incident upon glass?

12. Why are your answers for Questions 10 and 11 different?

13. How does the frequency of re-emitted light compare to the frequency of the light that stimulates its re-emission?

14. How does the average speed of light in glass compare to its speed in a vacuum?

Color

15. What is the relationship between the frequency of light and its color?

16. The visible color spectrum runs from red through violet. Compared to violet light, is red light high-frequency or low-frequency?

Selective Reflection

17. Distinguish between the white of this page and the black of this ink in terms of what happens to the white light that falls on both.

18. How does the color of an object differ when illuminated by candle light and by the light from a fluorescent lamp?

Selective Transmission

19. What color light is transmitted through a piece of red glass?

20. Which will warm quicker in sunlight, a clear or a colored piece of glass? Why?

Mixing Colored Light

21. What is the evidence for the statement that white light is a composite of all the colors of the spectrum?

22. What is the color of the peak frequency of solar radiation?

23. What color light are our eyes most sensitive to?

24. What frequency ranges of the radiation curve do red, green, and blue light occupy?

25. Why are red, green, and blue called the *additive primary colors*?

26. What is the resulting color of equal intensities of blue light and green light combined?

---

10 m high? This question does not have a straightforward numerical answer. Why?

26. In the hydraulic machine shown, it is observed that when the small piston is pushed down 10 cm, the large piston is raised 1 cm. If the small piston is pushed down with a force of 100 N, how much force is the large piston capable of exerting?

27. Consider the swinging-balls apparatus. If two balls are lifted and released, momentum is conserved as two balls pop out the other side with the same speed as the released balls at impact. But momentum would also be conserved if one ball popped out at twice the speed. Can you explain why this never happens?

28. Consider the inelastic collision between the two freight cars in Figure 3.8. The momentum before and after the collision is the same. The KE, however, is less after the collision than before the collision. How much less, and what becomes of this energy?

29. If an automobile had a 100 percent efficient engine, would it be warm to your touch? Would its exhaust heat the surrounding air? Would it make any noise? Would it vibrate? Would any of its fuel go unused?

30. We know more force is normally required to stop a moving truck than a moving skateboard. Make an argument that more force would be needed to stop a moving skateboard.

PROBLEMS

1. A railroad diesel engine weighs four times as much as a freightcar. If the diesel engine coasts at 5 km per hour into a freightcar that is initially at rest, how fast do the two coast after they couple together?

2. A 5-kg fish swimming at 1 m/s swallows an absent-minded 1-kg fish at rest. What is the speed of the larger fish after lunch?

3. This question is typical on some driver's license exams: A car moving at 50 km/h skids 15 m with locked brakes. How far will the car skid with locked brakes at 150 km/h?

4. How many kilometers per liter will a car obtain if its engine is 25 percent efficient and it encounters an average retarding force of 1000 N? Assume that the energy content of gasoline is 40 MJ/L.

5. A car with a mass of 1000 kg moves at 20 m/s. What braking force is needed to bring the car to a halt in 10 s?

6. What is the efficiency of a pulley system that will raise a 1000-N load a vertical distance of 1 m when 3000 J of effort are involved?

7. What is the efficiency of the body when a cyclist expends 1000 W of power to deliver mechanical energy to her bicycle at the rate of 100 W?

8. Your monthly electric bill is probably expressed in kilowatt-hours (kWh), a unit of energy delivered by the flow of 1 kW of electricity for 1 hr. How many joules of energy do you get when you buy 1 kWh?

SUMMARY OF TERMS

Summary of Terms facilitates review of chapter concepts.

REVIEW QUESTIONS

Grouped by chapter sections, **Review Questions** at the end of each chapter provide a structured review of chapter material.

PROBLEMS

In most chapters, several thought-provoking **Problems,** calling for simple computations, help to nail down central concepts.

EXERCISES

Chapter-end **Exercises** are designed to prompt the application of physical science to everyday situations and to encourage critical thinking.

ACKNOWLEDGMENTS

We are most grateful to the person who has been most helpful in shaping this book, and to whom this book is dedicated—Charlie Spiegel (photo facing page 1). Charlie, who long ago "retired" and is presently a research assistant at California State University, Dominguez Hills, has provided ideas and feedback that permeate the entire book. Thanks, Charlie!

We are also grateful for the input from geologist Bob Abrams, whose many ideas and suggestions add substance to the earth science chapters. Thanks go to Tracy Suchocki for critically reviewing the chemistry chapters, for helping to draw the periodic table keyboard insert, and for her loving support as the wife of the chemistry co-author. Thanks go to brother and uncle David Hewitt for his photography. We thank Meidor Hu for her photographs, and for her graphics help with the ancillaries. We are grateful to Ernie Brown for designing the lettering on the cover, and to Helen Yan for her hand lettering that adorns the figures throughout the entire book.

For input to the physics chapters, thanks go to friends Howie Brand, Paul Doherty, Marshall Ellenstein, John Hubisz, and Pablo Robinson. We are grateful to Peter Brancazio, Brooklyn College, Nick Brown, Cal Poly San Luis Obispo, and David Willey, University of Pittsburgh, Johnstown, for helpful physics suggestions. For chemistry suggestions we thank Leeward Community College colleagues Bob Asato, Manny Cabral, George Shiroma, and Pearl Takeuchi, and Ted Brattstrom, Pearl City High School. For classroom feedback we are grateful to Lucille Garmon, West Georgia College, and to Arnie Feldman, University of Hawaii at Manoa. Special thanks go to Albert Sneden and Everette May of Virginia Commonwealth University for their support and guidance during the early development of the chemistry chapters. For earth science suggestions we are grateful to Karen Grove and Lisa White, both of the Department of Geosciences at San Francisco State University. We also thank Dennis Tasa for his wonderful artwork. For astronomy suggestions we thank Richard Crowe and Bill Heacox of the University of Hawaii at Hilo. We are grateful to amateur astronomer Forrest Luke of Leilehua High School in Hawaii for valued ideas, and to author Paul Tipler for suggestions concerning relativity.

We are grateful to those whose own books were principal influences and references. For contributions to physics: Richard Feynman, *The Feynman Lectures on Physics—Volumes 1–3*; Kenneth Ford, *Basic Physics*; and Eric Rogers, *Physics for the Inquiring Mind*. For chemistry: John C. Kotz and Keith F. Purcell, *Chemistry and Chemical Reactivity*; Hugh W. Salzberg, *From Cavemen to Chemists: Circumstances and Achievements*; and Bassam Z. Shakhashiri, *Chemical Demonstration: A Handbook for Teachers of Chemistry*. For earth science: Edward J. Tarbuck and Frederick K. Lutgens, *The Earth—An Introduction to Physical Geology*; Brian J. Skinner and Stephen C. Porter, *Physical Geology*; and C. Donald Ahrens, *Meteorology Today*. For astronomy: William J. Kaufmann III, *Discovering the Universe*; and Michael Seeds, *Foundations of Astronomy*.

For reviewing portions of the manuscript we thank the following: Robert J. Backes, Pittsburg State University; Katherine Becker, Creighton University; Edward R. Borchardt, Mankato State University; Nicholas E. Brown, California Polytechnic

State University—San Luis Obispo; Douglas Cole, Confederation College; Leo Carson Davis, Southern Arkansas University; Dale J. DeGeeter, Richard J. Daley College; Hudson B. Eldridge, University of Central Arkansas; Abbas M. Faridi, Orange Coast College; Helene Gabelnick, Harold Washington College; Lucille Garmon, West Georgia College; Donald Greenberg, University of Alaska—Southeast; Allan Gubrud, Lane Community College; Sandra Harpole, Mississippi State University; Mani C. Jayaswal, Florida Community College at Jacksonville; Marylin J. Kouba, Harold Washington College; Paul D. Lee, Louisiana State University; Edward N. McCurry, Spartanburg Methodist College; James Merkel, University of Wisconsin—Eau Claire; Wayne R. Morgan, Jr., Hutchison Community College; J. Ronald Mowery, Harrisburg Area Community College; Charles W. O'Neill, Edison Community College; Roger A. Podewell, Olive-Harvey College; B.E. Powell, West Georgia College; Donald E. Rickard, Arkansas Technical University; Duane Sea, Bemidji State University; Malav M. Shah, Macon College; Stephen J. Shulik, Clarion University; Paul Tebbe, Johnson County Community College; Leonard W. Wall, California Polytechnic State University—San Luis Obispo; Larry Weaver, Kansas State University; Linda A. Wilson, Middle Tennessee State University; David C. Ziegler, Hannibal-LaGrange College.

Additional thanks go to the following for their feedback in the early planning stages of *Conceptual Physical Science:* Lawrence H. Adams, Polk Community College; Charles G. Aldrich, Solano Community College; Paul E. Beck, Clarion University; Louis R. Bedell, Northeast Louisiana University; Jayant Bhawalkar, University of North Texas; Don Bickard, Arkansas Tech University; Joseph E. Biron, Greater Hartford Community College; W. E. Blass, Univeristy of Tennessee; Stuart Bradford, Southern Vermont College; Steve Brandt, Henry Ford Community College; Treasure Brasher, West Texas State University; James L. Bray, Barstow College; William J. Brown, Montgomery College; Gary L. Buckwalter, Daytona Beach Community College; Mark M. Chamberlain, Glassboro State College; Do Ren Chang, Averett College; Austin Chappelle, Marion Military Institute; Michael Cherry, Louisiana State University; D. Fred Clark, Northeastern Bible College; Roy W. Clark, Middle Tennessee State University; Clyde W. Clendaniel, California University of Pennsylvania; Dennis Cravens, Vernon Regional Junior College; Don Cruikshank, Anderson University; Paul Deaton, Asbury College; Gerritt H. DeBoer, Rancho Santiago College; Dwight F. Decker, Community College of Rhode Island; Wayne Deckert, Pillsbury College; Benjamin deMayo, West Georgia College; Don DeYoung, Grace College; Philip DiLarore, Indiana State University; Joseph Di Rienzi, College of Notre Dame of Maryland; S. M. Lucia Dudziuski, College of Mt. St. Joseph; P. Elbert, Middle Georgia College; Gibril O. Fadika, Saint Paul's College; Donald L. Fick, Winona State University; Ronald L. Field, Morehead State University; Ronald Fietkau, Southeastern Oklahoma State University; Thomas J. Flair, Delgado Community College; Eugene Franks, Dyke College; Carl Gibbs, Hinds Community College; John Gieniec, Central Missouri State University; Peter Glanz, Rhode Island College; Frederick M. Glaser, University of Texas—Pan Am; John F. Goehl, Jr., Barry University; Richard Graham, Ricks College; Robert Graham, University of Southern Colorado; Robert E. Graves, East Texas Baptist University; Dave Green, Towson State University; Dennis M. Grev, Columbia College; Elbert C. Heath, Pennsylvania Valley Community College; James C. Healy, Florida Southern College; Carl F. Hedges, Jr., Panola College; Kenny Herbert, Carl Albert Junior College; Linda Hobart, Community College of Finger Lakes; Donald W. Hubbard, God's Bible School

& College; Michael R. Hudson, Ashland University; Thelma C. Ivery, Alabama State University; Norman Jacobs, Kennedy-King College; Sardari L. Khanna, York College of Pennsylvania; Lee Kleiss, Fayetteville State University; Ted Koehn, Southeast Community College; Diona Koerner, Marymount College; Kurtis Koll, Cameron University; Ken Ladner, Western New Mexico University; Gerhard Laule, Seminole Junior College; Bob MacKay, Clark College; James Mackey, Harding University; Bruce MacLaren, Eastern Kentucky University; M. P. McAdams, William Penn College; William E. McCorkle, West Liberty State College; M. L. McCurdy, Tarrant County Junior College, Northeast; Debra McDowell, Central Community College; John McGrath, College of St. Rose; Karen McGurk, Southwestern Oklahoma State University; Melvin E. McLester, Rockingham Community College; Ken McMurray, Hutchinson Community College; Arthur R. McRobbie, State University of New York—Potsdam; Lucy G. Merritt, Lawrence Technical University; Gundega Michel, Truman College; John B. Mudie, Modesto Junior College; Claire R. Olander, Appalachian State University; Bruce Oldfield, Broome Community College; Wayne Osborn, Central Michigan University; Albert E. Packard, Notre Dame College; Samuel T. Peavy, Francis Marion College; Nancy A. Perrin, Chapman College; Walter A. Placek, Wilkes College; Larry Rabideau, Kankakee Community College; Thomas F. Reed, Brevard Community College; W. S. Richardson, Auburn University at Montgomery; Heidi R. Ries, Norfolk State University; Wilfred Otaño Rivera, Cayey University College; James E. Roach, Bob Jones University; H. F. Robertson, Hendrix College; Ed Rocks, Los Medanos College; Suzanne Rudnick, Manhattan College; Neil Rudolph, Adams State College; Ross Sears, Lake City Community College; Gerald L. Seebach, Transylvania University; M. Shahriar, Southern Arkansas University; J. B. Shand, Berry College; Prasanta Sharma, Miles College; Yitzhak Sharon, Stockton State College; Daniel Sheeran, Wagner College; Sally Shelton, Austin Community College; Mel Sick, Iowa Central Community College; Pamela G. Simmerman, Gloucester County College; Mervin W. Smart, Spoon River College; W. H. Snedegar, Clarion University of Pennsylvania; Esther B. Sparberg, Hofstra University; William Standish, Skidmore College; Peter W. Stephens, State University of New York; Randall Stephens, Northeast State Community College; Andrew Stevenson, Morgan State University; Harold Stewart, Trevecca Nazarene College; Michael H. Suckley, Macomb Community College; Earl C. Swallow, Elmhurst College; Henry Teoh, SUNY College at Old Westbury; P. E. Thomas, Paine College; Aaron W. Todd, Middle Tennessee State University; Susan Todd, Brookhaven College; Hugo C. Tscharnack, University of Wisconsin—Whitewater; Stan Udd, Calvary Bible College; Lois Veath, Chadron State College; W. C. Wyatt, Wesleyan College.

We are very grateful to the staff at the HarperCollins Glenview office for their work and professional care: Jane Piro, Doug Humphrey, Louise Howe, Carol Zombo, Rachel Schneider, and Kelly Mountain. Special thanks go to copyeditor Suzanne Lyons for invaluable assistance and input.

Wow, Great Grandpa Charlie! Before this chickie exhausted its inner space resources and poked out of its shell, it must have thought it was at its last moments. But what seemed like its end was a new beginning. Are we like chickies, ready to poke through to a new environment and new beginning -- like humanizing outer space maybe?

PROLOGUE: ABOUT SCIENCE

Science is organized common sense about the physical world around us. It is a practical knowledge of nature that draws upon the collective efforts, observations, insights, and wisdom of the human race. Science aids human activity by identifying order in nature and finding explanations for this order. Science began before recorded history, when people first discovered regularities and relationships in nature, such as star patterns in the night sky and weather patterns—when the rainy season started or the days grew longer. From these regularities, people learned to make predictions that gave them some control over their surroundings.

Science made great headway in Greece in the seventh century BC and spread throughout the Mediterranean world. Scientific advances came to a halt in Europe when the Roman Empire fell in the fifth century AD. Barbarian hordes destroyed almost everything in their path as they overran Europe and ushered in the Dark Ages. During this time the Chinese were charting the stars and the planets and Arab nations were developing mathematics. Greek science was reintroduced to Europe by Islamic influences that penetrated into Spain during the tenth, eleventh, and twelfth centuries. Universities emerged in Europe in the thirteenth century, and the introduction of gunpowder changed the social and political structure of Europe in the fourteenth century. The fifteenth century saw art and science beautifully blended by Leonardo da Vinci. Scientific thought was bolstered in the sixteenth century with the advent of the printing press.

The sixteenth-century Polish astronomer Copernicus caused great controversy when he published a book proposing that the sun was stationary and the earth revolved around the sun. This model conflicted with the popular view that the earth was the center of the universe. It also conflicted with Church teachings and so was banned for 200 years. The Italian physicist Galileo was arrested for popularizing the Copernican theory and for other radical scientific ideas. Yet a century later the Copernican model was widely accepted.

This kind of cycle happens age after age. In the early 1800s geologists met with violent condemnation because they differed with the Genesis account of creation. Later in the same century geology was accepted. Charles Darwin's theory of the evolution of species was condemned and its teaching forbidden in the 1920s. Court battles were waged before teaching evolution was again made legal. Every age has its groups of intellectual rebels who are persecuted, condemned, or suppressed at the time but who later seem harmless and often essential to the elevation of human conditions. "At every crossway on the road that leads to the future, each progressive spirit is opposed by a thousand men appointed to guard the past."*

MATHEMATICS AND SCIENTIFIC MEASUREMENTS

Science and human conditions advanced dramatically after the discovery almost four centuries ago that nature could be analyzed and described mathematically. When the ideas of science are expressed in mathematical terms, they are unambiguous. They don't have the double meanings that so often confuse the discussion of ideas expressed in common language. When findings in nature are expressed mathematically, they are easier to verify or disprove by experiment. The methods of mathematics and experimentation have led to the enormous success of science.†

Measurements are also a hallmark of good science. How much you know about something is often related to how well you can measure it. This was well put by the famous physicist Lord Kelvin in the last century: "I often say that when you can measure something and express it in numbers, you know something about it. When

---

*From Count Maurice Maeterlinck's "Our Social Duty."

†The mathematical structure of physics is evident in the equations you will encounter throughout this book. You will see that equations are simply shortcut expressions of relationships that can be expressed in words. The focus of this book is on understanding concepts—in English. Equations are compact statements that help guide thinking, and are not used in this book as recipes for mathematical problem solving. A premature effort at mathematical problem solving often tends to obscure science, so emphasis on this kind of problem solving is best postponed until you understand the concepts—perhaps in a follow-up course. Conceptual Physical Science puts comprehension comfortably before computation.

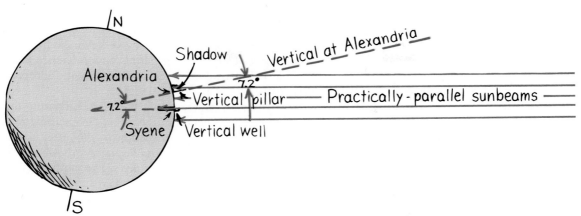

Figure 1 When the sun is directly overhead at Syene, it is not directly overhead in Alexandria, 800 km north. When the sun's rays shine directly down a vertical well in Syene, they cast a shadow of a vertical pillar in Alexandria. The verticals at both locations extend to the center of the earth, and make the same angle that the sun's rays make with the pillar at Alexandria. Eratosthenes measured this angle to be $\frac{1}{50}$ of a complete circle. Therefore the distance between Alexandria and Syene is $\frac{1}{50}$ the Earth's circumference. (Equivalently, the shadow cast by the pillar is $\frac{1}{8}$ the height of the pillar, which means the distance between the locations is $\frac{1}{8}$ the earth's radius.)

you cannot measure it, when you cannot express it in numbers, your knowledge is of a meager and unsatisfactory kind. It may be the beginning of knowledge, but you have scarcely in your thoughts advanced to the stage of science, whatever it may be." Mathematics and scientific measurements are the tools of good science. These tools are not something new, but go back to ancient times. In the third century BC, for example, fairly accurate measurements were made of the sizes of the earth, moon, sun, and the distances between them.

Size of the Earth

The size of the earth was first measured by the geographer and mathematician Eratosthenes in about 235 BC.* Eratosthenes calculated the circumference of the earth in the following way. He knew that the sun is highest in the sky at noon on June 22, the summer solstice. At this time a vertical stick casts its shortest shadow. If the sun is directly overhead, a vertical stick casts no shadow at all, which occurred in Syene, a city south of Alexandria (where the Aswan Dam stands today). Eratosthenes learned that the sun was directly overhead Syene from library information, which reported that at this unique time sunlight shines directly down a deep well there and is reflected back up again. Eratosthenes reasoned that if the sun's rays were extended into the earth at this point, they would pass through the center. Likewise, a vertical line extended into the earth at Alexandria (or anywhere else) would also pass through the earth's center. In Alexandria at this time, a vertical stick casts a shadow—which is very easy to measure.

At noon on June 22 Eratosthenes measured the shadow cast by a vertical pillar in Alexandria and found it to be $\frac{1}{8}$ the height of the pillar (Figure 1). This corresponds to a 7.2-degree angle between the sun's rays and the vertical pillar. Since 7.2 degrees is $\frac{7.2}{360}$, or $\frac{1}{50}$ of a circle, Eratosthenes reasoned the distance between

---

*Eratosthenes was second chief Librarian at the University of Alexandria in Egypt, founded by Alexander the Great. Eratosthenes was one of the foremost scholars of his time and wrote on philosophy, and scientific and literary matters. As a mathematician he invented a method for finding prime numbers. His reputation among his contemporaries was immense—Archimedes dedicated a book to him. As a geographer he wrote *Geography*, the first book to give geography a mathematical basis and to treat the earth as a globe divided into Frigid, Temperate, and Torrid zones. It long remained a standard work, and was used a century later by Julius Caesar. Eratosthenes spent most of his life in Alexandria and died there in 195 BC.

Alexandria and Syene must be $\frac{1}{50}$ the circumference of the earth. Thus the circumference of the earth becomes 50 times the distance between these two cities. This distance, quite flat and frequently traveled, was measured by surveyors to be 5000 stadia (800 kilometers). So Eratosthenes calculated the earth's circumference to be 50×5000 stadia = 250,000 stadia. This is within 5 percent of the currently accepted value of the earth's circumference.

We get the same result by bypassing degrees altogether and comparing the length of the shadow cast by the pillar to the height of the pillar. Geometrical reasoning shows, to a close approximation, that the ratio of *shadow length/pillar height* is the same as the ratio of *distance between Alexandria and Syene/earth's radius*. So just as the pillar is 8 times greater than its shadow, the radius of the earth must be 8 times greater than the distance between Alexandria and Syene.

Since the circumference of a circle equals 2π times its radius, the earth's radius is simply its circumference divided by 2π. In modern units the earth's radius is 6370 kilometers, and its circumference is 40,000 km.

Size of the Moon

Aristarchus was perhaps the first to suggest that the earth spins on a daily axis, and this accounts for the daily motion of the stars. He also hypothesized that the earth moves around the sun in a yearly orbit, and that the other planets do likewise.*

Figure 2 During a lunar eclipse the earth's shadow is observed to be 2.5 times wider than the moon's diameter. Because of the sun's large size, the earth's shadow must taper. The amount of taper is evident during a solar eclipse, where the moon's shadow tapers a whole moon diameter from moon to earth. So the earth's shadow must taper the same amount in the same distance. Therefore the earth's diameter must be 3.5 moon diameters.

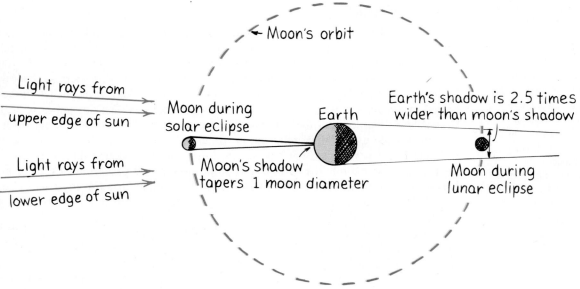

*Aristarchus was unsure of his heliocentric hypothesis, likely because the earth's seasons were not equal in length. This didn't support the idea of the earth circling the sun. More important, it was noted that the moon's distance from earth varies, clear evidence that it does not perfectly circle the earth. If the moon does not follow a circular path about the earth, it was hard to argue that the earth follows a circular path about the sun. The explanation, the elliptical paths of planets, was not discovered until centuries later by Kepler. In the meantime, the epicycles proposed by other astronomers accounted for these discrepancies. It is interesting to speculate about the course of astronomy if the moon didn't exist. Its irregular orbit would not have contributed to the early discrediting of the heliocentric theory, which may have taken hold centuries earlier!

Sun

Moon's orbit

Earth
(too small to see)

Figure 3 Correct scale of solar and lunar eclipses, which shows why eclipses are rare. (We will see in Chapter 27 that they are even rarer because the moon's orbit is tilted about 5° from the plane of the earth's orbit about the sun.)

He correctly measured the moon's diameter and its distance from the earth. He did all this in about 240 BC, seventeen centuries before his findings became fully accepted.

Aristarchus compared the size of the moon with the size of the earth by watching eclipses of the moon. The earth, like any body in sunlight, casts a shadow. An eclipse of the moon is simply the event wherein the moon passes into this shadow. Aristarchus carefully studied this event and found that the width of the earth's shadow out at the moon was 2.5 moon diameters. This would seem to indicate that the moon's diameter is 2.5 times smaller than the earth's. But the earth's shadow tapers, as evidenced during a solar eclipse (Figure 2 shows this in exaggerated scale). At that time the moon's shadow tapers almost to a point at the earth's surface, so the earth just barely falls in the moon's shadow. This means the taper of the moon at this distance is one moon diameter. So the earth's shadow, covering the same distance, must also taper one moon diameter. Taking the tapering of the sun's rays into account, the earth's diameter must be (2.5 + 1) times the moon's diameter. In this way Aristarchus showed that the moon's diameter is $\frac{1}{3.5}$ that of the earth's. The presently accepted diameter of the moon is 3640 km, within 5 percent of the value calculated by Aristarchus.

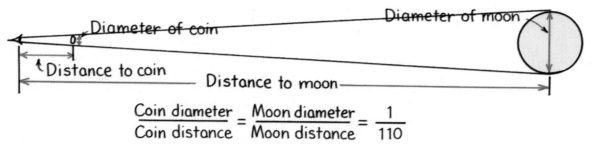

$$\frac{\text{Coin diameter}}{\text{Coin distance}} = \frac{\text{Moon diameter}}{\text{Moon distance}} = \frac{1}{110}$$

Figure 4 An exercise in ratio and proportion: When the coin barely "eclipses" the moon, then the diameter of the coin compared to the distance between you and the coin is equal to the diameter of the moon compared to the distance between you and the moon (not to scale here). Measurements give a ratio of $\frac{1}{110}$ for both.

Distance to the Moon

Tape a small coin like a dime to a window and view it with one eye so that it just blocks out the full moon. This occurs when your eye is about 110 coin diameters away. Then the ratio of *coin diameter/coin distance* is about $\frac{1}{110}$. Geometrical reasoning shows this is also the ratio of *moon diameter/moon distance* (Figure 4). So the distance to the moon is 110 times the moon's diameter. The early Greeks knew this. Aristarchus' measurement of the moon's diameter was all that was needed to calculate the distance between the earth and the moon. So the early Greeks knew both the size of the moon and its distance from earth.

With this information, Aristarchus made a measurement of the distance between the sun and the earth.

Distance to the Sun

Repeat the coin-on-the-window-and-moon exercise for the sun and guess what: The ratio of sun diameter/sun distance is also $\frac{1}{110}$. This is because the sun and moon both have the same size to the eye. They both taper the same angle (about

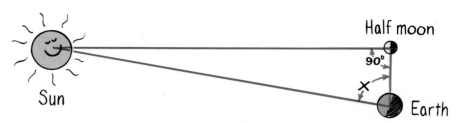

Figure 5 When the moon appears exactly half full, the sun, moon, and earth form a right triangle (not to scale). The hypotenuse is the earth-sun distance. By simple trigonometry, the hypotenuse of a right triangle can be found if you know the value of either non-right angle and the length of one side. The earth-moon distance is a known side. Measure angle X and you can calculate the earth-sun distance.

0.5 degrees). So although the ratio of diameter to distance was known to the early Greeks, diameter alone or distance alone would have to be determined by some other means. Aristarchus found a means of doing this and made a rough estimate. Here's what he did.

Aristarchus watched for the phase of the moon when it was *exactly* half full, with the sun still visible in the sky. Then the sunlight must be falling on the moon at right angles to the line of sight. The lines between earth and the moon, between earth and the sun, and between moon and the sun form a right triangle (Figure 5).

A rule of trigonometry states that if you know all the angles in a right triangle plus the length of any one of its sides, you can calculate the length of any other side. Aristarchus knew the distance from earth to moon. At the time of the half moon he also knew one of the angles, 90 degrees. All he had to do was measure the second angle between the line of sight to the moon and the line of sight to the earth. Then the third angle, a very small one, is 180 degrees minus the sum of the first two angles (the sum of the angles in any triangle = 180 degrees).

Measuring the angle between the lines of sight to moon and sun is difficult to do without a modern transit. For one thing, both the sun and moon are not points, but are relatively big. Aristarchus had to sight on their centers (or either edge) and measure the angle between—quite large, almost a right angle itself! By modern-day measure, his measurement was very crude. He measured 87 degrees, while the true value was 89.8 degrees. He figured the sun to be about 20 times more distant than the moon, when in fact it is about 400 times more distant. So although his method was ingenious, his measurements were not. Perhaps Aristarchus found it difficult to believe the sun was so far away, and he erred on the nearer side. We don't know.

Today we know the sun to be an average of 150,000,000 km away. It is somewhat closer in December (147,000,000 km), and somewhat farther in June (152,000,000 km).

$$\frac{d}{h} = \frac{D}{150,000,000 \text{ km}} = \frac{1}{110}$$

Figure 6 The round spot of light cast by the pinhole is an image of the sun. Its *diameter/distance* ratio is the same as the *sun's diameter/sun's distance* ratio; $\frac{1}{110}$. The sun's diameter is $\frac{1}{110}$ its distance from earth.

Size of the Sun

Once the distance to the sun is known, the $\frac{1}{110}$ ratio of diameter/distance enables a measure of the sun's diameter. Another way to measure the $\frac{1}{110}$ ratio, besides the method of Figure 5, is to measure the diameter of the sun's image cast through a pinhole opening. You may do this as a lab experiment, but if not, be sure to try this anyway. Poke a hole in a sheet of opaque cardboard and let sunlight shine on it. The round image that is cast on a surface below is actually an image of the sun. You'll see that the size of the image does not depend on the size of the pinhole, but on how

Figure 7 Renoir accurately painted the spots of sunlight that are actually images of the sun cast by relatively small openings in the leaves above.

Figure 8 The crescent-shaped spots of sunlight are images of the sun when it is partially eclipsed.

far away it is from the cardboard surface. Bigger holes make brighter images, not bigger ones. Of course if the hole is too big, the image is no longer evident. Careful measurement will show the ratio of image size to pinhole distance is $\frac{1}{110}$—the same ratio of *sun diameter/sun-earth distance* (Figure 6).

Interestingly enough, at the time of a partial solar eclipse, the image cast by the pinhole will be a crescent shape—the same as that of the partially covered sun! This provides an interesting way to view a partial eclipse without actually looking at the sun.

Have you noticed that the spots of sunlight you see on the ground beneath trees are perfectly round when the sun is overhead, and elliptical when the sun is low in the sky? These are pinhole images of the sun, where light shines through openings in the leaves that are small compared to the distance to the ground below. A round spot 10 cm in diameter is cast by an opening that is 110×10 cm above ground. Tall trees make large images; short trees make small images. And at the time of a partial solar eclipse, the images are crescents (Figure 8).

THE SCIENTIFIC METHOD

The methods of science usually are underscored by rational thinking and experimentation. In the sixteenth century the Italian physicist Galileo Galilei and the English philosopher Francis Bacon founded what is called the **scientific method**—a method that is extremely effective in gaining, organizing, and applying new knowledge. This method is essentially as follows:

1. Recognize a problem.
2. Make an educated guess—a **hypothesis**.
3. Predict the consequences of the hypothesis.

4. Perform experiments to test predictions.
5. Formulate the simplest general rule that organizes the three main ingredients—hypothesis, prediction, and experimental outcome.

Although this method has a certain appeal, it has not always been the key to the discoveries and advances in science. Many cases of trial and error, experimentation without guessing, and just plain accidental discovery have accounted for much of the progress in science. The success of science has more to do with an attitude common to scientists than with a particular method. This attitude is one of inquiry, experimentation, and humility before the facts.

THE SCIENTIFIC ATTITUDE

It is common to think of a fact as something that is unchanging and absolute. But in science, a **fact** is generally a close agreement by competent observers of a series of observations of the same phenomena. For example, where it was once a recognized fact in some cultures that the earth was flat, today it is a fact that the earth is round. A scientific hypothesis, on the other hand, is an educated guess that is only presumed to be factual until demonstrated by experiments. When a hypothesis has been tested over and over again and has not been contradicted, it may become known as a **law** or **principle**.

If a scientist believes a certain hypothesis, law, or principle is true but finds contradicting evidence, then ideally, in the scientific spirit, the hypothesis, law, or principle is changed or abandoned with little regard for the reputation or authority of the persons advocating it. For example, the greatly respected Greek philosopher Aristotle (384–322 BC) claimed that an object falls at a speed proportional to its weight. This false idea was held to be true for more than 2000 years because of Aristotle's compelling authority. In the scientific spirit, however, a single verifiable experiment to the contrary outweighs any authority, regardless of reputation or the number of followers or advocates. In modern science, argument by appeal to authority has little value.

Scientists must accept their experimental findings even when they would like them to be different. They must strive to distinguish between what they see and what they wish to see, for scientists, like most people, have a vast capacity for fooling themselves. People have always tended to adopt general rules, beliefs, creeds, ideas, and hypotheses without thoroughly questioning their validity and to retain them long after they have been shown to be meaningless, false, or at least questionable. The most widespread assumptions are often the least questioned. Too often, when an idea is adopted, particular attention is given to cases that seem to support it, while cases that seem to refute it are distorted, belittled, or ignored.

Scientists use the word *theory* in a way that differs from its usage in everyday speech. In everyday speech a theory is no different from a hypothesis—a supposition that has not been verified. A scientific **theory,** on the other hand, is a synthesis of a large body of information that encompasses well-tested and verified hypotheses about certain aspects of the natural world. Physicists, for example, speak of the quark theory of the atomic nucleus, chemists speak of the theory of metallic bonding in metals, geologists subscribe to the theory of plate tectonics, and astronomers speak of the theory of the Big Bang.

The theories of science are not fixed, but undergo change. Scientific theories evolve as they go through stages of redefinition and refinement. During the last hundred years, for example, the theory of the atom has been repeatedly refined as new evidence on atomic behavior has been gathered. Similarly, chemists have refined their view of the way atoms bond together to form molecules, geologists have refined the plate tectonics theory, and astronomers armed with new data from the Hubble telescope are presently sharpening their view of the universe. The refinement of theories is a strength of science, not a weakness. Many people feel that it is a sign of weakness to change one's mind. Competent scientists must be open to changing their minds. They change their minds, however, only when confronted with solid experimental evidence to the contrary or when a conceptually simpler hypothesis forces them to a new point of view. More important than defending beliefs is improving them. Better hypotheses are made by those who are honest in the face of fact.

Away from their profession, scientists may be no more honest or ethical than most other people. But in their profession they work in an arena that puts a high premium on honesty. The cardinal rule in science is that all hypotheses must be testable—they must be susceptible, at least in principle, to being proved *wrong*. In science, it is more important that there be a means of proving an idea wrong than that there be a means of proving it right. At first this may seem strange, for when we wonder about most things, we concern ourselves with ways of finding out whether they are true. This emphasis on determining possible wrongness distinguishes science from nonscience. If you want to distinguish whether a hypothesis is scientific or not, look to see if there is a test for proving it wrong. If there is no test for its possible wrongness, then the hypothesis is not scientific.

Consider Charles Darwin's hypothesis that life forms evolve from simpler to more complex forms. This could be proved wrong if paleontologists found that more complex forms of life appeared before their simpler counterparts. Einstein hypothesized that light is bent by gravity. This might be proved wrong if starlight that grazed the sun and could be seen during a solar eclipse were undeflected from its normal path. As it turns out, less complex life forms are found to precede their more complex counterparts and starlight is found to bend as it passes close to the sun, which support the claims. If and when a hypothesis or scientific claim is confirmed, it is regarded as useful and a stepping stone to additional knowledge.

Consider on the other hand the hypothesis that "intelligent life exists on other planets somewhere in the universe." At present, this hypothesis is not scientific. Reasonable or not, it is *speculation*. Although it can be proved correct by the verification of a single instance of intelligent life existing elsewhere in the universe, there is no way to prove the hypothesis wrong if no life is ever found. If we search the far reaches of the universe for eons and find no life, we would not prove that it doesn't exist around the next corner. A hypothesis that is capable of being proved right but not capable of being proved wrong is not a scientific hypothesis. Many such statements are quite reasonable and useful, but they lie outside the domain of science.

None of us has the time, energy, or resources to test every idea, so most of the time we take somebody's word. How do we know whose word to take? To reduce the likelihood of error, scientists accept the word only of those whose ideas, theories, and findings are testable—if not in practice, at least in principle. Speculations that cannot be tested are regarded as "unscientific." This has the long-run effect of compelling honesty—findings widely publicized among fellow scientists are generally subjected to further testing. Sooner or later, mistakes (or deception) are found out; wishful thinking is exposed. A discredited scientist does not get a second chance in the community of scientists. Honesty, so important to the progress of science, thus becomes a matter of self-interest to scientists. There is relatively little bluffing in a game where all bets are called. In fields of study where truth and falsehood are not so easily established, honesty is more difficult to enforce.

QUESTION

Which of these is a scientific hypothesis?

(a) Atoms are the smallest particles of matter that exist.
(b) Space is permeated with an essence that is undetectable.
(c) Albert Einstein is the greatest physicist of the twentieth century.

SCIENCE, ART, AND RELIGION

The search for order and meaning in the world around us has taken different forms: one is science, another is art, and another is religion. Although the roots of all three go back thousands of years, the traditions of science are relatively recent. More important, the domains of science, art, and religion are different, although they often overlap. Science is principally engaged with discovering and recording natural phenomena, the arts are an expression of human experience through sensory and emotional means, and religion addresses the source, purpose, and meaning of it all.

Science and the arts are comparable. In literature we find what is possible in human experience. We can learn about emotions ranging from anguish to love, even if we haven't yet experienced them. The arts do not necessarily give us those experiences, but describe them to us and suggest possibilities. A knowledge of science similarly tells us what is possible in nature. Scientific knowledge helps us predict possibilities in nature even before these possibilities have been experienced. It provides us with a way of connecting things, of seeing relationships between and among

ANSWER

Only *a* is scientific, because there is a test for its falseness. The statement is not only *capable* of being proved wrong, but it in fact *has* been proved wrong. Statement *b* has no test for possible wrongness and is therefore unscientific. Likewise for any principle or concept for which there is no means, procedure, or test whereby it can be shown to be wrong (if it is wrong). Some pseudoscientists and other pretenders of knowledge will not even consider a test for the possible wrongness of their statements. Statement *c* is an assertion that has no test for possible wrongness. If Einstein was not the greatest physicist, how could we know? It is important to note that because the name Einstein is generally held in high esteem, it is a favorite of pseudoscientists. So we should not be surprised that the name of Einstein, like that of Jesus and other highly respected sources, is cited often by charlatans who wish to bring respect to themselves and their points of view. In all fields it is prudent to be skeptical of those who wish to credit themselves by calling upon the authority of others.

them, and of making sense of the myriad of natural events around us. Science broadens our perspective of the natural environment of which we are a part. A knowledge of both the arts and the sciences makes for a wholeness that affects the way we view the world and the decisions we make about it and ourselves. A truly educated person is knowledgeable in both the arts and the sciences.

Science and religion have similarities also, but they are basically different—principally because their domains are different. Science is concerned with the physical realm; religion is concerned with the spiritual realm. Simply put, science asks *how;* religion asks *why.* The practices of science and religion are also different. Whereas scientists experiment to find nature's secrets, many religious practitioners worship God and work to build human community. In these respects, science and religion are as different as apples and oranges and do not contradict each other. Science and religion are two different yet complementary fields of human activity.

When we study the nature of light later in this book, we will treat light first as a wave and then as a particle. To the person who knows a little bit about science, waves and particles are contradictory; light can be only one or the other, and we have to choose between them. But to the enlightened person, waves and particles complement each other and provide a deeper understanding of light. In a similar way, it is mainly people who are either uninformed or misinformed about the deeper natures of both science and religion who feel that they must choose between believing in religion and believing in science. Unless one has a shallow understanding of either or both, there is no contradiction in being religious and being scientific in one's thinking.

SCIENCE AND TECHNOLOGY

Science and technology are also different from each other. Whereas science involves discovering evidence and relationships for observable phenomena in nature and establishing theories that organize and make sense of those phenomena, technology involves tools, techniques, and procedures for putting the findings of science to use.

Another difference between science and technology is how each affects human lives. Ideally, science excludes the human factor. Scientists who seek to comprehend the workings of nature cannot be influenced by their own or other people's likes or dislikes or by popular ideas about what is correct. What scientists discover may shock or anger people, as did Darwin's theory of evolution. But even an unpleasant truth is likely to be useful; besides, we have the option of refusing to believe it! But this is hardly so with technology once it is developed: We do not have the option of refusing to hear the sonic boom produced by a supersonic aircraft flying overhead, we do not have the option of refusing to breathe polluted air, and we do not have the option of living in a non-nuclear age. Unlike science, advances in technology must be measured in terms of the human factor, with the understanding that technology is our slave and not the reverse. The legitimate purpose of technology is to serve people—people in general, not just some people, and future generations, not just those who currently wish to gain advantage for themselves. Technology must be oriented to human well-being if it is to lead to a better world.

We are all familiar with the abuses of technology. Many people blame technology itself for widespread pollution, resource depletion, and even social decay in general—so much so that the promise of technology is obscured. That promise is a cleaner and healthier world. It is much wiser to combat the misuse of technology with knowledge than with ignorance. Wise applications of science and technology *can* lead to a better world.

PHYSICS, CHEMISTRY, GEOLOGY, AND ASTRONOMY

Science is the present-day equivalent of what used to be called *natural philosophy*. Natural philosophy was the study of unanswered questions about nature. As the answers were found, they became part of what is now called *science*. The study of science today branches into the study of living things and nonliving things: the life sciences and the physical sciences. The life sciences branch into such areas as biology, zoology, and botany. The *physical sciences* branch into such areas as physics, chemistry, geology, meteorology, and astronomy—the areas addressed in this book.

We begin with physics because it is basic to the other physical sciences. *Physics* is about the nature of basic things such as motion, force, energy, matter, heat, sound, light, and the insides of atoms. *Chemistry* builds on physics and tells us how matter is put together, how atoms combine to form molecules, and how the molecules combine to make the materials around us. Physics and chemistry applied to the earth and its processes make up the science of geology, and when applied to other planets and their stars we are speaking about *astronomy*.

Biology is more complex than physics and chemistry, for it involves matter that is alive. Underneath biology is chemistry, and underneath chemistry is physics. The concepts of physics reach up to the more complicated sciences. An understanding of science in general begins with an understanding of physics, which begins with this book.

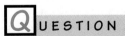

UESTION

Which of the following activities involves the utmost human expression of passion, talent, and intelligence?

(a) painting and sculpture   **(c)** music   **(e)** science
(b) literature   **(d)** religion

Ａ N S W E R

All of them! The human value of science, however, may be the least understood by most individuals in our society. The reasons are varied, ranging from the common notion that science is incomprehensible to people of average ability to the extreme view that science is a dehumanizing force in our society. Most of the misconceptions about science probably stem from the confusion between the *abuses* of science and science itself.

Science is an enchanting human activity shared by a wide variety of people who, with present-day tools and know-how, are reaching further and finding out more about themselves and their environment than people in the past were ever able to do. The more you know about science, the more passionate you feel toward your surroundings. There is physics and chemistry in everything you see, hear, smell, taste, and touch!

In Perspective

Only a few centuries ago the most talented and most skilled architects and artisans of the world directed their genius and effort to the construction of the great cathedrals, synagogues, temples, and mosques. Some of these architectural structures took centuries to build, which means that nobody witnessed both the beginning and the end of construction. Even the architects and early builders who lived to a ripe old age never saw the finished results of their labors. Entire lifetimes were spent in the shadows of construction that must have seemed without beginning or end. This enormous focus of human energy was inspired by a vision that went beyond worldly concerns—a vision of the cosmos. To the people of that time, the structures they erected were their "spaceships of faith," firmly anchored but pointing to the cosmos.

Today the efforts of many of our most skilled scientists, engineers, and artisans are directed to building the spaceships that already orbit the earth and others that will voyage beyond. The time required to build these spaceships is extremely brief compared to the time spent building the stone and marble structures of the past. Some people working on today's spaceships were alive before Charles Lindbergh made the first solo airplane flight across the Atlantic Ocean. Where will younger lives lead in a comparable time?

We seem to be at the dawn of a major change in human growth, for, as little Sarah suggests in the photo at the beginning of this book, we may be like the hatching chicken who has exhausted the resources of its inner-egg environment and is about to break through to a whole new range of possibilities. The earth is our cradle and has served us well. But cradles, however comfortable, are one day outgrown. So with the inspiration that in many ways is similar to the inspiration of those who built the early cathedrals, synagogues, temples, and mosques, we aim for the cosmos.

We live in an exciting time!

SUMMARY OF TERMS

Fact A phenomenon about which competent observers who have made a series of observations are in agreement.

Hypothesis An educated guess; a reasonable explanation of an observation or experimental result that is not fully accepted as factual until tested over and over again by experiment.

Law A general hypothesis or statement about the relationship of natural quantities that has been tested over and over again and has not been contradicted. Also known as a *principle*.

Scientific method An orderly method for gaining, organizing, and applying new knowledge.

Theory A synthesis of a large body of information that encompasses well-tested and verified hypotheses about certain aspects of the natural world.

MECHANICS

How neat! The number of balls I release into the array of balls is always the same number that emerge from the other side. But why? There's gotta be a reason -- mechanical rules of some kind. I'll know why the balls behave so predictably after I learn the rules of mechanics in the following five chapters. Best of all, learning these rules will provide a keener intuition in understanding the world around me!

1

MOTION

A systematic study of motion goes back to the time of Aristotle, the leading philosopher of the fourth century BC. In his view, every object in the universe had a proper place, determined by its "nature," and any object not in its proper place would "strive" to get there. An unsupported lump of clay, being of earth, properly fell to the ground; a puff of smoke, being of the air, properly rose. According to Aristotle's description of motion, the *distance* of an object from its natural place was the fundamentally important factor.

Galileo, a leading scientist of the seventeenth century AD, broke with this traditional concept in realizing that *time* was an important missing ingredient—motion was best described as a change in distance over a change in time. To measure motion Galileo used *rates*. A rate tells how fast something happens, or how much something changes in a certain amount of time. Rates that measure motion are *speed, velocity,* and *acceleration.*

Aristotle (384–322 BC)

Greek philosopher, scientist, and educator Aristotle was the son of a physician who personally served the king of Macedonia. At 17 he entered the Academy of Plato, where he worked and studied for 20 years until Plato's death. He then became the tutor of young Alexander the Great. Eight years later he formed his own school. Aristotle's aim was to systematize existing knowledge, just as Euclid had systematized geometry. Aristotle made critical observations, collected specimens, and gathered together, summarized, and classified almost all existing knowledge of the physical world. His systematic approach became the method from which Western science arose. After his death, his voluminous notebooks were preserved in caves near his home and were later sold to the library at Alexandria. Scholarly activity ceased in most of Europe through the Dark Ages, and the works of Aristotle were forgotten and lost. Scholarship continued in the Byzantine and Islamic empires, and various texts were reintroduced to Europe during the eleventh and twelfth centuries and translated into Latin. The Church, the dominant political and cultural force in Western Europe, first prohibited the works of Aristotle and then accepted and incorporated them into Christian doctrine. Any attack on Aristotle was an attack on the Church itself. It was in this climate that Galileo effectively challenged Aristotle's ideas on motion, and ushered in a new method of finding things out—experimentation.

1.1 SPEED

Things in motion cover certain distances in certain times. An automobile, for example, may travel so many kilometers in an hour. **Speed** is a measure of how fast something is moving. It is the rate at which distance is covered, and is always measured in terms of a unit of distance divided by a unit of time. In general,

$$\text{Speed} = \frac{\text{distance}}{\text{time}}$$

Any choice of distance unit and time unit can be used. For motor vehicles kilometers per hour (km/h) or miles per hour (mi/h or mph) are commonly used. For

Figure 1.1 A cheetah is the champion runner over distances less than 500 meters and can achieve peak speeds of 100 km/h.

TABLE 1.1 *Approximate Speeds in Different Units*

25	km/h	=	16	mi/h	=	7	m/s
60	km/h	=	37	mi/h	=	17	m/s
75	km/h	=	47	mi/h	=	21	m/s
100	km/h	=	62	mi/h	=	28	m/s
120	km/h	=	75	mi/h	=	33	m/s

shorter distances, meters per second (m/s) are often useful. The slash symbol (/) is read as *per* and means *divided by.* Table 1.1 shows speeds in various units.*

The speed something has at any instant is its instantaneous speed. It is the speed shown by the speedometer of a car. When we say that the speed of a car at some instant is 60 kilometers per hour, we are specifying its **instantaneous speed,** and we mean that if the car continued moving at that speed for an hour, it would travel 60 kilometers. If it continued at that speed for half an hour, it would go half the distance, 30 kilometers. If it continued for 1 minute, it would go 1 kilometer.

Figure 1.2 A speedometer gives readings in both miles per hour and kilometers per hour.

A car rarely moves at constant speed. On any trip the speed will likely vary. So we speak of the **average speed:**

$$\text{Average speed} = \frac{\text{total distance covered}}{\text{time interval}}$$

If we drive a distance of 80 kilometers, for example, in a time of 1 hour, we find our average speed is 80 kilometers per hour. Likewise, if we travel 320 kilometers in 4 hours,

$$\text{Average speed} = \frac{\text{total distance covered}}{\text{time interval}} = \frac{320 \text{ km}}{4 \text{ h}} = 80 \text{ km/h}$$

A distance in kilometers (km) divided by a time in hours (h), is an average speed in kilometers per hour (km/h). This doesn't show the speed variations that may occur, including possible stops when speed equals zero. We may undergo a variety of speeds on most trips, so the average speed is often different than the speed at each moment—the instantaneous speed. But whether we talk about average speed or instantaneous speed, we are talking about the rates at which distance is covered.

If we know average speed and time of travel, distance traveled is easy to find. A simple rearrangement of the above definition gives

$$\text{Total distance covered} = \text{average speed} \times \text{time}$$

If your average speed is 80 kilometers per hour on a 4-hour trip, for example, you cover a total distance of 320 kilometers.

---

*Conversion is based on 1 h = 3600 s, 1 mi = 1609.344 m.

Galileo Galilei (1564–1642)

Galileo was born in Pisa in the same year that Shakespeare was born and Michelangelo died. He studied medicine at the University of Pisa and then changed to mathematics. He developed an early interest in the mechanics of motion and was soon at odds with his contemporaries, who held to Aristotelian ideas on falling bodies. He left Pisa to teach at the University of Padua and became an advocate of the new Copernican theory of the solar system. He was one of the first to build a telescope, and he was the first to direct it to the nighttime sky and discover mountains on the moon and the moons of Jupiter. Because he published his findings in Italian instead of in the Latin expected of so reputable a scholar, and because of the recent invention of the printing press, his ideas reached a wide readership. He soon ran afoul of the Church and was warned not to teach and hold to Copernican views. He restrained himself publicly for nearly 15 years and then defiantly published his observations and conclusions, which were counter to Church doctrine. The outcome was a trial in which he was found guilty, and he was forced to renounce his discoveries. By then an old man broken in health and spirit, he was sentenced to perpetual house arrest. Nevertheless, he completed his studies on motion and his writings were smuggled from Italy and published in Holland. Earlier he damaged his eyes looking at the sun through a telescope, which led to blindness at the age of 74. He died 4 years later.

QUESTIONS

1. What is the average speed of a cheetah that sprints **(a)** 100 m in 4 seconds? **(b)** 50 m in 2 s?

2. A car moved at an average speed of 60 km/hr for 1 h. **(a)** How far did it go? **(b)** At this rate, how far would it go in 4 h? (c) in 10 h?

3. During this 1-hr trip, is it possible for the car to attain an average speed of 60 km/h and never exceed a reading of 60 km/h on the speedometer?

ANSWERS

(Are you reading this before you have a reasoned answer? If so, do you also exercise your body by watching others do push-ups? Exercise your thinking: When you encounter the questions throughout this book, think before you read the answer.)

1. In both cases the answer is 25 m/s:

$$\text{Average speed} = \frac{\text{distance covered}}{\text{time interval}} = \frac{100 \text{ meters}}{4 \text{ seconds}} = \frac{50 \text{ meters}}{2 \text{ seconds}} = 25 \text{ m/s}$$

2. The distance traveled is the average speed × time of travel, so

   (a) Distance = 60 km/h × 4 h = 240 km
   (b) Distance = 60 km/h × 10 hr = 600 km

3. No, not if the trip started from rest and ended at rest, because there would then be intervals with an instantaneous speed less than 60 km/h. Unless there is compensation of periods of speed greater than 60 km/h, it would not be possible to yield an average of 60 km/h. In practice, average speeds are usually appreciably less than peak instantaneous speeds.

Speed Is Relative

In a strict sense, everything moves—even things that appear to be at rest. They move relative to the sun and stars. A book that is at rest relative to the table it lies on is moving at about 30 kilometers per second relative to the sun. And it moves even faster relative to the center of our galaxy. When we discuss the speed of something, we describe its motion relative to something else. When we say an express train travels at 200 kilometers per hour, we mean relative to the track. When we say that a space shuttle moves at 8 kilometers per second, we mean relative to the earth below. Unless stated otherwise, when we discuss the speeds of things in our environment we mean relative to the surface of the earth.

1.2 VELOCITY

When we describe the speed and the *direction* of motion, we are specifying **velocity**. Loosely speaking, we can use the words *speed* and *velocity* interchangeably. Strictly speaking, however, there is a distinction between the two. When we say that something travels at 60 kilometers per hour, we are specifying its speed. But if we say that something travels at 60 kilometers per hour to the north, we are specifying its velocity. A racecar driver is concerned with his speed—how fast he is moving; an airplane pilot is concerned with her velocity—how fast and in what direction she is moving.

We distinguish between average velocity and instantaneous velocity as we do for speed. By custom, the word *velocity* alone is assumed to mean instantaneous velocity. Likewise for the word *speed* alone. If something moves at an unchanging or constant velocity, then, its average and instantaneous velocities will have the same value. The same is true for speed. When something moves at constant velocity or constant speed, then *equal distances* are covered in equal intervals of time. Constant velocity and constant speed, however, can be very different. Constant velocity means constant speed with no change in direction. A car that rounds a curve at a constant speed does not have a constant velocity—its velocity changes as its direction changes.

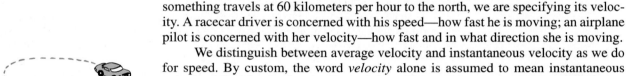

Figure 1.3 The car on the circular track may have a constant speed, but its velocity is changing every instant. Why?

Q UESTIONS

1. "She moves at a constant speed in a constant direction." Say the same sentence in fewer words.
2. The speedometer of a car moving to the east reads 100 km/h. It passes another car that moves to the west at 100 km/h. Do both cars have the same speed? Do they have the same velocity?
3. During a certain period of time, the speedometer of a car reads a constant 60 km/h. Does this indicate a constant speed? A constant velocity?

A NSWERS

1. "She moves at constant velocity."
2. Both cars have the same speed, but they have opposite velocities because they are moving in opposite directions.
3. The constant speedometer reading indicates a constant speed but not a constant velocity, because the car may not be moving along a straight-line path, in which case it is accelerating. (We'll see in Chapter 2 that whenever something accelerates, a force must be acting on it.)

Velocity Vectors

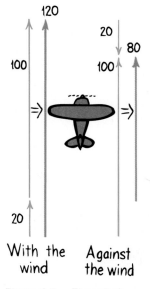

Figure 1.4 This vector, scaled so that 1 cm equals 20 km/h, represents 60 km/h to the right.

Pictures are often more descriptive than words. It is useful to represent velocity by an arrow. When the length of the arrow represents the speed and the direction of the arrow represents the direction of motion, the arrow is called a **velocity vector.** The vector in Figure 1.4 is scaled so that 1 centimeter represents 20 kilometers per hour; it is 3 centimeters long and points to the right, and therefore it represents a velocity of 60 kilometers per hour to the right.

Consider an airplane flying due north at 100 kilometers per hour relative to the surrounding air. Suppose there is a tailwind (wind from behind) that moves due north at 20 kilometers per hour. This example is represented with vectors in Figure 1.5 left. Here the velocity vectors are scaled so that 1 centimeter represents 20 kilometers per hour. Thus, the 100-kilometers per hour velocity of the airplane is shown by the 5-centimeter-long vector and the 20-kilometers per hour tailwind is shown by the 1-centimeter-long vector. You can see (with or without the vectors) that the resulting velocity is going to be 120 kilometers per hour relative to the ground. Without the tailwind, the airplane would travel 100 kilometers in one hour relative to the ground below. With the tailwind, it would travel 120 kilometers in one hour.

Suppose, instead, that the wind is a headwind (wind head-on), so the airplane flies into the wind rather than with the wind. Now the velocity vectors are in opposite directions (Figure 1.5 right). The result is 100 kilometers per hour minus 20 kilometers per hour that equals 80 kilometers per hour. Flying against a 20 kilometers per hour headwind, the airplane would travel only 80 kilometers relative to the ground in one hour.

QUESTION

> Consider a motor boat that travels 10 km/h relative to the water. If the boat travels in a river that flows also at a rate of 10 km/h, what will be its velocity relative to the shore when it heads directly upstream? When it heads directly downstream?

Combining vectors that act along parallel directions is simple: when in the same direction, they add; when in opposite directions, they subtract. The sum of two or more vectors is called their **resultant**. To find the resultant of two vectors that are not parallel to each other, we use the *parallelogram rule*.* Construct a parallelogram with the two vectors as adjacent sides; the diagonal of the parallelogram shows the resultant. This is shown in Figure 1.6, where the parallelogram is a rectangle (in this chapter we'll only treat vectors that are either parallel or that make rectangles. A more general treatment of vectors is in Appendix III at the end of the book).

Figure 1.5 The velocity of an airplane relative to the ground depends on its velocity relative to still air and the wind.

ANSWER

When the boat heads upstream its velocity is zero relative to the land (a velocity of +10 added to a velocity of −10 equals zero). When the boat heads directly downstream its velocity is 20 km/h in the direction of river flow (a velocity of +10 added to a velocity of +10 equals +20 in the same direction).

*A parallelogram ▱ is a four-sided plane figure having opposite sides parallel and equal.

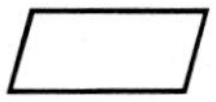

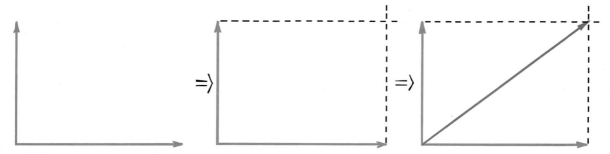

Figure 1.6 The pair of vectors at right angles to each other make two sides of a rectangle, the diagonal of which is their resultant.

With the vector technique we can correct for the effect of a crosswind (right angle) on the velocity of an airplane. Consider a slow-moving airplane that flies 80 km/h north and is caught in a strong easterly crosswind of 60 km/h. Figure 1.7 shows vectors for the airplane velocity and wind velocity. The scale here is 1 cm:20 km/h. The diagonal of the constructed parallelogram (rectangle in this case) measures 5 cm, which represents 100 km/h. So the airplane moves at 100 km/h relative to the ground, in a direction between north and northeast.*

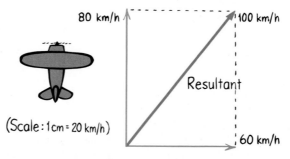

Figure 1.7 An 80-km/h airplane flying in a 60-km/h crosswind has a resultant speed of 100 km/h relative to the ground.

There is a special case of the parallelogram that often occurs. When two vectors that are equal in magnitude and at right angles to each other are to be added, the parallelogram becomes a square. We see this in Figure 1.8. Since for any square the length of a diagonal is $\sqrt{2}$, or 1.414, times one of the sides, the resultant is $\sqrt{2}$ times one of the vectors. For example, the resultant of two equal vectors of magnitude 100 acting at right angles to each other is 141.4.

---

*Whenever the vectors are at right angles to each other, their resultant can be found by the Pythagorean Theorem, a well-known tool of geometry. It states that the square of the hypotenuse of a right-angle triangle is equal to the sum of the squares of the other two sides. Note that two right triangles are present in the parallelogram (rectangle in this case) in Figure 1.12. From either one of these triangles we get:

$$\text{resultant}^2 = (60 \text{ km/h})^2 + (80 \text{ km/h})^2$$
$$= (3600 \text{ km}^2/\text{h}^2) + (6400 \text{ km}^2/\text{h}^2)$$
$$= (10{,}000 \text{ km}^2/\text{h}^2)$$

The square root of $(10{,}000 \text{ km}^2/\text{h}^2)$ is 100 km/h, as expected.

Figure 1.8 The diagonal of a square is $\sqrt{2}$ the length of one of its sides.

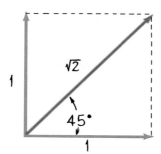

The kind of motion we have considered so far has primarily been steady motion—velocity that has not undergone change. Much of the motion around us, however, undergoes change. This is accelerated motion.

QUESTION

A boat travels between the two banks of a river at a speed of 10 km/hr relative to the water. The river itself, however, flows at a rate of 10 km/hr. What's the velocity of the boat relative to its starting point on shore?

1.3 ACCELERATION

When the velocity of a moving object changes, we say the object accelerates. We can change the velocity of something by changing its speed, by changing its direction, or by changing both its speed and its direction. We define the rate of change in velocity as **acceleration:**

$$\text{Acceleration} = \frac{\text{change of velocity}}{\text{time interval}}$$

We are all familiar with acceleration in an automobile. When driving, we call it "pickup" or "getaway"; we experience it when we tend to lurch toward the rear of the car. The key idea that defines acceleration is *change*. Suppose we are driving and in 1 second we steadily increase our velocity from 30 kilometers per hour to 35 kilometers per hour, and then to 40 kilometers per hour in the next second, to 45 in the next second, and so on. We change our velocity by 5 kilometers per hour each second. This change of velocity is what we mean by acceleration.

$$\text{Acceleration} = \frac{\text{change of velocity}}{\text{time interval}} = \frac{5 \text{ km/h}}{1 \text{ s}} = 5 \text{ km/h/s}$$

ANSWER

When the boat heads cross-stream (at right angles to the river flow) its velocity relative to its starting point is 14.14 km/h, 45 degrees downstream (in accord with the diagram in Figure 1.8).

In this case the acceleration is 5 kilometers per hour per second (abbreviated as 5 km/h/s). Note that a unit for time enters twice: once for the unit of velocity and again for the interval of time in which the velocity is changing. Also note that acceleration is not just the total change in velocity; it is the *time rate of change,* or *change per second,* of velocity.

QUESTIONS

1. A particular car can go from rest to 90 km/h in 10 s. What is its acceleration?

2. In 2.5 s a car increases its speed from 60 km/h to 65 km/h while a bicycle goes from rest to 5 km/h. Which undergoes the greater acceleration? What is the acceleration of each vehicle?

Figure 1.9 We say that an object undergoes acceleration when there is a *change* in its state of motion.

The term *acceleration* applies to decreases as well as to increases in velocity. We say the brakes of a car, for example, produce large retarding accelerations; that is, there is a large decrease per second in the velocity of the car. We often call this *deceleration,* or *negative acceleration.* We experience deceleration when we tend to lurch toward the front of the car.

We accelerate whenever we move in a curved path, even if we are moving at constant speed because our direction and hence our velocity is changing. We experience this acceleration as we tend to lurch toward the outer part of the curve. We distinguish speed and velocity for this reason and define *acceleration* as the rate at which velocity changes, thereby encompassing changes both in speed and in direction.

Anyone who has stood in a crowded bus has experienced the difference between velocity and acceleration. Except for the effects of a bumpy road, you can stand with no extra effort inside a bus that moves at constant velocity, no matter how fast it is going. You can flip a coin and catch it exactly as if the bus were at rest. It is only when the bus accelerates—speeds up, slows down, or turns—that you experience difficulty.

In much of this book we will be concerned primarily with motion along a straight line. When straight-line motion is being considered, it is common to use

ANSWERS

1. Its acceleration is 9 km/h/s. Strictly speaking, this would be its average acceleration, for there may have been some variation in its rate of picking up speed. Also, since acceleration like velocity is a vector quantity, its direction as well as magnitude should be specified.

2. The accelerations of both the car and the bicycle are the same: 2 km/h/s.

$$\text{Acceleration}_{car} = \frac{\text{change of velocity}}{\text{time interval}} = \frac{(65 \text{ km/h} - 60 \text{ km/h})}{2.5 \text{ s}} = \frac{5 \text{ km/h}}{2.5 \text{ s}} = 2 \text{ km/h/s}$$

$$\text{Acceleration}_{bike} = \frac{\text{change of velocity}}{\text{time interval}} = \frac{(5 \text{ km/h} - 0 \text{ km/h})}{2.5 \text{ s}} = \frac{5 \text{ km/h}}{2.5 \text{ s}} = 2 \text{ km/h/s}$$

Although the velocities involved are quite different, the rates of change of velocity are the same. Hence the accelerations are equal.

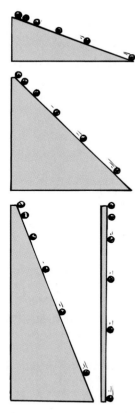

Figure 1.10 The greater the slope of the incline, the greater the acceleration of the ball. What is its acceleration if the incline is vertical?

speed and *velocity* interchangeably. When the direction is not changing, acceleration may be expressed as the rate at which *speed* changes.

$$\text{Acceleration (along a straight line)} = \frac{\text{change in speed}}{\text{time interval}}$$

Galileo developed the concept of acceleration in his experiments on inclined planes. His main interest was falling objects, and because he lacked suitable timing devices he used inclined planes to effectively slow down accelerated motion and investigate it more carefully.

Galileo found that a ball rolling down an inclined plane will pick up the same amount of speed in successive seconds; that is, the ball will roll with uniform or constant acceleration. For example, a ball rolling down a plane inclined at a certain angle might be found to pick up a speed of 2 meters per second for each second it rolls. This gain per second is its acceleration. Its instantaneous velocity at 1-second intervals, at this acceleration, is then 0, 2, 4, 6, 8, 10, and so forth, meters per second. We can see that the instantaneous speed or velocity of the ball at any given time after being released from rest is simply equal to its acceleration multiplied by the time:*

$$\text{Velocity acquired} = \text{acceleration} \times \text{time}$$

If we substitute the acceleration of the ball in this relationship, we can see that at the end of 1 second, the ball is traveling at 2 meters per second; at the end of 2 seconds, it is traveling at 4 meters per second; at the end of 10 seconds, it is traveling at 20 meters per second; and so on. After starting from rest, the instantaneous speed or velocity at any time is simply equal to the acceleration multiplied by the number of seconds it has been accelerating.

Galileo found greater accelerations for steeper inclines. The ball attains its maximum acceleration when the incline is tipped vertically. Then the acceleration is the same as that of a falling object (Figure 1.10). Regardless of the weight or size, Galileo discovered that when air resistance is small enough to be neglected, all material objects fall with the same constant acceleration.

1.4 FREE FALL

How Fast

Things fall because of the force of gravity. When a falling object is free of all restraints—no friction, air or otherwise, and falls under the influence of gravity alone, the object is in **free fall**. (We'll consider the effects of air resistance on falling in Chapter 2.) Table 1.2 shows the instantaneous speed of a freely falling object at 1 second intervals. The important thing to note in these numbers is the way the speed changes. *During each second of fall, the object gains a speed of 10 meters per second.* This gain per second is the acceleration. Free-fall acceleration is approximately equal to 10 meters per second each second, or, in shorthand notation, 10 m/s² (read as 10 meters per second squared). Note that the unit of time, the second, enters twice—once for the unit of speed and again for the time interval during which the speed changes.

---

*Note that this relationship follows from the definition of acceleration. From $a = (change\ in\ v)/t$, simple rearrangement (multiplying both sides of the equation by t) gives *change in v* $= at$.

TABLE 1.2 Free-Fall from Rest

Time of Fall (s)	Velocity Acquired (m/s)
0	0
1	10
2	20
3	30
4	40
5	50
.	.
.	.
.	.
t	$10t$

Figure 1.11 Pretend that a falling rock is equipped with a speedometer. In each succeeding second of fall, we would find that the rock's speed increases by the same amount: 10 m/s. (Table 1.2 shows the speed we would read at various seconds of fall.)

In the case of freely falling objects, it is customary to use the letter g to represent the acceleration (because the acceleration is due to *gravity*). Although the value of g varies slightly in different parts of the world, its average value is 9.8 meters per second each second, or, in shorter notation, 9.8 m/s^2. We round this off to 10 m/s^2 in our present discussion and in Table 1.2 to establish the ideas involved more clearly; multiples of 10 are more obvious than multiples of 9.8. Where precision is important, the value of 9.8 m/s^2 should be used.

Note in Table 1.2 that the instantaneous speed or velocity of an object falling from rest is consistent with the equation that Galileo deduced with his inclined planes:

$$\text{Velocity acquired} = \text{acceleration} \times \text{time}$$

The instantaneous velocity v of an object falling from rest* after a time t can be expressed in shorthand notation as

$$v = gt$$

The letter v symbolizes both speed and velocity. To see that this equation makes good sense, take a moment to check it with Table 1.2. Note that the instantaneous speed in meters per second is simply the acceleration $g = 10$ m/s^2 multiplied by the time t in seconds.

QUESTION

What would the speedometer reading on the falling rock shown in Figure 1.11 be 3.5 s after it drops from rest? How about 6 s after it is dropped? 100 s?

NSWER

The speedometer readings would be 35 m/s, 60 m/s, and 1000 m/s, respectively. You can reason this from Table 1.2 or use the equation $v = gt$, where g is replaced by 10 m/s^2.

---

*If instead of being dropped from rest the object is thrown downward at speed v_o, the speed v after any elapsed time t is $v = v_o + gt$. We will not be concerned with this added complication here, and will instead learn as much as we can from the most simple cases. That will be a lot!

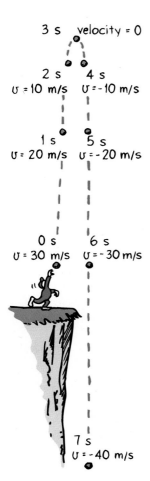

3 s   velocity = 0

2 s   4 s
$v = 10$ m/s $v = -10$ m/s

1 s   5 s
$v = 20$ m/s $v = -20$ m/s

0 s   6 s
$v = 30$ m/s $v = -30$ m/s

7 s
$v = -40$ m/s

Figure 1.12 The rate at which the velocity changes each second is the same.

TABLE 1.3 Distance Fallen in Free Fall

Time of Fall (s)	Distance of Fall (m)
0	0
1	5
2	20
3	45
4	80
5	125
.	.
.	.
.	.
t	$\frac{1}{2}10t^2$

So far we have been considering objects moving straight downward in the direction of gravity. How about an object thrown straight upward? Once released, it continues to move upward for a while and then comes back down. At the highest point, when it is changing its direction of motion from upward to downward, its instantaneous speed is zero. Then it starts downward just as if it had been dropped from rest at that height.

During the upward part of this motion, the object slows from its initial upward velocity to zero velocity. Its speed decreases—evidence of acceleration (or deceleration in the upward direction). How much does its speed decrease each second? It should come as no surprise that it decreases at the rate of 10 meters per second each second—the same acceleration it experiences on the way down. So interestingly enough, as Figure 1.12 shows, the instantaneous speed at points of equal elevation in the path is the same whether the object is moving upward or downward. The velocities are opposite, of course, because they are in opposite directions. During each second, the speed or the velocity changes by 10 meters per second. The acceleration is 10 meters per second squared the whole time, whether the object is moving upward or downward.

How Far

How *far* an object falls is altogether different from how *fast* it falls. With his inclined planes Galileo found that the distance a uniformly accelerating object travels is proportional to the *square of the time.* Your instructor may supply the details of this relationship. We will state here only the results. The distance traveled by a uniformly accelerating object starting from rest is

$$\text{Distance traveled} = \tfrac{1}{2}(\text{acceleration} \times \text{time} \times \text{time})$$

This relationship applies to the distance something falls. We can express it for the case of a freely falling object that starts from rest in shorthand notation as*

$$d = \tfrac{1}{2}gt^2$$

where d is the distance something falls when the time of fall in seconds is substituted for t and squared. If we use 10 m/s² for the value of g, the distance fallen for various times will be as shown in Table 1.3.

Note that an object falls a distance of only 5 meters during the first second of fall, although its speed at 1 second is 10 meters per second. This may be confusing, for we may think that the object should fall a distance of 10 meters. But for it to fall 10 meters in its first second of fall, it would have to fall at an *average* speed of 10 meters per second for the entire second. It starts its fall at 0 meters per second, and its speed is 10 meters per second only in the last instant of the 1-second interval.

_____

* $d = $ average velocity $\times$ time

$$d = \frac{\text{initial velocity} + \text{final velocity}}{2} \times \text{time}$$

$$d = \frac{0 + gt}{2} \times t = \frac{1}{2}gt^2$$

Hang Time

Figure 1.13 Michael Jordan can attain a hang time of 0.9 second.

Some ballet dancers, track and field stars, and basketball players have great jumping ability, and seem to "hang in the air" in defiance of gravity. Ask your friends to estimate the "hang time" of the great jumpers—the amount of time a jumper is airborne, with feet off the ground. Two or three seconds? Several seconds? Nope. Surprisingly, the hang time of the greatest jumpers is almost always less than 1 second. The seemingly longer time is one of many illusions we have about nature.

What vertical height do the greatest jumpers attain? Ask your friends this question and you'll likely win bets, for these high jumps are illusory. Michael Jordan's great leaps raise his center of gravity no more than 1 meter off the floor. We distinguish between how high one can reach and how high one can jump. Only very tall basketball players with long reaches can touch the top of the backboard. And we distinguish between the height of a horizontal bar and the vertical displacement of the jumper's center of gravity when clearing the bar. Jumpers often contort their bodies to clear bars that their centers of gravity pass beneath.

Jumping ability is best measured by a standing vertical jump. Stand facing a wall, and with feet flat on the floor and arms extended upward, make a mark on the wall at the top of your reach. Then make your jump and at the peak make another mark. The distance between these two marks measures your vertical leap. If it's more than 0.6 meters (2 feet), you're exceptional. Very few athletes can attain more than $2\frac{1}{2}$ feet.

Now let's look at the physics. When you leap upward, jumping force is applied only so long as your feet are still in contact with the ground. The greater the force, the greater your launch speed and the higher the jump. It is important to note that as soon as your feet leave the ground, whatever upward speed you attain immediately decreases at the steady rate of g—10 m/s^2. When your upward speed reaches zero, you're at maximum height. Then you begin to fall, gaining speed at exactly the same rate, g. Time rising equals time falling; hang time is the sum of time up and time down. While airborne, no amount of leg or arm pumping or other bodily motions can change your hang time.

The relationship between time up or down and vertical height is given by

$$d = \frac{1}{2}gt^2.$$

We can rearrange this for time in terms of vertical height

$$t = \sqrt{\frac{2d}{g}}.$$

For d, let's use 1.0 meter, Michael Jordan territory, and use the more precise 9.8 m/s^2 for g. Solving for t, half the hang time, we get

$$t = \sqrt{\frac{2d}{g}} = \sqrt{\frac{2[1.0]}{9.8}} = 0.45 \text{ second.}$$

Double this and we see that Michael's hang time is 0.9 second. Only rarely does anybody stay in the air for 1 second.

We've been talking about only vertical motion. How about running jumps? We'll learn in Chapter 4 that hang time depends only on the jumper's vertical speed at launch, and is independent of horizontal speed. While airborne the jumper moves horizontally at a constant speed while only vertical speed undergoes acceleration. Interesting physics!

Figure 1.14 Pretend that a falling rock is equipped with an odometer. The readings of distance fallen increase with time by $\frac{1}{2}gt^2$ and are shown in Table 1.3.

Its average speed during this interval is the average of its initial and final speeds, 0 and 10 meters per second. To find the average value of these or any two numbers, we simply add the two numbers and divide by 2. This equals 5 meters per second, which over a time interval of 1 second gives a distance of 5 meters. As the object continues to fall in succeeding seconds, it will fall through ever-increasing distances as its speed continuously increases.

QUESTIONS

A cat steps off a ledge and drops to the ground in $\frac{1}{2}$ second.

(a) What is its speed on striking the ground?
(b) What is its average speed during the $\frac{1}{2}$ second?
(c) How high is the ledge from the ground?

A common observation is that all objects do not fall with equal accelerations. A dry leaf, a feather, or a sheet of paper may flutter to the ground, while a stone or a coin fall rapidly. The fact that air resistance is responsible for those different accelerations can be shown very nicely with a closed glass tube containing light and heavy objects—a feather and a coin, for example. In the presence of air, the feather and coin fall with quite different accelerations. But if the air in the tube is removed by a vacuum pump and the tube is quickly inverted, the feather and coin fall with the same acceleration (Figure 1.15). Although air resistance appreciably alters the motion of things like falling feathers, the motion of heavier objects like stones and baseballs at ordinary low speeds is not appreciably affected by the air. The relationships $v = gt$ and $d = \frac{1}{2}gt^2$ can be used to a very good approximation for most objects falling from rest in air.

Please remember that it took people nearly 2000 years from the time of Aristotle to reach a clear understanding of motion, so be patient with yourself if you find that you require a few hours to achieve as much!

Figure 1.15 A feather and a coin fall at equal accelerations in a vacuum.

ANSWERS

If we round g off to 10 m/s², we find

(a) Speed: $v = gt = 10 \text{ m/s}^2 \times \frac{1}{2} \text{ s} = 5 \text{ m/s}$

(b) Average speed: $\bar{v} = \dfrac{\text{initial } v + \text{ final } v}{2} = \dfrac{0 \text{ m/s} + 5 \text{ m/s}}{2} = 2.5 \text{ m/s}$

We put a bar over the symbol to denote average speed—$\bar{v}$.

(c) Distance: $d = \bar{v}t = 2.5 \text{ m/s} \times \frac{1}{2} \text{ s} = 1.25 \text{ m}$
Or equivalently,

$$d = \tfrac{1}{2}gt^2 = \tfrac{1}{2} \times 10 \text{ m/s}^2 \times (\tfrac{1}{2} \text{ s})^2 = \tfrac{1}{2} \times 10 \text{ m/s}^2 \times \tfrac{1}{4} \text{ s}^2 = 1.25 \text{ m}$$

Notice that we can find the distance by either of these equivalent relationships. Doing it both ways is a good check!

Figure 1.16   Motion analysis.

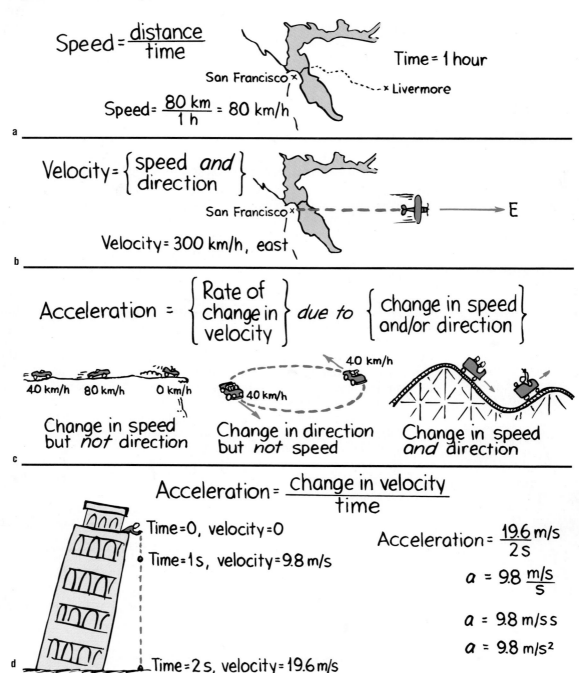

SUMMARY OF TERMS

Inertia The sluggishness or apparent resistance an object offers to changes in its state of motion.

Speed The distance traveled per time.

Velocity The speed of an object and specification of its direction of motion.

Acceleration The rate at which velocity changes with time; the change in velocity may be in magnitude or direction or both.

Free fall A state of fall free from air resistance and other forces except for gravity.

REVIEW QUESTIONS

*Each chapter in this book concludes with a set of review questions and exercises, and for some chapters, problems. The **Review Questions** are designed to help you fix ideas and catch the essentials of the chapter material. You'll notice that answers to the questions can be found within the chapters. The **Exercises** stress thinking rather than mere recall of information and call for an understanding of the definitions, principles, and relationships of the chapter material. In many cases the intention of particular exercises is to help you to apply the ideas of physics to familiar situations. Unless you cover only a few chapters in your course, you will likely be expected to tackle only a few exercises for each chapter. The large number of exercises is to allow your instructor a wide choice of assignments. **Problems** are found only for those chapters that feature concepts more clearly understood with numerical values and straightforward calculations. The problems are relatively few in number so as to avoid an undue emphasis on problem solving that could obscure the primary goal of* Conceptual Physical Science *—to develop a gut feel for the concepts of science in your everyday language. Calculations are to enhance the learning of concepts, and not the other way around.*

Speed

1. What two units of measurement are necessary for describing speed?

2. Distinguish speed in general from instantaneous speed. Give an example.

3. What is the average speed of a horse that gallops a distance of 15 km in a time of 30 min?

4. How far does a horse travel if it gallops at an average speed of 25 km/h for 30 min?

Velocity

5. Distinguish between speed and velocity.

6. If a car moves with a constant velocity, does it also move with a constant speed? Explain.

7. If a car moves with a constant speed, can you say that it also moves with a constant velocity? Give an example to support your answer.

Velocity Vectors

8. What two quantities are necessary for a vector quantity?

9. Why do we say velocity is a vector and speed is not?

Acceleration

10. Distinguish between velocity and acceleration.

11. What is the acceleration of a car that increases its velocity from 0 to 100 km/h in 10 s?

12. What is the acceleration of a car that maintains a constant velocity of 100 km/h for 10 s? (Why do many students who correctly answer the last question get this question wrong?)

13. When does one feel the effects of velocity in a moving vehicle when it is moving uniformly or when its motion changes?

14. Acceleration is generally defined as the time rate of change of velocity. When can it be defined as the time rate of change of speed?

Free Fall

How Fast

15. What exactly is meant by a "freely falling" object?

16. What is the velocity acquired by a freely falling object 5 s after being dropped from a rest position? What is it 6 s after?

17. The acceleration of free fall is about 10 m/s^2. Why does the seconds unit appear twice?

How Far

18. What relationship between distance traveled and time of travel did Galileo discover with his inclined planes?

19. What is the distance fallen for a "freely falling" object 5 s after being dropped from a rest position? What is it 6 s after?

20. Consider these measurements: 10 m, 10 m/s, and 10 m/s^2. Which is a measure of distance, which of speed, and which of acceleration?

EXERCISES

1. One airplane travels due north at 300 km/h while another travels due south at 300 km/h. Are their speeds the same? Are their velocities the same? Explain.

2. Charlie drove his car around the block at constant velocity. True or false?

3. Cite an example of something that undergoes acceleration while moving at constant speed. Can you also give an example of something that accelerates while traveling at constant velocity? Explain.

4. Can you give an example wherein the acceleration of a body is opposite in direction to its velocity? Do so if you can.

5. If you were standing in an enclosed car moving at constant velocity, would you have to lean in some special way to compensate for the car's motion? What if the car were moving with unchanging acceleration? Explain.

6. On which of these hills does the ball roll down with increasing speed and decreasing acceleration? (Use this example if you wish to explain to someone the difference between speed and acceleration.)

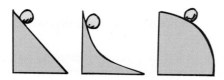

7. Why does vertically-falling rain make slanted streaks on the side windows of a moving automobile? If the streaks make an angle of 45°, what does this tell you about the relative speed of the car and the falling rain?

8. What is the ground speed of an airplane that has an airspeed of 120 km/h when it is in a 90-km/h crosswind?

9. What is the acceleration of a car that moves at a steady velocity of 100 km/h for 100 seconds? Explain your answer.

10. For a freely falling object dropped from rest, what is its acceleration at the end of the 5th second of fall? The 10th second? Defend your answer.

11. Suppose that a freely falling object were somehow equipped with a speedometer. By how much would its speed reading increase with each second of fall?

12. Suppose that the freely falling object in the preceding exercise were also equipped with an odometer. Would the readings of distance fallen each second indicate equal or different falling distances for successive seconds? Explain.

13. When a ball player throws a ball straight up, by how much does the speed of the ball decrease each second while ascending? In the absence of air resistance, by how much does it increase each second while descending? How much time is required for rising compared to falling?

14. Someone standing at the edge of a cliff (as in Figure 1.12) throws a ball straight up at a certain speed and another ball straight down with the same initial speed. If air resistance is negligible, which ball will have the greater speed when it strikes the ground below?

15. If you drop an object, its acceleration toward the ground is 9.8 m/s². If you throw it down instead, would its acceleration after throwing be greater than 9.8 m/s²? Why or why not?

16. In the preceding exercise can you think of a reason why the acceleration of the object thrown downward through the air would actually be less than 9.8 m/s²?

17. If it were not for air resistance, why would it be dangerous to go outdoors on rainy days?

18. Extend Tables 1.2 and 1.3 (which give values of from 0 to 5 s) to 0 to 10 s, assuming no air resistance.

PROBLEMS

1. What is the average speed of a jogger who jogs 2 km in 10 min?

2. What is the acceleration of a vehicle that changes its velocity from 100 km/h to a dead stop in 10 s?

3. A ball is thrown straight up with an initial speed of 30 m/s. How high does it go, and how long is it in the air (neglecting air resistance)?

4. A ball is thrown with enough speed straight up so that it is in the air several seconds. **(a)** What is the velocity of the ball when it gets to its highest point? **(b)** What is its velocity 1 s before it reaches its highest point? **(c)** What is the change in its velocity during this 1-s interval? **(d)** What is its velocity 1 s after it reaches its highest point? **(e)** What is the change in velocity during this 1-s interval? **(f)** What is the change in velocity during the 2-s interval? **(g)** What is the acceleration of the ball during any of these time intervals and when it passes through the zero velocity point?

5. What is the instantaneous velocity of a freely-falling object 10 s after it is released from a position of rest? What is its average velocity during this 10-s interval? How far will it go during this time?

6. A climber near the summit of a vertical dome accidentally knocks loose a large rock. She sees it shatter at the bottom of the cliff 8 s later. What was the speed of impact? How far did the rock fall?

7. A car goes from $v = 0$ to $v = 50$ m/s in 10 s. If you wish to find the distance traveled using the equation $d = \frac{1}{2}at^2$, what value should you use for a?

8. Consider a planet where the acceleration due to gravity is 20 m/s². How fast and how far will an object at rest freely fall in 1 s? Compared to earth, how much greater and farther are the speed and distance fallen on this planet at given times?

9. Surprisingly, few athletes can jump more than 2 feet (0.6 m) high. Use $d = \frac{1}{2}gt^2$ and solve for the time one spends moving upward (or downward) in a 2-foot vertical jump. Then double it for the "hang time"—the time one's feet are off the ground.

10. The Guinness Book of Records reports that basketball player Darrell Griffith in 1976 made a standing vertical jump of 4 feet (1.2 m). Calculate Darrell's hang time.

2

NEWTON'S LAWS
OF MOTION

The Greek philosopher Aristotle believed that natural laws could be understood by logical reasoning. One of his assertions was that heavy objects necessarily fall faster than lighter objects. Another was that moving objects must necessarily have forces exerted on them to keep them moving. These ideas were completely turned around two thousand years later by Galileo, who stated that experiment was superior to logic in uncovering natural laws. The idea that heavy things fall faster than lighter things was demolished by Galileo in his famous Leaning Tower of Pisa experiment, where he supposedly dropped objects of different weights and showed that except for the effects of air resistance they fell to the ground together. In his inclined plane experiments he showed that moving things, once moving, continued in motion *without* the application of forces. He called the property of objects to behave this way *inertia*. The stage was set for Isaac Newton, who was born shortly after Galileo died in 1642. By the time Newton was 23, he had developed his famous three laws of motion that completed the overthrow of Aristotelian ideas. These three important laws first

Figure 2.1 Galileo's famous demonstration.

appeared in one of the most important books of all time, Newton's *Philosophiae Naturalis Principia Mathematica.** The first law is a restatement of Galileo's concept of inertia; the second law relates acceleration to the cause that produces it, force; and the third is the law of action and reaction.

2.1 NEWTON'S FIRST LAW OF MOTION

The tendency of things to resist changes in motion was what Galileo called *inertia*. Newton refined Galileo's idea and made it his first law, appropriately called the **law of inertia**. From Newton's *Principia*:

> **LAW 1:   Every material object continues in its state of rest, or of uniform motion in a straight line, unless it is compelled to change that state by forces impressed upon it.**

The key word in this law is *continues:* An object *continues* to do whatever it happens to be doing unless a force is exerted upon it. If it is at rest, it *continues* in a state of rest. If it is moving, it *continues* to move without turning or changing its speed. In short, the law says that an object does not accelerate of itself; acceleration must be imposed against the tendency of an object to retain its state of motion. Things at rest tend to stay at rest—things moving tend to continue moving. **Inertia is the tendency of things to resist changes in motion.**

Mass

Every material object possesses inertia; how much depends on its amount of matter— the more matter, the more inertia. In speaking of how much matter something has, we use the term *mass*. The greater the mass of an object, the greater the amount of matter and the greater its inertia. **Mass is a measure of the inertia of a material object.**

**The Mathematical Principles of Natural Philosophy*

Loosely speaking, mass corresponds to our intuitive notion of **weight**. We say something has a lot of matter if it is heavy. That's because we are accustomed to measuring matter by gravitational attraction to the earth. But mass is more fundamental than weight; it is a fundamental quantity that completely escapes the notice of most people. There are times, however, when weight corresponds to our unconscious notion of inertia. For example, if you are trying to determine which of two small objects is the heavier, you might shake them back and forth in your hands or move them in some way instead of lifting them. In doing so, you are judging which of the two is more difficult to accelerate, seeing which is the more resistant to a *change* in motion. You are really making a comparison of the inertias of the objects.

It is easy to confuse the ideas of mass and weight, mainly because they are directly proportional to each other. If the mass of an object is doubled, its weight is also doubled; if the mass is halved, its weight is halved. But there is a distinction between the two.

We define each as follows:

Mass: The quantity of matter in a material object. More specifically, it is the measurement of the inertia or sluggishness that an object exhibits in response to any effort made to start it, stop it, or change in any way its state of motion.

Weight: The gravitational force exerted on an object by the nearest most-massive body (locally, by the earth).

In the United States, the quantity of matter in an object has commonly been described by the gravitational pull between it and the earth, or its weight. This has usually been expressed in *pounds*. In most of the world, however, the measure of matter is commonly expressed in a mass unit, the **kilogram**. At the surface of the earth, a 1-kilogram mass brick weighs 2.2 pounds. In the metric system of units, the unit of force is the **newton**, which is equal to a little less than a quarter pound (like the weight of a quarter-pounder hamburger *after* it is cooked). A 1-kilogram brick weighs 9.8 newtons (9.8 N).* Away from the earth's surface, where the influence of gravity is less, the same brick weighs less. It would also weigh less on the surface of planets with less gravity than the earth. On the moon's surface, for example, where the gravitational force on things is only $\frac{1}{6}$ as strong as on earth, 1 kilogram weighs about 1.6 newtons (or 0.37 pounds). On more massive planets it would weigh more. But the mass of the brick is the same everywhere. The brick offers the same resistance to speeding up or slowing down, regardless of whether the earth, moon, or anything at all is attracting it. In a space capsule located between the earth and the moon, where gravitational forces cancel each other, a brick still has mass. If placed on a scale, it wouldn't weigh anything, but its resistence to a change in motion is the same as on earth. Just as much force would have to be exerted by an astronaut in the space capsule to shake the brick back and forth as would be required to shake it back

---

*So 2.2 lb equal 9.8 N, or 1 N is approximately equal to 0.2 lb—about the weight of an apple. In the metric system it is customary to specify quantities of matter in units of mass (in grams or kilograms) and rarely in units of weight (in newtons). In the United States, however, quantities of matter have customarily been specified in units of weight (in pounds), and the unit of mass, the *slug,* is not well known. See Appendix I for more about systems of measurement.

Why will the coin drop into the glass when a force accelerates the card?

Why does the downward motion and sudden stop of the hammer tighten the hammerhead?

Why is it that a slow continuous increase in the downward force breaks the string above the massive ball, but a sudden increase breaks the lower string?

Figure 2.2   Examples of inertia.

and forth while on earth. It would take the same push to start a Cadillac limousine moving on a hard level surface on the moon as on earth. The difficulty of lifting it against gravity (weight) is something else. Mass and weight are different from each other (Figures 2.3 and 2.4).

It is also easy to confuse mass and **volume**. When we think of a massive object, we often think of a big object. An object's size, however, is not necessarily a good way to judge its mass. Which is easier to get moving: a car battery or an empty cardboard box of the same size? So, we find that mass is neither weight nor volume.

Although Galileo introduced the idea of inertia, Newton grasped its significance. The law of inertia defines natural motion and tells us what kinds of motion are the result of applied forces. Whereas Aristotle maintained that the forward motion of an arrow through the air required a force, Newton's law of inertia instead tells us that the behavior of the arrow is natural; constant speed along a straight line (or, simply, constant velocity) requires no force. Aristotle and his followers held that the circular motions of heavenly bodies were natural and moving without applied forces. The law of inertia, however, clearly states that in the absence of forces of some kind the planets would not move in the divine circles of ancient and medieval astronomy

Figure 2.3   An anvil in outer space, between the earth and moon, for example, may be weightless, but it is not massless.

Figure 2.4   The astronaut finds it just as difficult in space as it would be on earth to shake the "weightless" anvil. If the anvil is more massive than the astronaut, which shakes more—the anvil or the astronaut?

Isaac Newton (1642–1727)

Isaac Newton was born prematurely and barely survived on Christmas Day, 1642, the same year that Galileo died. Newton's birthplace was his mother's farmhouse in Woolsthorpe, England. His father died several months before his birth, and he grew up under the care of his mother and grandmother. As a child he showed no particular signs of brightness, and at the age of $14\frac{1}{2}$ he was taken out of school to work on his mother's farm. As a farmer he was a failure. He preferred to read books he borrowed from a neighboring druggist. An uncle sensed the scholarly potential in young Isaac and prompted him to study at the University of Cambridge, which he did for 5 years, graduating without particular distinction.

A plague swept through London, and Newton retreated to his mother's farm—this time to continue his studies. At the farm, at the age of 23, he laid the foundations for the work that was to make him immortal. Story has it that seeing an apple fall to the ground led him to consider the force of gravity extending to the moon and beyond, and he formulated the law of universal gravitation (which he later proved); he invented the calculus, an indispensable mathematical tool in science; he extended Galileo's work and formulated the three fundamental laws of motion; and he formulated a theory of the nature of light and showed with prisms that white light is composed of all colors of the rainbow. It was his experiments with prisms that first made him famous.

When the plague subsided, Newton returned to Cambridge and soon established a reputation for himself as a first-rate mathematician. His mathematics teacher resigned in his favor and Newton was appointed the Lucasian professor of mathematics, a post he held for 28 years. In 1672 he was elected to the Royal Society, where he exhibited the world's first reflector telescope.

It wasn't until Newton was 42 that he began to write his now famous *Principia Mathematica Philosophiae Naturalis*. He wrote it in Latin and completed it in 18 months. It appeared in print in 1687 and wasn't printed in English until 1729, 2 years after his death. When asked how he was able to make so many discoveries, Newton replied that he found his solutions to problems were not by sudden insight but by continually thinking very long and hard about them until he worked them out. He also said that he had stood on the shoulders of giants, acknowledging others like Galileo.

At the age of 46, his energies turned somewhat from science when he was elected a member of Parliament. He attended the sessions in Parliament for 2 years and never gave a speech. One day he rose and the House fell silent to hear the great man. Newton's "speech" was very brief; he simply requested that a window be closed because of a draft.

A further turn from his work in science was his appointment as warden and then as master of the mint, to the dismay of counterfeiters who flourished at that time. He maintained his membership in the Royal Society and was elected president, and re-elected each year for the rest of his life. At the age of 62, he wrote *Opticks,* which summarized his work on light. Nine years later he wrote a second edition to his *Principia.*

Although Newton's hair turned gray at 30, it remained full, long, and wavy all his life, and unlike others in his time he did not wear a wig. He was a modest man, very sensitive to criticism, and never married. He remained healthy in body and intellect into old age. At 80, he still had all his teeth, his eyesight and hearing were sharp, and his mind was alert. In his lifetime he was regarded by his countrymen as the greatest scientist who ever lived. In 1705 he was knighted by Queen Anne. Newton died at the age of 85 and was buried in Westminster Abbey along with England's kings and heroes.

but would move instead in straight-line paths off into space. Newton maintained that the curved motion of the planets was evidence of some kind of force. We shall see in the next chapter that his search for this force led to the law of gravity.

QUESTIONS

1. A hockey puck sliding across the ice finally comes to rest. How would Aristotle interpret this behavior? How would Galileo and Newton interpret it? How would you interpret it?

2. Does a 2-kg iron brick have twice as much *inertia* as a 1-kg iron brick? Twice as much *mass*? Twice as much *volume*? Twice as much *weight*?

3. Would it be easier to lift a Cadillac limousine on the earth or to lift it on the moon?

2.2 NEWTON'S SECOND LAW OF MOTION

Figure 2.5 The greater the mass, the greater the force must be for a given acceleration.

Every day we see things that do not continue in a constant state of motion: Objects initially at rest later may move; moving objects may follow paths that are not straight lines; things in motion may stop. Most of the motion we observe undergoes changes and is the result of one or more applied forces. The overall net force, whether it be from a single source or a combination of sources, produces acceleration. The relationship of acceleration to force and inertia is given in Newton's second law.

LAW 2:   The acceleration of an object is directly proportional to the net force acting on the object, is in the direction of the net force, and is inversely proportional to the mass of the object.

In summarized form, this is

$$\text{Acceleration} \sim \frac{\text{net force}}{\text{mass}}$$

ANSWERS

1. Aristotle would probably say that the puck slides to a stop because it seeks its proper and natural state, one of rest. Galileo and Newton would probably say that once in motion the puck would continue in motion, and that what prevents continued motion is not its nature or its proper rest state, but the friction between the puck and the ice. This friction is small compared to the friction between the puck and a wooden floor, which is why the puck slides so much farther on ice. Only you can answer the last question.

2. The answers to all parts are yes. A 2-kg iron brick has twice as many iron atoms and therefore twice the amount of matter and mass. In the same location, it also has twice the weight. And since both bricks have the same density (the same mass/volume), the 2-kg brick also has twice the volume.

3. A Cadillac limousine would be easier to lift on the moon because gravity is less on the moon. When you *lift* an object, you are contending with the force of gravity (its weight) in addition to its inertia (its mass). Although its mass is the same on the earth, the moon, or anywhere, its weight is only $\frac{1}{6}$ as much on the moon, so only $\frac{1}{6}$ as much effort is required to lift it there. To move it horizontally, however, you are not pushing against gravity. When mass is the only factor, equal forces will produce equal accelerations whether the object is on the earth or the moon.

Force of hand
accelerates
the brick

Twice as much force
produces twice as
much acceleration

Twice the force on
twice the mass gives
the same acceleration

Figure 2.6 Acceleration is directly proportional to force.

Force of hand
accelerates
the brick

The same force
accelerates 2 bricks
½ as much

3 bricks, ⅓ as
much acceleration

Figure 2.7 Acceleration is inversely proportional to mass.

In symbol notation, this is simply

$$a \sim \frac{F_{\text{net}}}{m}$$

We shall use the wiggly line $\sim$ as a symbol meaning "is proportional to." We say that acceleration a is directly proportional to the overall net force F and inversely proportional to the mass m. By this we mean that if F increases, a increases; but if m increases, a decreases. With appropriate units of F, m, and a, the proportionality may be expressed as an exact equation:

$$a = \frac{F_{\text{net}}}{m}$$

An object is accelerated in the direction of the force acting on it. Applied in the direction of the object's motion, a force will increase the object's speed. Applied in the opposite direction, it will decrease the speed of the object. Applied at right angles, it will deflect the object. Any other direction of application will result in a combination of speed change and deflection. *The acceleration of an object is always in the direction of the net force.*

A **force,** in the simplest sense, is a push or a pull. Its source may be gravitational, electrical, magnetic, or simply muscular effort. In the second law, Newton gives a more precise idea of force by relating it to the acceleration it produces. He says in effect that *force is anything that can accelerate an object.* Furthermore, he says that the larger the force, the more acceleration it produces. For a given object, twice the force results in twice the acceleration; three times the force, three times the acceleration; and so forth. Acceleration is directly proportional to net force (Figure 2.6).

We say *net* force because oftentimes more than a single force acts on an object. Let's examine more carefully what we mean by *net.* If you and a friend pull in the same direction with equal forces on an object, the forces add to produce a net force twice as great as your single force. The combination of forces produces twice the acceleration than if you pulled alone. If, however, you each pull with equal forces but in opposite directions, the object will not accelerate. Because they are oppositely directed, the forces on the object cancel each other. One of the forces can be considered to be the negative of the other, and they add algebraically to zero. The net force is zero.

Suppose you pull on an object with a force of 20 newtons and your friend pulls in the opposite direction with a force of 15 newtons. Then the force and resulting acceleration is the same as if you pulled alone with a force of 5 newtons. The resulting 5 newtons is the net force. If the forces are in the same direction they are added; if in opposite directions they are subtracted (Figure 2.8). It is the net force that accelerates the object.

We find that force involves both magnitude (the amount) and direction. Force is a *vector quantity.* When two or more forces are exerted on an object, the forces combine by vector rules. When the forces are parallel, in equal or opposite directions as just discussed, they simply add algebraically. But when two or more forces are exerted at angles to one another so they are neither in the same nor opposite directions, they combine via the parallelogram rule shown in Chapter 1. To simplify our study of physics, we will not treat forces at angles here.

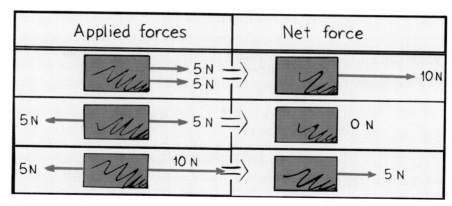

Applied forces	Net force
→ 5 N → 5 N ⇒	→ 10 N
5 N ← → 5 N ⇒	0 N
5 N ← 10 N → ⇒	→ 5 N

Figure 2.8   Net force.

Force and mass have opposite effects on acceleration. The more massive the object, the less its acceleration. For the same force, twice the mass results in half the acceleration; three times the mass, one-third the acceleration. Increasing the mass decreases the acceleration. For example, if we put identical Ford engines in a Cadillac and a Volkswagen, we would expect quite different accelerations even though the driving force in each car is the same. The Cadillac with its greater mass has a greater resistance to a change in velocity than does the Volkswagen. Consequently, the Cadillac requires a more powerful engine to achieve the same acceleration. For the same acceleration, a larger mass requires a correspondingly larger force. We say that acceleration is inversely proportional to mass.*

The acceleration of an object, then, depends on both the net force exerted on the object and the mass of the object.

Q UESTION

In Chapter 1 acceleration was defined to be the time rate of change of velocity; that is, a = (change in v)/time. Are we in this chapter saying that acceleration is instead the ratio of force to mass; that is, $a = F/m$? Which is it?

A NSWER

Acceleration is defined as the time rate of change of velocity and is *produced by* a force—how much force/mass (the cause) results in the rate change in v/time (the effect). So whereas we defined acceleration in Chapter 1, in this chapter we define the terms that cause acceleration.

---

*Mass is operationally defined as the proportionality constant between force and acceleration in Newton's second law, rearranged to read $m = F/a$. A 1-unit mass is that which requires 1 unit of force to produce 1 unit of acceleration. So 1 kg is the amount of matter that 1 N of force will accelerate 1 m/s². In British units, 1 slug is the amount of matter that 1 pound of force will accelerate 1 ft/s². We shall see later that mass is simply a form of concentrated energy.

Figure 2.9 The table pushes up on the book with as much force as the downward force of gravity on the book, so the net force on the book is zero.

When Acceleration Is Zero—Equilibrium

When the acceleration of an object is zero, we say the object is in **mechanical equilibrium**. Whatever forces may act on it are canceled out. The net force on an object in equilibrium is zero. A book lying motionless on a table is in equilibrium because it is not accelerating. And the same book sliding at constant velocity across a smooth surface is also in equilibrium, because it also is not accelerating. In both cases acceleration is zero. Let's consider the case without motion first. We call this *static* equilibrium.

A book lying motionless on a table has zero acceleration, so according to Newton's second law, the net force must also be zero. Weight acts downward, so another force must act upward on the book. The other force is the *support force* of the table (often called the *normal force*).* The table exerts an upward support force on the book that is equal in magnitude to the downward force of gravity on the book—its weight (Figure 2.9).

When you step on a bathroom scale, the downward pull of gravity and the upward support force of the floor compress a spring that is calibrated to give your weight. In effect, the scale shows the support force. Since you are not accelerating, the net force on you is zero, which means the support force and your weight are equal in magnitude. For any object that is not accelerating, either no force is acting on it or there is a combination of forces that cancel to zero. This idea that zero net force acts on things in equilibrium is often quite useful. For example, if you see that a force is exerted on something that doesn't accelerate, then you know there is another force acting—one that is equal in magnitude and opposite in direction. The net force on an object in equilibrium is always zero.

QUESTION

Suppose you stand on two bathroom scales with your weight evenly divided between the two scales. What will each scale read? How about if you stand with more of your weight on one foot than the other?

Consider a hockey puck that slides across the ice at constant velocity. It slides without accelerating because the net force on it is zero. Any object that moves without accelerating is said to be in *dynamic* equilibrium. A bowling ball rolling at constant velocity along a bowling alley is in dynamic equilibrium—until it interacts with the pins. When you push a crate at constant velocity across a factory floor, the crate

Figure 2.10 The sum of the upward pulls by the rings must equal her weight.

ANSWER

The reading on both scales adds up to your weight. This is because the sum of the scale readings, which equals the support force by the floor, must counteract your weight so the net force on you will be zero. If you stand equally on each scale, each will read half your weight. If you lean more on one scale than the other, more than half your weight will be read on that scale but less on the other, so they will still add up to your weight. For example, if one scale reads two-thirds your weight, the other scale will read one-third your weight. Get it?

---

*This force acts at right angles to the surface. When we say "normal to" we are saying "at right angles to," which is why this force is called a *normal force*.

Figure 2.11 The hockey puck has practically zero acceleration after it has been struck and slides across the ice.

is in dynamic equilibrium. In this case the force you apply is balanced by the force of friction between the crate and the floor. The net force is zero; hence the acceleration is zero. It's important to emphasize that zero acceleration does not mean zero velocity. Zero acceleration means that the object will maintain the velocity it happens to have, neither speeding up nor slowing down nor changing direction. Most things that are moved in our environment must be pushed or pulled to overcome friction.

When surfaces slide or tend to slide over one another, a force of **friction** occurs. Friction depends on the kinds of material and how much they are pressed together, and results from the mutual contact of irregularities in the surfaces.* The irregularities act as obstructions to motion. Even surfaces that appear to be very smooth have irregular surfaces when viewed microscopically.

The direction of friction force is always in a direction opposing motion. Thus, if an object is to move at constant velocity, a force equal to the opposing force of friction must be applied so that the two forces exactly cancel each other. The zero net force then results in zero acceleration.

75-N friction force   75-N applied force

Figure 2.12 The crate slides to the right because of an applied force of 75N. A friction force of 75N opposes motion and results in a zero net force on the crate, so the crate slides at constant velocity (zero acceleration).

Q UESTION

A jumbo jet cruises at constant velocity of 1000 km/h when the thrusting force of its engines is a constant 100,000 N. What is the acceleration of the jet? What is the force of air resistance on the jet?

When Acceleration Is g—Free Fall

Although Galileo founded the concepts of both inertia and acceleration, and was the first to measure the acceleration of falling objects, Galileo could not explain why objects of various masses fall with equal accelerations. Newton's second law provides the explanation.

A NSWER

The acceleration is zero because the velocity is constant. Since the acceleration is zero, it follows from Newton's second law that the net force is zero, which in turn means that the force of air resistance must just equal the thrusting force of 100,000 N and act in the opposite direction. So the air resistance on the jet is 100,000 N. (Note that we don't need to know the velocity of the jet to answer this question. We need only to know that it is constant, our clue that acceleration and therefore net force is zero.)

---

*Even though it may not seem so yet, most of the concepts in physics are not really complicated. But friction is different—it is a very complicated phenomenon. The findings are empirical (gained from a wide range of experiments) and the predictions approximate (also based on experiment).

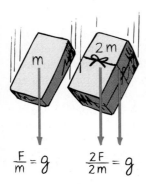

$$\frac{F}{m} = g \qquad \frac{2F}{2m} = g$$

Figure 2.13 The ratio of weight *F* to mass *m* is the same for all objects in the same locality; hence, their accelerations are the same in the absence of air drag.

We know that a falling object accelerates toward the earth because of the gravitational force of attraction between the object and the earth. We call this force the *weight* of the object.* When this is the only force—that is, when friction such as air resistance is negligible—we say that the object is in a state of **free fall**.

The attractive force between a more massive object and the earth is greater than that of a less massive object. The double brick in Figure 2.13, for example, has twice the attraction as the single brick. Why then, as Aristotle supposed, doesn't the double brick fall twice as fast? The answer is that the acceleration of an object depends not only on the force—in this case, the weight—but on the object's resistance to motion, its inertia. Whereas a force produces an acceleration, inertia is a *resistance* to acceleration. So twice the force exerted on twice the inertia produces the same acceleration as half the force exerted on half the inertia. Both accelerate equally. The acceleration due to gravity is symbolized by *g*. We use the symbol *g*, rather than *a*, to denote that acceleration is due to gravity alone.

The ratio of weight to mass for freely falling objects equals a constant—*g*. This is similar to the constant ratio of circumference to diameter for circles, which equals the constant *π*. The ratio of weight to mass is the same for both heavy and light objects, just as the ratio of circumference to diameter is the same for both large and small circles (Figure 2.14).

We now understand that the acceleration of free fall is independent of an object's mass. A boulder 100 times more massive than a pebble falls at the same acceleration as the pebble because although the force on the boulder (its weight) is 100 times greater than the force (or weight) on the pebble, its resistance to a change in motion (mass) is 100 times that of the pebble. The greater force offsets the equally greater mass.

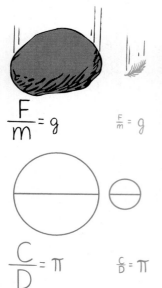

$$\frac{F}{m} = g \qquad \frac{F}{m} = g$$

$$\frac{C}{D} = \pi \qquad \frac{C}{D} = \pi$$

Figure 2.14 The ratio of weight *F* to mass *m* is the same for the large rock and the small feather; similarly the ratio of circumference *C* to diameter *D* is the same for the large and the small circle.

QUESTION

In a vacuum, a coin and a feather fall equally, side by side. Would it be correct to say that in a vacuum equal forces of gravity act on both the coin and the feather?

When Acceleration Is Less Than *g*—Nonfree Fall

Objects falling in a vacuum are one thing, but what of the practical cases of objects falling in air? Although a feather and a coin will fall equally fast in a vacuum, they fall quite differently in the presence of air. How do Newton's laws apply to objects falling in air? The answer is that Newton's laws apply for *all* objects,

ANSWER

No, no, no, a thousand times no! These objects accelerate equally not because the forces of gravity on them are equal (they aren't!), but because the ratios of their weights to masses are equal. Although air resistance is not present in a vacuum, gravity is (you'd know this if you stuck your hand into a vacuum chamber and a Mack truck rolled over it). If you answered yes to this question, let this be an alarm to be more careful when you think physics!

---

*Weight and mass are directly proportional to each other, and the constant of proportionality is *g*. We see that weight = *mg*, so 9.8 N = (1kg)(9.8 m/s^2).

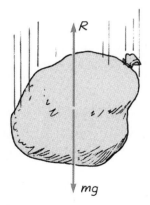

Figure 2.15 When weight *mg* is greater than air resistance *R*, the falling sack accelerates. At higher speeds, *R* increases. When *R* = *mg*, acceleration reaches zero, and the sack reaches its terminal velocity.

whether freely falling or falling in the presence of resistive forces. The accelerations, however, are quite different for the two cases. The important thing to keep in mind is the idea of *net force*. In a vacuum or in cases where air resistance can be neglected, the net force is the weight because it is the only force. In the presence of air resistance, however, the net force is less than the weight—it is the weight minus air drag, the force arising from air resistance.*

The force of air drag experienced by a falling object depends on two things. First, it depends on the frontal area of the falling object—that is, on the amount of air the object must plow through as it falls. Second, it depends on the speed of the falling object; the greater the speed, the greater the number of air molecules an object encounters per second and the greater the force of molecular impact. Air drag depends on the size and the speed of a falling object.

In some cases air drag greatly affects falling; in other cases it doesn't. Air drag is important for a falling feather. Since a feather has so much area compared to its small weight, it doesn't have to fall very fast before the upward-acting air drag cancels the downward-acting weight. The net force on the feather is then zero and acceleration terminates. When acceleration terminates, we say the object has reached its **terminal speed**. If we are concerned with direction, down for falling objects, we say the object has reached its **terminal velocity**. The same idea applies to all objects falling in air. Consider skydiving. As a falling skydiver gains speed, air drag may finally build up until it equals the weight of the skydiver. If and when this happens, the *net* force becomes zero and the skydiver no longer accelerates; she has reached her terminal velocity. For a feather, terminal velocity is a few centimeters per second, whereas for a skydiver it is about 200 kilometers per hour. A skydiver may vary this speed by varying position. Head or feet first is a way of encountering less air and thus less air drag and attaining maximum terminal velocity. A smaller terminal velocity is attained by spreading oneself out like a flying squirrel. Minimum terminal velocity is attained when the parachute is opened.

Consider a man and woman parachuting together from the same altitude (Figure 2.16). Suppose that the man is twice as heavy as the woman and that their same-sized chutes are initially opened. The same size chute means that at equal speeds the air resistance is the same on each. Who gets to the ground first—the heavy man or the lighter woman? The answer is the person who falls fastest gets to the ground first—that is, the person with the greatest terminal speed. At first we might think that because the chutes are the same, the terminal speeds for each would be the same, and therefore both would reach the ground together. This doesn't happen because air drag also depends on speed. The woman will reach her terminal speed when air drag against her chute equals her weight. When this happens, the air drag against

---

*In mathematical notation,

$$a = \frac{F_{\text{net}}}{m} = \frac{mg - R}{m}$$

where *mg* is the weight and *R* is the air resistance. Note that when *R* = *mg*, *a* = 0; then, with no acceleration, the object falls at constant velocity. With elementary algebra we can go another step and get

$$a = \frac{F_{\text{net}}}{m} = \frac{mg - R}{m} = g - \frac{R}{m}$$

We see that the acceleration *a* will always be less than *g* if air resistance *R* impedes falling. Only when *R* = 0 does *a* = *g*.

Figure 2.16 The heavier parachutist must fall faster than the lighter parachutist for air drag to cancel his greater weight.

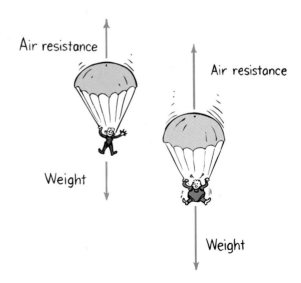

the chute of the man will not yet equal his weight. He must fall faster than she does for air drag to match his greater weight.* Terminal velocity is greater for the heavier person, with the result that the heavier person reaches the ground first.

QUESTION

A skydiver jumps from a high-flying helicopter. As she falls faster and faster through the air, does her acceleration increase, decrease, or remain the same?

Consider a pair of tennis balls, one a hollow regular ball and the other filled with iron pellets. Although they are the same size, the iron-filled ball is considerably heavier than the regular ball. If you hold them above your head and drop them simultaneously you'll see that they strike the ground at the same time. But if you drop them from a greater height, say from the top of a building, you'll note the heavier ball strikes the ground first. Why? In the first case the balls did not gain much speed

Figure 2.17 A stroboscopic study of a golf ball and a styrofoam ball falling in air. The air drag is negligible for the heavier golf ball, and it accelerates essentially at g. Air drag is not negligible for the lighter styrofoam ball, which reaches its terminal velocity sooner.

ANSWER

Acceleration decreases because the net force on her decreases. Net force is equal to her weight minus her air drag, and since air drag increases with increasing speed, net force and hence acceleration decrease. By Newton's second law,

$$a = \frac{F_{net}}{m} = \frac{mg - R}{m}$$

where mg is her weight and R is the air drag she encounters. As R increases, a decreases. Note that if she falls fast enough so that $R = mg$, $a = 0$, then with no acceleration she falls at constant speed.

---

*Terminal speed for the man will be somewhat less than double that of the half-as-heavy woman because the retarding force of air drag is not exactly proportional to the speed.

in their short fall. Air drag encountered was small compared to their weights, even for the regular ball. The tiny difference in their arrival time was not noticed. But when dropped from a greater height, the greater speeds of fall were met with greater air drag. At any given speed each ball encounters the same air drag because each has the same size. This same air drag may be a lot compared to the weight of the lighter ball, but only a little compared to the weight of the heavier ball (like the parachutists in Fig 2.16). For example, 1 N of air drag acting on a 2-N object will cut the acceleration to $g/2$, but 1 N of air drag on a 200-N object will only slightly diminish acceleration. So even with equal air drags the accelerations of each are different. There is a moral to be learned here. Whenever you consider the acceleration of something, use the equation of Newton's second law to guide your thinking: The acceleration is equal to the ratio of *net* force to mass. For the falling tennis balls, the net force on the hollow ball is appreciably reduced as air drag builds up, while comparably the net force on the iron-filled ball is only slightly reduced. Acceleration decreases as net force decreases, which in turn decreases as air drag increases. If and when the air drag builds up to equal the weight of the falling object, then the net force becomes zero and acceleration terminates.

2.3 NEWTON'S THIRD LAW OF MOTION

So far we have treated force in its simplest sense; as a push or pull that produces or tends to produce acceleration. In a broader sense, a force is not a thing in itself but makes up an *interaction* between one thing and another. We pull on a cart and it accelerates. A hammer hits a stake and drives it into the ground. One object interacts with another object. Which exerts the force and which receives the force? Newton's answer to this is that neither force has to be identified as "exerter" or "receiver"; he concluded that both objects must be treated equally. For example, when we pull the cart, the cart in turn pulls on us, as evidenced perhaps by the tightening of the rope wrapped around our hand. This pair of forces, our pull on the cart and the cart's pull on us, make up the single interaction between us and the cart. In the interaction between the hammer and the stake, the hammer exerts a force against the stake, but is itself brought to a halt in the process. Such observations led Newton to his third law—the law of action and reaction.

LAW 3:   Whenever one object exerts a force on a second object, the second object exerts an equal and opposite force on the first.

Newton's third law is often stated thus: "To every action there is always opposed an equal reaction." In any interaction there is an action and reaction pair of forces that are equal in magnitude and opposite in direction. Neither force exists without the other—forces come in *pairs,* one action and the other reaction. The action and reaction pair of forces make up one interaction between two things.

You interact with the floor when you walk on it. Your push against the floor is coupled to the floor's push against you. The pair of forces occur simultaneously. Likewise, the tires of a car push against the road while the road pushes back on the tires—the tires and road push against each other. In swimming you interact with the water that you push backward, while the water pushes you forward—you and the water push against each other. In each case there is a pair of forces, one action and the other reaction, that make up one interaction. The reaction forces are what account for our motion in these cases. These forces depend on friction; a person or

Figure 2.18 In the interaction between the hammer and the stake, each exerts the same amount of force on the other.

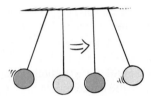

Figure 2.19 The impact force between the dark and light ball moves the light ball and stops the dark ball.

car on ice, for example, may not be able to exert the action force to produce the needed reaction force. Which force we call *action* and which we call *reaction* doesn't matter. The point is that neither exists without the other.

An interesting question often arises; since action and reaction forces are equal and opposite, why don't they cancel to zero? To answer this question we must consider the "system" involved. Consider the force pair between the apple and orange in Figure 2.21. We'll consider the "system" to be the orange. Then we see a force is indeed exerted on the orange and it accelerates. The fact that the orange simultaneously exerts a force on the apple, which is external to the system, may affect the apple but not the orange. In this case the interaction is between the system (the orange) and something external (the apple), so the action and reaction forces don't cancel. However, if we consider the system to be both the orange and apple together, the force pair is internal to the system. In this case the forces do cancel each other. Similarly, the many force pairs between molecules in a golf ball may hold the ball together into a cohesive solid, but they play no role at all in accelerating the ball. A force external to the ball is needed to accelerate the ball. Similarly, a force external to both the apple and the orange is needed to produce acceleration of both (like friction of the floor on the apple's feet). If action and reaction forces are internal to a system, they cancel each other and produce no acceleration of the system. Action and reaction forces do not cancel each other when either is external to the system being considered. If this is confusing, it may be well to note that Newton had difficulties with the third law himself.

Action: tire pushes on road   Reaction: road pushes on tire

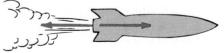

Action: rocket pushes on gas   Reaction: gas pushes on rocket

Action: man pulls on spring   Reaction: spring pulls on man

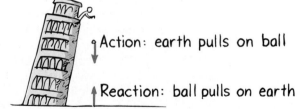

Action: earth pulls on ball

Reaction: ball pulls on earth

Figure 2.20 Action and reaction forces. Note that when action A exerts force on B, the reaction is simply B exerts force on A.

Figure 2.21 When an apple pulls on an orange, the orange accelerates—period! The fact that the orange pulls back on the apple affects the apple—not the orange. The force on the orange is not canceled by the force on the apple.

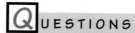

QUESTIONS

1. On a cold rainy day your car battery is dead and you must push the car to get it started. Why can't you push the car by remaining comfortably inside and pushing against the dashboard?

2. Does a speeding missile possess force?

3. We know that the earth pulls on the moon. Does it follow that the moon also pulls on the earth?

4. Can you identify the action and reaction forces in the case of an object falling in a vacuum?

When a rifle is fired there is an interaction between the rifle and the bullet (Figure 2.23). A pair of forces act on both the rifle and the bullet. The force exerted on the bullet is as great as the reaction force exerted on the rifle; hence, the rifle kicks. Since the forces are equal in magnitude, why doesn't the rifle recoil with the same speed as the bullet? In analyzing changes in motion, Newton's second law reminds us that we must also consider the masses involved. Suppose we let F represent both the action and reaction force, m the mass of the bullet, and m the mass of the more

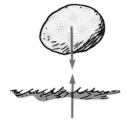

Figure 2.22 The gravitational interaction between the boulder and the earth consists of a pair of equal and opposite forces; one is the earth pulling down on the boulder, the other is the boulder pulling up on the earth.

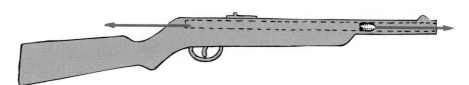

Figure 2.23 The force exerted against the recoiling rifle is just as great as the force that drives the bullet. Why, then, does the bullet accelerate more than the rifle?

ANSWERS

1. In this case the system to be accelerated is the car. If you remain inside and push on the dash, the force pair you produce acts and reacts within the system. These forces cancel out as far as any motion of the car is concerned. To accelerate the car there must be an interaction between the car and something else—you pushing outside against the road, for example.

2. No, a force is not something an object *has*, like mass, but is part of an interaction between one object and another. A speeding missile may possess the capability of exerting a force on another object when interaction occurs, but it does not possess force as a thing in itself. As we will see in the following chapter, a speeding missile possesses momentum and kinetic energy.

3. Yes, both pulls make up an action-reaction pair of forces that make up the gravitational interaction between the earth and the moon. We can say that (1) the earth pulls on the moon, and (2) the moon likewise pulls on the earth; but it is more insightful to think of this as a single interaction—the earth and moon simultaneously pull on each other, with the *same* amount of force.

4. To identify a pair of action-reaction forces in any situation, first identify the pair of interacting objects involved. Something is interacting with something. In this case the whole earth is interacting (gravitationally) with the falling object. So the earth pulls downward on the object (call it action), while the object pulls upward on the earth (reaction).

massive rifle. The accelerations of the bullet and the rifle are then found by taking the ratio of force to mass. The acceleration of the bullet is given by

$$\frac{F}{m} = a$$

while the acceleration of the recoiling rifle is

$$\frac{F}{m} = a$$

We see why the change in motion of the bullet is so huge compared to the change of motion of the rifle. A given force divided by a small mass produces a large acceleration, while the same force divided by a large mass produces a small acceleration. We use different-sized symbols to indicate the differences in relative masses and resulting accelerations.

   If we extend the idea of a rifle recoiling or "kicking" from the bullet it fires, we can understand rocket propulsion. Consider a machine gun recoiling each time a bullet is fired. If the machine gun is fastened so it is free to slide on a vertical wire (Figure 2.24), it accelerates upward as bullets are fired downward. A rocket accelerates the same way as it continually "recoils" from the ejected exhaust gas. Each molecule of exhaust gas is like a tiny bullet shot from the rocket (Figure 2.25).

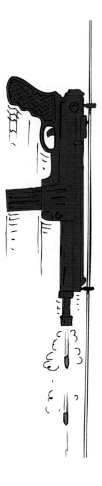

Figure 2.24 The machine gun recoils from the bullets it fires and climbs upward.

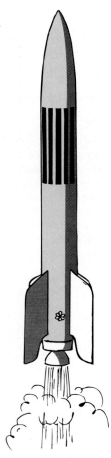

Figure 2.25 The rocket recoils from the "molecular bullets" it fires and climbs upward.

A common misconception is that a rocket is propelled by the impact of exhaust gases against the atmosphere. In fact, fifty years ago before the advent of rockets, it was commonly thought that sending a rocket to the moon was impossible because of the absence of an atmosphere for the rocket to push against. But this is like saying a gun wouldn't kick unless the bullet had air to push against. Not true! Both the rocket and recoiling gun accelerate not because of any pushes on the air, but because of the reaction forces by the "bullets" they fire—air or no air. A rocket works better, in fact, above the atmosphere where there is no air resistance to restrict its speed.

Using Newton's third law, we can understand how a helicopter gets its lifting force. The whirling blades are shaped to force air particles down (action), and the air forces the blades up (reaction). This upward reaction force is called *lift*. When lift equals the weight of the craft, the helicopter hovers in midair. When lift is greater, the helicopter climbs upward.

This is true for birds and airplanes. Birds fly by pushing air downward. The air in turn pushes the bird upward. When the bird is soaring, the wing must be shaped so that moving air particles are deflected downward. Slightly tilted wings that deflect oncoming air downward produce the lift on an airplane. Air must be pushed downward continuously to maintain lift. This supply of air is obtained by the forward motion of the aircraft, which results from jets that push air backward. When the jets push air backward, the air in turn pushes the jets forward. We will learn in Chapter 4 that the curved surface of a wing is an airfoil, which enhances the lifting force.

QUESTIONS

1. A car accelerates along a level road. Strictly speaking, what is the force that moves the car?

2. A high-speed bus and an innocent bug have a head-on collision. The force of the bus on the bug splatters the bug over the windshield. Is the corresponding force that the bug exerts against the windshield greater, less, or the same? Is the resulting deceleration of the bus greater than, less than, or the same as that of the bug?

We see Newton's third law at work everywhere. A fish pushes the water backward with its fins, and the water in turn pushes the fish forward. The wind pushes against the branches of a tree, and the branches in turn push back on the wind and we have whistling sounds. Forces are interactions between different things. Every

ANSWERS

1. The tires of the car interact with the road, so it is the road that pushes the car along. Really! Except for air resistance, only the road provides a horizontal force on the car. How does it do this? The rotating tires push back on the road (action). The road, in turn, pushes forward on the tires (reaction). How about that!

2. The magnitudes of both forces are the same, for they constitute an action-reaction force pair that makes up the interaction between the bus and the bug. The accelerations, however, are very different because the masses involved are different. The bug undergoes an enormous and lethal deceleration, while the bus undergoes a very tiny deceleration—so tiny that the very slight slowing of the bus is unnoticed by its passengers. But if the bug were more massive—as massive as another bus, for example—the slowing down would be quite evident!

Figure 2.26 You cannot touch without being touched—Newton's third law.

contact requires at least a twoness; there is no way that an object can exert a force on nothing. Forces, whether large shoves or slight nudges, always occur in pairs, each of which is opposite to the other. Thus, we cannot touch without being touched.

Summary of Newton's Three Laws

An object at rest tends to remain at rest; an object in motion tends to remain in motion at constant speed along a straight-line path. This tendency of objects to resist change in motion is called *inertia*. Mass is a measure of inertia. Objects will undergo changes in motion only in the presence of a net force.

When a net force is impressed upon an object, it will accelerate. The acceleration is directly proportional to the net force and inversely proportional to the mass. Symbolically, $a \sim \frac{F_{net}}{m}$. Acceleration is always in the direction of the net force. When objects fall in a vacuum, the net force is simply the weight, and the acceleration is g (the symbol g denotes that acceleration is due to gravity alone). When objects fall in air, the net force is equal to the weight minus the force of air drag, and the acceleration is less than g. If and when the force of air drag equals the weight of a falling object, acceleration terminates, and the object falls at constant speed.

Whenever one object exerts a force on a second object, the second object exerts an equal and opposite force on the first. Forces come in *pairs*, one action and the other reaction, both of which constitute the interaction between one thing and the other. Neither force exists without the other.

SUMMARY OF TERMS

Inertia The tendency of things to resist changes in motion.

Mass The quantity of matter in an object. More specifically, it is the measurement of the inertia or sluggishness that an object exhibits in response to any effort made to start it, stop it, or change in any way its state of motion.

Weight The gravitational force exerted on an object by the nearest most-massive body (locally, by the earth).

Kilogram The fundamental SI unit of mass. One kilogram (symbol kg) is this amount of mass in 1 liter (l) of water at 4°C.

Newton The SI unit of force. One newton (symbol N) is the force that will give an object of mass 1 kg an acceleration of 1 m/s^2.

Volume The quantity of space an object occupies.

Force Any influence that can cause an object to be accelerated, measured in newtons (in pounds in the British system).

Mechanical equilibrium The state of an object or system of objects for which any impressed forces cancel to zero and no acceleration occurs.

Friction The resistive forces that arise to oppose the motion or attempted motion of an object past another with which it is in contact.

Free fall Motion under the influence of gravitational pull only.

Terminal speed The speed at which the acceleration of a falling object terminates because friction balances the weight.

• •

REVIEW QUESTIONS

Newton's First Law of Motion

1. Is inertia the *reason* for objects to maintain their states of motion, or the *name* given to this property of matter?

Mass

2. Clearly distinguish among *mass, weight,* and *volume.*

3. Which is fundamental, *mass* or *weight?* (Does mass have weight, or does weight have mass?)

4. Does a 2-kg iron brick have twice as much *inertia* as a 1-kg block of wood? Twice as much *volume?* (Why are your answers different?)

5. What kind of path would the planets follow if suddenly no force acted on them?

Newton's Second Law of Motion

6. If we say that one quantity is *proportional* to another quantity, does this mean they are *equal* to each other? Explain briefly, using mass and weight as an example.

7. A cart is pulled to the left with a force of 100 N, and to the right with a force of 30 N. What is the net force on the cart?

8. Why do we say that force is a vector quantity?

9. If the net force impressed on a sliding block is tripled, by how much does the acceleration increase?

10. If the mass of a sliding block is tripled while a constant net force is applied, by how much does the acceleration decrease?

11. If the mass of a sliding block is tripled at the same time the net force on it is tripled, how does the resulting acceleration compare to the original accleration?

When Acceleration Is Zero—Equilibrium

12. What is the net force on something that is in mechanical equilibrium?

13. Consider a book that weighs 15 N at rest on a flat table. How many newtons of support force does the table provide? What is the net force on the book in this case?

14. Consider a woman weighing 500 N who stands with her weight evenly divided on a pair of bathroom scales. What is the reading on each scale? If she shifts her weight so one of the scales reads 300 N, what will the other scale read?

15. What is the acceleration of an object that moves at constant velocity? What is the net force on the object in this case?

16. Why is it more difficult to slide a crate from a position of rest across the floor than it is to keep it in motion once it is sliding?

17. In what direction does the force of friction act when a body slides?

18. If you push horizontally with a force of 50 N on a crate and make it slide at constant velocity, how much friction acts on the crate? If you increase your force, will the crate accelerate? Explain.

When Acceleration Is g—Free Fall

19. What is meant by *free fall?*

20. What is the net force that acts on a 10-N freely falling object?

21. Why doesn't a heavy object accelerate more than a lighter object when both are freely falling?

When Acceleration Is Less Than g—Nonfree Fall

22. What is the net force that acts on a 10-N falling object when it encounters 4 N of air resistance? 10 N of air resistance?

23. What two principal factors affect the force of air resistance on a falling object?

24. What is the acceleration of a falling object that has reached its terminal velocity?

25. Why does a heavy parachutist fall faster than a lighter parachutist who wears the same size parachute?

Newton's Third Law of Motion

26. First we said a force was a push or pull; now we say it is an interaction. Which is it, a push or pull or an interaction? And what does it mean to say *interaction?*

27. Consider hitting a baseball with a bat. If we call the force on the bat against the ball the *action* force, identify the *reaction* force.

28. When and when not do action and reaction pairs of forces cancel each another?

29. If the forces that act on a bullet and the recoiling gun from which it is fired are equal in magnitude, why do the bullet and gun have very different accelerations?

30. How does a helicopter get its lifting force?

HOME PROJECTS

1. Ask a friend to drive a small nail into a piece of wood placed on top of a pile of books on your head. Why doesn't this hurt you?

2. Drop a sheet of paper and a coin at the same time. Which reaches the ground first? Why? Now crumple the paper into a small, tight wad and again drop it with the coin. Explain the difference observed. Will they fall together if dropped from a second-, third-, or fourth-story window? Try it and explain your observations.

3. Drop a book and a sheet of paper and note that the book has a greater acceleration—g. Place the paper beneath the book and it is forced against the book as both fall, so both fall at g. How do the accelerations compare if you place the paper on top of the raised book and then drop both? You may be surprised, so try it and see. Then explain your observation.

4. Drop two balls of different weight from the same height, and at small speeds they practically fall together. Will they roll together down the same inclined plane? If each is suspended from an equal length of string, making a pair of pendulums, and displaced through the same angle, will they swing back and forth in unison? Try it and see; then explain using Newton's laws.

5. The net force acting on an object and the resulting acceleration are always in the same direction. You can demonstrate this with a spool. If the spool is pulled horizontally to the right, in which direction will it roll?

6. Hold your hand like a flat wing outside the window of a moving automobile. Then slightly tilt the front edge upward and notice the lifting effect. Can you see Newton's laws at work here?

EXERCISES

Please do not be intimidated by the large number of exercises in this and other meaty chapters. If your course work is to cover many chapters, your instructor will likely assign only a few exercises from each.

1. Your empty hand is not hurt when it bangs lightly against a wall. Why is it hurt if it does so while carrying a heavy load? Which of Newton's laws is most applicable here?

2. Why is a massive cleaver more effective for chopping vegetables than an equally sharp knife?

3. Each of the chain of bones forming your spine is separated from its neighbors by disks of elastic tissue. What happens, then, when you jump heavily on your feet from an elevated position? Can you think of a reason why you are a little taller in the morning than in the night? (Hint: Think about the hammerhead in Figure 2.2.)

4. Before the time of Galileo and Newton, it was thought by many learned scholars that a stone dropped from the top of a tall mast of a moving ship would fall vertically and hit the deck behind the mast by a distance equal to how far the ship had moved forward while the stone was falling. In light of your understanding of Newton's laws, what do you think about this?

5. Because the earth rotates once per 24 hours, the west wall in your room moves in a direction toward you at a linear speed that is probably more than 1000 km/h per hour (the exact speed depends on your latitude). When you stand facing the wall you are carried along at the same speed, so you don't notice it. But when you jump upward, with your feet no longer in contact with the floor, why doesn't the high-speed wall slam into you?

6. The chimney of a stationary toy train consists of a vertical spring gun that shoots a steel ball a meter or so straight into the air—so straight that the ball always falls back into the chimney. Suppose the train moves at constant speed along the straight track. Do you think the ball will still return to the chimney if shot from the moving train? How about if the train accelerates along the straight track? How about if it moves at a constant speed on a circular track? Why are your answers different?

7. When a junked car is crushed into a compact cube, does its mass change? Its weight? Explain.

8. Gravitational force on the moon is only $\frac{1}{6}$ that of the gravitational force on the earth. What would be the weight of a 10-kg object on the moon and on the earth? What would its mass be on the moon and on the earth?

9. What is your own mass in kilograms? Your weight in newtons?

10. A rocket becomes progressively easier to accelerate as it travels through space. Why is this so? (Hint: About 90 percent of the mass of a newly launched rocket is fuel.)

11. If an object has no acceleration, can you conclude that no forces are exerted on it? Explain.

12. We say that the acceleration of an object is always in the direction of the net force. Does this mean that the velocity of an object is in the direction of the net force also? Defend your answer.

13. As you stand on a floor, does the floor exert an upward force against your feet? How much force does it exert? Why are you not moved upward by this force?

14. The little girl hangs at rest from the ends of the rope as shown. How does the reading on the scale compare to her weight?

15. Harry the painter swings year after year from his bosun's chair. His weight is 500 N and the rope, unknown to him, has a breaking point of 300 N. Why doesn't the rope break when he is supported as shown to the left below? One day Harry is painting near a flagpole, and, for a change, he ties the free end of the rope to the flagpole instead of to his chair as shown to the right. Why did Harry end up taking his vacation early?

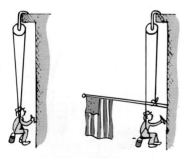

16. Consider the two forces acting on the person who stands still, namely, the downward pull of gravity and the upward support of the floor. Are these forces equal and opposite? Do they form an action-reaction pair? Why or why not?

17. Two 100-N weights are attached to a spring scale as shown. Does the scale read 0, 100, or 200 N, or give some other reading? (Hint: Would it read any differently if one of the ropes were tied to the wall instead of to the hanging 100-N weight?)

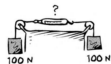

100 N   100 N

18. When your car moves along the highway at constant velocity, the net force on it is zero. Why, then, do you continue running your engine?

19. A "shooting star" is usually a grain of sand from outer space that burns up and gives off light as it enters the atmosphere. What exactly causes this burning?

20. What is the net force on an apple that weighs 1 N when you hold it at rest above your head? What is the net force on it when you release it?

21. Does a stick of dynamite contain force?

22. Can a dog wag its tail without the tail in turn "wagging the dog"? (Consider a dog with a relatively massive tail.)

23. When the athlete holds the barbell overhead, the reaction force is the weight of the barbell on his hand. How does this force vary for the case where the barbell is accelerated upward? Downward?

24. Why can you exert greater force on the pedals of a bicycle if you pull up on the handlebars?

25. If the earth exerts a gravitational force of 1000 N on an orbiting communications satellite, how much force does the satellite exert on the earth?

26. Your weight is the result of a gravitational force of the earth on your body. What is the corresponding reaction force?

27. The strong man will push the two initially stationary freight cars of equal mass apart before he himself drops straight to the ground. Is it possible for him to give either of the cars a greater speed than the other? Why or why not?

28. Suppose two carts, one twice as massive as the other, fly apart when the compressed spring that joins them is released. How fast does the heavier cart roll compared to the lighter cart?

29. If you exert a horizontal force of 200 N to slide a crate across a factory floor at constant velocity, how much friction is exerted by the floor on the crate? Is the force of friction equal and oppositely directed to your 200-N push? Does the force of friction make up the reaction force to your push? Why not?

30. If a Mack truck and motorcycle have a head-on collision, upon which vehicle is the impact force greater? Which vehicle undergoes the greater change in its motion? Explain your answers.

31. Two people of equal mass attempt a tug-of-war with a 12-m rope while standing on frictionless ice. When they pull on the rope, they each slide toward each other. How do their accelerations compare, and how far does each person slide before they meet?

32. Suppose in the preceding exercise that one person has twice the mass of the other. How far does each person slide before they meet?

33. A horse pulls a heavy wagon with a certain force. The wagon, in turn, pulls back with an opposite but equal force on the horse. Doesn't this mean the forces cancel one another, making acceleration impossible? Why or why not? (Hint: Consider the role of the ground.)

34. How does the weight of a falling body compare to the air resistance it encounters just before it reaches terminal velocity? After?

35. You tell your friend that the acceleration of a skydiver decreases as falling progresses. Your friend then asks if this means the skydiver is slowing down? What is your response?

36. Why is it that a tennis ball dropped from the top of a 50-story building will hit the ground no faster than if it were dropped from the 20th story?

37. As an object falls faster and faster through the air, where air resistance is a factor, does its acceleration increase, decrease, or remain constant?

38. If and when Galileo dropped two balls from the top of the Leaning Tower of Pisa, air resistance was not really negligible. Assuming both balls were the same size yet one much heavier than the other, which ball struck the ground first? Why?

39. If you simultaneously drop a pair of tennis balls from the top of a building, they will strike the ground at the same time. If one of the tennis balls is filled with lead pellets, will it fall faster and hit the ground first? Which of the two will encounter more air resistance? Defend your answers.

40. What is the acceleration of a rock at the top of its trajectory when thrown straight upward? (Is your answer consistent with Newton's second law?)

. . . .

PROBLEMS

1. Find the net force produced by a 30-N and 20-N force in each of the following cases:

    (a) Both forces act in the same direction.
    (b) Both forces act in opposite directions.
    (c) Both forces act at right angles to each other.

2. When two horizontal forces are exerted on a cart, 600 N forward and 400 N backward, the cart undergoes acceleration. What additional force is needed to produce nonaccelerated motion?

3. A horizontal force of 100 N is required to push a box across a floor at a constant speed.

    (a) What is the net force acting on the box?
    (b) What is the force of friction acting on the box?

4. If a mass of 1 kg is accelerated 1 m/s² by a force of 1 N, what would be the acceleration of 2 kg acted on by a force of 2 N?

5. How much acceleration does a 747 jumbo jet of mass 30,000 kg experience in takeoff when the thrust for each of four engines is 30,000 N?

6. You push with 20-N horizontal force on a 2-kg mass on a horizontal surface against a horizontal friction force of 12 N. What is the acceleration?

7. What horizontal force must be applied to produce an acceleration of g for a 1-kg puck on a horizontal friction-free air table?

8. What is the weight in newtons of a 100-kg person?

9. Suzie Skydiver with parachute and all has a mass of 50 kg.

    (a) Before opening her chute, what force of air resistance will she encounter when she reaches terminal velocity?
    (b) What force of air resistance will she encounter when she reaches terminal velocity after the chute is open?
    (c) Discuss why your answers are the same or different.

10. Suzie Skydiver with parachute and all has a mass of 50 kg. How much air drag will be acting on her when she accelerates at $g/2$?

3

MOMENTUM
AND ENERGY

Moving objects have a quantity that objects at rest don't have. More than a hundred years ago this quantity was called *impedo*. A boulder at rest had no impedo, while the same boulder rolling down a steep incline had much impedo. The faster it moved, the greater the impedo. The change in impedo depended on force, and importantly, on how long the force acted. Apply a force to a cart and you give it impedo. Apply a long force and you give it more impedo. But what do we mean by "how long"? Does "how long" refer to time, or to distance? When this distinction was made, the term impedo gave way to two new terms—*momentum* and *kinetic energy*. This chapter is about both these important concepts.

3.1 MOMENTUM

We all know that a heavy truck is harder to stop than a small car moving at the same speed. We state this fact by saying that the truck has more momentum than the car. By **momentum** we mean inertia in motion, or, more specifically, the product of the

Figure 3.1 The boulder, unfortunately, has more momentum than the runner.

mass of an object and its velocity; that is,

$$\text{Momentum} = \text{mass} \times \text{velocity}$$

Or, in shorthand notation,

$$\text{Momentum} = mv$$

When direction is not an important factor, we can say momentum $=$ mass $\times$ speed, which we still abbreviate mv.

We can see from the definition that a moving object can have a large momentum if either its mass or its velocity is large, or if both its mass and its velocity are large. The truck has more momentum than a car moving at the same velocity because it has a greater mass. We can see that a huge ship moving at a small velocity can have a large momentum, as can a small bullet moving at a high velocity. A massive truck rolling down a steep hill with no brakes has a large momentum, whereas the same truck at rest has no momentum at all.

3.2 IMPULSE CHANGES MOMENTUM

To change the momentum of something requires force. The size of the force and how long it acts determine how much the momentum is changed. "How long" refers to *time*. Apply a force briefly to a stalled automobile, and you produce a small change in its momentum. Apply the same force over an extended time interval, and a greater change in momentum results. We name the product of force and this time interval, **impulse**. The relationship of impulse to momentum is a rearrangement of Newton's second law ($a = F/m$). The time interval of impulse is "buried" in the term for acceleration (change in velocity/time interval). Rearrangement of Newton's second law gives*

$$\text{Force} \times \text{ time interval} = \text{change in (mass} \times \text{velocity)}$$

We can express all terms in this relationship in shorthand notation and introduce the delta symbol Δ (Greek letter D), which stands for "change in" (or "difference in"):

$$Ft = \Delta mv$$

which reads, "force multiplied by the time-during-which-it-acts equals change in momentum."

This rearrangement of Newton's second law explains why "follow through" is important in increasing the momentum of things. You apply the largest force you can, and you apply this force for as long a time as you can to impart maximum velocity to a given mass. We also see why the long barrels of cannons increase the velocity of emerging cannonballs. The force of exploding gunpowder in a long barrel acts on the cannonball for a longer time. The longer the time that the force acts, the greater will be the increase in momentum.

---

*If we equate the cause of acceleration $\left(\dfrac{F}{m} = a\right)$ to what acceleration is $\left(a = \dfrac{\text{change in } v}{t}\right)$,

$$\frac{F}{m} = \frac{\text{change in } v}{t}$$

simple rearrangement gives

$$Ft = \text{change in } (mv); \quad \text{or,} \quad Ft = \Delta mv.$$

Figure 3.2 A large change in momentum which occurs over a long time requires a small force.

Figure 3.3 A large change in momentum in a short time requires a large force.

Correspondingly, if we decrease momentum over a long time, a smaller force is required to bring about the momentum change. A car out of control is better off hitting a haystack than a brick wall (Figure 3.2). By hitting the haystack the car may extend its time of impact 100 times. Then the force of impact is reduced by a factor of 100. So whenever you wish the force of impact to be small, extend the time of impact. A safety net used by acrobats provides an obvious example. The net provides a small impact force over a long time in bringing the acrobat's momentum to zero after a fall. The impulse-momentum equation also explains why a boxer rides or rolls with the punch to reduce the force of impact (Figure 3.4).

The converse idea of short time of contact explains how a karate expert can sever a stack of bricks with the blow of her bare hand (Figure 3.5). By swift execution she makes the time of contact very brief and correspondingly makes the force of impact huge.

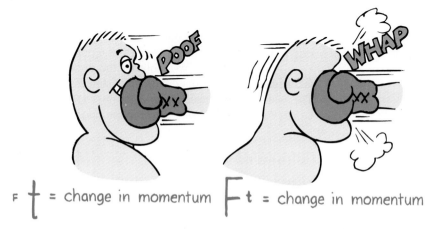

Figure 3.4 In both cases the impulse by the boxer's jaw reduces the momentum of the punch. (a) The boxer is moving away when the glove hits, thereby extending the time of contact. Much of the impulse therefore involves time. (b) The boxer is moving into the glove, thereby lessening the time of contact. Much of the impulse therefore involves force.

Figure 3.5 He imparts a large impulse to the bricks in a short time and produces a considerable force.

QUESTIONS

1. If the boxer in Figure 3.4 is able to make the duration of impact three times as long by riding with the punch, by how much will the force of impact be reduced?

2. If the boxer instead moves into the punch such as to decrease the duration of impact by half, by how much will the force of impact be increased?

3. A boxer being hit with a punch contrives to extend time for best results, whereas a karate expert delivers a force in a short time for best results. Isn't there a contradiction here?

Bouncing

You know that if a flower pot falls from a shelf onto your head, you may be in trouble. And whether you know it or not, if it bounces from your head, you're certainly in trouble. Impulses are greater when bouncing takes place. This is because the impulse required to bring something to a stop and then, in effect, "throw it back again" is greater than the impulse required merely to bring something to a stop. Suppose, for example, that you catch the falling pot with your hands. Then you provide an impulse to catch it and reduce its momentum to zero. If you were to then throw the pot upward, you would have to provide additional impulse. So it would take more impulse to catch it and throw it back up than merely to catch it. The same greater impulse is supplied by your head if the pot bounces from it.

An interesting application of the greater impulse that occurs when bouncing takes place was employed with great success in California during the gold rush days. The water wheels used in gold-mining operations were inefficient. A man named Lester A. Pelton saw that the problem had to do with their flat paddles. He designed

ANSWERS

1. The force of impact will be three times less than if he didn't pull back.

2. The force of impact will be two times greater than if he were hit at rest. This force is further increased because of the additional impulse produced when his momentum of approach is stopped. The increased impulse and short time of impact result in forces that account for many knockouts.

3. There is no contradiction because the best results for each are quite different. The best result for the boxer is reduced force, accomplished by maximizing time, and the best result for the karate expert is increased force delivered in minimum time.

curved-shape paddles that would cause the incident water to make a U-turn upon impact—to "bounce." In this way the impulse exerted on the water wheels was greatly increased. Pelton patented his idea and made more money from his invention, the Pelton wheel, than any of the gold miners.

Figure 3.6 The Pelton wheel. The curved blades cause water to bounce and make a U-turn, which produces a greater impulse to turn the wheel.

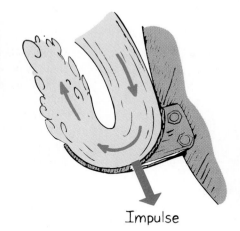

Impulse

Bungee Jumping

Would you like to put the impulse-momentum relationship to a thrilling test? Then consider bungee jumping. Simply dive head-first off a tall tower, high bridge, or even from a high-flying balloon. Never mind a parachute, but remember to securely fasten one end of a big rubber band around your ankles—and be sure the other end is firmly anchored. That big rubber band for today's bungee jumpers is a wrist-thick cord made up of bundles of rubber strands wrapped in woven nylon. You'll be glad the cord stretches, for whatever momentum you gain in your fall, the cord will have to supply the same impulse to bring you to a halt—hopefully above ground level.

We see how $Ft = \Delta mv$ applies here. The mv you wish to change is the momentum you have gained before the cord does its thing. The Ft is the impulse the cord will supply to reduce your mv to zero. Because the cord takes so long stretching, a large t insures a correspondingly small force F on your ankles. (Would you like to try such a jump with a non-stretchable cord? Goodbye feet!) Today's cords will stretch to about twice their length, depending on your momentum.

Bungee jumping is not something new, but began in ancient times as a rite of passage to maturity on the Pentecost Island of Vanuatu in the South Pacific. In the spring, platforms were built on high towers constructed of eucalyptus trees. Liana vines were wound into a long cord with just the right length, strength, and elasticity so that a jumper remains in free fall most of the way down. Then the cord starts to stretch, stopping our hero just short of disaster. The danger is twofold: hitting the ground if the cord breaks or stretches too much, and losing your feet if it stretches too little.

Bungee jumping became a modern sport on April Fool's Day in 1979 when members of the Oxford Dangerous Sports Club dived 245 feet from Clifton Bridge, Bristol, England. Wearing top hats and tuxedos, they did it again from the Golden Gate Bridge in San Francisco. It's one of the new rages, and a thrilling way to demonstrate the laws of physics!

3.3 CONSERVATION OF MOMENTUM

Newton's second law tells us that if we want to accelerate an object, we must apply a force. The force must be an external force. If you want to accelerate a car, you must push from outside. Inside forces don't count—sitting inside an automobile pushing against the dashboard has no effect in accelerating the automobile. Likewise, if we want to change the momentum of an object, the force must be external. In the impulse-momentum concept, $Ft = \Delta mv$, internal forces don't count. If no external force is present, then no change in momentum is possible.

Consider a rifle being fired. Both the force that drives the bullet and the force that makes the rifle recoil are equal and opposite (Newton's third law). To the system comprised of the rifle and the bullet, they are internal forces. No external net force acts on the rifle-bullet system so the total momentum of the system undergoes no net change (Figure 3.7). Before the firing, the momentum is zero; after firing, the *net* momentum is still zero. Like velocity, momentum is a *vector quantity*. The momentum gained by the bullet is equal and opposite to the momentum gained by the rifle. They cancel. No momentum is gained and no momentum is lost.

Whenever a physical quantity remains unchanged during a process, that quantity is said to be *conserved*. We say momentum is conserved. The conservation of momentum is especially useful in collisions, where the forces involved are internal forces. In any collision, we can say

Net momentum before collision = net momentum after collision.

When a moving billiard ball hits another billiard ball at rest head on, the first ball comes to rest and the second ball moves with the initial velocity of the first ball. We see that momentum is transferred from the first ball to the second. Momentum is conserved. When objects collide without lasting deformation, or without the generation of heat, the collision is said to be an **elastic collision**. The collisions between molecules in a gas are perfectly elastic. Billiard balls approximate perfectly elastic collisions (Figure 3.8).

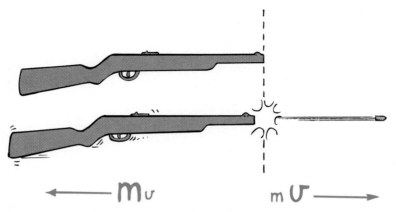

Figure 3.7 The momentum before firing is zero. After firing, the net momentum is still zero, because the momentum of the rifle is equal and opposite to the momentum of the bullet.

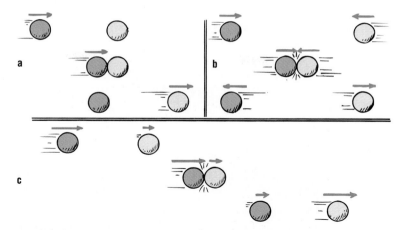

Figure 3.8   Elastic collisions. (a) A dark ball strikes a ball at rest. (b) A head-on collision. (c) A collision of balls moving in the same direction. In each case, momentum is transferred from one ball to the other.

Momentum is conserved even when the colliding objects become entangled during the collision. This is an **inelastic collision**, characterized by deformation or the generation of heat or both. Consider, for example, the case of a freight car moving along a track and colliding with another freight car at rest (Figure 3.9). If the freight cars are of equal mass and are coupled by the collision, can we predict the velocity of the coupled cars after impact?

Suppose the single car is moving at 10 m/s, and we consider the mass of each car to be m. Then, from the conservation of momentum,

$$(\text{total } mv)_{\text{before}} = (\text{total } mv)_{\text{after}}$$
$$(m \times 10)_{\text{before}} = (2m \times V)_{\text{after}}$$

By simple algebra, $V = 5$ m/s. This makes sense, for since twice as much mass is moving after the collision, the velocity must be half as much as the velocity before collision. Both sides of the equation are then equal.

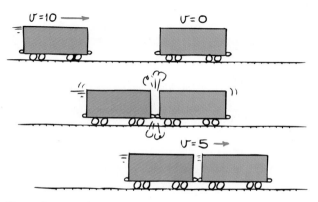

Figure 3.9   Inelastic collision. The momentum of the freight car on the left is shared with the freight car on the right after collision.

UESTION

Consider the air track in Figure 3.10. Suppose a gliding cart bumps into and sticks to a stationary cart that has three times the mass. Compared to the initial speed of the gliding cart, how fast will the coupled carts glide after collision?

Figure 3.10 An air track. Blasts of air from tiny holes provide a friction-free air cushion for the carts to glide upon.

So we see that changes in an object's motion depend on both force and how long the force acts. When "how long" means time, we call the quantity "force × time" *impulse*. But "how long" can mean distance also. When we consider the quantity force × *distance*, we are talking about something entirely different—the concept of *energy*.

3.4 ENERGY

Energy is perhaps the most central concept to all of science. The combination of energy and matter makes up the universe: Matter is substance, and energy is the mover of substance. The idea of matter is easy to grasp. Matter is stuff that we can

ANSWER

According to momentum conservation, momentum of cart of mass m and velocity v before = momentum of both carts stuck together after.

$$mv_{before} = (m + 3m)V_{after}$$
$$mv = 4mV$$
$$v = \frac{mv}{4m} = \frac{v}{4}$$

This makes sense, for four times as much mass will be moving after collision, so the coupled carts will glide more slowly. In keeping the momentum equal, four times the mass glides $\frac{1}{4}$ as fast.

see, smell, and feel. It has mass and occupies space. Energy, on the other hand, is abstract. We cannot see, smell, or feel most forms of energy. Surprisingly, the idea of energy was unknown to Isaac Newton, and its existence was still being debated in the 1850s. Although energy is familiar to us, it is difficult to define, because it is not only a "thing" but also a process—as if it were both a noun and a verb. Persons, places, and things have energy, but we usually observe energy only when it is happening— only when it is being transformed. It comes to us in the form of electromagnetic waves from the sun and we feel it as heat; it is captured by plants and binds atoms of matter together; it is in the food we eat and we receive it by digestion. We begin our study of energy by observing a related concept: work.

3.5 WORK

Figure 3.11 Work is done in lifting the barbell. If the weight lifter were taller, he would have to expend proportionally more energy to press the barbell over his head.

We have seen that force × time is impulse, which changes momentum. Force × distance, however, changes energy. We call the quantity "force × distance" **work**. When we lift a load against earth's gravity, work is done on the load. The heavier the load or the higher we lift the load, the more work is done. Two things enter into every case where work is done: (1) the exertion of a force and (2) the movement of something by that force. We define the work done on an object as the product of the applied force and the distance through which the object is moved:*

Work = force × distance

If we lift two loads one story up, we do twice as much work as in lifting one load, because the *force* needed to lift twice the weight is twice as much. Similarly, if we lift a load two stories instead of one story, we do twice as much work because the *distance* is twice as much.

A weight lifter who holds a barbell weighing 1000 newtons overhead does no work on the barbell. He may get really tired doing so, but if the barbell is not moved by the force he exerts, he does no work on the barbell. Work may be done on his muscles by stretching and contracting, which is force times distance on a biological scale, but this work is not done on the barbell. *Lifting* the barbell, however, is a different story. When the weight lifter raises the barbell from the floor, he does work on the barbell.

The unit of measurement for work combines a unit of force (N) with a unit of distance (m); the unit of work is the newton-meter (N·m), also called the *joule* (J) (rhymes with *pool*). One joule of work is done when a force of 1 newton is exerted

Figure 3.12 He may expend energy when he pushes on the wall, but if it doesn't move, no work is performed on the wall.

---

*More specifically, work is the product of the component of force that acts in the direction of motion and the distance moved.

over a distance of 1 meter, as in lifting an apple over your head. For larger values we speak of kilojoules (kJ), thousands of joules, or megajoules (MJ), millions of joules. The weight lifter in Figure 3.11 does work in kilojoules. The energy released by 1 kilogram of gasoline is rated in megajoules.

3.6 WORK CHANGES ENERGY

Work is done in lifting the heavy ram of a pile driver, and, as a result, the ram acquires the property of being able to do work on a piling beneath it when it falls. When work is done by an archer in drawing a bow, the bent bow has the ability of being able to do work on the arrow. When work is done to wind a spring mechanism, the spring acquires the ability to do work on various gears to run a clock, ring a bell, or sound an alarm. In each case, something has been acquired. This "something" that is given to the object enables the object to do work. This "something" may be a compression of atoms in the material of an object; it may be a physical separation of attracting bodies; it may be a rearrangement of electric charges in the molecules of a substance. This "something" that enables an object to do work is *energy*.* Like work, energy is measured in joules. It appears in many forms, which will be discussed in the following chapters. For now we focus on mechanical energy—the form of energy due to position (potential energy) or to the movement of mass (kinetic energy). Mechanical energy may be in the form of either potential energy or kinetic energy.

Potential Energy

An object may store energy because of its position. The energy that is stored and held in readiness is called **potential energy** (PE), because in the stored state it has the potential to do work. For example, a stretched or compressed spring has the potential for doing work. When a bow is drawn, energy is stored in the bow. A stretched rubber band has potential energy because of its position, for if it is part of a slingshot, it is capable of doing work.

The chemical energy in fuels is potential energy, for it is energy of position from a microscopic point of view. This energy is available when the positions of electric charges within and between molecules are altered, that is, when a chemical change takes place. Any substance that can do work through chemical action possesses potential energy. Potential energy is found in fossil fuels, electric batteries, and the food we eat.

Work is required to elevate objects against earth's gravity. The potential energy due to elevated positions is called *gravitational potential energy*. Water in an

Figure 3.13 The potential energy of Tenny's drawn bow equals the work (average force × distance) she did in drawing the arrow into position. When released, the potential energy of the drawn bow will become the kinetic energy of the arrow.

---

*Strictly speaking, that which enables an object to do work is its available energy, for not all the energy in an object can be transformed to work—some unavoidably dissipates as heat.

elevated reservoir and the elevated ram of a pile driver has gravitational potential energy. The amount of gravitational potential energy possessed by an elevated object is equal to the work done against gravity in lifting it. The work done equals the force required to move it upward times the vertical distance it is moved ($W = Fd$). The upward force equals the weight mg of the object. So the work done in lifting it through a height h is given by the product mgh:

$$\text{Gravitational potential energy} = \text{weight} \times \text{height}$$
$$PE = mgh$$

Note that the height h is the distance above some reference level, such as the ground or the floor of a building. The potential energy mgh is relative to that level and depends only on mg and the height h. You can see in Figure 3.14 that the potential energy of the ball at the top of the structure does not depend on the path taken to get it there.

QUESTIONS

1. How much work is done on a 75-N bowling ball when you carry it horizontally across a 10-m-wide room?
2. How much work is done on it when you lift it 1 m?
3. What is its gravitational potential energy in the lifted position?

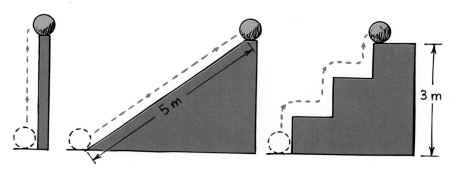

Figure 3.14   The potential energy of the 10-N ball is the same (30 J) in all three cases because the work done in elevating it 3 m is the same whether it is (a) lifted with 10 N of force, (b) pushed with 6 N of force up the 5-m incline, or (c) lifted with 10 N up each 1-m stair. No work is done in moving it horizontally (neglecting friction).

ANSWERS

1. You do no work on the ball moved horizontally, for you apply no force (except for the tiny bit to start it) in its direction of motion. It has no more PE across the room than it had initially.
2. You do 75 J of work when you lift it 1 m ($Fd = 75\,\text{N} \cdot \text{m} = 75\,\text{J}$).
3. It depends. With respect to its starting position its PE is 75 J; with respect to some other reference level, it would be some other value.

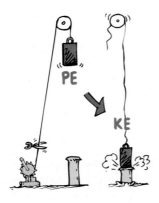

Figure 3.15 The potential energy of the elevated ram is converted to kinetic energy when released.

Kinetic Energy

If we push on an object, we can set it in motion. More specifically, if we do work on an object, we can change the energy of motion of that object. If an object is moving, then by virtue of that motion it is capable of doing work. We call energy of motion **kinetic energy** (KE). The kinetic energy of an object depends on mass and speed. It is equal to half the mass multiplied by the square of the speed.

$$\text{Kinetic energy} = \tfrac{1}{2}\text{mass} \times \text{speed}^2$$
$$\text{KE} = \tfrac{1}{2}mv^2$$

When a ball is thrown, work is done on it, giving it kinetic energy. The moving ball can then hit something and push against it, doing work on what it hits. The kinetic energy of a moving object is equal to the work done in bringing it from rest to that speed or to the work a moving object can do in being brought to rest:*

$$\text{Net force} \times \text{distance} = \text{kinetic energy}$$
$$Fd = \tfrac{1}{2}mv^2$$

Notice that speed is squared, so if the speed of an object is doubled, its kinetic energy is quadrupled ($2^2 = 4$), and the object can do four times as much work. An object

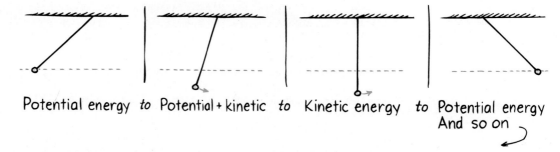

Potential energy *to* Potential + kinetic *to* Kinetic energy *to* Potential energy
And so on

Figure 3.16 Energy transitions in a pendulum.

---

*This can be derived as follows: If we multiply both sides of $F = ma$ (Newton's second law) by d, we get $Fd = mad$. Recall from Chapter 1 that $d = \tfrac{1}{2}at^2$, so we can say $Fd = ma(\tfrac{1}{2}at^2) = \tfrac{1}{2}m(at)^2$; and substituting $v = at$, we get $Fd = \tfrac{1}{2}mv^2$.

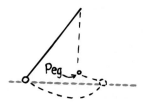

Figure 3.17 The pendulum bob will swing to its original height whether or not the peg is present.

moving twice as fast as another takes four times as much work to stop. Accident investigators are well aware that an automobile going 100 kilometers per hour has four times the kinetic energy it would have at 50 kilometers per hour. This means a car going 100 kilometers per hour will skid four times as far when its brakes are locked as it would going 50 kilometers per hour. This is because speed is squared for kinetic energy.

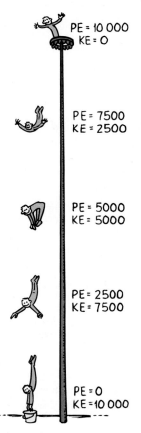

PE = 10 000
KE = 0

PE = 7500
KE = 2500

PE = 5000
KE = 5000

PE = 2500
KE = 7500

PE = 0
KE = 10 000

QUESTIONS

1. When the brakes of a car going 90 km/h are locked, how much farther will it skid than if the brakes lock at 30 km/h?
2. Can an object have energy?
3. Can an object have work?

Kinetic energy underlies other seemingly different forms of energy such as heat, sound, and light. Random molecular motion is sensed as heat, molecules vibrating in rhythmic patterns are perceived as sound, and electrons in motion make electric currents. Even light energy originates from the motion of electrons within atoms. We will find there is much in common among the various forms of energy that we will investigate.

Figure 3.18 A circus diver at the top of a pole has a potential energy of 10,000 J. As he dives, his potential energy converts to kinetic energy. Note that at successive positions one-fourth, one-half, three-fourths, and all the way down, the total energy is constant. (Adapted from K. F. Kuhn and J. S. Faughn, *Physics in Your World,* Philadelphia: Saunders, 1980.)

ANSWERS

1. Nine times farther. The car has nine times as much energy when it travels three times as fast: $\frac{1}{2}m(3v)^2 = \frac{1}{2}m9v^2 = 9(\frac{1}{2}mv^2)$. The friction force will ordinarily be the same in either case; therefore, to do nine times the work requires nine times as much sliding distance.

2. Yes, but in a relative sense. For example, an elevated object may possess PE relative to the ground below, but none relative to a point at the same elevation. Similarly, the KE of an object is relative to a frame of reference, usually taken to be the earth's surface. We will see later that material objects have energy of being—the congealed energy that makes up mass—$E = mc^2$. Read on!

3. No, unlike momentum or energy, work is not something that an object *has.* Work is something that an object *does* to some other object. An object can *do* work only if it has energy.

3.7 KINETIC ENERGY AND MOMENTUM COMPARED

Kinetic energy and momentum are both concepts concerning the motion of an object. But they are different. Like velocity, momentum is a vector quantity and is therefore directional and capable of being cancelled entirely. But kinetic energy is a positive non-vector (*scalar*) quantity, like mass, and so can never be cancelled.* The momenta of two cars just before a head-on collision may cancel to zero, and the combined wreck after collision will have the same zero value for the momentum; but the energies add, as evidenced by the deformation and heat after collision. Or the momenta of two firecrackers approaching each other may cancel, but when they explode, there is no way their energies can cancel. Energies transform to other forms; momenta do not. The vector quantity momentum is different from the scalar quantity kinetic energy.

Another difference is the velocity dependence of the two. Whereas momentum depends on velocity (mv), kinetic energy depends on the square of velocity ($\frac{1}{2}mv^2$).† An object that moves with twice the velocity of another object of the same mass has twice the momentum but four times the kinetic energy. So when a car going twice as fast crashes, it crashes with four times the energy.

The difference between momentum and kinetic energy is illustrated with momentum conservation. Recall that when a bullet is fired from a gun, the momentum of the emerging bullet is equal and opposite to the momentum of the recoiling gun. But the KE of the bullet is enormously larger than the KE of the recoiling gun. For example, if the bullet's speed is a hundred times greater than the gun's recoil speed (which means the bullet must have one-hundredth the mass of the gun) then the KE of the bullet is a hundred times greater than the KE of the gun.††

Momentum and kinetic energy are properties of moving things, but we see they are different from each other. If this distinction is not really clear to you, you're in good company. Failure to make this distinction, when impedo was in vogue, resulted in disagreements and arguments between the best British and French physicists for two centuries.

3.8 CONSERVATION OF ENERGY

More important than being able to state *what energy is* is understanding how it behaves—*how it transforms*. We can better understand processes or changes that occur in nature if we analyze them in terms of transformations of energy from one form to another or of transfers from one place to another.

---

*If you're into mathematics, you may know that when a vector quantity (velocity) is multiplied by a scalar quantity (mass), the product is also a vector (momentum). But a vector quantity squared (velocity²) is a scalar. So KE is a scalar quantity. It can't be cancelled. Although momenta can be combined in such a way to cancel to zero, there is no way to combine KEs to equal zero.

†Interestingly enough, kinetic energy is equal to the momentum squared divided by twice the mass. To see this, let momentum mv be p. Then KE $= \frac{1}{2}mv^2 = \frac{1}{2}mvv = \frac{1}{2}pv$. Multiply by m/m, and we see KE $= \frac{1}{2}m/m(v) = \frac{1}{2}mp^2$.

††Let m be the mass of the bullet, and $100m$ the mass of the gun. Then by momentum conservation if the speed of the gun is v, the speed of the bullet is $100v$. Then KE$_{gun} = \frac{1}{2}(100m)v^2 = 50mv^2$. KE$_{bullet} = \frac{1}{2}m(100v)^2 = \frac{1}{2}m(10000)v^2 = 5000mv^2$. So the bullet's KE is 100 times greater than the recoiling gun's KE.

The Swinging Wonder

Figure 3.19 The Swinging Wonder. The number of balls that are raised, released, and that make impact with the array of identical balls always equals the number of balls that emerge. Two conservation principles explain why.

Momentum conservation is nicely demonstrated with the swinging wonder, the novel device shown in Figure 3.19. When a single ball is raised and allowed to swing into the array of other identical balls, a single ball from the other side pops out. When two balls are similarly raised and released, presto—two balls on the other side pop out. The number of balls incident on the array is always the same as the number of balls that emerge. We can see that *momentum before = momentum after*. That is, $mv = mv$, or $2mv = 2mv$, or $3mv = 3mv$, and so on. The intriguing question arises: When a single ball is raised, released, and makes impact, why cannot two balls emerge with half the speed? Or if two balls make impact, why cannot one ball emerge with twice the speed? If either of these cases occurred, the momentum before would still be equal to the momentum after: $mv = 2m(\frac{1}{2}v)$; or $2mv = m(2v)$. Intriguingly, this never happens. Nor can it happen.

Why? Because something besides momentum must be conserved in this interaction—energy. Since the collisions are quite elastic, with very little energy transforming to heat and sound, to a good approximation the kinetic energy before equals the kinetic energy after. That is, $\frac{1}{2}mv^2_{\text{before}} = \frac{1}{2}mv^2_{\text{after}}$. Consider dropping two balls with one emerging at twice the speed. Then will $\frac{1}{2}2mv^2 = \frac{1}{2}m(2v)^2$? The answer is no! If this case were to occur, there would be more energy after the collision than before (we'll leave it to you to figure how much more). Give this some thought and you'll see there is a reason why, for identical balls, the number of balls that make impact will always equal the number of balls that emerge.

In any collision, elastic or inelastic, momentum before collision equals momentum after collision. In the special case of a perfectly elastic collision, where no energy is transformed to other forms, kinetic energy before collision equals kinetic energy after collision.

Why is this device called the swinging wonder? Because the unequal-balls situation and its impossibility has left many people wondering—and wondering—and wondering. But you know the reason why the number of incident and emerging balls must be the same. It's nice to know some physics!

As we draw back the arrow in a bow, we do work in bending the bow; we give the arrow and bow potential energy. When released, this potential energy is transferred to the arrow as kinetic energy. The arrow in turn transfers this energy to its target, perhaps a rigid wooden fence post. The slight distance the arrow penetrates multiplied by the average force of impact doesn't quite match the kinetic energy of the arrow. The energy score doesn't balance. But if we investigate further, we'll find that both the arrow and fence post are a bit warmer. By how much? By the energy difference. Energy changes from one form to another. It transforms without net loss or net gain.

The study of various forms of energy and their transformations from one form into another has led to one of the greatest generalizations in physics—the law of **conservation of energy:**

> **Energy cannot be created or destroyed; it may be transformed from one form into another, but the total amount of energy never changes.**

When we consider any system in its entirety, whether it be as simple as a swinging pendulum or as complex as an exploding galaxy, there is one quantity that doesn't change: energy. It may change form or it may simply be transferred from one place to another, but, as far as we can tell, the total energy score stays the same. This energy score takes into account the fact that the atoms that make up matter are themselves concentrated bundles of energy. When the nuclei (cores) of atoms rearrange themselves, enormous amounts of energy can be released. The sun shines because some of this energy is transformed into radiant energy. In nuclear reactors much of this energy is transformed into heat. Enormous gravitational forces in the deep hot interior of the sun crush the cores of hydrogen atoms together to form helium atoms. This welding together of atomic cores is called *thermonuclear fusion*. This process releases radiant energy, some of which reaches the earth. Part of this energy falls on plants, and part of this in turn later becomes coal. Another part supports life in the food chain that begins with plants, and part of this energy later becomes oil. Part of the energy from the sun goes into the evaporation of water from the ocean, and part of this returns to the earth as rain that may be trapped behind a dam. By virtue of its position, the water in a dam has energy that may be used to power a generating plant below, where it will be transformed to electric energy. The energy travels through wires to homes, where it is used for lighting, heating, cooking, and operating electric gadgets. How nice that energy is transformed from one form to another!

QUESTION

Does an automobile consume more fuel when its air conditioner is turned on? When its lights are on? When its radio is on while it is sitting in the parking lot?

3.9 POWER

Energy can be transformed quickly or it can be transformed slowly. The rate at which energy is changed, or the rate at which work is done, is called **power**. We say power is equal to the amount of work done per time it takes to do it:

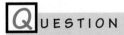

$$\text{Power} = \frac{\text{work done}}{\text{time interval}}$$

An engine of great power can do work rapidly. An automobile engine with twice the power of another does not necessarily produce twice as much work or go twice as fast as the less powerful engine. Twice the power means it will do the same amount of work in half the time or twice the work in the same time. The main advantage of a powerful automobile engine is the acceleration it can produce. It can get the automobile up to a given speed in less time than less powerful engines.

Here's another way to look at power: A liter (L) of gasoline can do a certain amount of work, but the power produced when we burn it can be any amount, de-

ANSWER

The answer to all three questions is yes, for the energy they consume ultimately comes from the fuel. Even the energy from the battery must be given back to the battery by the alternator, which is turned by the engine, which runs from the energy of the fuel. There's no free lunch!

Figure 3.20 The three main engines of a space shuttle can develop 33,000 MW of power when fuel is burned at the enormous rate of 3400 kg/s. This is like emptying an average-size swimming pool in 20 s.

pending on how *fast* it is burned. The liter may produce 50 units of power for a half hour in an automobile or 90,000 units of power for one second in a Boeing 747.

The unit of power is the joule per second (J/s), also known as the watt (in honor of James Watt, the eighteenth-century developer of the steam engine). One watt (W) of power is expended when 1 joule of work is done in 1 second. One kilowatt (kW) equals 1000 watts. One megawatt (MW) equals 1 million watts. In the United States we customarily rate engines in units of horsepower and electricity in kilowatts, but either may be used. In the metric system of units, automobiles are rated in kilowatts. (One horsepower is the same as three-fourths of a kilowatt, so an engine rated at 134 horsepower is a 100-kW engine.)

3.10 MACHINES

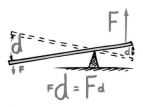

Figure 3.21 The lever.

A *machine* is a device for multiplying forces or simply changing the direction of forces. Underlying every machine is the *conservation of energy and machines* concept. Consider one of the simplest machines, the **lever** (Figure 3.21). At the same time we do work on one end of the lever, the other end does work on the load. We see that the direction of force is changed, for if we push down, the load is lifted up. If the heat from friction forces is small enough to neglect, the work input will be equal to the work output.

$$\text{Work input} = \text{work output}$$

Since work equals force times distance, input force × input distance = output force × output distance.

$$(\text{Force} \times \text{distance})_{\text{input}} = (\text{force} \times \text{distance})_{\text{output}}$$

If the pivot point, *fulcrum,* of the lever is relatively close to the load, then a small input force will produce a large output force. This is because the input force is exerted through a large distance and the load is moved over a correspondingly short distance. In this way, a lever can multiply forces. But no machine can multiply work or multiply energy. That's a conservation of energy no-no!

A child uses the principle of the lever in jacking up the front end of an automobile. By exerting a small force through a large distance, she is able to provide a large force acting through a small distance. Consider the ideal example illustrated in Figure 3.22. Every time she pushes the jack handle down 25 centimeters, the car rises only a hundredth as far but with 100 times the force.

5000 N

25 cm

$_F d = F d$
$50 \times 25 = 5000 \times 0.25$

Figure 3.22 Applied force $\times$ applied distance = output force $\times$ output distance.

A block and tackle, or system of pulleys, is a simple machine that multiplies force at the expense of distance. One can exert a relatively small force through a relatively large distance and lift a heavy load through a relatively short distance. With the ideal pulley system such as that shown in Figure 3.23, the man pulls 10 meters of rope with a force of 50 newtons and lifts 500 newtons through a vertical distance of 1 meter. The energy the man expends in pulling the rope is numerically equal to the increased potential energy of the 500-newton block.

Any machine that multiplies force does so at the expense of distance. Likewise, any machine that multiplies distance, such as the construction of your forearm and elbow, does so at the expense of force. No machine or device can put out more energy than is put into it. No machine can create energy; it can only transform it from one form to another.

Efficiency

The three previous examples were of ideal machines; 100 percent of the work input was transformed to work output. An ideal machine would operate at 100 percent efficiency. In practice, this doesn't happen and we can never expect it to happen. In any transformation some energy is dissipated to molecular kinetic energy—heat. This makes the machine warmer.

Even a lever rocks about its fulcrum and converts a small fraction of the input energy into heat. We may do 100 joules of work and get out 98 joules of work. The lever is then 98 percent efficient, and we waste only 2 joules of work input on heat. If the girl in Figure 3.22 puts in 100 joules of work and increases the potential energy of the car by 60 joules, the jack is 60 percent efficient; 40 joules of work have been used to do the work in overcoming the friction force, and these appear as heat. In a pulley system, a larger fraction of input energy goes into heat. If we do 100 joules of work, the forces of friction acting through the distances through which the pulleys turn and rub about their axles may dissipate 60 joules of energy as heat. So the work output is only 40 joules and the pulley system has an efficiency of 40 percent. The lower the efficiency of a machine, the greater the amount of energy wasted as heat.

Inefficiency exists whenever energy is transformed from one form to another. **Efficiency** can be expressed by the ratio

$$\text{Efficiency} = \frac{\text{work done}}{\text{energy used}}$$

500 N

$_F d = F d$

Figure 3.23 Applied force $\times$ applied distance = output force $\times$ output distance.

QUESTION

Suppose a miracle car has a 100 percent efficient engine and burns fuel that has an energy content of 40 MJ/L. If the air drag and overall frictional forces on the car traveling at highway speed is 1000 N, what is the upper limit in distance per liter the car could go at this speed?

An automobile engine is a machine that transforms chemical energy stored in fuel into mechanical energy. The bonds between the molecules in the petroleum fuel break up when the fuel burns by reacting with the oxygen in the air. Carbon atoms then bond with oxygen to form carbon monoxide and carbon dioxide, in which the bonds have less energy stored in them than they do in the original bonds. Some of the remaining energy goes into running the engine. We'd like all this energy converted into mechanical energy; that is, we'd like an engine that is 100 percent efficient. This is impossible, because some of the energy goes out in the hot exhaust gases, and nearly half is wasted in the friction of the moving engine parts. In addition to these inefficiencies, the fuel doesn't even burn completely and a certain amount of fuel energy goes unused.

Look at the inefficiency that accompanies transformations of energy this way: In any transformation there is a dilution of available *useful energy* to a state of wasted

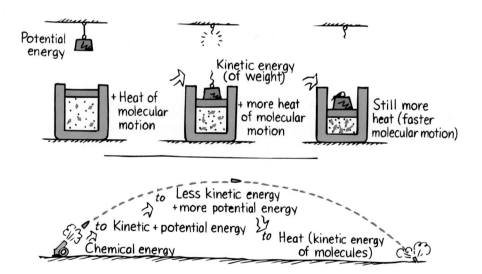

Figure 3.24 Energy transitions. The graveyard of kinetic energy is heat.

ANSWER

From the definition work = force × distance, simple rearrangement gives distance = work/force. If all 40 million J of energy in 1 L were used to do the work of overcoming the air drag and frictional forces, the distance would be:

$$\text{Distance} = \frac{\text{work}}{\text{force}} = \frac{40{,}000{,}000 \text{ J/L}}{1000 \text{ N}} = 40{,}000 \text{ m/L} = 40 \text{ km/L}$$

The important point here is that even with a perfect engine, there is an upper limit of fuel economy dictated by the conservation of energy.

energy. The amount of usable energy decreases with each transformation and ultimately becomes heat. When we study thermodynamics, we'll see that the energy in the form of heat is useless for doing work unless it can be transformed to a lower temperature. Heat is the graveyard of useful energy.

3.11 SOURCES OF ENERGY

The source of practically all of our energy is the sun. This includes the energy we obtain from the combustion of petroleum, coal, natural gas, and wood, for these materials are the result of photosynthesis, a biological process that incorporates the sun's radiant energy. There are many other ways of using the sun as a source of energy. Sunlight, for example, can be directly transformed into electricity by way of photovoltaic cells, like those found in solar powered calculators. Solar radiation can also be used indirectly to generate electricity. Sunlight evaporates water, which later falls as rain; rainwater flows into rivers and turns generator turbines as it returns to the sea. Using mirrors, solar radiation can be concentrated to heat water to steam, which can also be used to turn generator turbines. Furthermore, wind is solar energy that has already been converted into mechanical energy. The mechanical energy of wind can be used to turn generator turbines within specially equipped windmills.

An important non-solar source of energy is nuclear energy. Presently, over 19 percent of the electricity in the United States is produced by nuclear-fission power plants. Nuclear energy in the earth's interior heats the earth. This is geothermal energy, which can convert water to steam for the generation of electricity.

All methods for obtaining energy have disadvantages. The combustion of fossil fuels, for example, leads to increased atmospheric concentrations of carbon dioxide, sulfur dioxide and other pollutants. Methods using solar energy are limited in that they require proper atmospheric conditions.* Nuclear fission energy, though it is an efficient source of energy, remains controversial because of the nuclear wastes that are generated. Geothermal energy is predominantly limited to areas of volcanic activity, such as Iceland, New Zealand, and Japan.

As the world population increases, so does our need for energy. Our best bet is to optimize all possible sources of energy, and to make sure that we use energy efficiently and wisely.

*One interesting way to get around this problem is to collect solar energy in space and beam it down to the surface in the form of microwaves, which can penetrate the worst of weather conditions. This and other similar topics are discussed in Chapter 7—Heat Transfer.

SUMMARY OF TERMS

Momentum The product of the mass of an object and its velocity.
Impulse The product of the force acting on an object and the time during which it acts.

Relationship of impulse and momentum Impulse is equal to the change in the momentum of the object that the impulse acts on. In symbol notation,

$$Ft = \Delta mv$$

Conservation of momentum When no external net force acts on an object or a system of objects, no change of momentum takes place. Hence, the momentum before an event involving only internal forces is equal to the momentum after the event:

$$mv \text{ (before event)} = mv \text{ (after event)}$$

Elastic collision A collision in which colliding objects rebound without lasting deformation or the generation of heat.

Inelastic collision A collision in which the colliding objects become distorted and generate heat during the collision.

Work The product of the force and the distance through which the force moves:

$$W = Fd$$

Power The time rate of work:

$$\text{Power} = (\text{work/time})$$

Energy The property of a system that enables it to do work.

Potential energy The stored energy that a body possesses because of its position.

Kinetic energy Energy of motion, described by the relationship: Kinetic energy $= \frac{1}{2}mv^2$.

Conservation of energy Energy cannot be created or destroyed; it may be transformed from one form into another, but the total amount of energy never changes.

Conservation of energy and machines The work output of any machine cannot exceed the work input. In an ideal machine, where no energy is transformed into heat, Work$_{\text{input}}$ = work$_{\text{output}}$ and $(Fd)_{\text{input}} = (Fd)_{\text{output}}$.

Efficiency The percent of the work put into a machine that is converted into useful work output.

. . . .

REVIEW QUESTIONS

Momentum

1. Which has a greater momentum, a heavy truck at rest or a moving skateboard?

Impulse Changes Momentum

2. How does impulse differ from force?

3. What are the two ways that the impulse exerted on something can be increased or decreased?

4. What is the relationship of the impulse-momentum relationship to Newton's second law?

5. Why is it incorrect to say that impulse equals momentum?

6. For the same force, which cannon imparts the greatest speed to a cannonball—a long cannon or a short one? Explain.

7. Why is it best to extend your hand forward when you catch a fast-moving baseball with your bare hand?

Bouncing

8. Which undergoes the greater change in momentum: (1) a moving object brought to rest, (2) the same object projected from rest to the speed it had before, or (3) the same moving object brought to rest and then projected backward to its original speed?

9. In the preceding question, in which case is the greatest impulse required?

Conservation of Momentum

10. Can you produce a net impulse to an automobile by sitting inside and pushing on the dashboard? Can the internal forces within a baseball produce an impulse on the baseball that will change its momentum?

11. Is it correct to say that if no impulse is exerted on an object, then no change in the momentum of the object will occur?

12. What does it mean to say that a quantity is *conserved?*

13. Distinguish between an *elastic* collision and an *inelastic* collision. Why is momentum conserved for both types of collisions?

14. Railroad car A rolls at a certain speed against car B of the same mass, which is at rest. Compare the motions of the cars after collision if the collision is perfectly elastic.

15. In the preceding question, compare the motions of the cars before and after collision if the collision is inelastic.

Energy

16. When is energy most evident?

Work

17. A force sets an object in motion. When the force is multiplied by the time of its application, we call the quantity *impulse,* which changes the *momentum* of that object. What do we call the quantity *force × distance,* and what quantity does this change?

18. Which requires more work—lifting a 50-kg sack a vertical distance of 2 m or lifting a 25-kg sack a vertical distance of 4 m?

Work Changes Energy

19. In what units are work and energy measured?

Potential Energy

20. A car is lifted a certain distance in a service station and therefore has potential energy with respect to the floor. If it were lifted twice as high, how much potential energy would it have?

21. Two cars are lifted to the same elevation in a service station. If one car is twice as heavy as the other, how do their potential energies compare?

Kinetic Energy

22. How many joules of kinetic energy does a 1-kg book have when it is tossed across the room at a speed of 2 m/s? How much energy is imparted to the wall it accidentally encounters?

23. A moving car has kinetic energy. If it speeds up by four times, how much kinetic energy does it have in comparison? Compared to its original speed, how much work must the brakes supply to stop the four-times-as-fast car?

24. Which has the greater kinetic energy—a car traveling at 30 km/h or a half-as-heavy car traveling at 60 km/h?

Kinetic Energy and Momentum Compared

25. Can momenta cancel? Can energies cancel?

26. If a moving object doubles its speed, how much more momentum does it have? Energy?

27. If a moving object doubles its speed, how much more impulse does it provide to whatever it bumps into (how much more wallop)? How much more work (how much more damage)?

Conservation of Energy

28. What will be the kinetic energy of an arrow shot from a bow having a potential energy of 40 J?

29. An apple hanging from a limb has potential energy. If it falls, what becomes of this energy just before it hits the ground? After it hits the ground?

Machines

30. Can a machine multiply input force? Input distance? Input energy? (If your three answers are the same, seek help, for the last question is especially important.)

31. A force of 50 N is applied to the end of a lever, which is moved a certain distance. If the other end of the lever is moved half as far, how much force can it exert?

Efficiency

32. Compared to work input, how much work is put out by a machine that operates at 30 percent efficiency?

33. Is a machine physically possible that has an efficiency greater than 100 percent? Discuss.

Sources of Energy

34. Explain how the energy that operates an electric toothbrush is actually the energy of sunlight.

HOME PROJECT

When you get a bit ahead in your studies, cut classes some afternoon and visit your local pool or billiards parlor and bone up on momentum conservation. Note that no matter how complicated the collision of balls, the momentum along the line of action of the cue ball before impact is the same as the combined momentum of all the balls along this direction after impact and that the components of momenta perpendicular to this line of action cancels to zero after impact, the same value as before impact in this direction. You'll see both the vector nature of momentum and its conservation more clearly when rotational skidding—English—is not imparted to the cue ball. When English is imparted by striking the cue ball off center, rotational momentum, which is also conserved, somewhat complicates analysis. But regardless of how the cue ball is struck, in the absence of external forces, both linear and rotational momentum are always conserved. Pool or billiards offers a first-rate exhibition of momentum conservation in action.

EXERCISES

1. Why might a wine glass survive a fall onto a carpeted floor but not onto a concrete floor?

2. To bring a super tanker to a stop, its engines are typically cut off about 25 km from port. Why is it so difficult to stop or turn a super tanker?

3. In terms of impulse and momentum, why are nylon ropes, which stretch considerably under stress, favored by mountain climbers?

4. It is generally much more difficult to stop a heavy truck than a skateboard when they move at the same speed. State a case where the moving skateboard could require more stopping force. (Consider relative times.)

5. If a ball is projected upward from the ground with 10 units of momentum, what is the momentum of recoil of the world? Why do we not feel this?

6. Why is a punch more forceful with a bare fist than with a boxing glove?

7. A boxer can punch a heavy bag for more than an hour without tiring, but will tire quickly when boxing with an opponent for a few minutes. Why? (Hint: When aimed at the bag, what supplies the impulse to stop the punches? When aimed at the opponent, what or

who supplies the impulse to stop the punches that are missed?)

8. Railroad cars are loosely coupled so that there is a noticeable time delay from the time the first car is moved and last cars are moved from rest by the locomotive. Discuss the advisability of this loose coupling and slack between cars from the point of view impulse and momentum.

9. A fully dressed person is at rest in the middle of a frozen pond on perfectly frictionless ice and must get to shore. How can this be accomplished?

10. If you throw a ball horizontally while standing on roller skates, you roll backward with a momentum that matches that of the ball. Will you roll backward if you go through the motions of throwing the ball, but instead hold on to it? Explain.

11. Using examples, show that situations explained by Newton's third law can also be explained by the conservation of momentum.

12. Why is it difficult for a fire fighter to hold a hose that ejects large amounts of water at a high speed?

13. If a Mack truck and a motorcycle have a head-on collision, which vehicle will experience the greater force of impact? The greater impulse? The greater change in its momentum? The greater acceleration?

14. Would a head-on collision between two cars be more damaging to the occupants if the cars stuck together or if the cars rebounded upon impact?

15. Suppose there are three astronauts outside a spaceship, and two of them decide to play catch with the third man. All the astronauts weigh the same on earth and are equally strong. The first astronaut throws the second one toward the third one and the game begins. Describe the motion of the astronauts as the game proceeds. How long will the game last?

16. To determine the potential energy of Tenny's drawn bow (Figure 3.12), would it be an underestimate or an overestimate to multiply the force with which she holds the arrow in its drawn position by the distance she pulled it? Why do we say the work done is the *average* force × distance?

17. When a rifle with a long barrel is fired, the force of expanding gases acts on the bullet for a longer distance. What effect does this have on the velocity of the emerging bullet? (Do you see why long-range cannons have such long barrels?)

18. You and a flight attendant toss a ball back and forth in an airplane in flight. Does the KE of the ball depend on the speed of the airplane? Carefully explain.

19. Can something have energy without having momentum? Explain. Can something have momentum without having energy? Defend your answer.

20. At what point in its motion is the KE of a pendulum bob a maximum? At what point is its PE a maximum? When its KE is half its maximum value, how much PE does it have?

21. A physics instructor demonstrates energy conservation by releasing a heavy pendulum bob, as shown in the sketch, allowing it to swing to and fro. What would happen if in his exuberance he gave the bob a slight shove as it left his nose? Why?

22. Discuss the design of the roller coaster shown in the sketch in terms of the conservation of energy.

23. Strictly speaking, does a car burn more gasoline when its lights are turned on? Does the overall consumption of gasoline depend on whether or not the engine is running? Defend your answer.

24. You tell your friend that no machine can possibly put out more energy than is put into it, and your friend states that a nuclear reactor puts out more energy than is put into it. What do you say?

25. This may seem like an easy question for a physics type to answer: With what force does a rock that weighs 10 N strike the ground if dropped from a rest position

10 m high? This question does not have a straightforward numerical answer. Why?

26. In the hydraulic machine shown, it is observed that when the small piston is pushed down 10 cm, the large piston is raised 1 cm. If the small piston is pushed down with a force of 100 N, how much force is the large piston capable of exerting?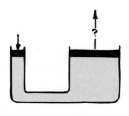

27. Consider the swinging-balls apparatus. If two balls are lifted and released, momentum is conserved as two balls pop out the other side with the same speed as the released balls at impact. But momentum would also be conserved if one ball popped out at twice the speed. Can you explain why this never happens?

28. Consider the inelastic collision between the two freight cars in Figure 3.8. The momentum before and after the collision is the same. The KE, however, is less after the collision than before the collision. How much less, and what becomes of this energy?

29. If an automobile had a 100 percent efficient engine, would it be warm to your touch? Would its exhaust heat the surrounding air? Would it make any noise? Would it vibrate? Would any of its fuel go unused?

30. We know more force is normally required to stop a moving truck than a moving skateboard. Make an argument that more force would be needed to stop a moving skateboard.

PROBLEMS

1. A railroad diesel engine weighs four times as much as a freightcar. If the diesel engine coasts at 5 km per hour into a freightcar that is initially at rest, how fast do the two coast after they couple together?

2. A 5-kg fish swimming at 1 m/s swallows an absentminded 1-kg fish at rest. What is the speed of the larger fish after lunch?

3. This question is typical on some driver's license exams: A car moving at 50 km/h skids 15 m with locked brakes. How far will the car skid with locked brakes at 150 km/h?

4. How many kilometers per liter will a car obtain if its engine is 25 percent efficient and it encounters an average retarding force of 1000 N? Assume that the energy content of gasoline is 40 MJ/L.

5. A car with a mass of 1000 kg moves at 20 m/s. What braking force is needed to bring the car to a halt in 10 s?

6. What is the efficiency of a pulley system that will raise a 1000-N load a vertical distance of 1 m when 3000 J of effort are involved?

7. What is the efficiency of the body when a cyclist expends 1000 W of power to deliver mechanical energy to her bicycle at the rate of 100 W?

8. Your monthly electric bill is probably expressed in kilowatt-hours (kWh), a unit of energy delivered by the flow of 1 kW of electricity for 1 hr. How many joules of energy do you get when you buy 1 kWh?

4

GRAVITY AND SATELLITE MOTION

From the time of Aristotle, the circular motions of heavenly bodies were regarded as natural. The ancients believed that the stars, planets, and moon moved in divine circles, free from any impressed forces. As far as the ancients were concerned, this circular motion required no explanation. Isaac Newton, however, recognized that a force of some kind must be acting on the planets; otherwise, their paths would be straight lines. And whereas others of his time, influenced by Aristotle, would say any force would be directed along the planet's motion, Newton reasoned a force on the planets must be perpendicular to their motion, directed toward the center of their curved paths—toward the sun. This was the force of gravity, the same force that pulls apples off trees. Newton's stroke of intuition, that the force between the earth and apples is the same force that pulls moons and planets and everything else in our universe, was a revolutionary break with the prevailing notion that there were two sets of natural laws, one for earthly events and another altogether for motions in the heavens.

4.1 THE UNIVERSAL LAW OF GRAVITY

Figure 4.1 If the moon did not fall, it would follow the straight-line path. Because of its attraction to the earth, it falls along a curved path.

According to popular legend, Newton was sitting under an apple tree when he got the idea that gravity extended beyond the earth. Perhaps he looked up through tree branches toward the origin of a falling apple and noticed the moon. In any event, Newton had the insight to see that the force between the earth and a falling apple is the same force that pulls the moon in an orbital path around the earth, a path similar to a planet's path around the sun.

To test this hypothesis, Newton compared the fall of an apple with the fall of the moon. He realized that the moon falls in the sense that *it falls away from the straight line it would follow if there were no forces acting on it.* Because of its horizontal speed, it "falls around" the round earth. By simple geometry the moon's distance of fall per second could be compared to the distance that an apple or anything at that distance would fall in 1 second. Newton's calculations didn't check. Disappointed, but recognizing that brute fact must always win over a beautiful hypothesis, he placed his papers in a drawer where they remained for nearly 20 years. During this period he founded and developed the field of geometric optics for which he first became famous.

Newton's interest in mechanics was rekindled with the advent of a spectacular comet in 1680 and another two years later. He returned to the moon problem at the prodding of his astronomer friend, Edmond Halley, for whom the second comet was later named. He made corrections in the experimental data used in his earlier method and obtained excellent results. Only then did he publish what is one of the most far-reaching generalizations of the human mind: the **law of universal gravitation.***

Everything pulls on everything else in a beautifully simple way that involves only mass and distance. According to Newton, every mass attracts every other mass with a force that is directly proportional to the product of the masses involved and inversely proportional to the square of the distance separating them.

This statement can be expressed symbolically as

$$F \sim \frac{m_1 m_2}{d^2}$$

where m_1 and m_2 are the masses, and d is the distance between their centers. Thus, the greater the masses m_1 and m_2, the greater the force of attraction between them.†
The greater the distance of separation d, the weaker the force of attraction, but weaker as the inverse square of the distance between their centers of mass.

---

*This is a dramatic example of the painstaking effort and cross-checking that go into the formulation of a scientific theory. Contrast Newton's approach with the failure to "do one's homework," the hasty judgements, and the absence of cross-checking that so often characterize the pronouncements of less-than-scientific theories.

†Note the different role of mass here. Thus far we have treated mass as a measure of inertia, which is called *inertial mass*. Now we see mass as a measure of gravitational force, which in this context is called *gravitational mass*. It is experimentally established that the two are equal, and, as a matter of principle, the equivalence of inertial and gravitational mass is the foundation of Einstein's general theory of relativity.

The Universal Gravitational Constant, G

The proportionality form of the universal law of gravitation can be expressed as an exact equation when the constant of proportionality G, called the *universal gravitational constant,* is introduced. Then the equation is

$$F = G\frac{m_1 m_2}{d^2}$$

In words, the force of gravity between two objects is found by multiplying their masses, dividing by the square of the distance between their centers, and then multiplying this result by the constant G. The magnitude of G is the same as the magnitude of the force between two masses of 1 kilogram each, 1 meter apart: 0.0000000000667 newton. G has this magnitude, indicating an extremely weak force. The units of G are such as to make the force come out in newtons. In scientific notation,*

$$G = 6.67 \times 10^{-11} \text{ N·m}^2/\text{kg}^2$$

G was first measured long after the time of Newton by an English physicist, Henry Cavendish, in the eighteenth century. He accomplished this by measuring the tiny force between lead masses with an extremely sensitive torsion balance. A simpler method was later developed by Philipp von Jolly, who attached a spherical flask of mercury to one arm of a sensitive balance (Figure 4.2). After the balance was put in equilibrium, a 6-ton lead sphere was rolled beneath the mercury flask. The gravitational force between the two masses was equal to the weight that had to be placed on the opposite end of the balance to restore equilibrium. All the quantities m_1, m_2, F, and d were known, from which the ratio G was calculated:

$$\frac{F}{m_1 m_2/d^2} = 6.67 \times 10^{-11} \text{ N/kg}^2/\text{m}^2 = 6.67 \times 10^{-11} \text{ N·m}^2/\text{kg}^2$$

The value of G tells us that the force of gravity is a very weak force. It is the weakest of the presently known four fundamental forces. (The other three are the electromagnetic force and two kinds of nuclear forces.) We sense gravitation only

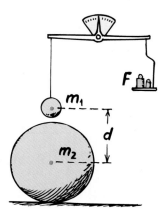

Figure 4.2 Jolly's method of measuring G. Balls m_1 and m_2 attract each other with a force F equal to the weights needed to restore balance.

---

*The numerical value of G depends entirely on the units of measurements we choose for mass, distance, and time. The international system of choice is: for mass, the kilogram; for distance, the meter; and for time, the second. Scientific notation is discussed in Appendix I at the end of this book.

when masses like that of the earth are involved. The force of attraction between you and a large ship on which you stand is too weak for ordinary measurement. The force of attraction between you and the earth, however, can be measured. It is your weight.

In addition to your mass, your weight also depends on your distance from the center of the earth. At the top of a mountain your mass is no different than it is anywhere else, but your weight is slightly less than at ground level because your distance from the center of the earth is greater.

Once the value of G was known, the mass of the earth was easily calculated. The force that the earth exerts on a mass of 1 kilogram at its surface is 9.8 newtons. The distance between the 1-kilogram mass and the center of mass of the earth is the earth's radius, 6.4×10^6 meters. Therefore, from $F = G(m_1 m_2/d^2)$, where m_1 is the mass of the earth,

$$9.8 \text{ N} = 6.67 \times 10^{-11} \text{ N·m}^2/\text{kg}^2 \frac{1 \text{ kg} \times m_1}{(6.4 \times 10^6 m)^2}$$

from which the mass of the earth $m_1 = 6 \times 10^{24}$ kilograms.

QUESTION

If there is an attractive force between all objects, why do we not feel ourselves gravitating toward massive buildings in our vicinity?

Gravity and Distance: The Inverse-Square Law

We can better understand how gravity is diluted with distance by considering how paint from a paint gun spreads with increasing distance (Figure 4.3). Suppose we position a paint gun at the center of a sphere with a radius of 1 meter, and a burst

Figure 4.3 The inverse-square law. Paint spray travels radially away from the nozzle of the can in straight lines. Like gravity, the "strength" of the spray obeys the inverse-square law.

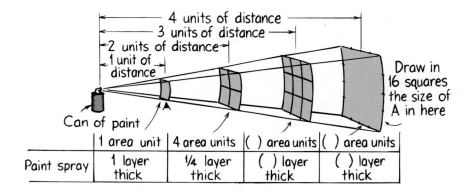

Paint spray	1 area unit	4 area units	( ) area units	( ) area units
	1 layer thick	¼ layer thick	( ) layer thick	( ) layer thick

ANSWER

Gravity pulls us to massive buildings and everything else in the universe. Physicist Paul A. M. Dirac, winner of the 1933 Nobel Prize, put it this way: "Pick a flower on earth and you move the farthest star!" How much we are influenced by buildings or how much interaction there is between flowers and stars is another story. The forces between us and buildings are relatively small because the masses are small compared to the mass of the earth. The forces due to the stars are small because of their great distances. These tiny forces escape our notice when they are overwhelmed by the overpowering attraction to the earth.

Figure 4.4 According to Newton's equation, her weight (not mass) decreases as she increases her distance from the earth's center (not surface).

of paint spray travels 1 meter to produce a square patch of paint that is 1 millimeter thick. How thick would the patch be if the experiment were done in a sphere with twice the radius? If the same amount of paint travels in straight lines for 2 meters, it will spread to a patch twice as tall and twice as wide. The paint would be spread over an area four times as big, and its thickness would be only $\frac{1}{4}$ millimeter. Can you see from the figure that for a sphere of radius 3 meters the thickness of the paint patch would be only $\frac{1}{9}$ millimeter? Can you see the thickness of the paint decreases as the square of the distance increases? This is known as the **inverse-square law**. The inverse-square law holds for gravity and for all phenomena wherein the effect from a localized source spreads uniformly throughout the surrounding space: the electric field about an isolated electron, light from a match, radiation from a piece of uranium, and sound from a cricket.

In using Newton's equation for gravity, it is important to emphasize that the distance term d is the distance between the centers of masses of the objects that are attracted to each other. Note in Figure 4.5 that the apple that normally weighs 1 newton at the earth's surface weighs only $\frac{1}{4}$ as much when it is twice the distance from the earth's center. The greater the distance from the earth's center, the less the weight of an object. A child who weighs 300 newtons at sea level will weigh only 299 newtons atop Mt. Everest. But no matter how great the distance, the earth's gravitational force approaches, but never reaches, zero. Even if you were transported to the far reaches of the universe, the gravitational influence of home would still be with you. It may be overwhelmed by the gravitational influences of nearer and/or more massive bodies, but it is there. The gravitational influence of every material object, however small or however far, is exerted through all of space.

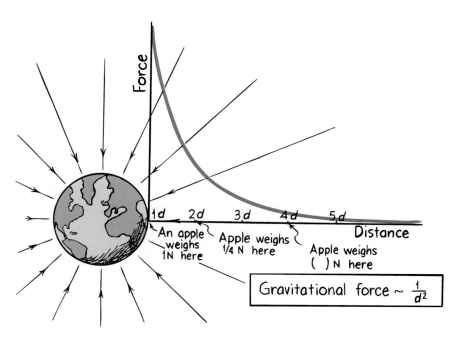

Figure 4.5 If an apple weighs 1 N at the earth's surface, it weighs only $\frac{1}{4}$ N twice as far from the center of the earth. At three times the distance, it weighs only $\frac{1}{9}$ N. What would it weigh at four times the distance? Five times? Gravity versus distance is plotted in color.

Force Fields

We know there is an attraction between the earth and the moon. This is *action at a distance,* because the earth and moon interact with each other without being in contact. But we can look at this in a different way: We can regard the moon as in contact with the *gravitational field* of the earth. A gravitational field is the space surrounding a massive body in which another mass experiences a force of attraction. A gravitational field is an example of a **force field,** for any mass in

for these fields. The field concept plays an in-between role in our thinking about the forces between different masses.

A familiar force field is the *magnetic field*. Iron filings sprinkled over a sheet of paper on top of a magnet reveal the shape of the magnet's magnetic field (Figure 4.6). The pattern of filings shows the strength and direction of the magnetic field at different points in the space around the magnet. Where the filings are close together, the field is strong. The dire of the filings shows the direction of the field at each point. Planet Earth is a giant magnet, and like all magnets is surrounded by a magnetic field. Evidence of the field is easily seen by the orientation of a magnetic compass. The pattern of the earth's gravitational field can be represented by field lines (Figure 4.7). Like the iron filings around a magnet, the field lines are closer together where the gravitational field is stronger. At each point on a field line, the direction of the field at that point is along the line. Arrows show the field direction. A particle, astronaut, spaceship, or any mass in the vicinity of the earth will be accelerated in the direction of the field line at that location. The strength of the earth's gravitational field, like the strength of its force on objects, follows the inverse-square law. The field is strongest near the earth's surface and weakens with increased distance from the earth.

Another example of a force field is the one that surrounds electrical charges; the electric field, which we shall study in Chapter 9. In Part 6 we'll learn that atoms align with the electric fields of other atoms to form molecules. In Chapter 10 we'll learn how magnets align with the magnetic field of the earth to become compasses. Then in Chapter 27 we'll learn how the moon similarly aligns with the earth's gravitational field, which is why only one side of the moon faces us. Force fields have far-reaching effects.

Figure 4.6 Top view of the magnetic field pattern of a bar magnet. Iron filings have been sprinkled on a sheet of paper that covers the magnet.

Figure 4.7 Field lines represent the gravitational field about the earth. Where the field lines are closer together, the field is stronger. Farther away, where the field lines are farther apart, the field is weaker.

the field space experiences a force. We think of rockets and distant space probes as interacting with the gravitational fields rather than with the masses responsible

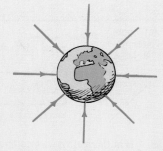

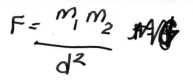

$$F = \frac{m_1 m_2}{d^2} \cancel{M/G}$$

Q UESTIONS

1. By how much does the gravitational force between two objects decrease when the distance between them is doubled? Tripled? Increased tenfold?

2. Consider an apple at the top of a tree that is pulled by earth gravity with a force of 1 N. If the tree were twice as tall, would the force of gravity be only $\frac{1}{4}$ as strong? Defend your answer.

4.2 TIDES

People near the seashore have always known there was a connection between the ocean tides and the moon. Newton was the first to show that tides are caused by *differences* in the gravitational pull between the moon and the earth on opposite sides of the earth. Gravitational force between the moon and earth is stronger on the side of the earth nearer to the moon, and weaker on the side of the earth farthest from the moon.

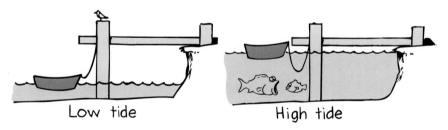

Low tide   High tide

Figure 4.8 Ocean tides.

To understand why the difference in pulls produces tides, consider a spherical ball of Jell-O. If you exert the same force on every part of the ball, it would remain as a sphere as it accelerates. But if you pulled harder on one side than the other, there would be a difference in accelerations and the ball would become elongated (Figure 4.9). That's what's happening to this big ball we're living on. The different pulls of the moon produce a stretching of the earth, the ocean bulges that extend nearly 1 meter above the average surface level of the ocean. The earth spins once per day, so a fixed point on earth passes beneath both of these bulges each day. This produces two sets of ocean tides per day—two high tides and two low tides. It turns out that

---

A NSWERS

1. It decreases to one-fourth, one-ninth, and one-hundredth.

2. No, because the twice-as-tall apple tree is not twice as far from the earth's center. The taller tree would have to have a height equal to the radius of the earth (6,370 km) before the weight of the apple reduces to $\frac{1}{4}$ N. Before its weight decreases by 1 percent, an apple or any object must be raised 32 km—nearly four times the height of Mt. Everest. So as a practical matter we disregard the effects of everyday changes in elevation.

Figure 4.9 A ball of Jell-O remains spherical when all parts of it are pulled equally in the same direction. When one side is pulled more than the other, however, the shape is elongated.

while the earth spins, the moon moves in its orbit and appears at the same position in our sky every 24 hours and 50 minutes, so the two-high-tide cycle is actually at 24-hour-and-50-minute intervals. This means tides do not occur at the same time every day.

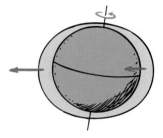

Figure 4.10 Two tidal bulges remain relatively fixed with respect to the moon while the earth spins daily beneath them.

The sun also contributes to ocean tides, about half as effectively as the moon—even though it pulls 180 times more on the earth than the moon. Why aren't tides due to the sun 180 times greater than lunar tides? Because the *difference* in gravitational pulls on opposite sides of the earth is very small (only about 0.017 percent, compared to 6.7 percent across the earth by the moon). When the sun, earth, and moon are all lined up, the tides due to the sun and the moon coincide and we have higher-than-average high tides and lower-than-average low tides. These are called **spring tides**. (Spring tides have nothing to do with the spring season.) Spring tides occur at the times of a new or full moon.

When the moon is half way between a new moon and a full moon, in either direction, the solar and lunar tides partly cancel each other. Then high tides are lower than average and the low tides are not as low as average low tides. These are called **neap tides.**

Because much of the earth is molten we have earth tides, though less pronounced than ocean tides. There are also atmospheric tides, which regulate the cosmic rays that reach the earth's surface. Our brief treatment of tides is quite simplified, for the tilt of the earth's axis, interfering land masses, friction with the ocean bottom, and other factors complicate tidal motions. Tides are interesting!

4.3 WEIGHT AND WEIGHTLESSNESS

When you step on a spring balance such as a bathroom scale, you compress a spring inside. When the pointer stops, the strong electrical forces between the molecules inside the spring material balance the gravitational attraction between you and the earth—nothing moves, for you and the scale are in static equilibrium. The pointer is calibrated to show your weight. If you stood on a bathroom scale in an accelerating elevator, you'd find variations in your weight. If the elevator accelerated upward,

Figure 4.11 The sensation of weight (your apparent weight) equals the force with which you press against the supporting floor. If the floor accelerates up or down, your apparent weight varies (while your weight *mg* remains the same).

Normal weight

Greater than normal weight

Less than normal weight

zero weight

the springs inside the bathroom scale would be more compressed and your weight reading would increase. If the elevator accelerated downward, the springs inside the scale would be less compressed and your weight reading would decrease. If the elevator cable broke and the elevator fell freely, the reading on the scale would go to zero. According to the reading, you would be weightless. Would you really be weightless? We can answer this question only if we agree on what we mean by *weight*.

Recall in Chapter 2 we defined **weight** as the gravitational force exerted on an object by the nearest most massive body. According to this definition, you would have weight whether or not you were falling, for you are still gravitationally attracted to the earth. So your weight and the weight you experience, your *apparent weight,* can be very different. We define **apparent weight** as the force an object exerts against the supporting floor or the weighing scales. According to this definition, you are as heavy as you feel; so in an elevator that accelerates downward, the supporting force of the floor is less and your apparent weight is less. If the elevator is in free fall, your apparent weight is zero (Figure 4.11). Even in this weightless condition, however, there is still a gravitational force acting on you, causing your downward acceleration. But gravity now is not felt as weight because there is no support force.

Consider an astronaut in orbit. She feels weightless because she is not supported by anything. There would be no compression in the springs of a bathroom scale placed beneath her feet because the bathroom scale is falling as fast as she is. (We'll see shortly that a satellite is an object in free fall; its sideways motion is great enough to insure it falls around the earth rather than into it.) If she drops a couple of objects in her vicinity, she drops with them and they remain in her vicinity, unlike what happens on the ground. Local effects of gravity seem to be eliminated. The body organs respond as though gravity forces were absent, and this gives the sensation of weightlessness. The astronaut experiences the same sensation in orbit that she would feel in a falling elevator—a state of free fall.

Note that the astronaut whose apparent weight is zero is still under the influence of gravitational force, which keeps her in orbit. To be truly weightless, she would have to be far out in space, well away from the earth, sun, and other attracting bodies, where gravitational forces are negligible. In this truly weightless environment, any motion would be in a straight-line path rather than the curved path of closed orbit.

Figure 4.12 Both are "weightless."

Figure 4.13 The half-dozen or so inhabitants in this proposed laboratory and docking facility will continually experience weightlessness in their zero-*g* environment.

4.4 PROJECTILE MOTION

Without gravity, toss a rock skyward and it follows a straight-line path. But because of gravity the path curves. A tossed rock, a cannonball, or any object that is projected by some means and continues in motion by its own inertia is called a **projectile.** To the early cannoneers, the curved paths of projectiles seemed hopelessly complicated. But now we see these paths surprisingly simple when we look at the horizontal and vertical components of motion separately.

The horizontal component of motion for a projectile is no more complex than the horizontal motion of a bowling ball rolling freely along a level bowling alley. If the retarding effect of friction can be ignored, the bowling ball moves at constant velocity. Since there is no force acting horizontally on the ball, it rolls of its own inertia and covers equal distances in equal intervals of time. It rolls without accelerating. The horizontal component of a projectile's motion is just like the bowling ball's motion along the alley (Figure 4.14b).

The vertical component of motion for a projectile following a curved path is just like the motion described in Chapter 1 for a freely falling object. Like a ball dropped in mid-air, the projectile moves in the direction of earth gravity and accelerates downward (Figure 4.14a). The increase in speed in the vertical direction causes successively greater distances to be covered in each successive equal-time interval.

The curved path of a projectile is a combination of horizontal and vertical motion. The horizontal component of motion for a projectile is completely independent of the vertical component of motion. Unless air drag or some other horizontal force acts, the constant horizontal velocity component is not affected by the vertical force of gravity. It is important to understand that each component acts independently of the other. Their combined effects produce the curved paths that projectiles follow.

Figure 4.14 (a) Drop a ball, and it accelerates downward and covers a greater vertical distance each second. (b) Roll it along a level surface, and its velocity is constant because no component of gravitational force acts horizontally.

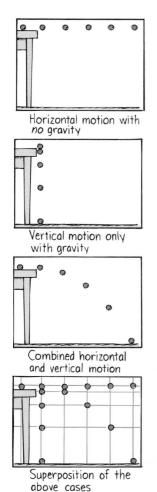

Horizontal motion with *no* gravity

Vertical motion only with gravity

Combined horizontal and vertical motion

Superposition of the above cases

Figure 4.15 Simulated photographs of a moving ball illuminated with a strobe light.

These ideas are neatly illustrated in the simulated multiple-flash exposure in Figure 4.15, which shows equally timed successive positions for a ball rolled off a horizontal table. Investigate the photo carefully, for there's a lot of good physics there. The curved path of the ball is best analyzed by considering the horizontal and vertical components of motion separately. There are two important things to notice. The first is that the ball's horizontal component of motion doesn't change as the falling ball moves sideways. The ball travels the same horizontal distance in the equal times between each flash. That's because there is no component of gravitational force acting horizontally. Gravity acts only downward, so the only acceleration of the ball is downward. The second thing to note from the photo is that the vertical positions become farther apart with time. The distances traveled vertically are the same as if the projected ball were simply dropped. Figure 4.16 is an actual strobe-light photograph of two balls being released simultaneously. Note the curvature of the ball's path. This curvature results from the combination of a horizontal velocity, due to an initial sideways kick, and a vertical acceleration, due to the constant force of gravity.

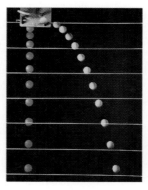

Figure 4.16 A strobe-light photograph of two golf balls released simultaneously from a mechanism that allows one ball to drop freely while the other is projected horizontally.

QUESTION

At the instant a horizontally held rifle is fired over a level range, a bullet held at the side of the rifle is released and drops to the ground. Which bullet, the one fired downrange or the one dropped from rest, strikes the ground first?

ANSWER

Both bullets fall the same vertical distance with the same acceleration *g* due to gravity and therefore strike the ground at the same time. Can you see that this is consistent with our analysis of Figures 4.15 and 4.16? We can reason this another way by asking which bullet would strike the ground first if the rifle were pointed at an upward angle. In this case, the bullet that is simply dropped would hit the ground first. Now consider the case where the rifle is pointed downward. The fired bullet hits first. So upward, the dropped bullet hits first; downward, the fired bullet hits first. There must be some angle at which there is a dead heat—where both hit at the same time. Can you see what it would be when the rifle is neither pointing upward nor downward—when it is horizontal?

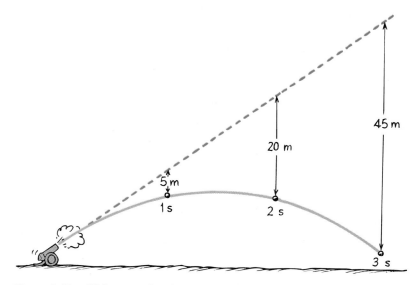

Figure 4.17 With no gravity, the projectile would follow a straight-line path (dashed line). But because of gravity, it falls beneath this line the same vertical distance it would fall if released from rest. Compare the distances fallen with those given in Table 1.3 in Chapter 1. (With $g = 9.8$ m/s^2, these distances are more accurately 4.9 m, 19.6 m, and 44.1 m.)

Consider a cannonball shot at an upward angle (Figure 4.17). Pretend for a moment that there is no gravity; according to the law of inertia, the cannonball would follow the straight-line path shown by the dashed line. But there is gravity, so this doesn't happen. What really happens is that the cannonball continually falls beneath the imaginary line until it finally strikes the ground. Get this: The vertical distance it falls beneath any point on the dashed line is the same vertical distance it would fall if it were dropped from rest and had been falling for the same amount of time. This distance, as introduced in Chapter 1, is given by $d = \frac{1}{2}gt^2$, where t is the elapsed time.

We can put it another way: Shoot a projectile skyward at some angle and pretend there is no gravity. After so many seconds t, it should be at a certain point along a straight-line path. But because of gravity, it isn't. Where is it? The answer is that it's directly below this point. How far below? The answer in meters is $5t^2$ (or, more accurately, $4.9t^2$). How about that!

Note another thing from Figure 4.17. The cannonball moves equal horizontal distances in equal time intervals. That's because no acceleration takes place horizontally. The only acceleration is vertical, in the direction of earth's gravity. The vertical distance it falls below the imaginary straight-line path during equal time intervals continually increases with time.

QUESTIONS

1. Suppose the cannonball in Figure 4.17 were fired faster. How many meters below the dashed line would it be at the end of the 5 s?

2. If the horizontal component of the cannonball's velocity were 20 m/s, how far downrange would the cannonball be at the end of 5 s?

Figure 4.18 Vertical and horizontal components of a projectile's velocity.

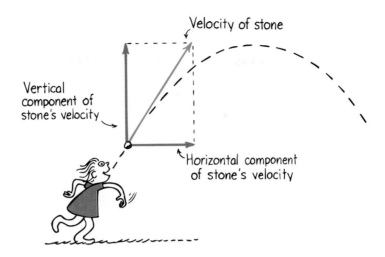

In Figure 4.18 we see vectors representing both horizontal and vertical components of velocity for a projectile. Likewise for Figure 4.19, where we see that the horizontal component is everywhere the same, and only the vertical component changes. Note also that the actual velocity is represented by the vector that forms the diagonal of the rectangle formed by the vector components. At the top of the trajectory the vertical component vanishes to zero, so the actual velocity there is the horizontal component of velocity at all other points. Everywhere else the magnitude of velocity is greater (just as the diagonal of a rectangle is greater than either of its sides).

Figure 4.20 shows the paths of several projectiles in the absence of air resistance, all with the same initial speed but different projection angles. Notice that these projectiles reach different *altitudes,* or heights above the ground. They also have dif-

Figure 4.19 The velocity of a projectile at various points along its path. Note that the vertical component changes and the horizontal component is the same everywhere.

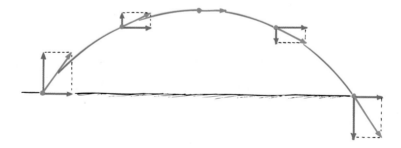

Ⓐ N S W E R S ·

1. Assuming $g = 10$ m/s^2, the vertical distance beneath the dashed line at the end of 5 s would be 125 m [$d = 5t^2 = 5(5)^2 = 5(25) = 125$]. (If we use $g = 9.8$ m/s^2, then this distance is 122.5 m/s). Interestingly enough, this distance doesn't depend on the angle of the cannon. If air resistance is neglected, any projectile will fall $5t^2$ meters below where it would have reached if there were no gravity.

2. In the absence of air resistance, the cannonball will travel a horizontal distance of 100 m [$d = vt = (20)(5) = 100$]. Note that since gravity acts only vertically and there is no acceleration in the horizontal direction, the cannonball travels equal horizontal distances in equal times. This distance is simply its horizontal component of velocity multiplied by the time (and not $5t^2$, which applies only to vertical motion under the acceleration of gravity).

Figure 4.20 Ranges of a projectile shot at the same speed at different projection angles.

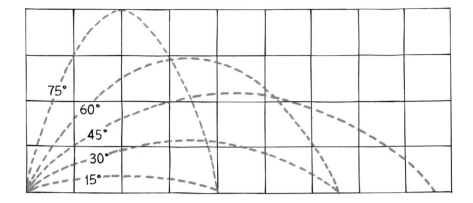

ferent *horizontal ranges,* or distances traveled horizontally. The remarkable thing to note from Figure 4.20 is that the same range is obtained from two different projection angles—angles that add up to 90 degrees! An object thrown into the air at an angle of 60 degrees, for example, will have the same range as if it were thrown at the same speed at an angle of 30 degrees. For the smaller angle, of course, the object remains in the air for a shorter time.

Figure 4.21 If the batted ball's speed were the same for all angles and air resistance weren't a factor, maximum range would be attained when a ball is batted at an angle of 45°. Actually a more forceful ball-bat contact is made lower on the bat, so a batted ball's speed is greater at a lower angle. This combined with air resistance makes about 35° the favored angle for maximum range.

QUESTIONS

1. A projectile is shot at an angle into the air. If air resistance is negligible, what is the acceleration of its vertical component of motion? Of its horizontal component of motion?

2. At what part of its trajectory does a projectile have minimum speed?

ANSWERS

1. The acceleration in the vertical direction is g because the force of gravity is along the vertical direction. (Recall from Chapter 2 that acceleration is always in the direction of the force that acts on an object.) The acceleration is zero in the horizontal direction because no horizontal force acts on the projectile.

2. The speed of a projectile is minimum at the top of its path. If it is launched vertically, its speed at the top is zero. If it is projected at an angle, the vertical component of speed is zero at the top, leaving only the horizontal component. So the speed at the top is equal to the horizontal component of the projectile's velocity at any point.

Figure 4.22 In the presence of air resistance, the trajectory of a high-speed projectile falls short of a parabolic path.

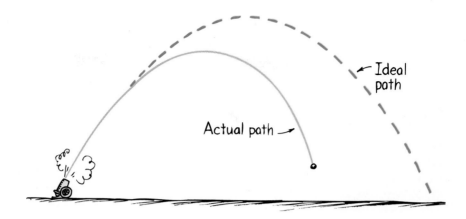

We have emphasized the special case of projectile motion without air resistance. When there is air resistance, the range of the projectile is shorter and the altitude of the projectile less (Figure 4.22).

When air resistance is small enough to be negligible, a projectile will rise to its maximum height in the same time it takes to fall from that height to the ground (Figure 4.23). This is because its deceleration by gravity while going up is the same as its acceleration by gravity while coming down. The speed it loses while going up is therefore the same as the speed it gains while coming down. So the projectile arrives at the ground with the same speed it had when it was projected from the ground.

Baseball games normally take place on level ground. For the short-range projectile motion on the playing field, the earth can be considered to be flat because the flight of the baseball is not affected by the earth's curvature. For very long range projectiles, however, the curvature of the earth's surface must be taken into account. We'll now see that if an object is projected fast enough, it will fall all the way around the earth and be an earth satellite.

Figure 4.23 Without air resistance, speed lost while going up equals speed gained while coming down; time going up equals time coming down.

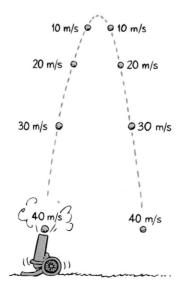

Hang Time Revisited

Figure 4.24 Michael Jordan's hang time depends only on the vertical component of his lift-off velocity when he makes his jump.

Recall our discussion of hang time back in Chapter 1. We stated that the time one is airborne during a jump is independent of horizontal speed. We see now why this is so—horizontal and vertical components of motion are independent of each other. We can apply the rules of projectile motion to jumping. Whatever maneuverings are employed to attain maximum launching velocity, once the feet leave the ground, the athlete can be considered to be a projectile. Barring any air resistance effects, no amount of leg or arm pumping or other bodily motions can change the time in the air—the hang time. Hang time depends only on the vertical component of lift-off velocity. Unless you are able to muster a greater upward jumping force by running, your hang time will not be increased by horizontal motion. The illusion of a prolonged hang time, however, is en-

hanced by horizontal motion, for at the top of the jump horizontal motion continues while vertical motion ceases. Your hang time in a bus at rest and in a bus that's moving is the same.

Can time in the air be increased by the transformation of kinetic energy to potential energy? Not unless there is a mechanism for doing so, like a rope or a pole. Tarzan could grab a vine while running and swing higher than he could jump, providing the vine didn't break. Similarly for a pole vaulter: A pole vaulter transforms much of the kinetic energy of running into bending the pole and increasing its elastic potential energy, which then is transformed into gravitational potential energy. But without a means of converting kinetic energy into potential energy, horizontal motion has no effect on hang time.

QUESTION

The boy on the tower throws a ball 20 m downrange as shown in Figure 4.25. What is his pitching speed?

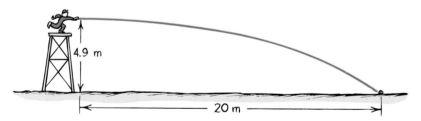

Figure 4.25 How fast is the ball thrown?

ANSWER

The ball is thrown horizontally, so the pitching speed is horizontal distance divided by time. A horizontal distance of 20 m is given, but the time is not stated. However, while the ball is moving horizontally at constant speed, it is falling under gravity a vertical distance of 4.9 m, which takes 1 s. So pitching speed $v = d/t = (20 \text{ m})/(1 \text{ s}) = 20 \text{ m/s}$. It is interesting to note that consideration of the equation for constant speed, $v = d/t$, guides thinking about the crucial factor in this problem—the time.

4.5 SATELLITE MOTION

Consider the stone thrower in Figure 4.26. If gravity did not act on the stone, the stone would follow a straight-line path shown by the dashed line. But there is gravity, so the stone falls below this straight-line path. In fact, 1 second after the stone leaves the thrower's hand it will have fallen a vertical distance of 4.9 meters below the dashed line—whatever the throwing speed. It is important to understand this, for it is the crux of satellite motion.

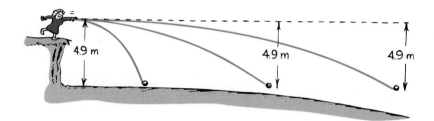

Figure 4.26 Throw a stone at any speed and one second later it will have fallen 4.9 m below where it would have been without gravity.

An earth satellite is simply a projectile that falls *around* the earth rather than *into* it. The speed of the satellite must be great enough to ensure that its falling distance matches the earth's curvature.* A geometrical fact about the curvature of our earth is that its surface drops a vertical distance of 4.9 meters for every 8000 meters tangent to the surface (Figure 4.27).† This means that if you were floating in a calm ocean, you would be able to see only the top of a 4.9-meter mast on a ship 8 kilometers away. So if a baseball could be thrown fast enough to travel a horizontal distance of 8 kilometers during the time (1 second) it takes to fall 4.9 meters, then it would follow the curvature of the earth. A little thought will show that this speed is 8 kilometers per second. If this doesn't seem fast, convert it to kilometers per hour and you get an impressive 29,000 kilometers per hour (or 18,000 miles per hour)!

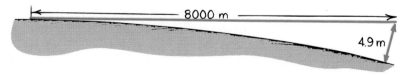

Figure 4.27 Earth's curvature—not to scale!

Figure 4.28 If the speed of the stone and the curvature of its trajectory are great enough, the stone may become a satellite.

---

*The conventional definition of *to fall* is "to get closer to the earth"; satellites such as the moon do not do this. In science we will find many cases where the technical definition differs from the conventional. For example, we say "the sun rises" and "the moon sets," but technically they do not.

†A tangent to a circle or to the earth's surface is a straight line that touches the circle or surface at one place, so it is parallel to the circle at the point of contact.

At this speed, atmospheric friction would burn the baseball or even a piece of iron to a crisp. This is the fate of grains of sand and other meteorites that graze the earth's atmosphere and burn up, appearing as "falling stars." That is why satellites such as the space shuttles are launched to altitudes of 150 kilometers or so. A common misconception is that satellites orbiting at high altitudes are free from gravity. Nothing could be further from the truth. The force of gravity on a satellite 150 kilometers above the earth's surface is nearly as great as at the surface. The high altitude is to put the satellite beyond the earth's atmosphere, not beyond the earth's gravity.

Figure 4.29 The space shuttle is a projectile in a constant state of free fall. Because of its tangential velocity, it falls around the earth rather than vertically into it.

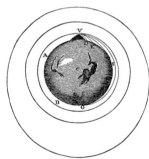

Figure 4.30 "The greater the velocity . . . with which [a stone] is projected, the farther it goes before it falls to the earth. We may therefore suppose the velocity to be so increased, that it would describe an arc of 1, 2, 5, 10, 100, 1000 miles before it arrived at the earth, till at last, exceeding the limits of the earth, it should pass into space without touching."—Isaac Newton, *System of the World.*

Satellite motion was understood by Newton, who reasoned that the moon was simply a projectile circling the earth under the attraction of gravity. This concept is illustrated in a drawing by Newton, shown in Figure 4.30. He compared motion of the moon to a cannonball fired from the top of a high mountain. He imagined that the mountain top was above the earth's atmosphere, so air resistance would not impede the motion of the cannonball. If a cannonball were fired with a small horizontal speed, it would follow a curved path and soon hit the earth below. If it were fired faster, its path would be less curved and it would hit the earth farther away. If the cannonball were fired fast enough, Newton reasoned, the curved path would become a circle and the cannonball would circle indefinitely. It would be in orbit.

Both cannonball and moon have "sideways" velocity, or *tangential velocity*—the velocity parallel to the Earth's surface—sufficient to insure motion *around* the earth rather than *into* it. If there is no resistance to reduce its speed, the moon "falls" around and around the earth indefinitely. Similarly with the planets that continually fall around the sun in closed paths. Why don't the planets crash into the sun? They don't because of their tangential velocities. What would happen if their tangential

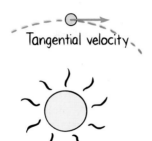

Tangential velocity

Figure 4.31 The tangential velocity of the earth about the sun allows it to fall around the sun rather than directly into it. If this tangential velocity were reduced to zero, what would be the fate of the earth?

velocities were reduced to zero? The answer is simple enough: Their motion would be straight toward the sun and they would indeed crash into it. Any objects in the solar system with insufficient tangential velocities have long ago crashed into the sun. What remains is the harmony we observe.

Circular Orbits

So the tangential velocity a projectile needs to orbit the earth is 8 kilometers per second. That's because in the 1 second the projectile travels 8 kilometers horizontally, it falls a vertical distance of 4.9 meters—the same vertical distance the earth curves for each 8-kilometer tangent. So an 8-kilometers-per-second cannonball fired horizontally from Newton's mountain would follow the earth's curvature and coast around the earth again and again (provided the cannoneer and the cannon got out of the way). Fired slower, the cannonball would strike the earth's surface; fired faster it would overshoot a circular orbit as we will discuss shortly. Newton calculated the speed for circular orbit, and since such a cannon-muzzle velocity was clearly impossible, he did not foresee people launching satellites (he did not foresee multistage rockets).

Note that in circular orbit the speed of a satellite is not changed by gravity; only the direction changes. We can understand this by comparing a satellite in circular orbit with a bowling ball rolling along a bowling alley. Why doesn't the gravity that acts on the bowling ball change its speed? Because gravity is not pulling forward or backward; gravity pulls straight downward. The bowling ball has no component of gravitational force along the direction of the alley (Figure 4.32).

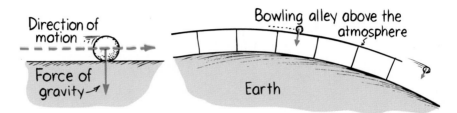

Direction of motion

Force of gravity

Bowling alley above the atmosphere

Earth

Figure 4.32 (a) The force of gravity on the bowling ball is at 90° to its direction of motion, so it has no components of force to pull it forward or backward, and the ball rolls at constant speed. (b) The same is true even if the bowling alley is larger and remains "level" with the curvature of the earth. (c) If the ball moves at 8 km/s with no air resistance, would it need the alley?

The same is true for a satellite in circular orbit. In circular orbit a satellite is always moving in a direction perpendicular to the force of gravity that acts on it. The satellite does not move in the direction of the force, which would increase its speed, nor does it move in a direction against the force, which would decrease its speed. Instead, the satellite moves at right angles to the gravitational force that acts on it. With no component of motion along this force, no change in speed occurs—only change in direction. A satellite in circular orbit around the earth coasts parallel to the surface of the earth at constant speed.

QUESTIONS

Consider a ball rolling along a bowling alley that completely circles the earth, and is elevated high enough so air resistance can be neglected.

1. Why would the force of gravity not change the speed of the ball?

2. If a section of the alley were cut away to leave a large gap, how fast must the ball travel to clear the gap and continue its motion as usual? At this speed, what would be the maximum gap for unchanged motion?

For a satellite close to the earth, the period (the time for a complete orbit about the earth) is about 90 minutes. For higher altitudes, the orbital speed is less and the period is longer. For example, communication satellites located in orbit 5.5 earth radii above the surface of the earth have a period of 24 hours. This period matches the period of daily earth rotation. For an orbit around the equator, these satellites stay above the same point on the ground. The moon is even farther away and has a period of 27.3 days. The higher the orbit of a satellite, the less its speed and the longer its period.*

ANSWERS

1. A change in speed requires a force or component of force along the direction of travel. In this case the alley and gravitational force on the ball are everywhere perpendicular to each other, so there is no force component in the direction of motion.

2. To clear the gap without bumping into the edge of the alley, the ball must have orbital speed. Then its curved path matches that of the alley's surface. In this case the alley can be completely removed to have in effect a 360° gap, because the ball would be in earth orbit anyway!

---

*The speed of a satellite in circular orbit is given by $v = \sqrt{GM/d}$ and the period of satellite motion is given by $T = 2\pi\sqrt{d^3/GM}$, where G is the universal gravitational constant, M is the mass of the earth (or whatever body the satellite orbits), and d is the altitude of the satellite measured from the center of the earth or parent body.

QUESTIONS

1. One of the beauties of physics is that there are usually different ways to view and explain a given phenomenon. Is the following explanation valid? Satellites remain in orbit instead of falling to the earth because they are beyond the main pull of earth's gravity.

2. Satellites in close circular orbit fall about 4.9 m during each second of orbit. Why doesn't this distance accumulate and send satellites crashing into the earth's surface?

Elliptical Orbits

If a projectile just above the drag of the atmosphere is given a horizontal speed somewhat greater than 8 kilometers per second, it will overshoot a circular path and trace an oval-like path, an **ellipse.**

An ellipse is a specific curve: the closed path taken by a point that moves in such a way that the sum of its distances from two fixed points (called *foci*) is constant. For a satellite orbiting a planet, one focus is at the center of the planet; the other focus is empty. An ellipse can be easily constructed by using a pair of tacks, one at each focus, a loop of string, and a pencil (Figure 4.33). The closer the foci are to each other, the closer the ellipse is to a circle. When both foci are together, the ellipse is a circle. So we see that a circle is a special case of an ellipse.

Figure 4.33 A simple method for constructing an ellipse.

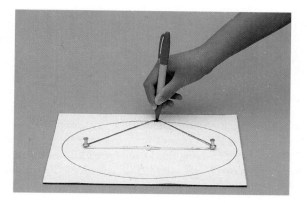

ANSWERS

1. No, no, a thousand times no! If any moving object were beyond the pull of gravity, it would move in a straight line and would not curve around the earth. Satellites remain in orbit because they are being pulled by gravity, not because they are beyond it. For the altitudes of most earth satellites, the force of gravity is only a few percent weaker than at the earth's surface.

2. In each second, the satellite falls about 4.9 m below the straight-line tangent it would have taken if there were no gravity. The earth's surface also curves 4.9 m beneath a straight-line 8-km tangent. The process of falling with the curvature of the earth continues from tangent line to tangent line, so the curved path of the satellite and the curve of the earth's surface "match" all the way around the earth. Satellites in fact do crash to the earth's surface from time to time, but this is principally because they encounter air resistance in the upper atmosphere that decreases their orbital speed.

Figure 4.34 The shadows cast by the ball are all ellipses, one for each lamp in the room. The point at which the ball makes contact with the table is the common focus of all three ellipses.

Unlike the constant speed of a satellite in a circular orbit, speed varies in an elliptical orbit. When the initial speed is greater than 8 kilometers per second, the satellite overshoots a circular path and moves away from the earth, against the force of gravity. It therefore loses speed. Like a rock thrown into the air, it slows to a point where it no longer recedes and then begins to fall back toward the earth. The speed it loses in receding is regained as it falls back toward the earth, and it finally crosses its original path with the same speed it had initially (Figure 4.35). The procedure repeats over and over, and an ellipse is traced each cycle.

Figure 4.35 Elliptical orbit. An earth satellite that has a speed somewhat greater than 8 km/s overshoots a circular orbit (a) and travels away from the earth. Gravitation slows it to a point where it no longer leaves the earth (b). It falls toward the earth gaining the speed it lost in receding (c) and overshoots as before in a repetitious cycle.

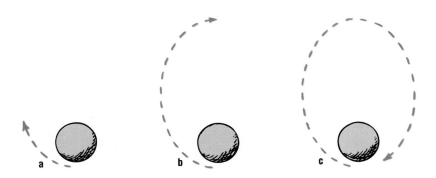

It's interesting to note that the curved path of a projectile such as a tossed baseball or a cannonball is actually a tiny segment of an ellipse that extends within and just beyond the center of the earth (Figure 4.36a). In Figure 4.36b, we see several paths of cannonballs fired from Newton's mountain. All ellipses have the center of the earth as one focus. As muzzle velocity is increased, the ellipses are less eccen-

Figure 4.36 (a) The parabolic path of the cannonball is part of an ellipse that extends within the earth. The earth's center is the far focus. (b) All paths of the cannonball are ellipses. For less than orbital speeds, the center of the earth is the far focus; for circular orbit, both foci are the earth's center; for greater speeds, the near focus is the earth's center.

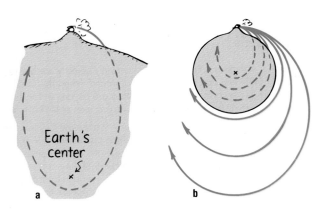

Earth's center

Figure 4.37 The initial thrust of the rocket pushes it up above the atmosphere. Another thrust to a horizontal speed of at least 8 km/s is required if the rocket is to fall around rather than into the earth.

tric (wider); and when muzzle velocity reaches 8 kilometers per second, the ellipse rounds into a circle and does not intercept the earth's surface. The cannonball coasts in circular orbit. At greater muzzle velocities, the orbiting cannonball traces the familiar external ellipse.

Putting a payload into earth orbit requires control over the speed and direction of the rocket that carries it above the atmosphere. A rocket initially fired vertically is intentionally tipped from the vertical course; then, once above the drag of the atmosphere, it is aimed *horizontally,* whereupon the payload is given a final thrust to 8 kilometers per second or more. This is shown in Figure 4.37, where for the sake of simplicity the payload is the entire single-stage rocket. We see that with the proper tangential velocity it falls around the earth, rather than into it, and becomes an earth satellite.

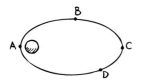

QUESTION

The orbital path of a satellite is shown in the sketch at the left. In which of the marked positions A through D does the satellite have the greatest speed? Lowest speed?

4.6 ENERGY CONSERVATION AND SATELLITE MOTION

Recall from Chapter 3 that an object in motion possesses kinetic energy (KE) by virtue of its motion. An object above the earth's surface possesses potential energy (PE) by virtue of its position. Everywhere in its orbit, a satellite has both KE and PE with respect to the body it orbits. The sum of the KE and PE will be a constant all through the orbit. The simplest case occurs for a satellite in circular orbit.

In circular orbit the distance between the body's center and the satellite does not change, which means the PE of the satellite is the same everywhere in orbit.

ANSWER

The satellite has its greatest speed as it whips around A and has its lowest speed at position C. Beyond C it gains speed as it falls back to A to repeat its cycle.

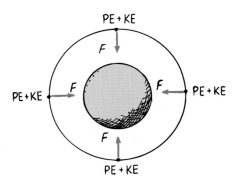

Figure 4.38 The force of gravity on the satellite is always toward the center of the body it orbits. For a satellite in circular orbit, no component of force acts along the direction of motion. The speed and, thus, the KE do not change.

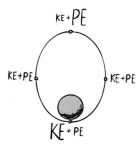

Figure 4.39 The sum of KE and PE for a satellite is constant at all points along its orbit.

This component of force does work on the satellite

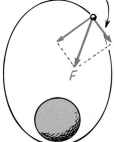

Figure 4.40 In elliptical orbit, a component of force exists along the direction of the satellite's motion. This component changes the speed and, thus, the KE. (The perpendicular component changes only the direction.)

Then, by the conservation of energy, the KE must also be constant. So a satellite in circular orbit coasts at an unchanging PE, KE, and speed (Figure 4.38).

In elliptical orbit the situation is different. Both speed and distance vary. PE is greatest when the satellite is farthest away (at the apogee) and least when the satellite is closest (at the perigee). Note in Figure 4.39 that the KE will be least when the PE is most, and the KE will be most when the PE is least. At every point in the orbit, the *sum* of KE and PE is the same.

At all points along the elliptical orbit there is a component of gravitational force in the direction of motion of the satellite (with two exceptions, the apogee and perigee, where the force is perpendicular only). The component of force in the direction of motion changes the speed of the satellite. Or by the work-energy relationship we can say (this component of force) × (distance moved) = ΔKE. Either way, when the satellite gains altitude and moves against this component, its speed and KE decrease. The decrease continues to the apogee. Once past the apogee, the satellite moves in the same direction as the component, and the speed and KE increase. The increase continues until the satellite whips past the perigee and repeats the cycle.

QUESTION

Why does the force of gravity change the speed of a satellite when it is in an elliptical orbit, but not when it is in a circular orbit?

ANSWER

At any point on its path, the direction of motion of a satellite is always tangent to its path. If a component of force exists along this tangent, then the acceleration of the satellite will involve a change in speed as well as direction. In circular orbit the gravitational force is always perpendicular to the direction of motion of the satellite, just as every part of the circumference of a circle is perpendicular to the radius. So there is no component of gravitational force along the tangent, and only the direction of motion changes, not the speed. But when the satellite moves in directions that are not perpendicular to the force of gravity, as in an elliptical path, there is a component of force along the direction of motion that changes the speed of the satellite. From a work-energy point of view, a component of force along the distance the satellite moves does work to change its KE.

4.7   ESCAPE SPEED

We know that a cannonball fired horizontally at 8 kilometers per second from New-ton's mountain would find itself in orbit. But what would happen if the cannonball were instead fired at the same speed *vertically*? It would rise to some maximum height, reverse direction, and then fall back to earth. Then the old saying "What goes up must come down" would hold true, just as surely as a stone tossed skyward will be returned by gravity (unless, as we shall see, its speed is too great).

In today's spacefaring age, it is more accurate to say "What goes up may come down," for there is a critical speed at which a projectile is able to outrun gravity and escape the earth. This critical speed is called **escape speed** or, if direction is involved, *escape velocity*. From the surface of the earth, escape speed is 11.2 kilometers per second.* Launch a projectile at any speed greater than that and it will leave the earth, traveling slower and slower, never stopping due to earth gravity. Gravitational interaction with the earth becomes weaker and weaker with increased distance, its speed becomes less and less, though both are never reduced to zero. The payload outruns the gravity of the earth. It escapes.

So the escape speed from the surface of Planet Earth is 11.2 km/s.† The escape speeds of other bodies in the solar system are shown in Table 4.1. Note that the escape

TABLE 4.1   Escape Speeds at the Surface of Bodies in the Solar System

Astronomical Body	Mass (earth masses)	Radius (earth radii)	Escape Speed (km/s)
Sun	330 000	109	620
Sun (at a distance of the Earth's orbit)	23 000		42.5
Jupiter	318	11	61.0
Saturn	95.2	9	37.0
Neptune	17.3	3.4	25.4
Uranus	14.5	3.7	22.4
Earth	1.00	1.00	11.2
Venus	0.82	0.96	10.4
Mars	0.11	0.53	5.2
Mercury	0.05	0.38	4.3
Moon	0.01	0.27	2.4

*From an energy point of view, as it continues outward, its PE increases and its KE decreases. By energy conservation, the PE of a 1-kg mass infinitely far from earth is 60 million joules. So to put a payload that far from the earth's surface requires an initial KE of at least 60 MJ/kg. This corresponds to a speed of 11.2 km/s, whatever the mass involved.

Escape speed, from any planet or any body, is given by $v = \sqrt{2GM/d}$, where G is the universal gravitational constant, M is the mass of the attracting body, and d is the distance from its center. (At the surface of the body, d would simply be the radius of the body.)

†This might well be called the *maximum falling speed*. Any object, however far from earth, released from rest and allowed to fall to earth only under the influence of the earth's gravity would not exceed 11.2 km/s.

Figure 4.41 Pioneer 10, launched from earth in 1972, escaped from the solar system in 1984 and is wandering in interstellar space.

speed from the sun is 620 kilometers per second at the surface of the sun. Even at a distance equaling that of the earth's orbit, the escape speed from the sun is 42.5 kilometers per second, considerably more than the escape speed from the earth. An object projected from the earth at a speed greater than 11.2 kilometers per second but less than 42.5 kilometers per second will escape the earth but not the sun. Rather than recede forever, it will take up an orbit around the sun.

The first probe to escape the solar system, Pioneer 10, was launched from earth in 1972 with a speed of only 15 kilometers per second. The escape was accomplished by directing the probe into the path of oncoming Jupiter. It was whipped about by Jupiter's great gravitation, picking up speed in the process—similar to the increase in the speed of a ball encountering an oncoming bat when it departs from the bat. Its speed of departure from Jupiter was increased enough to exceed the sun's escape speed at the distance of Jupiter. Pioneer 10 passed the orbit of Pluto in 1984. Unless it collides with another body, it will wander indefinitely through interstellar space. Like a note in a bottle cast into the sea, Pioneer 10 contains information about the earth that might be of interest to extraterrestrials, in hopes that it will one day wash up and be found on some distant "seashore."

It is important to point out that the escape speeds for different bodies refer to the initial speed given by a brief thrust, after which there is no force to assist motion. One could escape the earth at any sustained speed more than zero, given enough time. For example, suppose a rocket is launched to a destination such as the moon. If the rocket engines burn out when still close to the earth, the rocket needs a minimum speed of 11.2 kilometers per second. But if the rocket engines can be sustained for long periods of time, the rocket could go to the moon without ever attaining 11.2 kilometers per second.

Interestingly enough, the accuracy with which an unmanned rocket reaches its destination is not accomplished by staying on a preplanned path or by getting back on that path if it strays off course. No attempt is made to return the rocket to its original path. Instead, the control center in effect asks, "Where is it now with respect to where it ought to go? What is the best way to get there from here, given its present situation?" With the aid of high-speed computers, the answers to these questions are used in finding a new path. Corrective thrusters put the rocket on this new path. This process is repeated over and over again all the way to the goal.*

---

*Is there a lesson to be learned here? Suppose you find that you are off course. You may, like the rocket, find it more fruitful to take a course that leads to your goal as best plotted from your present position and circumstances, rather than try to get back on the course you plotted from a previous position and under, perhaps, different circumstances.

SUMMARY OF TERMS

The Universal Law of Gravity Every mass in the universe attracts every other mass with a force that for two masses is directly proportional to the product of their masses and inversely proportional to the square of the distance separating them:

$$F = G\frac{m_1 m_2}{d^2}$$

Inverse-square law A law relating the intensity of an effect to the inverse square of the distance from the cause:

$$\text{Intensity} \sim \frac{1}{\text{distance}^2}$$

Gravity follows an inverse-square law, as do the effects of electric, magnetic, light, sound, and radiation phenomena.

Weightlessness A condition wherein gravitational pull appears to be lacking.

Spring tide A high or low tide that occurs when the sun, earth, and moon are all lined up so that the tides due to the sun and moon coincide, making the high tides higher than average and the low tides lower than average.

Neap tide A tide that occurs when the moon is midway between new and full, in either direction. Tides due to the sun and moon partly cancel, making the high tides lower than average and the low tides higher than average.

Satellite A projectile or small celestial body that orbits a larger celestial body.

Ellipse The closed oval-like curve wherein the sum of the distances from any point on the curve to both foci is a constant. When the foci are together at one point, the ellipse is a circle. The farther apart the foci, the more eccentric the ellipse.

Escape speed The speed that a projectile, space probe, or similar object, must reach to escape the gravitational influence of the earth or celestial body to which it is attracted.

. . . .

REVIEW QUESTIONS

The Universal Law of Gravity

1. In Newton's insight, what did a falling apple have in common with the moon?

2. How can the moon "fall" without getting closer to the earth?

3. How does the force of gravity between two objects depend on their masses?

4. How does the force of gravity depend on the distance between two objects?

The Universal Gravitational Constant, G

5. How was G measured?

6. What is the magnitude of the gravitational force between the earth and your body?

7. What is the magnitude of the gravitational force between the earth and a 1-kg mass?

Gravity and Distance: The Inverse-Square Law

8. Why is the force of gravity between a pair of objects reduced to one-fourth instead of one-half when the distance between them is doubled?

9. Hanging from a tree is a certain apple that weighs 1 N. Twice as high above the ground on the same tree is another apple of the same mass. Why is its weight practically the same and not $\frac{1}{4}$ N?

Tides

10. Why do both the sun and the moon exert a greater gravitational force on one side of the earth than the other?

11. Do tides depend more on the strength of gravitational pull or on the *difference* in strengths? Explain.

12. Why are all tides greatest at the time of a full moon or new moon?

Weight and Weightlessness

13. Would the springs inside a bathroom scale be more compressed or less compressed if you weighed yourself in an elevator that accelerated upward? Downward?

14. Would the springs inside a bathroom scale be more compressed or less compressed if you weighed yourself in an elevator that moved upward at constant velocity? Downward at s constant velocity?

15. Distinguish between *weightlessness* and *apparent weightlessness*.

Projectile Motion

16. Why does the horizontal component of a projectile's motion remain constant?

17. Why does the vertical component of a projectile's motion undergo acceleration?

18. How does the vertical distance a projectile falls below an otherwise straight-line path compare with the vertical distance it would fall from rest in the same time?

19. What angle from the horizontal will give the greatest range for a projectile launched from ground level?

20. How does air resistance affect the range of projectiles?

Satellite Motion

21. How far does a baseball fall beneath a straight-line tangent to its motion in 1 s?

22. How far does an earth satellite in close orbit fall beneath a straight-line tangent to its motion in 1 s?

23. What exactly is *tangential velocity*?

24. Why don't the satellites like the space shuttle simply fall to earth like tossed baseballs?

Circular Orbits

25. Why doesn't the force of gravity change the speed of a satellite in circular orbit?

26. How much time is taken for a complete revolution of a satellite in close orbit about the earth?

27. For orbits of greater altitude, is the period greater or less?

Elliptical Orbits

28. Why does the force of gravity change the speed of a satellite in an elliptical orbit?

29. At what part of an elliptical orbit does a satellite have the greatest speed? The least speed?

Energy Conservation and Satellite Motion

30. Why is kinetic energy a constant for a satellite in circular orbit?

31. Why is kinetic energy a variable for a satellite in an elliptical orbit?

32. With respect to the apogee and perigee of an elliptical orbit, where is the gravitational potential greatest? Least?

33. Is the sum of kinetic and potential energies a constant for satellites in circular orbits, elliptical orbits, or both?

Escape Speed

34. What is the minimum speed for orbiting the earth in close orbit? The maximum speed? What happens above this speed?

35. How was Pioneer 10 able to escape the solar system at a speed less than escape speed?

HOME PROJECTS

1. Hold up your thumb and first two fingers and make a V sign. Place a strong rubber band across your thumb and first finger. This represents the force of gravity between the sun and the earth. Place a medium strength rubber band across your thumb and second finger to represent the force of gravity between the sun and moon. Then place a weak rubber band across your first two fingers to represent the force of gravity between the moon and earth. Note how all fingers pull on each other. Likewise for the gravitational pulls between the sun, earth, and the moon.

2. Hold your hands outstretched, one twice as far from your eyes as the other, and make a casual judgement as to which hand looks bigger. Most people see them to be about the same size, while many see the nearer hand as slightly bigger. Almost nobody upon casual inspection sees the nearer hand as four times as big. But by the inverse-square law, the nearer hand should appear twice as tall and twice as wide and therefore occupy four times as much of your visual field as the farther hand. Your belief that your hands are the same size is so strong that you likely overrule this information. Now if you overlap your hands slightly and view them with one eye closed, you'll see the nearer hand as clearly bigger. This raises an interesting question: What other illusions do you have that are not so easily checked?

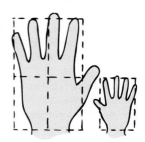

EXERCISES

1. Gravitational force acts on all bodies in proportion to their masses. Why, then, doesn't a heavy body fall faster than a light body?

2. Which weighs more, a sheet of aluminum foil or the same sheet crumpled into a tight wad?

3. Which planets, those closer to the sun than the earth or those farther from the sun than the earth, have a period greater than 1 earth year?

4. What are the magnitude and direction of the gravitational force that acts on a man who weighs 700 N at the surface of the earth?

5. The earth and the moon are attracted to each other by gravitational force. Does the more massive earth attract the less massive moon with a force that is greater, smaller, or the same as the force with which the moon attracts the earth? (With an elastic band stretched between your thumb and forefinger, which is pulled more strongly by the band, your thumb or your forefinger?)

6. If the mass of the earth somehow increased, with all other factors remaining the same, would your weight also increase? (Hint: Let the equation for gravitational force guide your thinking.)

7. A small light source located 1 m in front of a 1-m² opening illuminates a wall behind. If the wall is 1 m behind the opening (2 m from the light source), the illuminated area covers 4 m². How many square meters will be illuminated if the wall is 3 m from the light source? 5 m? 10 m?

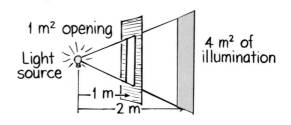

1 m² opening
Light source
4 m² of illumination
1 m
2 m

8. The planet Jupiter is more than 300 times as massive as earth, so it might seem that a body on the surface of Jupiter would weigh 300 times as much as on earth. But it so happens that a body would scarcely weigh three times as much on the surface of Jupiter as it would on the surface of the earth. Can you think of an explanation for why this is so? (Hint: Let the terms in the equation for gravitational force guide your thinking.)

9. From the data in the preceding exercise, estimate how much bigger in size Jupiter is compared to earth.

10. Why do the passengers of high-altitude jet planes feel the sensation of weight while passengers in an orbiting space vehicle such as the space shuttle do not?

11. If the earth made one revolution each 90 min instead of each 24 h, would you press against the earth's surface if you were at the equator? At the poles? In the middle of the United States? Explain.

12. If you were in a car that drove off the edge of a cliff, why would you feel weightless? Would gravity still be acting on you in this state?

13. If you were in a freely falling elevator and you dropped a pencil, you'd see the pencil hovering. Is the pencil falling? Explain.

14. Explain why the following reasoning is wrong. "The sun attracts all bodies on the earth. At midnight, when the sun is directly below, it pulls on an object in the same direction as the pull of the earth on that object; at noon, when the sun is directly overhead, it pulls on an object in a direction opposite to the pull of the earth. Therefore, all objects should be somewhat heavier at midnight than they are at noon." (Hint: Relate this to the preceding two exercises.)

15. If the mass of the earth increased, your weight would correspondingly increase. But if the mass of the sun increased, your weight would not be affected at all. Why?

16. Most people today know that the ocean tides are caused by the gravitational influence of the moon. And most people therefore think that the gravitational pull of the moon on the earth is greater than the gravitational pull of the sun on the earth. What do you think?

17. Would ocean tides exist if the gravitational pull of the moon (and sun) were somehow equal on all parts of the world? Explain.

18. Why aren't high ocean tides exactly 12 h apart?

19. With respect to spring and neap ocean tides, when are the lowest tides? That is, when is it best for digging clams?

20. Whenever the ocean tide is unusually high, will the following low tide be unusually low? Defend your answer in terms of "conservation of water." (If you slosh water in a tub so it is extra deep at one end, will the other end be extra low?)

21. The Mediterranean Sea has very little sediment churned up and suspended in its waters, mainly because of the absence of any substantial ocean tides. Why do you suppose the Mediterranean Sea has practically no tides? Similarly, are there tides in the Black Sea? Great Salt Lake? Your county reservoir? A glass of water? Explain.

22. The value of g at the earth's surface is about 9.8 m/s². What is the value of g at a distance of twice the earth's radius?

23. Which requires more fuel—a rocket going from the earth to the moon or a rocket coming from the moon to the earth? Why?

24. If a cannonball is fired from a tall mountain, gravity changes its speed all along its trajectory. But if it is fired fast enough to go into circular orbit, gravity does not change its speed at all. Explain.

25. Since the moon is gravitationally attracted to the earth, why doesn't it simply crash into the earth?

26. Does the speed of a falling object depend on its mass? Does the speed of a satellite in orbit depend on its mass? Defend your answers.

27. If you have ever watched the launching of an earth satellite, you may have noticed that the rocket departs from a vertical course and continues its climb at an angle. Why?

28. Why are satellites normally sent into orbit by firing them in an easterly direction (the direction in which the earth spins)?

29. In the sketch on the left, a ball gains KE when rolling down a hill because work is done by the component of weight (F) that acts in the direction of motion. Sketch in the similar component of gravitational force that does work to change the KE of the satellite on the right.

30. Why is work done by the force of gravitation on a satellite when it is in an elliptical orbit, but not when it is in a circular orbit?

31. What is the shape of the orbit when the velocity of the satellite is everywhere perpendicular to the force of gravity?

32. If a pair of satellites at different altitudes but moving in the same direction are seen from the earth to pass overhead, which of the two overtakes the other?

33. If a flight mechanic drops a wrench from a high-flying jumbo jet, it crashes to earth. If an astronaut on the orbiting space shuttle drops a wrench, does it crash to earth also? Defend your answer.

34. The orbiting space shuttle travels at 8 km/s with respect to the earth. Suppose it projects a capsule rearward at 8 km/s with respect to the shuttle. Describe the path of the capsule with respect to the earth.

35. Would the speed of a satellite in close circular orbit about Jupiter be greater than, equal to, or less than 8 km/s?

36. The orbital velocity of the earth about the sun is 30 km/s. If the earth were suddenly stopped in its tracks, it would simply fall radially into the sun. Devise a plan whereby a rocket loaded with radioactive wastes could be fired into the sun for permanent disposal. How fast and in what direction with respect to the earth's orbit should the rocket be fired?

37. Escape speed from the surface of the earth is 11.2 km/s, but a space vehicle could escape from the earth at half this speed and less. Explain.

38. What is the maximum possible speed of impact upon the surface of the earth for a faraway body initially at rest that falls to earth by virtue of the earth's gravity only?

39. At which of the indicated positions does the satellite in elliptical orbit experience the greatest gravitational force? The greatest speed? The greatest velocity? The greatest momentum? The greatest kinetic energy? The greatest gravitational potential energy? The greatest total energy? The greatest angular momentum? The greatest acceleration?

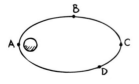

40. Many schools and hospitals could have been constructed with the funds used in the last decade to probe and explore the solar system. How many of these schools and hospitals do you speculate would in fact have been constructed if the various space missions were not funded?

• •

PROBLEMS

1. The value of g at the earth's surface is about 9.8 m/s^2. What is the value of g at a distance from the earth's center that is 4 times the earth's radius?

2. A 1-kg mass at the earth's surface is gravitationally attracted to earth with a force of 9.8 N. Calculate the force of gravity with which the 1-kg mass on earth is attracted to the moon. (The moon's mass is 7.4×10^{22} kg and its distance is 3.8×10^8 m).

3. The moon is about 3.8×10^5 km from the earth. Find its average orbital speed about the earth.

4. Calculate the speed in m/s at which the earth revolves about the sun. You may assume the orbit is nearly circular.

5. Calculate the force of gravity between the earth (mass = 6×10^{24} kg) and the sun (mass = 2×10^{30} kg, distance = 1.5×10^{11} m).

6. Suppose the force of gravity between the sun and earth vanishes. Instead a steel cable between the sun and earth keeps the earth in orbit. Estimate the thickness (diameter) of such a cable. You need to know that the strength of steel is 2×10^{11} N/m^2. That is, a steel cable with a cross-sectional area of 1 m^2 will support a force of 2×10^{11} N.

5

FLUID MECHANICS

Liquids and gases have the ability to flow; hence, they are called *fluids*. Because they are both fluids we find that they obey similar mechanical laws.* How is it that iron boats don't sink in water or that helium balloons don't sink from the sky? Why is it impossible to breathe through a snorkel when you're under more than a meter of water? Why do your ears pop when riding an elevator? How do hydrofoils and airplanes attain lift? To discuss fluids it is important to introduce two concepts—*density* and *pressure*.

5.1 DENSITY

An important property of materials, whether in the solid, liquid, or gaseous phase is the measure of compactness: **density**. We think of density as the "lightness"

---

*The phases of matter—solid, liquid, gas, and plasma—are described from a chemical point of view in Chapter 19.

TABLE 5.1   Densities of Some Substances

Material	Density	
	Grams per Cubic Centimeter	Kilograms per Cubic Meter
Liquids		
Mercury	13.6	1,360
Glycerin	1.26	1,260
Seawater	1.03	1,025
Water at 4°C	1.00	1,000
Benzene	0.90	899
Ethyl alcohol	0.81	806
Solids		
Osmium	22.5	22,480
Platinum	21.5	21,450
Gold	19.3	19,320
Uranium	19.0	19,050
Lead	11.3	11,344
Silver	10.5	10,500
Copper	8.9	8,920
Brass	8.6	8,560
Iron	7.8	7,800
Tin	7.3	7,280
Aluminum	2.7	2,702
Ice	0.92	917

or "heaviness" of materials of the same size. It is a measure of how much mass is squeezed into a given space; it is the amount of matter per unit volume:

$$\text{Density} = \frac{\text{mass}}{\text{volume}}$$

The densities of a few materials are given in Table 5.1. Mass is measured in grams or kilograms and volume in cubic centimeters (cm^3) or cubic meters (m^3).* A gram of material is equal to the mass of 1 cubic centimeter of water at a temperature of 4°C. Therefore, water has a density of 1 gram per cubic centimeter. Mercury, density 13.6 grams per cubic centimeter, is therefore 13.6 times as massive as an equal volume of water. Osmium, a hard, bluish-white metallic element, is the densest solid substance on earth.

A quantity known as weight density, commonly used when discussing liquid pressure, can be expressed by the amount of weight a body per unit volume:†

$$\text{Weight density} = \frac{\text{weight}}{\text{volume}}$$

Figure 5.1 When the volume of the bread is reduced, its density increases.

---

*A cubic meter is a sizable volume and contains a million cubic centimeters, so there are a million grams of water in a cubic meter (or, equivalently, a thousand kilograms of water in a cubic meter). Hence, 1 g/cm^3 = 1000 kg/m^3.

†Weight density is common to U.S. Customary units (formerly British or English), in which one cubic foot of fresh water (almost 7.5 gallons) weighs 62.4 pounds. So fresh water has a weight density of 62.4 lb/ft^3. Salt water is a bit denser, 64 lb/ft^3.

$\boxed{\text{Q}}$UESTIONS

1. Which has the greater density—1 kg of water or 10 kg of water?
2. Which has the greater density—5 kg of lead or 10 kg of aluminum?

5.2 PRESSURE

Figure 5.2 Although the weight of both books is the same, the upright book exerts greater pressure against the table.

Place a book on a bathroom scale and whether you place it on its back, on its side, or balanced on a corner, it still exerts the same force. The weight reading is the same. Now balance the book on the palm of your hand and you sense a difference—the *pressure* of the book depends on the area over which the force is distributed. There is a difference between force and pressure. **Pressure** is defined as the force exerted over a unit of area, such as a square meter or square foot:*

$$\text{Pressure} = \frac{\text{force}}{\text{area}}$$

A dramatic illustration of pressure is shown in Figure 5.3. The physics author applies appreciable force when he breaks the cement block with a sledge hammer. Yet his friend, who is sandwiched between two beds of sharp nails, is unharmed. This is because the force is distributed over more than 200 nails that make contact with his body. The combined surface area of the nails results in a tolerable pressure that does not puncture the skin. Force and pressure are different from each other.

Pressure in a Liquid

When you swim under water, you can feel the water pressure acting against your eardrums. The deeper you swim, the greater the pressure. What causes this pressure? It is simply the weight of the water above pushing against you. If you swim twice as deep, there is twice the weight of water above, and you therefore feel twice the

$\boxed{\text{A}}$NSWERS

1. The density of any amount of water is the same: 1 g/cm³ or equivalently, 1000 kg/m³, which means that the mass of water that would exactly fill a thimble of volume 1 cubic centimeter would be 1 gram; or the mass of water that would fill a 1-cubic-meter tank would be 1000 kg. 1 kg of water would fill a tank only a thousandth as much, 1 liter. 10 kg would fill 10 liters. Nevertheless, the important concept is that the ratio of mass/volume is the same for any amount of water.

2. Density is a *ratio* of weight or mass per volume, and this ratio is greater for any amount of lead than for any amount of aluminum—see Table 5.1.

---

*Pressure may be measured in any unit of force divided by any unit of area. The standard international (SI) unit of pressure, the newton per square meter, is called the *pascal* (Pa), after the seventeenth-century theologian and scientist, Blaise Pascal. A pressure of 1 Pa is very small and approximately equals the pressure exerted by a dollar bill resting flat on a table. Science types more often use kilopascals (1 kPa = 1000 Pa).

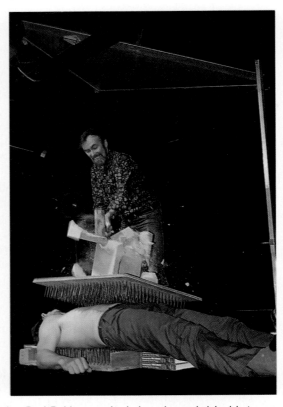

Figure 5.3 Paul Hewitt applies a force to fellow physics teacher Paul Robinson, who is bravely sandwiched between beds of sharp nails. The driving force per nail is not enough to puncture the skin. From an inertia point of view, is Robinson better off that the block is massive? From the point of view of energy, is he better off that the block breaks?

water pressure. At three times the depth, you feel three times the water pressure, and so on. Liquid pressure depends also on the weight density of the liquid. If you were submerged in a liquid more dense than water, the pressure would be proportionally greater.*

Liquid pressure = weight density × depth

It is important to note that pressure does not depend on amount of liquid. You feel the same pressure a meter deep in a small pool as you do a meter deep in the middle of the ocean. This is illustrated by the connecting vases shown in

---

*This is derived from the definitions of pressure and density. Consider an area at the bottom of a vessel of liquid. The weight of the column of liquid directly above this area produces pressure. From the definition Weight density = weight/volume, we can express this weight of liquid as Weight = weight density × volume, where the volume of the column is simply the area multiplied by the depth. Then we get

$$\text{Pressure} = \frac{\text{force}}{\text{area}} = \frac{\text{weight}}{\text{area}} = \frac{\text{weight density} \times \text{volume}}{\text{area}}$$

$$= \frac{\text{weight density} \times \cancel{\text{area}} \times \text{depth}}{\cancel{\text{area}}} = \text{weight density} \times \text{depth}$$

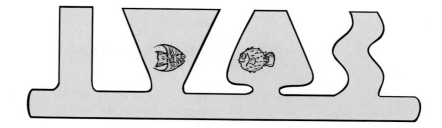

Figure 5.4 Liquid pressure is the same for any given depth below the surface, regardless of the shape of the containing vessel.

Figure 5.4. If the pressure at the bottom of a large vase were greater than the pressure at the bottom of a neighboring, narrower vase, the greater pressure would force water sideways and then up the narrower vase to a higher level. We find, however, that this doesn't happen. Pressure is dependent upon depth, not volume. That a liquid tends to seek its own level can be demonstrated by filling a garden hose with water and holding the two ends upright. The water levels will be the same even when the two ends are held far apart. Pressure is depth dependent and not volume dependent, so we see there is a reason why water seeks its own level.

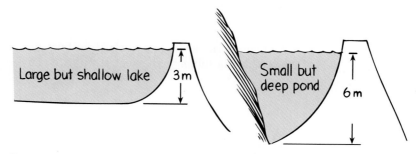

Figure 5.5 The average water pressure acting against the dam depends on the average depth of the water and not on the volume of water held back. The large shallow lake exerts only one-half the pressure that the small deep pond exerts.

Figure 5.6 The forces in a liquid that produce pressure against a surface add up to a net force that is perpendicular to the surface.

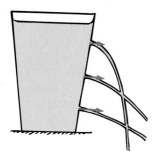

Figure 5.7 Water pressure pushes perpendicularly to the sides of its container and increases with increasing depth.

Another interesting fact about liquid pressure is that it is exerted equally in all directions. For example, if we are submerged in water, no matter which way we tilt our heads we feel the same amount of water pressure on our ears. Because a liquid can flow, the pressure isn't only downward. We know pressure acts upward when we try to push a beach ball beneath the water surface. The bottom of a boat is certainly pushed upward by water pressure. And we know water pressure acts sideways when we see water spurting sideways from a leak in an upright can. Pressure in a liquid at any point is exerted in equal amounts in all directions.

When liquid presses against a surface, there is a net force directed perpendicular to the surface (Figure 5.6). If there is a hole in the surface, the liquid spurts at right angles to the surface before curving downward due to gravity (Figure 5.7). At greater depths the pressure is greater and the speed of the escaping liquid is greater.*

---

*The speed of liquid out of the hole is $\sqrt{2gh}$, where h is the depth below the free surface. Interestingly enough, this is the same speed the water or anything else would have if freely falling the same distance h.

5.3 BUOYANCY IN A LIQUID

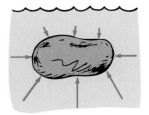

Figure 5.8 The greater pressure against the bottom of a submerged object produces an upward buoyant force.

Anyone who has ever lifted a submerged object out of water is familiar with buoyancy, the apparent loss of weight of submerged objects. For example, lifting a large boulder off the bottom of a riverbed is a relatively easy task as long as the boulder is below the surface. When it is lifted above the surface, however, the force required to lift it is considerably more. This is because when the boulder is submerged, the water exerts an upward force on it that is exactly opposite in direction to gravity. This upward force is called the **buoyant force** and is a consequence of pressure increasing with depth. Figure 5.8 shows why the buoyant force acts upward. Pressure is exerted everywhere against the object in a direction perpendicular to its surface. The arrows represent the magnitude and directions of points of pressure. Forces that produce pressures against the sides are at equal depths and cancel one another. Pressure is greatest against the bottom of the boulder because the bottom of the boulder is at a greater depth. Since this upward pressure is greater than the downward pressure acting against the top, the pressures do not cancel, and there is a net pressure upward. This net pressure distributed over the bottom of the boulder produces a net upward force, which is the buoyant force.

If the weight of the submerged object is greater than the buoyant force, the object will sink. If the weight is equal to the buoyant force on the submerged object, it will remain at any level, like a fish. If the buoyant force is greater than the weight of the completely submerged object, it will rise to the surface and float.

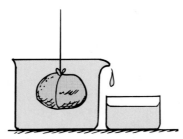

Figure 5.9 When a stone is submerged, it displaces a volume of water equal to the volume of the stone.

Understanding buoyancy requires understanding the meaning of the expression "volume of water displaced." If a stone is placed in a container brimful of water, some water will overflow (Figure 5.9). Water is *displaced* by the stone. A little thought will tell us that the *volume*—that is, the amount of space taken up or the number of cubic centimeters—*of water displaced* is equal to the *volume of the stone*. Place any object into a container partly filled with water, and the level of the surface rises (Figure 5.10). By how much? Exactly the same as if a volume of water were poured in that was equal to the volume of the submerged object. This is a good method for determining the volume of irregularly shaped objects: *A completely submerged object always displaces a volume of liquid equal to its own volume.*

Water displaced

Figure 5.10 The raised level equals that which would occur if water equal to the stone's volume were poured in.

5.4 ARCHIMEDES' PRINCIPLE

Figure 5.11 A liter of water occupies a volume of 1000 cm³, has a mass of 1 kg, and weighs 9.8 N. Its density may therefore be expressed as 1kg/L and its weight density as 9.8 N/L. (Seawater is slightly denser, 10 N/L).

The relationship between buoyancy and displaced liquid was first discovered in the third century BC by the Greek philosopher Archimedes. It is stated as follows:

An immersed body is buoyed up by a force equal to the weight of the fluid it displaces.

This relationship is called **Archimedes' principle**. It is true of liquids and gases, which are both fluids. If an immersed body displaces 1 kilogram of fluid, the buoyant force acting on it is equal to the weight of 1 kilogram.* By *immersed*, we mean either *completely* or *partially submerged*. If we immerse a sealed 1-liter container halfway into the water, it will displace a half-liter of water and be buoyed up by the weight of a half-liter of water. If we immerse it completely (submerge it), it will be buoyed up by the weight of a full liter or 1 kilogram of water. Unless the container is compressed, the buoyant force will equal the weight of 1 kilogram at *any* depth, as long as it is completely submerged. This is because at any depth it can displace no greater volume of water than its own volume. And the weight of this volume of water (not the weight of the submerged object!) is equal to the buoyant force.

If a 25-kilogram object displaces 20 kilograms of fluid upon immersion, its apparent weight will be equal to the weight of 5 kilograms. Note that in Figure 5.12 the 3-kilogram block has an apparent weight equal to the weight of 1 kilogram when submerged. The apparent weight of a submerged object is its weight in air minus the buoyant force.

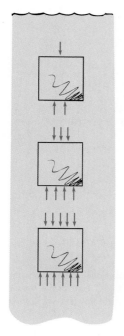

Figure 5.13 The difference in the upward and downward force acting on the submerged block is the same at any depth.

Figure 5.12 A 3-kg block weighs more in air than in water. When submerged, its loss in weight is the buoyant force, which is equal to the weight of water displaced.

Perhaps your instructor will summarize Archimedes' principle by way of a numerical example to show that the difference between the upward-acting and the downward-acting forces due to the similar pressure differences on a submerged cube is numerically identical to the weight of fluid displaced. It makes no difference how deep the cube is placed, for although the pressures are greater with increasing depths, the *difference* in pressure up against the bottom of the cube and down against the top of the cube is the same at any depth (Figure 5.13). Whatever the shape of the submerged body, the buoyant force is equal to the weight of fluid displaced.

---

*A kilogram is not a unit of force but a unit of mass. So, strictly speaking, the buoyant force is not 1 kg, but the *weight* of 1 kg, which is 9.8 N. We could as well say that the buoyant force is 1 *kilogram weight*, not simply 1 kg.

Archimedes and the Gold Crown

According to popular legend, Archimedes (287–212 BC) had been given the task of determining whether a crown made for King Hieron II was of pure gold or whether it contained some cheaper metals such as silver. Archimedes' problem was to determine the density of the crown without destroying it. He could weigh the gold, but determining its volume was a problem. Story has it that Archimedes came to the solution while bathing in the public baths of Syracuse and immediately rushed naked through the streets shouting "Eureka, Eureka" ("I have found it, I have found it").

Archimedes' insight preceded Newton's law of motion, from which Archimedes' Principle can be derived, by almost 2000 years. What Archimedes had discovered was a simple and accurate way of finding the volume of an irregular object—the displacement method of determining volumes. Once he knew both the weight and volume, he could calculate the density. Then the density of the crown could be compared to the density of gold.

QUESTIONS

1. Does Archimedes' principle tell us that if an immersed object displaces 10 N of fluid, the buoyant force on the object is 10 N?

2. A 1-L container completely filled with lead has a mass of 11.3 kg and is submerged in water. What is the buoyant force acting on it?

3. A boulder is thrown into a deep lake. As it sinks deeper and deeper into the water, does the buoyant force increase? Decrease?

4. Since buoyant force is the upward force that a fluid exerts on a body and we learned in Chapter 2 that forces produce accelerations, why doesn't a submerged body accelerate?

ANSWERS

1. Yes. Looking at it another way, the immersed object pushes 10 N of fluid aside. The displaced fluid reacts by pushing back on the immersed object with 10 N.

2. 9.8 N. The buoyant force is equal to the weight of 1 kg (9.8 N) because the volume of water displaced is 1 L, which has a mass of 1 kg and a weight of 9.8 N. The 11.3 kg of the lead is irrelevant; 1 L of anything submerged will displace 1 L of water and be buoyed upward with a force of 1 kg weight, or 9.8 N.

3. Buoyant force does not change as the boulder sinks because the boulder displaces the same volume of water at any depth. Since water is practically incompressible, its density is the same at all depths; hence, the weight of water displaced, or the buoyant force, is the same at all depths.

4. It does accelerate if the buoyant force is not balanced by other forces that act on it—the force of gravity and fluid resistance. The net force on a submerged body is the result of the force the fluid exerts (buoyant force), the weight of the body, and, if moving, the force of fluid friction.

5.5 FLOTATION

Iron is much denser than water and therefore an iron chunk sinks, but an iron ship floats. Why is this so? Consider a solid 1-ton block of iron. Iron is nearly eight times as dense as water, so when it is submerged it will displace only $\frac{1}{8}$ ton of water, which is hardly enough to keep it from sinking. Suppose we reshape the same iron block into a bowl, as shown in Figure 5.14. It still weighs 1 ton. When we place it in the water, it settles into the water, displacing a greater volume of water than before. The deeper it is immersed, the more water it displaces and the greater the buoyant force acting on it. When the buoyant force equals 1 ton, it will sink no further.

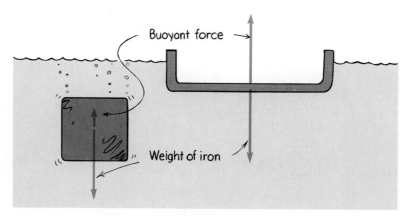

Figure 5.14 An iron block sinks, while the same block shaped like a bowl floats.

When the iron boat displaces a weight of water equal to its own weight, it floats. This is sometimes called the **principle of flotation,** which states:

A floating object displaces a weight of fluid equal to its own weight.

Every ship, every submarine, and every dirigible must be designed to displace a weight of fluid equal to its own weight. Thus, a 10,000-ton ship must be built wide enough to displace 10,000 tons of water before it sinks too deep in the water. The same holds true in air, which is a fluid. A dirigible or balloon that weighs 100 tons displaces at least 100 tons of air. If it displaces more, it rises; if it displaces less, it falls. If it displaces exactly its weight, it hovers at constant altitude.

Figure 5.15 The weight of a floating object equals the weight of the water displaced by the submerged part.

Figure 5.16   A floating object displaces a weight of fluid equal to its own weight.

Since the buoyant force upon a body equals the weight of the fluid it displaces, denser fluids will exert a greater buoyant force upon a body than less dense fluids of the same volume. A ship therefore floats higher in salt water than in fresh water because salt water is slightly denser than fresh water. In the same way, a solid chunk of iron will sink in water but float in mercury.

Q UESTIONS

Fill in the blanks for these statements:

1. The volume of a submerged body is equal to the _____ of the fluid displaced.
2. The weight of a floating body is equal to the _____ of the fluid displaced.

Figure 5.17   The same ship empty and loaded. How does the weight of its load compare to the weight of extra water displaced?

Did you notice in our discussion of liquids that Archimedes' Principle and the Law of Flotation were in terms of *fluids*, not liquids? That's because although liquids and gases are different physical phases of matter, they are both fluids and we find that much of the mechanics of liquids and gases are essentially the same. We now turn our attention to the mechanics of gases in particular.

5.6 PRESSURE IN A GAS

There are similarities and there are differences between gases and liquids. The primary difference between a gas and a liquid is the distance between molecules. In a gas, the molecules are far apart and free from the cohesive forces that dominate their motions when in the liquid and solid phases. Their motions are less restricted. A gas expands, fills all space available to it, and exerts a pressure against its container. Only when the quantity of gas is very large, such as the earth's atmosphere or a star, do the gravitational forces limit the size or determine the shape of the mass of gas.

A NSWERS

1. volume.
2. weight.

Boyle's Law

We know that the air pressure inside the inflated tires of an automobile is considerably greater than the atmospheric pressure outside. The density of air inside is also more than that of the air outside. To understand the relation between pressure and density, think of the molecules of air (primarily nitrogen and oxygen) inside the tire. Inside the tire, the molecules behave like tiny Ping-Pong balls, perpetually moving helter skelter and banging against the inner walls. Their impacts on the inner surface of the tire produce a jittery force that appears to our coarse senses as a steady push. This pushing force averaged over a unit of area provides the pressure of the enclosed air.

Suppose there are twice as many molecules in the same volume (Figure 5.18). Then the air density is doubled. If the molecules move at the same average speed—or, equivalently, if they have the same temperature—then, to a close approximation, the number of collisions will be doubled. This means the pressure is doubled. So pressure is proportional to density.

Figure 5.18 When the density of gas in the tire is increased, pressure is increased.

We could instead double the density by simply compressing the air to half its volume. Consider the cylinder with the movable piston in Figure 5.19. If the piston is pushed downward so that the volume is half the original volume, the density of molecules will be doubled, and the pressure will correspondingly be doubled. Decrease the volume to a third its original value, and the pressure will be increased by three, and so forth.

Notice in these examples that the product of pressure and volume is the same. For example, a doubled pressure multiplied by a halved volume gives the same value as a tripled pressure multiplied by a one-third volume. In general, we can say that the product of pressure and volume for a given mass of gas is a constant as long as the temperature does not change. "Pressure × volume" for a quantity of gas at one time is equal to any "different pressure × different volume" at any other time. In shorthand notation,

$$P_1 V_1 = P_2 V_2$$

Figure 5.19 When the volume of gas is decreased, density and therefore pressure are increased.

where P_1 and V_1 represent the original pressure and volume, respectively, and P_2 and V_2 the second, or final, pressure and volume. This relationship is called **Boyle's Law,** after Robert Law, the seventeenth-century physicist who is credited with its discovery.*

Boyle's law applies to ideal gases. An ideal gas is one in which the disturbing effects of the intermolecular forces and molecular volume are so small that they are negligible. Air and other gases under normal pressures approach ideal gas conditions.

UESTIONS

1. A piston in an airtight pump is withdrawn so that the volume of the air chamber is increased three times. What is the change in pressure?

2. A scuba diver 10.3 m deep breathes compressed air. If she holds her breath while returning to the surface, by how much does the volume of her lungs tend to increase?

5.7 ATMOSPHERIC PRESSURE

We live at the bottom of an ocean of air. The atmosphere, much like the water in a lake, exerts a pressure. One of the most celebrated experiments demonstrating the pressure of the atmosphere was conducted in 1654 by Otto von Gueicke, burgomaster of Magdeburg and inventor of the vacuum pump. Von Gueicke placed together two copper hemispheres about $\frac{1}{2}$ meter in diameter to form a sphere, as shown in Figure 5.20. He set a gasket made of a ring of leather soaked in oil and wax between them to make an airtight joint. When he evacuated the sphere with his vacuum pump, two teams of eight horses each were unable to pull the hemispheres apart.

When the air pressure inside a cylinder like that shown in Figure 5.21 is reduced, there is an upward force on the piston. This force is large enough to lift a heavy weight. If the inside diameter of the cylinder is 12 centimeters or greater, a person can be lifted by this force.

ANSWERS

1. The pressure in the piston chamber is reduced to one-third. This is the principle that underlies a mechanical vacuum pump.

2. Atmospheric pressure can support a column of water 10.3 m high, so the pressure in water due to the weight of the water alone equals atmospheric pressure at a depth of 10.3 m. Taking the pressure of the atmosphere at the water's surface into account, the total pressure at this depth is twice atmospheric pressure. Unfortunately for the scuba diver, her lungs tend to inflate to twice their normal size if she holds her breath while rising to the surface. A first lesson in scuba diving is not to hold your breath when ascending. To do so can be fatal.

---

*Humor aside, Boyle's law is named after Robert Boyle. A general law that takes temperature changes into account is $P_1 V_1/T_1 = P_2 V_2/T_2$, where T_1 and T_2 represent the initial and final *absolute* temperatures, measured in SI units called *kelvins* (Chapter 8).

Figure 5.20 The famous "Magdeburg hemispheres" experiment of 1654, demonstrating atmosphere pressure. Two teams of horses couldn't pull the evacuated hemisphere apart. Were the hemispheres sucked together or pushed together? By what?

What do the experiments of Figures 5.20 and 5.21 demonstrate? Do they show that air exerts pressure or that there is a "force of suction"? If we say there is a force of suction, then we assume that a vacuum can exert a force. But what is a vacuum? It is an absence of matter; it is a condition of nothingness. How can nothing exert a force? The hemispheres are not sucked together, nor is the piston holding the weight sucked upward. The hemispheres and the piston are being pushed against by the pressure of the atmosphere.

Just as water pressure is caused by the weight of water, **atmospheric pressure** is caused by the weight of air. We have adapted so completely to the invisible air that we sometimes forget it has weight. Perhaps a fish "forgets" about the weight of water in the same way. The reason we don't feel this weight crushing against our bodies is that the pressure inside our bodies equals that of the surrounding air. There is no net force for us to sense.

At sea level, 1 cubic meter of air at 20° has a mass of about 1.2 kilograms. Estimate the number of cubic meters in your room, multiply by 1.2 kg/m^3, and you'll have the mass of air in your room. Don't be surprised if it's heavier than your kid sister. If your kid sister doesn't believe air has weight, maybe its because she's always surrounded by air. Hand her a plastic bag of water and she'll tell you it has weight. But hand her the same bag of water while she's submerged in a swimming pool, and she won't feel the weight. Similarly with air; we don't notice its weight because we're submerged in air.

Unlike the constant density of water in a lake, the density of air in the atmosphere decreases with altitude. At 10 kilometers 1 cubic meter of air has a mass of about 0.4 kilograms. To compensate for this, airplanes are pressurized; the additional air needed to fully pressurize a 747 jumbo jet, for example, is more than 1000 kilograms. Air is heavy if you have enough of it.

Consider the mass of air in an upright 30-kilometer-tall bamboo pole that has an inside cross-sectional area of 1 square centimeter. If the density of air inside the pole matches the density of air outside, the enclosed mass of air would be about

To vacuum
← pump

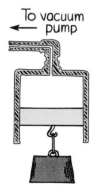

Figure 5.21 Is the piston pulled up or pushed up?

Figure 5.22 You don't notice the weight of a bag of water while you're submerged in water. Similarly, you don't notice the weight of air.

one kilogram. The weight of this much air is about 10 newtons. So air pressure at the bottom of the bamboo pole would be about 10 newtons per square centimeter (10 N/cm²). Of course, the same is true without the bamboo pole. There are 10,000 square centimeters in 1 square meter, so a column of air 1-square meter in cross section that extends up through the atmosphere has a mass of about 10,000 kilograms. The weight of this air is about 100,000 newtons (10^5 N). This weight produces a pressure of 100,000 newtons per square meter—or, equivalently, 100,000 pascals, or 100 kilopascals. To be more exact, the average atmospheric pressure at sea level is 101.3 kilopascals (101.3 kPa).*

The pressure of the atmosphere is not uniform. Besides altitude variations, there are variations in atmospheric pressure at any one locality due to moving air currents and storms. Measurement of changing air pressure is important to meteorologists in predicting weather.

Figure 5.23 The mass of air that would occupy a bamboo pole that extends to the "top" of the atmosphere is about 1 kg. This air has a weight of 10 N.

*The pascal or kilopascal is the SI unit of measurement. Before the advent of SI units, the average pressure at sea level was called 1 atmosphere. This term is still commonly used. In British units, the average atmospheric pressure at sea level is 14.7 lb/in² (psi).

Figure 5.24 The weight of air that bears down on a one-square-meter surface at sea level is about 100,000 newtons. So atmospheric pressure is about 10^5 N/m², or about 100 kPa.

Barometers

Instruments used for measuring the pressure of the atmosphere are called **barometers**. A simple mercury barometer is illustrated in Figure 5.25. A glass tube, longer than 76 centimeters and closed at one end, is filled with mercury and tipped upside down in a dish of mercury. The mercury in the tube runs out of the submerged open bottom until the level falls to about 76 centimeters. The empty space trapped above, except for some mercury vapor, is a pure vacuum. The vertical height of the mercury column remains constant even when the tube is tilted, unless the top of the tube is less than 76 centimeters above the level in the dish—in which case the mercury completely fills the tube.

Why does mercury behave this way? The explanation is similar to the reason a simple see-saw will balance when the weights of people at its two ends are equal. The barometer "balances" when the weight of liquid in the tube exerts the same pressure as the atmosphere outside. Whatever the width of the tube, a 76-centimeter column of mercury weighs the same as the air that would fill a vertical 30-kilometer tube of the same width. If the atmospheric pressure increases, then it pushes the mercury column higher than 76 centimeters. The mercury is literally pushed up into the tube of a barometer by atmospheric pressure.

Could water be used to make a barometer? The answer is yes, but the glass tube would have to be much longer—13.6 times as long, to be exact. You may recognize this number as the density of mercury compared to that of water. A volume of water 13.6 times that of mercury is needed to provide the same weight as the mercury in

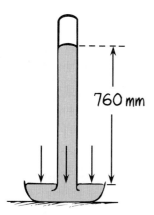

760 mm

Figure 5.25 A simple mercury barometer.

Figure 5.26 Strictly speaking, they do not suck the soda up the straw. They instead reduce pressure in the straw and allow the weight of the atmosphere to press the liquid up into the straws. Could they drink a soda this way on the moon?

ANSWERS

1. The mass of air is 1000 kg. The volume of air is 200 m² × 4 m = 800 m³; each cubic meter of air has a mass of about 1.25 kg, so 800 m³ × 1.25 kg/m³ = 1000 kg.

2. Atmospheric pressure is exerted on both sides of a window, so no net force is exerted on the window. If for some reason the pressure is reduced or increased on one side only, then watch out!

Figure 5.27   The atmosphere pushes water from below up into a pipe that is evacuated of air by the pumping action.

the tube (or in the imaginary tube of air outside). So the height of the tube would have to be at least 13.6 times taller than the mercury column. A water barometer would have to be 13.6 × 0.76 meter, or 10.3 meters high—too tall to be practical.

What happens in a barometer is similar to what happens during the process of drinking through a straw. By sucking, you reduce the air pressure in the straw that is placed in a drink. Atmospheric pressure on the drink pushes liquid up into the reduced-pressure region. Strictly speaking, the liquid is not sucked up; it is pushed up by the pressure of the atmosphere. If the atmosphere is prevented from pushing on the surface of the drink, as in the party trick bottle with the straw through the air-tight cork stopper, one can suck and suck and get no drink.

If you understand these ideas, you can understand why there is a 10.3-meter limit on the height water can be lifted with vacuum pumps. The old fashioned farm-type pump, Figure 5.27, operates by producing a partial vacuum in a pipe that extends down into the water below. The atmospheric pressure exerted on the surface of the water simply pushes the water up into the region of reduced pressure inside the pipe. Can you see that even with a perfect vacuum, the maximum height to which water can be lifted is 10.3 meters?

Q︎UESTION

What is the maximum height to which water could be drunk through a straw?

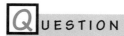NSWER

However strong your lungs may be, or whatever device you use to make a vacuum in the straw, at sea level the water could not be pushed up by the atmosphere higher than 10.3 m.

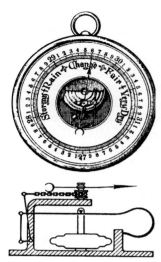

Figure 5.28 The aneroid barometer.

A small portable instrument that measures atmospheric pressure is the aneroid barometer (Figure 5.28). It uses a small metal box that is partially exhausted of air and has a slightly flexible lid that bends in or out with changes in atmospheric pressure. Motion of the lid is indicated on a scale by a mechanical spring-and-lever system. Since the atmospheric pressure decreases with increasing altitude, a barometer can be used to determine the elevation. An aneroid barometer calibrated for altitude is called an altimeter (altitude meter). Some of these instruments are sensitive enough to indicate a change in elevation less than a meter.*

The pressure (or vacuum) inside a television picture tube is about 1 ten-thousandth pascal (10^{-4} Pa). At an altitude of about 500 kilometers, artificial satellite territory, gas pressure is about one ten-thousandth of this (10^{-8} Pa). This is a pretty good vacuum by earthbound standards. Still greater vacuums exist in the wakes of satellites orbiting at this distance, and they can reach 10^{-13} pascals. This is a hard vacuum.

Vacuums on the earth are produced by pumps, which work by virtue of a gas tending to fill its container. If a space with less pressure is provided, gas will flow from the region of higher pressure to the one of lower pressure. A vacuum pump simply provides a region of lower pressure into which the normally fast-moving gas molecules randomly fill. The air pressure is repeatedly lowered by piston and valve action (Figure 5.29). The best vacuums attainable with mechanical pumps are about a pascal. Better vacuums, down to 10^{-8} pascals, are attainable with vapor diffusion pumps. Greater vacuums are very difficult to attain. Technologists requiring hard vacuums are looking more to the prospects of orbiting laboratories in space.

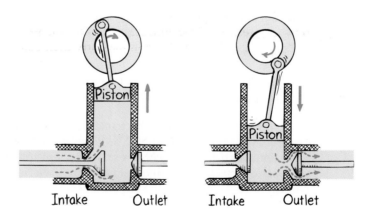

Figure 5.29 A mechanical vacuum pump. When the piston is lifted, the intake valve opens and air moves in to fill the empty space. When the piston is moved downward, the outlet valve opens and the air is pushed out. What changes would you make to convert this pump into an air compressor?

*Evidence of a noticeable pressure difference over a 1-m or less difference in elevation is any small helium-filled balloon that rises in air. The atmosphere really does push harder against the lower bottom than against the higher top.

● ●

5.8 BUOYANCY IN A GAS

A crab lives at the bottom of its ocean of water and looks upward at jellyfish and other lighter-than-water objects floating above it. Similarly, we live at the bottom of our ocean of air and look upward to balloons and other lighter-than-air objects floating above us. A balloon floats in air and a jellyfish "floats" in water for the same reason: each is buoyed upward by a displaced weight of fluid equal to its own weight. In one case the displaced fluid is air, and in the other case it is water. In water, immersed objects are buoyed upward because the pressure acting up against the bottom of the object exceeds the pressure acting down against the top. Likewise, air pressure acting up against an object immersed in air is greater than the pressure above pushing down. The buoyancy in both cases is numerically equal to the weight of fluid displaced. We can state **Archimedes' principle for air** as

> **An object surrounded by air is buoyed up by a force equal to the weight of the air displaced.**

Figure 5.30 All bodies are buoyed up by a force equal to the weight of air they displace. Why, then, don't all objects float like this balloon?

We know that a cubic meter of air at ordinary atmospheric pressure and room temperature has a mass of about 1.2 kilograms, so its weight is about 12 newtons. Therefore any 1-cubic-meter object in air is buoyed up with a force of 12 newtons. If the mass of the 1-cubic-meter object is greater than 1.2 kilograms (so that its weight is greater than 12 newtons), it falls to the ground when released. If this size object has a mass less than 1.2 kilograms, it rises in the air. Any object that has a mass less than the mass of an equal volume of air will rise in air. Another way to say this is that any object less dense than air will rise in air. Gas-filled balloons that rise in air are less dense than air.

The gas used in balloons prevents the atmosphere from collapsing the balloon. Helium is commonly used because its mass is small enough that the combined weight of helium, balloon, and whatever the cargo happens to be is less than the weight of air it displaces.* Helium is used in a balloon for the same reason cork is used in a swimmer's life preserver. The cork possesses no strange tendency to be drawn toward the surface of water, and helium possesses no strange tendency to rise. Both are buoyed upward like anything else. They are simply light enough for the buoyancy to be significant.

Unlike water, there is no sharp surface at the "top" of the atmosphere. Furthermore, unlike water, the atmosphere becomes less dense with altitude. Whereas cork will float to the surface of water, a released helium-filled balloon does not rise to any atmospheric surface. Will a lighter-than-air balloon rise indefinitely? How high will a balloon rise? We can state the answer in several ways. A balloon will rise only so long as it displaces a weight of air greater than its own weight. Since air becomes less dense with altitude, a lesser weight of air is displaced per given volume as the balloon rises. When the weight of displaced air equals the total weight of the balloon, upward acceleration of the balloon will cease. We can also say that when the buoyant force on the balloon equals its weight, the balloon will cease rising. Equivalently, when the density of the balloon equals the density of the surrounding air, the balloon will cease rising. Helium-filled toy balloons usually break when released

---

*Hydrogen is the least dense gas but is highly flammable, so it is seldom used.

in the air because as the balloon rises to regions of less pressure, the helium in the balloon expands, increasing the volume and stretching the rubber until it ruptures.

QUESTIONS

1. Is there a buoyant force acting on you? If there is, why are you not buoyed up by this force?
2. Two balloons that have the same weight and volume contain equal amounts of helium. One is rigid and the other is free to expand as the pressure outside decreases. When released, which balloon will rise higher?

Large dirigible airships are designed so that when loaded they will slowly rise in air; that is, their total weight is a little less than the weight of air displaced. When in motion, the ship may be raised or lowered by means of horizontal rudders or "elevators."

Thus far we have treated pressure only as it applies to stationary fluids. Motion produces an additional influence.

5.9 BERNOULLI'S PRINCIPLE

Mark this statement true or false: Atmospheric pressure increases in a gale, tornado, or hurricane. If you answered true, sorry; the statement is false. High-speed winds may blow the roof off your house, but the pressure within the winds is actually less than for still air of the same density inside the house. As strange as it may first seem, when the speed of a fluid increases, the internal pressure decreases proportionally. This is true whether the fluid is a gas or liquid.

Daniel Bernoulli, a Swiss scientist of the eighteenth century, studied the relationship of fluid speed and pressure. When a fluid flows through a narrow constriction, its speed increases. This is easily noticed by the increased speed of a brook when it flows through the narrow parts. The fluid must speed up in the constricted region if the flow is to be continuous.

Bernoulli wondered how the fluid got this extra speed. He reasoned that it is acquired at the expense of a lowered internal pressure. His discovery, now called **Bernoulli's principle**, states:

The pressure in a fluid decreases as the speed of the fluid increases.

Bernoulli's principle is a consequence of the conservation of energy. For a steady flow of fluid there are three kinds of energy; kinetic energy due to motion,

ANSWERS

1. There is a buoyant force acting on you, and you are buoyed upward by it. You don't notice it only because your weight is so much greater.
2. The balloon that is free to expand will displace more air as it rises than the balloon that is restrained. Hence, the balloon that is free to expand will experience more buoyant force and rise higher.

Figure 5.31 The pressure in the spout reduces when the plug is removed.

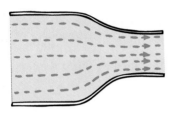

Figure 5.32 Water speeds up when it flows into the narrower pipe. The constricted streamlines indicate increased speed and decreased internal pressure.

Figure 5.33 The paper rises when air is blown across its top surface.

potential energy due to pressure, and gravitational potential energy due to elevation. In a steady fluid flow where no energy is added or taken away, the sum of these forms of energy remains constant.* If the elevation of the flowing fluid doesn't change, then an increase in speed means a decrease in the pressure, and vice versa.

The decrease of fluid pressure with increasing speed may at first seem surprising, particularly if we fail to distinguish between the pressure in the fluid and the pressure by the fluid on something that interferes with its flow. The pressure within fast-moving water in a fire hose is relatively low, whereas the pressure that the water can exert on anything in its path to slow it down may be huge.

In steady flow, the paths taken by each little region of fluid do not change as time passes. The motion of a fluid in steady flow follows streamlines, which are represented by dashed lines in Figure 5.31 and later figures. Streamlines are the smooth paths, or trajectories, of the neighboring regions of fluid. The lines are closer together in the narrower regions, where the flow speed is greater and the pressure within the fluid is less.

If the flow speed is too great, the flow may become turbulent and follow changing, curling paths known as eddies. Then Bernoulli's principle will not hold.

Applications of Bernoulli's Principle

Hold a sheet of paper in front of your mouth, as shown in Figure 5.33. When you blow across the top surface, the paper rises. This is because the moving air pushes against the top of the paper with less pressure than the air that pushes against the lower surface, which is at rest.

We began our discussion of Bernoulli's principle by stating that atmospheric pressure decreases in a strong wind, tornado, or hurricane. As it turns out, an unvented building with airtight closed windows is in more danger of losing its roof than a well-vented building. This is because the air pressure inside may be appreciably greater than the reduced atmospheric pressure outside, and the roof is more likely to be pushed off by the relatively compressed air in the building than blown off by the wind. When the wind is blowing over a peaked roof as shown in Figure 5.34, the effect is even more pronounced. The crowding of the streamlines shows this. The difference in outside and inside pressure need not really be very much. A small pressure difference over a large area can be formidable. So if you're ever caught in an unvented building in a tornado or hurricane, consider opening the windows a bit so that the pressures inside and outside are more nearly equal.†

If we think of the blown-off roof as an airplane wing, we can better understand the lifting force that supports a heavy airliner. In both cases a greater pressure below pushes the roof and wing into a region of lesser pressure above. A cambered (arched) roof is more apt to be blown off than a flat roof. Similarly, a wing with

---

*In mathematical form: $\frac{1}{2}mv^2 + pV + mgy =$ constant, where m is the mass of a unit volume V, v its speed, p its pressure, g the acceleration due to gravity, and y its elevation. If mass m is expressed in terms of density ρ, where $\rho = m/V$, and each term is divided by V, Bernoulli's equation takes the form: $\frac{1}{2}\rho v^2 + p + \rho gy =$ constant. Then all three terms have units of pressure. If y does not change, then an increase in v means a decrease in p, and vice versa.

†A word of caution: In a building that has venting adequate to tolerate the sudden pressure drop without the need to open windows, opening a window may actually increase damage.

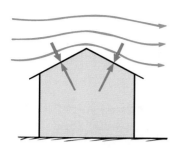

Figure 5.34 Air pressure above the roof is less than air pressure beneath the roof.

more curvature on the top surface has greater lift than a wing with flat surfaces (like the wings of those balsa-wood gliders you used to play with). Whether it has flat or curved wings, an airplane will fly by virtue of air impact against the lower surface of wings that are tilted back slightly to deflect oncoming air downward (as discussed briefly in Chapter 2). The airfoil of a curved wing, however, adds considerably to lift and results in a greater difference in pressure between the lower and upper wing surfaces. This net upward pressure multiplied by the surface area of the wing gives the net lifting force. The lift is greater when there is a large wing area and when the plane is traveling fast. Gliders have a very large wing area so that they do not have to be going very fast for sufficient lift. At the other extreme, fighter planes designed for high speed have very small wing areas. Consequently, they must take off and land at relatively high speeds.

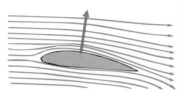

Figure 5.35 The arrow represents the net upward force (lift) that results from less air pressure above the wing than below the wing.

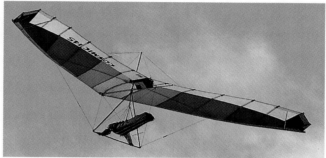

Figure 5.36 Where is air pressure greater—on the top or bottom surface of the hang glider?

We all know that a baseball pitcher can throw a ball in such a way that it will curve off to one side of its trajectory. This is accomplished by imparting a large spin to the ball. Similarly, a tennis player can hit a ball that will curve. A thin layer of air is dragged around the spinning ball by friction, which is enhanced by the baseball's threads or the tennis ball's fuzz. The moving layer produces a crowding of streamlines on one side. Note in Figure 5.37 (right) that the streamlines are more crowded at B than at A for the direction of spin shown. Air pressure is greater at A, and the ball curves as shown.

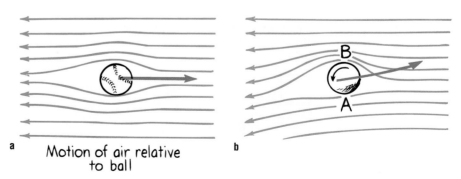

Motion of air relative to ball

Figure 5.37 (a) The streamlines are the same on either side of a nonspinning baseball. (b) A spinning ball produces a crowding of streamlines and curves as shown by the thick arrow.

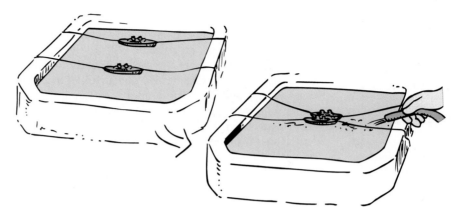

Figure 5.38 Try this in your sink. Loosely moor a pair of toy boats side by side. Then direct a stream of water between them. The boats will draw together and collide. Why?

Bernoulli's principle accounts for the fact that passing ships run the risk of a sideways collision. Water flowing between the ships travels faster than water flowing past the outer sides. The streamlines are more compressed between the ships than outside. Water pressure acting against the hulls is therefore reduced between the ships. Unless the ships are steered to compensate for this, the greater pressure against the outer sides of the ships then forces them together. Figure 5.38 shows how this can be demonstrated in your kitchen sink or bathtub.

You can demonstrate Bernoulli's principle quite interestingly in your kitchen sink (Figure 5.39). Tape a table-tennis ball to a string and allow the ball to swing into a stream of running water. You'll see that it will remain in the stream even when tugged slightly to the side, as shown. This is because the pressure of the stationary air next to the ball is greater than the pressure of the moving water. The ball is pushed into the region of reduced pressure by the atmosphere.

The same thing happens to a bathroom shower curtain when the shower water is turned on full blast. The pressure in the shower stall is reduced, and the relatively greater pressure outside the curtain pushes it inward. The next time you're taking a shower and the curtain swings in against your legs, think of Daniel Bernoulli!

Figure 5.39 Pressure is greater in the stationary fluid (air) than in the moving fluid (water stream). The atmosphere pushes the ball into the region of reduced pressure.

QUESTION

Will the mercury level in a barometer be affected on a windy day?

ANSWER

Atmospheric pressure is reduced on a windy day, so the barometer level will fall from its normal position.

Figure 5.40 The curved shape of an umbrella can be disadvantageous on a windy day.

SUMMARY OF TERMS

Density The amount of matter per unit volume.

$$\text{Density} = \frac{\text{mass}}{\text{volume}}$$

Weight density is expressed as weight per unit volume.
Pressure The ratio of force to the area over which that force is distributed:

$$\text{Pressure} = \frac{\text{force}}{\text{area}}$$

Liquid pressure = weight density × depth

Buoyant force The net force that a fluid exerts on an immersed object.
Archimedes' principle An immersed body is buoyed up by a force equal to the weight of the fluid it displaces.
Principle of flotation A floating object displaces a weight of fluid equal to its own weight.
Pascal's principle The pressure applied to a fluid confined in a container is transmitted undiminished throughout the fluid and acts in all directions.
Barometer Any device that measures atmospheric pressure.
Boyle's law The product of pressure and volume is a constant for a given mass of confined gas regardless of changes in either pressure or volume individually, so long as temperature remains unchanged:

$$P_1 V_1 = P_2 V_2$$

Archimedes' principle for air An object surrounded by air is buoyed up with a force equal to the weight of displaced air.
Bernoulli's principle The pressure in a fluid decreases with an increase in fluid velocity.

REVIEW QUESTIONS

Density

1. Distinguish between the "heaviness" of a material and its density.

2. Distinguish between *mass density* and *weight density*. What is the mass density and weight density of water?

Pressure

3. Distinguish between *force* and *pressure*.

Pressure in a Liquid

4. What is the relationship between liquid pressure and the depth of a liquid? Between liquid pressure and density?

5. If you swim twice as deep in water, how much more water pressure is exerted on your ears? If you swim in salt water, will the pressure at the same depth be greater than in fresh water? Why or why not?

6. How does water pressure one meter below the surface of a small pond compare to water pressure one meter below the surface of a huge lake?

7. If you punch a hole in a container filled with water, in what direction does the water initially flow outward from the container?

Buoyancy in a Liquid

8. Why does buoyant force act upward for an object submerged in water?

9. How does the volume of a submerged object compare to the volume of water displaced?

Archimedes' Principle

10. How does the buoyant force that acts on a fish compare to the weight of the fish?

11. If a 1-L container is immersed halfway into water, what is the volume of water displaced? What is the buoyant force on the container?

12. How does the volume of a submerged object compare to the volume of water displaced?

13. What is the mass of 1 L of water? What is its weight in newtons?

14. Does the buoyant force on a submerged object depend on the weight of the object itself or on the weight of the fluid displaced by the object? Does it depend on the weight of the object itself or on its volume? Defend your answer.

Flotation

15. How much water is displaced by a 100-ton ship? What is the buoyant force that acts on a 100-ton ship?

16. Does the buoyant force on a floating object depend on the weight of the object itself or on the weight of the fluid displaced by the object? Or are these both the same for the special case of floating? Defend your answer.

Pressure in a Gas

17. Cite differences in liquids and gases.

Boyle's Law

18. By how much does the density of air increase when it is compressed to half its volume?

19. What happens to the air pressure inside a balloon when it is squeezed to half its volume?

20. What is an ideal gas?

Atmospheric Pressure

21. What is the cause of atmospheric pressure?

22. What is the mass of a cubic meter of air at room temperature (20°C)?

23. What is the approximate mass of a column of air 1 cm^2 in area that extends from sea level to the upper atmosphere? What is the weight of this amount of air?

24. What is the SI unit of atmospheric pressure?

Barometers

25. How does the downward pressure of the 76-cm column of mercury in a barometer compare to the pressure due to the weight of the atmosphere?

26. How does the weight of mercury in a barometer compare to the weight of an equal cross-section of air from sea level to the top of the atmosphere?

27. Why is a water barometer 13.6 times taller than a mercury barometer?

28. When you drink liquid through a straw, is it more accurate to say the liquid is pushed up the straw rather than sucked up the straw? What exactly does the pushing? Defend your answer.

29. Why will a vacuum pump not operate for a well that is more than 10.3 m deep?

Buoyancy in a Gas

30. How much buoyant force acts on a floating balloon that weighs 1 N? What happens if the buoyant force decreases? If it increases?

31. Does the air exert buoyant force on all objects in air or only on very light objects such as balloons that are in the air? Why is buoyancy effective in floating only things with a very low density?

Bernoulli's Principle

32. What happens to the internal pressure in a fluid flowing in a pipe when its speed increases?

33. What are streamlines? Is pressure greater or less in regions where streamlines are crowded?

Applications of Bernoulli's Principle

34. Is Bernoulli's principle the sole explanation for the flight of airplanes?

35. Why does a spinning ball curve in its flight?

36. Does Bernoulli's principle hold for liquids? Defend your answer.

. . . .

HOME PROJECTS

1. Try to float an egg in water. Then dissolve salt in the water until the egg floats. How does the density of an egg compare to that of tap water? To salt water?

2. Punch a couple of holes in the bottom of a water-filled container, and water will spurt out because of water pressure. Now drop the container, and as it freely falls note that the water no longer spurts out! If your friends don't understand this, could you figure it out and then explain it to them?

3. You can find the weight of a car by measuring the area of contact that its four tires make with the road and then multiplying this area by the air pressure in the tires. The area can be closely approximated by tracing the edges of tire contact on sheets of paper marked with 1-cm^2 squares

beneath each tire. If you actually did this, your calculation would be an exercise in converting units of measurement, especially if area is in square centimeters and air pressure in pounds per square inch.

4. Try this in the bathtub or when you're washing dishes. Lower a drinking glass, mouth downward, over a small floating object. What do you observe? How deep will the glass have to be pushed in order to compress the enclosed air to half its volume? (You won't be able to do this in your bathtub unless it's 10.3 m deep!)

5. You ordinarily pour water from a full glass to an empty glass by simply placing the full glass above the empty glass and tipping. Have you ever poured air from one glass to the other? The procedure is similar. Lower two glasses in water, mouths  downward. Let one fill with water by tilting its mouth upward. Then hold the water-filled glass mouth downward above the air-filled glass. Slowly tilt the lower glass and let the air escape, filling the upper glass. You will be pouring air from one glass to another!

6. Raise a filled glass of water above the waterline, but with its mouth beneath the surface. Why does the water not run out? How tall would a glass have to be before water began to run out? (You won't be able to do this indoors unless you have a 10.3-m ceiling.)

7. Place a card over the open top of a glass filled to the brim with water, and invert it. Why does the card stay intact? Try it sideways.

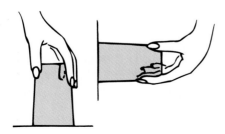

8. Invert a water-filled pop bottle or small-necked jar. Notice that the water doesn't simply fall out, but gurgles out of the container—a result of the weight of the many kilometers of air above pushing down and forcing air up into the bottle and forcing the water out. How would an inverted, water-filled bottle empty if this were done on the moon?

9. Heat a small amount of water to boiling in an aluminum soda-pop can and invert it quickly into a dish of cold water. Surprisingly dramatic!

10. Make a small hole near the bottom of an open tin can. Fill it with water, which proceeds to spurt from the hole. Cover the top of the can firmly with the palm of your hand and the flow stops. Explain.

11. Lower a narrow glass tube or drinking straw in water and place your finger over the top of the tube. Lift the tube from the water and then lift your finger from the top of the tube. What happens? (You'll do this often in chemistry experiments.)

12. Fold the ends of a filing card down so that you make a little bridge. Stand it on the table and blow through the arch as shown. No matter how hard you blow, you will not succeed in blowing the card off the table (unless you blow against the side of it). Try this with your nonphysics friends. Then explain it to them!

13. Push a pin through a small card sheet and place it in the hole of a thread spool. Try to blow the card from the spool by blowing through the hole. Try it in all directions.

14. Hold a spoon in a stream of water as shown and feel the effect of the differences in pressure.

EXERCISES

1. You know that a sharp knife cuts better than a dull knife. Do you know why this is so?

2. If water faucets upstairs and downstairs are turned fully on, will more water per second come out of the upstairs or downstairs faucet?

3. Which do you suppose exerts more pressure on the ground—an elephant or a woman standing on spike heels? (Which will be more likely to make dents in a linoleum floor?) Can you approximate a rough calculation for each?

4. The sketch shows the reservoir that supplies water to a farm. It is made of wood and reinforced with metal hoops. (a) Why is it elevated? (b) Why are the hoops closer together near the bottom part of the tank?

5. A block of aluminum with a volume of 10 cm^3 is placed in a beaker of water filled to the brim. Water overflows. The same is done in another beaker with a 10-cm^3 block of lead. Does the lead displace more, less, or the same amount of water?

6. A block of aluminum with a mass of 1 kg is placed in a beaker of water filled to the brim. Water overflows. The same is done in another beaker with a 1-kg block of lead. Does the lead displace more, less, or the same amount of water?

7. A block of aluminum with a weight of 10 N is placed in a beaker of water filled to the brim. Water overflows. The same is done in another beaker with a 10-N block of lead. Does the lead displace more, less, or the same amount of water? (Why are your answers to this exercise and Exercise 10 different from your answer to Exercise 9?)

8. There is a legend of a Dutch boy who bravely held back the whole Atlantic Ocean by plugging a hole in a dike with his finger. Is this possible and reasonable?

9. Why does water "seek its own level"?

10. Suppose you wish to lay a level foundation for a home on hilly and bushy terrain. How can you use a garden hose filled with water to determine equal elevations for distant points?

11. If liquid pressure were the same at all depths, would there be a buoyant force on an object submerged in the liquid? Explain.

12. How much force is needed to push a nearly weightless but rigid 1-L carton beneath a surface of water?

13. The density of ice is 0.9 g/cm^3, 0.9 percent the density of water. What proportion of a floating chunk of ice is above water level?

14. The Himalayan Mountains are slightly less dense than the mantle material upon which they "float." Do you suppose that, like floating icebergs, they are deeper than they are high?

15. Consider the plug in your bathtub the next time you take a bath. When the tub is full, is there a buoyant force on the plug when it is in the drain? When it's pulled out and lying submerged on the tub bottom beside the drain? Defend your answer.

16. Why is it inaccurate to say that heavy objects sink and that light objects float? Give exaggerated examples to support your answer.

17. Compared to an empty ship, would a ship loaded with a cargo of styrofoam sink deeper into water or rise in water? Defend your answer.

18. A barge filled with scrap iron is in a canal lock. If the iron is thrown overboard, does the water level at the side of the lock rise, fall, or remain unchanged? Explain.

19. Would the water level in a canal lock go up or down if a battleship in the lock sank?

20. A balloon is weighted so that it is barely able to float in water. If it is pushed beneath the surface, will it come back to the surface, stay at the depth to which it is pushed, or sink? Explain. (Hint: What change in density, if any, does the balloon undergo?)

21. The density of a rock doesn't change when it is submerged in water, but your density changes when you are submerged. Why is this so?

22. In answering the question of why bodies float higher in salt water than fresh water, your friend replies that the reason is that salt water is denser than fresh water. (Does your friend often answer questions by reciting only factual statements that relate to the answers but don't provide any concrete reasons?) How would you answer the same question?

23. Suppose you wear two life preservers that are identical in size, first a light one filled with styrofoam and then a very heavy one filled with lead pellets. If you wear these life preservers in the water, upon which will the buoyant force be greater? Upon which will the buoyant force be ineffective? Why are your answers different?

24. When an ice cube in a glass of water melts, does the water level in the glass rise, fall, or remain unchanged?

25. Count the tires on a large tractor trailer that is unloading food at your local supermarket, and you may be surprised to count 18 or so tires. Why so many tires? (Hint: Consider Home Project 3.)

26. Why do your ears pop when you ascend to higher altitudes?

27. Two teams of eight horses each were unable to pull the Magdeburg hemispheres apart (Figure 5.20). Why? Suppose two teams of nine horses each could pull them apart. Then would one team of nine horses succeed if the other team were replaced with a strong tree? Defend your answer.

28. Before boarding an airplane, you buy a roll of camera film or any item packaged in an airtight foil package, and while in flight you notice that it is puffed up. Explain why this happens.

29. We can understand how pressure in water depends on depth by considering a stack of bricks. The pressure at the bottom of the first brick corresponds to the weight of the entire stack. Halfway up the stack, the pressure is half because the weight of the bricks above is half. To explain atmospheric pressure, we should consider compressible bricks, like foam rubber. Why is this so?

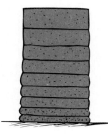

30. Would it be slightly more difficult to draw soda through a straw at sea level or on top of a very high mountain? Explain.

31. The pressure exerted against the ground by an elephant's weight distributed evenly over its four feet is less than 1 atmosphere. Why, then, would you be crushed beneath the foot of an elephant, while you're unharmed by the pressure of the atmosphere?

32. Your friend says that the buoyant force of the atmosphere on an elephant is significantly greater than the buoyant force of the atmosphere on a small helium-filled balloon. What do you say?

33. Why is it so difficult to breathe when snorkeling at a depth of 1 m, and practically impossible at a 2-m depth? Why can't a diver simply breathe through a hose that extends to the surface?

34. A block of wood and a block of iron on weighing scales each weigh 1 ton. Which has the greater mass?

35. Why does the weight of an object in air differ from its weight in a vacuum? Cite an example where this would be an important consideration.

36. Estimate the buoyant force that acts on you. (To do this, you can estimate your volume by knowing your weight and by assuming that your weight density is a bit less than that of water.)

37. Two identical balloons filled with air to the same volumes are suspended on the ends of a stick that is horizontally balanced. One of the balloons is then punctured. Is the balance of the stick upset? If so, which way does it tip?

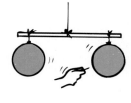

38. Two balloons that have the same weight and volume are filled with equal amounts of helium. One is rigid and the other is free to expand as the pressure outside decreases. When released, which will rise higher? Explain.

39. Estimate the *force* that the atmosphere exerts on you, and then compare this force with the weight of something familiar in your environment. To do this, estimate your surface area from the area of your clothes, and multiply by 10^5 N/m^2. Why is your answer to this exercise so different from your answer to Exercise 36?

40. The force of the atmosphere at sea level against the outside of a 10-m^2 store window is about a million N. Why does this not shatter the window? Why might the window shatter in a strong wind?

41. In a department store, an airstream from a hose connected to the exhaust of a vacuum cleaner blows

upward at an angle and supports a beach ball in midair. Does the air blow under or over the ball to provide support?

42. What provides the lift to keep a Frisbee in flight?

43. Why is it that when passing an oncoming truck on the highway, your car tends to lurch toward the truck?

44. Why is it that the canvas roof of a convertible automobile bulges upward when the car is traveling at high speeds?

45. Why is it that the windows of older trains sometimes break when a high-speed train passes by on the next track?

46. A steady wind blows over the waves of an ocean. Why does the wind increase the humps and troughs of the waves?

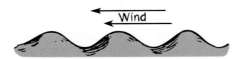

47. In answering the question of why a flag flaps in the wind, your friend replies that it flaps because of Bernoulli's principle. (Not really a convincing explanation, is it?) How would you answer the same question?

48. Wharves are made with pilings that permit the free passage of water. Why would a solid-walled wharf be disadvantageous to ships attempting to pull alongside?

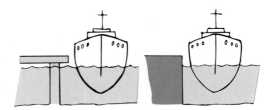

49. Why does the fire in a fireplace burn more briskly on a windy day?

50. It is often said that fast-moving air has lower pressure than air at rest or slower-moving air. Can you make an argument that the opposite is the case; that fast-moving air is the *result*, not the *cause of*, lower pressure?

HEAT

Although the temperature of these sparks exceeds 2000°C, the heat they impart in striking my skin is very small-- which illustrates that temperature and heat are different concepts. Learning to distinguish between closely related concepts is the essence of *Conceptual Physical Science*.

6

THERMAL ENERGY

All matter is composed of continually jiggling atoms or molecules. Whether the atoms and molecules combine to form solids, liquids, gases, or plasmas depends on the rate of molecular vibrations. By virtue of this vibratory motion, the molecules or atoms in matter possess kinetic energy. The average kinetic energy of the individual particles is directly related to a property you can sense: how hot something is. Whenever something becomes warmer, the kinetic energy of its particles increases. Strike a solid penny with a hammer and it becomes warm because the hammer's blow causes the atoms in the metal to jostle faster. Put a flame to a liquid and it too becomes warmer. Rapidly compress air in a tire pump and the air becomes warmer. When a solid, liquid, or gas gets warmer, its atoms or molecules move faster. The atoms or molecules have more kinetic energy. We say the substance made up of these atoms or molecules has more *thermal energy.*

6.1 TEMPERATURE

The quantity that tells how warm or cold an object is with respect to some standard is called **temperature.** We express the temperature of matter by a number which corresponds to the degree of hotness on some chosen scale. A common thermometer measures temperature by means of the expansion and contraction of a liquid, usually mercury or colored alcohol.

Figure 6.1 Can we trust our sense of hot and cold? Will both fingers feel the same temperature when they are put in the warm water?

Most common in the world is the *Celsius thermometer* in honor of the Swedish astronomer Anders Celsius (1701–1744), who first suggested the scale of 100 degrees between the freezing point and boiling point of water. The number 0 is assigned to the temperature at which water freezes and the number 100 to the temperature at which water boils (at standard atmospheric pressure), with 100 equal parts called *degrees* between. In the United States, the number 32 is assigned to the temperature at which water freezes, and the number 212 is assigned to the temperature at which water boils. Such a scale makes up a Fahrenheit thermometer, named after its originator, the German physicist G. D. Fahrenheit (1686–1736). The Fahrenheit scale will become obsolete if and when the United States goes metric.*

Still another temperature scale, favored by scientists, is the Kelvin scale, named after the British physicist Lord Kelvin (1824–1907). This scale is calibrated not in terms of the freezing and boiling points of water, but in terms of energy itself. The number 0 is assigned to the lowest possible temperature—**absolute zero,** where a substance has absolutely no kinetic energy to give up. Absolute zero corresponds to $-273°C$ on the Celsius scale. Units on the Kelvin scale are the same size as degrees on the Celsius scale, so the temperature of melting ice is $+273$ kelvins. There are no negative numbers on the Kelvin scale. We won't treat this scale further until we return to it when we study thermodynamics in Chapter 8.

Arithmetic formulas are used for converting from one temperature scale to the other and are popular in classroom exams. Such arithmetic exercises are not really physics, and the probability of your having the occasion to do this task elsewhere is small, so we will not be concerned with it here. Besides, this conversion can be very closely approximated by simply reading the corresponding temperature from the side-by-side scales in Figure 6.2.

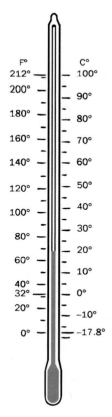

Figure 6.2 Fahrenheit and Celsius scales on a thermometer.

---

*Americans are slow to convert, perhaps because the Fahrenheit scale seems much better suited to everyday use. For example, its degrees are smaller ($1°F = \frac{5}{9}°C$), which gives greater accuracy when reporting the weather in whole-number temperature readings. Then, too, people somehow attribute a special significance to numbers increasing by an extra digit, so that when the temperature of a hot day is reported to reach 100°F, the idea of heat is conveyed more dramatically than by saying it is 37.7°C. Like so much of the British system of measure, the Fahrenheit scale is geared to human beings.

Temperature is related to the random motion of the molecules in a substance. In the simplest case of an ideal gas, temperature is proportional to the average kinetic energy of molecular translational motion (that is, molecular motion along a straight or curved path). In solids and liquids, where molecules are more constrained and have potential energy, temperature is more complicated. Nonetheless, temperature is normally related to the average kinetic energy of translational motion of molecules. It is important to understand that temperature is not a measure of the total kinetic energy of molecules in a substance. For example, there is twice as much molecular kinetic energy in 2 liters of boiling water as in 1 liter of boiling water—but the temperatures of both amounts of water are the same because the average kinetic energy per molecule in each is the same.

Interestingly enough, a thermometer actually registers its own temperature. When a thermometer is in thermal contact with whatever is being measured, energy will flow between the two until their temperatures are equal; thermal equilibrium is established. To know the temperature of the thermometer is to know also the temperature of whatever is being measured. A thermometer should be small enough that it doesn't appreciably alter the temperature of that being measured. If you are measuring the room air temperature, then your thermometer is small enough. But if you are measuring a drop of water, its temperature after thermal contact may be quite different from its initial temperature.

6.2 HEAT

If you touch a hot stove, thermal energy enters your hand because the stove is warmer than your hand. When you touch a piece of ice, on the other hand, thermal energy passes out of your hand and into the colder ice. The direction of energy transfer is always from a warmer thing to a neighboring cooler thing. A physicist defines **heat** as the transfer of thermal energy from one thing to another because of a temperature difference.

According to this definition, matter does not *contain* heat. Matter contains thermal energy, *not heat*. Heat is thermal energy in transit. Once heat has been transferred to an object or substance, it ceases to be heat. It becomes thermal energy.*

Figure 6.3 The temperature of the sparks is very high, more than 2000°C. That's a lot of energy per molecule of spark. But because of the few molecules in a spark, the total amount of thermal energy is safely small. Temperature is one thing; transfer of thermal energy is another.

---

*In place of *thermal energy,* physicists use the term *internal energy*—the grand total of all energies inside a substance. This includes potential energy due to the forces between molecules or atoms and kinetic energy due to movements of atoms within molecules.

When a substance absorbs or gives off heat, thermal energy in the substance changes. Thus, as a substance absorbs heat, this energy may or may not make the molecules or atoms jostle faster. In some cases, as when ice is melting, a substance absorbs thermal energy without an increase in molecular kinetic energy. The substance undergoes a change of phase, which we will cover in detail in Chapter 8.

For things in thermal contact, heat will flow from the substance at a higher temperature into a substance at a lower temperature, but it will not necessarily flow from a substance with more thermal energy into a substance with less thermal energy. There is more thermal energy in a bowl of warm water than there is in a red-hot thumbtack; if the tack is immersed in the water, thermal energy will not flow from the warm water to the tack. Instead, thermal energy will flow from the hot tack to the relatively cooler water. Thermal energy never flows of itself from a low-temperature substance into a higher-temperature substance. More about this in Chapter 8.

QUESTIONS

1. Suppose you apply a flame to 1 L of water and its temperature rises by 2°C. If you apply the same flame to 2 L of water, by how much will its temperature rise?

2. If a fast marble hits a random scatter of slow marbles, does the fast marble usually speed up or slow down? Which lose(s) kinetic energy and which gain(s) kinetic energy, the initially fast-moving marble or the initially slow ones? How do these questions relate to the direction of heat flow?

Figure 6.4 Just as water in the pipes seeks a common level (where the pressures at the bottom are the same), the thermometer and its immediate surroundings reach a common temperature (where the average molecular KE for both is the same).

ANSWERS

1. Its temperature will rise by only 1°C, because there are twice as many molecules in 2 L of water and each molecule receives only half as much energy on the average. So the average kinetic energy, and thus the temperature, increases by half as much.

2. A fast-moving marble slows when it hits slower-moving marbles. It gives up some of its kinetic energy to the slower ones. Likewise with the flow of heat. Molecules with more kinetic energy that are in contact with less energetic molecules give up some of their kinetic energy to the less energetic ones. The direction of energy transfer is from hot to cold. For both the marbles and the molecules, however, the total energy before and after contact is the same.

6.3 QUANTITY OF HEAT

So heat is the thermal energy that transfers from one thing to another by virtue of a temperature difference. The quantity of heat involved in such a transfer is measured by some change that accompanies the process. For example, in determining the energy value in food, the amount of thermal energy that is released as heat is measured by burning. Fuels are rated on how much energy a certain amount of the fuel will produce when burned.

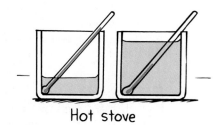

Figure 6.5 Although the same quantity of heat is added to both containers, the temperature increases more in the container with the smaller amount of water.

Hot stove

The unit of heat is defined as the energy necessary to produce some standard, agreed-on change. The most commonly used unit for heat is the calorie. The calorie is defined as the amount of heat required to change the temperature of 1 gram of water by 1 Celsius degree.*

The kilocalorie is 1000 calories (the heat required to change 1 kilogram of water by 1°C). The heat unit used in rating foods is actually a kilocalorie. To distinguish this unit from the smaller calorie, the food unit is sometimes called a *Calorie* (written with a capital *C*). It is important to remember that the calorie and Calorie are units of energy. These names are historical carry-overs from the early idea that heat was an invisible fluid called *caloric*. This view persisted almost to the nineteenth century. We now know that heat is a form of energy, and we are presently in a transition period to the International System (SI) of units, in which the quantity of heat is measured in joules, the SI unit for all forms of energy. (The relationship between calories and joules is that 1 calorie = 4.187 joules.)

Figure 6.6 To the weight watcher, the peanut contains 10 Calories; to the physicist, it releases 10,000 calories (or 41,870 joules) of energy when burned or digested.

6.4 SPECIFIC HEAT CAPACITY

Figure 6.7 The filling of hot apple pie may be too hot to eat, whereas the crust is not.

You've likely noticed that some foods remain hotter much longer than others—like the filling of hot apple pie that can burn your tongue while the crust will not, even when the pie has just been taken out of the oven. Or a piece of toast may be comfortably eaten a few seconds after coming from the hot toaster, whereas you must wait several minutes before eating soup from a stove as hot as the toaster.

Different substances have different capacities for storing thermal energy. If we heat a soup pot of water on a stove, we might find that it requires 15 minutes to raise it from room temperature to its boiling temperature. But if we put an equal mass of iron on the same flame, we would find that it would rise through the same temperature

---

*The SI unit of heat is the SI unit of energy, the joule. There are 4.187 joules in 1 calorie. Another common unit of heat is the British thermal unit (Btu). The Btu is defined as the amount of heat required to change the temperature of 1 lb of water by 1 Fahrenheit degree.

range in only about 2 minutes. For silver, the time would be less than a minute. We find that different materials require different quantities of heat to raise the temperature of a given mass of the material by a specified number of degrees. This is because different materials absorb energy in different ways. The energy may increase the to-and-fro vibrational motion of molecules, which raises the temperature; or it may increase the amount of internal vibration or rotation within the molecules and go into potential energy, which does not raise the temperature. Generally there is a combination of both.

A gram of water requires 1 calorie of energy to raise the temperature 1 Celsius degree. It takes only about one-eighth as much energy to raise the temperature of a gram of iron by the same amount. Water absorbs more heat than iron for the same change in temperature. We say water has a higher **specific heat capacity** (sometimes simply called *specific heat*).*

> **The specific heat capacity of any substance is defined as the quantity of heat required to change the temperature of a unit mass of the substance by 1 degree.**

We can think of specific heat capacity as thermal inertia. Recall that inertia is a term used in mechanics to signify the resistance of an object to a change in its state of motion. Specific heat capacity is like a thermal inertia since it signifies the resistance of a substance to a change in temperature.

Water has a much higher capacity for storing energy than all but a few uncommon materials. A relatively small amount of water absorbs a great deal of heat for a correspondingly small temperature rise. Because of this, water is a very useful cooling agent and is used in the cooling system of automobiles and other engines. If a liquid of lower specific heat capacity were used in cooling systems, its temperature would rise higher for a comparable absorption of heat.

QUESTION

Which has a higher specific heat capacity, water or sand?

Water also takes a long time to cool, a fact that explains the wide use of hot-water bottles by old timers on cold winter nights. (Better blankets, including electric ones, have since taken their place.) This tendency on the part of water to resist changes in temperature improves the climate in many places. The next time you're looking at a world globe, notice the high latitude of Europe. If water did not have a high specific heat capacity, the countries of Europe would be as cold as the northeastern regions of Canada, for both Europe and Canada get about the same amount of

ANSWER

Water has the higher specific heat capacity. The temperature of water increases less than the temperature of sand in the same sunlight. Sand's low specific heat capacity, as evidenced by how quickly the surface warms in the morning sun and how quickly it cools at night, affects local climates.

---

*If we know the specific heat capacity c, the formula for the quantity of heat Q involved when a mass m of a substance undergoes a change in temperature ΔT is $Q = mc\Delta T$. In words, heat transferred = mass $\times$ specific heat capacity $\times$ temperature change.

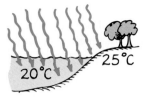

Figure 6.8 Water has a high specific heat and is transparent, so it takes more energy to heat up than land. Solar energy incident upon land is concentrated at the surface, but upon water it extends and dilutes beneath the surface.

sunlight per square kilometer. The Atlantic current known as the Gulf Stream carries warm water northeast from the Caribbean. It holds much of its thermal energy long enough to reach the North Atlantic off the coast of Europe, where it then cools. The energy released, 1 calorie per degree for each gram of water that cools, is carried by the westerly winds over the European continent. A similar effect occurs in the United States. The winds in the latitudes of North America are westerly. On the West Coast, air moves from the Pacific Ocean to the land. Because of water's high specific heat capacity, ocean temperatures do not vary much from summer to winter. The water is warmer than the air in the winter and cooler than the air in the summer. In winter the water warms the air, which then moves over and warms the coastal regions of North America. In summer, the water cools the air and the coastal regions are cooled. The East Coast does not benefit from the moderating effects of water because air moves from the land to the Atlantic Ocean. Land, with a lower specific heat capacity, gets hot in the summer but cools rapidly in the winter. Because of water's high specific heat capacity and westerly winds, the West Coast city of San Francisco is warmer in the winter and cooler in the summer than the East Coast city of Washington, D.C., which is at about the same latitude.

Islands and peninsulas that are more or less surrounded by water do not have the same extremes of temperatures that are observed in the interior of a continent. The high summer and low winter temperatures common in Manitoba and the Dakotas, for example, are largely due to the absence of large bodies of water. Europeans, islanders, and people living near ocean air currents should be glad that water has such a high specific heat capacity. San Franciscans are!

Life at the Extremes

Some deserts such as those on the plains of Spain, the Sahara in Africa, and the Gobi Desert in central Asia reach surface temperatures of 60°C (140°F). Too hot for life? Not for a species of ant (*Cataglyphis*) that thrives at this searing temperature. At this extremely high temperature the desert ants can forage for food without the presence of lizards who otherwise prey upon them. Resilient to heat, these ants can withstand higher temperatures than any other creatures in the desert. How they are able to do this is currently being researched. They scavenge the desert surface for corpses of those who did not find cover in time, touching the hot sand as little as possible while often sprinting on four legs with two high in the air. Although their foraging paths zig zag over the desert floor, their return paths are almost straight lines to their nest holes. They attain speeds of 100 body lengths per second. During an average six day life, most of these ants retrieve 15 to 20 times their weight in food.

From deserts to glaciers, a variety of creatures have invented ways to survive the harshest corners of the world. A species of worm thrives in the glacial ice in the Arctic. There are insects in the Antarctic ice that pump their bodies full of antifreeze to ward off becoming frozen solid. Some fish beneath the ice are able to do the same. Then there are bacteria that thrive in boiling hot springs as a result of heat-resistant proteins.

An understanding of how creatures survive at the extremes of temperature can provide clues for practical solutions for physical challenges to humans. Astronauts who venture from our nest, for example, will need all the techniques available for coping with unfamiliar environments.

6.5 THERMAL EXPANSION

When the temperature of a substance is increased, its atoms or molecules jiggle faster and on the average tend to move farther apart. The result is an expansion of the substance. With few exceptions, all phases of matter—solids, liquids, gases, and plasmas—generally expand when they are heated and contract when they are cooled.

In many cases the changes in the size of substances are not very noticeable, but careful observation will usually detect them. Telephone wires are longer and sag more on a hot summer day than they do on a cold winter day. Metal lids on glass fruit jars can often be loosened by heating them under hot water. If one part of a piece of glass is heated or cooled more rapidly than adjacent parts, the expansion or contraction that results may break the glass. This is especially true with thick glass. Pyrex glass is an exception because it is specially formulated to expand very little with increasing temperature.

The expansion of substances must be allowed for in the construction of structures and devices of all kinds. A dentist uses filling material that has the same rate of expansion as teeth. The aluminum pistons of some automobile engines are just smaller enough in diameter than the steel cylinders to allow for the much greater expansion rate of aluminum. A civil engineer uses reinforcing steel of the same expansion rate as concrete. Long steel bridges commonly have one end fixed while the other rests on rockers (Figure 6.9). The roadway itself is segmented with tongue-and-groove type gaps called *expansion joints* (Figure 6.10). Similarly, concrete roadways and sidewalks are intersected by gaps, sometimes filled with tar, so that the concrete can expand freely in summer and contract in winter.

Figure 6.9 One end of the bridge is fixed, while the end shown rides on rockers to allow for thermal expansion.

Figure 6.10 This gap is called an expansion joint; it allows the bridge to expand and contract.

Different substances expand at different rates. When two strips of different metals, say one of brass and the other of iron, are welded or riveted together, the greater expansion of one metal results in the bending shown in Figure 6.11. Such a

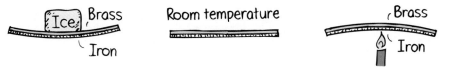

Figure 6.11 A bimetallic strip. Brass expands (or contracts) more when heated (or cooled) than iron does, so the strip bends as shown.

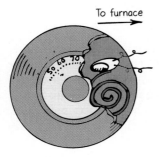

To furnace

Figure 6.12 A thermostat. When the coil expands, the mercury rolls away from the electrical contacts and breaks the circuit. When the coil contracts, the mercury rolls against the contacts and completes the electrical circuit.

compound thin bar is called a *bimetallic strip.* When the strip is heated, one side of the double strip becomes longer than the other, causing the strip to bend into a curve. On the other hand, when the strip is cooled, it tends to bend in the opposite direction, because the metal that expands more also shrinks more. The movement of the strip may be used to turn a pointer, regulate a valve, or close a switch.

A practical application of this is the thermostat (Figure 6.12). The back-and-forth bending of the bimetallic coil opens and closes an electric circuit. When the room becomes too cold, the coil bends toward the brass side, and in so doing activates an electrical switch that turns on the heat. When the room becomes too warm, the coil bends toward the iron side, which activates an electrical contact that turns off the heating unit. Refrigerators are equipped with special thermostats to prevent them from becoming either too warm or too cold. Bimetallic strips are used in oven thermometers, electric toasters, and various other devices.

Liquids expand appreciably with increases in temperature. In most cases the expansion of liquids is greater than the expansion of solids. The gasoline overflowing a car's tank on a hot day is evidence of this. If the tank and contents expanded at the same rate, they would expand together and no overflow would occur.

QUESTION

The Concorde supersonic airplane is 20 cm longer when in flight. Offer an explanation.

6.6 CHANGE OF PHASE

The matter in our environment exists in four common phases.* Ice, for example, is the *solid* phase of H_2O. Add energy, and you add motion to the rigid molecular structure, which breaks down to form H_2O in the *liquid* phase, water. Add more energy, and the liquid changes to the *gaseous* phase. Add still more energy, and the molecules break into ions and electrons, giving the *plasma* phase. Plasma, an incandescent gas found in fluorescent and other vapor lamps, is actually the predominant phase of matter in the universe. It makes up the sun and other stars, and much of the matter between them. We'll postpone discussion of this phase of matter until Part 8—Astronomy. In Chapter 8 we'll discuss the energies that accompany the phase changes in solids, liquids, and gases. In this chapter we'll see how changes in molecular motion, temperature, and pressure affect changes of phase.

ANSWER

At cruising speed (faster than the speed of sound), air friction against the Concorde raises its temperature dramatically, resulting in this significant thermal expansion.

*Some texts use the word *state* in place of *phase.* In this context matter is made up of the four states of matter. Either may be used.

Evaporation

Water in an open container will eventually evaporate, or dry up. The liquid that disappears becomes water vapor in the air. **Evaporation** is a change of phase from liquid to gas that takes place at the surface of a liquid. Molecules in the liquid phase move about in all directions and bump into one another while moving at different speeds. Some gain kinetic energy while others lose kinetic energy. Molecules at the surface that gain kinetic energy by being bumped from below may have enough energy to break free of the liquid. They can leave the surface and fly into the space above the liquid. In this way they become gas molecules.

The increased kinetic energy of molecules bumped free comes from molecules remaining in the liquid. This is "billiard-ball physics": When balls bump into one another and some gain kinetic energy, the others lose the same amount. Molecules going from liquid to gas are the gainers, while the losers of energy remain in the liquid. Thus the average kinetic energy of the molecules remaining in the liquid is lowered—evaporation is a cooling process.

Unless heated by the surroundings, the temperature of a container of water will decrease as evaporation proceeds. As the water cools, so does the rate of evaporation. A higher rate of evaporation can be maintained if the water is in contact with a relatively warm surface, such as your skin. Body heat then flows from you into the water. In this way the water maintains a higher temperature and evaporation continues at a relatively high rate. This is why you feel cool as you dry off after getting wet—you are losing your body heat to the energy-requiring process of evaporation.

When our bodies tend to overheat, our sweat glands produce perspiration. This is part of nature's thermostat, for the evaporation of perspiration cools us and helps us maintain a stable body temperature. Many animals do not have sweat glands and must cool themselves by other means (Figures 6.13 and 6.14).

Figure 6.13 Dogs have no sweat glands (except between the toes). They cool themselves by panting. In this way evaporation occurs in the mouth and within the bronchial tract.

Figure 6.14 Pigs have no sweat glands and therefore cannot cool by the evaporation of perspiration. Instead, they wallow in the mud to cool themselves.

QUESTION

Would evaporation be a cooling process if each molecule at the surface of a liquid had the same kinetic energy before and after bumping into other molecules?

The rate of evaporation is greater for higher temperatures, because there are more molecules with enough kinetic energy to break free of the liquid. We will learn in Chapter 19 that the hydrogen bonds between water molecules are broken when evaporation occurs. Water evaporates at lower temperatures too, but at a lower rate. A puddle of water, for example, may evaporate to dryness on a cool day. Even frozen water "evaporates." This form of evaporation in which molecules jump directly from a solid to a gaseous phase is called **sublimation.** Since water molecules are so tightly held in the solid phase, frozen water does not evaporate (sublime) as readily as liquid water. Sublimation, however, does account for the loss of significant portions of snow and ice, especially on sunny and dry mountain tops.

Condensation

The process that is the opposite of evaporation is **condensation**—the changing of a gas to a liquid. When gas molecules near the surface of a liquid are attracted to the liquid, they strike the surface with increased kinetic energy and become part of the liquid. This kinetic energy is absorbed by the liquid, and increases its temperature. Condensation is a warming process.

A dramatic example of warming that results from condensation is the energy given up by steam when it condenses—a painful experience if it condenses on you.

ANSWER

No. If there were no transfer of kinetic energy during molecular bumping, there would be no change in temperature. The liquid cools only when there is a lowering of average kinetic energy of molecules in the liquid. This occurs when some molecules (like billiard balls) gain speed at the expense of others that lose speed. Those that leave (evaporate) are gainers while losers remain behind to effectively lower the temperature of the water.

Figure 6.15 Heat is given up by steam when it condenses inside the radiator.

That's why a steam burn is much more damaging than a burn from boiling water of the same temperature; the steam gives up considerable energy when it condenses to a liquid and wets the skin. This energy release by condensation is utilized in steam-heating systems.

Water vapor need not be as hot as steam to give up energy when it condenses. You are warmed by condensation after you have taken a shower—even a cold shower—if you remain in the moist shower area. You quickly sense the difference if you step outside. Away from the moisture, net evaporation takes place quickly and you feel chilly. When you remain in the shower stall, even with the water off, the warming effect of condensation counteracts the cooling effect of evaporation. If as much moisture condenses as evaporates, you feel no change in body temperature. If condensation exceeds evaporation, you are warmed. If evaporation exceeds condensation, you are cooled. So now you know why you can dry yourself with a towel much more comfortably if you remain in the shower area. To dry yourself thoroughly, you can finish the job in a less moist area.

Spend a July afternoon in dry Tucson or Phoenix where evaporation is appreciably greater than condensation. The result of this pronounced evaporation is a much cooler feeling than we would experience in a same-temperature July afternoon in New York City or New Orleans. In these humid locations, condensation noticeably counteracts evaporation, and you feel the warming effect as vapor in the air condenses on your skin. We are quite literally being "tattooed" by the impact of H_2O molecules in the air that slam into us. Put more mildly, we are warmed by the condensation of vapor in the air upon our skin. We will explore condensation in the atmosphere when we study meteorology in Chapter 26.

Figure 6.16 If you're chilly outside the shower stall, step back inside and be warmed by the condensation of the excess water vapor there.

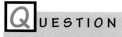UESTION

If the water level in a dish of water remains unchanged from one day to the next, can you conclude that no evaporation or condensation is taking place?

A NSWER

Not at all, for there is much activity taking place at the molecular level. Both evaporation and condensation occur continuously. The fact that the water level remains constant simply indicates equal rates of evaporation and condensation. As many molecules leave the surface by evaporation as return by condensation, so no *net* evaporation or condensation takes place. The two processes are in equilibrium, and effectively cancel each other.

Boiling

Evaporation occurs beneath the surface of a liquid in the process called **boiling.** Bubbles of vapor form in the liquid and are buoyed to the surface where they escape. Bubbles can form only when the pressure of the vapor within the bubbles is great enough to resist the pressure at the depth of the surrounding water, which includes atmospheric pressure. This occurs at the boiling temperature. At lower temperatures the vapor pressure in the bubbles is not enough, and the surrounding pressure serves to collapse any bubbles that form.

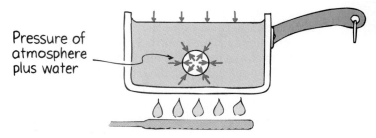

Pressure of atmosphere plus water

Figure 6.17   The motion of molecules in the bubble of steam (much enlarged) creates a gas pressure that counteracts the atmospheric and water pressure against the bubble.

As the atmospheric pressure is increased, the molecules in the vapor are required to move faster to exert increased pressure within the bubble to counteract the additional atmospheric pressure. So increasing the pressure on the surface of a liquid raises the boiling temperature of the liquid. Conversely, lowered pressure (as at high altitudes) decreases the boiling temperature. So we see that boiling depends not only on temperature, but on pressure also.

The role of pressure is evident in a pressure cooker (Figure 6.18). Vapor pressure builds up inside and prevents boiling, which results in higher water temperature. It is important to note that it is the high temperature of the water that cooks the food, not the boiling process itself. At high altitudes, water boils at a lower temperature. In Denver, Colorado, the "mile-high city," for example, water boils at 95°C, instead of the 100°C boiling temperature characteristic of sea level. If you try to cook food in boiling water of a lower temperature, you must wait a longer time for proper cooking. A "three-minute" boiled egg in Denver is yukky. If the temperature of the boiling water were very low, food would not cook at all.

Figure 6.18   The tight lid of a pressure cooker holds pressurized vapor above the liquid surface that inhibits boiling; in this way the boiling temperature is above 100°C.

Boiling, like evaporation, is a cooling process. At first thought, this may seem surprising—perhaps because we usually associate boiling with heating. But heating water is one thing; boiling is another. When 100°C water at atmospheric pressure is boiling, it is in thermal equilibrium. It is being cooled by boiling as fast as it is being heated by energy from the heat source (Figure 6.19). If cooling did not take place, continued application of heat to a pot of boiling water would raise its temperature. The reason a pressure cooker reaches higher temperatures is because boiling is forestalled, which in effect prevents cooling.

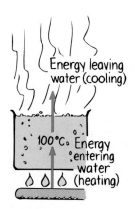

Figure 6.19 Heating warms the water from below, and boiling cools it from above.

Q UESTION

Since boiling is a cooling process, would it be a good idea to cool your hot and sticky hands by dipping them into boiling water?

We usually boil water by the application of heat. But water can be boiled by the reduction of atmospheric pressure. We can dramatically show the cooling effect of evaporation and boiling when a shallow dish of room-temperature water is placed in a vacuum jar (Figure 6.20). If the pressure in the jar is slowly reduced by a vacuum pump, the water will start to boil. The boiling process takes heat away from the water, which cools to a lower temperature. As the pressure is further reduced, more and more of the slower-moving molecules boil away. Continued boiling results in a lowering of temperature until the freezing point of approximately 0°C is reached. Continued cooling by boiling causes ice to form over the surface of the bubbling water. Boiling and freezing are taking place at the same time! This must be witnessed to be appreciated. Frozen bubbles of boiling water are a remarkable sight.

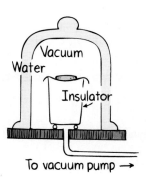

Figure 6.20 Apparatus to demonstrate that water will freeze and boil at the same time in a vacuum. A gram or two of water is placed in a dish that is insulated from the base by a polystyrene cup.

Spray some drops of coffee into a vacuum chamber, and they, too, will boil until they freeze. Even after they are frozen, the water molecules will continue to evaporate into the vacuum until little crystals of coffee solids are left. This is how freeze-dried coffee is made. The low temperature of this process tends to keep the chemical structure of coffee solids from changing. When hot water is added, more of the original flavor of the coffee is retained. Boiling really is a cooling process!

A NSWER

No, no, no! When we say boiling is a cooling process, we mean that the water (not your hands!) is being cooled relative to the higher temperature it would attain otherwise. Because of cooling, it remains at 100°C instead of getting hotter. A dip in 100°C water would be most uncomfortable for your hands!

Melting and Freezing

To visualize the processes of a phase change from solid to liquid, pretend you held hands with someone, and each of you started jumping around randomly. The more violently you jumped, the more difficult keeping hold would be. If you jumped violently enough, keeping hold would be impossible. Something like this happens to the molecules of a solid when it is heated. As heat is absorbed, the molecules vibrate more and more violently. If enough heat is absorbed, the attractive forces between the molecules will no longer be able to hold them together. The solid melts.

Freezing is the converse of this process. As energy is withdrawn from a liquid, molecular motion diminishes until finally the molecules, on the average, are moving slowly enough so that the attractive forces between them are able to cause cohesion. The molecules then vibrate about fixed positions and form a solid.

Ice has a crystalline structure. The crystals of most solids are arranged in such a way that the solid phase occupies a smaller volume than the liquid phase. Ice, however, has open-structured crystals (Figure 6.21). This structure results from the angular shape of the water molecules and the fact that the forces binding them together are strongest at certain angles. This is hydrogen bonding, which will be treated in detail in Chapter 19. Water molecules in this open structure occupy a greater volume than they do in the liquid phase. Consequently, ice is less dense than water.

At atmospheric pressure, ice forms at 0°C. At this temperature, water molecules on the average are sufficiently slowed for crystal formation to take place—unless substances such as sugar or salt are dissolved in the water. Then the freezing point is lower. In the case of salt, chlorine ions grab electrons from the hydrogen atoms in H_2O and impede crystal formation. The result of this impedance by "foreign" ions is that slower motion is required for the formation of the six-sided ice-crystal structures. As ice crystals do form, the impedance is intensified because the proportion of "foreign" molecules or ions among nonfused water molecules increases. Connections become more and more difficult. Only when the water molecules move

Snowballs and Ice Skating

The open structure of H_2O molecules in the solid phase has some interesting applications. The application of pressure to the ice crystals lowers their melting point. Simply put, the crystals are crushed to the liquid phase. The effect is small, for the melting point is only lowered 0.007°C for each additional atmosphere of added pressure. Nonetheless, when the pressure is removed, molecules crystallize and refreezing occurs. This phenomenon of melting under pressure and freezing again when the pressure is reduced is called *regelation*. It is one of the properties of water that make it different from other materials.

When we make a snowball we are using this principle. We compress the snow with our hands and cause a slight melting of the ice crystals; when pressure is removed, refreezing occurs and binds the snow together. Making snowballs is difficult in very cold weather because the pressure we can apply may not be enough to melt the snow.

Regelation enables ice skating. Strictly speaking, we don't skate on ice; we skate on a thin film of water between the blade and the ice, which is produced by blade pressure and friction. Only melting ice is slippery, not non-moist ice, as anyone who has handled ice colder than 0°C can attest. As soon as the pressure is released, the film of water refreezes.

Figure 6.21 Crystals of water in ice. The open-structured hexagonal arrangement results in the expansion of water upon freezing. Ice therefore is less dense than water.

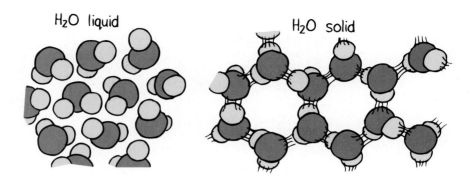

slowly enough for attractive forces to play an unusually large part in the process can freezing be completed. The ice first formed is almost always pure H_2O. In general, adding anything to water lowers the freezing temperature. Antifreeze is a practical application of this process.

Expansion of Water

When ice melts, whether pure H_2O or otherwise, not all the six-sided crystals collapse. Some crystals remain in the ice water, making up a sort of microscopic slush that slightly "bloats" the water. The result is that 0°C ice water is less dense than slightly warmer water. As the temperature of ice water is increased, more of the remaining ice crystals collapse and the volume of water decreases. This is most unusual, for we have learned that increasing the temperature of a liquid results in expansion. Water near its freezing temperature is an exception. Ice water *contracts* when its temperature is raised. This contraction continues until it reaches a temperature of 4°C. With further increase in temperature, the water undergoes a net expansion that continues all the way to the boiling point, 100°C (Figure 6.22).

Figure 6.22 The expansion of water with increasing temperature.

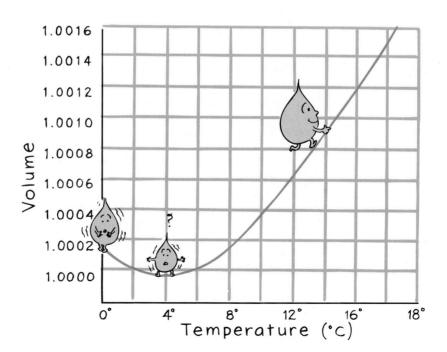

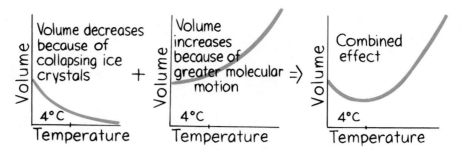

Figure 6.23 The collapsing of ice crystals plus increased molecular motion with increasing temperature produce the overall effect of water being most dense at 4°C.

We say *net* expansion because the near-freezing water has been expanding due to greater thermal motion while it has been contracting due to crystal collapse. Both processes occur together. Expansion simply overtakes contraction at the temperature of 4°C. Ice crystals are pretty well gone by about 10°C and expansion thereafter has free reign. This odd behavior is shown graphically in Figure 6.23.

So a given amount of water has its smallest volume, and thus its greatest density, at 4°C. The same amount of water has its largest volume, and smallest density, in its solid form, ice. (The fact that ice floats in water is evidence that it is less dense than water.) The volume of ice at 0°C is not shown in Figure 6.22. (If it were plotted to the same exaggerated scale, the graph would extend far beyond the top of the page.) After water has turned to ice, further cooling causes it to contract.

QUESTION

What was the precise temperature at the bottom of Lake Michigan on New Year's Eve in 1901?

This behavior of water is of great importance in nature. Suppose that the greatest density of water were at its freezing point, as is true of most liquids. Then the coldest water would settle to the bottom, and ponds would freeze from the bottom up. Pond organisms would then be destroyed in winter months. Fortunately, this doesn't happen. The densest water, which settles at the bottom of a pond, is 4 degrees above the freezing temperature. Water at the freezing point, 0°C, is less dense and "floats," so ice forms at the surface while the pond remains liquid below the ice.

Let's examine this in more detail. Most of the cooling in a pond takes place at its surface when the surface air is colder than the water. As the surface water is cooled, it becomes more dense and sinks to the bottom. Water will "float" at the surface for further cooling only if it is equally as dense or less dense than the water below.

ANSWER

The temperature at the bottom of any body of water that has 4°C water in it is 4°C at the bottom, for the same reason that rocks are at the bottom. Both 4°C water and rocks are more dense than water at any other temperature. Water is a poor heat conductor, so if the body of water is deep and in a region of long winters and short summers, the water at the bottom is likely a constant 4°C year round.

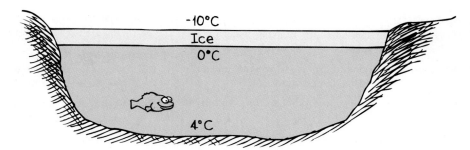

Figure 6.24 As water is cooled, it sinks until the entire pond is 4°C. Only then can surface cooling to the freezing point take place without further sinking.

Consider a pond that is initially at, say, 10°C. It cannot possibly be cooled to 0°C without first being cooled to 4°C. And water at 4°C cannot remain at the surface for further cooling unless all the water below has at least an equal density—that is, unless all the water below is at 4°C. If the water below the surface is any temperature other than 4°C, any surface water at 4°C will be denser and will sink before it can be further cooled. So before any ice can form, all the water in a pond must be cooled to 4°C. Only when this condition is met can the surface water be cooled to 3°, 2°, 1°, and 0°C without sinking. Then ice can form.

So we see that the water at the surface is first to freeze. Continued cooling of the pond results in the freezing of the water next to the ice, so a pond freezes from the surface downward. In a cold winter the ice will be thicker than in a milder winter. Very deep bodies of water are not ice-covered even in the coldest of winters. This is because all the water in a lake must be cooled to 4°C before lower temperatures can be reached. For deep lakes, the winter is not long enough for this to occur. If only some of the water is 4°C, it lies on the bottom. Because of water's high specific heat and poor ability to conduct heat, the bottom of deep lakes in cold regions is a constant 4°C the year round. Fish should be glad that this is so.

· ·

SUMMARY OF TERMS

Temperature A measure of the hotness and coldness of substances, related to the average kinetic energy per molecule in a substance, measured in degrees Celsius or Fahrenheit or in kelvins.

Absolute zero The lowest possible temperature that a substance may have—the temperature at which molecules of a substance have their minimum kinetic energy.

Thermal energy The total of all molecular energies, kinetic plus potential energy, internal to a substance.

Heat The thermal energy that flows from a substance of higher temperature to a substance of lower temperature, commonly measured in calories or joules.

Specific heat capacity The quantity of heat per unit mass required to raise the temperature of a substance by 1 Celsius degree.

Evaporation The change of phase at the surface of a liquid as it passes to the gaseous phase. This is caused

by the random motion of molecules that occasionally escape from the liquid surface. Cooling of the liquid results.

Condensation The change of phase from gas to liquid; the opposite of evaporation. Warming of the liquid results.

Boiling A rapid state of evaporation that takes place within the liquid as well as at its surface. As with evaporation, cooling of the liquid results.

· ·

REVIEW QUESTIONS

1. Why does a penny become warmer when it is struck by a hammer?

2. What happens to the temperature of air when it is rapidly compressed?

Temperature

3. What are the temperatures for freezing water on the Celsius and Fahrenheit scales? For boiling water?

4. What are the temperatures for freezing water and boiling water on the Kelvin temperature scale?

5. Why are there no negative numbers on the Kelvin scale?

6. What is meant by the statement that a thermometer measures its own temperature?

Heat

7. When you touch a cold surface, does cold travel from the surface to your hand or does energy travel from your hand to the cold surface? Explain.

8. Distinguish between temperature and heat.

9. Distinguish between heat and thermal energy.

10. What determines the direction of heat flow?

Quantity of Heat

11. How is the energy value of foods determined?

12. Distinguish between a calorie and a Calorie.

13. Distinguish between a calorie and a joule.

Specific Heat Capacity

14. Which warms up faster when heat is applied—iron or silver?

15. Name three ways in which a material can absorb energy.

16. Does a substance that heats up quickly have a high or a low specific heat?

17. How does the specific heat of water compare to the specific heats of other common materials?

18. Northeastern Canada and much of Europe receive about the same amount of sunlight per unit area. Why then is Europe generally warmer in the winter?

19. By energy conservation: If ocean water cools, then does something else warm? If so, what?

20. Why is the temperature fairly constant for land masses surrounded by large bodies of water?

Thermal Expansion

21. Why will hot water poured into a drinking glass be more likely break the glass if the glass is thick?

22. How can a bimetallic strip be used to regulate temperature?

23. Which generally expands more for increases in temperature—solids or liquids?

Change of Phase

24. What are the four common phases of matter?

Evaporation

25. Do all the molecules or atoms in a liquid have about the same speed or much different speeds?

26. What is evaporation, and why is it a cooling process? Exactly what is it that cools?

27. Why does a hot dog pant?

28. Give an example of a solid bypassing the liquid phase and changing directly to the gaseous phase.

Condensation

29. What is condensation, and why is it a warming process? Exactly what is it that warms?

30. Why is a steam burn more damaging than a burn from boiling water of the same temperature?

31. Why do we feel uncomfortably warm on a hot and humid day?

Boiling

32. Distinguish between evaporation and boiling.

33. Why does water not boil at 100°C when it is under higher pressure?

34. Why is the boiling point of water higher in a pressure cooker?

35. Is it the boiling process or the higher temperature that cooks food faster in a pressure cooker?

36. What cools when boiling occurs—the water that leaves in the form of vapor or the water that remains in the liquid phase?

37. We observe that the temperature of boiling water doesn't increase with continued heat input. Explain how this observation is evidence that boiling is a cooling process.

38. What condition permits water to boil at a temperature less than 100°C?

39. We can add heat to water and boil it to form steam. Is it a contradiction to say we can boil water to form ice? Explain.

Melting and Freezing

40. Why does increasing the temperature of a solid make it melt?

41. Why does decreasing the temperature of a liquid make it freeze?

42. Why does freezing of water not occur at 0°C when molecules other than H_2O are present?

43. Which freezes at the lower temperature, pure water or salt water?

Expansion of Water

44. When the temperature of ice-cold water is increased slightly, does it undergo a net expansion or net contraction?

45. What is the reason for ice being less dense than water?

46. Does "microscopic slush" in water tend to make it more dense or less dense?

47. What happens to the amount of "microscopic slush" in cold water when its temperature is increased?

48. What happens to the amount of molecular motion in water when its temperature is increased?

49. At what temperature do the effects of contraction and expansion produce the smallest volume for water?

50. Why does ice form at the surface of a body of water instead of at the bottom?

HOME PROJECTS

1. Place a Pyrex funnel mouth down in a saucepan full of water so the straight tube of the funnel sticks above the water. Rest a part of the funnel on a nail or coin so water can get under it. Place the pan on a stove and watch the water as it begins to boil. Where do the bubbles form first? Why? As the bubbles rise, they expand rapidly and push water ahead of them. The funnel confines the water, which is forced up the tube and driven out at the top. Now do you know how a geyser and a coffee percolator work?

2. Watch the spout of a tea kettle of boiling water. Notice that you cannot see the steam that issues from the spout. The cloud that you see farther away from the spout is not steam, but condensed water droplets. Now hold a candle in the cloud of condensed steam. Can you explain your observations?

3. Compare the temperature of boiling water with the temperature of a boiling solution of salt and water.

EXERCISES

1. Why can't you establish whether you are running a high temperature by touching your own forehead?

2. Which has the greater amount of thermal energy, an iceberg or a cup of hot coffee? Explain.

3. Would a common mercury thermometer be feasible if glass and mercury expanded at the same rates for changes in temperature? Explain.

4. Does it make sense to talk about the temperature of a vacuum?

5. What is *temperature* a measurement of?

6. If you drop a hot rock into a pail of water, the temperature of the rock and the water will change until both are equal. The rock will cool and the water will warm. Does this hold true if the hot rock is dropped into the Atlantic Ocean? Explain.

7. Would you expect the temperature of water at the bottom of Niagara Falls to be slightly higher than the temperature at the top of the falls? Why?

8. Why does the pressure of gas enclosed in a rigid container increase as the temperature increases?

9. Bermuda is close to North Carolina, but unlike North Carolina it has a tropical climate year round. Why is this so?

10. If the winds at the latitude of San Francisco and Washington, D.C., were from the east rather than from the west, why might San Francisco be able to grow only cherry trees and Washington, D.C., only palm trees?

11. San Francisco is warmer in winter than Washington, D.C. Why isn't it warmer in summer?

12. In addition to the to-and-fro vibrations of a molecule that are associated with temperature, some molecules can absorb large amounts of energy in the form of internal vibrations and rotations of the molecule itself. Would you expect materials composed of such molecules to have a high or a low specific heat? Explain.

13. The desert sand is very hot in the day and very cool at night. What does this tell you about its specific heat?

14. Would you or the gas company gain by having gas warmed before it passed through your gas meter?

15. A metal ball is just able to pass through a metal ring. When the ball is heated, however, it will not pass through the ring. What would happen if the ring, rather than the ball, were heated? Does the size of the hole increase, stay the same, or decrease?

16. After a machinist very quickly slips a hot, snugly fitting iron ring over a very cold brass cylinder, there is no way that the two can be separated intact. Can you explain why this is so?

17. Suppose you cut a small gap in a metal ring. If you heat the ring, will the gap become wider or narrower?

18. When a mercury thermometer is warmed, the mercury level momentarily goes down before it rises. Can you give an explanation for this?

19. One of the reasons the first light bulbs were expensive was that the electrical lead wires into the bulb were made of platinum, which expands at about the same rate as glass when heated. Why is it important that the metal leads and the glass have the same coefficient of expansion?

20. If you measure a plot of land with a steel tape on a hot day, will your measurements of the plot be larger or smaller than they actually are?

21. You can determine wind direction by wetting your finger and holding it up in the air. Explain.

22. Can you give two reasons why blowing over hot soup cools the soup?

23. Can you give two reasons why pouring a cup of hot coffee into the saucer results in faster cooling?

24. Porous canvas bags filled with water are used by travelers in hot weather. When the bags are slung on the outside of a fast-moving car, the water inside is cooled considerably. Explain.

25. What is the function of the cloth that commonly covers metal canteens? Should this cloth be wet or dry, and why?

26. Why will wrapping a bottle in a wet cloth at a picnic often produce a cooler bottle than placing the bottle in a bucket of cold water?

27. The human body can maintain its customary temperature of 37°C on a day when the temperature is above 40°C. How is this done?

28. Why does the temperature of boiling water remain the same as long as the heating and boiling continue?

29. Water will boil spontaneously in a vacuum—on the moon, for example. Could you cook an egg in this boiling water? Explain.

30. Your inventor friend proposes a design of cookware that will allow boiling to take place at a temperature less than 100°C so that food can be cooked with less energy. Comment on this idea.

31. Your instructor hands you a closed flask of room-temperature water. When you hold it, the heat from your bare hands causes the water to boil. Quite impressive! How is this accomplished?

32. When you boil potatoes, will your cooking time be reduced with vigorously boiling water compared to gently boiling water? (Directions for cooking spaghetti call for vigorously boiling water—not to lessen cooking time, but to prevent something else. If you don't know what it is, ask a cook.)

33. Why does putting a lid over a pot of water on a stove shorten the time it takes for the water to come to a boil, whereas after the water is boiling, use of the lid only slightly shortens the cooking time?

34. What was the precise temperature at the bottom of Lake Superior at 12:01 AM on October 31, 1894?

35. Suppose that water is used in a thermometer instead of mercury. If the temperature is at 4°C and then changes, why can't the thermometer indicate whether the temperature is rising or falling?

36. How does the combined volume of the billions and billions of hexagonal open spaces in the structures of ice crystals in a piece of ice compare to the portion of ice that floats above the water line?

37. How would the shape of the curve in Figure 6.22 differ if density were plotted against temperature instead of volume? Make a rough sketch.

38. Why is it important to protect water pipes so they don't freeze?

39. If cooling occurred at the bottom of a pond instead of at the surface, would a lake freeze from the bottom up? Explain.

40. If water had a lower specific heat, would ponds be more likely to freeze or less likely to freeze?

. . . .

PROBLEMS

1. Suppose a bar 1 m long expands $\frac{1}{2}$ cm when heated. By how much will a bar 100 m long of the same material expand when similarly heated?

2. Steel expands 1 part in 100 000 (10^{-5}) for each Celsius degree increase in temperature. Suppose the 1.3-km main span of the Golden Gate Bridge had no expansion joints. How much longer would it be for an increase in temperature of 10°?

3. Consider a 40 000-km steel pipe that forms a ring to fit snugly all around the circumference of the world. Suppose people along its length breathe on it so as to raise its temperature 1 Celsius degree. The pipe gets longer. It also is no longer snug. How high does it stand above ground level? (To simplify, consider only the expansion of its radial distance from the center of the earth, and apply the geometry formula that relates circumference C and radius r, $C = 2\pi r$. The result is surprising!)

7

···

HEAT TRANSFER

Heat transfers from warmer to cooler things. If several things of different temperatures are in contact, those that are warm become cooler and those that are cool become warmer. They tend to reach a common temperature. This equalizing of temperature occurs in three ways: *conduction, convection,* and *radiation.*

···

7.1 CONDUCTION

Hold one end of an iron nail in a flame. It will quickly become too hot to hold. The heat enters the metal nail at the end kept in the flame and is transmitted along its whole length. The transmission of heat in this manner is called **conduction**. The fire causes the molecules at the heated end of the nail to move more rapidly. Because of this increased motion, these molecules and free electrons collide with their neighbors, and so on. This process continues until the increased motion has been transmitted to all the molecules and the entire body has become hot. Heat conduction occurs by electron and molecular collisions.

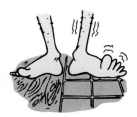

Figure 7.1 The tile floor feels colder than the wooden floor, even though both floor materials are the same temperature. This is because tile is a better conductor than wood, and heat is more readily conducted from the foot that makes contact with the tile.

How well an object conducts heat depends on the electrical bonding of the molecular structure. Solids whose molecules have "loose" outer electrons conduct heat (and electricity) well. Metals have the "loosest" outer electrons and are the best conductors of heat and electricity for this reason. Silver is the best, copper is next, and, among the common metals, aluminum and then iron are next in order. Wood, wool, straw, paper, cork, and styrofoam are poor conductors of heat. The outer electrons in the molecules of these materials are firmly attached.

Interestingly enough, wood is a poor conductor even when it's hot—red hot. This is demonstrated by co-author John Suchocki walking on red-hot coals of wood in front of his physical science class (Figure 7.2). Because of the low conductivity of the coals, energy from the interior of the coal is slow to move out to the surface and into John's feet. (Using red-hot pieces of metal would be a painful mistake!) John's success is enhanced by first walking across wet grass. Like the moisture on your finger when you touch a hot iron, energy that goes into vaporization is energy that doesn't go to the feet.

Liquids and gases, in general, are poor conductors. Air is a very poor conductor. Porous substances, which have many small air spaces, are poor conductors and good insulators. The good insulating properties of such things as wool, fur, and feathers are largely due to the air spaces they contain. Be glad that air is a poor conductor; if it weren't, you'd feel quite chilly on a 20°C (68°F) day! Poor conductors are called *insulators.*

Snow is a poor conductor (good insulator), about the same as dry wood, and hence is popularly said to keep the earth warm. Its flakes are formed of crystals, which collect into feathery masses, imprisoning air and thereby interfering with the escape of heat from the earth's surface. Eskimo winter dwellings are shielded from the cold by their snow covering. Animals in the forest find shelter from the cold in

Figure 7.2 John Suchocki walks barefoot on red-hot wooden coals, demonstrating that wood is a poor conductor of heat, even when red hot.

Figure 7.3 Snow patterns on the roof of a house reveal the conduction, or lack of conduction, of heat through the roof.

snowbanks and in holes in the snow. The snow doesn't provide them with heat; it simply prevents the heat they generate from escaping.

Heat is transmitted from a higher to a lower temperature. We often hear people say they wish to keep the cold out of their homes. A better way to put this is to say that they want to prevent the heat from escaping. There is no "cold" that flows into a warm home. If the home becomes colder, it is because heat flows out. Homes are insulated with rock wool or spun glass to prevent heat from escaping rather than to prevent cold from entering. Interestingly enough, insulation of whatever kind does not actually prevent heat from getting through it; it simply slows the rate at which heat penetrates. Even a well-insulated warm home in winter will gradually cool. Insulation delays the transfer of heat.

Q UESTION

In desert regions that are hot in the daytime and cold at nighttime, the walls of houses are usually made of mud. Why is it important that the mud walls be thick?

7.2 CONVECTION

Liquids and gases transmit heat mainly by **convection,** which is heat transfer by the actual motion of the fluid—currents. Convection may occur in all fluids, whether liquids or gases. Whether we heat water in a pan or warm air in a room, the process is the same (Figure 7.4). If the fluid is heated from below, its molecules increase in speed; the fluid becomes less dense and is pushed up by the denser cooler fluid that takes its place at the bottom. In this way, convection currents keep the fluid stirred up as it heats. Convection currents occur in the atmosphere, and to understand this we must first understand why warm air rises—then why it cools.

A NSWER

A wall of appropriate thickness provides a proper amount of time for heat flow from outer to inner wall surfaces to produce maximum interior temperature during the sleeping hours, when it is cold outside, and minimum interior temperature during midday, when it is hot outside.

Figure 7.4 (*Top*) Convection currents in air. (*Bottom*) Convection currents in liquid.

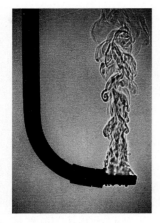

Figure 7.5 A heater at the tip of the submerged J-tube produces convection currents, which are revealed as shadows as light is deflected in water of different temperatures.

Why Warm Air Rises

We all know that warm air rises. From our study of buoyancy we understand why this is so. Warm air expands, becomes less dense than the surrounding air, and is buoyed upward like a balloon. The buoyancy is in an upward direction because the air pressure below a region of warmed air is greater than the air pressure above. And the warmed air rises because the buoyant force is greater than its weight.

We can understand the rising of warm air from a different point of view—by considering the motion of individual molecules. Consider a fairly large region of identical gas molecules. Because of gravity we would find more molecules near the bottom of our region than near the top; the gas would be slightly denser toward the ground. Suppose the region is of uniform temperature; then each molecule, on the average, has the same kinetic energy and the same average velocity. Each molecule, therefore, has the same tendency to migrate throughout the region. Suppose now that we introduce a faster-moving molecule; a "hot" one. Until it gives up its excess energy to slower-moving molecules, it will migrate farther in a shorter time than any of its neighbors. If our sample molecule is placed in the middle of our region, it will bump into and rebound from molecules in all directions. A little thought will show it will rebound from a greater number of molecules whenever it happens to be moving downward rather than upward. This is because the density of molecules is greater below; there is more opposition to a downward migration than to an upward migration where the air is less dense. Furthermore, when our "hot" molecule moves in an upward direction, it travels farther before making a collision than when it travels downward. We say it has a longer "mean-free path" when moving upward. So our faster-moving molecule will tend to bumble upward in its random jostling.

We have simplified the idea of rising warm air by considering the behavior of a single molecule. A single fast-moving molecule would, of course, soon share its excess energy and momentum with its less energetic neighbors and would not rise very high.* However, if we start with a large cluster of energetic molecules, many of these will rise to appreciable heights before their energy and momentum dissipate.

Why Expanding Air Cools

As warm air rises, it expands and cools. It expands because it moves up into a region of less pressure. But why does it cool? We can understand this if we do two things: (1) consider the converse situation where compressing air increases temperature, and (2) think of air molecules as tiny Ping-Pong balls.

If you've ever slammed a Ping-Pong ball with a paddle, you know the speed of the ball is increased; and if you've ever pumped air into a bicycle tire with a hand pump, you probably noticed that the pump soon became hot. Your pumping action

---

*Interestingly enough, a single helium atom will rise in the atmosphere for exactly the reasons stated in the previous paragraph, because even with no excess kinetic energy its average speed will always be considerably greater than the speeds of neighboring heavier molecules of nitrogen and oxygen. Can you see that in a mixture of gaseous particles of various masses at the same temperature, the particles of least mass have the greatest speeds? (Recall KE $= \frac{1}{2}mv^2$; if m is small, then v must be large compared to an equally energetic more-massive molecule.) And can you see that lightweight helium with its greater speed will migrate more than its heavier neighbors, bumble its way to the top of the atmosphere, and escape into outer space? Can you see why helium, which in fact is the seventh most common gas in the earth's atmosphere, doesn't normally exist in the lower atmosphere?

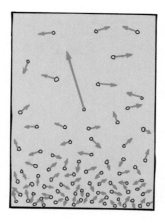

Figure 7.6 A fast-moving molecule tends to migrate toward the region of least obstruction—upward. Warm air therefore rises.

served to increase the speed of the molecules of air, just as a paddle slamming against Ping-Pong balls increases the speed of the balls. The act of compressing air warms it. The converse is true when air expands.

Let's see why. Like Ping-Pong balls, when molecules collide with others that are approaching at greater speeds, their rebound speeds are increased; the converse is that when molecules collide with others that are *receding,* their rebound speeds are reduced. This is easy to see: Although a Ping-Pong ball picks up speed when struck by an approaching paddle, it slows down when it rebounds from a receding paddle. In a region of air that is expanding, molecules will collide, on the average, with more molecules that are receding than are approaching. (This is just the opposite of what happens when air is compressed.) Thus, in expanding air, the average speed of the molecules decreases and the air cools. Where does the energy go in this case? It actually goes into the work done on the surrounding air as the expanding air pushes outward. In this way thermal energy is diluted. So we see that the energy per volume and the temperature are less. (We'll return to this concept in the next chapter.)

A common misconception about temperature and molecular motion is that heat is produced by the *number* of collisions molecules make with one another; that the more frequently molecules collide, the higher the temperature of the gas will be. This is not true. Although individual molecules may gain or lose speed in approaching and receding collisions, the total number of molecules bouncing off one another have

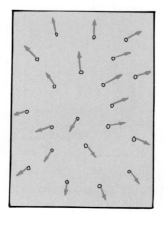

Figure 7.7 A molecule in a region of expanding air collides primarily with receding molecules, not approaching ones. Its velocity of rebound therefore lessens with each collision and results in a cooling of expanding air.

Figure 7.8 With your mouth wide open, blow air on your hand and note the air temperature. Now close your lips so the air expands when you blow. Do you notice a diffence in air temperature? (Try it now!)

the same total energy and momentum before and after a collision. Temperature is a measure of kinetic energy, not of collision rates. The temperature of gas would be no different if all the molecules were able to move without colliding with one another. Put another way, when a gas is heated, the molecules collide more often. We can say they collide more often *because* they're heated. But we can *not* say that they are heated because they collide more often. ("Cause produces effect" is a one-way street!)

Thus, compression heats air; expansion cools it. But once compressed or expanded, it soon comes to a temperature equilibrium with its surroundings. Put your hand on a tank of compressed air and you will find that it has the same temperature as its surroundings. Put your hand in the path of the same air as it escapes and expands from a nozzle and you will find a considerably lower temperature. Do the following right now: blow air on your hand with your mouth wide open. Now close your lips so the air expands when it leaves your mouth and blow on your hand again (Figure 7.8). Notice its coolness. A more dramatic example occurs with steam that expands through the nozzle of a pressure cooker (Figure 7.9). The cooling effect of both expansion and rapid mixing with cooler air allows you to hold your hand comfortably in the jet of condensed vapor. (Caution: If you try this, be sure to place your

Figure 7.9 The hot steam expands from the pressure cooker and is cool to Millie's touch.

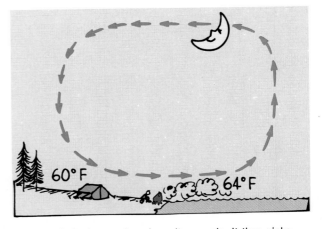

Figure 7.10 Convection currents produced by unequal heating. (a) Warmed air rises and cools as it expands. It then sinks and flows toward the warmed area to replace the air that has risen. The land is warmer than the water in the day (lower specific heat) and (b) cooler than the water at night, so the direction of air flow reverses.

hand high above the nozzle at first and then lower it to a comfortable distance; if you put your hand at the nozzle where no steam appears, watch out! Steam is invisible, and that's where it is before it has sufficiently expanded and cooled. The cloud of "steam" you see is actually condensed water vapor—much cooler.)

Convection currents stirring the atmosphere result in winds. Some parts of the earth's surface absorb heat from the sun more readily than others, and as a result the air near the surface is heated unevenly and convection currents form. This is evident at the seashore. In the daytime the shore warms more easily than the water; air over the shore is pushed up (we say it rises) by cooler air from above the water taking its place. The result is a sea breeze. At night the process reverses because the shore cools off more quickly than the water, and then the warmer air is over the sea (Figure 7.10). Build a fire on the beach and you'll notice that the smoke sweeps inward during the day and seaward at night.

7.3 RADIATION

Heat from the sun passes through space before it passes through the atmosphere to warm the earth's surface. Neither convection nor conduction is possible in the empty space between our atmosphere and the sun, so we can see that heat must be transmitted some other way—by radiation.* The energy so radiated is called *radiant energy.*

Although the sun is an obvious source of radiant energy, all objects continually radiate energy in a mixture of wavelengths. Objects at low temperatures emit long waves, just as long lazy waves are produced when you shake a rope with little energy (Figure 7.11). Higher-temperature objects emit waves of shorter wavelengths. Objects of everyday temperatures emit waves mostly in the long-wavelength infrared region, between radio and light waves. It is the shorter-wavelength infrared radiation that our skin commonly experiences as heat. So it is common to refer to infrared radiation as *heat radiation.*

Figure 7.11 Shorter wavelengths are produced when the rope is shaken more vigorously.

Objects that are hot enough radiate some of their energy in the range of visible light. At a temperature of about 600°C an object begins to emit the longest waves we can see, red light. Higher temperatures produce a yellowish light. At 1200°C all the different waves to which the eye is sensitive are emitted and we see an object as "white hot." At the same time, however, the object continues to emit the long waves,

---

*Do not confuse radiation with radioactivity—reactions that involve the atomic nucleus and are characteristic of nuclear power and nuclear power plants and the like. Radiation here is electromagnetic radiation, "heat" waves of low-frequency light, which we will treat in detail in Parts 2 and 4.

Figure 7.12　Types of radiant energy (electromagnetic waves).

thus radiating a wide variety of wavelengths. When any kind of matter absorbs these waves, the waves set the electrons in the matter into vibration, and temperature increases. If radiation in the wave range of infrared falls on our skin, it may excite the sensation of warmth. In this way, heat is transmitted from the radiating object to the absorbing object.

Emission, Absorption, and Reflection of Radiation

All objects continually radiate energy. Why, then, doesn't the temperature of all objects continually decrease? The answer is that all objects also continually absorb radiant energy. If an object is radiating more energy than it is absorbing, its temperature does decrease; but if an object is absorbing more energy than it is emitting, its temperature increases. An object that is warmer than its surroundings emits more energy than it receives, and therefore it cools; an object colder than its surroundings is a net gainer of energy, and its temperature therefore increases. An object whose temperature is constant, then, emits as much radiant energy as it receives. If it receives none, it will radiate away all its available energy, and its temperature will approach absolute zero.

The rate at which an object radiates or absorbs radiant energy depends on the nature of the object and the difference between its temperature and the surrounding temperature. Emission and absorption occur at the surface of an object. A rough surface is a better absorber and emitter than a smooth surface because its irregularities give it more area, and its microscopic hills and valleys enable multiple scattering of radiation at the surface to occur. This is true both for long-wavelength heat radiation and shorter-wavelength light radiation. So we can see that a very good absorber of radiation reflects very little light and therefore appears black. Things that reflect little visible light appear black to us. A perfect absorber reflects no radiation and appears perfectly black.

A body that absorbs all the radiation incident upon it is called a *blackbody*. Holes and cavities are practical examples of blackbodies. A keyhole in a closet door, for example, appears perfectly black. Small holes appear black because the radiation that enters is reflected from the inside walls many times and is partly absorbed at each reflection until none remains (Figure 7.14). For the same reason, the pupil of the eye is black (except when illuminated directly—with a flash camera, for example—in which case it appears pink).

If the walls of the hole or cavity are heated, the radiation given off in all directions is more than the incoming radiation, and there is a net escape of radiation from the hole. So we find that good absorbers are also good emitters. And poor absorbers are poor emitters. For example, a radio antenna constructed to be a good emitter of

Figure 7.13 The hole looks perfectly black and indicates a black interior, when in fact the interior has been painted a bright white.

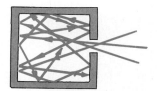

Figure 7.14 Radiation that enters the cavity has little chance of leaving before it is completely absorbed.

Figure 7.15 When the containers are filled with hot (or cold) water, the blackened one cools (or warms) faster.

radio waves will also by its very design be a good receiver of radio waves. And a poorly designed transmitting antenna will also be a poor receiver. The same holds true for the atomic and molecular construction of matter. A blackbody is both an ideal absorber and an ideal radiator.

Whether a surface plays the role of net radiator or net absorber depends on whether its temperature is above or below the temperature of its surroundings. If the surface is hotter than the surrounding air, for example, it will be a net radiator and will cool. If the surface is colder than the surrounding air, it will be a net absorber and will become warmer.

The heat radiation that falls on a surface and is absorbed obviously cannot also be reflected; likewise, the heat radiation that falls on a surface and is reflected cannot be absorbed. Hence, a good absorber is a poor reflector of radiation, and a good reflector is a poor absorber.

Polished or mirrorlike surfaces are poor absorbers of both visible and heat radiation. This can be illustrated by placing thermometers in a pair of metal containers of the same size and shape, one having a brightly polished surface and the other a blackened surface (Figure 7.15). If they are filled with hot water, we find that the container with the blackened surface cools faster. The blackened surface is a better radiator. Coffee will stay hot longer in a polished pot than in a blackened one. If we repeat the same experiment, but this time fill each container with ice water, we will find that the container with the blackened surface warms up faster. The blackened surface is also a better absorber of radiant energy.

Clean snow is a good reflector and therefore does not melt rapidly in sunlight. If the snow is dirty, it absorbs radiant energy from the sun and melts faster. Dropping black soot by aircraft on snowed-in mountainsides is a technique sometimes used in flood control. Controlled melting at favorable times rather than a sudden runoff of melted snow is thereby accomplished.

Light-colored buildings stay cooler in summer because they reflect much of the incoming radiant energy. They are also poor emitters, so they retain more of their thermal energy, which is beneficial in the winter. So paint your house a light color.

Cooling at Night by Radiation

Bodies that radiate more energy than they receive become cooler. This happens at night when solar radiation is absent. Objects out in the open radiate energy into the night and, because of the absence of warmer bodies, may receive very little energy in return. They give out more energy than they receive and become cooler. If the object is a good conductor of heat—like metal, stone, or concrete—heat from the ground will be conducted to it, somewhat stabilizing its temperature. But materials such as wood, straw, and grass are poor conductors, and little heat is conducted into them from the ground. These insulating materials are net radiators and get *colder than the air.* It is common for frost to form on these kinds of materials even when the temperature of the air does not go down to freezing. Have you ever seen a frost-

Figure 7.16 Patches of frost crystals betray the hidden entrances to mouse burrows. Each cluster of crystals is frozen mouse breath!

covered lawn or field on a chilly but above-freezing morning before the sun is up? The next time you see this, notice that the frost forms only on the grass, straw, or other poor conductors, while none forms on the cement, stone, or other good conductors.

Snow is a good example. During the day the snow gains very little heat energy from the sun because of its reflectivity, and at night it loses energy rapidly by radiating infrared radiation to space. Because of snow's poor conductivity, the ground conducts very little heat to it, and the surface of the snow becomes cooler than the surrounding air. Snow therefore has a significant cooling effect on the earth and atmosphere. The bottom of the snow remains at about ground temperature. That's why Midwestern wheat farmers like deep snows in winter—to protect their wheat fields from the harsh, cold weather.

7.4 NEWTON'S LAW OF COOLING

Objects at different temperatures from their surroundings will eventually come to a common temperature with the surroundings. A hot object will cool as it warms the surroundings, and an object cooler than its surroundings will warm as the surroundings cool. The cooling *rate* of an object depends on how much hotter the object is than the surroundings. The temperature change per second of a hot apple pie will be more if the hot pie is put in a cold freezer than if put on the kitchen table. When the pie cools in the freezer, the temperature difference is greater. A warm home will leak heat to the cold outside at a greater rate when there is a large difference between the temperature inside and outside. Keeping the inside of your home at a high temperature on a cold day is more costly than keeping it at a lower temperature; the smaller the temperature difference, the lower the cooling rate. At ordinary temperatures, the cooling rate—by conduction, convection, or radiation—is approximately proportional to the temperature difference, ΔT, between the object and its surroundings.

$$\text{Rate of cooling} \sim \Delta T$$

This is known as *Newton's law of cooling*. (Guess who was the first to establish this?)

A similar law holds for heating. If an object is cooler than its surroundings, its rate of warming up is also proportional to ΔT. Frozen food will warm up faster in a warm room than in a cold room.

Q UESTION

Since a hot cup of coffee loses heat more rapidly than a lukewarm cup of coffee, would it be correct to say that a hot cup of coffee will cool to room temperature before a not-so-hot cup of coffee?

A NSWER

No! Although the rate of cooling is greater for the hotter cup, it has farther to cool to reach thermal equilibrium. The extra time is equal to the time it takes to cool to the initial temperature of the lukewarm cup of coffee. Cooling *rate* and cooling *time* are not the same thing.

7.5 THE GREENHOUSE EFFECT

Park your car with its windows closed tightly in the bright sun on a hot day and the temperature of air inside the car soon climbs significantly higher than the outside air temperature. This is the greenhouse effect, so named for the same temperature-raising effect in florists' glass greenhouses. There are two things we need to know to understand the greenhouse effect. The first is what we previously stated: that all things radiate, and the wavelength of radiation depends on the temperature of the thing emitting the radiation—high-temperature things radiate short waves; low-temperature things radiate long waves. The second thing we need to know is that the transparency of things such as the air, glass, or water, depends on the wavelength of radiation. Air is transparent to both infrared waves and visible waves, unless there are nominal amounts of CO_2 in it, in which case it is opaque to infrared. Glass is transparent to the radiations that make up visible light, but is opaque to infrared waves. Water is transparent to visible light, but not to infrared. Why the transparencies and opacities of different materials vary is discussed in Chapter 12.

Now to why that car gets so hot in bright sunlight: Compared to the car, the temperature of the sun is very high. This means the waves it emits are very short. These short waves easily pass through both the earth's atmosphere and the glass windows of the car. So energy from the sun gets into the car interior, where, except for reflection, it is absorbed. The interior of the car warms up. Like the sun, it emits its own waves, but unlike the sun, its waves are longer. This is because its temperature is lower. The reradiated long waves encounter opaque glass windows. So reradiated energy remains in the car, which gets warmer still. As hot as the interior becomes, it won't be hot enough to emit waves that can pass though glass (unless it glows red or white hot!)

A similar story occurs in the earth's atmosphere, which is transparent to solar radiation. The surface of the earth absorbs this energy, and reradiates part of this at longer wavelengths. Atmospheric gases (mainly carbon dioxide and water vapor) absorb and re-emit much of this long-wavelength radiation back to earth. So the long-wave radiation that cannot escape the earth's atmosphere warms the earth. This process is very nice, for the earth would be a frigid $-18°C$ otherwise. Our present environmental concern is that increased levels of carbon dioxide and other

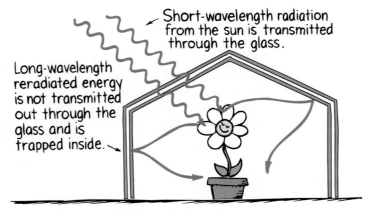

Figure 7.17 The glass window acts as a one-way valve, letting short waves in and preventing the outgo of long waves. So the thermal energy in the interior builds up.

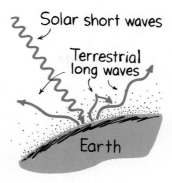

Figure 7.18 The greenhouse effect of the earth's atmosphere. Carbon dioxide in the atmosphere acts to absorb and retain heat that would otherwise be radiated from earth into space.

atmospheric gases in the atmosphere will make the earth too warm. We will return to the greenhouse effect as it affects the earth in Chapter 26.

Interestingly enough, in the florist's greenhouse, heating is mainly due to the ability of glass to prevent convection currents from mixing the cooler outside air with the warmer inside air. So the greenhouse effect plays a bigger role in the warming of the earth than it does in the warming of greenhouses.

QUESTION

What does it mean to say that the greenhouse effect is like a one-way valve?

7.6 SOLAR POWER

Step from the shade into the sunshine and you're noticeably warmed. The warmth you feel isn't so much because the sun is hot, for its surface temperature of 6000°C is no hotter than the flames of some welding torches; you're warmed principally because the sun is so big. We'll treat the sun in detail in Chapter 27, and discuss the effect of the sun's energy on the earth when we discuss meteorology in Chapter 26. For now we will briefly discuss the harvesting of solar energy here on earth.

The amount of solar energy per square meter that spreads across the earth's surface per day is about twice the power input needed to heat or cool the average American house.* That's why we see more and more homes using solar power for domestic heating and cooling. Solar heating needs a distribution system to move solar

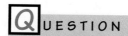

ANSWER

The transparent material (atmosphere for the earth and glass for the florist's greenhouse) passes only incoming short waves and blocks outgoing long waves. As a result, radiant energy is trapped within the "greenhouse."

---

*The solar energy received from the sun each second at the top of the earth's atmosphere is 1400 J/m². This is called the *solar constant*; expressed in terms of power the solar constant is 1.4 kW/m².

Figure 7.19 Solar water heaters are covered with glass to provide a greenhouse effect, which further heats the water. Why are the collectors painted black?

energy from the collector to the storage or living space. When the distribution system requires external energy to operate fans or pumps, we have an active system. When the distribution is by natural means (conduction, convection, or radiation), we have a passive system. Even in the northern states, solar homes with either active or passive systems are essentially problem-free and economical. Solar space heating and cooling are presently feasible; solar electricity is not yet here.

The problems of generating electricity by solar power are greater than those of solar heating and cooling. First, there is the fact that no energy arrives at night. This calls for a solar power plant to have supplemental sources of energy or efficient solar-energy storage devices. Variations in weather, particularly cloud cover, produce a variable energy supply from day to day and from season to season. Even in clear daylight hours, the sun is high in the sky only part of the day. Solar-energy collecting and concentration systems, whether arrays of mirrors or photovoltaic cells, at this writing are not yet competitive with the costs of electrical power generation by conventional power sources. Projections indicate the story may be different at the beginning of the next century.

7.7 GLOBAL WARMING

A radiative equilibrium exists that balances the amount of radiation the earth receives from the sun with the radiation the earth emits into space. This equilibrium results in an average temperature that supports the development of life as we know it. The power generation in the present century is a new factor in the radiative equilibrium. We are all familiar with the problem of thermal pollution that is a by-product of power generation plants. The earth's environment is the heat sink for wasted

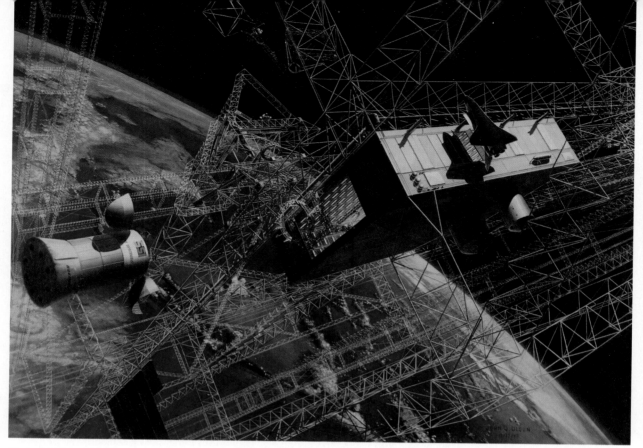

Figure 7.20 A Boeing conception of a photovoltaic power satellite being constructed in low earth orbit. A space-shuttle orbiter (upper right) docks at the facility's assembly bay. To the left, an upper stage of a massive-lift launch vehicle approaches the facility to discharge its cargo of construction materials. The weightlessness in orbit allows the use of large weblike structures of a kind that would be crushed if used on earth. The satellite would be deployed in geosynchronous orbit after completion. The electricity produced would be converted to microwaves and beamed to earth.

energy. Power plants in orbit will not solve the problem. The amount of heat put into the earth's environment has to do not only with the power production but also with energy consumption.

Suppose, for example, that enough power to supply all our needs was beamed to earth by orbiting solar power stations, and even suppose that all thermally polluting facilities—from steel mills to manufacturing plants—were also located in space. Then the heat sink for these facilities would be outer space rather than the planet earth. This would significantly cut down the buildup of heat on earth that would otherwise match continued growth, but as long as more and more energy is consumed on earth, more and more heat is the end product. When you make your toast in the morning, the extra heat your kitchen or breakfast nook receives has little to do with the nature of the power plant that produced the electricity. All the energy we consume, whether from making toast, operating a television set, or running a power saw, ultimately becomes heat.

However you look at it, increased energy consumption on earth results in increased heating of the earth. This increase in heat must be carried away by terrestrial radiation, and to do this, the temperature of the radiator, the earth, increases—and climate changes. Warmer oceans result in increased evaporation and increased snowfalls in polar regions, which in turn result in increased glacial growth. The fraction of the earth that is presently beneath glaciers is about equal to the total area used for farmlands. Glacial growth and the corresponding larger areas of white snow reflect

Satellite Solar Power

Solar power plants in orbit will not be in vogue during the remainder of this century, but are serious considerations for the future. Solar collectors in orbit can harvest solar energy undiminished by the atmosphere at a constant 24 hours per day. Because of weightlessness and the absence of wind and rain, huge structures of low mass could be built. These power plants could consist of giant banks of photovoltaic cells in geosynchronous orbits (similar to the orbits of communications satellites), which would render them always stationary with respect to any desired location on earth. These cells could collect solar energy continuously and convert it to electricity that would be fed to microwave generators aboard the satellites. The microwaves would be beamed to earth and picked up by antennae that could be located just about anywhere—on land, at sea, or in the desert.

The receiving antenna for such an arrangement might be a network of wires with about 80 percent open space. Sunlight would pass through it, but the microwaves would be absorbed by the antenna, making it relatively safe for occupants beneath it (providing present studies confirm the safety of microwave exposure to humans). If we ever have cities covered by huge geodesic domes of the type proposed by Buckminster Fuller, these domes might be effective microwave-receiving antennae, since they would be transparent to sunlight and opaque to microwaves.

Whether large-scale solar plants are better located on earth or in space is debatable. The technological proficiency is now available for doing either, however, and is beyond debate. The essential considerations involve economic priorities rather than technology.

Figure 7.21   A climate-controlled city beneath a microwave-receiving antenna.

Figure 7.22 All the energy we consume ultimately becomes heat.

more solar radiation, which may well lead to a significant drop in global temperature. So overheating the earth might well trigger the next ice age! Or it might not. We don't know.

We can speculate about the long-term effects of worldwide climate changes. On one hand, it may turn out that overall changes will be tolerable and that civilizations will adapt and continue to function. On the other hand, the effects may be intolerable and be the final push to the colonization of space, where communities in orbit could easily control their temperatures by radiating excess energy into the enormous heat sink of space. Would space colonies thermally pollute the solar system in time? No, the energy used in a space facility is intercepted from a tiny fraction of the energy radiated from the sun, and whatever is discharged partially fills in what was taken away. It simply goes back to where it came from in the first place. Regulating the temperature of a colony in the vacuum of space should be an achievable task; doing the same on earth surrounded by its insulating atmosphere is a different story.

Back here on earth we must seriously question the idea of continued growth. (Please take time to read Appendix II, "Exponential Growth and Doubling Time"—very important stuff.)

7.8 THE THERMOS BOTTLE

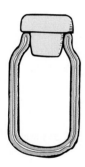

Figure 7.23
A Thermos bottle.

We can briefly summarize the ways in which heat is transferred by considering a device that inhibits these three methods—the common vacuum, or Thermos, bottle. The Thermos bottle is double-walled glass, with a vacuum between the walls. The two inner glass surfaces facing each other are silvered. When a hot liquid is poured into such a bottle, it remains at very nearly the same temperature for many hours. This is because the transfer of heat by conduction, convection, and radiation is severely inhibited.

1. Heat transfer by *conduction* through the vacuum is impossible. Some heat escapes by conduction through the glass and stopper, but this is a slow process as glass and plastic or cork are poor conductors.
2. The vacuum also prevents heat loss through the walls by *convection*.
3. Heat loss by *radiation* is prevented by the silvered surfaces of the walls, which reflect heat waves back into the bottle.

SUMMARY OF TERMS

Conduction The transfer and distribution of heat energy that moves from molecule to molecule within a substance.

Convection The transfer of heat energy in a gas or liquid by means of currents in the heated fluid. The fluid moves, carrying energy with it.

Radiation The transfer of energy at the speed of light by means of electromagnetic waves.

Newton's law of cooling The rate of loss of heat with time from an object is proportional to the excess temperature of the substance over the temperature of its surroundings.

Greenhouse effect The heating effect of a medium such as glass or the earth's atmosphere that is transparent to the short-wavelength radiation of sunlight but opaque to long-wavelength terrestrial radiation. Energy of sunlight that enters the glass of a florist's greenhouse or the atmosphere of the earth is absorbed and reradiated at a longer wavelength that is consequently trapped, which produces heating.

Solar power Energy per unit time obtained from the sun.

REVIEW QUESTIONS

1. What are the three common ways in which heat is transferred?

Conduction

2. What is the role of "loose" electrons in heat conductors?

3. Distinguish between a conductor and an insulator.

4. Why does room-temperature tile feel cooler to the bare feet than the wooden floor?

5. Why are materials such as wood, fur, feathers, and even snow good insulators?

6. How does a blanket keep you warm on a cold night, even though it is not really a source of energy?

7. Why do we say that cold is not a tangible thing?

Convection

8. How is heat transferred from one place to another by convection?

Why Warm Air Rises

9. How does buoyancy relate to convection?

10. Why are fast-moving air molecules more likely to migrate upward than downward?

Why Expanding Air Cools

11. What happens to the pressure of air as it rises? What happens to its volume? What happens to its temperature?

12. How are the speeds of molecules of air affected when they are compressed by the action of a tire pump?

13. How are the speeds of molecules of air affected when a volume of air undergoes rapid expansion?

14. Does the rate at which molecules in a gas collide affect temperature, or is it the other way around? Explain.

15. Why is Millie's hand not burned when she places it above the escape valve of the pressure cooker (Figure 7.7)?

16. Why does the direction of coastal winds change from day to night?

Radiation

17. What exactly is radiant energy?

18. How do the wavelengths of radiant energy vary with the temperature of the radiating source?

Emission, Absorption, and Reflection of Radiation

19. Since all objects are absorbing energy from their surroundings, why doesn't the temperature of all objects continually increase?

20. Since all objects are emitting energy to their surroundings, why doesn't the temperature of all objects continually decrease?

21. What determines whether an object is a net absorber or net emitter?

22. Why does the pupil of the eye appear black?

23. Which will normally cool faster, a black pot of hot water or a silvered pot of hot water? Explain.

24. Which will normally warm faster, a black pot of cold water or a silvered pot of cold water? Explain.

25. Is a good absorber of radiation a good emitter or a poor emitter?

Cooling at Night by Radiation

26. What happens to the temperature of something that radiates energy without absorbing the same amount in return?

27. Will a good conductor that is in contact with the relatively warm earth become significantly colder than the earth when it radiates energy? Why or why not?

28. Will a good insulator that is in contact with the relatively warm earth become significantly colder than the earth when it radiates energy? Why or why not?

Newton's Law of Cooling

29. Why will a can of beverage cool faster in the freezer compartment than in the main part of a refrigerator?

30. Which will undergo the greater rate of cooling, a red-hot poker in a warm oven or a red-hot poker in a cold room (or do both cool at the same rate)?

31. Does Newton's law of cooling apply to warming as well as cooling?

The Greenhouse Effect

32. What is meant by terrestrial radiation?

33. Why is radiant energy from the sun composed of short waves and terrestrial radiation composed of relatively longer waves?

Solar Power

34. Distinguish between *active* and *passive* solar heating systems.

35. When will the technology exist for solar power stations in orbit?

36. How can solar energy from satellites be transferred to earth?

Global Warming

37. How does worldwide energy consumption relate to the average temperature of the world?

38. Will space colonies that get their energy from solar power not thermally pollute the solar system? Defend your answer.

The Thermos Bottle

39. What is the function of the silver surfaces of the Thermos bottle?

40. Heat cannot readily escape a Thermos bottle, so hot things inside stay hot. Will cold things inside a Thermos bottle likewise stay cold? Explain.

. . . .

HOME PROJECTS

1. Hold the bottom end of a test tube full of cold water in your hand. Heat the top part in a flame until it boils. The fact that you can still hold the bottom shows that water is a poor conductor of heat. This is even more dramatic when you wedge chunks of ice at the bottom; then the water above can be brought to a boil without melting the ice. Try it and see.

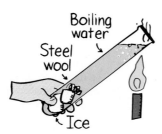

2. If you live where there is snow, do as Benjamin Franklin did nearly two centuries ago and lay samples of light and dark cloth on the snow. Note the difference in the rate of melting beneath the cloths.

3. Wrap a piece of paper around a thick metal bar and place it in a flame. Note that the paper will not catch fire. Can you figure out why? (Paper generally will not ignite until its temperature reaches 233°C.)

. . . .

EXERCISES

1. Why is it difficult to estimate the temperature of things by touching them?

2. If 70°F air feels warm and comfortable to us, why does 70°F water feel cool when we swim in it?

3. At what common temperature will a block of wood and a block of metal both feel neither hot nor cold to the touch?

4. If you hold one end of a metal nail against a piece of ice, the end in your hand soon becomes cold. Does cold flow from the ice to your hand? Explain.

5. Silver is a very good conductor of heat. Is this quality favorable or unfavorable for silverware? Explain.

6. Why do restaurants serve baked potatoes wrapped in aluminum foil?

7. Many tongues have been injured by licking a piece of metal on a very cold day. Why would no harm result if a piece of wood were licked on the same day?

8. Wood is a better insulator than glass. Yet fiberglass is commonly used as an insulator in wooden buildings. Explain.

9. Visit a snow-covered cemetery and note the snow does not slope upward against the gravestones, but instead forms depressions as shown. Can you think of a reason for this?

10. If you were caught in freezing weather with only a candle for a heat source, would you be warmer in an Eskimo igloo or in a wooden shack?

11. When it is daytime on the moon, the moon-rock surface is hot enough to melt solder. Why is this not so on earth?

12. If you wish to cool something by placing it in contact with ice, should you put it on top of the ice block or put the ice block on top of it?

13. You can bring water in a paper cup to a boil by placing it in a hot flame. Why doesn't the paper cup burn?

14. Why can you comfortably hold your fingers close beside a candle flame, but not very close above the flame?

15. Why is it that you can safely hold your bare hand in a hot pizza oven for a few seconds, but if you momentarily touch the metal insides you'll burn yourself?

16. Wood conducts heat very poorly—it has a very low conductivity. Does wood still have a low conductivity if it is hot? Could you safely grab the wooden handle of a pan from a hot oven with your bare hand? Although the pan handle is hot, does much heat conduct from it to your hand if you do it quickly? Could you do the same with an iron handle? Explain.

17. Wood has a very low conductivity. Does it still have a low conductivity if it is very hot—that is, in the stage of smoldering red hot coals? Could you safely walk across a bed of red-hot wooden coals with bare feet? Although the coals are hot, does much heat conduct from them to your feet if you step quickly? Could you do the same on red-hot iron coals? Explain.

18. In a still room, smoke from a cigarette will sometimes rise and then settle in the air before reaching the ceiling. Explain why.

19. Why would you expect a single helium atom to continually rise in an atmosphere of nitrogen and oxygen? Why doesn't it "settle off" like the smoke in the preceding question?

20. In a mixture of hydrogen and oxygen gases at the same temperature, which molecules move faster? Why?

21. One container is filled with argon gas and the other with krypton gas. If both gases have the same temperature, in which container are the atoms moving faster? Why?

22. If we warm a volume of air, it expands. Does it then follow that if we expand a volume of air, it warms? Explain.

23. How could you change the drawing in Figure 7.6 to make it illustrate the heating of air when it is compressed? Make a sketch of this case.

24. A snow-making machine used for ski areas consists of a mixture of compressed air and water blown through a nozzle. The temperature of the mixture may initially be well above the freezing temperature of water, yet crystals of snow are formed as the mixture is ejected from the nozzle. Explain how this happens.

25. What does the high specific heat of water have to do with convection currents in the air at the seashore?

26. What would be the most efficient color for steam radiators?

27. Why does a good *emitter* of heat radiation appear black at room temperature?

28. A number of bodies at different temperatures placed in a closed room will ultimately come to the same temperature. Would this thermal equilibrium be possible if good absorbers were poor emitters and if poor absorbers were good emitters? Explain.

29. From the rules that a good absorber of radiation is a good radiator and a good reflector is a poor absorber, state a rule relating the reflecting and radiating properties of a surface.

30. Suppose at a restaurant you are served coffee before you are ready to drink it. In order that it be hottest when you are ready for it, would you be wiser to add cream to it right away or when you are ready to drink it?

31. Even though metal is a good conductor, frost can be seen on parked cars in the early morning even when the air temperature is above freezing. Can you explain this?

32. Why is whitewash sometimes applied to the glass of florists' greenhouses in the summer?

33. On a very cold sunny day you wear a black coat and a transparent plastic coat. Which should be worn on the outside for maximum warmth?

34. If the composition of the upper atmosphere were changed so that it permitted a greater amount of terrestrial radiation to escape, what effect would this have on the earth's climate? How about if the atmosphere reduced the escape of terrestrial radiation?

35. Is it important to convert temperatures to the Kelvin scale when we use Newton's law of cooling? Why or why not?

36. If you wish to save fuel and you're going to leave your warm house for a half hour or so on a very cold day, should you turn your thermostat down a few degrees, turn it off altogether, or let it remain at the room temperature you desire?

37. If you wish to save fuel and you're going to leave your cool house for a half hour or so on a very hot day, should you turn your air conditioning thermostat up a bit, turn it off altogether, or let it remain at the room temperature you desire?

38. As more energy is consumed on earth, the overall temperature of the earth tends to rise. Regardless of the increase in energy, however, the temperature does not rise indefinitely. By what process is an indefinite rise prevented? Explain your answer.

8

THERMODYNAMICS

The study of heat and its transformation to mechanical energy is called **thermodynamics** (which stems from Greek words meaning "movement of heat"). The science of thermodynamics was developed at the beginning of the last century, before the atomic and molecular theory of matter was understood. Whereas in previous chapters we described heat in terms of the microscopic behavior of jiggling atoms and molecules, in this chapter we invoke only macroscopic notions—such as mechanical work, pressure, and temperature—and their roles in energy transformations. Thermodynamics is a powerful theoretical science that bypasses the molecular details of the system altogether. Its foundation is the conservation of energy and the fact that heat flows from hot to cold and not the other way around. It provides the basic theory of heat engines, from steam turbines to fusion reactors, and the basic theory of refrigerators and heat pumps. We begin our study of thermodynamics with a look at one of its early concepts—a lowest limit of temperature.

8.1 ABSOLUTE ZERO

In principle, there is no upper limit of temperature. As thermal motion increases, a solid object first melts and then vaporizes; as the temperature is further increased,

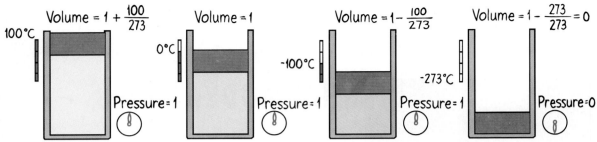

Figure 8.1 The volume of a gas changes by $\frac{1}{273}$ its volume at 0°C with each 1°C change in temperature when the pressure is held constant. (a) At 100°C the volume is $\frac{100}{273}$ greater than it is at (b) 0°C. (c) When the temperature is reduced to −100°C, the volume is reduced by $\frac{100}{273}$. (d) At −273°C the volume of the gas would be reduced by $\frac{273}{273}$ and so would be zero.

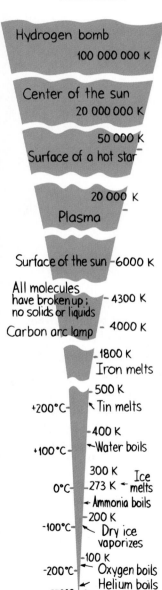

Figure 8.2 Some absolute temperatures.

molecules break up into atoms, and atoms lose some or all of their electrons, thereby forming a cloud of electrically charged particles—a plasma. This situation exists in stars, where the temperature is many millions of degrees Celsius.

In contrast, there is a definite limit at the other end of the temperature scale. Gases expand when heated and contract when cooled. Nineteenth-century experiments found that all gases, regardless of their initial pressures or volumes, change by $\frac{1}{273}$ of their volume at 0°C for each degree Celsius change in temperature, provided the pressure is held constant. So if a gas at 0°C were cooled down by 273°C, it would contract $\frac{273}{273}$ volumes and be reduced to zero volume. Clearly, we cannot have a substance with zero volume.

It was also found that the pressure of any gas in any container of *fixed* volume would change by $\frac{1}{273}$ for each degree Celsius change in temperature. So a gas in a container of fixed volume cooled 273°C below zero would have no pressure whatsoever. In practice, every gas liquefies before it gets this cold. Nevertheless, these decreases by $\frac{1}{273}$ increments suggested the idea of a lowest temperature: −273°C. So there is a limit to coldness. When atoms and molecules lose all available kinetic energy, they reach the **absolute zero** of temperature.* At absolute zero, no more energy can be extracted from a substance and no further lowering of its temperature is possible. This limiting temperature is actually 273.15° below zero on the Celsius scale (and 459.69° below zero on the Fahrenheit scale).

On the absolute temperature scale, the Kelvin scale,† absolute zero is 0K (short for "0 Kelvin," rather than "0 degrees Kelvin"). There are no minus numbers in the Kelvin scale. Degrees on the Kelvin scale are calibrated with the same-sized divisions as on the Celsius scale. The melting point of ice thus is 273.15K, and the boiling point of water is 373.15K.

---

*Even at absolute zero, molecules still have a small kinetic energy, called the *zero-point energy*. Helium, for example, has enough motion at absolute zero to keep it from freezing. The explanation for this involves quantum theory.

†Named after the famous British physicist Lord Scale, who coined the word *thermodynamics* and was the first to suggest this thermodynamic temperature scale.

QUESTIONS

1. Which is larger, a Celsius degree or a Kelvin?

2. A piece of metal has a temperature of 0°C. If someone says a second, identical piece of metal is twice as hot, what do they mean? What is its temperature?

8.2 THERMAL ENERGY

We all know that there is a vast amount of energy locked in all materials—this book, for example. The pages of this book are composed of molecules that are in constant motion. They have kinetic energy. Due to interactions with neighboring molecules, they also have potential energies. The pages can be easily burned, so we know they store chemical energy, which is really electric potential energy at the molecular level. We know there are vast amounts of energy associated with atomic nuclei. Then there is the "energy of being," described by the celebrated equation $E = mc^2$ (mass energy). Energy within a substance is found in these and other forms, which, when taken together, are called **thermal energy** (or to the physicist, *internal energy*).* The thermal energy in even the simplest substance can be quite complex. But in our study of heat changes and heat flow, we will be concerned only with the *changes* in the thermal energy of a substance. Changes in temperature will indicate these changes in thermal energy.

8.3 FIRST LAW OF THERMODYNAMICS

More than a hundred years ago heat was thought to be an invisible fluid called *caloric*, which flowed like water from hot objects to cold objects. Caloric was conserved in its interactions, and this discovery was the forerunner of the law of conservation of energy. In the mid-1880s, it became apparent that the flow of heat was

ANSWERS

1. Neither. They are equal.

2. To say a piece of metal is twice as hot means it has twice the thermal energy. This means it has twice the absolute temperature, or two times 273K. This would be 546K, or 273°C.

---

*The physicist uses the term *internal energy* to emphasize that thermal energy is *internal* to the system. If, for example, this book were poised at the edge of a table and ready to fall, it would possess gravitational potential energy; if it were tossed into the air, it would possess kinetic energy. But these are external to the book and not part of the book's thermal energy. To include the gravitational interactions with the earth and motion with respect to the earth, we must do so in terms of a larger "system"—a system enlarged to include both the book and the earth.

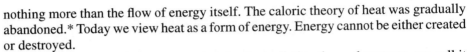

nothing more than the flow of energy itself. The caloric theory of heat was gradually abandoned.* Today we view heat as a form of energy. Energy cannot be either created or destroyed.

When the law of energy conservation is applied to thermal systems, we call it the **first law of thermodynamics**. We state it generally in the following form:

Whenever heat is added to a system, it transforms to an equal amount of some other form of energy.

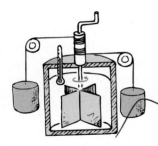

Figure 8.3 Paddle-wheel apparatus first used to compare thermal energy with mechanical energy. As the weights fall, they give up potential energy and warm the water accordingly. This was first demonstrated by James Joule, for whom the unit of energy is named.

By *system,* we mean a well-defined group of atoms, molecules, particles, or objects. The system may be the steam in a steam engine or it may be the whole earth's atmosphere. It can even be the body of a living creature. The important point is that we must be able to define what is contained within the system and what is outside it. If we add heat energy to the steam in a steam engine, to the earth's atmosphere, or to the body of a living creature, we will increase the thermal energies of these things. And because of this added energy, these things may do work on external things. This added energy does one or both of two things: (1) it increases the thermal energy of the system if it remains in the system or (2) it does work on things external to the system if it leaves the system. So, more specifically, the first law states:

Heat added = increase in thermal energy + external work done by the system.

Note that the first law is an overall principle that is not concerned with the inner workings of the system itself. Whatever the details of molecular behavior in the system, the added heat energy serves only two functions: It increases the thermal energy of the system or it enables the system to do external work (or both). Our ability to describe and predict the behavior of systems that may be too complicated to analyze in terms of atomic and molecular processes is one of the beauties of thermodynamics. Thermodynamics bridges the microscopic and macroscopic worlds.

Put an airtight can of air on a hot stove and heat it up. Since the can is of fixed size, no work is done on it, and all the heat that goes into the can increases the thermal energy of the enclosed air. Its temperature rises. If the can is fitted with a movable piston, then the heated air can do work as it expands and pushes the piston outward. Can you see that the temperature of the enclosed air will be less than if no work were done on the piston? If heat is added to a system that does no external work, then the amount of heat added is equal to the increase in the thermal energy of the system. If the system does external work, then the increase in thermal energy is correspondingly less. The first law of thermodynamics is simply the thermal version of the law of conservation of energy.

Consider a given amount of heat supplied to a steam engine. The amount supplied will be evident in the increase in the thermal energy of the steam and in the mechanical work done. The sum of the increase in thermal energy and the work done will equal the heat input. In no way can energy output in any form exceed energy input.

Figure 8.4 Do work on the pump by pressing down on the piston and you compress the air inside. What happens to the temperature of the enclosed air? What happens to its temperature if the air expands and pushes the piston outward?

*Popular ideas, when proved wrong, are seldom suddenly discarded. People tend to identify with the ideas that characterize their time; hence, it is often the young who are more prone to discover and accept new ideas and push the human adventure forward.

QUESTIONS

1. If 100 J of energy is added to a system that does no external work, by how much is the thermal energy of that system raised?

2. If 100 J of energy is added to a system that does 40 J of external work, by how much is the thermal energy of that system raised?

Adding heat to a system so that it can do mechanical work is only one application of the first law of thermodynamics. If instead of adding heat to a system we do mechanical work on it, the first law tells us what we can expect: an increase in thermal energy. A bicycle pump provides a good example. When we pump on the handle, the pump becomes hot. Why? Because we are primarily putting mechanical work into the system and raising its thermal energy. If the process happens quickly enough so that very little heat is conducted from the system during compression, then most of the work input will go into increasing the thermal energy, and the system becomes hotter.

Adiabatic Processes

The process of compression or expansion of a gas so that no heat enters or leaves a system is said to be *adiabatic* (from the Greek for "impassable"). **Adiabatic processes** can be achieved either by thermally insulating a system from its surroundings (with Styrofoam, for example) or by performing the process rapidly so that heat has little time to enter or leave. If we set the "heat added" part of the first law to zero, we see that changes in thermal energy are equal to the work done on or by the system.* If work is done on a system—compressing it, for example—the thermal energy increases. We raise its temperature. If work is done by the system—expanding against its surroundings, for example—the thermal energy decreases. It cools.

A common example of an adiabatic process is the compression and expansion of the gases in the cylinders of an automobile engine (Figure 8.5). Compression and expansion occur in only hundredths of a second, too short a time for any appreciable amount of energy to leave the combustion chamber. For very high compressions, like those which occur in diesel engines, the temperatures achieved are sufficient to ignite a fuel mixture without the use of spark plugs. Diesel engines have no spark plugs.

ANSWERS

1. 100 J.

2. 60 J. We see from the first law that 100 J = 60 J + 40 J.

---

*ΔHeat = Δthermal energy + work

        0 = Δthermal energy + work

Then we can say

−Work = Δthermal energy

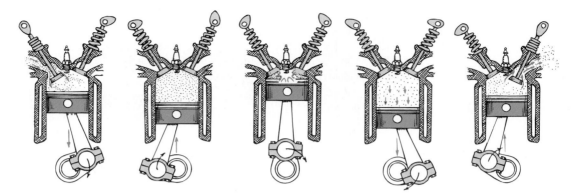

Figure 8.5 A four-cycle internal-combustion engine. (a) A fuel-air mixture from the carburetor fills the cylinder as the piston moves down. (b) The piston moves upward and compresses the mixture—adiabatically, since no heat transfer occurs. (c) The spark plug fires, ignites the mixture, and raises it to a high temperature. (d) Adiabatic expansion pushes the piston downward, the power stroke. (e) The burned gases are pushed out the exhaust pipe. Then the intake valve opens and the cycle repeats.

So when work is done on a gas by adiabatically compressing it, the gas gains thermal energy and becomes warmer. We note this by the warmth of a bicycle pump when air is compressed. When a gas adiabatically expands, it does work on its surroundings and gives up thermal energy and becomes cooler. This can be noted by an experiment you can do right now: With your mouth wide open, blow on your hand. Your breath is warm. Now repeat, but pucker your lips to make a small hole so your breath must expand as it leaves your mouth. Your breath is noticeably cooler! Expanding air cools.

8.4 SECOND LAW OF THERMODYNAMICS

Suppose we place a hot brick next to a cold brick in a thermally insulated region. We know that the hot brick will cool as it gives heat to the colder brick, which will warm. They will arrive at a common temperature: thermal equilibrium. No energy will be lost, in accordance with the first law of thermodynamics. But pretend the hot brick extracts heat from the cold brick and becomes hotter. Would this violate the first law of thermodynamics? Not if the cold brick becomes correspondingly colder so that the combined energy of both bricks remains the same. If this were to happen, it would not violate the first law. But it would violate the **second law of thermodynamics**. The second law distinguishes the direction of energy transformation in natural processes. The second law of thermodynamics can be stated in many ways, but most simply it is this:

Heat will never of itself flow from a cold object to a hot object.

Heat flows one way, downhill from hot to cold. In winter, heat flows from inside a warm heated home to the cold air outside. In summer, heat flows from the hot air outside into the cooler interior. The direction of heat flow is from hot to cold. Heat can be made to flow the other way, but only by imposing external effort—as occurs with heat pumps that increase the temperature of air, or air conditioners that reduce air temperature. The huge amount of thermal energy in the ocean cannot be used to light a single flashlight lamp without external effort. Energy will not of itself flow from the lower temperature ocean to the higher temperature lamp filament. Without external effort, the direction of heat flow is from hot to cold.

Heat Engines

It is easy to change work completely into heat—simply rub your hands together briskly. Or push a crate at constant speed along a floor. All the work you do in over-coming friction is completely converted to heat. But the reverse process, changing heat completely into work, can never occur. The best that can be done is the conversion of some heat to mechanical work. The first heat engine to do this was the steam engine, invented in about 1700.

A **heat engine** is any device that changes thermal energy into mechanical work. The basic idea behind a heat engine, whether a steam engine, internal combustion engine, or jet engine, is that mechanical work can be obtained only when heat flows from a high temperature to a low temperature. In every heat engine only some of the heat can be transformed into work.

Every heat engine—whether it be a steam engine, gasoline or diesel engine, gas turbine, or jet engine—will (1) absorb thermal energy from a reservoir of higher temperature, (2) convert some of this energy into mechanical work, and (3) expel the remaining energy to some lower-temperature reservoir, usually called a sink (Figure 8.6). In a gasoline engine, for example, (1) the burning fuel in the combustion chamber is the high-temperature reservoir, (2) mechanical work is done on the piston, and (3) the expelled energy goes out as exhaust. The second law tells us that no heat engine can convert all the heat supplied into mechanical energy. Only some of the heat can be transformed into work, with the remainder expelled in the process. Applied to heat engines, the second law may be stated:

> **When work is done by a heat engine running between two temperatures, T_{hot} and T_{cold}, only some of the input energy at T_{hot} can be converted to work, and the rest is expelled as heat at T_{cold}.**

There is always heat exhaust, which may be desirable or undesirable. Hot steam expelled in a laundry on a cold winter day may be quite desirable, while the same steam on a hot summer day is something else. When expelled heat is undesirable, we call it *thermal pollution*.

Before the second law was understood, it was thought that a very low friction heat engine could convert nearly all the input energy to useful work. But not so. In 1824 the French engineer Sadi Carnot carefully analyzed the cycles of compression and expansion in a heat engine and made a fundamental discovery. He showed that

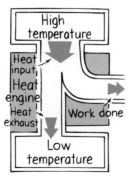

Figure 8.6 When heat flows in any heat engine from a high-temperature place to a low-temperature place, part of this heat can be turned into work. (If work is put into a heat engine, the flow of energy may go from a low-temperature to a high-temperature place, as in a refrigerator or air conditioner.)

the upper fraction of heat that can be converted to useful work, even under ideal conditions, depends on the temperature difference between the hot reservoir and the cold sink. His equation is

$$\text{Ideal efficiency} = \frac{(T_{\text{hot}} - T_{\text{cold}})}{T_{\text{hot}})}$$

where T_{hot} is the temperature of the hot reservoir and T_{cold} the temperature of the cold.* Ideal efficiency depends only on the temperature difference between input and exhaust. Whenever ratios of temperatures are involved, the absolute temperature scale must be used. So T_{hot} and T_{cold} are expressed in Kelvins. For example, when the hot reservoir in a steam turbine is 400K (127°C), and the sink is 300K (27°C), the ideal efficiency is

$$\frac{(400 - 300)}{400} = \frac{1}{4}$$

This means that even under ideal conditions, only 25 percent of the thermal energy of the steam can be converted into work, while the remaining 75 percent is expelled as waste. This is why steam is superheated to high temperatures in steam

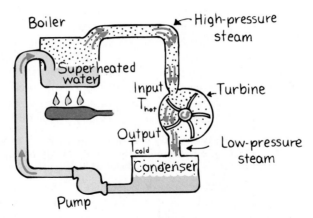

Figure 8.7   A simplified steam turbine. The turbine turns because high-temperature steam from the boiler exerts more pressure on the front side of the turbine blades than the low-temperature steam exerts on the back side of the blades. Without a pressure difference, the turbine would not turn and deliver energy to an external load (an electric generator, for example). The presence of steam pressure on the back side of the blades, even in the absence of friction, prevents a perfectly efficient engine.

------

*Efficiency $= \frac{\text{work output}}{\text{heat input}}$.

Heat input = work output + heat that flows out at low temperature (see Figure 8.6). So we can say work output = heat input − heat out.

So efficiency $= \frac{(\text{heat input} - \text{heat out})}{(\text{heat input})}$.

In the ideal case, it can be shown that the ratio $\frac{(\text{heat out})}{(\text{heat in})} = \frac{T_{\text{cold}}}{T_{\text{hot}}}$.

Then we can say

Ideal efficiency $= \frac{(T_{\text{hot}} - T_{\text{cold}})}{T_{\text{hot}}}$.

engines and power plants. The higher the steam temperature driving a motor or turbogenerator, the higher the efficiency of power production. (Increasing operating temperature in the example to 600K yields an efficiency $\frac{(600-300)}{600} = \frac{1}{2}$; twice the efficiency at 400K.)

Carnot's equation states the upper limit of efficiency for all heat engines. The higher the operating temperature (compared to exhaust temperature) of any heat engine, whether in an ordinary automobile, nuclear-powered ship, or jet aircraft, the higher the efficiency of that engine. In practice, friction is always present in all engines, and efficiency is always less than ideal.* So whereas friction is solely responsible for the inefficiencies of many devices, in the case of heat engines the overriding concept is the second law of thermodynamics; only some of the heat input can be converted to work—even without friction.

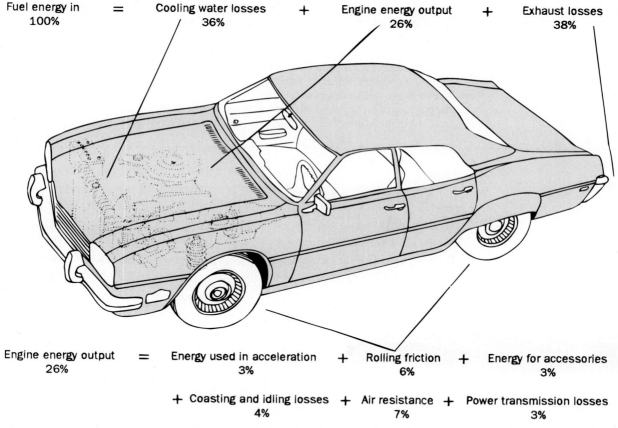

Fuel energy in = Cooling water losses + Engine energy output + Exhaust losses
100% 36% 26% 38%

Engine energy output = Energy used in acceleration + Rolling friction + Energy for accessories
26% 3% 6% 3%

+ Coasting and idling losses + Air resistance + Power transmission losses
4% 7% 3%

Figure 8.8 Only 26 percent of the heat energy produced in burning the gasoline in an automobile becomes useful mechanical energy, and most of that is lost in friction and in overcoming air resistance. The losses shown here are for a typical American automobile and are averages over different driving situations.

*The ideal efficiency of an ordinary automobile engine is about 56%, but in practice the actual efficiency is about 25%. Engines of higher operating temperatures (compared to sink temperatures) would be more efficient, but the melting point of engine materials limits the upper temperatures at which they can operate. Higher efficiencies await engines made with new materials with higher melting points. Watch for ceramic engines! (See the box on ceramic engines.)

Ceramic Engines

More than 10,000 years ago someone cooked a hunk of clay, and it hardened into pottery. Today we call it ceramics. Compared to metals, ceramics can be harder, lighter, stiffer, and more resistant to heat and corrosion—ideal for an automobile engine? The Carnot efficiency of an engine goes up with higher operating temperatures. Today's metal automobile engines are relatively inefficient because their operating temperatures must be kept lower than the melting point of the metals. Costly radiators are installed to get rid of valuable heat that otherwise would raise efficiency. So why aren't today's engines made of ceramics?

Ceramics, like pottery, are brittle and shatter and crack when dropped or hit. Unlike metals, ceramics cannot bend and deform to absorb impacts. Intense research is presently underway to combat the problem of brittleness, with some success. Ceramic engines are already employed in Isuzu's Ceramic Research Institute in Yokohama, Japan, where a high-powered sedan labeled CERAMIC features a small engine and no radiator. This car uses heat instead of rejecting it. Likewise with the prototype turbine ceramic engines developed in the United States by the Department of Energy. In place of pistons, the United States' versions feature two gas turbines of silicon nitride. With other ceramic parts sharing the same high-temperature area, the cars will run at a hot 1600K, with higher efficiency and cleaner emissions. Optimistically, these engines may be on the market before the year 2000.

Ceramics are being used to coat metal to extend life, and metals are being used to coat ceramics to help lubricate the normal high-friction ceramic surface. But don't expect too much too soon. The ceramic engine will likely require a long, continued, and coordinated international effort.

Q UESTIONS

1. What would be the ideal efficiency of an engine if its hot reservoir and exhaust were the same temperature—say 400K?
2. What would be the ideal efficiency of a machine having a hot reservoir at 400K and a cold reservoir at absolute zero, 0K?

8.5 ENERGY AND CHANGES OF PHASE

Recall from Chapter 6 that matter in our everyday environment exists in four common phases, solid, liquid, gas, and plasma. When a substance changes phase, a transfer of energy occurs. To melt ice, for example, heat is applied. The phase change from ice to water requires energy. Likewise for boiling water, energy is required to change water to steam. The steam in turn can be changed into a glowing incandescent gas, a plasma, by additional energy. We won't treat plasma until Part 8, when we discuss its role in the sun and stars.

A NSWERS

1. Zero efficiency; $\frac{400 - 400}{400} = 0$. So no work output is possible for any heat engine unless a temperature difference exists between the reservoir and the sink.
2. $\frac{400 - 0}{400} = 1$; only in this idealized case is an ideal efficiency of 100% possible.

Figure 8.9 Energy changes with change of state.

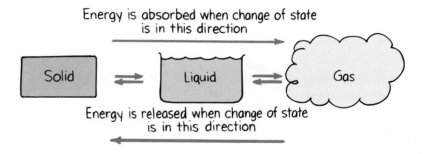

Energy is absorbed when change of state is in this direction

Solid ⇌ Liquid ⇌ Gas

Energy is released when change of state is in this direction

The energy required to melt ice is energy that breaks the bonds between ice crystals, and energy that turns water to steam is energy that breaks the molecular bonds between water molecules. We will treat these molecular bonds in great detail in Chapter 19. For now we will simply say that energy is required for both liquefaction of a solid and vaporization of a liquid. Conversely, energy must be extracted from a substance to change its phase in the direction from gas to liquid to solid (Figure 8.9).

The cooling cycle of a refrigerator neatly employs the concepts shown in Figure 8.9. A liquid of low boiling point, usually Freon, is pumped into the cooling unit, where it turns into a gas and draws heat from the things stored there. The gas with its added energy is directed outside the cooling unit to coils located in the back, appropriately called condensation coils, where heat is given off to the air as the gas condenses to a liquid again. A motor pumps the fluid through the system, where it is made to undergo the cyclic process of vaporization and condensation. The next time you're near a refrigerator, place your hand near the condensation coils in the back and you'll feel the heat that has been extracted from inside.

An air conditioner employs the same principles and simply pumps heat energy from one part of the unit to another. When the roles of vaporization and condensation are reversed, the air conditioner becomes a heater.

So we see that a solid must absorb energy to melt, and a liquid must absorb energy to vaporize. Conversely, a gas must release energy to liquefy, and a liquid must release energy to solidify.

QUESTIONS

1. A pot of water containing soaked hickory chips is placed in the oven with a turkey that is to be roasted for a holiday dinner. Why does the turkey take longer than expected to cook?

2. When H_2O in the vapor state condenses, is the surrounding air warmed or cooled?

ANSWERS

1. Much of the heat from the oven is consumed in changing the phase of the water. As long as water remains in the liquid phase, the temperature of the oven will not rise much higher than the boiling temperature of water—100°C.

2. The surrounding air is warmed because energy in the form of heat is released by the gaseous water molecules as they turn into liquid water molecules. This release of heat results from the formation of the chemical interactions between liquid water molecules. This is analogous to the collision of two magnets. As magnets collide and stick, they too release energy, in the form of heat—the same amount of energy that was originally required to pull them apart!

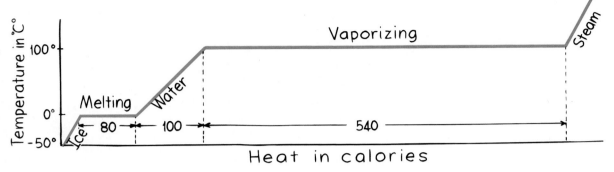

Figure 8.10   A graph showing the energy involved in the heating and the change of state of 1 g of H_2O.

The general behavior of many substances can be illustrated by a description of the changes of phase of H_2O. To make the numbers simple, consider a 1-gram piece of ice at a temperature of $-50°C$ in a closed container, put on a stove to heat. A thermometer in the container reveals a slow increase in temperature up to $0°C$. At $0°C$, the temperature stops rising, even though heat is continually added. The added heat breaks crystal bonds in the ice, effectively increasing their potential energies rather than their kinetic energies. So energy that goes into the water doesn't make the water molecules move faster; they simply break apart from one another, and the temperature remains the same. This process of melting 1 gram of ice requires 80 calories. Only when all the ice melts will the temperature rise. Then each calorie absorbed by the water increases its temperature by $1°C$ until the boiling temperature, $100°C$, is reached. Again, as heat is added, the temperature remains constant while more and more of the gram of water is boiled away and becomes steam. The water must absorb 540 calories of thermal energy to vaporize the whole gram. Finally, when all the water has become steam at $100°C$, the temperature begins to rise once more. It will continue to rise as long as heat is added. This process is graphed in Figure 8.10.

QUESTIONS

1. How much energy is transferred when a gram of steam at $100°C$ condenses to water at $100°C$?

2. How much energy is transferred when a gram of boiling water at $100°C$ cools to ice water at $0°C$?

3. How much energy is transferred when a gram of ice water at $0°C$ freezes to ice at $0°C$?

4. How much energy is transferred when a gram of steam at $100°C$ turns to ice at $0°C$?

ANSWERS

1. One gram of steam at $100°C$ transfers 540 calories of energy when it condenses to become water at the same temperature.

2. One gram of boiling water transfers 100 calories when it cools $100°C$ to become ice water.

3. One gram of ice water at $0°C$ transfers 80 calories to become ice at $0°C$.

4. One gram of steam at $100°C$ transfers to the surroundings a grand total of the above values, 720 calories, to become ice at $0°C$.

The 540 calories required to vaporize a gram of water is a relatively large amount of energy—much more than would be required to bring a gram of ice at absolute zero to boiling water at 100°C! Although the molecules in steam and boiling water at 100°C have the same average kinetic energy, steam has more potential energy because the molecules are relatively free of one another and are not bound together as in the liquid phase.

This 540 calories of energy per gram of steam can be released when the steam condenses to water. Then the potential energy of the far-apart water molecules transforms to heat. This is like the potential energy of two attracting magnets separated from each other; when released their potential energy is converted into kinetic energy, and then into heat as the magnets strike each other.

So, we see that the energies required to both melt ice and boil water are the same amounts of energy released when the phase changes are in the opposite direction. The processes are reversible. The amount of energy to change phase from solid to liquid (and vice versa) is called **heat of fusion**. For water this is 80 calories per gram. The amount of energy required to change phase from liquid to gas (and vice versa) is called **heat of vaporization**. For water this is a whopping 540 calories per gram.* We will see later in Chapter 19 that these relatively high values are due to the strong forces between water molecules—hydrogen bonds.

Some people who live in cold climates take advantage of water's high heat of fusion by placing large bathtubs of water in their basements. Outside, the temperature may drop to well below freezing. Downstairs in the basement, however, as the water freezes, millions of calories of heat are released to keep the basement considerably warmer.

Water's high heat of fusion allows you to briefly touch your wetted finger to a hot skillet on a hot stove without harm. You can even touch it a few times in succession as long as your finger remains wet. This is because energy that ordinarily would go into burning your finger goes instead into changing the phase of the moisture on your finger. The energy converts the moisture to a vapor. Similarly, you are able to judge the hotness of a hot clothes iron.

QUESTION

Suppose 4 g of boiling water is spread over a large surface so 1 g rapidly evaporates. If evaporation takes 540 calories from the remaining 3 g of water and no other heat transfer takes place, what will be the temperature of the remaining 3 g?

ANSWER

The remaining 3 g will turn to 0°C ice under conditions where all 540 calories are taken from the remaining water (for example when the surroundings are below freezing and don't contribute energy). 540 calories from 3 g means each gram gives up 180 calories. 100 calories from a gram of boiling water reduces its temperature to 0°C, and 80 more calories taken away turns it to ice. This is why hot water so quickly turns to ice in a freezing-cold environment.

---

*In SI units, the heat of vaporization of water is 2.26 megajoules per kilogram (MJ/kg), and the heat of fusion of water is 0.335 MJ/kg.

Figure 8.11 Paul Ryan tests the hotness of molten lead by dragging his wetted finger through it.

Supervisor Paul Ryan of the Department of Public Works in Malden, Massachusetts, has for years used molten lead to seal pipes in certain plumbing operations. He startles onlookers by dragging his finger through molten lead to judge its hotness (Figure 8.11). He is sure that the lead is very hot and his finger is thoroughly wet before he does this. (Do not try this on your own, for if the lead is not hot enough it will stick to your finger—ouch!)

8.6 ORDER TENDS TO DISORDER

We have seen that the first law of thermodynamics states that energy can be neither created nor destroyed. The second law qualifies this by adding that the form energy takes in transformations "deteriorates" to less useful forms. Energy becomes more diffuse and ultimately degenerates into waste. Another way to say this is organized energy (concentrated and therefore useable energy) degenerates into disorganized energy (nonusable energy). The energy of gasoline is organized and usable energy. When gasoline burns in a car engine, part of its energy does useful work, part heats the engine and surroundings, and part goes out the exhaust. Useful energy degenerates to nonuseful forms and is unavailable for doing the same work again, such as driving another car. Heat is the graveyard of useful energy.

The quality of energy is lowered with each transformation as energy of an organized form tends to disorganized forms. In this broader regard, the second law can be stated another way:

Natural systems tend to proceed toward a state of greater disorder.

Figure 8.12 Push a heavy crate across a rough floor and all your work will go into heating the floor and crate. Work against friction turns into thermal energy, which cannot in turn do work on the crate even to begin to push it back to its starting place.

Consider a system consisting of a stack of pennies on your table, all heads up. Somebody walks by and accidentally bumps into the table and the pennies topple to the floor below, certainly not all heads up. Order becomes disorder. Molecules of gas all moving in harmony make up an orderly state—and also an unlikely state. On the other hand, molecules of gas moving in haphazard directions and speeds make up a disorderly state—a disorderly and a more probable state. If you remove the lid of a bottle containing some gas, the gas molecules escape into the room and make up a more disorderly state (Figure 8.13). Relative order becomes disorder. You would not expect the reverse to happen; that is, you would not expect the molecules to spontaneously order themselves back into the bottle and thereby return to the more ordered containment. Such processes in which disorder goes to order are simply not observed to happen.

Disordered energy can be changed to ordered energy only at the expense of some organizational effort or work input. For example water in a refrigerator freezes and becomes more ordered because work is put into the refrigeration cycle; gas can be ordered into a small region if work input is supplied to a compressor. But without some imposed work input, processes in which the net effect is an increase in order are not observed in nature.

8.7 ENTROPY

Figure 8.13 Molecules of gas go from the bottle to the air and not vice versa.

The idea of lowering the "quality" of energy is embodied in the idea of **entropy**.* Entropy is the measure of the *amount of disorder*. The second law states that, in the long run, entropy always increases. Gas molecules escaping from a bottle move from a relatively orderly state to a disorderly state. Disorder increases; entropy increases. Whenever a physical system is allowed to distribute its energy freely, it always does so in a manner such that entropy increases while the available energy of the system for doing work decreases.

Consider the old riddle, "How do you unscramble an egg?" The answer is simple: "Feed it to a chicken." But even then you won't get all your original egg back—egg making has its inefficiencies too. All living organisms, from bacteria to trees to human beings, extract energy from their surroundings and use it to increase their own organization. In living organisms, entropy decreases. But the order in life forms is maintained by increasing entropy elsewhere; life forms plus their waste

Figure 8.14 Why is the motto of this contractor—"Increasing entropy is our business"—so appropriate?

---

*Entropy can be expressed as a mathematical equation, stating that the increase in entropy ΔS in an ideal thermodynamic system is equal to the amount of heat added to a system ΔQ divided by the temperature T of the system: $\Delta S = \Delta Q/T$.

products have a net increase in entropy.* Energy must be transformed into the living system to support life. When it isn't, the organism soon dies and tends toward disorder.

The first law of thermodynamics is a universal law of nature for which no exceptions have been observed. The second law, however, is a probabilistic statement. Given enough time, even the most improbable states may occur; entropy may sometimes decrease. For example, the haphazard motions of air molecules could momentarily become harmonious in a corner of the room, just as a barrelful of pennies dumped on the floor could all come up heads. These situations are possible—but not probable. The second law tells us the most probable course of events, not the only possible one.

The laws of thermodynamics are often stated this way: You can't win (because you can't get any more energy out of a system than you put into it), you can't break even (because you can't even get as much energy out as you put in), and you can't get out of the game (entropy in the universe is always increasing).

---

* Interestingly enough, the American writer Ralph Waldo Emerson, who lived during the time the second law of thermodynamics was the new science topic of the day, philosophically speculated that not everything becomes more disordered with time and cited the example of human thought. Ideas about the nature of things grow increasingly refined and better organized as they pass through the minds of succeeding generations. Human thought is evolving toward more order.

SUMMARY OF TERMS

Thermodynamics The study of heat and its transformation to other forms of energy.

Absolute zero The lowest possible temperature that a substance may have; the temperature at which molecules of a substance have their minimum kinetic energy.

Thermal energy (or *internal energy*) The total of all molecular energies—kinetic energy and potential energy—internal to a substance. *Changes* in thermal energy are of principal concern in thermodynamics.

First law of thermodynamics A restatement of the law of energy conservation, usually as it applies to systems involving changes in temperature: The heat added to a system equals an increase in thermal energy plus external work done by the system.

Adiabatic process A process, usually of expansion or compression, wherein no heat enters or leaves a system.

Second law of thermodynamics Heat will never spontaneously flow from a cold object to a hot object. Also, no machine can be completely efficient in converting energy to work; some input energy is dissipated as heat. And finally, all systems tend to become more and more disordered as time goes by.

Heat Engine A device that changes thermal energy to mechanical work.

Evaporation The change of phase at the surface of a liquid as it passes to the gaseous phase.

Condensation The change of phase from gas to liquid; the opposite of evaporation.

Entropy A measure of the disorder of a system. Whenever energy freely transforms from one form to another, the direction of transformation is toward a state of greater disorder and therefore toward one of greater entropy.

REVIEW QUESTIONS

1. The word *thermodynamics* stems from which Greek words?

2. Is the study of thermodynamics concerned primarily with microscopic or macroscopic processes?

Absolute Zero

3. By how much does a volume of gas at 0°C reduce in size for each 1 Celsius degree decrease when the pressure is held constant?

4. By how much does the pressure of gas at 0°C reduce for each 1 Celsius degree decrease when the volume is held constant?

5. If we assume that the gas does not condense to a liquid, what volume is approached for a gas at 0° cooled by 273 Celsius degrees?

6. What is the lowest possible temperature on the Celsius scale? On the Kelvin scale?

Thermal Energy

7. Which is always greater, the molecular kinetic energy in a substance or the thermal energy in a substance?

8. In the study of thermodynamics, is the principal concern the amount of thermal energy in a substance or the changes in thermal energy in a substance?

First Law of Thermodynamics

9. How does the law of the conservation of energy relate to the first law of thermodynamics?

10. What happens to the thermal energy of a system when mechanical work is done on it? What happens to its temperature?

11. What is the relationship between heat added to a system and the thermal energy and external work done by the system?

Adiabatic Processes

12. What condition is necessary for a process to be adiabatic?

13. If work is done *on* a system, does the thermal energy of the system increase or decrease? If work is done *by* a system, does the thermal energy of the system increase or decrease?

Second Law of Thermodynamics

14. How does the second law of thermodynamics relate to the direction of heat flow?

Heat Engines

15. What three processes occur in every heat engine?

16. What exactly is thermal pollution?

17. If all friction could be removed from a heat engine, would it be 100 percent efficient? Explain.

18. What does it mean to say that energy becomes more diffuse when it transforms?

Energy and Changes of Phase

19. Does a liquid give off or absorb energy when it evaporates? When it solidifies?

20. Does a gas give off or absorb energy when it condenses?

21. How many calories are required to change the temperature of 1 g of water by 1°C? To melt 1 g of ice at 0°C? To vaporize 1 g of boiling water at 100°C?

22. Why does the temperature of melting ice not rise when heat is applied?

23. Why does the temperature of boiling water not rise when heat is applied?

24. Is the food compartment in a refrigerator cooled by vaporization or by condensation of the refrigerating fluid?

25. Why is it important that your finger be wet when you touch it briefly to a hot clothes iron?

Order Tends to Disorder

26. Give at least two examples to distinguish between organized energy and disorganized energy.

27. How much of the electrical energy transformed by a common light bulb becomes heat energy?

28. With respect to orderly and disorderly states, what do natural systems tend to do? Can a disorderly state ever transform to an orderly state? Explain.

Entropy

29. What is the physicist's term for *measure of messiness?* What is its relationship to the second law of thermodynamics?

30. Distinguish between the first and second laws of thermodynamics in terms of whether or not exceptions occur.

. . . .

HOME PROJECT

How much energy is in a nut? Burn it and find out. The heat of the flame is energy released upon the formation of chemical bonds (carbon dioxide, CO_2, and water, H_2O). Pierce a nut (pecan or walnut halves work best) with a bent paper clip that holds the nut above the table surface. Above this secure a can of water so you can measure its temperature change when the nut burns. Use about 10 cubic centimeters (10 milliliters) of water and a Celsius thermometer. As soon as you ignite the nut with a match place the can of water above it and record the increase in water temperature as soon as the flame has extinguished. The number of calories released by the burning nut can be calculated by the formula, $Q = mc\Delta T$, where m is the mass of water, c its specific heat (1) and ΔT the change in temperature. The energy in food is expressed in terms of the dietetic Calorie, which is 1000 of the calories you'll measure. So to find the number of dietetic Calories, divide your result by 1000.

. . . .

EXERCISES

1. Consider a piece of metal that has a temperature of 10°C. If it is heated until it is twice as hot (has twice the thermal energy), what would its temperature be?

2. On a chilly 10°C day, your friend who likes cold weather says she wishes it were twice as cold. Taking this literally, what temperature would this be?

3. A friend says the temperature inside a certain oven is 600, and the temperature inside a certain star is 60,000. You're not sure if your friend means degrees Celcius or Kelvins. How much difference does it make in each case?

4. If you shake a can of liquid back and forth, will the temperature of the liquid increase? (Try it and see.)

5. When you pump a tire with a bicycle pump, the cylinder of the pump becomes hot. Give two reasons why this is so.

6. Is it possible to wholly convert a given amount of mechanical energy into heat? Is it possible to wholly convert a given amount of heat into mechanical energy? Cite examples to illustrate your answers.

7. Why do diesel engines need no spark plugs?

8. When warm air rises, what happens to its temperature, and why?

9. Imagine a giant dry-cleaner's bag full of air floating like a balloon with a string hanging from it high above the ground. What would be the effect on its temperature if it were yanked suddenly to ground?

10. We say that energy "deteriorates" in energy transformations. Does this contradict the first law of thermodynamics? Defend your answer.

11. The combined molecular kinetic energies of molecules in a very large container of cold water are greater than the combined molecular kinetic energies in a cup of hot tea. Pretend you partially immerse the teacup in the cold water and that the tea absorbs 10 units of energy from the water and becomes hotter, while the water that gives up 10 units of energy becomes cooler. Would this energy transfer violate the first law of thermodynamics? The second law of thermodynamics? Explain.

12. Why is *thermal pollution* a relative term?

13. Is it possible to construct a heat engine that produces no thermal pollution? Defend your answer.

14. Why is it advantageous to use steam as hot as possible in a steam-driven turbine?

15. How does the ideal efficiency of an automobile relate to the temperature of the engine and the temperature of the environment in which it runs? Be specific.

16. What happens to the efficiency of a heat engine when the temperature of the reservoir into which heat energy is rejected is lowered?

17. To increase the efficiency of a heat engine, would it be better to increase the temperature of the reservoir while holding the temperature of the sink constant or to decrease the temperature of the sink while holding the temperature of the reservoir constant? Explain.

18. What is the ideal efficiency of an automobile engine in which gasoline is heated to 2700K and the outdoor air is 300K?

19. Under what conditions would a heat engine be 100 percent efficient?

20. Suppose one wishes to cool a kitchen by leaving the refrigerator door open and closing the kitchen door and windows. What will happen to the room temperature? Why?

21. Why will a refrigerator with a fixed amount of food consume more energy in a warm room than in a cold room?

22. People who live where snowfall is common will tell you that air temperatures are higher on snowy days than on clear days. Some misinterpret this by stating that snowfall can't occur on very cold days. Explain this misinterpretation.

23. What effect does melting ice have on the temperature of the surrounding air?

24. What effect does freezing water have on the temperature of the surrounding air?

25. Why does dew form on the surface of an ice-cold soft drink can?

26. Why is it that in cold winters a tub of water placed in a farmer's canning cellar helps prevent canned food from freezing?

27. Why will spraying fruit trees with water before a frost help to protect the fruit from freezing?

28. In buildings that are being electrically heated, is it at all wasteful to turn all the lights on? Is turning all the lights on wasteful if the building is being cooled by air conditioning?

29. Using both the first and the second law of thermodynamics, defend the statement that 100 percent of the electrical energy that goes into lighting a lamp is converted to thermal energy.

30. Is the total energy of the universe becoming more unavailable with time? Explain.

31. Water evaporates from a salt solution and leaves behind salt crystals that have a higher degree of molecular order than the more randomly moving molecules in the salt water. Has the entropy principle been violated? Why or why not?

32. Water put into a freezer compartment in your refrigerator goes to a state of less molecular disorder when it freezes. Is this an exception to the entropy principle? Explain.

PART 3

ELECTRICITY AND MAGNETISM

Just think, this magnet outpulls the whole world when it lifts these nails. The pull between the nails and the earth I call a **gravitational force,** and the pull between the nails and the magnet I call a **magnetic force.** I can *name* these forces, but I don't yet *understand* them. My learning begins by realizing there's a big difference in *knowing the names* of things and really *understanding* those things.

9

ELECTRICITY

Electricity underlies just about everything around us. It's in the lightning from the sky, it's in the spark beneath our feet when we scuff across a rug, and it's what holds atoms together to form molecules. The control of electricity is evident in technological devices of many kinds, from lamps to computers. In this technological age it is important to have an understanding of the basics of electricity and of how these basics can be manipulated to produce a prosperity unknown before recent times.

9.1 ELECTRIC FORCE AND CHARGE

What if there were a universal force like gravity, which varies inversely as the square of the distance but which is billions upon billions of times stronger? If there were such a force and if it were attractive like gravity, the universe would be pulled together into a tight sphere with all the matter in the universe pulled as

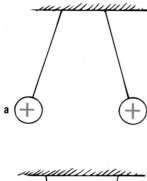

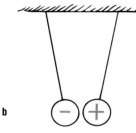

Figure 9.1 (a) Like charges repel. (b) Unlike charges attract.

close together as possible. But suppose this force were a repelling force, where every particle repelled every other particle. What then? The universe would be an ever-expanding gaseous cloud. Suppose, however, that the universe consisted of a strong force that could be both attractive *and* repulsive, and that this force was manifested by two different kinds of particles, one that was of a positive sign and the other of negative sign. Suppose that like signs repelled and unlike signs attracted (Figure 9.1). Suppose there were equal numbers of each so that this strong force were perfectly balanced! What would the universe be like? The answer is simple: It would be like the one we are living in. For there is such a force. We call it *electrical force.*

The terms *positive* and *negative* refer to electric *charge*, the fundamental quantity that underlies all electrical phenomena. The positively charged particles are *protons*, and the negatively charged particles are *electrons*. The attractive force between these particles causes them to lump together into incredibly small units, which we call atoms. More of the interesting details about atoms are presented in Chapter 14 as well as in subsequent chapters. In order to understand the basic principles of electricity, however, it is important to be aware of some fundamental facts about atoms:

1. Every atom is composed of a positively charged nucleus, surrounded by negatively charged electrons.
2. The electrons of all atoms are identical. Each has the same quantity of negative charge and the same mass.
3. Protons and neutrons compose the nucleus. (The common form of hydrogen that has no neutrons is the only exception.) Protons are about 1800 times more massive than electrons but carry an amount of positive charge equal to the negative charge of electrons. Neutrons have slightly more mass than protons and have no net charge.

Normally, an atom has as many electrons as protons. When an atom loses one or more electrons it has a positive net charge, and when it gains one or more electrons it has a negative net charge. A charged atom is called an *ion*. A *positive ion* has a net positive charge. A *negative ion,* with one or more extra electrons, has a net negative charge.

Material objects are made of atoms, which means they are composed of electrons and protons (and neutrons as well). Objects ordinarily have equal numbers of electrons and protons and are therefore electrically neutral. But if there is a slight imbalance in the numbers, the object is electrically charged. An imbalance comes about when electrons are added or removed from an object. Although the innermost electrons in an atom are bound very tightly to the oppositely charged atomic nucleus, the outermost electrons of many atoms are bound very loosely and can be easily dislodged. How much work is required to tear an electron away from an atom varies

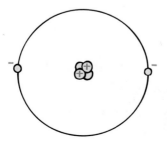

Figure 9.2 Model of a helium atom. The atomic nucleus is made up of two protons and two neutrons. The positively charged protons attract two negative electrons. What is the net charge of this atom?

Figure 9.3 Electrons are transferred from the fur to the rod. The rod is then negatively charged. Is the fur charged? How much compared to the rod? Positively or negatively?

for different substances. The electrons are held more firmly in rubber or plastic than in your hair, for example. Thus, when a comb is rubbed through your hair, electrons transfer from the hair to the comb. The comb then has an excess of electrons and is said to be *negatively charged.* Your hair, in turn, has a deficiency of electrons and is said to be positively charged. If you rub a glass or plastic rod with silk, you'll find that the rod becomes positively charged. The silk has a greater affinity for electrons than the glass or plastic rod. Electrons are rubbed off the rod and onto the silk.

So we see that an object having unequal numbers of electrons and protons is electrically charged. If it has more electrons than protons, it is negatively charged. If it has fewer electrons than protons, it is positively charged.

It is important to note that when we charge something, no electrons are created or destroyed. Electrons are simply transferred from one material to another. Charge is conserved. In every event, whether large-scale or at the atomic and nuclear level, the principle of *conservation of charge* has always been found to apply. No case of the creation or destruction of net electric charge has ever been found. The conservation of charge ranks with the conservation of energy and momentum as a significant fundamental principle in physics.

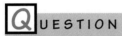UESTION

If you walk across a rug and scuff electrons from your feet, are you negatively or positively charged?

· ·
Ⓐ N S W E R

You have fewer electrons after you scuff your feet, so you are positively charged (and the rug is negatively charged).

Electrostatic Charge

Static charge can be dangerous. Two hundred years ago young boys called "powder monkeys" ran below the decks of warships to bring sacks of black gunpowder to the cannons above. It was ship law that this task be done barefoot. Why? Because it was important that no static charge build up on their bodies as they ran to and fro. Bare feet assured no charge buildup that might result in an igniting spark and an explosion.

Static charge is a danger in many industries today—not because buildings may blow up, but because delicate electronic circuits may be destroyed by static charge. Some circuit components are sensitive enough to be "fried" by static electric sparks. Electronics technicians often wear clothing of special fabrics with ground wires between their sleeves and their socks. Some wear special wrist bands that are connected to a grounded surface so that static charge will not build up, when moving a chair, for example. The smaller the electronic circuit, the more hazardous are sparks that may short-circuit the circuit elements.

Coulomb's Law

The electrical force, like gravitational force, decreases inversely as the square of the distance between charges increases. This relationship was discovered by Charles Coulomb in the eighteenth century and is called **Coulomb's law**. It states that for charged objects that are much smaller than the distance between them, the force between two charges varies directly as the product of the charges and inversely as the square of the separation distance. The force acts along a straight line from one charge to the other. Coulomb's law can be expressed as

$$F = k\frac{q_1 q_2}{d^2}$$

where d is the distance between the charged particles, q_1 represents the quantity of charge of one particle, q_2 represents the quantity of charge of the other particle, and k is the proportionality constant.

The unit of charge is called the **coulomb**, abbreviated C. It turns out that a charge of 1 C is the charge associated with 6.25 billion billion electrons. This might seem like a great number of electrons, but it represents only the amount of charge that passes through a common 100-watt light bulb in a little over a second.

The proportionality constant k in Coulomb's law is similar to G in Newton's law of gravity. Instead of being a very small number like G (6.67×10^{-11}), the electrical proportionality constant k is a very large number. It is approximately

$$k = 9,000,000,000 \text{ N·m}^2/\text{C}^2$$

or, in scientific notation, $k = 9 \times 10^9$ N·m^2/C^2. The units N·m^2/C^2 are not central to our interest here. What is important is the large magnitude of k. If, for example, a pair of like charges of 1 coulomb each were 1 meter apart, the force of repulsion between the two charges would be 9 billion newtons.* That would be about ten times the weight of a battleship! Obviously, such amounts of net charge do not usually exist in our everyday environment.

So Newton's law of gravitation for masses is similar to Coulomb's law for electric charges.† Whereas the gravitational force of attraction between particles such as an electron and a proton is extremely small, the electrical force between these particles is relatively enormous. The greatest difference between gravitational and electrical forces is that although gravitational forces are only attractive, electrical forces may be either attractive or repulsive.

---

*Contrast this to the gravitational force of attraction between two unit masses (kilograms) 1 m apart: 6.67×10^{-11} N. This is an extremely small force. For the force to be 1 N, the masses at 1 m apart would have to be nearly 123,000 kg each! Gravitational forces between ordinary objects are exceedingly small, and electrical forces (noncancelled) between ordinary objects are exceedingly huge. We don't sense them because the positives and negatives normally balance out, and even for highly charged objects, the imbalance of electrons to protons is normally less than one part in a trillion trillion.

†The similarities between these two forces have made some people think they may be different aspects of the same thing. Albert Einstein was one of these people, and he spent the latter part of his life searching with little success for a "unified field theory." More recently, the electrical force has been unified with one of the two nuclear forces, the *weak force,* which plays a role in radioactive decay. Physicists are still looking for a way to unify electrical and gravitational forces.

Balloon Stick

Charge an inflated balloon by rubbing it on your hair. Then place the balloon against a wall and it sticks. This is because the charge on the balloon alters the charge distribution in the atoms or molecules in the wall, effectively inducing an opposite charge on the wall. The molecules cannot move from their relatively fixed positions, but their "centers of charge" are moved. The positive part of the atom or molecule is attracted toward the balloon while the negative part is repelled. This has the effect of distorting the atom or molecule (Figure 9.4). The atom or molecule is said to be *electrically polarized*. We will treat electrical polarization further in Chapter 19.

We know that after you rub the balloon on your hair, the balloon will stick to the wall. Question: If you put your head to the wall, will that stick too?

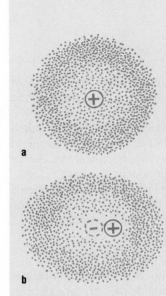

a

b

Figure 9.4 (a) The center of the negative "cloud" of electrons coincides with the center of the positive nucleus in an atom. (b) When an external negative charge is brought nearby to the left, as on a charged balloon, the electron cloud is distorted so the centers of negative and positive charge no longer coincide. The atom is electrically polarized.

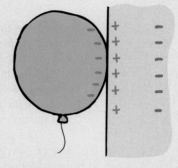

Figure 9.5 The negatively charged balloon polarizes molecules in the wooden wall and creates a positively charged surface, so the balloon sticks to the wall.

QUESTIONS

1. The proton that makes up the nucleus of the hydrogen atom attracts the electron that orbits it. Relative to this force, does the electron attract the proton with less force, more force, or the same amount of force?

2. If a proton at a particular distance from a charged particle is repelled with a given force, by how much will the force decrease when the proton is three times as distant from the particle? Five times as distant?

3. What is the sign of charge of the particle in this case?

ANSWERS

1. The same amount of force, in accord with Newton's third law—basic mechanics! Recall that a force is an interaction between two things, in this case between the proton and the electron. They pull on each other—equally.

2. It decreases to $\frac{1}{9}$ its original value; to $\frac{1}{25}$.

3. Positive.

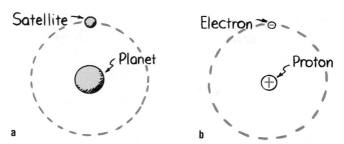

Figure 9.6 A gravitational force holds the satellite in orbit about the planet (a), and an electrical force holds the electron in orbit about the proton (b). In both cases there is no contact between the bodies. We say that the orbiting bodies interact with the *force fields* of the planet and proton and are everywhere in contact with these fields. Thus, the force that one electric charge exerts on another can be described as the interaction between one charge and the field set up by the other.

9.2 ELECTRIC FIELD

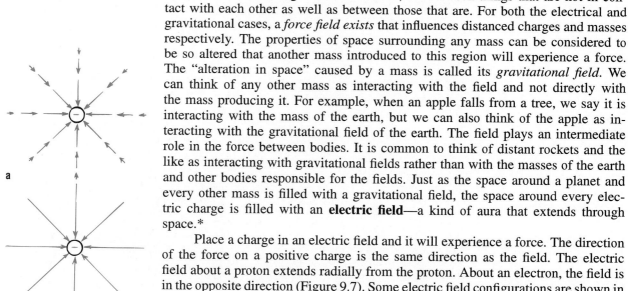

Figure 9.7 Electric field representations about a negative charge. (a) A vector representation. (b) A lines-of-force representation.

Electrical forces, like gravitational forces, act between things that are not in contact with each other as well as between those that are. For both the electrical and gravitational cases, a *force field exists* that influences distanced charges and masses respectively. The properties of space surrounding any mass can be considered to be so altered that another mass introduced to this region will experience a force. The "alteration in space" caused by a mass is called its *gravitational field*. We can think of any other mass as interacting with the field and not directly with the mass producing it. For example, when an apple falls from a tree, we say it is interacting with the mass of the earth, but we can also think of the apple as interacting with the gravitational field of the earth. The field plays an intermediate role in the force between bodies. It is common to think of distant rockets and the like as interacting with gravitational fields rather than with the masses of the earth and other bodies responsible for the fields. Just as the space around a planet and every other mass is filled with a gravitational field, the space around every electric charge is filled with an **electric field**—a kind of aura that extends through space.*

Place a charge in an electric field and it will experience a force. The direction of the force on a positive charge is the same direction as the field. The electric field about a proton extends radially from the proton. About an electron, the field is in the opposite direction (Figure 9.7). Some electric field configurations are shown in

---

*An electric field has both magnitude (strength) and direction. The magnitude of the field at any point is simply the force per unit of charge. If a charge q experiences a force F at some point in space, then the electric field E at that point is

$$E = \frac{F}{q}$$

Figure 9.8 Some electric field configurations.
(a) Lines of force about a single positive charge.
(b) Lines of force for a pair of equal but opposite charges. Note that the lines emanate from the positive charge and terminate on the negative charge.
(c) Uniform lines of force between two oppositely charged parallel plates.

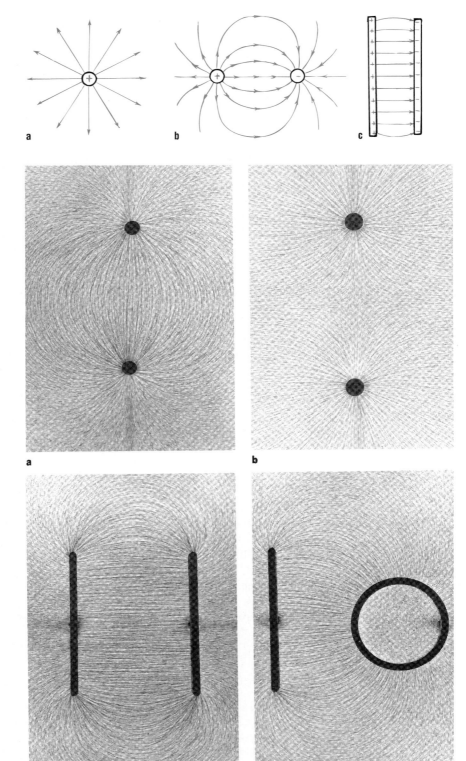

Figure 9.9 Bits of thread suspended in an oil bath surrounding charged conductors line up end-to-end along the direction of the field.
(a) Equal and opposite charges. (b) Equal like charges. (c) Oppositely charged plates.
(d) Oppositely charged cylinder and plate.

Figure 9.10 Both Meidor and the spherical dome of the Van de Graaff generator are electrically charged. Why does her hair stand out?

Figure 9.8, and photographs of field patterns are shown in Figure 9.9. In the next chapter we'll see how bits of iron similarly align with magnetic fields.

Perhaps your instructor will demonstrate the effects of the electric field that surrounds the charged dome of a Van de Graaff generator (Figure 9.10). Charged objects in the field of the dome are either attracted or repelled, depending on their sign of charge.

Q UESTION

The woman in Figure 9.10, like the dome of the Van de Graaff generator, is charged (because she was in contact with the dome when charging took place). Why does her hair stand out?

━━━ ••

9.3 ELECTRIC POTENTIAL

When we studied energy in Chapter 3, we learned that an object may have gravitational potential energy because of its location in a gravitational field. Similarly, a charged object can have potential energy by virtue of its location in an electric field. Just as work is required to lift a massive object against the gravitational field of the earth, work is required to push a charged particle against the electric field of a charged body. This work increases the electric potential energy of the

A NSWER ••

Meidor and her hair are charged. Each hair is repelled by others around it—evidence that *like charges repel*. Even a small charge produces an electrical force greater than hair weight. Fortunately, the electrical force is not great enough to make her arms stand out!

charged particle.* Consider the small positive charge located at some distance from a positively charged sphere in Figure 9.11a. If you push the small charge closer to the sphere (Figure 9.11b), you will expend energy to overcome electrical repulsion; that is, you will do work in pushing the charge against the electric field of the sphere. This work done in moving the small charge to its new location increases the energy of the small charge. We call the energy the charge now possesses by virtue of its location **electric potential energy**. If the charge is released, it accelerates in a direction away from the sphere, and its electric potential energy changes to kinetic energy.

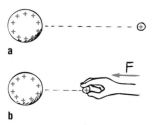

Figure 9.11 The small charge has more PE in (b) than in (a) because work was required to move it to the closer location.

If we push two charges instead, we do twice as much work, and the two charges in the same location have twice the electric potential energy as before; three charges have three times as much potential energy; ten charges ten times the potential energy; and so on. Rather than dealing with the total potential energy of a charged body, it is convenient when working with electricity to consider the electric potential energy *per charge*. We simply divide the amount of energy in any case by the amount of charge. For example, ten charges at a specific location will have ten times as much energy as one, but will also have ten times as much charge; so we see that the energy per charge remains the same at a particular field location whatever the amount of charge. The concept of potential energy per charge is called **electric potential**; that is,

$$\text{Electric potential} = \frac{\text{electric potential energy}}{\text{amount of charge}}$$

The unit of measurement for electric potential is the volt, so electric potential is often called *voltage*. A potential of 1 volt (V) equals 1 joule (J) of energy per 1 coulomb (C) of charge.

$$1 \text{ volt} = \frac{1 \text{ joule}}{\text{coulomb}}$$

Thus a 1.5-volt battery gives 1.5 joules of energy to every 1 coulomb of charge passing through the battery. Both the names *electric potential* and *voltage* are common, so either may be used. In this book the names will be used interchangeably.

The significance of voltage is that a definite value for it can be assigned to a location, whether or not a charge exists at that location. We can speak about the voltages at different locations in an electric field whether or not charges occupy those locations. Likewise with voltages at various locations in an electric circuit. Later in this chapter we will see that the location of the positive terminal of a 12-volt battery is maintained at a voltage 12 volts higher than the location of the negative terminal. When a conducting medium connects this voltage difference, charges in the medium will move between these voltage-differing locations.

Figure 9.12 The sketch on top shows the PE (potential energy) of a mass held in a gravitational field. The sketch on the bottom shows the PE of a charge held in an electric field. When released, how does the KE (kinetic energy) acquired by each compare to the decrease in PE?

---

*This work is positive if it increases the electric potential energy of the charged particle and negative if it decreases it.

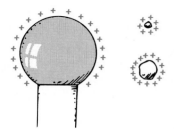

Figure 9.13 The larger test charge has more PE in the field of the charged dome, but the *electric potential* of any amount of charge at the same location is the same.

QUESTIONS

1. If there were twice as many coulombs in the test charge near the charged sphere in Figure 9.13, would the electric potential energy of the test charge with respect to the charged sphere be the same or would it be twice as great? Would the electric potential of the test charge be the same or would it be twice as great?

2. What does it mean to say that an automobile battery is "12 volts"?

Rub a balloon on your hair, and the balloon becomes negatively charged—perhaps to several thousand volts! That would be several thousand joules of energy if the charge were 1 coulomb. However, 1 coulomb is a fairly respectable amount of charge. The charge on a balloon rubbed on hair is more typically much less than a millionth of a coulomb. Therefore, the amount of energy associated with the charged balloon is very, very small. A high voltage means a lot of energy only if a lot of charge is involved.* Here we see there is a difference between electric potential energy and electric potential.

Now we will see what happens when an electric potential is applied to one end of a piece of metal wire, and a different electric potential is applied to the other end. This difference in electric potential, or voltage, acts like an "electric pressure" that produces an electric current—a flow of electric charge.

Figure 9.14 Although the voltage of the charged balloon is high, the electric potential energy is low because of the small amount of charge.

ANSWERS

1. Twice as many coulombs would find the test charge with twice as much electric potential energy (because it takes twice as much work to put the charge there). But the electric potential would be the same. This is because the electric potential is total electric potential energy divided by total charge. For example, ten times the energy divided by ten times the charge will give the same value as two times the energy divided by two times the charge. Electric potential is not the same thing as electric potential energy. Be sure you understand this before you study further.

2. It means that one of the battery terminals is 12 V higher in potential than the other battery terminal. Soon we'll see that it also means that when a circuit is connected across these terminals, each coulomb of charge that makes up the resulting current will be given 12 J of energy as it passes through the battery.

---

*This is very similar to the harmless high-temperature sparks emitted by a fireworks sparkler. Temperature is the ratio of energy/molecule. A high temperature means a lot of energy only if a lot of molecules are involved.

9.4 VOLTAGE SOURCES

When the ends of a conductor of heat are at different temperatures, heat energy flows from the higher temperature to the lower temperature. The flow ceases when both ends reach the same temperature. Similarly, when the ends of an electrical conductor are at different electric potentials—when there is a **potential difference**—charges in the conductor flow from the higher potential to the lower potential. The flow of charge persists until both ends reach the same potential. Without a potential difference, no flow of charge will occur. Connect one end of a wire to a charged Van de Graaff generator, for example, and the other end to the ground, and a surge of charge will flow through the wire. The flow will be brief, however, for the sphere will quickly reach a common potential with the ground.

To attain a sustained flow of charge in a conductor, some arrangement must be provided to maintain a difference in potential while charge flows from one end to the other. The situation is analogous to the flow of water from a higher reservoir to a lower one (Figure 9.15a). Water will flow in a pipe that connects the reservoirs only as long as a difference in water level exists. The flow of water in the pipe, like the flow of charge in the wire that connects the Van de Graaff generator to the ground, will cease when the pressures at each end are equal (we imply this when we say that water seeks its own level). A continuous flow is possible if the difference in water levels—hence water pressures—is maintained with the use of a suitable pump (Figure 9.15b).

Figure 9.15 Water flows from the reservoir of higher pressure to the reservoir of lower pressure. (a) The flow will cease when the difference in pressure ceases. (b) Water continues to flow because a difference in pressure is maintained with the pump.

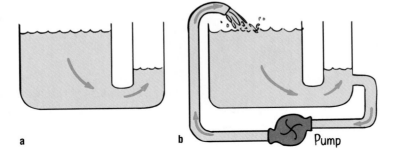

a   b   Pump

A sustained electric current requires a suitable pumping device to provide a difference in electric potential—to provide a **voltage**. Chemical batteries or generators are "electrical pumps" that can maintain a steady flow of charge. These devices do work to pull negative charges away from positive ones. In chemical batteries, this work is done by the chemical disintegration of zinc or lead in acid, and the energy stored in the chemical bonds is converted to electric potential energy.* Generators separate charge by electromagnetic induction, a process we will describe in the next chapter. The work done by whatever means in separating the opposite charges is available at the terminals of the battery or generator. This energy per charge provides the difference in potential (voltage) that provides the "electrical pressure" to move electrons through a circuit joined to these terminals.

Figure 9.16 An unusual source of voltage. The electric potential between the head and tail of the electric eel (*Electrophorus electricus*) can be up to 600 V.

---

*The chemical nature of batteries is described in Chapter 20—Chemical Reactions.

The unit of electric potential (voltage) is the *volt*. A common automobile battery will provide an electrical pressure of 12 volts to a circuit connected across its terminals. Then 12 joules of energy are supplied to each coulomb of charge that is made to flow in the circuit.

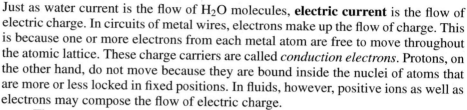

9.5 ELECTRIC CURRENT

Just as water current is the flow of H_2O molecules, **electric current** is the flow of electric charge. In circuits of metal wires, electrons make up the flow of charge. This is because one or more electrons from each metal atom are free to move throughout the atomic lattice. These charge carriers are called *conduction electrons*. Protons, on the other hand, do not move because they are bound inside the nuclei of atoms that are more or less locked in fixed positions. In fluids, however, positive ions as well as electrons may compose the flow of electric charge.

The *rate* of electrical flow is measured in **amperes**. An ampere is the rate of flow of 1 coulomb of charge per second. (Recall that 1 coulomb, the standard unit of charge, is the electric charge of 6.25 billion billion electrons.) In a wire that carries 5 amperes, for example, 5 coulombs of charge pass any cross section in the wire each second. In a wire that carries 10 amperes, twice as many coulombs pass any cross section each second.

It is interesting to note that a current-carrying wire is not electrically charged. Under ordinary conditions, negative conduction electrons swarm through the atomic lattice made up of positively charged atomic nuclei. So there are as many electrons as protons in the wire. Whether a wire carries a current or not, the net charge of the wire is normally zero at every moment.

There is often some confusion about charge flowing *through* a circuit and voltage placed, or impressed, *across* a circuit. We can distinguish between these ideas by considering a long pipe filled with water. Water will flow *through* the pipe if there is a difference in pressure *across* or between its ends. Water flows from the high-pressure end to the low-pressure end. Only the water flows, not the pressure. Similarly, electric charge flows because of the differences in electrical pressure (voltage). You say that charges flow *through* a circuit because of an applied voltage *across* the circuit.* You don't say that voltage flows through a circuit. Voltage doesn't go anywhere, for it is the charges that move. Voltage produces current (if there is a complete circuit).

Direct Current and Alternating Current

Electric current may be dc or ac. By *dc*, we mean **direct current**, which refers to the flowing of charges in *one direction*. A battery produces direct current in a circuit because the terminals of the battery always have the same sign. Electrons move from the repelling negative terminal toward the attracting positive terminal, always moving through the circuit in the same direction.

Alternating current (ac) acts as the name implies. Electrons in the circuit are moved first in one direction and then in the opposite direction, alternating to and fro

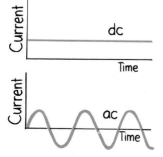

Figure 9.17 Each coulomb of charge that is made to flow in a circuit that connects the ends of this 1.5-V flashlight cell is energized with 1.5 J.

Figure 9.18 A time graph of dc and ac.

---

*It is conceptually simpler to say that current flows through a circuit, but don't say this around somebody who is picky about grammar, for the expression "current flows" is redundant. More properly, charge flows—which *is* current.

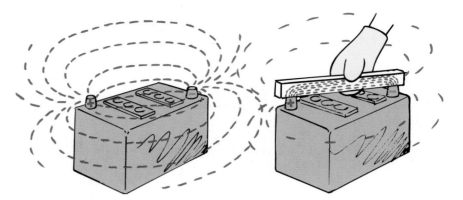

Figure 9.19   The electric field lines between the terminals of a battery are directed through a conductor, which joins the terminals. A metal bar is shown here, but the conductor is usually an electric circuit. (If you do this, you won't be shocked, but the bar will heat quickly and may burn your hand!)

about relatively fixed positions. This can be accomplished by periodically switching the sign at the terminals. Nearly all commercial ac circuits involve currents that alternate back and forth at a frequency of 60 cycles per second. This is 60-hertz current (a cycle per second is called a *hertz*). In some countries, 25-hertz, 30-hertz, or 50-hertz current is used. Throughout the world, most residential and commercial circuits are ac because electric energy in the form of ac can be transmitted great distances. Why this is so is quite interesting, and will be touched on in the next chapter.

9.6   ELECTRICAL RESISTANCE

A battery or generator of some kind is the prime mover and source of voltage in an electric circuit. How much current there is depends not only on the voltage but also on the **electrical resistance** the conductor offers to the flow of charge. This is similar to the rate of water flow in a pipe, which depends not only on the pressure behind the water but also on the resistance offered by the pipe itself. The resistance of a wire depends on the conductivity of the material and also on its thickness and length. Electrical resistance is less in thick wires. The longer the wire, of course, the greater the resistance. In addition, electrical resistance depends on temperature. The greater the jostling about of atoms within the conductor, the greater resistance the conductor offers to the flow of charge. For most conductors, increased temperature means increased resistance.* The resistance of some materials reaches zero at very low temperatures. These are the superconductors discussed in the *Special Interest* of Chapter 10.

---

*Carbon is an interesting exception. As temperature increases, more carbon atoms are able to shake loose an electron. This increases the electric current. The resistance of carbon, in effect, lowers with increasing temperature. This is one reason for the use of carbon in arc lamps (the primary reason is its high melting point).

Figure 9.20 Analogy between a simple hydraulic circuit (a) and an electrical circuit (b).

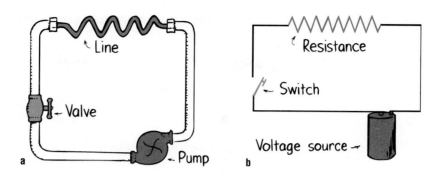

Electrical resistance is measured in units called *ohms*. The Greek letter *omega*, Ω, is commonly used as the symbol for the ohm. This unit is named after Georg Simon Ohm, a German physicist who in 1826 discovered a simple and very important relationship among voltage, current, and resistance.

Ohm's Law

The relationship among voltage, current, and resistance is summarized by a statement called **Ohm's law**. Ohm discovered that the amount of current in a circuit is directly proportional to the voltage established across the circuit, and is inversely proportional to the resistance of the circuit. In short,

$$\text{Current} = \frac{\text{voltage}}{\text{resistance}}$$

Or, in units form,

$$\text{Amperes} = \frac{\text{volts}}{\text{ohms}}$$

So for a given circuit of constant resistance, current and voltage are proportional to each other.* This means we'll get twice the current for twice the voltage. The greater the voltage, the greater the current. But if the resistance is doubled for a circuit, the current will be half what it would be otherwise. The greater the resistance, the smaller the current. Ohm's law makes good sense.

Ohm's law tells us that a potential difference of 1 volt established across a circuit that has a resistance of 1 ohm will produce a current of 1 ampere. If 12 volts are impressed across the same circuit, the current will be 12 amperes. The resistance of a typical lamp cord is much less than 1 ohm, while a typical light bulb has a resistance of about 100 ohms. An iron or electric toaster has a resistance of 15 to 20 ohms. The low resistance permits a large current, which produces considerable heat. Inside electrical devices such as radio and television receivers, current is regulated by circuit elements called *resistors,* whose resistance may be a few ohms or millions of ohms.

Figure 9.21 Resistors. The symbol of resistance in an electric circuit is $-\!\!\mathsf{W\!W\!V}\!\!-$.

---

*Many texts us *V* for voltage, *I* for current, and *R* for resistance, and express Ohm's law as $V = IR$. It then follows that $I = V/R$, or $R = V/I$, so if any two variables are known, the third can be found. Units are abbreviated V for volts, A for amperes, and Ω for ohms.

Superconductors

In ordinary conductors such as house wiring, moving electrons that flow as electric current often collide with atomic nuclei in the wire, transferring their kinetic energy to the wire as heat. Early in this century, certain metals in a bath of 4K liquid helium lost all electrical resistance. The electrons in these conductors traveled pathways that avoided these collisions, permitting them to flow indefinitely. These materials are called **superconductors,** having zero electrical resistance. Current does not decrease and no heat is generated in superconductivity. Until recently, it was generally thought that zero electrical resistance could be brought about only in certain metals near absolute zero. Then in 1986 superconductivity was achieved at 30K, which spurred hopes of finding superconductity above 77K, the point at which nitrogen boils.

Nitrogen is easier to handle and much less expensive because it is a byproduct of producing liquid oxygen from air. The historic leap came in the following year with an yttrium compound that lost its resistance at 90K.

Various ceramic oxides have since been found to be superconducting at temperatures above 100K. Once electric current is established in a superconductor, the current will flow indefinitely, even without an electric field. Steady currents have been observed to persist for years in certain superconductors without apparent loss. There is presently enormous interest in the physics community as to exactly why certain materials acquire superconducting properties. Explanations have generally to do with the wave nature of matter (quantum mechanics) and are being vigorously researched.

QUESTIONS

1. How much current will flow through a lamp that has a resistance of 60 Ω when 12 V are impressed across it?

2. What is the resistance of an electric frying pan that draws 12 A when connected to a 120-V circuit?

Electric Shock

What causes electric shock in the human body—current or voltage? The damaging effects of shock are the result of current passing through the body. From Ohm's law, we can see that this current depends on the voltage that is applied, and also on the electrical resistance of the human body. The resistance of one's body depends on its condition and ranges from about 100 ohms if soaked with salt water to about

ANSWERS

1. $\frac{1}{5}$ A. This is calculated from Ohm's law: $0.2\ A = \frac{12\ V}{60\Omega}$.

2. 10 Ω. Rearrange Ohm's law to read:

$$\text{Resistance} = \frac{\text{voltage}}{\text{current}} = \frac{120\ V}{12\ A} = 10\ \Omega$$

TABLE 9.1   Effect of Electric Currents on the Body

Current (A)	Effect
0.001	Can be felt
0.005	Is painful
0.010	Causes involuntary muscle contractions (spasms)
0.015	Causes loss of muscle control
0.070	Goes through the heart; serious disruption, probably fatal if current lasts for more than 1 s

500,000 ohms if the skin is very dry. If we touch the two electrodes of a battery with dry fingers, the resistance our body normally offers to the flow of charge is about 100,000 ohms. We usually cannot feel 12 volts, and 24 volts just barely tingles. If our skin is moist, 24 volts can be quite uncomfortable. Table 9.1 describes the effects of different amounts of current on the human body.

QUESTIONS

1. At 100,000 Ω, how much current will flow through your body if you touch the terminals of a 12-V battery?

2. If your skin is very moist—so your resistance is only 1000 Ω—and you touch the terminals of a 12-V battery, how much current will you receive?

Figure 9.22   The bird can stand harmlessly on one wire of high potential, but it had better not reach over and grab a neighboring wire! Why not?

Figure 9.23   The third prong connects the body of the appliance directly to ground. Any charge that builds up on an appliance is therefore conducted to the ground.

To receive a shock, there must be a *difference* in electric potential between one part of your body and another part. Most of the current will pass along the path of least electrical resistance connecting these two points. Suppose you fell from a bridge and managed to grab onto a high-voltage power line, halting your fall. So long as you touch nothing else of different potential, you will receive no shock at all. Even if the wire is a few thousand volts above ground potential and even if you hang by it with two hands, no appreciable charge will flow from one hand to the other. This is because there is no appreciable difference in electric potential between your hands. If, however, you reach over with one hand and grab onto a wire of different potential . . . zap! We have all seen birds perched on high-voltage wires. Every part of their bodies is at the same high potential as the wire, so they feel no ill effects.

Most electric plugs and sockets today are wired with three, instead of two, connections. The principal two flat prongs on a plug are for the current-carrying double wire, one part of which is "live" and the other neutral, while the third, round prong is connected directly to the earth (Figure 9.23). Appliances such as irons, stoves, washing machines, and dryers are connected with these three wires. If the live wire accidentally comes in contact with the metal surface of the appliance, the current will be directed to ground and won't shock anyone who handles it.

ANSWERS

1. The current through your body will be $\frac{12}{100\,000}$ A; (0.00012 A).
2. You will receive $\frac{12}{1000}$ A; (0.012 A). Ouch!

Safety and Electricity

Many people are killed each year by current from common 120-volt electric circuits. If you touch a faulty 120-volt light fixture with your hand while your feet are on the ground, there may be a 120-volt "electrical pressure" between your hand and the ground. Resistance to current flow is usually greatest between your feet and the ground, so the current is usually not enough to do serious harm. But if your feet and the ground are wet, there is a low-resistance electrical path between you and the ground. The 120 volts across this lowered resistance may produce a current greater than your body can withstand.

Pure water is not a good conductor. But the ions normally found in water make it a fair conductor. More dissolved materials, especially small amounts of salt, reduce the resistance even more. There is usually a layer of salt left from perspiration on your skin, which when wet lowers your skin resistance to a few hundred ohms or less. Handling electrical devices while taking a bath is a definite no-no.

Injury by electric shock comes in three forms: (1) burning of tissues by heating, (2) galvanic muscle contraction, and (3) disruption of cardiac rhythm. These conditions are caused by too much power delivered in critical body volumes for too long a time.

Electric shock can upset the nerve center that controls breathing. In rescuing shock victims, the first thing to do is clear them from the electric supply with a dry wooden stick or some other nonconductor so that you don't get electrocuted yourself. Then apply artificial respiration. It is important to continue artificial respiration, for there have been cases of lightning victims who did not breathe for several hours but were eventually revived and completely regained good health.

QUESTION

What causes electric shock, current or voltage?

9.7 ELECTRIC CIRCUITS

Any path along which electrons can flow is a *circuit*. For a continuous flow of electrons, there must be a complete circuit with no gaps. A gap is usually provided by an electric switch that can be opened or closed to either cut off or allow energy flow. Most circuits have more than one device that receives electric energy. These devices are commonly connected in a circuit in one of two ways, *series* or *parallel*. When connected in series, they form a single pathway for electron flow between the terminals of the battery, generator, or wall socket (which is simply an extension of these terminals). When connected in parallel, they form branches, each of which is a separate path for the flow of electrons. Both series and parallel connections have their own distinctive characteristics. We shall briefly treat circuits using these two types of connections.

ANSWER

Electric shock *occurs* when current is produced in the body, which is *caused* by an impressed voltage.

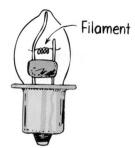

Filament

Figure 9.24 The conduction electrons that surge to and fro in the filament of the lamp do not come from the voltage source. They are in the filament to begin with. The voltage source simply provides them with surges of energy.

Series Circuits

A simple **series circuit** is shown in Figure 9.25. Three lamps are connected in series with a battery. The same current exists almost immediately in all three lamps when the switch is closed. The current does not "pile up" in any lamp but flows *through* each lamp. Electrons that make up this current leave the negative terminal of the battery, pass through each of the resistive filaments in the lamps in turn, and then return to the positive terminal of the battery (the same amount of current passes through the battery). This is the only path of the electrons through the circuit. A break anywhere in the path results in an open circuit, and the flow of electrons ceases. Burning out one of the lamp filaments or simply opening the switch could cause such a break. The circuit shown in Figure 9.25 illustrates the following important characteristics of series connections:

1. Electric current has but a single pathway through the circuit. This means that the current passing through the resistance of each electrical device is the same.
2. This current is resisted by the resistance of the first device, the resistance of the second, and that of the third also, so the total resistance to current in the circuit is the sum of the individual resistances along the circuit path.
3. The current in the circuit is numerically equal to the voltage supplied by the source divided by the total resistance of the circuit. This is in accord with Ohm's law.
4. The total voltage impressed across a series circuit divides among the individual electrical devices in the circuit so that the sum of the "voltage drops" across the resistance of each individual device is equal to the total voltage supplied by the source. This follows from the fact that the amount of

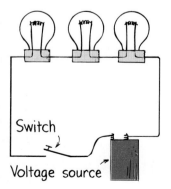

Switch

Voltage source

Figure 9.25 A simple series circuit. The 6-V battery provides 2 V across each lamp.

energy given to the total current is equal to the sum of energies given to each device.

5. The voltage drop across each device is proportional to its resistance. This follows from the fact that more energy is dissipated when a current passes through a large resistance than when the same current passes through a small resistance.

QUESTIONS

1. What happens to current in other lamps if one lamp in a series circuit burns out?
2. What happens to the light intensity of each lamp in a series circuit when more lamps are added to the circuit?

It is easy to see the main disadvantage of a series circuit: If one device fails, current in the whole circuit ceases. Some cheap Christmas tree lights are connected in series. When one bulb burns out, it's fun and games (or frustration) trying to find which one to replace.

Most circuits are wired so that it is possible to operate several electrical devices, each independently of the other. In your home, for example, a lamp can be turned on or off without affecting the operation of other lamps or electrical devices. This is because these devices are connected not in series, but in parallel with one another.

Parallel Circuits

A simple **parallel circuit** is shown in Figure 9.26. Three lamps are connected to the same two points A and B. Electrical devices connected to the same two points of an electrical circuit are said to be *connected in parallel*. Electrons leaving the negative terminal of the battery need travel through only *one* lamp filament before returning to the positive terminal of the battery. In this case, current branches into three separate pathways from A to B. A break in any one path does not interrupt the flow of charge in the other paths. Each device operates independently of the other devices. The circuit shown in Figure 9.26 illustrates the following major characteristics of parallel connections:

1. Each device connects the same two points A and B of the circuit. The voltage is therefore the same across each device.
2. The total current in the circuit divides among the parallel branches. Since the voltage across each branch is the same, the amount of current in each branch is inversely proportional to the resistance of the branch.
3. The total current in the circuit equals the sum of the currents in its parallel branches.

ANSWERS

1. If one of the lamp filaments burns out, the path connecting the terminals of the voltage source will break and current will cease. All lamps will go out.
2. The addition of more lamps in a series circuit results in a greater circuit resistance. This decreases the current in the circuit and therefore in each lamp, which causes dimming of the lamps. Energy is divided among more lamps, so the voltage drop across each lamp is less.

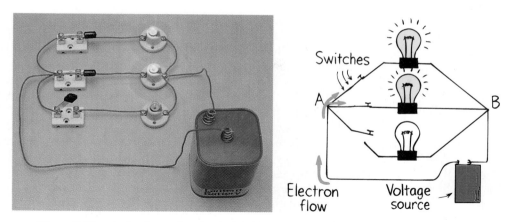

Figure 9.26   A simple parallel circuit. A 6-V battery provides 6 V across each lamp.

4. As the number of parallel branches is increased, the overall resistance of the circuit is decreased. Overall resistance is lowered with each added path between any two points of the circuit. This means the overall resistance of the circuit is less than the resistance of any one of the branches.

QUESTIONS

1. What happens to the current in other lamps if one of the lamps in a parallel circuit burns out?
2. What happens to the light intensity of each lamp in a parallel circuit when more lamps are added in parallel to the circuit?

Parallel Circuits and Overloading

Electricity is usually fed into a home by way of two lead wires called *lines*. These lines are very low in resistance and are connected to wall outlets in each room. About 110–120 volts are impressed on these lines by generators at the power utility. This voltage is applied to appliances and other devices that are connected in parallel by

ANSWERS

1. If one lamp burns out, the other lamps will be unaffected. The current in each branch, according to Ohm's law, is equal to voltage/resistance, and since neither voltage nor resistance is affected in the other branches, the current in those branches is unaffected. The total current in the overall circuit (the current through the battery), however, is decreased by an amount equal to the current drawn by the lamp in question before it burned out. But the current in any other single branch is unchanged.

2. The light intensity for each lamp is unchanged as other lamps are introduced (or removed). Only the total resistance and total current in the total circuit changes, which is to say the current in the battery changes. (There is resistance in a battery also, which we assume is negligible here.) As lamps are introduced, more paths are available between the battery terminals, which effectively decreases total circuit resistance. This decreased resistance is accompanied by an increased current, the same increase that feeds energy to the lamps as they are introduced. Although changes of resistance and current occur for the circuit as a whole, no changes occur in any individual branch in the circuit.

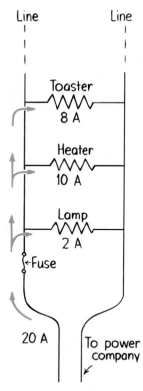

Figure 9.27 Circuit diagram for appliances connected to a household supply line.

plugs to these lines. As more devices are connected to the lines, more pathways for current result in lowering of the combined resistance of the circuit. Therefore, a greater amount of current occurs in the lines. Lines that carry more than a safe amount of current are said to be *overloaded*.

We can see how overloading occurs by considering the circuit in Figure 9.27. The supply line is connected to an electric toaster that draws 8 amperes, to an electric heater that draws 10 amperes, and to an electric lamp that draws 2 amperes. When only the toaster is operating and drawing 8 amperes, the total line current is 8 amperes. When the heater is also operating, the total line current increases to 18 amperes (8 amperes to the toaster and 10 amperes to the heater). If you turn on the lamp, the line current increases to 20 amperes. Connecting any more devices increases the current still more. Connecting too many devices into the same line results in overheating that may cause a fire.

Safety Fuses

To prevent overloading in circuits, fuses are connected in series along the supply line. In this way the entire line current must pass through the fuse. The fuse shown in Figure 9.28 is constructed with a wire ribbon that will heat up and melt at a given current. If the fuse is rated at 20 amperes, it will pass 20 amperes, but no more. A current above 20 amperes will melt the fuse, which "blows out" and breaks the circuit. Before a blown fuse is replaced, the cause of overloading should be determined and remedied. Often, insulation that separates the wires in a circuit wears away and allows the wires to touch. This effectively shortens the path of the circuit and is called a *short circuit*. Circuits may also be protected by circuit breakers, which use magnets or bimetallic strips to open the switch. Utility companies use circuit breakers to protect their lines all the way back to the generators.

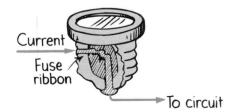

Figure 9.28 A safety fuse.

9.8 ELECTRIC POWER

The moving charges in an electric current do work. This work, for example, can heat a circuit or turn a motor. The rate at which work is done—that is, the rate at which electric energy is converted into another form such as mechanical energy, heat, or light—is called **electric power**. Electric power is equal to the product of current and voltage.*

---

*Recall from Chapter 3 that Power = work/time; 1 Watt = 1 J/s. Note that the units for mechanical power and electrical power check (work and energy are both measured in joules):

$$\text{Power} = \frac{\text{charge}}{\text{time}} \times \frac{\text{energy}}{\text{charge}} = \frac{\text{energy}}{\text{time}}$$

Power = current × voltage

If the voltage is expressed in volts and the current in amperes, then the power is expressed in watts. So, in units form,

Watts = amperes × volts

If a lamp rated at 120 watts operates on a 120-volt line, you can see that it will draw a current of 1 ampere (120 watts = 1 ampere × 120 volts). A 60-watt lamp draws $\frac{1}{2}$ ampere on a 120-volt line. This relationship becomes a practical matter when you wish to know the cost of electrical energy, which is usually a small fraction of a dollar per kilowatt-hour depending on locality. A kilowatt is 1000 watts, and a kilowatt-hour represents the amount of energy consumed in 1 hour at the rate of 1 kilowatt.* Therefore, in a locality where electric energy costs 25 cents per kilowatt-hour, a 100-watt electric light bulb can be run for 10 hours at a cost of 25 cents, or a half nickel for each hour. A toaster or iron, which draws much more current and therefore much more energy, costs several times as much to operate.

Figure 9.29 The power and voltage on the light bulb read "60 W 120 V." How many amperes will flow though the bulb?

 UESTIONS

1. If a 120-V line to a socket is limited to 15 A by a safety fuse, will it operate a 1200-W hair dryer?

2. At 30¢/kWh, what does it cost to operate the 1200-W hair dryer for 1 h?

ANSWERS

1. From the expression Watts = amperes × volts, we see that A = 1200 W/120 V = 10 A, so the hair dryer will operate when connected to the circuit. But two hair dryers on the same socket will blow the fuse.

2. 36¢ (1200 W = 1.2 kW; 1.2 kW × 1 h × 30¢/1 kWh = 36¢).

---

*Since Power = energy/time, simple rearrangement gives Energy = power × time; thus, energy can be expressed in the unit *kilowatt-hours* (kWh).

SUMMARY OF TERMS

Electrostatics The study of electric charges at rest relative to one another (not *in motion,* as in electric currents).

Coulomb's law The relationship among electrical force, charge, and distance:

$$F = k\frac{q_1 q_2}{d^2}$$

If the charges are alike in sign, the force is repelling; if the charges are unlike, the force is attractive.

Coulomb The SI unit of electrical charge. One coulomb (symbol C) is equal to the total charge of $6.25 × 10^{18}$ electrons.

Conductor Any material through which charge easily flows when subject to an external electrical force.

Superconductor Any material with zero electrical resistance, wherein electrons flow without losing energy and without generating heat.

Insulator Any material that resists charge flow through it when subject to an external electrical force.

Electrically polarized Term applied to an atom or molecule in which the charges are aligned so that one side is slightly more positive or negative than the opposite side.

Electric field The energetic region of space surrounding a charged object. About a charged point, the field decreases with distance according to the inverse-square law, like a gravitational field. Between oppositely charged parallel plates, the electric field is uniform. A charged object placed in the region of an electric field experiences a force.

Electric potential energy The energy a charge possesses by virtue of its location in an electric field.

Electric potential The electric potential energy per amount of charge, measured in volts, and often called *voltage:*

$$\text{Voltage} = \frac{\text{electric energy}}{\text{amount of charge}}$$

Potential difference The difference in voltage between two points, measured in volts. It can be compared to the difference in water pressure between two containers: If two containers having different water pressures are connected by a pipe, water will flow from the one with the higher pressure to the one with the lower pressure until the two pressures are equalized. Similarly, if two points with a difference in potential are connected by a conductor, charge will flow from the one with the greater potential to the one with the smaller potential until the potentials are equalized.

Electric current The flow of electric charge that transports energy from one place to another. Measured in amperes, where 1 A is the flow of 6.25×10^{18} electrons per second.

Electrical resistance The property of a material that resists the flow of an electric current through it. Measured in ohms (Ω).

Ohm's law The statement that the current in a circuit varies directly with the potential difference or voltage and inversely with resistance:

$$\text{Current} = \frac{\text{voltage}}{\text{resistance}}$$

A potential difference of 1 V across a resistance of 1 Ω produces a current of 1 A.

Direct current (dc) An electric current flowing in one direction only.

Alternating current (ac) Electric current that repeatedly reverses its direction; the electric charges vibrate about relatively fixed points. In the United States the vibrational rate is 60 Hz.

Electric power The rate of energy transfer, or the rate of doing work; the amount of energy per unit time, which electrically can be measured by the product of current and voltage:

$$\text{Power} = \text{current} \times \text{voltage}$$

Measured in watts (or kilowatts), where 1 A $\times$ 1 V = 1 W.

Series circuit An electric circuit with devices having resistances arranged in such a way that the same electric current flows through each of them.

Parallel circuit An electric circuit with two or more resistances arranged in branches in such a way that any single one completes the circuit independently of all the others.

REVIEW QUESTIONS

Electrical Force and Charge

1. How strong are electrical forces between an electron and a proton compared to gravitational forces between charged particles?

2. Why does gravitational force predominate over electrical force for astronomical bodies?

3. What part of an atom is *positively* charged and what part is *negatively* charged?

4. How does the charge of one electron compare to that of another electron?

5. How do the masses of electrons compare to the masses of protons? Neutrons?

6. How do the numbers of protons in the atomic nucleus normally compare to the number of electrons that orbit the nucleus?

7. What kind of charge does an object acquire when electrons are stripped from it?

8. What is meant by saying charge is *conserved?*

Coulomb's Law

9. How does a *coulomb* of charge compare to the charge of a *single* electron?

10. How is Coulomb's law similar to Newton's law of gravitation? How is it different?

11. How does the magnitude of electrical force between a pair of charges change when the charges are moved twice as far apart? Three times as far apart?

12. Why are metals good conductors of heat and electricity?

13. Why are materials such as glass and rubber good insulators?

14. How does an electrically *polarized* object differ from an electrically *charged* object?

15. A piece of paper becomes polarized in the presence of, say, a negative charge. The positive side of the paper is attracted to the negative charge, and the negative side of the paper is repelled by the negative charge. So why don't these forces cancel out?

Electric Field

16. Give two examples of common force fields.

17. How is the direction of an electric field defined?

18. How is the magnitude of an electric field defined?

Electric Potential

19. Distinguish between *electric potential energy* and *electric potential*.

20. A balloon may easily be charged to several thousand volts. Does that mean it has several thousand joules of energy? Explain.

21. Why does Meidor's hair stand out in Figure 9.29?

22. What condition is necessary for the flow of heat energy to occur from one end of a metal bar to the other? For the flow of electric charge?

23. For electric circuits, what is analogous to the statement "Water seeks its own level"?

24. What condition is necessary for a sustained flow of electric charge through a conducting medium?

Voltage Sources

25. How much energy is given to each coulomb of charge passing through a 6-V battery?

26. Does current flow *across* a circuit or *through* a circuit? Does voltage *flow* across a circuit or is it *established* across a circuit? Explain.

27. Does voltage produce current or does current produce voltage? Which is the cause and which is the effect?

Electric Current

28. Why do electrons rather than protons make up the flow of charge in a metal wire?

29. Does voltage flow through a circuit? Explain.

Direct Current and Alternating Current

30. Distinguish between *dc* and *ac*.

31. Which produces dc, a battery or a simple generator? Which produces ac?

Electrical Resistance

32. Which has the greater resistance, a thick wire or a thin wire of the same length?

Ohm's Law

33. What is the effect on current through a circuit of steady resistance when the voltage is doubled? What if both voltage and resistance are doubled?

34. How much current will flow through a radio speaker of 8 Ω resistance when 12 V are impressed across it?

Electric Shock

35. Which has the greater electrical resistance, wet skin or dry skin?

36. High voltage by itself does not produce electric shock. What does?

37. What is the function of the third prong on the plug of an electric appliance?

38. From where do the electrons originate that produce an electric shock when you touch a charged conductor?

Electric Circuits

39. What exactly is an electric circuit, and what is the role of a gap in an electric circuit?

Series Circuit

40. In a circuit of two lamps in series, if the current through one lamp is 1 A, what is the current through the other lamp?

41. If 6 V are impressed across the above circuit and the voltage across the first lamp is 2 V, what is the voltage across the second lamp?

Parallel Circuits

42. In a circuit of two lamps in parallel, if there are 6-V across one lamp, what is the voltage across the other lamp?

43. If the current through each of the two branches of a parallel circuit is the same, what does this tell you about the resistance of the two branches?

44. How does the total current through the branches of a simple parallel circuit compare to the current that flows through the voltage source?

45. As more doors are opened in a crowded room, the resistance to the motion of people trying to leave the room is reduced. How is this similar to what happens when more branches are added to a parallel circuit?

46. Are household circuits normally wired in series or in parallel?

Parallel Circuits and Overloading

47. How does the amount of line current in a home circuit differ from the amount of current that lights the lamp you are likely reading this by?

48. Why will too many electrical devices operating at one time often blow a fuse?

Electric Power

49. What is the relationship among electric power, current, and voltage?

50. Which draws more current, a 40-W bulb or a 100-W bulb?

HOME PROJECTS

1. Demonstrate charging by friction and discharging from points with a friend who stands at the far end of a carpeted room. With leather shoes, scuff your way across the rug until your noses are close together. This can be a delightfully tingling experience, depending on how dry the air is and how pointed your noses are.

2. Briskly rub a comb on your hair or a woolen garment and bring it near a small but smooth stream of running water. Is the stream of water charged? (Before you say yes, note the behavior of the stream when an opposite charge is brought nearby.)

3. An electric cell is made by placing two plates of different materials that have different affinities for electrons in a conducting solution. You can make a simple 1.5-V cell by placing a strip of copper and a strip of zinc in a tumbler of salt water. The voltage of a cell depends on the materials used and the solution they are placed in, not the size of the plates. A battery is actually a series of cells.

    An easy cell to construct is the citrus cell. Stick a paper clip and a piece of copper wire into a lemon. Hold the ends of the wire close together, but not touching, and place the ends on your tongue. The slight tingle you feel and the metallic taste you experience result from a slight current of electricity pushed by the citrus cell through the wires when your moist tongue closes the circuit.

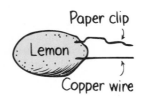

Paper clip

Lemon

Copper wire

4. Examine the electric meter in your house. It is probably in the basement or on the outside of your house. You will see that in addition to the clocklike dials in the meter, there is a circular aluminum disk that spins between the poles of magnets when electric current goes into the house. The more electric current, the faster the disk turns. The speed of the disk is directly proportional to the number of watts used; for example, it will spin five times as fast for 500 W as for 100 W.

    You can use the meter to determine how many watts an electrical device uses. First, see that all electrical devices in your home are disconnected (you may leave electric clocks connected, for the 2 watts they use will hardly be noticeable). The disk will be practically stationary. Then connect a 100-W bulb and note how many seconds it takes for the disk to make five complete revolutions. The black spot painted on the edge of the disk makes this easy. Disconnect the 100-W bulb and plug in a device of unknown wattage. Again, count the seconds for five revolutions. If it takes the same time, its a 100-W device; if it takes twice the time, its a 50-W device; half the time, a 200-W device; and so forth. In this way you can estimate the power consumption of devices fairly accurately.

EXERCISES

1. We do not feel the gravitational forces between ourselves and the objects around us because these forces are extremely small. Electrical forces, in comparison, are extremely huge. Since we and the objects around us are composed of charged particles, why don't we usually feel electrical forces?

2. Why do clothes often cling together after tumbling in a clothes dryer?

3. When combing your hair, you scuff electrons from your hair onto the comb. Is your hair then positively or negatively charged? How about the comb?

4. An electroscope is a simple device consisting of a metal ball that is attached by a conductor to two fine gold leaves that are protected from air disturbances in a jar, as shown. When the ball is touched by a charged body, the leaves that normally hang straight down spread apart. Why? (Electroscopes are useful not only as charge detectors, but also for measuring the quantity of charge: the more charge transferred to the ball, the more the leaves diverge.)

5. The leaves of a charged electroscope collapse in time. At higher altitudes they collapse more readily. Why is this true? (Hint: The existence of cosmic rays was first indicated by this observation.)

6. It is relatively easy to strip the outer electrons from a heavy atomic nucleus like that of uranium (which is then a uranium ion) but very difficult to strip the inner electrons. Why do you suppose this is so?

7. How does the magnitude of electric force compare between a pair of charged particles when they are brought to half their original distance of separation? To one-

quarter their original distance? To four times their original distance? (What law guides your answers?)

8. Why is a good conductor of electricity also a good conductor of heat?

9. If you put in 10 J of work to push a unit charge against an electric field, what will be its voltage with respect to its starting position? When released, what will be its kinetic energy if it flies past its starting position?

10. If you rub an inflated balloon against your hair and bring it in the vicinity of uncharged bits of paper, the bits of paper will be attracted to the balloon. Why?

11. You are not harmed by contact with a charged balloon, even though its voltage may be very high. Is the reason similar to why you are not harmed by the greater-than-1000°C sparks from a 4th-of-July type sparkler? Defend your answer by comparing electric potential to temperature.

12. What is the voltage at the location of a 0.0001 C charge that has an electric potential energy of 0.5 J?

13. If a conductor is placed in contact with two separated objects charged to different electrical potential energies, can you say for certain which way charge will flow in the conductor? How about if the objects are charged to different electric potentials?

14. One example of a water system is a garden hose that waters a garden. Another is the cooling system of an automobile. Which of these exhibits behavior more analogous to that of an electric circuit? Why?

15. What happens to the brightness of light emitted by a lightbulb when the current that flows in it increases?

16. Your tutor tells you that an *ampere* and a *volt* really measure the same thing, and the different terms only serve to make a simple concept seem confusing. Why should you consider getting a different tutor?

17. A simple lie detector consists of an electric circuit, one part of which is part of your body—like between your fingers. A sensitive meter shows the current that flows when a small voltage is applied. How does this technique indicate that a person is lying? (And when does this technique not tell when someone is lying?)

18. Only a small percentage of the electric energy fed into a common light bulb is transformed into light. What happens to the rest?

19. Why are thick wires rather than thin wires usually used to carry large currents?

20. Will a lamp with a thick or thin filament of the same length draw the most current?

21. Will the current in a light bulb connected to a 220-V source be greater or less than when the same bulb is connected to a 110-V source?

22. Which will do less damage—plugging a 110-V appliance into a 220-V circuit or plugging a 220-V appliance into a 110-V circuit? Explain.

23. Would the resistance of a 110-W bulb be greater or less than the resistance of a 60-W bulb? Assuming the filaments in each bulb are of the same length, which bulb has the thicker filament?

24. Would you expect to find dc or ac in the filament of a light bulb in your home? How about in an automobile?

25. The wattage marked on a light bulb is not an inherent property of the bulb but depends on the voltage to which it is connected, usually 110 or 120 V. How many amperes flow through a 60-W bulb connected in a 120-V circuit?

26. The damaging effects of electric shock result from the amount of current that flows in the body. Why, then, do we see signs that read "Danger—High Voltage" rather than "Danger—High Current"?

27. If electrons flow very slowly through a circuit, why does it not take a noticeably long time for a lamp to glow when you turn on a distant switch?

28. If a glowing light bulb is jarred and oxygen leaks inside, the bulb will momentarily brighten considerably before burning out. Putting excess current through a light bulb will also burn it out. What physical change occurs when a light bulb burns out?

29. Consider a pair of flashlight bulbs connected to a battery. Will they each glow brighter connected in series or in parallel? Will the battery run down faster if they are connected in series or in parallel?

30. In the circuit shown, how do the brightnesses of the identical light-bulbs compare? Which lightbulb draws the most current? What will happen if bulb A is unscrewed? If C is unscrewed?

31. As more and more bulbs are connected in series to a flashlight battery, what happens to the brightness of each bulb? Assuming heating inside the battery is negligible, what happens to the brightness of each bulb when more and more bulbs are connected in parallel?

32. What changes occur in the line current when more devices are introduced in a series circuit? In a parallel circuit? Why are your answers different?

33. It so happens that if too great a load is placed on a battery, the internal resistance of the battery is increased. This lowers the voltage that is supplied to the external circuit. If too many lamps are connected in parallel across a battery, will their brightness diminish? Explain.

34. Why are devices in household circuits almost never connected in series?

35. If a 60-W bulb and a 100-W bulb are connected in series in a circuit, through which bulb will there be the greater voltage drop? How about if they are connected in parallel?

36. Your friend says that power companies sell energy, not power. Do you agree or disagree with your friend?

• •

PROBLEMS

1. Rearrange the equation: Current = voltage/resistance to express *resistance* in terms of current and voltage. Then solve the following: A certain device in a 120-V circuit has a current rating of 20 A. What is the resistance of the device (how many ohms)?

2. Using the formula: Power = current × voltage, find the current drawn by a 1200-W hair dryer connected to 120 V. Then using the method you used in the previous exercise, find the resistance of the hair dryer.

3. The useful life an automobile battery has without being recharged is given in terms of ampere-hours. A typical 12-V battery has a rating of 60 ampere-hours, which means that a current of 60 A can be drawn for 1 h, 30 A can be drawn for 2 h, and so forth. Suppose you forget to turn off the headlights in your parked automobile. If each of the two headlight draws 3 A, how long will it be before your battery is "dead"?

4. How much does it cost to operate a 100-W lamp continuously for 1 month if the power utility rate is 20¢/kWh?

5. A 4-W night light is plugged into a 120-V circuit and operates continuously for 1 year. Find the following: (a) the current it draws, (b) the resistance of its filament, (c) the energy consumed in a year, and (d) the cost of its operation for a year at the utility rate of 20¢/kWh.

10

MAGNETISM

The term *magnetism* comes from the region of Magnesia, an island in the Aegean Sea, where certain stones were found by the Greeks more than 2000 years ago. These stones, called *lodestones,* had the unusual property of attracting pieces of iron. Magnets were first fashioned into compasses and used for navigation by the Chinese in the twelfth century.

In the sixteenth century, William Gilbert, Queen Elizabeth's physician, made artificial magnets by rubbing pieces of iron against lodestone, and suggested that a compass always points north and south because the earth itself has magnetic properties. Later, in 1750, John Michell in England found that magnetic poles obey the inverse-square law, and his results were confirmed by Charles Coulomb. The subjects of magnetism and electricity developed independently of each other until 1820, when a Danish physicist named Hans Christian Oersted discovered in a classroom demonstration that an electric current affects a magnetic compass.*

---

*We can only speculate about how often such relationships become evident when they "aren't supposed to" and are dismissed as "something wrong with the apparatus." Oersted, however, was keen enough to see that nature was revealing another of its secrets.

He saw that magnetism was related to electricity. Shortly thereafter, the French physicist André-Marie Ampere proposed that electric currents are the source of all magnetism.

10.1 MAGNETIC POLES

Figure 10.1 A horseshoe magnet.

Anyone who has played around with magnets knows that magnets exert forces on one another. Magnetic forces are similar to electrical forces, for they can both attract and repel without touching, depending on which ends of the magnets are held near one another. And similar to electrical forces, the strength of their interaction depends on the distance the two magnets are separated. Whereas electric charges produce electrical forces, regions called *magnetic poles* produce magnetic forces.

Suspend a bar magnet at its center by a piece of string and you've got a compass. One end points northward, called the *north-seeking pole,* and the opposite end points southward, called the *south-seeking pole.* More simply, these are called the *north* and *south poles.* All magnets have both a north pole and a south pole. In a simple bar magnet these are located at the two ends. A common horseshoe magnet is simply a bar magnet that has been bent into a U shape. Its poles are also at its two ends.

When the north pole of one magnet is brought near the north pole of another magnet, they repel.* The same is true of a south pole near a south pole. If opposite poles are brought together, however, attraction occurs. We find that

Like poles repel; opposite poles attract.

This rule is similar to the rule for the forces between electric charges, where like charges repel one another and unlike charges attract. But there is a very important difference between magnetic poles and electric charges. Whereas electric charges can be isolated, magnetic poles cannot. Electrons and protons are entities by themselves. A cluster of electrons need not be accompanied by a cluster of protons and vice versa. But a north magnetic pole never exists without the presence of a south pole and vice versa. The north and south poles of a magnet are like the head and tail of the same coin.

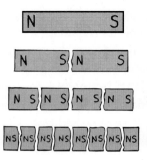

Figure 10.2 Break a magnet in half and you have two magnets. Break these in half and you have four magnets, each with a north and south pole. Keep breaking the pieces further and further and you find the same results. Magnetic poles exist in pairs.

---

*The force of interaction between magnetic poles is given by $F \sim \frac{p_1 p_2}{d^2}$, where p_1 and p_2 represent magnetic pole strengths and d represents the separation distance between the poles. Note the similarity of this relationship to Coulomb's law.

If you break a bar magnet in half, each half still behaves as a complete magnet. Break the pieces in half again, and you have four complete magnets. You can continue breaking the pieces in half and never isolate a single pole. Even when your piece is one atom thick, there are two poles. This suggests that atoms themselves are magnets.

QUESTIONS

1. Does every magnet necessarily have a north and south pole?
2. Is there a limit to the number of times you can break a magnet in half only to get more magnets?

10.2 MAGNETIC FIELDS

Sprinkle some iron filings on a sheet of paper placed on a magnet and you'll see that the filings trace out an orderly pattern of lines that surround the magnet. The space around the magnet contains a magnetic field. The shape of the field is revealed by magnetic field lines that spread out from one pole and return to the other pole. It is interesting to compare the field patterns in Figures 10.3 and 10.4 with the electric field patterns in Figure 9.7 in the previous chapter.

Figure 10.3 Top view of iron filings sprinkled on a sheet of paper on top of a magnet. The filings trace out a pattern of *magnetic field lines* in the surrounding space. Interestingly enough, the magnetic field lines continue inside the magnet (not revealed by the filings) and form closed loops.

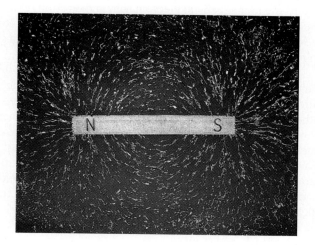

ANSWERS

1. Yes, just as every coin has two sides, a head and a tail. Some "trick" magnets may have more than one pair of poles, but nevertheless poles occur in pairs.

2. The limit is when you get down to the actual particles responsible for magnetism, which are the protons and electrons of the atom. Whether protons and electrons themselves are divisible is a good question and the focus of much scientific research.

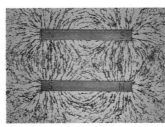 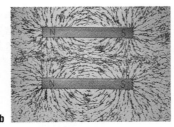

a b

Figure 10.4 The magnetic field patterns for a pair of magnets. (a) Opposite poles are parallel to each other, and (b) like poles are parallel to each other.

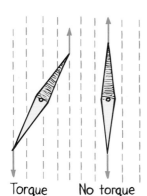

Torque No torque

Figure 10.5 When the compass needle is not aligned with the magnetic field, the oppositely directed forces produce a pair of troques (called a *couple*) that twist the needle into alignment.

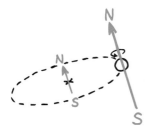

Figure 10.6 Both the spinning motion and the orbital motion of every electron in an atom produce magnetic fields. These fields combine constructively or destructively to produce the magnetic field of the atom. The resulting field is greatest for iron atoms.

The direction of the field outside the magnet is from the north to the south pole. Where the lines are closer together, the field is stronger. We see the magnetic field strength is greater at the poles. If we place another magnet or a small compass anywhere in the field, its poles will tend to line up with the magnetic field.

Magnetism is very much related to electricity. Just as an electric charge is surrounded by an electric field, the same charge is also surrounded by a magnetic field if it is moving. This is due to the "distortions" in the electric field caused by motion, and was explained by Albert Einstein in 1905 in his special theory of relativity. We won't go into the details, except to acknowledge that a magnetic field is a relativistic byproduct of the electric field. Charges in motion have associated with them both an electric and a magnetic field. A magnetic field is produced by the motion of electric charge.*

If the motion of electric charges produces magnetism, where is this motion in a common bar magnet? The answer is: in the electrons of atoms that make up the magnet. These electrons are in constant motion. Two kinds of electron motion contribute to magnetism: electron spin and electron revolution. Electrons spin about their own axes like tops, and they revolve about the atomic nucleus like planets revolve about the sun. In most common magnets, electron spin is the chief contributor to magnetism.

Every spinning electron is a tiny magnet. A pair of electrons spinning in the same direction makes up a stronger magnet. A pair of electrons spinning in opposite directions, however, has the opposite effect. The magnetic fields of each cancel the other. This is why most substances are not magnets. In most atoms, the various fields cancel one another because the electrons spin in opposite directions. In materials such as iron, nickel, and cobalt, however, the fields do not cancel each other entirely. Each iron atom has four electrons whose spin magnetism is uncancelled. Each iron atom, then, is a tiny magnet. The same is true to a lesser extent for the atoms of nickel and cobalt. Most common magnets are therefore made from alloys containing iron, nickel, cobalt, and aluminum in various proportions.

---

*Interestingly enough, since motion is relative, the magnetic field is relative. For example, when a charge moves by you, there is a definite magnetic field associated with the moving charge. But if you move along with the charge, so there is no motion relative to you, you will find no magnetic field associated with the charge. Magnetism is relativistic, as first explained by Albert Einstein when he published his first paper on special relativity, "On the Electrodynamics of Moving Charges." (More on relativity in Chapter 30.)

10.3 MAGNETIC DOMAINS

Figure 10.7 A microscopic view of magnetic domains in a crystal of iron. Each domain consists of billions of aligned iron atoms.

The magnetic field of an individual iron atom is so strong that interaction among adjacent atoms causes large clusters of them to line up with one another. These clusters of aligned atoms are called **magnetic domains.** Each domain is perfectly magnetized and is made up of billions of aligned atoms. The domains are microscopic (Figure 10.7), and there are many of them in a crystal of iron.

Not every piece of iron, however, is a magnet. This is because the domains in ordinary iron are not aligned. Consider a common iron nail: The domains in the nail are randomly oriented. They can be induced into alignment, however, when a magnet is brought nearby. (It is interesting to listen with an amplified stethoscope to the clickety-clack of domains undergoing alignment in a piece of iron when a strong magnet approaches.) The domains align themselves much as electrical charges in a piece of paper align themselves in the presence of a charged rod. When you remove the nail from the magnet, ordinary thermal motion causes most or all of the domains in the nail to return to a random arrangement.

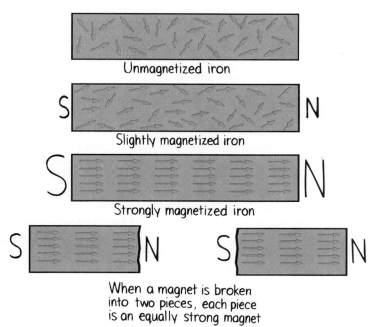

Unmagnetized iron

Slightly magnetized iron

Strongly magnetized iron

When a magnet is broken into two pieces, each piece is an equally strong magnet

Figure 10.8 Pieces of iron in successive stages of magnetism. The arrows represent domains; the head is a north pole and the tail a south pole. Poles of neighboring domains neutralize each other's effects, except at the ends.

Figure 10.9 The iron nails become induced magnets.

Permanent magnets can be made by placing pieces of iron or paramagnetic materials in strong magnetic fields. Alloys of iron differ; soft iron is easier to magnetize than steel. It helps to tap the material to nudge any stubborn domains into alignment. Another way is to stroke the material with a magnet. The stroking motion aligns the domains. If a permanent magnet is dropped or heated outside of the strong magnetic field from which it was made, some of the domains are jostled out of alignment and the magnet becomes weaker.

Magnetism in Your Home

Bring a magnetic compass near the tops of iron or steel objects in your home (radiators, refrigerators, stoves, lamps, etc.). You'll find that the north pole of the compass needle points to the tops of these objects, and the south pole of the compass needle points to the bottoms. The iron objects have been in one place in the earth's magnetic field and have become slightly magnetized. They have become magnets, having a south pole on top and a north pole on the bottom. (If you live in the southern hemisphere, the opposite will be the case.) You must do this to appreciate it!

Even more interesting, place your compass alongside cans of stored food on your kitchen or pantry shelf. You'll find that cans are magnetized. Turn one over and test to see how many days it takes to lose its magnetism, and then to reverse its polarity.

UESTION

How can a magnet attract a piece of iron that is not magnetized?

10.4 ELECTRIC CURRENTS AND MAGNETIC FIELDS

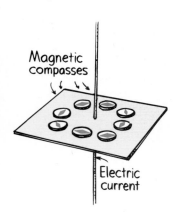

Magnetic compasses

Electric current

Figure 10.10 The compasses show the circular shape of the magnetic field surrounding the current-carrying wire.

A moving charge produces a magnetic field. A current of charges, then, also produces a magnetic field. The magnetic field that surrounds a current-carrying wire can be demonstrated by arranging an assortment of compasses around the wire (Figure 10.10). The magnetic field about the current-carrying wire makes up a pattern of concentric circles. When the current reverses direction, the compass needles turn around, showing that the direction of the magnetic field changes also.*

If the wire is bent into a loop, the magnetic field lines become bunched up inside the loop (Figure 10.11). If the wire is bent into another loop, overlapping the first, the concentration of magnetic field lines inside the loops are doubled. It follows that the magnetic field intensity in this region is increased as the number of loops is increased. The magnetic field intensity is appreciable for a current-carrying coil of many loops.

A NSWER

Like the compass needle in Figure 10.4, domains in the unmagnetized piece of iron are induced into alignment by the magnetic field of the nearby magnet. One domain pole is attracted to the magnet and the other domain pole is repelled. Does this mean the net force is zero? No, because the domain pole that is attracted is closer than the pole repelled. Closeness wins, and there is a net attraction. In this way a magnet will attract non-magnetized pieces of iron (Figure 10.9). (This induction of pole alignment is similar to the way a charged comb attracts bits of uncharged paper.)

---

*Earth's magnetism is thought to be the result of electric currents that accompany thermal convection in the molten parts of the earth's interior. Earth scientists have found evidence that the earth's poles periodically reverse places—more than 20 reversals in the past 5 million years. This is perhaps the result of changes in the direction of electric currents within the earth. More about this in Chapter 27.

Figure 10.11 Magnetic field lines about a current-carrying wire crowd up when the wire is bent into a loop.

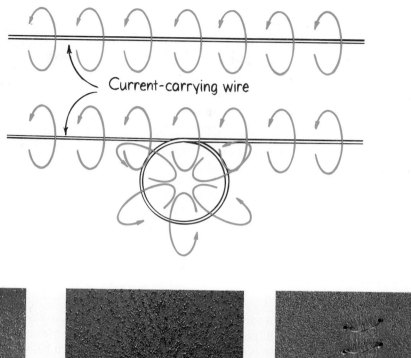

Current-carrying wire

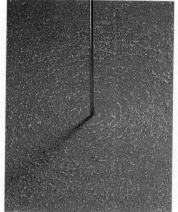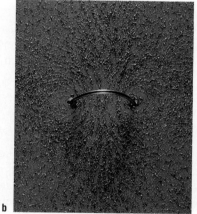

a   b   c

Figure 10.12 Iron filings sprinkled on paper reveal the magnetic field configurations about (a) a current-carrying wire, (b) a current-carrying loop, and (c) a coil of loops.

Electromagnets

If a piece of iron is placed in a current-carrying coil of wire, the magnetic domains in the iron are induced into alignment. This further increases the magnetic field intensity, and we have an **electromagnet!** The strength of an electromagnet is increased by simply increasing the current. Strong electromagnets are used to control charged particle beams in high-energy accelerators such as the supercollider. They also levitate and propel high-speed trains (Figure 10.13).

    Electromagnets powerful enough to lift automobiles are a common sight in junkyards. The strength of these electromagnets is limited by overheating of the

Figure 10.13 Conventional trains vibrate as they ride on rails at high speeds. This Japanese magnetically levitated train is capable of vibration-free high speeds, even in excess of 200 km/h.

Superconducting Electromagnets

The relatively new ceramic superconductors (see the box in previous chapter) have the interesting property of expelling magnetic fields. Because magnetic fields cannot penetrate the surface of a superconductor, magnets will levitate above them (Figure 10.14). The reasons for this behavior are presently being researched, and seem explainable only in terms of quantum mechanics. One of the hot applications of superconducting electromagnets is the levitation of high-speed trains for transportation. Prototype trains have already been demonstrated in Japan and Germany (Figure 10.13). Watch for the growth of this relatively new technology.

Figure 10.14 A permanent magnet levitates above a superconductor because its magnetic field cannot penetrate the superconducting material.

current-carrying coils (due to electrical resistance) and saturation of magnetic domain alignment in the core. The most powerful electromagnets omit the iron core and use superconducting coils through which large electrical currents easily flow.

10.5 MAGNETIC FORCES ON MOVING CHARGES

A charged particle at rest will not interact with a static magnetic field. But if the charged particle moves in a magnetic field, the magnetic character of a charge in motion becomes evident. It experiences a deflecting force.* The force is greatest when the particle moves in a direction perpendicular to the magnetic field lines. At other angles, the force is less and becomes zero when the particle moves parallel to the field lines. In any case, the direction of the force is always perpendicular to the magnetic field lines and the velocity of the charged particle (Figure 10.15). So a moving charge is deflected when it crosses through a magnetic field, but when it travels parallel to the field no deflection occurs.

This sideways deflection is very different from the forces that occur in other interactions like the gravitation between masses, the electrostatic forces between charges, and the forces between magnetic poles. The force that acts on a moving charged particle does not act along the line that joins the sources of interaction, but instead acts perpendicularly to both the magnetic field and the electron beam.

We are fortunate that charged particles are deflected by magnetic fields. This fact is employed to spread electrons onto the inner surface of a TV tube and provide a picture. More interesting, charged particles from outer space are deflected by the

---

*When particles of electric charge q and velocity v move perpendicularly into a magnetic field of strength B, the force F on each particle is simply the product of the three variables: $F = qvB$. For nonperpendicular angles, v in this relationship must be the component of velocity perpendicular to B.

Figure 10.15   A beam of electrons is deflected by a magnetic field.

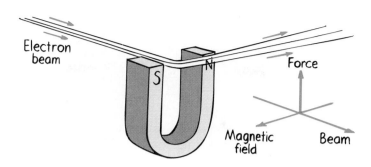

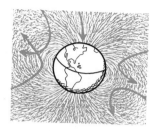

Figure 10.16   The magnetic field of the earth deflects many charged particles that make up cosmic radiation.

earth's magnetic field. The intensity of harmful cosmic rays bombarding the earth's surface would be more intense otherwise.

Magnetic Force on Current-Carrying Wires

Simple logic tells you that if a charged particle moving through a magnetic field experiences a deflecting force, then a current of charged particles moving through a magnetic field experiences a deflecting force also. If the particles are trapped inside a wire when they respond to the deflecting force, the wire will also move (Figure 10.17).

If we reverse the direction of current, the deflecting force acts in the opposite direction. The force is strongest when the current is perpendicular to the magnetic field lines. The direction of force is not along the magnetic field lines nor along the direction of current. The force is perpendicular to both field lines and current. It is a sideways force.

We see that just as a current-carrying wire will deflect a magnetic compass, as discovered by Oersted in a high-school classroom in 1820, a magnet will deflect a current-carrying wire. Both cases show different effects of the same phenomenon. This discovery created much excitement, for almost immediately people began harnessing this force for useful purposes—with great sensitivity in electric meters and with great force in electric motors.

Electric Meters

The simplest meter to detect electric current is simply a magnetic compass. The next most simple meter is a compass in a coil of wires (Figure 10.18). When an electric current passes through the coil, each loop produces its own effect on the needle, so a very small current can be detected. A current-indicating instrument is called a *galvanometer.*

Figure 10.17
A current-carrying wire experiences a force in a magnetic field. (Can you see this is a simple extension of Figure 10.15?)

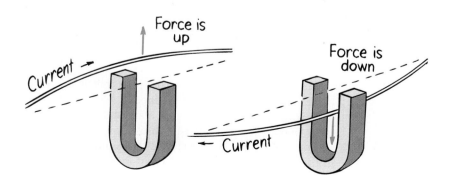

Figure 10.18 A very simple galvanometer.

A more common design is shown in Figure 10.19. It employs more loops of wire and is therefore more sensitive. The coil is mounted for movement and the magnet is held stationary. The coil turns against a spring, so the greater the current in its windings, the greater its deflection. A galvanometer may be calibrated to measure current (amperes), in which case it is called an *ammeter*. Or it may be calibrated to measure electric potential (volts), in which case it is called a *voltmeter*.

Figure 10.19 A common galvanometer design.

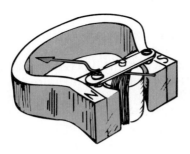

Electric Motors

If we modify the design of the galvanometer slightly, we have an electric motor. The principal difference is that the current is made to change direction every time the coil makes a half rotation. After being forced to turn one half rotation, it overshoots just in time for the current to reverse, whereupon it is forced to continue another half rotation, and so on in cyclic fashion to produce continuous rotation.

Figure 10.20 Both the ammeter and the voltmeter are basically galvanometers. (The electrical resistance of the instrument is made to be very low for the ammeter, and very high for the voltmeter.)

In Figure 10.21 we see the principle of the electromagnetic motor in bare outline. A permanent magnet produces a magnetic field in a region where a rectangular loop of wire is mounted to turn about the axis shown. When a current passes through the loop, it flows in opposite directions in the upper and lower sides of the loop (it has to do this because if charge flows into one end of the loop, it must flow out the other end). If the upper portion of the loop is forced to the left, then the lower portion is forced to the right, as if it were a galvanometer. But unlike a galvanometer, the

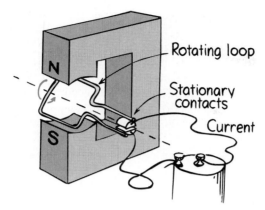

Figure 10.21 A simplified motor.

current is reversed during each half revolution by means of stationary contacts on the shaft. The parts of the wire that brush against these contacts are called *brushes*. In this way, the current in the loop alternates so that the forces in the upper and lower regions do not change directions as the loop rotates. The rotation is continuous as long as current is supplied.

We have described here only a very simple dc motor. Larger motors, dc or ac, are usually made by replacing the permanent magnet by an electromagnet that is energized by the power source. Of course, more than a single loop is used. Many loops of wire are wound about an iron cylinder, called an *armature*, which then rotates when energized with electric current.

The advent of electric motors, needless to say, brought about the replacement of enormous human and animal toil, the world over. Electric motors have greatly changed the way people live.

UESTION

What is the major similarity between a galvanometer and a simple electric motor? What is the major difference?

10.6 ELECTROMAGNETIC INDUCTION

In the early 1800s, the only current-producing devices were voltaic cells, which produced small currents by dissolving metals in acids. These were the forerunners of our present-day batteries. The question arose as to whether electricity could be produced from magnetism. The answer was provided in 1831 by two physicists, Michael

ANSWER

A galvanometer and a motor are similar in that they both employ coils positioned in a magnetic field. When a current passes through the coils, forces on the wires rotate the coils. The fundamental difference is that the maximum rotation of the coil in a galvanometer is one half turn, whereas in a motor the coil (armature) rotates through many complete turns. This is accomplished by alternating this current with each half turn of the armature.

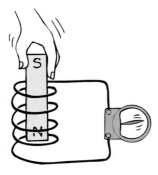

Figure 10.22 When the magnet is plunged into the coil, charges in the coil are set in motion; voltage is induced in the coil.

Faraday in England and Joseph Henry in the United States—each working without knowledge of the other. Their discovery changed the world by making electricity commonplace—powering industries by day and lighting up cities at night.

Faraday and Henry both discovered that electric current could be produced in a wire by simply moving a magnet in or out of a coil of wire (Figure 10.22). No battery or other voltage source was needed—only the motion of a magnet in a wire loop. They discovered that voltage is caused or *induced* by the relative motion between a wire and a magnetic field. Voltage is induced whether the magnetic field of a magnet moves near a stationary conductor or the conductor moves in a stationary magnetic field (Figure 10.23). The results are the same whether either or both move.

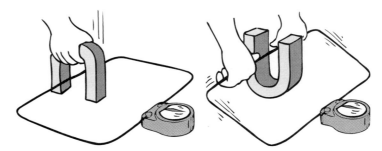

Figure 10.23 Voltage is induced in the wire loop whether the magnetic field moves past the wire or the wire moves through the magnetic field.

The greater the number of loops of wire that move in a magnetic field, the greater the induced voltage (Figure 10.24). Pushing a magnet into twice as many loops will induce twice as much voltage; pushing into ten times as many loops will induce ten times as much voltage; and so on. It may seem that we get something (energy) for nothing by simply increasing the number of loops in a coil of wire. But we don't: We find it is more difficult to push the magnet into a coil with more loops. This is because the induced voltage makes a current, which makes an electromagnet,

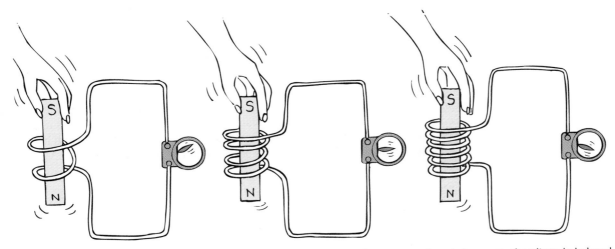

Figure 10.24 When a magnet is plunged into a coil of twice as many loops as another, twice as much voltage is induced. If the magnet is plunged into a coil with three times as many loops, then three times as much voltage is induced.

Figure 10.25 It is more difficult to push the magnet into a coil with many loops because the magnetic field of each current loop resists the motion of the magnet.

which repels the magnet in our hand. So we do more work to induce more voltage (Figure 10.25). The amount of voltage induced depends on how fast the magnetic field lines are entering or leaving the coil. Very slow motion produces hardly any voltage at all. Quick motion induces a greater voltage. This phenomenon of inducing voltage by changing the magnetic field in a coil of wire is called **electromagnetic induction.**

Faraday's Law

Electromagnetic induction is summarized by **Faraday's law,** which states,

> **The induced voltage in a coil is proportional to the number of loops × the rate at which the magnetic field changes within those loops.**

The amount of *current* produced by electromagnetic induction depends on the resistance of the coil and the circuit that it connects, as well as the induced voltage.* For example, we can plunge a magnet in and out of a closed rubber loop and in and out of a closed loop of copper. The voltage induced in each is the same, providing each intercepts the same number of magnetic field lines. But the current in each is quite different. The electrons in the rubber sense the same voltage as those in the copper, but their bonding to the fixed atoms prevents the movement of charge that so freely occurs in the copper.

Q UESTION

If you push a magnet into a coil, as shown in Figure 10.25, you'll feel a resistance to your push. Why is this resistance greater in a coil with more loops?

Voltage can be induced in a loop of wire in three different ways: by moving the loop near a magnet, by moving a magnet near the loop, or by changing a current in a nearby loop. In each case we have the important ingredient—a change in the amount of magnetic field in the loop.

We see electromagnetic induction all around us. On the road we see it operate when a car drives over buried coils of wire to activate a nearby traffic light; when iron parts of a car move over the coils the earth's magnetic field in the coils is changed, which induces voltage to trigger the changing of the traffic lights. Similarly, when one walks through the upright coils in the security system at airports, metal will

A NSWER

Simply put, more work is required to induce the greater voltage in more loops. You can also look at it this way: when two magnets (electro- or permanent) are close to each other, the two magnets are either forced together or forced apart. In cases where one of the magnets is induced by motion of the other, the polarity of the fields is always such as to force the magnets apart. This is the resistive force you feel. Inducing more current in more coils simply increases the induced magnetic field strength and hence the resistive force.

---

*Current also depends on the "reactance" of the coil. Reactance is similar to resistance and is important in ac circuits; it depends on the number of loops in the coil and on the frequency of the ac source, among other things. We will not treat this complication in this book.

slightly alter the magnetic field in the coils, induce voltage, and sound an alarm. When the magnetic strip on the back of a credit card is scanned, changing magnetic fields induce voltages that identify the card. Similarly with the recording head of a tape recorder, when domains are aligned as blank tape moves by a current-carrying coil. Variations in the current are recorded as variations in domain alignment in the tape. On playback, the variations of magnetic field strength produce changes in the magnetic field that move by the playback coil. The variations in induced voltage match the variations that produced the domain alignment upon recording. The result is beautiful music, if that's what you started with. Of course the voltages induced are tiny, but even they are amplified with transformers, which themselves are based on electromagnetic induction. Electromagnetic induction is everywhere. As we will see in Chapter 12, it even underlies the electromagnetic waves we call light.

10.7 GENERATORS AND ALTERNATING CURRENT

When a magnet is plunged into and back out of a coil of wire, the direction of the induced voltage alternates. As the magnetic field strength inside the coil is increased (magnet entering), the induced voltage in the coil is directed one way. When the magnetic field strength diminishes (magnet leaving), the voltage is induced in the opposite direction. The greater the frequency of field change, the greater the induced voltage. The frequency of the alternating voltage induced is equal to the frequency of the changing magnetic field within the loop.

It is more practical to induce voltage by moving the coil rather than moving the magnet. This can be done by rotating the coil in a stationary magnetic field (Figure 10.26). This arrangement is called a **generator.** The construction of a generator is in principle identical to that of a motor. Only the roles of input and output are reversed. In a motor, electric energy is the input and mechanical energy the output; in a generator, mechanical energy is the input and electric energy the output. Both devices simply transform energy from one form to another.

We can see the details of electromagnetic induction in Figure 10.27. Note that when the loop of wire is rotated in the magnetic field, there is a change in the number of magnetic field lines within the loop. In (a) the loop has the largest number of lines inside it. As the loop rotates, (b), it encircles fewer of the field lines until at (c) the loop lies along the field lines and encloses none at all. As rotation continues, it encloses more field lines (d) and reaches a maximum of lines when

Figure 10.26 A simple generator. Voltage is induced in the loop when it is rotated in the magnetic field.

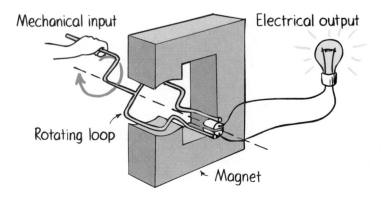

Mechanical input

Electrical output

Rotating loop

Magnet

Figure 10.27 As the loop rotates, there is a change in the number of magnetic field lines it encloses. It varies from a maximum at (a) to a minimum at (c) and back to a maximum again at (a). Induced voltage occurs at (c), where the greatest rate of change occurs.

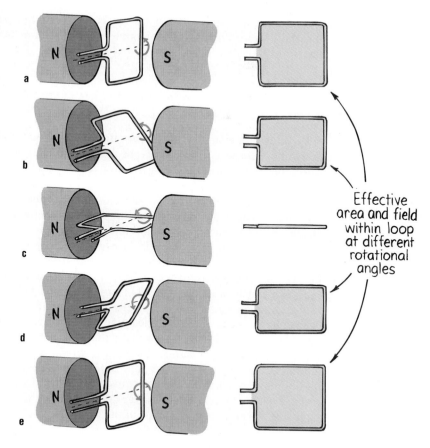

Effective area and field within loop at different rotational angles

it has made a half turn (e). As rotation continues, the magnetic field inside the loop changes in cyclic fashion, with the greatest rate of change of field lines as it goes through the zero point. Hence the induced voltage is greatest at these points (Figure 10.28). Because the voltage induced by the generator alternates, the current produced is ac, an alternating current.* It changes magnitude and direction periodically.

Figure 10.28 As the loop rotates, the magnitude and direction of the induced voltage (and current) changes. One complete rotation of the loop produces one complete cycle in voltage (and current).

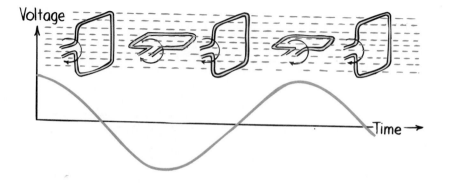

*With appropriate brushes and by other means, the ac in the loop(s) can be taken off as dc to make a dc generator.

The alternating current in our homes is produced by generators standardized so that the current changes its magnitude and direction 60 cycles per second—60 hertz.

10.8 POWER PRODUCTION

Fifty years after Faraday and Henry discovered electromagnetic induction, Nikola Tesla and George Westinghouse put their findings to practical use and showed the world that electricity could be generated reliably and in sufficient quantities to light entire cities.

Tesla built generators like those of today—quite a bit more complicated than the simple model we have discussed. Tesla's generators had armatures consisting of bundles of copper wires that were made to spin within strong magnetic fields by means of a turbine, which in turn was spun by the energy of falling water or steam. The rotating loops of wire in the armature cut through the magnetic field of the surrounding electromagnets, thereby inducing alternating voltage and current.

We can look at this process from an atomic point of view. When the wires in the spinning armature cut through the magnetic field, oppositely directed electromagnetic forces act on the negative and positive charges. Electrons respond to this force by momentarily swarming relatively freely in one direction throughout the crystalline copper lattice; the copper atoms, which are actually positive ions, are forced in the opposite direction. But the ions are anchored in the lattice, so they hardly move at all. Only the electrons move, sloshing back and forth in alternating fashion with each rotation of the armature. The energy of this electronic sloshing is tapped at the electrode terminals of the generator.

Generators, of course, don't produce energy—they simply convert energy from some other form to electric energy. As we discussed in Chapter 8, energy from the source, usually some type of fuel, is converted to mechanical energy to drive the turbine, and the generator converts most of this to electrical energy. The electricity produced simply carries this energy to distant places. Some people think that electricity is a primary source of energy. It is not. It is a form of energy that must have a source.

Figure 10.29 Steam drives the turbine, which is connected to the armature of the generator.

Steam

Boosting or Lowering Voltage—the Transformer

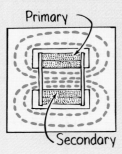

Figure 10.30 A simple transformer.

When changes in the magnetic field of a current-carrying coil of wire are intercepted by a second coil of wire, voltage is induced in the second coil. This is the principle of the **transformer**—a simple electromagnetic-induction device consisting of an input coil of wire (the primary) and an output coil of wire (the secondary). The coils need not physically touch each other, but are commonly wound on a common iron core so that the magnetic field of the primary passes through the secondary. The primary is powered by an ac voltage source, and the secondary is connected to some external circuit. Changes in primary current produce changes in its magnetic field. These changes extend to the secondary, and by electromagnetic induction, voltage is induced in the secondary. If the number of turns of wire in both coils is the same, voltage input and voltage output will be the same. Nothing is gained. But if the secondary has more turns than the primary, then greater voltage will be induced in the secondary. This is a *step-up transformer*. If the secondary has fewer turns than the primary, the ac voltage induced in the secondary will be lower than that in the primary. This is a *step-down transformer*.

The relationship between primary and secondary voltages relative to the number of turns is:

$$\frac{\text{Primary voltage}}{\text{Number of primary turns}}$$
$$= \frac{\text{secondary voltage}}{\text{number of secondary turns}}$$

It might seem we get something for nothing with a transformer that steps up the voltage. But we don't. When voltage is stepped up, the primary draws more current. The transformer actually transfers energy from one coil to the other. The rate of transferring energy is *power*. The power used in the secondary is supplied by the primary. The primary gives no more than the secondary uses, in accord with the law of energy conservation. If slight power losses due to heating of the core are neglected, then

Power into primary
= power out of secondary

Electric power is equal to the product of voltage and current, so we can say

$(\text{Voltage} \times \text{current})_{\text{primary}}$
$= (\text{voltage} \times \text{current})_{\text{secondary}}$

The ease with which voltages can be stepped up or down with a transformer is the principal reason that most electric power is ac rather than dc.

Figure 10.31 A practical transformer. The iron coil guides the changing magnetic field lines.

10.9 FIELD INDUCTION

Electromagnetic induction has thus far been discussed as the production of voltages and currents. Actually, the more fundamental *fields* underlie both voltages and currents. The modern view of electromagnetic induction holds that electric and magnetic fields are induced, which in turn give rise to the voltages we have considered. Induction takes place whether or not a conducting wire or any material medium is present. In this more general sense, Faraday's law states:

An electric field is induced in any region of space in which a magnetic field is changing with time. The magnitude of the induced electric field is propor-

tional to the rate at which the magnetic field changes. The direction of the induced electric field is at right angles to the changing magnetic field.

There is a second effect, which is the counterpart to Faraday's law. It is the same as Faraday's law, except that the roles of electric and magnetic fields are interchanged. It is one of the many symmetries in nature. This effect was advanced by the British physicist James Clerk Maxwell in about 1860, and is known as **Maxwell's counterpart to Faraday's law:**

> **A magnetic field is induced in any region of space in which an electric field is changing with time. The magnitude of the induced magnetic field is proportional to the rate at which the electric field changes. The direction of the induced magnetic field is at right angles to the changing electric field.**

These statements are two of the most important statements in physics. They underlie an understanding of the nature of light and of electromagnetic waves in general.

SUMMARY OF TERMS

Magnetic force (1) Between magnets, it is the attraction of unlike magnetic poles for each other and the repulsion between like magnetic poles. (2) Between a magnetic field and a moving charge, it is a deflecting force due to the motion of the charge: the deflecting force is perpendicular to the motion of the charge and perpendicular to the magnetic field lines. This force is greatest when the charge moves perpendicular to the field lines and is smallest (zero) when moving parallel to the field lines.

Magnetic field The region of magnetic influence around a magnetic pole or a moving charged particle.

Magnetic domains Clustered regions of aligned magnetic atoms. When these regions themselves are aligned with one another, the substance containing them is a magnet.

Electromagnet A magnet whose field is produced by an electric current. Usually in the form of a wire coil with a piece of iron inside the coil.

Electromagnetic induction The induction of voltage when a magnetic field changes with time. If the magnetic field within a closed loop changes in any way, a voltage is induced in the loop:

$$\text{Voltage induced} = -\text{ number of loops} \times \frac{\text{magnetic field change}}{\text{time}}$$

This is a statement of Faraday's law. The induction of voltage is actually the result of a more fundamental phenomenon: the induction of an electric *field*, as defined for the more general case below.

Faraday's law An electric field is induced in any region of space in which a magnetic field is changing with time. The magnitude of the induced electric field is proportional to the rate at which the magnetic field changes. The direction of the induced field is at right angles to the changing magnetic field.

Generator An electromagnetic induction device that produces electric current by rotating a coil within a stationary magnetic field.

Transformer A device for transferring electric power from one coil of wire to another by means of electromagnetic induction.

Maxwell's counterpart to Faraday's law A magnetic field is induced in any region of space in which an electric field is changing with time. The magnitude of the induced magnetic field is proportional to the rate at which the electric field changes. The direction of the induced magnetic field is at right angles to the changing electric field.

REVIEW QUESTIONS

Magnetic Poles

1. Where are the magnetic poles located on a common bar magnet?

2. In what way is the rule for the interaction between magnetic poles similar to the rule for the interaction between electric charges?

3. In what way are *magnetic poles* very different from *electric charges?*

Magnetic Fields

4. An electric field surrounds an electric charge. What additional field surrounds a moving electric charge?

5. Why is *motion* a key word for magnetism?

6. What two kinds of motion are exhibited by electrons in an atom?

Magnetic Domains

7. What is a magnetic domain?

8. Why is iron magnetic and wood not?

9. Why will dropping an iron magnet on a hard floor make it a weaker magnet?

Electric Currents and Magnetic Fields

10. How do the directions of magnetic field lines about a current-carrying wire differ from the directions of electric field lines about a charge?

11. What happens to the direction of the magnetic field about an electric current when the direction of the current is reversed?

12. Why is the magnetic field strength greater inside a current-carrying loop of wire than about a straight section of wire?

Magnetic Force on Moving Charges

13. In what direction relative to a magnetic field does a charged particle move in order to experience maximum deflecting force? Minimum deflecting force?

14. Both gravitational and electrical forces act in a direction along and parallel to the force fields. How is the direction of the magnetic force on a moving charge different?

15. What effect does the magnetic field about the earth have on cosmic-ray bombardment?

16. Since a magnetic force acts on a moving charged particle, does it make sense that a magnetic force also acts on a current-carrying wire? Defend your answer.

17. What relative direction between a magnetic field and a current-carrying wire results in greatest deflection? Smallest deflection?

18. What happens to the direction of deflection when the current in a wire is reversed?

19. What is a galvanometer called when calibrated to read current? Voltage?

Electromagnetic Induction

20. When a magnet is thrust into a coil of wire, voltage is induced in the wire, which in turn produces a current in the wire. Is each current-carrying loop of wire in the coil then an electromagnet?

21. Exactly what is it that must change for electromagnetic induction to occur?

22. What are the three ways that voltage can be induced in a wire?

Generators and Alternating Current

23. How does the frequency of induced voltage compare to how frequently a magnet is plunged in and out of a coil of wire?

24. What is the basic *difference* between a generator and an electric motor?

25. What is the basic *similarity* between a generator and an electric motor?

26. Where in the rotation cycle of a simple generator is the greatest rate of change of field lines? Where, then, is induced voltage at a maximum?

27. Why does the voltage induced in a generator alternate?

Power Production

28. What commonly supplies the energy input to a turbine?

Field Induction

29. What is induced by the rapid alternation of a *magnetic field?*

30. What is induced by the rapid alternation of an *electric field?*

· ·

HOME PROJECTS

1. Find the direction and dip of the earth's magnetic field lines in your locality. Magnetize a large steel needle or straight piece of steel wire by stroking it a couple of dozen times with a strong magnet. Run the needle through a cork and float it in a plastic or wooden container of water. The needle will point to the magnetic pole. Then remove both the needle and cork and press an unmagnetized common pin into each side of the cork. Rest the pins on the rims of a pair of drinking glasses so that the needle points to the magnetic pole. It should dip in line with the earth's magnetic field.

2. An iron bar can be easily magnetized by aligning it with the magnetic field lines of the earth and striking it lightly a few times with a hammer. The hammering jostles the domains so they can better fall into alignment with the earth's field. The bar can be demagnetized by striking it when it is in an east-west direction.

· ·

EXERCISES

1. In what sense are all magnets electromagnets?

2. Since every iron atom is a tiny magnet, why aren't all iron materials themselves magnets?

3. "An electron always experiences a force in an electric field, but not necessarily in a magnetic field." Defend this statement.

4. The core of the earth is probably composed of iron and nickel, excellent metals for making permanent magnets. Why is it unlikely that the earth's core is a permanent magnet?

5. Why will a magnet attract an ordinary nail or paper clip, but not a wooden pencil?

6. One way to make a compass is to stick a magnetized needle into a piece of cork and float it in a glass bowl full of water. The needle will align itself with the magnetic field of the earth. Since the north pole of this compass is attracted northward, will the needle float toward the northward side of the bowl? Defend your answer.

7. Your friend says that when a compass is taken across the equator, it turns around and points in the opposite direction. Your other friend says this is not true, that people living in the Southern Hemisphere use the south pole of the compass to find direction. You're on; what do you say?

8. Why will a magnet placed in front of a television picture tube distort the picture? (*Note:* Do NOT try this with a color set. If you succeed in magnetizing the metal mask in back of the glass screen, you will have picture distortion even when the magnet is removed!)

9. Magnet A has twice the magnetic field strength of magnet B and at a certain distance pulls on magnet B with a force of 50 N. With how much force, then, does magnet B pull on magnet A?

10. A strong magnet attracts a paper clip to itself with a certain force. Does the paper clip exert a force on the strong magnet? If not, why not? If so, does it exert as much force on the magnet as the magnet exerts on it? Defend your answers.

11. When iron naval ships are built, the location of the shipyard and the orientation in the ship while in the shipyard are recorded on a brass plaque permanently fixed to the ship. Why?

12. A cyclotron is a device for accelerating charged particles in ever-increasing circular orbits to high speeds. The charged particles are subjected to both an electric field and a magnetic field. One of these fields increases the speed of the charged particles, and the other field holds them in a circular path. Which field performs which function?

13. A magnetic field can deflect a beam of electrons, but it cannot do work on the electrons to speed them up. Why?

14. Two charged particles are projected into a magnetic field that is perpendicular to their velocities. If the charges are deflected in opposite directions, what does this tell you about them?

15. Why does an iron core increase the magnetic induction of a coil of wire?

16. Why is a generator armature harder to rotate when it is connected to and supplying electric current to a circuit?

17. Will a cyclist coast farther if the lamp connected to his generator is turned off? Explain.

18. If your metal car moves over a wide, closed loop of wire embedded in a road surface, will the magnetic field of the earth within the loop be altered? Will this produce a current pulse? Can you think of a practical application for this at a traffic intersection?

19. At the security area of an airport, you walk through a weak magnetic field inside a coil of wire. What is the result of a small piece of iron on your person that slightly alters the magnetic field in the coil?

20. A certain earthquake detector consists of a little box that contains a massive magnet suspended by sensitive springs. The magnet is surrounded by stationary coils of wire that are fastened to the box, which is firmly anchored to the earth. Explain how this device works, using two important principles of physics—one studied in Chapter 2, and the other in this chapter.

21. What is the primary difference between an electric *motor* and an electric *generator?*

22. Does the voltage output increase when a generator is made to spin faster? Explain.

23. If you place a metal ring in a region where a magnetic field is rapidly alternating, the ring may become hot to your touch. Why?

24. In what sense can a transformer be thought of as an electrical lever?

25. How could the motor in Figure 10.20 be used as a generator?

26. How could the generator in Figure 10.25 be used as a motor?

27. When a bar magnet is dropped through a vertical length of copper pipe, it falls noticeably more slowly than when dropped through a vertical length of plastic pipe. Why?

28. What is wrong with this scheme? To generate electricity without fuel, arrange a motor to run a generator that will produce electricity that is stepped up with transformers so that the generator can run the motor and simultaneously furnish electricity for other uses.

29. Would electromagnetic waves exist if changing magnetic fields could produce electric fields, but changing electric fields could not in turn produce magnetic fields? Explain.

PART 4

SOUND AND LIGHT

This disc is the pits -- billions of them, carefully inscribed in an array that is scanned by a laser beam millions of pits per second. Digitized music! But the beauty of CD recordings is not confined to sound -- just look at the brilliant spectrum of colors diffracted by the evenly spaced rows of pits. I find this CD even more beautiful when I know why it's so colorful and why it sounds so great. That's physical science!

11

SOUND WAVES

Many things in nature wiggle and jiggle—the surface of a bell, a string on a guitar, the reed in a clarinet, lips on the mouthpiece of a trumpet, the vocal chords of your larynx when you speak or sing. All these things **vibrate.** When they vibrate in air they make the air molecules they touch wiggle and jiggle too, exactly the same way, and these vibrations spread out in all directions, getting weaker, losing energy as heat, until they die out. If these vibrations reach an ear they are transmitted to a part of our brain, and we hear **sound.**

11.1 VIBRATIONS AND WAVES

In a general sense, anything that moves back and forth, to and fro, side to side, in and out, or up and down is vibrating. A **vibration** is a wiggle in time. A wiggle in space and time is a **wave.** A wave extends from one place to another. Light and sound are both vibrations that propagate throughout space as waves, but as two very different kinds. Sound is the propagation of vibrations through a material medium— a solid, liquid, or a gas. If there is no medium to vibrate, then no sound is possible. Sound cannot travel in a vacuum. But light can, for as we will study in the following chapter, light is a vibration of nonmaterial electric and magnetic fields—a vibration

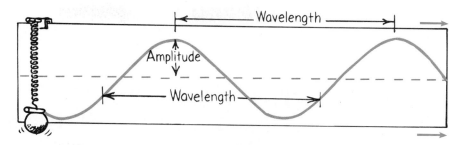

Figure 11.1 When the bob vibrates up and down, a marking pen traces out a sine curve on the paper that is moved horizontally at constant speed.

of pure energy. Although light can pass though many materials, it needs none. This is evident when it propagates through the vacuum between the sun and the earth.

The relationship between a vibration and a wave is shown in Figure 11.1. A marking pen on a bob attached to a vertical spring vibrates up and down and traces a wave form on a sheet of paper that is moved horizontally at constant speed. The wave form is actually a *sine curve,* a pictorial representation of a wave. Like a water wave, the high points are called *crests,* and the low points are the *troughs.* The straight dashed line represents the "home" position, or midpoint of the vibration. The term **amplitude** refers to the distance from the midpoint to the crest (or trough) of the wave. So the amplitude equals the maximum displacement from equilibrium.

The **wavelength** of a wave is the distance from the top of one crest to the top of the next one, or equivalently, the distance between successive identical parts of the wave. The wavelengths of waves at the beach are measured in meters, the wavelengths of ripples in a pond in centimeters, and the wavelengths of light in millionths of a meter (micrometers).

How frequently a vibration occurs is described by its **frequency.** The frequency of a vibrating pendulum, or object on a spring, specifies the number of to-and-fro vibrations it makes in a given time (usually one second). A complete to-and-fro oscillation is one vibration. If it occurs in one second, the frequency is one vibration per second. If two vibrations occur in one second, the frequency is two vibrations per second.

The unit of frequency is called the **hertz** (Hz), after Heinrich Hertz, who demonstrated radio waves in 1886. We call one vibration per second 1 hertz; two vibrations per second is 2 hertz, and so on. Higher frequencies are measured in kilohertz (kHz), and still higher frequencies in megahertz (MHz). AM radio waves are broadcast in kilohertz, while FM radio waves are broadcast in megahertz. A station at 960 kHz on the AM radio dial, for example, broadcasts radio waves that have a frequency of 960,000 vibrations per second. A station at 101 MHz on the FM dial broadcasts radio waves with a frequency of 101,000,000 hertz. These radio-wave frequencies are the frequencies at which electrons are forced to vibrate in the antenna of a radio station's transmitting tower. The source of all waves is something that vibrates. The frequency of the vibrating source and the frequency of the wave it produces are the same.

The **period** of a wave or vibration is the time it takes for a complete vibration. Period can be calculated from frequency, and vice versa. Suppose, for example, that a pendulum makes two vibrations in one second. Its frequency is 2 Hz. The time needed to complete one vibration—that is, the period of vibration—is $\frac{1}{2}$ second. Or if the vibration frequency is 3 Hz, then the period is $\frac{1}{3}$ second. The frequency and period are the inverse of each other:

Figure 11.2 Electrons in the transmitting antenna vibrate 940,000 times each second and produce 940-kHz radio waves.

$$\text{Frequency} = \frac{1}{\text{period}}$$

or vice versa:

$$\text{Period} = \frac{1}{\text{frequency}}$$

QUESTIONS

1. An electric vibrator completes 60 cycles every second. What is (a) its frequency? (b) its period?
2. Gusts of wind make the Sears Building in Chicago sway back and forth, completing a cycle every ten seconds. What is (a) its frequency? (b) its period?

11.2 WAVE MOTION

Figure 11.3 Water waves.

Drop a stone into a quiet pond and waves will travel outward in expanding circles. Note that the waves move, not the water. This can be seen by a floating leaf that the waves encounter. The leaf bobs up and down, but doesn't travel with the waves. The waves move, not the water. Likewise with waves of wind over a field of tall grass on a gusty day. Waves travel across the grass, while individual stems of grass do not leave their places; instead, they swing to and fro between definite limits but go nowhere. When you speak, wave motion through the air travels across the room at about 340 meters per second. The air itself doesn't travel across the room at this speed. In these examples, when the wave motion stops, the water, the grass, and the air return to their initial positions. It is characteristic of wave motion that the medium carrying the wave returns to its initial condition after the disturbance has passed.

Wave Speed

The speed of periodic wave motion is related to the frequency and wavelength of the waves. We can understand this by considering the simple case of water waves (Figures 11.3 and 11.4). If we fix our eyes at a stationary point on the water surface and count the number of crests of water passing this point each second (the frequency) and also observe the distance between crests (the wavelength), we can calculate the horizontal distance that a particular crest travels each second. We will then know how frequently a distance equal to the wavelength is traveled—that is, the **wave speed.**

For example, if two crests pass a stationary point each second, and if the wavelength is 10 meters, then 2×10 meters of waves pass by in 1 second. The waves therefore travel at 20 meters per second. We can say

Wave speed = frequency × wavelength

This relationship holds true for all kinds of waves, whether they are water waves, sound waves, or light waves.

Wavelength

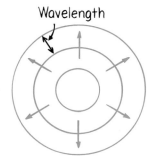

Figure 11.4 A top view of water waves.

ANSWERS

1. (a) 60 cycles per second or 60 Hz; (b) $\frac{1}{60}$ second.
2. (a) $\frac{1}{10}$ Hz. (b) 10 s.

Figure 11.5 If the wavelength is 1 m, and one wavelength per second passes the pole, then the speed of the wave is 1 m/s.

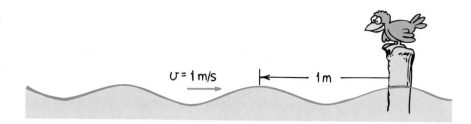

$v = 1$ m/s

1m

QUESTIONS

1. If a train of freight cars, each 10 m long, rolls by you at the rate of three cars each second, what is the speed of the train?

2. If a water wave oscillates up and down three times each second and the distance between wave crests is 2 m, (a) what is its frequency? (b) its wavelength? (c) its wave speed?

3. A 60-Hz vibrator produces air waves that spread out at 340 meters per second. What is (a) the frequency of the waves? (b) their period? (c) their speed? (d) their wavelength?

11.3 TRANSVERSE AND LONGITUDINAL WAVES

Fasten one end of a Slinky to a wall and hold the free end in your hand. Shake it up and down and you produce vibrations that are at right angles to the direction of wave travel. The right-angled, or sideways, motion is called *transverse motion*. This type of wave is called a **transverse wave.** Waves in the stretched strings of musical

Figure 11.6 Both waves transfer energy from left to right. When the end of the Slinky is shaken up and down, a transverse wave is produced. When it's shaken back and forth, a longitudinal wave is produced.

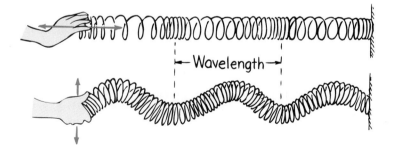

Wavelength

ANSWERS

1. 30 m/s. We can see this two ways, the Chapter 1 way and the Chapter 11 way. We have $v = d/t = (3 \times 10 \text{ m})/1 \text{ s} = 30$ m/s, where d is the distance of train that passes you in t seconds. If we compare our train to wave motion, where wavelength corresponds to 10 m and frequency is 3 Hz, then speed = frequency × wavelength = 3×10 m/s = 30 m/s.

2. (a) 3 Hz; (b) 2 m; (c) Wave speed = frequency × wavelength = 3×2 = 6 m/s. It is customary to express this as the equation $v = f\lambda$ where v is wave speed, f is wave frequency, and λ (the Greek letter lambda) is wavelength.

3. (a) 60 Hz; (b) $\frac{1}{60}$ second; (c) 340 m/s; (d) 5.7 m.

instruments and upon the surfaces of liquids are transverse. We will see later that electromagnetic waves, some of which are radio waves and light, are also transverse.

A **longitudinal wave** is produced by shaking the end of the Slinky back and forth. In this case the vibrations are parallel to the direction of energy transfer. Part of the Slinky is compressed, and a wave of *compression* travels along the spring. In between successive compressions is a stretched region, called a *rarefaction*. Both compressions and rarefactions travel in the same direction along the Slinky. Sound waves are longitudinal waves. The wavelength of a longitudinal wave is the distance between successive compressions or equivalently, the distance between successive rarefactions. In the case of sound, each molecule in the air vibrates to and fro about some equilibrium position as the waves move by.

11.4 SOUND WAVES

Figure 11.7 Vibrate a Ping-Pong paddle in the midst of a lot of Ping-Pong balls, and you make the balls vibrate also.

Think of the air molecules in the room as tiny Ping-Pong balls, all haphazardly moving. Vibrate a Ping-Pong paddle in the midst of the Ping-Pong balls and you set them vibrating also. They will vibrate to and fro in rhythm with your vibrating paddle. Like shaking the Slinky back and forth, in some regions the balls are momentarily bunched up (compressions) and spread out in between (rarefactions). The vibrating prongs of a tuning fork do the same to air molecules. Vibrations of compressions and rarefactions spread from the tuning fork throughout the air. It is important to again point out that the vibrations travel through the air, not the air molecules themselves. We hear these vibrations as sound.

We describe our subjective impression about the frequency of sound by pitch. A high-pitch sound like that from a tiny bell has a high-vibration frequency. Sound from a large bell has a low pitch, because its vibrations are of a low frequency.

The human ear can normally hear pitches corresponding to the range of frequencies between about 20 to 20,000 hertz. As we age, the limits of this range shrink. So by the time you can afford to trade in your old radio for an expensive hi-fi set, you may not be able to tell the difference. Sound waves with frequencies below 20 hertz are called infrasonic, and waves with frequencies above 20,000 hertz are called ultrasonic. We cannot hear infrasonic or ultrasonic sound waves. But dogs and some other animals can.

Most sound is transmitted through air, but any elastic substance—solid, liquid, or gas—can transmit sound.* Air is a poor conductor of sound compared to

Figure 11.8
Compressions and rarefactions travel (both at the same speed in the same direction) from the tuning fork through the air in the tube.

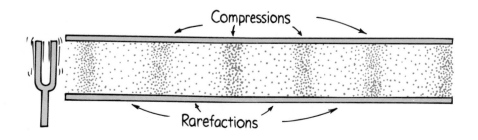

———

*An elastic substance is "springy," has resilience, and can transmit energy with little loss. Steel, for example is elastic, while lead or putty is not.

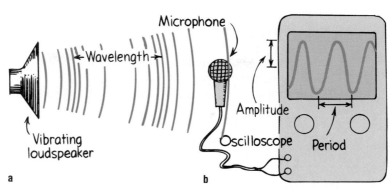

Figure 11.9 (a) The radio loudspeaker is a paper cone that vibrates in rhythm with an electric signal. The sound that is produced sets up similar vibrations in the microphone, which are displayed on an oscilloscope. (b) The shape of the waveform on the screen of the oscilloscope reveals information about the sound.

solids and liquids. You can hear the sound of a distant train clearly by placing an ear against the rail. When swimming, have a friend at a distance click two rocks together beneath the surface of water while you are submerged. Observe how well water conducts the sound. Sound cannot travel in a vacuum because there is nothing to compress and expand. The transmission of sound requires a medium.

Pause to reflect on the physics of sound while you are quietly listening to your radio sometime. The radio loudspeaker is a paper cone that vibrates in rhythm with an electrical signal. Air molecules that constantly impinge on the vibrating cone of the speaker are themselves set into vibration. These in turn vibrate against neighboring molecules, which in turn do the same, and so on. As a result, rhythmic patterns of compressed and rarefied air emanate from the loudspeaker, showering the whole room with undulating motions. The resulting vibrating air sets your eardrum into vibration, which in turn sends cascades of rhythmic electrical impulses along the cochlea nerve canal and into the brain. And you listen to the sound of music.

Speed of Sound

If we watch a person at a distance chopping wood or hammering, we can easily see that the blow takes place an appreciable time before its sound reaches our ears. Thunder is heard after a flash of lightning. These common experiences show that sound requires time to travel from one place to another. The speed of sound depends on wind conditions, temperature, and humidity. It does not depend on the loudness or the frequency of the sound; all sounds travel at the same speed in a given medium. The speed of sound in dry air at 0°C is about 330 meters per second, nearly 1200 kilometers per hour. Water vapor in the air increases this speed slightly. Sound travels faster through warm air than cold air. This is to be expected because the faster-moving molecules in warm air bump into each other more often and therefore can transmit a pulse in less time.* For each degree rise in temperature above 0°C, the speed of sound in air increases by 0.6 meter per second. So in air at a normal room temperature of about 20°C, sound travels at about 340 meters per second. In water, sound speed is about four times its speed in air; in steel it's about fifteen times.

Figure 11.10 Waves of compressed and rarefied air, produced by the vibrating cone of the loudspeaker, make up the pleasing sound of music.

---

*The speed of sound in a gas is about $\frac{3}{4}$ the average speed of molecules.

QUESTIONS

1. Do compressions and rarefactions in a sound wave travel in the same direction or in opposite directions from one another?

2. What is the approximate distance of a thunderstorm when you note a 3-s delay between the flash of lightning and the sound of thunder?

11.5 REFLECTION OF SOUND

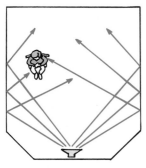

Figure 11.11 The angle of incident sound is equal to the angle of reflected sound.

We call the reflection of sound an *echo*. The fraction of sound energy reflected from a surface is large if the surface is rigid and smooth, and less if the surface is soft and irregular. Sound energy that is not reflected is transmitted or absorbed.

Sound reflects from a smooth surface the same way that light does—the angle of incidence is equal to the angle of reflection (Figure 11.11). Sometimes when sound reflects from the walls, ceiling, and floor of a room, the surfaces are too reflective and the sound becomes garbled. This is due to multiple reflections called **reverberations.** On the other hand, if the reflective surfaces are too absorbent, the sound level is low and the hall may sound dull and lifeless. Reflected sound in a room makes it sound lively and full, as you have probably found out while singing in the shower. In the design of an auditorium or concert hall, a balance must be found between reverberation and absorption. The study of sound properties is called *acoustics*.

It is often advantageous to place highly reflective surfaces behind the stage to direct sound out to an audience. Above the stage in some concert halls are suspended

ANSWERS

1. They travel in the same direction.

2. Assuming the speed of sound in air is about 340 m/s, in 3 s it will travel (340 × 3) 1020 m. We can assume no time delay for the light, so the storm is slightly more than 1 km away.

Figure 11.12 The plastic plates above the orchestra reflect both light and sound. Adjusting them is quite simple: what you see is what you hear.

reflecting surfaces. The ones in the opera hall in San Francisco are large shiny plastic surfaces that also reflect light (Figure 11.12). A listener can look up at these reflectors and see the reflected images of the members of the orchestra (the plastic reflectors are somewhat curved, which increases the field of view). Both sound and light obey the same law of reflection, so if a reflector is oriented so that you can see a particular musical instrument, rest assured that you will hear it also. Sound from the instrument will follow the line of sight to the reflector and then to you.

11.6 REFRACTION OF SOUND

Sound waves bend when parts of the wave fronts travel at different speeds. This happens in uneven winds or when sound is traveling through air of uneven temperatures. This bending of sound is called **refraction.** On a warm day, the air near the ground may be appreciably warmer than air above, so the speed of sound near the ground increases. Sound waves therefore tend to bend away from the ground, resulting in sound that does not seem to carry well (Figure 11.13).

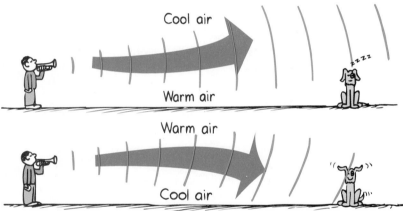

Figure 11.13 The wave fronts of sound are bent in air of uneven temperatures.

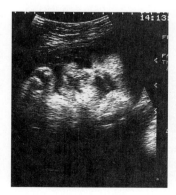

Figure 11.14 The 38-week-old fetus of Ian Suchocki displayed on a viewing screen by ultrasound.

The refraction of sound occurs under water, where the speed of sound varies with temperature. This poses a problem for surface vessels that bounce ultrasonic waves off the bottom of the ocean to chart its features. This poses a blessing to submarines that wish to escape detection. Because of thermal gradients and layers of water at different temperatures, the refraction of sound leaves gaps or "blind spots" in the water. This is where submarines hide. If it weren't for refraction, submarines would be easy to detect.

The multiple reflections and refractions of ultrasonic waves are used by physicians in a technique for harmlessly "seeing" inside the body without the use of X rays. When high-frequency sound (ultrasound) enters the body, it is reflected more strongly from the outside of organs than from their interior, and a picture of the outline of the organs is obtained (Figure 11.14). The ultrasound echo technique may be relatively new to humans, but not to bats who emit ultrasonic squeaks and locate objects by their echoes, or to dolphins who do this and more.

Dolphins and Acoustical Imaging

The primary sense of the dolphin is acoustic, for vision is not a very useful sense in the often murky and dark depths of the ocean. Whereas sound is a passive sense for us, it is an active sense for the dolphin who sends out sounds and then perceives its surroundings on the basis of the echoes that come back. The ultrasonic waves emitted by a dolphin enable it to "see" through the bodies of other animals and people. Skin, muscle, and fat are almost transparent to dolphins, so they "see" a thin outline of the body—but the bones, teeth, and gas-filled cavities are clearly apparent. Physical evidence of cancers, tumors, and heart attacks can all be "seen" by the dolphin—as humans have only recently been able to do with ultrasound.

What's more interesting, the dolphin can reproduce the sonic signals that paint the mental image of its surroundings; thus the dolphin probably communicates its experience to other dolphins by communicating the full acoustic image of what is "seen," placing it directly in the minds of other dolphins. It needs no word or symbol for "fish," for example, but communicates an image of the real thing—perhaps with emphasis highlighted by selective filtering, as we similarly communicate a musical concert to others via various means of sound reproduction. Small wonder that the language of the dolphin is very unlike our own!

Figure 11.15   A dolphin emits ultrahigh-frequency sound to locate and identify objects in its environment. Distance is sensed by the time delay between sending sound and receiving its echo, and direction is sensed by differences in time for the echo to reach its two ears. A dolphin's main diet is fish and, since hearing in fish is limited to fairly low frequencies, they are not alerted to the fact they are being hunted.

$\boxed{\text{Q}}$UESTION

An oceanic depth-sounding vessel surveys the ocean bottom with ultrasonic sound that travels 1530 m/s in seawater. How deep is the water if the time delay of the echo from the ocean floor is 2 s?

11.7 FORCED VIBRATIONS

If we strike an unmounted tuning fork, its sound is rather faint. Do the same when the fork is held against a table and the sound is louder. This is because the table is forced to vibrate, and with its larger surface it will set more air in motion. The table will be forced into vibration by a fork of any frequency. This is a case of **forced vibration.** The vibration of a factory floor caused by the running of heavy machinery is an example of forced vibration. A more pleasing example is given by the sounding boards of stringed instruments.

Drop a wrench and a baseball bat on a concrete floor and you can tell the difference in their sounds. This is because each vibrates differently when striking the floor. They are not forced to vibrate at a particular frequency, but instead each vibrates at its own characteristic frequency. Any object composed of an elastic material when disturbed will vibrate at its own special set of frequencies, which together form its special sound. We speak of an object's **natural frequency,** which depends on factors such as the elasticity and shape of the object. Bells and tuning forks, of course, vibrate at their own characteristic frequencies. And interestingly enough, most things from planets to atoms and almost everything else in between have a springiness to them and vibrate at one or more natural frequencies. A natural frequency is one at which the least amount of energy is required to produce forced vibrations.

11.8 RESONANCE

When the frequency of forced vibrations on an object matches the object's natural frequency, a dramatic increase in amplitude occurs. This phenomenon is called **resonance.** Literally, *resonance* means "resounding," or "sounding again." Putty doesn't resonate because it isn't elastic, and a dropped handkerchief is too limp. In order for something to resonate, it needs a force to pull it back to its starting position and enough energy to keep it vibrating.

A common experience illustrating resonance occurs on a swing. When pumping a swing, we pump in rhythm with the natural frequency of the swing. More important than the force with which we pump is the timing. Even small pumps or small pushes from someone else, if delivered in rhythm with the frequency of the swinging motion, produce large amplitudes. A common classroom demonstration of

$\boxed{\text{A}}$NSWER

1530 m

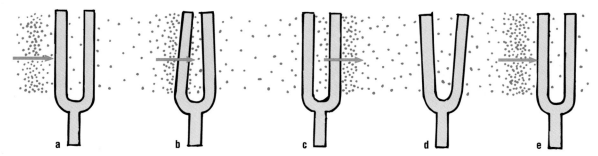

Figure 11.16 Stages of resonance. (a) The first compression meets the fork and gives it a tiny and momentary push; (b) the fork bends and then (c) returns to its initial position just at the time a rarefaction arrives and (d) overshoots in the opposite direction. Just when it returns to its initial position (e) the next compression arrives to repeat the cycle. Now it bends farther because it is moving.

resonance is illustrated with a pair of tuning forks adjusted to the same frequency and spaced a meter or so apart. When one of the forks is struck, it sets the other fork into vibration. This is a small-scale version of pushing a friend on a swing—it's the timing that's important. When a series of sound waves impinge on the fork, each compression gives the prong of the fork a tiny push. Since the frequency of these pushes corresponds to the natural frequency of the fork, the pushes will successively increase the amplitude of vibration. This is because the pushes occur at the right time and repeatedly occur in the same direction as the instantaneous motion of the fork.

If the forks are not adjusted for matched frequencies, the timing of pushes is off, and resonance will not occur. When you tune your radio set, you are similarly adjusting the natural frequency of the electronics in the set to match one of the many surrounding signals. The set then resonates to one station at a time instead of playing all stations at once.

Resonance is not restricted to wave motion. It occurs whenever successive impulses are applied to a vibrating object in rhythm with its natural frequency. Troops marching across a footbridge near Manchester, England, in 1831 inadvertently caused the bridge to collapse when they marched in rhythm with the bridge's natural frequency. Since then, it is customary to order troops to "break step" when crossing bridges. A more recent bridge disaster was caused by wind-generated resonance (Figure 11.17).

Figure 11.17 In 1940, four months after being completed, the Tacoma Narrows Bridge in the state of Washington was destroyed by wind-generated resonance. The mild gale produced a fluctuating force in resonance with the natural frequency of the bridge, steadily increasing the amplitude until the bridge collapsed.

11.9 INTERFERENCE

One of the most interesting properties of all waves is the phenomenon of **interference.** Consider transverse waves. When the crest of one wave overlaps the crest of another, their individual effects add together. The result is a wave of increased amplitude. This is *constructive interference* (Figure 11.18). When the crest of one wave overlaps the trough of another, their individual effects are reduced. The high part of one wave simply fills in the low part of another. This is called *destructive interference.*

Wave interference is easiest to see in water. In Figure 11.19 we see the interference pattern made when two vibrating objects touch the surface of water. We can see the regions where a crest of one wave overlaps the trough of another to produce regions of zero amplitude. At points along these regions, the waves arrive out of step. We say they are *out of phase* with one another.

Interference is characteristic of all wave motion, whether the waves are water waves, sound waves, or light waves. We see a comparison of interference for both transverse and longitudinal waves in Figure 11.20. In either case, when crests overlap crests, increased amplitude results. Or when crests overlap troughs, decreased amplitude results. In the case of sound, the crest of a wave corresponds to a compression, and the trough of a wave corresponds to a rarefaction.

Destructive sound interference is a useful property in *anti-noise technology.* Noisy devices such as jackhammers are being equipped with microphones that send

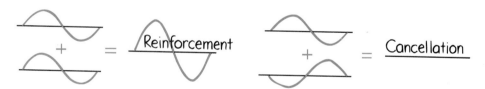

Figure 11.18 Constructive and destructive interference in a transverse wave.

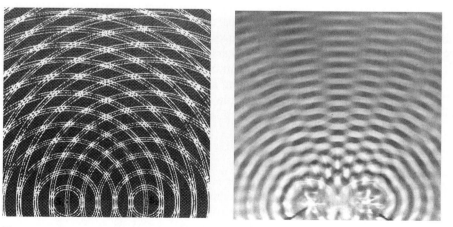

Figure 11.19 Two sets of overlapping water waves produce an interference pattern.

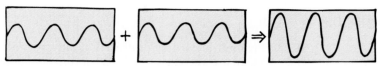

The superposition of two identical transverse waves in phase produces a wave of increased amplitude.

The superposition of two identical longitudinal waves in phase produces a wave of increased intensity.

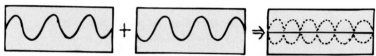

Two identical transverse waves that are out of phase destroy each other when they are superimposed.

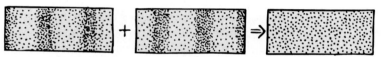

Two identical longitudinal waves that are out of phase destroy each other when they are superimposed.

Figure 11.20   Wave interference for transverse and longitudinal waves.

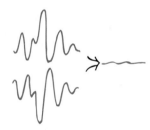

Figure 11.21   When a mirror image of a sound signal combines with the sound itself, the sound is canceled.

the sound of the device to electronic microchips, which create mirror-image wave patterns of the sound signals. For the jackhammer, this mirror-image sound signal is fed to earphones worn by the operator. Sound compressions (or rarefactions) from the hammer are canceled by mirror-image rarefactions (or compressions) in the earphones. The combination of signals cancels the jackhammer noise. Watch for this principle applied to electronic mufflers in cars—the anti-noise is blasted through loudspeakers, canceling about 95 percent of the original noise.

Sound interference is dramatically illustrated when monaural sound is played by stereo speakers that are out of phase. Speakers are out of phase when the input wires to one speaker are interchanged (plus and negative wire inputs reversed). For a monaural signal this means that when one speaker is sending a compression of sound, the other is sending a rarefaction. The sound produced is not as full and not as loud as from speakers properly connected in phase, because the longer waves are being canceled by interference. Shorter waves are canceled as the speakers are brought closer together, and when the pair of speakers are brought face to face, against each other, very little sound is heard! Only the highest frequencies survive cancellation. You must try this to appreciate it.

Beats

When two tones of slightly different frequencies are sounded together, a fluctuation in the loudness of the combined sounds is heard; the sound is loud, then faint, then loud, then faint, and so on. This periodic variation in the loudness of sound is called **beats** and is due to interference. Strike two slightly mismatched tuning forks, and because one fork vibrates at a different frequency than the other, the vibrations of

Figure 11.22 The plus and negative wire inputs to one of the stereo speakers has been interchanged, resulting in speakers that are out of phase. When far apart, monaural sound is not as loud as from properly-phased speakers. When brought face to face, very little sound is heard. Interference is nearly complete as the compressions of one speaker fill in the rarefactions of the other!

the forks will be momentarily in step, then out of step, then in again, and so on. When the combined waves reach our ears in step—say when a compression from one fork overlaps a compression from the other—the sound is a maximum. A moment later, when the forks are out of step, a compression from one fork is met with a rarefaction from the other, resulting in a minimum. The sound that reaches our ears throbs between maximum and minimum loudness and produces a tremolo effect.

Figure 11.23 The interference of two sound sources of slightly different frequencies produces beats.

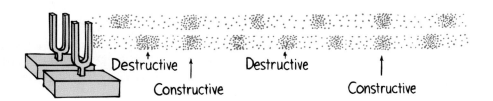

Destructive ↑   Destructive ↑

Constructive   Constructive

Beats can occur with any kind of wave and provide a practical way to compare frequencies. To tune a piano, for example, a piano tuner listens for beats produced between a standard tuning fork and those of a particular string on the piano. When the frequencies are identical, the beats disappear. The members of an orchestra tune up by listening for beats between their instruments and a standard tone produced by a piano or some other instrument.

Standing Waves

Another interesting effect of interference is *standing waves.* Tie a rope to a wall and shake the free end up and down. The wall is too rigid to shake, so the waves are reflected back along the rope. By shaking the rope just right, we can cause the incident and reflected waves to interfere and form a **standing wave,** where parts of the rope, called the *nodes,* are stationary. You can hold your fingers on either side of the rope at a node, and the rope will not touch them. Other parts of the rope would make contact with your fingers. The positions on a standing wave with the largest displacements are known as *antinodes.* Antinodes occur halfway between nodes.

Figure 11.24 The incident and reflected waves interfere to produce a standing wave.

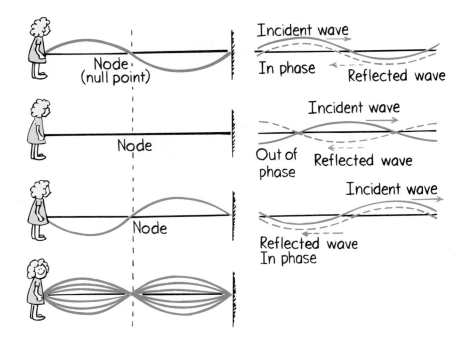

Standing waves are produced when two sets of waves of equal amplitude and wavelength pass through each other in opposite directions. Then the waves are steadily in and out of phase with each other and produce stable regions of constructive and destructive interference (Figure 11.24).

It is easy to make standing waves yourself. Tie a rope, or better, a rubber tube between two firm supports. Shake the tube from side to side with your hand near one of the supports. If you shake the tube with the right frequency, you will set up a standing wave as shown in Figure 11.25. Shake the tube with twice the frequency, and a standing wave of half the previous wavelength, two loops, will result. (The distance between successive nodes is a half wavelength; two loops make up a full wavelength.) Triple the frequency, and a standing wave with one-third the original wavelength, three loops, results, and so forth.

Figure 11.25 (a) Shake the rope until you set up a standing wave of one segment ($\frac{1}{2}$ wavelength). (b) Shake with twice the frequency and produce a wave with two segments (1 wavelength). (c) Shake with three times the frequency and produce three segments (1-$\frac{1}{2}$ wavelengths).

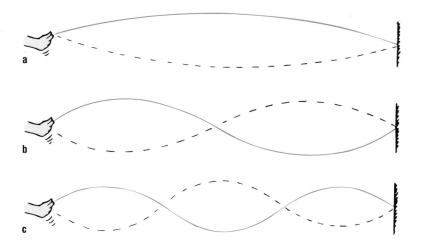

Standing waves are set up in the strings of musical instruments when plucked, bowed, or struck. They are set up in the air in an organ pipe and the air of a soda-pop bottle when air is blown over the top. Standing waves can be set up in a tub of water or a cup of coffee by sloshing it back and forth with the right frequency. Standing waves can be produced in either transverse or longitudinal waves.

QUESTIONS

1. Is it possible for one wave to cancel another wave so that no amplitude remains?

2. Suppose you set up a standing wave of three segments, as shown in Figure 11.25c. If you shake with twice as much frequency, how many wave segments will occur in your new standing wave? How many wavelengths?

11.10 DOPPLER EFFECT

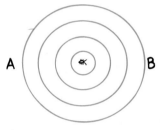

Figure 11.26 Top view of water waves made by a stationary bug jiggling in still water.

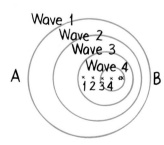

Figure 11.27 Water waves made by a bug swimming in still water.

A pattern of water waves produced by a bug jiggling its legs and bobbing up and down in the middle of a quiet puddle is shown in Figure 11.26. The bug is not going anywhere but is merely treading water in a fixed position. The waves it makes are concentric circles, because wave speed is the same in all directions. If the bug bobs in the water at a constant frequency, the distance between wave crests (the wavelength) is the same for all successive waves. Waves encounter point A as frequently as they encounter point B. This means that the frequency of wave motion is the same at points A and B, or anywhere in the vicinity of the bug. This wave frequency is the same as the bobbing frequency of the bug.

Suppose the jiggling bug moves across the water at a speed less than the wave speed. In effect, the bug chases part of the waves it has produced. The wave pattern is distorted and is no longer concentric (Figure 11.27). The center of the outer wave was made when the bug was at the center of that circle. The center of the next smaller wave was made when the bug was at the center of that circle, and so forth. The centers of the circular waves move in the direction of the swimming bug. Although the bug maintains the same bobbing frequency as before, an observer at B would encounter the waves more often. The observer would encounter a higher frequency. This is because each successive wave has a shorter distance to travel and therefore arrives at B more frequently than if the bug weren't moving toward B. An observer at A, on the other hand, encounters a lower frequency because of the longer time between wave-crest arrivals. This occurs because each successive wave travels farther to A as a result of the bug's motion. This change in frequency due to the motion of the source (or receiver) is called the **Doppler effect.**

ANSWERS

1. Yes. This is called destructive interference. In a standing wave in a rope, for example, parts of the rope have no amplitude—the nodes.

2. If you impart twice the frequency to the rope, you'll produce a standing wave with twice as many segments. You'll have six segments. Since a full wavelength has two segments, you'll have three complete wavelengths in your standing wave.

Water waves spread over the flat surface of the water. Sound and light waves, on the other hand, travel in three-dimensional space in all directions like an expanding balloon. Just as circular waves are closer together in front of the swimming bug, spherical sound or light waves ahead of a moving source are closer together and encounter a receiver more frequently. The Doppler effect holds for all types of waves.

The Doppler effect is evident when you hear the changing pitch of a car horn as the car drives by. When it approaches, the pitch is higher than normal. This is because the sound waves are encountering you more frequently. And when the car passes and moves away, you hear a drop in pitch because the waves are encountering you less frequently.

The Doppler effect also occurs for light. When a light source approaches, there is an increase in its measured frequency; and when it recedes, there is a decrease in its frequency. An increase in frequency is called a *blue shift,* because the increase is toward the high frequency, or blue end of the color spectrum. A decrease in frequency is called a *red shift,* referring to the lower-frequency, or red, end of the color spectrum. The galaxies, for example, show a red shift in the light they emit. A measurement of this shift permits a calculation of their speeds of recession. A rapidly spinning star shows a red shift on the side turning away from us and a relative blue shift on the side turning toward us. This enables a calculation of the star's spin rate.

Figure 11.28   The pitch of sound increases when the source moves toward you and decreases when the source moves away.

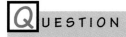UESTION

When a source moves toward you, do you measure an increase or decrease in wave speed?

11.11   WAVE BARRIERS AND BOW WAVES

When the speed of a source is as great as the speed of the waves it produces, a "wave barrier" is produced. Consider the bug in our previous example when it swims as fast as the wave speed. Can you see that the bug will keep up with the waves it produces?

ANSWER

Neither! It is the *frequency* of a wave that undergoes a change where there is motion of the source, not the wave speed. Be clear about the distinction between frequency and speed. How frequently a wave vibrates is altogether different from how fast it moves from one place to another.

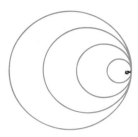

Figure 11.29 Wave pattern made by a bug swimming at wave speed.

Instead of the waves getting ahead of the bug, they pile up and superimpose on one another directly in front of the bug (Figure 11.29). The bug encounters a wave barrier. Much effort is required of the bug to swim over this barrier before it can swim faster than wave speed.

The same thing happens when an aircraft travels at the speed of sound. The waves overlap to produce a barrier of compressed air on the leading edges of the wings and other parts of the craft. Considerable thrust is required for the aircraft to push through this barrier (Figure 11.30). Once through, the craft can fly faster than the speed of sound without similar opposition. The craft is *supersonic.* It is like the bug, which once over its wave barrier finds the water ahead relatively smooth and undisturbed.

When the bug swims faster than wave speed, ideally it produces a wave pattern as shown in Figure 11.31. It outruns the waves it produces. The waves overlap at the edges, and the pattern made by these overlapping waves is a V shape, called a **bow wave,** which appears to be dragging behind the bug. The familiar bow wave generated by a speedboat knifing through the water is a non-periodic wave produced by the overlapping of many periodic circular waves.

Figure 11.30 The aircraft has just cracked the wave barrier. Condensation of water vapor by rapid expansion of air can be seen in the rarefied region behind the wall of compressed air.

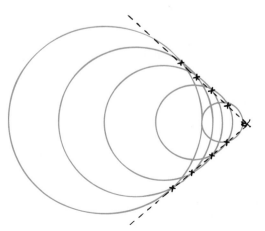

Figure 11.31 Wave pattern made by a bug swimming faster than wave speed.

Some wave patterns made by sources moving at various speeds are shown in Figure 11.32. Note that after the speed of the source exceeds wave speed, increased speed produces a narrower V shape.*

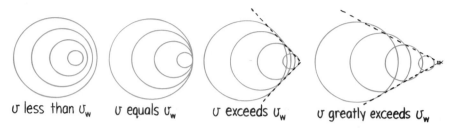

υ less than υ$_w$ υ equals υ$_w$ υ exceeds υ$_w$ υ greatly exceeds υ$_w$

Figure 11.32   Patterns made by a bug swimming at successively greater speeds. Overlapping at the edges occurs only when it travels faster than wave speed.

11.12   SHOCK WAVES AND THE SONIC BOOM

A speedboat knifing through the water generates a two-dimensional bow wave. A supersonic aircraft similarly generates a three-dimensional **shock wave.** Just as a bow wave is produced by overlapping circles that form a V, a shock wave is produced by overlapping spheres that form a cone. And just as the bow wave of a speedboat spreads until it reaches the shore of a lake, the conical wake generated by a supersonic craft spreads until it reaches the ground.

The bow wave of a speedboat that passes by can splash and douse you if you are at the water's edge. In a sense, you can say that you are hit by a "water boom." In the same way, when the conical shell of compressed air that sweeps behind a supersonic aircraft reaches listeners on the ground below, the sharp crack they hear is described as a **sonic boom.**

We don't hear a sonic boom from slower-than-sound, or subsonic, aircraft because the sound waves reach our ears one at a time and are perceived as one continuous tone. Only when the craft moves faster than sound do the waves overlap to encounter the listener in a single burst. The sudden increase in pressure is much the same in effect as the sudden expansion of air produced by an explosion. Both processes direct a burst of high-pressure air to the listener. The ear is hard pressed to distinguish between the high pressure from an explosion and the high pressure from many overlapping waves.

A water skier is familiar with the fact that next to the high hump of the V-shaped bow wave is a V-shaped depression. The same is true of a shock wave, which actually consists of two cones: a high-pressure cone generated at the bow of the supersonic aircraft and a low-pressure cone that follows at the tail of the craft. The edges of these cones are visible in the photograph of the supersonic bullet in

---

*Bow waves generated by boats in water are more complex than is indicated here. Our idealized treatment serves as an analogy for the production of the less complex shock waves in air.

Figure 11.33 Shock waves of a bullet piercing a sheet of Plexiglas. Light is deflected as it passes through the compressed air that makes up the shock waves, which makes them visible.

Figure 11.33. Between these two cones the air pressure rises sharply to above atmospheric pressure, then falls below atmospheric pressure before sharply returning to normal beyond the inner tail cone (Figure 11.34). This overpressure suddenly followed by underpressure intensifies the sonic boom.

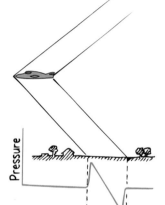

Figure 11.34 A shock wave.

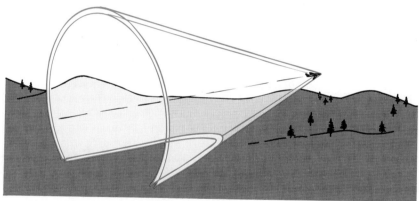

Figure 11.35 The shock wave is actually made up of two cones—a high-pressure cone with the apex at the bow and a low-pressure cone with the apex at the tail. A graph of the air pressure at ground level between the cones takes the shape of the letter N.

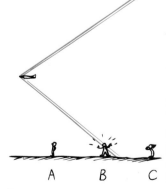

Figure 11.36 The shock wave has not yet reached listener A, but is now reaching listener B and has already reached listener C.

A common misconception is that sonic booms are produced when an aircraft breaks through the sound barrier—that is, just as the aircraft surpasses the speed of sound. This is the same as saying that a boat produces a bow wave when it overtakes its own waves. This is not so. The fact is that a shock wave and its resulting sonic boom are swept continuously behind an aircraft traveling faster than sound, just as a bow wave is swept continuously behind a speedboat. In Figure 11.36, listener B is in the process of hearing a sonic boom. Listener C has already heard it, and listener A will hear it shortly. The aircraft that generated this shock wave may have broken through the sound barrier hours ago!

It is not necessary that the moving source emit sound to produce a shock wave. Once an object is moving faster than the speed of sound, it will *make* sound. A supersonic bullet passing overhead produces a crack, which is a small sonic boom. If the bullet were larger and disturbed more air in its path, the crack would be more boomlike. When a lion tamer cracks a circus whip, the cracking sound is actually a sonic boom produced by the tip of the whip when it travels faster than the speed of sound. Both the bullet and the whip are not in themselves sound sources, but when traveling at supersonic speeds they produce their own sound as waves of air are generated to the sides of the moving objects.

11.13 MUSICAL SOUNDS

Most of the sounds we hear are noises. The impact of a falling object, the slamming of a door, the roaring of a motorcycle, and most of the sounds from traffic in city streets are noises. Noise corresponds to an irregular vibration of the eardrum produced by some irregular vibration. If we make a diagram to indicate the pressure of the air on the eardrum as it varies with time, the graph corresponding to a noise might look like that shown in Figure 11.37a. The sound of music has a different character, having more or less periodic tones—or musical "notes." (Musical instruments can make noise as well!) The graph representing a musical sound has a shape that repeats itself over and over again (Figure 11.37b). Such graphs can be displayed on the screen of an oscilloscope when the electrical signal from a microphone is fed into the input terminal of this useful device.

Figure 11.37 Graphical representation of noise and music.

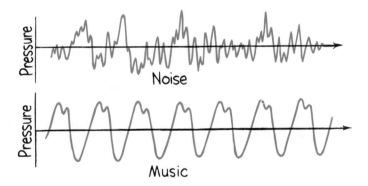

We have no trouble distinguishing between the tone from a piano and a like-pitched tone from a clarinet. Each of these tones has a characteristic sound that differs in **quality**, or timbre. Most musical sounds are composed of a superposition of many frequencies called **partial tones,** or simply *partials.* The lowest frequency, called the *fundamental frequency,* determines the pitch of the note. Partial tones that are whole multiples of the fundamental frequency are called **harmonics.** A tone that has twice the frequency of the fundamental is the second harmonic, a tone with three times the fundamental frequency is the third harmonic, and so on (Figure 11.38).* It is the variety of partial tones that give a musical note its characteristic quality.

---

*Not all partial tones present in a complex tone are integer multiples of the fundamental. Unlike the harmonics of woodwinds and brasses, stringed instruments such as a piano produce "stretched" partial tones that are nearly, but not quite, harmonics. This is an important factor in tuning pianos and happens because the stiffness of the strings adds a little bit of restoring force to the tension.

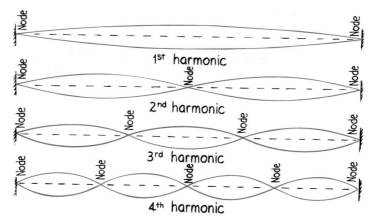

Figure 11.38 Modes of vibration of a guitar string.

Thus, if we strike middle C on the piano, we produce a fundamental tone with a pitch of about 262 hertz and also a blending of partial tones of two, three, four, five, and so on times the frequency of middle C. The number and relative loudness of the partial tones determine the quality of sound associated with the piano. Sound from practically every musical instrument consists of a fundamental and partials. Pure tones, those having only one frequency, can be produced electronically. Electronic synthesizers, for example, produce pure tones and mixtures of these to give a vast variety of musical sounds.

Figure 11.39 A composite vibration of the fundamental mode and the third harmonic.

The quality of a tone is determined by the presence and relative intensity of the various partials. The sound produced by a certain tone from the piano and the one produced by one of the same pitch from a clarinet have different qualities that the ear recognizes because their partials are different. A pair of tones of the same pitch with different qualities have either different partials or a difference in the relative intensity of the partials.

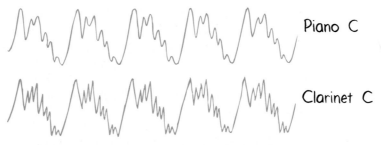

Figure 11.40 Sounds from the piano and clarinet differ in quality.

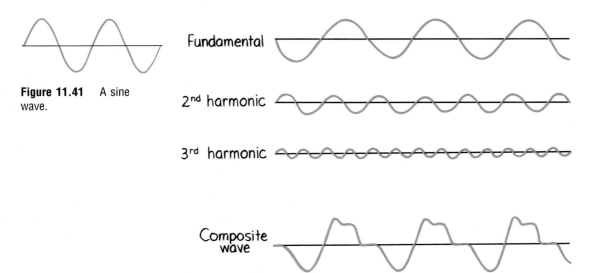

Figure 11.41 A sine wave.

Figure 11.42 Sine waves combine to produce a composite wave.

Musical Instruments

Conventional musical instruments can be grouped into one of three classes: those in which the sound is produced by vibrating strings, those in which the sound is produced by vibrating air columns, and those in which the sound is produced by *percussion*—the vibrating of a two-dimensional surface.

In a stringed instrument, the vibration of the strings is transferred to a sounding board and then to the air, with considerable dissipation of energy. Stringed instruments are low-efficiency producers of sound, so to compensate for this, we find relatively large string sections in orchestras. A smaller number of the high-efficiency wind instruments sufficiently balances a much larger number of violins.

In a wind instrument, the sound is a vibration of an air column in the instrument. There are various ways to set the air columns into vibration. In brass instruments such as trumpets, French horns, and trombones, vibrations of the players lips interact with standing waves that are set up by acoustic energy reflected within the instrument by the flared bell. The lengths of the vibrating air columns are manipulated by valves that add or subtract extra segments. In woodwinds such as clarinets, oboes, and saxophones, a stream of air produced by the musician sets a reed vibrating, whereas in fifes, flutes, and piccolos, the musician blows air against the edge of a hole to produce a fluttering stream that sets the air columns into vibration.

In percussion instruments such as drums and cymbals, a two-dimensional membrane or elastic surface is struck to produce sound. The fundamental tone produced depends on the geometry, the elasticity, and, in some cases, the tension of the surface. Changes in pitch result from changing the tension in the vibrating surface; depressing the edge of a drum membrane with the hand is one way of accomplishing this. Different modes of vibration can be set up by striking the surface in different places. In the kettledrum, the shape of the kettle changes the frequency of the drum. As in all musical sounds, the quality depends on the number and relative loudness of the partial tones.

Electronic musical instruments differ markedly from conventional musical instruments. Instead of strings that must be bowed, plucked, or struck, or reeds over which air must be blown, or diaphragms that must be tapped to produce sounds,

Fourier Analysis

One of the most interesting discoveries about music was made in 1822 by the French mathematician Joseph Fourier. He discovered that wave motion could be broken down into simple sine waves. A sine wave is the simplest of waves, having a single frequency, Figure 11.41. All periodic waves, however complicated, can be broken down into constituent sine waves of different amplitudes and frequencies. The mathematical operation for doing this is called **Fourier analysis.** We will not explain the mathematics here but simply point out that by such analysis one can find the pure sine tones that compose the tone of, say, a violin. When these pure tones are sounded together, as by striking a number of tuning forks or by selecting the proper keys on an electric organ, they combine to give the tone of the violin. The lowest-frequency sine wave is the fundamental and determines the pitch of the note. The higher-frequency sine waves are the partials that give the characteristic quality. Thus, the wave form of any musical sound is no more than a sum of simple sine waves.

Since the wave form of music is a multitude of various sine waves, to duplicate sound accurately by radio, tape recorder, or CD player, we should have as large a range of frequencies as possible. The notes of a piano keyboard range from 27 hertz to 4200 hertz, but to duplicate the music of a piano composition accurately, the sound system must have a range of frequencies up to 20,000 hertz. The greater the range of the frequencies of an electrical sound system, the closer the musical output approximates the original sound, hence the wide range of frequencies in a high-fidelity sound system.

Our ear performs a sort of Fourier analysis automatically. It sorts out the complex jumble of air pulsations that reach it and transforms them into pure tones. And we recombine various groupings of these pure tones when we listen. What combinations of tones we have learned to focus our attention on determines what we hear when we listen to a concert. We can direct our attention to the sounds of the various instruments and discern the faintest tones from the loudest; we can delight in the intricate interplay of instruments and still detect the extraneous noises of others around us. This is a most incredible feat.

Figure 11.43   Does each hear the same music?

some electronic instruments use electrons to generate the signals that make up musical sounds. Others start with sound from an acoustical instrument and then modify it. Electronic music demands of the composer and player an expertise beyond the knowledge of musicology. It brings a powerful new tool to the hands of the musician.

SUMMARY OF TERMS

Sine curve A wave form traced by simple harmonic motion, like the wavelike path traced on a moving conveyor belt by a pendulum swinging at right angles above the moving belt.

Amplitude For a wave or vibration, the maximum displacement on either side of the equilibrium (midpoint) position.

Wavelength The distance between successive crests, troughs, or identical parts of a wave.

Frequency For a body undergoing simple harmonic motion, the number of vibrations it makes per unit time. For a series of waves, the number of waves that pass a particular point per unit time.

Hertz The SI unit of frequency. One hertz (symbol Hz) equals one vibration per second.

Period The time required for a vibration or a wave to make a complete cycle; equal to 1/frequency.

Wave speed The speed with which waves pass by a particular point:

$$\text{Wave speed } = \text{ frequency } \times \text{ wavelength}$$

Transverse wave A wave in which the individual particles of a medium vibrate from side to side in a direction perpendicular (transverse) to the direction in which the wave travels. Light consists of transverse waves.

Longitudinal wave A wave in which the individual particles of a medium vibrate back and forth in a direction parallel (longitudinal) to the direction in which the wave travels. Sound consists of longitudinal waves.

Compression Condensed region of the medium through which a longitudinal wave travels.

Rarefaction Rarefied region, or region of lessened pressure, of the medium through which a longitudinal wave travels.

Interference pattern The pattern formed by superposition of different sets of waves that produces mutual reinforcement in some places and cancellation in others.

Standing wave A stationary wave pattern formed in a medium when two sets of identical waves pass through the medium in opposite directions.

Doppler effect The change in frequency of wave motion resulting from motion of the sender or receiver.

Bow wave The V-shaped wave made by an object moving across a liquid surface at a speed greater than the wave velocity.

Shock wave The cone-shaped wave made by an object moving at supersonic speed through a fluid.

Sonic boom The loud sound resulting from the incidence of a shock wave.

Infrasonic Describes a sound of frequency too low to be heard by the normal human ear—below 20 hertz.

Ultrasonic Describes a sound of frequency too high to be heard by the normal human ear—above 20,000 hertz.

Reverberation Reechoed sound.

Refraction The bending of a wave through either a nonuniform medium or from one medium to another, caused by differences in wave speed.

Forced vibration The setting up of vibrations in an object by a vibrating force.

Natural frequency A frequency at which an elastic object naturally tends to vibrate, so that minimum energy is required to produce a forced vibration or to continue vibration at that frequency.

Fundamental Frequency The lowest frequency of vibration, or first harmonic. In a string the vibration makes a single segment.

Resonance The result of forced vibrations in a body when an applied frequency matches the natural frequency of the body.

Beats A series of alternate reinforcements and cancellations produced by the interference of two sets of superimposed waves of different frequencies, heard as a throbbing effect in sound waves.

Quality The characteristic timbre of a musical sound, governed by the number and relative intensities of partial tones.

Partial tone One of the frequencies present in a complex tone. When a partial tone is an integer multiple of the lowest frequency, it is a harmonic.

Harmonic A partial tone that is an integer multiple of the fundamental. The vibration that begins with the fundamental vibrating frequency is the first harmonic, twice the fundamental is the second harmonic, and so on in sequence.

· ·

REVIEW QUESTIONS

Vibrations and Waves

1. What is a *wiggle in time* called? A *wiggle in space and time?*

2. What is the source of all waves?

3. How do *frequency* and *period* relate to each other?

Wave Motion

4. Exactly what is it that moves from source to receiver in wave motion?

5. Does the medium in which a wave moves travel along with the wave itself? Give examples to support your answer.

Wave Speed

6. What is the relationship among frequency, wavelength, and wave speed?

7. As the frequency of a wave of constant speed is increased, does the wavelength increase or decrease?

Transverse and Longitudinal Waves

8. In what direction are the vibrations compared to the direction of wave travel in a transverse wave?

9. In what direction are the vibrations compared to the direction of wave travel in a longitudinal wave?

10. How do compressions and rarefactions of longitudinal waves compare to crests and troughs of transverse waves?

Sound Waves

11. Distinguish between a *compression* and a *rarefaction*.

12. Do compressions and rarefactions travel in the same or opposite directions from one another in a wave? Cite evidence to support your answer.

13. How does the paper cone of a radio loudspeaker emit sound?

14. Compared to solids and liquids, how does air rank as a conductor of sound?

15. Why will sound not travel in a vacuum?

Speed of Sound

16. What common factors does the speed of sound depend upon? What common things does it *not* depend upon?

17. Does sound travel faster in warm air than in cold air? Defend your answer.

18. How does the speed of sound in water and steel compare to the speed of sound in air?

Reflection of Sound

19. What is the law of reflection for sound?

20. What exactly is a *reverberation*?

Refraction of Sound

21. What is the cause of refraction?

22. Does sound tend to bend upward or downward when its speed is less near the ground?

23. There is a difference between the way we passively see our surroundings by daylight and the way we actively probe our surroundings with a searchlight in the darkness. Which of these ways of perceiving our surroundings is most like the way a dolphin perceives its environment?

Forced Vibrations

24. Why will a struck tuning fork sound louder when it is held against a table?

25. Give at least three examples of forced vibration.

26. Give at least two factors that determine the natural frequency of an object.

Resonance

27. Distinguish between *forced vibrations* and *resonance*.

28. What is required to make an object resonate?

29. How does a radio select one station at a time instead of playing all stations at once?

30. Why do troops "break step" when crossing a bridge?

Interference

31. Is it possible for one wave to cancel another? Defend your answer.

32. What kind of waves exhibit interference?

33. Distinguish between *constructive interference* and *destructive interference*.

34. What is responsible for "dead spots" in poorly designed theaters or concert halls?

Beats

35. What physical phenomenon underlies the production of beats?

36. What does it mean to say one wave is out of phase with another?

Standing Waves

37. What kind of waves are characterized by interference?

38. What causes a standing wave?

39. What is a *node*?

Doppler Effect

40. Is it the frequency of a wave, the wave speed, or both that changes in the Doppler effect?

41. Is the Doppler effect characteristic of longitudinal waves, transverse waves, or both?

Wave Barriers and Bow Waves

42. How do the speed of a source of waves and the speed of the waves themselves compare for the production of a wave barrier?

43. How do the speed of a source of waves and the speed of the waves themselves compare for the production of a bow wave?

44. How does the V shape of a bow wave depend on the speed of the source?

Shock Waves and the Sonic Boom

45. How does the V shape of a shock wave depend on the speed of the source?

46. True or false: A sonic boom occurs only when an aircraft breaks through the sound barrier.

47. True or false: In order for an object to produce a sonic boom, it must be an emitter of sound.

Musical Sounds

48. Distinguish between a musical sound and noise.

Musical Instruments

49. What are the three principal classes of musical instruments?

50. Why do orchestras generally have a greater number of stringed instruments than wind instruments?

HOME PROJECTS

1. Tie a rubber tube, a spring, or a rope to a fixed support and produce standing waves. See how many nodes you can produce.

2. Test to see which ear has the better hearing by covering one ear and finding how far away your open ear can hear the ticking of a clock; repeat for the other ear. Notice also how the sensitivity of your hearing improves when you cup your ears with your hands.

3. Make the lowest-pitched sound you are capable of; then keep doubling the pitch to see how many octaves your voice can span.

EXERCISES

1. What kind of motion should you impart to the nozzle of a garden hose so that the resulting stream of water approximates a sine curve?

2. What kind of motion should you impart to a stretched coiled spring (or Slinky) to provide a transverse wave? A longitudinal wave?

3. If a gas tap is turned on for a few seconds, someone a couple of meters away will hear the gas escaping long before she smells it. What does this indicate about the way in which sound waves travel?

4. If we double the frequency of a vibrating object, what happens to its period?

5. You dip your finger repeatedly into a puddle of water and make waves. What happens to the wavelength if you dip your finger more frequently?

6. How does the frequency of vibration of a small object floating in water compare to the number of waves passing it each second?

7. In terms of wavelength, how far does a wave travel during one period?

8. The wave patterns seen in Figure 11.4 are composed of circles. What does this tell you about the speed of waves moving in different directions?

9. A rock is dropped in water, and waves spread over the flat surface of the water. What becomes of the energy in these waves when they die out?

10. Why is lightning seen before thunder is heard?

11. Why is it undesirable for a radio loudspeaker to have resonant frequencies in the audio range? (Some cheap AM radios have this problem.)

12. Would there be a Doppler effect if the source of sound were stationary and the listener in motion? Why or why not? In which direction should the listener move to hear a higher frequency? A lower frequency?

13. When you blow your horn while driving toward a stationary listener, an increase in frequency of the horn is heard by the listener. Would the listener hear an increase in horn frequency if he were also in a car traveling at the same speed in the same direction as you are? Explain.

14. How does the Doppler effect aid police in detecting speeding motorists?

15. How does the phenomenon of interference play a role in the production of bow or shock waves?

16. Does the conical angle of a shock wave open wider, narrow down, or remain constant as a supersonic aircraft increases its speed?

17. Does a sonic boom occur at the moment when an aircraft exceeds the speed of sound? Explain.

18. Why is it that a subsonic aircraft, no matter how loud it may be, cannot produce a sonic boom?

19. Imagine a super-fast fish that is able to swim faster than the speed of sound in water. Would such a fish produce a "sonic boom"?

20. A weight suspended from a spring is seen to bob up and down over a distance of 20 centimeters twice each second. What is its frequency? Its period? Its amplitude?

21. Why do flying bees buzz?

22. A cat can hear sound frequencies up to 70,000 Hz. Bats send and receive ultrahigh-frequency squeaks up to 120,000 Hz. Which hears shorter wavelengths, cats or bats?

23. At the stands of a race track you notice smoke from the starter's gun before you hear it fire. Explain.

24. When a sound wave moves past a point in air, are there changes in the density of air at this point? Explain.

25. At the instant that a high-pressure region is created just outside the prongs of a vibrating tuning fork, what is being created inside between the prongs?

26. Why is it so quiet after a snowfall?

27. If a bell is ringing inside a bell jar, we can no longer hear it if the air is pumped out, but we can still see it. What differences in the properties of sound and light does this indicate?

28. Why is the moon described as a "silent planet"?

29. As you pour water into a glass, you repeatedly tap the glass with a spoon. As the tapped glass is being filled,

does the pitch of the sound increase or decrease? (What should you do to answer this question?)

30. If the speed of sound depended on its frequency, how would distant music sound?

31. If the frequency of sound is doubled, what change will occur in its speed? In its wavelength?

32. Why does sound travel faster in warm air?

33. Why does sound travel faster in moist air? (Hint: At the same temperature, water vapor molecules have the same average kinetic energy as the heavier nitrogen and oxygen molecules in the air. How, then, do the average speeds of H_2O molecules compare with those of N_2 and O_2 molecules?)

34. Would the refraction of sound be possible if the speed of sound were unaffected by wind, temperature, and other conditions? Defend your answer.

35. Why can the tremor of the ground from a distant explosion be felt before the sound of the explosion can be heard?

36. What kinds of wind conditions would make sound more easily heard at long distances? Less easily heard at long distances?

37. Ultrasonic waves have many applications in technology and medicine. One advantage is that large intensities can be used without danger to the ear. Cite another advantage of their short wavelength. (Hint: Why do microscopists use blue light rather than white light to see detail?)

38. Why is an echo weaker than the original sound?

39. How can you estimate the distance of a distant thunderstorm?

40. A rule of thumb for estimating the distance in kilometers between an observer and a lightning stroke is to divide the number of seconds in the interval between the flash and the sound by 3. Is this rule correct?

41. If a single disturbance some unknown distance away sends out both transverse and longitudinal waves that travel with distinctly different speeds in the medium, such as in the ground during earthquakes, how could the origin of the disturbance be located?

42. Why will marchers at the end of a long parade following a band be out of step with marchers near the front?

43. If the handle of a tuning fork is held solidly against a table, the sound from the tuning fork becomes louder. Why? How will this affect the length of time the fork keeps vibrating? Explain.

44. A special device can transmit sound out of phase from a noisy jackhammer to its operator using earphones. Over the noise of the jackhammer, the operator can easily hear your voice while you are unable to hear his. Explain.

45. A certain note has a frequency of 1000 Hz. What is the frequency of a note one octave above it? Two octaves above it? One octave below it? Two octaves below it?

46. How many octaves does normal human hearing span? How many octaves are on a common piano keyboard?

47. If a guitar string vibrates in two segments, where can a tiny piece of folded paper be supported without flying off? How many pieces of folded paper could similarly be supported if the wave form were of three segments?

48. If the wavelength of a vibrating string is reduced, what effect does this have on the frequency of vibration and on the pitch?

PROBLEMS

1. You watch a distant woman driving nails into her front porch at a regular rate of 1 stroke per second. You hear the sound of the blows exactly synchronized with the blows you see. And then you hear one more blow after you see her stop hammering. How far away is she?

2. A skipper on a boat notices wave crests passing his anchor chain every 5 s. He estimates the distance between wave crests to be 15 m. He also correctly estimates the speed of the waves. What is this speed?

3. Radio waves travel at the speed of light—300 000 km/s. What is the wavelength of radio waves received at 100 MHz on your radio dial?

4. What is the wavelength of a 340-Hz tone in air? What is the wavelength of a 34,000-Hz ultrasonic wave in air?

5. An oceanic depth-sounding vessel surveys the ocean bottom with ultrasonic sound that travels 1530 m/s in seawater. How deep is the water if the time delay of the echo from the ocean floor is 6 s?

6. A bat flying in a cave emits a sound and receives its echo 1 s later. How far away is the cave wall?

7. What is the shortest wavelength you can hear?

8. What frequency of sound produces a wavelength of 1 meter in room-temperature air?

12

....

LIGHT WAVES

Light is the only thing we can really see. But what *is* light? We know that during the day the primary source of light is the sun, and the secondary source is the brightness of the sky. Other common sources are flames, white-hot filaments in light bulbs, and glowing gases in glass tubes. In all cases we find that light originates from the accelerated motion of electrons. Light is an electromagnetic phenomenon, and it is only a tiny part of a larger whole—a wide range of electromagnetic waves called the *electromagnetic spectrum*. We begin our study of light by investigating its electromagnetic properties, how it interacts with materials, and its appearance—color. We see its wave nature in the way it diffracts, interferes, and how it aligns to become polarized.

....

12.1 ELECTROMAGNETIC SPECTRUM

Shake the end of a stick back and forth in still water, and you'll produce waves on the water surface. Similarly shake an electrically charged rod to and fro in empty space, and you'll produce electromagnetic waves in space. This is because the moving charge is actually an electric current. Recall from Chapter 10 that a magnetic field surrounds an electric current. The magnetic field is changing as the charges accelerate. Recall also from Chapter 10 that a changing magnetic field induces an electric field—electromagnetic induction. And what does the changing electric field do? The changing electric field, in accordance with Maxwell's counterpart to Faraday's law, will induce a changing magnetic field. The vibrating electric and magnetic fields regenerate each other to make up an **electromagnetic wave.**

Figure 12.1 Shake an electrically charged object to and fro, and you produce an electromagnetic wave.

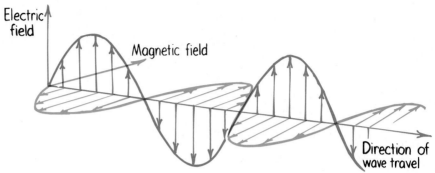

Figure 12.2 The electric and magnetic fields of an electromagnetic wave are perpendicular to each other and to the direction of motion of the wave.

In a vacuum, all electromagnetic waves move at the same speed and differ from one another in their frequency. The classification of electromagnetic waves according to frequency is the **electromagnetic spectrum** (Figure 12.3). Electromagnetic waves have been detected with a frequency as low as 0.01 hertz (Hz). Electromagnetic waves with frequencies on the order of several thousand hertz (kHz) are classified as radio waves. The very high frequency (VHF) television band of waves starts at about 50 million hertz (MHz). Still higher frequencies are called microwaves; these are followed by infrared waves, often called "heat waves." Further still is visible light, which makes up less than a millionth of 1 percent of the electromagnetic spectrum. The lowest frequency of light we can see with our eyes appears red. The highest visible frequencies are nearly twice the frequency of red and appear violet. Still higher frequencies are ultraviolet. These higher-frequency waves are more energetic and cause sunburns. Higher frequencies beyond ultraviolet extend into the X-ray and gamma-ray regions. There is no sharp boundary between these regions, which actually overlap each other. The spectrum is broken up into these arbitrary regions for classification.

Figure 12.3 The electromagnetic spectrum is a continuous range of waves extending from radio waves to gamma rays. The descriptive names of the sections are merely a historical classification, for all waves are the same in nature, differing principally in frequency and wavelength; all have the same speed.

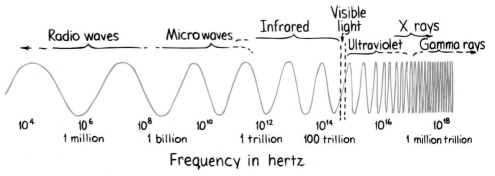

The frequency of the electromagnetic wave as it vibrates through space is identical to the frequency of the oscillating electric charge that generates it. Different frequencies result in different wavelengths—low frequencies produce long wavelengths and high frequencies produce short wavelengths. The higher the frequency of the vibrating charge, the shorter the wavelength of radiation.*

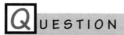UESTION

Is it correct to say that a radio wave is a low-frequency light wave? Is a radio wave also a sound wave?

12.2 TRANSPARENT AND OPAQUE MATERIALS

Light is energy carried in an electromagnetic wave emitted by vibrating electrons in atoms. When light is incident upon matter, some of the electrons in the matter are forced into vibration. In this way, vibrations in the emitter are transmitted to vibrations in the receiver. This is similar to the way that sound is transmitted (Figure 12.4).

Figure 12.4 Just as a sound wave can force a sound receiver into vibration, a light wave can force electrons in materials into vibration.

Thus the way a receiving material responds when light is incident upon it depends on the frequency of the light and the natural frequency of the electrons in the material. Visible light vibrates at a very high rate, some 100 trillion times per second (10^{14} hertz). If a charged object is to respond to these ultra-fast vibrations, it must have very, very little inertia. Electrons are light enough to vibrate at this rate.

[A]NSWER

Both a radio wave and light wave are electromagnetic waves, which originate in the vibrations of electrons. Radio waves have lower frequencies than light waves, so a radio wave may be considered to be a low-frequency light wave (and a light wave a high-frequency radio wave). But a sound wave is a mechanical vibration of matter and is not electromagnetic. A sound wave is fundamentally different from an electromagnetic wave. So a radio wave is definitely not a sound wave.

---

*The relationship is $c = f\lambda$, where c is the wave's speed (constant), f is the frequency, and λ is the wavelength. It is common to describe sound and radio by frequency and light by wavelength. In this book, however, we'll favor the single concept of frequency in describing light.

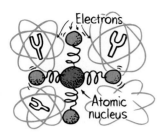

Figure 12.5 The outer electrons in an atom vibrate as if they were attached to the nucleus by springs. As a result, atoms and molecules behave somewhat like optical tuning forks.

Materials such as glass and water allow light to pass through in straight lines. We say they are **transparent** to light. To understand how light gets through a transparent material, visualize the electrons in an atom as if they were connected by springs (Figure 12.5).* When a light wave is incident upon them, they are set into vibration.

Materials that are springy (elastic) respond more to vibrations at some frequencies than others. Bells ring at a particular frequency, tuning forks vibrate at a particular frequency, and so do the electrons of atoms and molecules. The natural vibration frequencies of an electron depend on how strongly it is attached to its atom or molecule. Different atoms and molecules have different "spring strengths." Electrons in glass have a natural vibration frequency in the ultraviolet range. When ultraviolet rays shine on glass, resonance occurs as the wave builds and maintains a large amplitude of vibration between the electron and the atomic nucleus, just as pushing someone at the resonant frequency on a swing builds a large amplitude. The energy the atom receives may be passed on to neighboring atoms by collisions, or it may be re-emitted. Resonating atoms in the glass can hold onto the energy of the ultraviolet light for quite a long time (about 100 millionths of a second). During this time the atom makes about 1 million vibrations, and it collides with neighboring atoms and gives up its energy as heat. Thus, glass is not transparent to ultraviolet.

At lower wave frequencies, like those of visible light, electrons in the glass are forced into vibration, but at less amplitude. The atom or molecule holds the energy for less time, with less chance of collision with neighboring atoms and molecules, and less energy transformed to heat. The energy of vibrating electrons is re-emitted as light. Glass is transparent to all the frequencies of visible light. The frequency of the re-emitted light that is passed from molecule to molecule is identical to the frequency of the light that produced the vibration in the first place. The principal difference is a slight time delay between absorption and re-emission.

It is this time delay that results in a lower average speed of light through a transparent material (Figure 12.6). Light travels at different average speeds through different materials. We say *average speeds,* for the speed of light in a vacuum, whether in interstellar space or in the space between molecules in a piece of glass, is a constant 300,000 kilometers per second. We call this speed of light c.† Light travels a slight bit less than this in the atmosphere, which is usually rounded off as c. In water light travels at 75 percent of its speed in a vacuum, or $0.75\,c$. In glass light travels about $0.67\,c$, depending on the type of glass. In a diamond light travels at less than half its speed in a vacuum, only $0.41\,c$. When light emerges from these materials into the air, it travels at its original speed, c.

---

*Electrons, of course, are not really connected by springs. We are simply presenting a visual "spring model" of the atom to help us understand the interaction of light with matter. Physicists devise such conceptual models to understand nature, particularly at the submicroscopic level. The worth of a model lies not in whether it is "true," but whether it is useful. A good model not only is consistent with and explains observations, but also predicts what may happen. If predictions of the model are contrary to what happens, the model is usually either refined or abandoned. The simplified model that we present here—of an atom whose electrons vibrate as if on springs, with a time interval between absorbing energy and re-emitting energy—is quite useful for understanding how light passes through transparent material.

†The presently accepted value is 299,792 km/s, rounded to 300,000 km/s. (This corresponds to 186,000 mi/s).

Figure 12.6 A light wave incident upon a pane of glass sets up vibrations in the molecules that produce a chain of absorptions and re-emissions, which in turn pass the light energy through the material and out the other side. Because of the time delay between absorptions and re-emissions, the light travels more slowly in the glass.

Infrared waves, with frequencies lower than visible light, vibrate not only the electrons, but entire molecules in the structure of the glass. This vibration increases the internal energy and temperature of the structure, which is why infrared waves are often called *heat waves*. Glass is transparent to visible light, but not to ultraviolet and infrared light.

Q︎UESTIONS

1. Why is glass transparent to visible light but opaque to ultraviolet and infrared?
2. Pretend that while you walk across a room, you make several momentary stops along the way to greet people who are "on your wavelength." How is this analogous to light traveling through glass?
3. In what way is it not analogous?

A︎NSWERS

1. The natural frequency of vibration for electrons in glass is the same as the frequency of ultraviolet light, so resonance in the glass occurs when ultraviolet waves shine on it. The energetic vibrations of electrons generate heat instead of wave re-emission, so the glass is opaque at higher frequencies. In the range of visible light, the forced vibrations of electrons in the glass are at smaller amplitudes— vibrations are more subtle, re-emission of light rather than the generation of heat occurs, and the glass is transparent. Lower-frequency infrared causes whole molecules, rather than electrons, to resonate, and again heat is generated and the glass is opaque.

2. Your average speed across the room would be less because of the time delays associated with your momentary stops. Likewise, the speed of light in glass is less because of the time delays in interactions with atoms along its path.

3. In the case of walking across the room, it is you who begin the walk and you who complete the walk. This is not analogous to the similar case of light, for according to our model for light passing through a transparent material, the light that is absorbed by an electron made to vibrate is not the same light that is re-emitted—even though the two, like identical twins, are indistinguishable.

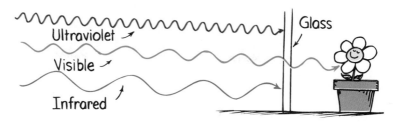

Figure 12.7 Glass blocks both infrared and ultraviolet, but is transparent to all the frequencies of visible light.

Most things around us are **opaque**—they absorb light without re-emission. Books, desks, chairs, and people are opaque. Vibrations given by light to their atoms and molecules are turned into random kinetic energy—into internal energy. They become slightly warmer.

Metals are opaque. The outer electrons of atoms in metals are not bound to any particular atom. They are free to wander with very little restraint throughout the material (which is why metal conducts electricity and heat so well). When light shines on metal and sets these free electrons into vibration, their energy does not "spring" from atom to atom in the material, but is instead reflected. That's why metals are shiny.

The earth's atmosphere is transparent to some ultraviolet, all visible light, some infrared, but is opaque to high-frequency ultraviolet waves. The small amount of ultraviolet that does get through is responsible for sunburns. If it all got through we would be fried to a crisp. Clouds are semitransparent to ultraviolet, which is why you can get a sunburn on a cloudy day. Ultraviolet rays are not only harmful to the skin, but are also damaging to tar roofs. Now you know why tarred roofs are covered with gravel.

So we see that when light encounters material it is either transmitted, absorbed, or reflected. Usually it is some combination of these.

Figure 12.8 Metals are shiny because light that shines on them forces into vibration free electrons, which then emit their "own" light waves as reflection.

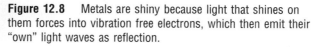

12.3 COLOR

Roses are red and violets are blue; colors intrigue artists and physics types too. To the physicist, the colors of objects are not in the substances of the objects themselves or even in the light they emit or reflect. Color is a physiological experience and is in the eye of the beholder. So when we say that light from a rose is red, in a stricter sense we mean that it appears red. Many organisms, including people with defective color vision, will not see the rose as red at all.

The colors we see depend on the frequency of the light we see. Different frequencies of light are perceived as different colors; the lowest frequency we detect appears to most people as the color red, and the highest as violet. Between them range the infinite number of hues that make up the color spectrum of the rainbow. By convention these hues are grouped into the seven colors of red, orange, yellow,

Figure 12.9 The colors of things depend on the colors of light that illuminate them.

green, blue, indigo, and violet. These colors together appear white. The white light from the sun is a composite of all the visible frequencies.

Except for light sources such as lamps, lasers, and gas discharge tubes (which we will treat in Chapter 14), most of the objects around us reflect rather than emit light. They reflect only part of the light that is incident upon them, the part that gives them their color.

Selective Reflection

A rose, for example, doesn't emit light; it reflects light (Figure 12.9). If we pass sunlight through a prism and then place a deep-red rose in various parts of the spectrum, the rose will appear brown or black in all parts of the spectrum except in the red. In the red part of the spectrum, the petal also will appear red, but the green stem and leaves will appear black. This shows that the red rose has the ability to reflect red light, but it cannot reflect other kinds of light; the green leaves have the ability to reflect green light and likewise cannot reflect other kinds of light. When the rose is held in white light, the petals appear red and the leaves green, because the petals reflect the red part of the white light and the leaves reflect the green part of the white light. To understand why objects reflect specific colors of light, we must turn our attention to the atom.

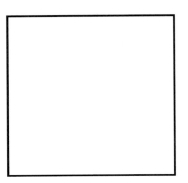 

Figure 12.10 The square on the left *reflects* all the colors illuminating it. In sunlight it is white. When illuminated with blue light, it is blue. The square on the right *absorbs* all the colors illuminating it. In sunlight it is warmer than the white square.

Light is reflected from objects in a manner similar to the way sound is "reflected" from a tuning fork when another that is nearby sets it into vibration. A tuning fork can be made to vibrate even when the frequencies are not matched, although at significantly reduced amplitudes. The same is true of atoms and molecules. Electrons can be forced into vibration (oscillation) by the vibrating (oscillating) electric fields of electromagnetic waves.* Once vibrating, these electrons send out their own electromagnetic waves just as vibrating acoustical tuning forks send out sound waves.

---

*We use the words *oscillate* and *vibrate* interchangeably. Also, the words *oscillators* and *vibrators* have the same meaning.

Usually a material will absorb light of some frequencies and reflect the rest. If a material absorbs most of the light and reflects red, for example, the material appears red. If it reflects light of all the visible frequencies, like the white part of this page, it will be the same color as the light that shines on it. If a material absorbs all the light that shines on it, it reflects none and is black.

When white light falls on a flower, light of some frequencies is absorbed by the cells in the flower and some is reflected. Cells that contain chlorophyll absorb most of the light and reflect the green part that falls on it. The petals of a red rose, on the other hand, reflect primarily red light, with a lesser amount of blue. Interestingly enough, the petals of most yellow flowers, like daffodils, reflect red and green as well as yellow. Yellow daffodils reflect a broad band of frequencies. The reflected colors of most objects are not pure single-frequency colors, but are composed of a spread of frequencies.

An object can only reflect frequencies that are present in the illuminating light. The appearance of a colored object therefore depends on the kind of light used. A candle flame emits light that is deficient in blue; its light is yellowish. An incandescent lamp emits light that is richer toward the lower frequencies, enhancing the reds. In a fabric with a little bit of red in it, for example, the red will be more apparent under an incandescent lamp than when illuminated with a fluorescent lamp. Fluorescent lamps are richer in the higher frequencies, so blues are enhanced under them. With various kinds of illumination, it is difficult to tell the true color of objects. Colors appear different in daylight from how they appear when illuminated by either kind of lamp (Figure 12.11).

Figure 12.11 Color depends on the light source.

Selective Transmission

The color of a transparent object depends on the color of the light it transmits. A red piece of glass appears red because it absorbs all the colors that compose white light, except red, which it *transmits*. Similarly, a blue piece of glass appears blue because it transmits primarily blue and absorbs the other colors that illuminate it. The piece of glass contains dyes or *pigments*—fine particles that selectively absorb certain frequencies and selectively transmit others. From an atomic point of view, electrons in the pigment molecules are set into vibration by the illuminating light. Some of the frequencies are absorbed by the pigments, and others are re-emitted from molecule to molecule in the glass. The energy of the absorbed frequencies of light increases the kinetic energy of the molecules and the glass is warmed. Ordinary window glass is colorless because it transmits light of all visible frequencies equally well.

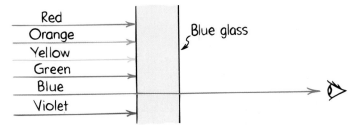

Figure 12.12 Only energy having the frequency of blue light is transmitted; energy of the other frequencies is absorbed and warms the glass.

Mixing Colored Lights

The fact that white light from the sun is a composite of all the visible frequencies is easily demonstrated by passing sunlight through a prism and observing the rainbow-colored spectrum. The distribution of solar frequencies (Figure 12.13) is uneven, being most intense in the yellow-green part of the spectrum. It is interesting to note that our eyes have evolved to have maximum sensitivity in this range. That's the reason for new fire engines to be painted yellow-green, particularly at airports where visibility is vital. This also explains why at night we see better under the illumination of yellow sodium-vapor lamps than under common tungsten-filament lamps of the same brightness.

Figure 12.13 The radiation curve of sunlight is a graph of brightness versus frequency. Sunlight is brightest in the yellow-green region, in the middle of the visible range.

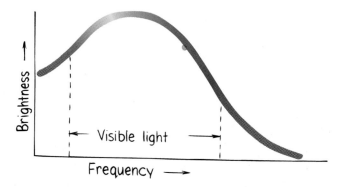

All the colors together make white. Interestingly enough, the perception of white also results from the combination of only red, green, and blue light. We can understand this by greatly simplifying the solar radiation curve by dividing it into three regions as in Figure 12.14. Three types of cone-shaped receptors in our eyes perceive color. Light in the lowest third of the spectral distribution stimulates the cones

ANSWERS

1. The leaves absorb rather than reflect red light, so the leaves become warmer.
2. The petals absorb rather than reflect the green light. Since green is the only color illuminating the rose, and green contains no red to be reflected, the rose reflects no color at all and appears black.
3. Reflection from the top surface is white, because the light doesn't go far enough into the colored glass for absorption of nonred light. The back surface has only red to reach it, so red is what you see reflected.

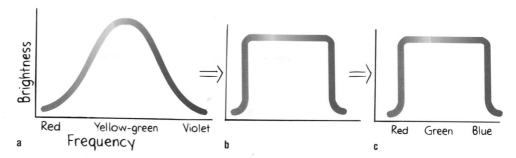

Figure 12.14 (a) Radiation curve of sunlight (b) simplified and (c) divided into three regions.

sensitive to low frequencies and appears red; light in the middle third stimulates the mid-frequency-sensitive cones and appears green; light in the high-frequency third stimulates the higher-frequency-sensitive cones and appears blue. When all three types of cones are stimulated equally, we see white.

Project red, green, and blue lights on a screen, and where they all overlap white is produced. If two of the three colors overlap, or are added, then another color sensation will be produced (Figure 12.15). By adding various amounts of red, green, and blue, the colors to which each of our three types of cones are sensitive, we can produce any color in the spectrum. For this reason, red, green, and blue are called the **additive primary colors.** A close examination of the picture on most color television tubes will reveal that the picture is an assemblage of tiny spots, each less than a millimeter across. When the screen is lit, some of the spots are red,

Figure 12.15 The white golf ball appears white because it is illuminated with red, green, and blue lights. Why are the shadows of the ball the complementary colors of red, green, and blue?

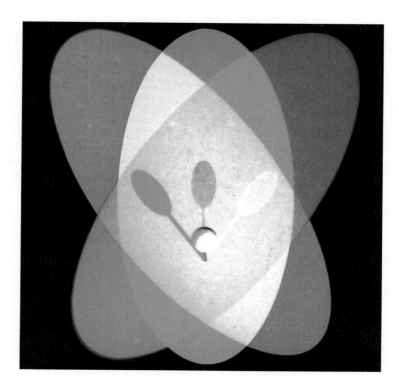

Color Vision

Light from the world around us focuses upon the retina in our eyes, and we see. The *retina* is composed of tiny antennae of two kinds that resonate to the incoming light—the rods and the cones. As the names imply, the rods are rod-shaped and the cones cone-shaped. Rods perceive only intensity of light, while cones perceive color. We see color because of the three types of cones—those sensitive to red, those sensitive to green, and those sensitive to blue. Cones are denser toward the region of distinct vision—the *fovea*. The rods are sensitive to intensity rather than frequency, and predominate away from the fovea, toward the periphery of the retina. Primates and a species of ground squirrel are the only mammals that have the three types of cones and experience full color vision. The retinas of other mammals consist primarily of rods, which are sensitive only to lightness or darkness, like a black-and-white photograph or movie.

Compared to rods, cones require more energy to "fire" an impulse through the nervous system. If the intensity of light is very low, the things we see have no color. We see low intensities with our rods. That's why its difficult to tell the color of a car by moonlight. Dark-adapted vision is almost entirely due to the rods, while vision in bright light is due to the cones. Stars, for example, look white to us. Yet most stars are actually brightly colored. A time exposure of the stars with a camera reveals reds and red-oranges for the "cooler" stars and blues and blue-violets for the "hotter" stars. The starlight is too weak, however, to fire the color-perceiving cones in the retina. So we see the stars with our rods and perceive them as white or, at best, as only faintly colored. Females have a slightly lower threshold of firing for the cones, however, and can see a bit more color than males. So if she says she sees colored stars and he says she doesn't, she is probably right!

What goes on in the eye seems to be quite complex. Some color sensations depend on intensity with both rods and cones responding. As intensity increases, orange seems to get yellower and violet seems to get bluer—with no change in frequency. Yellow, green, and blue, however, are independent of intensity and are called "psychological primaries." The eye is indeed fascinating.

some green, some blue; the mixtures of these primary colors at a distance provide a complete range of colors, plus white.*

Note carefully in Figure 12.15 the colors produced where only two primaries overlap. The sum of blue and red is **magenta**; the sum of green and blue is the greenish-blue color **cyan**; and the sum of red and green is **yellow** (which makes sense when you realize yellow is between red and green in the spectrum). We say that magenta is opposite green; cyan is opposite red; and yellow is opposite blue. A color plus its opposite appear white. We call any two colors that add together to produce white **complementary colors.**

So we see that when an additive primary and its opposite are added they produce white. Blue plus yellow, for example, produces white. (This follows logically: since yellow is (red + green), then blue + (red + green) = white.) From the figure we also see that red plus cyan produces white; green plus magenta produces white also.

---

*It's interesting to note that the "black" you see on the darkest scenes on a black-and-white TV tube is simply the color of the tube face itself, which is more a light gray than black. Because our eyes are sensitive to the contrast with the illuminated parts of the screen, we see this gray as black.

Q|UESTIONS

1. From Figure 12.15, find the complements of cyan, of yellow, and of red.

2. Red + cyan = _____.

3. White − cyan = _____.

4. White − red = _____.

Figure 12.16 Seen through a magnifying glass, the color green on a printed page is made up of blue and yellow dots.

The fact that a color and its complement combine to produce white light is nicely used in lighting stage performances. Blue and yellow lights shining on performers, for example, produce the effect of white light alone—except where one of the two colors is absent, like in the shadows. The shadow of one lamp, say the blue, will be illuminated by the yellow lamp and appear yellow. Similarly, the shadow cast by the yellow lamp will appear blue. This is a most interesting effect.

We see this effect in Figure 12.15, where red, green, and blue light shine on the golf ball. Note the shadows cast by the ball. The middle shadow is cast by the green spotlight and is not dark because it is illuminated by the red and blue lights, which overlap and add to produce magenta. Similarly, the shadow cast by the blue light appears yellow because it is illuminated by red and green light. Can you see why the shadow cast by the red light appears cyan?

Mixing Colored Pigments

Every artist knows that if you mix red, green, and blue paint, the result will not be white, but a muddy dark brown. Red and green paint certainly do not mix to form yellow, as is the rule for adding colored lights. The mixing of pigments in paints and dyes is entirely different than mixing lights. Pigments are tiny particles that absorb specific colors. For example, pigments that produce the color red absorb the complementary color cyan. So something painted red absorbs cyan, which is why it reflects red. In effect, cyan has been *subtracted* from white light. Something painted blue absorbs yellow, so it reflects all the colors except yellow. Take yellow away from white and you've got blue. The colors magenta, cyan, and yellow are the *subtractive primaries*. The variety of colors you see in the colored photographs in this or any book are the result of magenta, cyan, and yellow dots. Light illuminates the book, and light of some frequencies is subtracted from the light reflected. The rules of color subtraction differ from the rules of light addition. We leave this topic to the recommended reading.

A|NSWERS

1. Red, blue, cyan.

2. White.

3. Red.

4. Cyan. Interestingly enough, the cyan color of seawater is the result of red taken away from white sunlight. The natural frequency of water molecules coincides with the frequency of infrared, so infrared is strongly absorbed by water. To a lesser extent, red light is absorbed by water—enough to give it the greenish-blue or cyan color.

Why the Sky Is Blue

Not all colors are the result of the addition of light or the subtraction of light. Some colors, like the blue of the sky, are the result of selective scattering. Consider the analogous case of sound: If a beam of a particular frequency of sound is directed to a tuning fork of similar frequency, the tuning fork will be set into vibration and will effectively redirect the beam in many directions. The tuning fork *scatters* the sound. A similar process occurs with the scattering of light from atoms and particles that are far apart from one another, as in the atmosphere.*

We know that atoms behave like tiny optical tuning forks and re-emit light waves that shine on them. Very tiny particles do the same. The tinier the particle, the higher the frequency of light it will scatter. This is similar to the way small bells ring with higher notes than larger bells. The nitrogen and oxygen molecules and the tiny particles that make up the atmosphere are like tiny bells that "ring" with high frequencies when energized by sunlight. Like sound from the bells, the re-emitted light is sent in all directions. It is scattered.

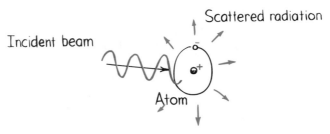

Figure 12.17 A beam of light falls on an atom and causes the electrons in the atom to vibrate. The vibrating electrons, in turn, re-emit light in various directions. Light is scattered.

Most of the ultraviolet light from the sun is absorbed by a thin protective layer of ozone gas in the upper atmosphere. The remaining ultraviolet sunlight that passes through the atmosphere is scattered by atmospheric particles and molecules. Of the visible frequencies, violet is scattered the most, followed by blue, green, yellow, orange, and red, in that order. Red is scattered only a tenth as much as violet. Although violet light is scattered more than blue, our eyes are not very sensitive to violet light. The lesser amount of blue predominates in our vision, so we see a blue sky!

The blue of the sky varies in different places under different conditions. A principal factor is the water-vapor content of the atmosphere. On clear dry days the sky is a much deeper blue than on clear days with high humidity. Places where the upper air is exceptionally dry, such as Italy and Greece, have beautifully blue skies that have inspired painters for centuries. Where there are a lot of particles of dust and other particles larger than oxygen and nitrogen molecules, light of the lower frequencies is scattered more. This makes the sky less blue, and it takes on a whitish appearance. After a heavy rainstorm when the particles have been washed away, the sky becomes a deeper blue.

---

*This type of scattering is called *Rayleigh scattering* (after Lord Scattering) and occurs whenever the scattering particles are much smaller than the wavelength of incident light and have resonances at frequencies higher than the scattered light. The shorter the wavelength of light, the more light is scattered.

Figure 12.18 In clean air the scattering of high-frequency light gives us a blue sky. When the air is full of particles larger than molecules, light of lower frequencies is also scattered, which adds to give a whitish sky.

The grayish haze in the skies of large cities is the result of particles emitted by internal combustion engines (cars, trucks, and industrial plants). Even when idling, a typical automobile engine emits more than 100 billion particles per second. Most are invisible and provide a framework to which other particles adhere. These are the primary scatterers of lower frequency light. For the larger of these particles, absorption rather than scattering takes place and a brownish haze is produced. Yuk!

Why Sunsets Are Red

Light of lower frequencies is scattered the least by nitrogen and oxygen molecules. Therefore red, orange, and yellow light are transmitted through the atmosphere much more than violet and blue. Red, which is scattered the least, passes through more atmosphere than any other color. Therefore, when white light passes through a thick atmosphere, higher-frequency light is scattered most while light of the lower frequencies is transmitted with minimal scattering. Such a thicker atmosphere is presented to sunlight at sunset.

At noon the sunlight travels through the least amount of atmosphere to reach the earth's surface. Only a small amount of high-frequency light is scattered from sunlight, enough to make the sun somewhat yellow. As the day progresses and the sun is lower in the sky (Figure 12.19), the path through the atmosphere is longer, and more blue is scattered from the sunlight. The removal of blue leaves the transmitted light more reddish in appearance. The sun becomes progressively redder, going from yellow to orange and finally to a red-orange at sunset. Sunsets and sunrises are unusually colorful following volcanic eruptions, for particles larger than atmospheric molecules are then more abundant in the air.

The colors of the sunset are consistent with our rules for color mixing. When blue is subtracted from white light, the complementary color that is left is yellow. When higher-frequency violet is subtracted, the resulting complementary color is

Figure 12.19 A sunbeam must travel through more kilometers of atmosphere at sunset than at noon. As a result, more blue is scattered from the beam at sunset than at noon. By the time a beam of initially white light gets to the ground, only light of lower frequencies survives to produce a red sunset.

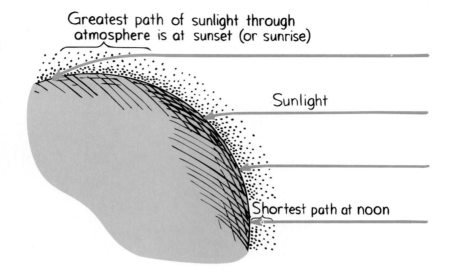

orange. When medium-frequency green is subtracted, magenta is left. The combinations of resulting colors vary with atmospheric conditions, which change from day to day and give us a variety of sunsets.

Why Clouds Are White

Clusters of water molecules in a variety of sizes make up clouds. The different-size clusters result in a variety of scattered frequencies: the tiniest, blue; slightly larger clusters, say, green; and still larger clusters, red. The overall result is a white cloud. Electrons close to one another in a cluster vibrate together and in step, which results in a greater intensity of scattered light than from the same number of electrons vibrating separately. Hence, clouds are bright!

Absorption occurs for larger droplets, and the scattered intensity is less. Clouds composed of larger droplets are darker. Further increase in the size of the drops causes them to fall to earth, and we have rain.

The next time you find yourself admiring a crisp blue sky, or delighting in the shapes of bright clouds, or watching a beautiful sunset, think about all those ultra-tiny optical tuning forks vibrating away—you'll appreciate these everyday wonders of nature even more!

Figure 12.20 A cloud is composed of a variety of particle sizes. The tiniest scatter blue, slightly larger scatter green, and still larger scatter reds. The overall result is a white cloud.

Q UESTIONS

1. If molecules in the sky scattered low-frequency light more than high-frequency light, how would the colors of the sky and sunsets appear?

2. Distant dark mountains are bluish in color. What is the source of this blueness? (Hint: Exactly what is between us and the mountains we see?)

3. Distant snow-covered mountains reflect a lot of light and are bright. But they look yellowish, depending on how far away they are. Why do they appear yellowish? (Hint: What happens to the reflected white light in traveling from the mountain to us?)

Figure 12.21 Water is cyan because it absorbs red. The froth in the waves is white because, like clouds, it is composed of a variety of tiny clusters of water droplets that scatter all the visible frequencies.

A NSWERS

1. If light of low frequencies were scattered, the noontime sky would appear reddish orange. At sunset more reds would be scattered by the longer path of the sunlight, and the sunlight would be predominantly blue and violet. So sunsets would appear blue!

2. If we look at distant dark mountains, very little light from them reaches us, and the blueness of the atmosphere between us and the mountains predominates. The blueness is of the low-altitude "sky" between us and the mountains. That's why distant mountains look blue!

3. The reason that bright snow-covered mountains appear yellow is because the blue in the white light from the snowy mountains is scattered on its way to us. So by the time the light gets to us, it is weak in the high frequencies and strong in the low frequencies—hence it is yellowish. For greater distances, farther away than mountains are usually seen, they would appear orange for the same reason a sunset appears orange.

   Why do we see the scattered blue when the background is dark, but not when the background is bright? Because the scattered blue is faint. A faint color will show itself against a dark background, but not against a bright background. For example, when we look from the earth's surface at the atmosphere against the darkness of space, the atmosphere is sky blue. But astronauts above who look below through the same atmosphere to the bright surface of the earth do not see the same blueness.

Figure 12.22   The oscillating meter stick makes plane waves in the tank of water. Waves diffract through the opening.

12.4   DIFFRACTION

Plane waves can be generated in water by successively dipping a horizontally held straightedge such as a meter stick into the surface (Figure 12.22). The photographs in Figure 12.23 are top views of a ripple tank, where plane waves are produced in water by an oscillating straightedge. The straightedge is not shown, but the plane waves produced are shown incident upon openings of various sizes. In (a), where the opening is wide, we see the plane waves continue through the opening without change—except at the corners, where the waves are bent into the shadow region. This bending is **diffraction.** Any bending of light by means other than reflection and refraction is called *diffraction*. As the width of the opening is narrowed as in (b), the spreading of waves into the shadow region is somewhat more pronounced. When the opening is small compared to the wavelength of the incident wave as in (c), diffraction is much more pronounced. We say the waves are *diffracted* as they spread into the shadow region. Diffraction occurs for all kinds of waves, including sound and light waves.

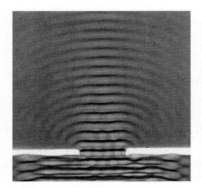
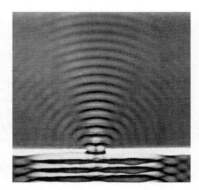
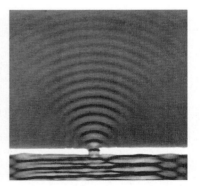

Figure 12.23   Plane waves passing through openings of various sizes. The smaller the opening, the greater the bending of the waves at the edges.

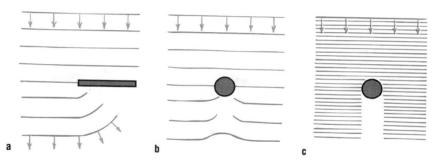

Figure 12.24 (a) Waves tend to spread into the shadow region. (b) When the wavelength is about the size of the object, the shadow is soon filled in. (c) When the wavelength is short compared to the object, a sharp shadow is cast.

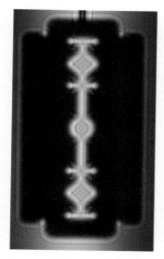

Figure 12.25 Diffraction fringes are evident in the shadows of monochromatic (single-frequency) laser light.

Diffraction is not confined by narrow slits or to openings in general but can be seen for all shadows. On close examination, even the sharpest shadow is blurred slightly at the edge (Figure 12.25).

The amount of diffraction depends on the wavelength of the wave compared to the size of the obstruction that casts the shadow. Long waves are better at filling in shadows, which is why foghorns emit low-frequency sound waves—to fill in any "blind spots." Likewise for radio waves of the standard AM broadcast band, which are very long compared to the size of most objects in their path. The wavelength of AM radio waves ranges from 180 to 6000 meters, and the waves readily bend around buildings and other objects that might otherwise obstruct them. A long-wavelength radio wave doesn't "see" a relatively small building in its path—but a short-wavelength radio wave does. The radio waves of the FM band range from 2.7 to 3.7 meters and don't bend very well around buildings. This is one of the reasons that FM reception is often poor in localities where AM comes in loud and clear. In the case of radio reception, we don't wish to "see" objects in the path of radio waves, so diffraction is nice.

Diffraction is not so nice for viewing very small objects. If the size of the object is about the same as the wavelength of light, the image of the object will be blurred by diffraction. If the object is smaller than the wavelength of light, no structure can be seen. The entire image is lost due to diffraction. No amount of magnification or perfection of microscope design can defeat this fundamental diffraction limit.

UESTION

Why does a microscopist use blue light rather than white light to illuminate the objects viewed?

A NSWER

For the objects viewed, less diffraction results from the short waves of blue light compared to other, longer waves. So the microscopist sees more detail with short-wave blue light, just as a dolphin beautifully investigates fine detail in its environment by the echoes of ultra-short wavelengths of sound.

To minimize this problem, microscopists illuminate tiny objects with the shorter wavelengths of electron beams rather than light. They use *electron microscopes,* which take advantage of the fact that all matter has wave properties: A beam of electrons has a wavelength smaller than visible light. In an electron microscope, electric and magnetic fields, rather than optical lenses, are used to focus and magnify images.

The fact that smaller details can be better seen with smaller wavelengths is neatly employed by the dolphin in scanning its environment with ultrasound. The echoes of long-wavelength sound give the dolphin an overall image of objects in its surroundings. To examine more detail, the dolphin emits sound of shorter wavelengths. The dolphin has always done naturally what physicians have only recently been able to do with an ultrasonic imaging device.

12.5 INTERFERENCE

Note that the diffracted light in Figure 12.25 shows fringes. These fringes are produced by **interference,** which we discussed in the previous chapter. Constructive and destructive interference is reviewed in Figure 12.26. We see that the adding, or superposition, of a pair of identical waves in phase with each other produces a wave of the same frequency but with twice the amplitude. If the waves are exactly one-half wavelength out of phase, their superposition results in complete cancellation. If they are out of phase by other amounts, partial cancellation occurs.

Figure 12.26
Wave interference.

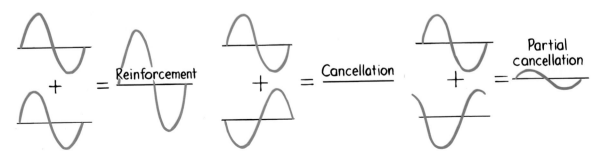

In 1801 the wave nature of light was convincingly demonstrated when the British physicist and physician Thomas Young performed his now-famous interference experiment.* Young found that light directed through two closely spaced pinholes recombined to produce fringes of brightness and darkness on a screen behind. The bright fringes of light resulted from light waves from the two holes arriving crest to crest, while the dark areas resulted from light waves arriving trough to crest. Figure 12.27 shows Young's drawing of the pattern of superimposed waves from the two sources. His experiment is now done with two closely spaced slits instead of pinholes, so the fringe patterns are straight lines (Figure 12.28).

We see in Figures 12.29 and 12.30 how the series of bright and dark lines results from the different path lengths from the slits to the screen. For the central

*Thomas Young read fluently at the age of 2; by 4, he had read the Bible twice; by 14, he knew eight languages. In his adult life he was a physician and scientist, contributing to an understanding of fluids, work and energy, and the elastic properties of materials. He was the first person to make progress in deciphering Egyptian hieroglyphics. No doubt about it—Thomas Young was a bright guy!

Figure 12.27 Thomas Young's original drawing of a two-source interference pattern. Letters C, D, E, and F mark regions of destructive interference.

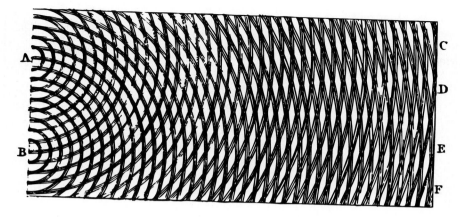

bright fringe, the paths from each slit are the same length, and the waves arrive in phase and reinforce each other. The dark fringes on either side of the central fringe result from one path being longer (or shorter) by one-half wavelength, where the waves arrive half a wavelength, or 180°, out of phase. The other sets of dark fringes occur where the paths differ by odd multiples of one-half wavelength: $\frac{3}{2}$, $\frac{5}{2}$, and so on.

Figure 12.28 When monochromatic light passes through two closely spaced slits, a striped interference pattern is produced.

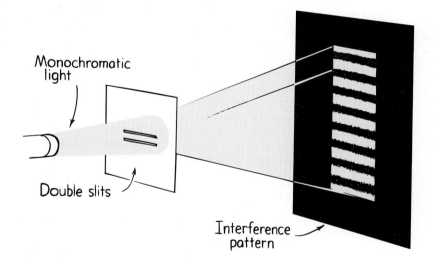

Interference patterns are not limited to one or two slits. A multitude of closely spaced slits makes up a *diffraction grating*. These devices, like prisms, disperse white light into colors. These are used in devices called *spectrometers,* which we will discuss in Chapter 14. The feathers of some birds act as diffraction gratings and disperse colors. Likewise for the microscopic pits on the reflective surface of compact discs.

Figure 12.29 Bright fringes occur when waves from both slits arrive in phase; dark areas result from the overlapping of waves that are out of phase.

Figure 12.30 Light from O passes through slits M and N and produces an interference pattern on the screen S.

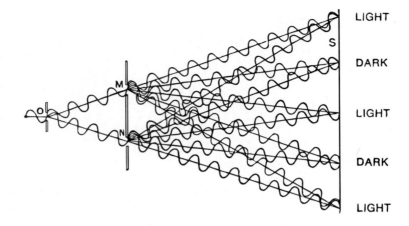

LIGHT

DARK

LIGHT

DARK

LIGHT

Star-Shaped Stars

Have you wondered why stars are represented with spikes? The stars on the American flag have five spikes, and the Jewish Star of David has six spikes. All through the ages stars have been drawn with spikes. The reason for this has not to do with the stars, which are point sources of light in the night sky, but rather with poor eyesight.

The surface of our eyes, the cornea, becomes scratched by a variety of causes. These scratches make up a diffraction grating of sorts. A scratched cornea is not a very good diffraction grating, but its effects are evident if you look at a bright point source against a dark background—like a star in the night sky. Instead of seeing a point of light, you see a spiky shape. The spikes will even shimmer and twinkle if there are some temperature differences in the atmosphere to produce some refraction. And if you live in a windy desert region where sandstorms are frequent, your cornea will be even more scratched and you'll see more vivid star spikes.

So stars don't have spikes. They appear spiked because of scratches that behave as diffraction gratings on the surfaces of our eyes.

QUESTIONS

1. If the double slits were illuminated with monochromatic red light, would the fringes be more widely or more closely spaced than if illuminated with monochromatic blue light?

2. Why is it important that monochromatic (single-frequency) light is used?

ANSWERS

1. More widely spaced. Can you see in Figure 12.29 that a slightly longer—and therefore a slightly more displaced—path from the entrance slit to the screen would result for the longer waves of red light?

2. If light of various wavelengths were diffracted by the slits, dark fringes for one wavelength would be filled in with bright fringes for another, resulting in no distinct fringe pattern. If you haven't seen this, be sure to ask your instructor to demonstrate it.

Interference Colors by Reflection from Thin Films

We have all noticed the beautiful spectrum of colors reflected from a soap bubble or from gasoline on a wet street. These colors are produced by the interference of light waves. This phenomenon is often called *iridescence* and is observed in thin films.

A soap bubble appears iridescent in white light when the thickness of the soap film is about the same as the wavelength of light. Light waves reflected from the outer and inner surfaces of the film travel different distances. When illuminated by white light, the film may be just the right thickness at one place to cause the destructive interference of, say, yellow light. When yellow light is subtracted from white light, the mixture left will appear as the complementary color—blue. At another place where the film is thinner, a different color may be canceled by interference, and the light seen will be its complementary color. The same thing happens to gasoline on a wet street (Figure 12.32). Light reflects from both the upper gasoline surface and the lower gasoline-water surface. If the thickness of the gasoline is such to cancel blue, as the figure suggests, then the gasoline surface appears yellow to the eye. Blue subtracted from white leaves yellow. The variety of colors seen in the thin film of gasoline, then, correspond to different film thicknesses, providing a vivid "contour map" of microscopic differences in surface "elevations."

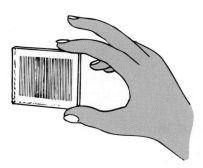

Figure 12.31   A diffraction grating disperses light into colors by interference. It may be used in place of a prism in a spectrometer.

View the thin film of gasoline at a lower angle and you see different colors. The apparent thickness of the film changes. Rays transmitted to the gasoline's lower surface travel a longer distance. Longer waves are canceled in this case, and different colors appear.*

Dishes washed in soapy water and poorly rinsed have a thin film of soap on them. Hold such a dish up to a light source so that interference colors can be seen. Then turn the dish to a new position, keeping your eye on the same part of the dish, and the color will change. Light reflecting from the bottom surface of the transparent

---

*Phase shifts at some reflecting surfaces also contribute to interference. For simplicity and brevity, our concern with this topic will be limited to this footnote: In short, when light in a medium is reflected at the surface of a second medium in which the speed of transmitted light is less (where there is a greater index of refraction), there is a 180° phase shift. However, no phase shift occurs when the second medium is one that transmits light at a higher speed (and there is a lower index of refraction). For example, in a soap bubble, light reflects from the first surface 180° out of phase. Light reflects from the second surface without a phase change. If the thickness of the soap film is very small compared to the wavelength of light so the distance through the film is negligible, the parts of the wave reflected from the two surfaces are out of phase and cancel for all frequencies. This is why parts of a soap film that are extremely thin appear black. Waves of all frequencies are canceled.

Figure 12.32 The thin film of gasoline is just the right thickness to cancel the reflections of blue light from the top and bottom surfaces. If the film were thinner, perhaps shorter-wavelength violet would be canceled.

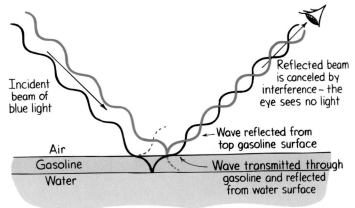

soap film is canceling light reflecting from the top surface of the soap film. Different wavelengths of light are canceled for different angles.

The wavelengths of light and other regions of the electromagnetic spectrum are measured with interference techniques. The principle of interference provides a means of measuring extremely small distances with great accuracy. Instruments called *interferometers*, which use the principle of interference, are the most accurate instruments known for measuring small distances.

QUESTIONS

1. What color will appear to be reflected from a soap bubble in sunlight when its thickness is such that green light is canceled?
2. In the left column are the colors of certain objects. In the right column are various ways in which colors are produced. Match the right column to the left.

   (a) yellow daffodil   (1) interference
   (b) blue sky   (2) diffraction
   (c) rainbow   (3) selective reflection
   (d) peacock feathers   (4) refraction
   (e) soap bubble   (5) scattering

12.6 POLARIZATION

Interference and diffraction provide the best evidence that light is wavelike in nature. Waves can be either longitudinal or transverse. Sound is longitudinal, where the vibratory motion is *along* the direction of wave travel. When we shake a taut rope, the vibratory motion that travels along the rope is perpendicular, or *transverse,* to the rope. Are light waves longitudinal or transverse? The fact that light waves can be **polarized** demonstrates that they are transverse.

ANSWERS

1. The composite of all the visible wavelengths except green results in the complementary color, magenta. See Figure 12.15.
2. a–3; b–5; c–4; d–2; e–1.

Figure 12.33 A vertically polarized plane wave and a horizontally polarized plane wave.

Shake a taut rope up and down as in Figure 12.33, and you produce a transverse wave along the rope in a plane. We say that such a wave is *plane-polarized.** Shake it up and down vertically, and the wave is vertically plane-polarized; that is, the waves traveling along the rope are confined to a vertical plane. If we shake the rope from side to side, we produce a horizontally plane-polarized wave.

A single vibrating electron emits an electromagnetic wave that is also plane-polarized. The plane of polarization will be in the same vibrational direction of the electron. A vertically accelerating electron, then, emits light that is vertically polarized, while a horizontally accelerating electron emits light that is horizontally polarized (Figure 12.34).

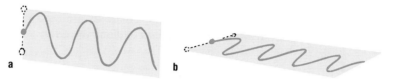

Figure 12.34 (a) A vertically polarized plane wave from a charge vibrating vertically. (b) A horizontally polarized plane wave from a charge vibrating horizontally.

A common light source such as an incandescent lamp, a fluorescent lamp, or a candle flame, emits light that is nonpolarized. This is because there is no preferred vibrational direction for the accelerating electrons emitting the light. The number of planes of vibration might be as numerous as the accelerating electrons producing them. A few planes are represented in Figure 12.35a. We can represent all these planes by radial lines (Figure 12.35b) or, more simply, by vectors in two mutually perpendicular directions (Figure 12.35c), as if we had resolved all the vectors of Figure 12.35b into horizontal and vertical components. This simpler schematic represents nonpolarized light. Polarized light would be represented by a single vector.

All transparent crystals of a noncubic natural shape have the property of transmitting light through two mutually perpendicular planes. Certain crystals not only produce two internal beams polarized at right angles to each other but also strongly

Figure 12.35 Representations of planes of plane-polarized waves.

*Light may also be circularly polarized and elliptically polarized, which are also transverse polarizations. But we will not study these cases.

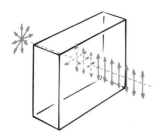

Figure 12.36 One component of the incident nonpolarized light is absorbed, resulting in emerging polarized light.

absorb one beam while transmitting the other (Figure 12.36). Such a crystal is herapathite. Microscopic crystals of herapathite are embedded between cellulose sheets in uniform alignment and are used in making Polaroid filters. Some Polaroid sheets consist of certain aligned molecules rather than tiny crystals.*

If you look at nonpolarized light through a Polaroid filter, you can rotate the filter in any direction, and the light will appear unchanged. But if this light is polarized, then as you rotate the filter, you can progressively cut off more and more of the light until it is blocked out. An ideal Polaroid will transmit 50 percent of incident nonpolarized light. That 50 percent is, of course, polarized. When two Polaroids are arranged so that their polarization axes are aligned, light will be transmitted through both (Figure 12.37). If their axes are at right angles to each other, no light will penetrate the pair. Actually, some of the shorter wavelengths do get through, but not to any significant degree. When Polaroids are used in pairs like this, the first one is called the *polarizer* and the second one the *analyzer*.

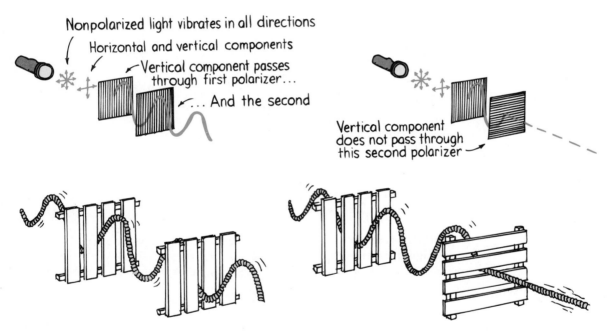

Figure 12.37 A rope analogy illustrates the effect of crossed polaroids.

Much of the light reflected from nonmetallic surfaces is polarized. The glare from glass or water is a good example. Except for normal incidence, the reflected ray contains more vibrations parallel to the reflecting surface, while the transmitted beam contains more vibrations at right angles to the surface (Figure 12.38). Skipping flat rocks off the surface of a pond is analogous. When the rocks hit parallel to the surface, they easily reflect; but if they hit with their faces at right angles to the surface, they "refract" into the water. The glare from reflecting surfaces can be appreciably diminished with the use of Polaroid sunglasses. The polarization axes of the lenses are vertical, as most glare reflects from horizontal surfaces.

---

*The molecules are polymeric iodine in a sheet of polyvinyl alcohol or polyvinylene.

Figure 12.38 Polaroid sunglasses block out horizontally vibrating light. When the lenses overlap at right angles, no light gets through.

Q UESTION

Which pair of glasses is best suited for automobile drivers? (The polarization axes are shown by the straight lines.)

A   B   C

Beautiful colors similar to interference colors can be seen when certain materials are placed between crossed polaroids. Cellophane works wonderfully. Why these colors are produced is another story, one left to the recommended reading at the back of the book.

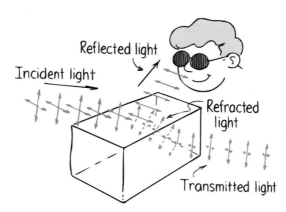

Figure 12.39 Most glare from nonmetallic surfaces is polarized. Here we see that the components of incident light that are parallel to the surface are reflected, while the components that are perpendicular to the surface are refracted into the medium. Since most of the glare we encounter is from horizontal surfaces, the polarization axes of Polaroid sunglasses are vertical.

a   b   c

Figure 12.40 Light is transmitted when the axes of the Polaroids are aligned (a), but absorbed when she rotates one so that the axes are at right angles to each other (b). When she inserts a third Polaroid at an angle between the crossed Polaroids, light is again transmitted (c). Why?

A NSWER

Glasses A are best suited because the vertical axis blocks horizontally polarized light, which composes much of the glare from horizontal surfaces. Glasses C are suited for viewing 3-D movies.

SUMMARY OF TERMS

Electromagnetic wave An energy-carrying wave emitted by vibrating electrons that is composed of oscillating electric and magnetic fields that regenerate one another.

Electromagnetic spectrum The range of electromagnetic waves extending in frequency from radio waves to gamma rays.

Transparent The term applied to materials through which light can pass in straight lines.

Additive primary colors The three colors—red, blue, and green—that when added in certain proportions will produce any color in the spectrum.

Subtractive primary colors The three colors of absorbing pigments—magenta, yellow, and cyan—that when mixed in certain proportions will reflect any color in the spectrum.

Complementary colors Any two colors that when added produce white light.

Diffraction The bending of light around an obstacle or through a narrow slit in such a way that fringes of light and dark or colored bands are produced.

Interference The superposition of waves producing regions of reinforcement and regions of cancellation. Constructive interference refers to regions of reinforcement; destructive interference refers to regions of cancellation. The interference of selected wavelengths of light produces colors known as *interference colors*.

Polarization The alignment of the electric vectors that make up electromagnetic radiation. Such waves of aligned vibrations are said to be *polarized*.

REVIEW QUESTIONS

The Electromagnetic Spectrum

1. Does light make up a relatively large part or a relatively small part of the electromagnetic spectrum?

2. What is the principal difference between a *radio wave* and *light*?

3. What is the principal difference between *light* and an *X ray*?

4. How much of the measured electromagnetic spectrum does light occupy?

5. What color do the lowest visible frequencies appear? The highest?

6. How does the frequency of light compare to the frequency of the vibrating electron that produces it?

7. How does the wavelength of light compare to its frequency?

Transparent and Opaque Materials

8. One tuning fork can force another to vibrate. How is this similar to light?

9. In what region of the electromagnetic spectrum is the resonant frequency of electrons in glass?

10. What is the fate of the energy in ultraviolet light that is incident upon glass?

11. What is the fate of the energy in visible light that is incident upon glass?

12. Why are your answers for Questions 10 and 11 different?

13. How does the frequency of re-emitted light compare to the frequency of the light that stimulates its re-emission?

14. How does the average speed of light in glass compare to its speed in a vacuum?

Color

15. What is the relationship between the frequency of light and its color?

16. The visible color spectrum runs from red through violet. Compared to violet light, is red light high-frequency or low-frequency?

Selective Reflection

17. Distinguish between the white of this page and the black of this ink in terms of what happens to the white light that falls on both.

18. How does the color of an object differ when illuminated by candle light and by the light from a fluorescent lamp?

Selective Transmission

19. What color light is transmitted through a piece of red glass?

20. Which will warm quicker in sunlight, a clear or a colored piece of glass? Why?

Mixing Colored Light

21. What is the evidence for the statement that white light is a composite of all the colors of the spectrum?

22. What is the color of the peak frequency of solar radiation?

23. What color light are our eyes most sensitive to?

24. What frequency ranges of the radiation curve do red, green, and blue light occupy?

25. Why are red, green, and blue called the *additive primary colors*?

26. What is the resulting color of equal intensities of blue light and green light combined?

27. What is the resulting color of equal intensities of red light and cyan light combined?

Mixing Colored Pigments

28. What are the subtractive primary colors?

29. If you look with a magnifying glass at pictures printed in full color in magazines, you'll notice three colors of ink plus black. What are these colors?

30. Why are red and cyan called *complementary colors*?

Why the Sky Is Blue

31. Which interacts more with high-pitched sounds, small bells or large bells?

32. Which interacts more with high-frequency light, small particles or large particles?

33. Why does the sky sometimes appear whitish?

34. Why is the sky a deeper blue after a heavy rainstorm?

Why Sunsets Are Red

35. Why does the sun look reddish at sunrise and sunset, but not at noon?

36. Why does the color of sunsets vary from day to day?

Why Clouds Are White

37. What is the evidence for a variety of particle sizes in a cloud?

38. What is the evidence for extra big particles in a rain cloud?

Diffraction

39. Is diffraction more pronounced through a small opening or through a large opening?

40. What is the relationship between the wavelength of a wave and the size of the opening through which diffraction occurs?

41. What are some of the assets and liabilities of diffraction?

Interference

42. Is interference restricted to only some types of waves or does it occur for all types of waves?

43. What is monochromatic light?

44. What produces iridescence?

Interference Colors by Reflection from Thin Films

45. What causes the spectrum of colors seen in gasoline splotches on a wet street? Why are these not seen on a dry street?

46. What accounts for the different colors in either a soap bubble or a layer of gasoline on water?

47. If you look at a soap bubble from different angles so you're viewing different apparent thicknesses of soap film, will you see different colors? Explain.

Polarization

48. What phenomenon distinguishes between longitudinal and transverse waves?

49. How does the plane of polarization of light compare to the plane of vibration of the electron that produced it?

50. Why will light pass through a pair of Polaroids when the axes are aligned but not when the axes are at right angles to each other?

51. How much ordinary light will an ideal Polaroid transmit?

52. When *ordinary* light is incident at a grazing angle upon water, what can you say about the *reflected* light?

. . . .

HOME PROJECTS

1. Which eye do you use more? To test which you favor, hold a finger up at arm's length. With both eyes open, look past it at a distant object. Now close your right eye. If your finger appears to jump to the right, then you use your right eye more.

2. Stare at a piece of colored paper for 45 seconds or so. Then look at a plain white surface. The cones in your retina receptive to the color of the paper become fatigued, so you see an afterimage of the complementary color when you look at a white area. This is because the fatigued cones send a weaker signal to the brain. All the colors produce white, but all the colors minus one produce the complementary to the missing color. Try it and see!

3. Simulate your own sunset: Add a few drops of milk to a glass of water and look through it at a light bulb. The bulb appears to be red or pale orange, while light scattered to the side appears blue. Try it and see.

4. With a razor blade, cut a slit in a card and look at a light source through it. You can vary the size of the opening by bending the card slightly. See the interference fringes? Try it with two closely spaced slits.

5. Next time you're in the bathtub, froth up the soapsuds and notice the colors of highlights from the illuminating light overhead on each tiny bubble. Notice that different bubbles reflect different colors, due to the different thicknesses of soap film. If a friend is bathing with you, compare the different colors that you each see reflected from the same bubbles. You'll see that they're different—for what you see depends on your point of view!

6. Do this one at your kitchen sink. Dip a dark-colored coffee cup (dark colors make the best background for viewing interference colors) in dishwashing detergent, and then hold it sideways and look at the reflected light from the soap film that covers its mouth. Swirling colors appear as the soap runs down to form a wedge that grows thicker at the bottom with time. The top becomes thinner, so thin that it appears black. This tells us that its thickness is less than one-fourth the thickness of the shortest waves of visible light. Whatever its wavelength, light reflecting from the inner surface reverses phase, rejoins light reflecting from the outer surface, and cancels. The film soon becomes so thin it pops.

7. When you're wearing Polaroid sunglasses, view the glare from nonmetallic surfaces such as a road or body of water, tip your head from side to side, and see how the glare intensity changes as you vary the number of electric vector components aligned with the polarization axis of the glasses. Also notice the polarization of different parts of the sky when you hold the sunglasses in your hand and rotate them.

8. Place a bottle of corn syrup between two sheets of Polaroid. Place a white light source behind the syrup. Then look through the Polaroids and syrup and view spectacular colors as you rotate one of the Polaroids.

9. See spectacular interference colors with a polarized-light microscope. Any microscope, including an inexpensive toy microscope, can be converted into a polarized-light microscope by fitting a piece of Polaroid inside the eyepiece and taping another onto the stage of the microscope. Mix drops of naphthalene and benzene on a slide and watch the growth of crystals. Rotate the eyepiece and change the colors.

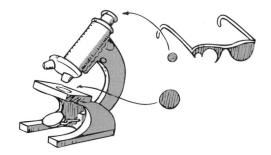

10. Make some slides for a slide projector by sticking some crumpled cellophane onto pieces of slide-sized Polaroid. (Also try strips of cellophane tape overlapped at different angles and experiment with different brands of transparent tape.) Project them onto a large screen or white wall and rotate a second, slightly larger piece of Polaroid in front of the projector lens in rhythm with your favorite music. You'll have your own light show.

EXERCISES

1. Which waves have longer wavelengths: light waves, X rays, or radio waves?

2. In about 1675 the Danish astronomer Olaus Roemer found that light from eclipses of Jupiter's moon took an extra 1000 s to travel 300,000,000 km across the diameter of the earth's orbit around the sun. Show how this finding provided the first reasonably accurate measurement for the speed of light.

3. The sun is 1.50×10^{11} meters from the earth. How long does it take for the sun's light to reach the earth?

4. The nearest star beyond the sun is Alpha Centauri, 4.2×10^{16} meters away. If we received a radio message from this star today, how long ago would it have been sent?

5. What evidence can you cite to support the idea that light can travel in a vacuum?

6. Why would you expect the speed of light to be slightly less in the atmosphere than in a vacuum?

7. If you fire a bullet through a tree, it will slow down inside the tree and emerge at a speed less than the speed at which it entered. Does light, then, similarly slow down when it passes through glass and also emerge at a lower speed? Defend your answer.

8. Is glass transparent, or is it opaque, to frequencies of light that match its own natural frequencies? Explain.

9. What determines whether a material is transparent or opaque?

10. You can get a sunburn on a cloudy day, but you can't get a sunburn even on a sunny day if you are behind glass. Explain.

11. Suppose that sunlight falls on both a pair of reading glasses and a pair of dark sunglasses. Which pair of glasses would you expect to be warmer in sunlight? Defend your answer.

12. In a clothing shop with only fluorescent lighting, a customer insists on taking garments into the daylight at the doorway to check their color. Is the customer being reasonable? Explain.

13. If the sunlight were somehow green instead of white, what color garment would be most advisable on an uncomfortably hot day? On a very cold day?

14. What color would red cloth appear if it were illuminated by sunlight? By red light from a neon sign? By cyan light?

15. Why does a white piece of paper appear white in white light, red in red light, blue in blue light, and so on for every color?

16. A spotlight with a white-hot filament is coated so that it won't transmit yellow light. What color is the emerging beam of light?

17. How could you use the spotlights at a play to make the yellow clothes of the performers suddenly change to black?

18. Suppose two flashlight beams are shone on a white screen, one beam through a pane of blue glass and the other through a pane of yellow glass. What color appears on the screen where the two beams overlap? Suppose, instead, that the two panes of glass are placed in the beam of a single flashlight. What then?

19. Does a color television work by color addition or by color subtraction? Which dots must be struck by electrons to create the color yellow? Magenta? White?

20. Complete the following equations:

    Yellow light + blue light = _____ light.
    Green light + _____ light = white light.
    Magenta + yellow + cyan = _____ light.

21. Check to see if the following three statements are accurate. Then fill in the last statement. (All colors are combined by the addition of light.)

    Red + green + blue = white.
    Red + green = yellow = white − blue.
    Red + blue = magenta = white − green.
    Green + blue = cyan = white − _____ .

22. On a photographic print, your friend is seen wearing a red sweater. What color is the sweater on the negative?

23. If the sky were composed of atoms that predominantly scattered orange light rather than blue, what color would sunsets be?

24. Tiny particles, like tiny bells, scatter high-frequency waves more than low-frequency waves. Large particles, like large bells, mostly scatter low frequencies. Intermediate-size particles and bells mostly scatter intermediate frequencies. What does this have to do with the whiteness of clouds?

25. Very big particles, like droplets of water, absorb more radiation than they scatter. What does this have to do with the darkness of rain clouds?

26. If the atmosphere of the earth were about fifty times thicker, would ordinary snowfall still seem white or would it be some other color? What color?

27. The atmosphere of Jupiter is more than 1000 km thick. From the surface of this planet, would you expect to see a white sun?

28. A lunar eclipse occurs when the moon passes into the earth's shadow. Rather than being black, the moon appears a copper red. Is there a connection between the reddish moon color and the ring of sunsets and sunrises that circle the earth? Explain.

29. Why do *radio waves* diffract around buildings, while *light waves* do not?

30. A pattern of fringes is produced when monochromatic light passes through a pair of thin slits. Would such a pattern be produced by three parallel thin slits? By thousands of such slits? Give an example to support your answer.

31. The colors of peacocks and hummingbirds are the result not of pigments, but of ridges in the surface layers of their feathers. By what physical principle do these ridges produce colors?

32. The colored wings of many butterflies are due to pigmentation, but in others, such as the Morpho butterfly, the colors do not result from any pigmentation. When the wing is viewed from different angles, the colors change. How are these colors produced?

33. Why do the iridescent colors seen in some seashells (such as abalone shells) change as the shells are viewed from different positions?

34. When dishes are not properly rinsed after washing, different colors are reflected from their surfaces. Explain.

35. Why are interference colors more apparent for thin films than for thick films?

36. Why do Polaroid sunglasses reduce glare, whereas nonpolarized sunglasses simply cut down on the total amount of light reaching the eyes?

37. How can you determine the polarization axis for a single sheet of Polaroid?

38. Most of the glare from nonmetallic surfaces is polarized, the axis of polarization being parallel to that of the reflecting surface. Would you expect the polarization axis of Polaroid sunglasses to be horizontal or vertical? Why?

39. How can a single sheet of Polaroid film be used to show that the sky is partially polarized? (Interestingly enough, unlike humans, bees and many insects can discern polarized light and use this ability for navigation.)

40. What percentage of light would be transmitted by two ideal Polaroids sandwiched with their polarization axes aligned? With their axes at right angles to each other?

13

PROPERTIES
OF LIGHT

Most of the things we see around us do not emit their own light. They are visible because they re-emit most of the light reaching their surface from a primary source, such as the sun or a lamp, or from a secondary source, such as the illuminated sky. When light falls on the surface of a material it is usually either re-emitted without change in frequency or is absorbed in the material and turned into heat. Usually both of these processes occur in varying degrees. When the re-emitted light is returned into the medium from which it came, it is *reflected*, and the process is **reflection**. When the re-emitted light bends from its original course and proceeds in straight lines from molecule to molecule into a transparent material, it is *refracted*, and the process is **refraction**.

13.1 REFLECTION

When this page is illuminated with sunlight or lamplight, a general electron vibration is set up in the normal energy states of its atoms; whole electron clouds vibrate in response to the oscillating electric fields of the illuminating light. Even under bright

Figure 13.1 The law of reflection.

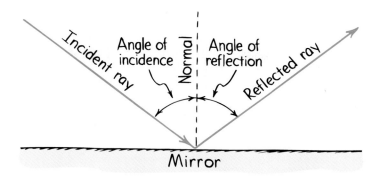

sunlight, the amplitudes of these vibrations are less than 1 percent of the radius of the atomic nucleus. It is these tiny electron vibrations that re-emit the light by which we see the page. When the page is illuminated by white light, it appears white, which reveals the fact that the electrons are set into vibrations at all the visible frequencies. Very little absorption occurs. The ink on the page is a different story. Except for a bit of reflection, it absorbs all the visible frequencies and therefore appears black.

So on the surfaces of all the objects around us, the electron clouds of the atoms undergo slight vibrations under the influence of illuminating light. These tiny vibrations over a wide frequency range reflect the various colors of light by which we see these objects. Simply put, we say that we see these objects by the light they reflect.

Law of Reflection

Anyone who has played pool or billiards knows that when a ball bounces from a surface, the angle of incidence is equal to the angle of rebound. Likewise for light. This is the **law of reflection**, and it holds for all angles:

The angle of incidence equals the angle of reflection.

The law of reflection is illustrated with arrows representing light rays in Figure 13.1. Instead of measuring the angles of incidence and reflected rays from the reflecting surface, it is customary to measure them from a line perpendicular to the plane of the reflecting surface. This imaginary line is called the *normal*. The incident ray, the normal, and the reflected ray all lie in the same plane.

A practical case of the law of reflection is an ordinary plane mirror. Suppose a candle flame is placed in front of a plane mirror. Rays of light are sent from the flame in all directions. Figure 13.2 shows only four of the infinite number of rays leaving one of the infinite number of points on the candle. When these rays encounter the mirror, they are reflected at angles equal to their angles of incidence. The rays diverge from the flame and on reflection diverge from the mirror. These divergent rays appear to emanate from behind the mirror, from a point located where the rays seem to diverge (dashed lines). An observer sees an image of the candle at this point. The light rays do not actually come from this point, so the image is called a *virtual image*. The image is as far behind the mirror as the object is in front of the mirror, and image and object have the same size. When you view yourself in a mirror, for example, the size of your image is the same as the size your twin would appear if located as far behind the mirror as you are in front—as long as the mirror is flat.

When the mirror is curved, the sizes and distances of object and image are no longer equal. We will not get into curved mirrors in this text, except to say that the

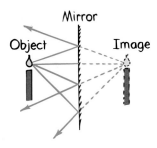

Figure 13.2 A virtual image is formed behind the mirror and is located at the position where the extended reflected rays (dashed lines) converge.

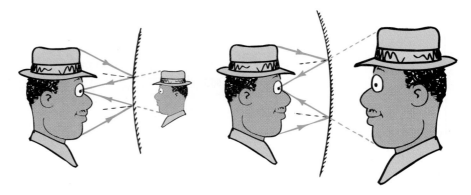

Figure 13.3 (a) The virtual image formed by a *convex* mirror (a mirror that curves outward) is smaller and closer to the mirror than the object. (b) When the object is close to a *concave* mirror (a mirror that curves inward like a "cave"), the virtual image is larger and farther away than the object. In either case the law of reflection holds for each ray.

law of reflection still holds. A curved mirror behaves as a succession of flat mirrors, each at a slightly different angular orientation from the one next to it. At each point, the angle of incidence is equal to the angle of reflection (Figure 13.3). Note that in a curved mirror, unlike in a plane mirror, the normals (shown by the dashed black lines) at different points on the surface are not parallel to one another.

Whether the mirror is plane or curved, the eye-brain system cannot ordinarily tell the difference between an object and its reflected image. So the illusion that an object exists behind a mirror (or in some cases in front of a concave mirror) is merely due to the fact that the light from the object enters the eye in exactly the same manner, physically, as it would have entered if the object really were at the image location.

Q UESTIONS

1. What evidence can you cite to support the claim that the frequency of light does not change upon reflection?
2. If you wish to take a picture of your image while standing 5 m in front of a plane mirror, for what distance should you set your camera to provide sharpest focus?

Only part of the light that strikes a surface is reflected. On a surface of clear glass, for example, only about 4 percent is reflected from each surface, while on a clean and polished aluminum or silver surface, about 90 percent of incident light is reflected.

A NSWERS

1. Simply stand in front of a mirror and compare the color of your shirt with the color of its image. The fact that the color is the same is evidence that the frequency of light doesn't change upon reflection.
2. You should set your camera for 10 m; the situation is equivalent to your standing 5 m in front of an open window and viewing your twin standing 5 m in back of the window.

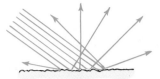

Figure 13.4 Diffuse reflection. Although the reflection of each single ray obeys the law of reflection, the many different surface angles that light rays encounter in striking a rough surface cause reflection in many directions.

Diffuse Reflection

When light is incident on a rough surface, it is reflected in many directions. This is called **diffuse reflection** (Figure 13.4). If the surface is so smooth that the distances between successive elevations on the surface are less than about one-eighth the wavelength of the light, there is very little diffuse reflection, and the surface is said to be *polished*. A surface therefore may be polished for radiation of long wavelength but not polished for light of short wavelength. The wire-mesh "dish" shown in Figure 13.5 is very rough for light waves and is hardly mirrorlike. But for long-wavelength radio waves it is "polished" and is an excellent reflector.

Light reflecting from this page is diffuse. The page may be smooth to a long radio wave, but to a fine light wave it is rough. Rays of light hitting on this page encounter millions of tiny flat surfaces facing in all directions. The incident light therefore is reflected in all directions. This is a desirable circumstance. It enables us to see objects from any direction or position. Most of our environment is seen by diffuse reflection.

An undesirable circumstance related to diffuse reflection is the ghost image that occurs on a TV set when the TV signal bounces off buildings and other obstructions. For antenna reception, this difference in path lengths for the direct signal and the reflected signal produces a slight time delay. The ghost image is normally displaced to the right, the direction of scanning in the TV tube, because the reflected signal arrives at the receiving antenna later than the direct signal. Multiple reflections may produce multiple ghosts.

Figure 13.5 The open-mesh parabolic dish is a diffuse reflector for short-wavelength light but polished for long-wavelength radio waves.

Figure 13.6 A magnified view of the surface of ordinary paper.

13.2 REFRACTION

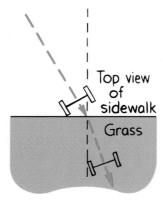

Figure 13.7 The direction of the rolling wheels changes when one part slows down before the other part.

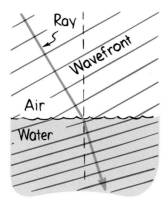

Figure 13.8 The direction of the light waves changes when one part of the wave slows down before the other part.

Recall from the previous chapter that the average speed of light is lower in glass and other transparent materials than in empty space. Light travels at different speeds in different materials.* It travels at 300 000 kilometers per second in a vacuum, at a slightly lower speed in air, and at about three-fourths that speed in water. In a diamond, light travels at about 40 percent of its speed in a vacuum. When light bends as it passes obliquely from one medium to another, we call the process *refraction*.

The cause of refraction is the changing of the average speed of light in going from one transparent medium to another. We can understand this by considering the action of a pair of toy cart wheels connected to an axle as the wheels roll from a smooth sidewalk onto a grass lawn. If the wheels meet the grass at some angle (Figure 13.7), they will be deflected from their straight-line course. The direction of the rolling wheels is shown by the dashed line. Note that on meeting the lawn, where the wheels roll slower owing to interaction with the grass, the left wheel slows down first. This is because it meets the grass while the right wheel is still on the smooth sidewalk. The faster-moving right wheel tends to pivot about the slower-moving left wheel. It travels farther during the same time the left wheel travels a lesser distance in the grass. This action bends the direction of the rolling wheels toward the "normal," the lightly dashed line perpendicular to the grass-sidewalk border.

A light wave bends in a similar way (Figure 13.8). Note the direction of light shown by the solid arrow (the light ray) and wave fronts at right angles to the ray. (If the light source were close, the wave fronts would appear as segments of circles; but if we assume the distant sun is the source, the waves form practically straight lines.) In any case, we see that the wave fronts are everywhere perpendicular to the light rays. In the figure the wave meets the water surface at an angle, so the left portion of the wave slows down in the water while the part still in the air travels at speed c. The ray or beam of light remains perpendicular to the wave front and bends

---

*Just how much the speed of light differs from its speed in a vacuum is given by the index of refraction, n, of the material; n is the ratio of the speed of light in a vacuum to the speed of light in the material:

$$n = \frac{\text{speed of light in vacuum}}{\text{speed of light in material}}$$

For example, the speed of light in a diamond is $(\frac{1}{2.4})c$, so $n = 2.4$. For a vacuum, $n = 1$.

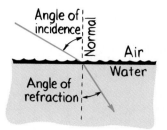

Figure 13.9   Refraction.

Figure 13.10   When light slows down in going from one medium to another, like going from air to water, it bends toward the normal. When it speeds up in traveling from one medium to another, like going from water to air, it bends away from the normal.

at the surface, just as the wheels bend to change direction when they roll from the sidewalk into the grass. In both cases the bending is caused by a change in speed.*

Figure 13.10 shows a beam of light entering water at the left and exiting at the right. The path would be the same if the light entered from the right and exited at the left. The light paths are reversible for both reflection and refraction. If you see someone by way of a reflective or refractive device, such as a mirror or a prism, then that person can see you by the device also.

QUESTION

If the speed of light were the same in all media, would refraction still occur when light passes from one medium to another?

The refraction of light is responsible for many illusions; one of them is the apparent bending of a stick partly immersed in water. The submerged part seems closer to the surface than it really is. Likewise when you view a fish in the water. The fish appears nearer to the surface and closer (Figure 13.11). Because of refraction, submerged objects appear to be magnified. If we look straight down into water, an object submerged 4 meters beneath the surface will appear to be 3 meters deep.

Refraction occurs in the earth's atmosphere. Whenever we watch a sunset, we see the sun for several minutes after it has sunk below the horizon (Figure 13.12). The earth's atmosphere is thin at the top and dense at the bottom. Since light travels faster in thin air than it does in dense air, parts of the wavefronts of sunlight higher up travel faster than parts of the wavefronts closer to ground. Since the density of the atmosphere changes gradually, the light path bends gradually and produces a curved

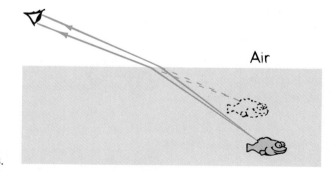

Figure 13.11   Because of refraction, a submerged object appears to be nearer to the surface than it actually is.

ANSWER

No.

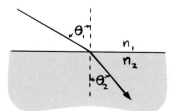

*The quantitative law of refraction, called *Snell's law*, was first worked out in 1621 by W. Snell, a Dutch astronomer and mathematician: $n_1 \sin \theta_1 = n_2 \sin \theta_2$, where n_1 and n_2 are the indices of refraction of the media on either side of the surface, and θ_1 and θ_2 are the respective angles of incidence and refraction. If three of these values are known, the fourth can be calculated from this relationship.

Figure 13.12 Because of atmospheric refraction, when the sun is near the horizon it appears higher in the sky.

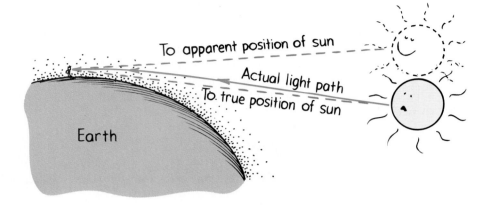

Figure 13.13 The sun is distorted by differential refraction.

Figure 13.14 Air is warmer and less dense near the ground and the wavefronts pick up speed as they travel downward and bend upward. The observer sees a mirage.

path. So we get a slightly longer period of daylight each day. Furthermore, when the sun (or moon) is near the horizon, the rays from the lower edge are bent more than the rays from the upper edge. This produces a shortening of the vertical diameter, causing the sun to appear elliptical (Figure 13.13).

We are all familiar with the mirage we see while driving on a hot road. The sky appears to be reflected from water on the distant road, but when we get there, the road is dry. Why is this so? The air is very hot just above the road surface and cooler above. Light travels faster through the less dense and thinner hot air than in the cool region. Wavefronts near the ground travel faster than above, so the light is refracted upward (Figure 13.14). So we see an inverted view as if reflection were occurring from a water surface. We see a mirage. A mirage is not, as many people mistakenly believe, a "trick of the mind." A mirage is formed by real light and can be photographed, as shown in Figure 13.15.

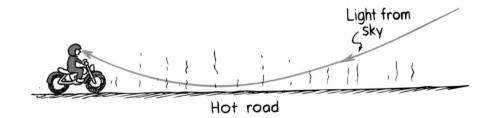

Figure 13.15 A mirage. The apparent wetness of the road is not reflection of the sky by water but refraction of sky light through the warmer and less-dense air near the road surface.

When we look at an object over a hot stove or over a hot pavement, we see a wavy, shimmering effect. This is due to varying temperatures and therefore varying densities of air. The twinkling of stars results from similar phenomena in the sky, where light passes through unstable layers in the atmosphere.

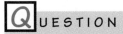UESTION

If the speed of light were the same in the various temperatures and densities of air, would there still be slightly longer daytimes, twinkling stars at night, mirages, and slightly squashed suns at sunset?

13.3 DISPERSION

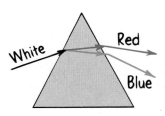

Figure 13.16 Dispersion through a prism makes the components of white light visible.

We know that the average speed of light is less than c in a transparent medium; how much less depends on the nature of the medium and on the frequency of light. The speed of light in a transparent medium depends on its frequency. Recall from Chapter 12 that frequencies of light that match the natural or resonant frequencies of the electron oscillators in the atoms and molecules of the transparent medium are absorbed, and frequencies near the resonant frequencies interact more often in the absorption/re-emission sequence and therefore travel more slowly. Since the natural or resonant frequency of most transparent materials is in the ultraviolet part of the spectrum, the higher frequencies of visible light travel more slowly than the lower frequencies. Violet travels about 1 percent slower in ordinary glass than does red light. The colors between red and violet travel at their own respective speeds.

Different frequencies of light travel at different speeds in transparent materials; because they travel at different speeds, they refract differently and bend by different amounts. When light is bent twice, as in a prism, the separation of the different colors of light is quite noticeable. This separation of light into colors arranged according to their frequency is called *dispersion* (Figure 13.16).

Rainbows

A most spectacular illustration of dispersion is the rainbow. The conditions for seeing a rainbow are a low sun shining in one part of the sky and rain falling in the opposite part of the sky. When we turn our backs toward the sun, we see the spectrum of colors in a bow. Seen high enough from an airplane, the bow forms a complete circle. All rainbows would be completely round if the ground were not in the way.

The beautiful colors of rainbows are dispersed from the sunlight by thousands of tiny spherical drops that act like prisms. We can better understand this by considering an individual raindrop, as shown in Figure 13.17. Follow the ray of sunlight as it enters the drop near its top surface. Some of the light here is reflected (not shown), and the remainder is refracted into the water. At this first refraction, the light is

---

Answer

No.

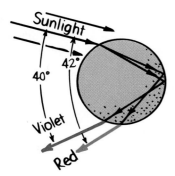

Figure 13.17 Dispersion of sunlight by a single raindrop.

dispersed into its spectrum colors, violet being deviated the most and red the least. Reaching the opposite side of the drop, each color is partly refracted out into the air (not shown) and partly reflected back into the water. Arriving at the lower surface of the drop, each color is again reflected (not shown) and refracted into the air. This second refraction is similar to that of a prism, where refraction at the second surface increases the dispersion already produced at the first surface.

Although each drop disperses a full spectrum of colors, an observer is in a position to see only a single color from any one drop (Figure 13.18). If violet light from a single drop reaches the eye of an observer, red light from the same drop is incident elsewhere toward the feet. To see red light, one must look to a drop higher in the sky. The color red will be seen when the angle between a beam of sunlight and the dispersed light is 42°. The color violet is seen when the angle between the sunbeams and dispersed light is 40°.

Why does the light dispersed by the raindrops form a bow? The answer to this involves a bit of geometry. First of all, a rainbow is not the flat two-dimensional arc it appears to be. It appears flat for the same reason a spherical burst of fireworks high in the sky appears as a disc—because of a lack of distance cues. The rainbow you see is actually a three-dimensional cone with the apex at your eye. Consider a glass cone, the shape of those paper cones you sometimes see at drinking fountains. If you held the tip of such a glass cone against your eye, what would you see? You'd see the glass as a circle. Likewise with a rainbow. All the drops that disperse the rainbow's light toward *you* lie in the shape of a cone—a cone of different layers with drops that disperse red to your eye on the outside, orange beneath the red, yellow beneath the orange, and so on all the way to violet on the inner conical surface (Figure 13.19). The thicker the region containing water drops, the thicker the conical edge you look through, and the more vivid the rainbow.

To see this further, consider only the dispersion of red light. You see red when the angle between the incident rays of sunlight and dispersed rays make a 42° angle. Of course beams are dispersed 42° from drops all over the sky in all directions—up, down, or sideways. But the only red light *you* see is from drops that lie on a cone with a side-to-axis angle of 42°. Your eye is at the apex of this cone as we see in Figure

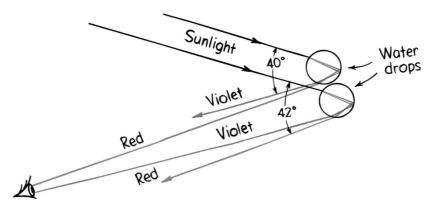

Figure 13.18 Sunlight is incident on two sample raindrops as shown and emerges from them as dispersed light. The observer sees the red light from the upper drop and the violet light from the lower drop. Millions of drops produce the whole spectrum.

Figure 13.19 When your eye is located between the sun (not shown off to the left) and the water-drop region, the rainbow you see is the edge of a 3-dimensional cone that extends through the water-drop region. Violet is dispersed by drops that form a 40° conical surface; red is seen from drops along a 42° conical surface, with other colors in between. (Innumerable layers of drops form innumerable 2-dimensional arcs like the four suggested here.)

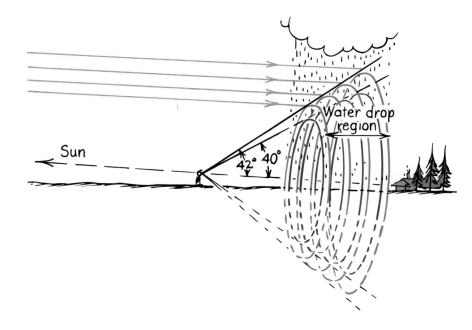

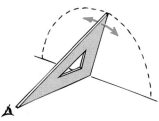

Figure 13.20 Only raindrops along the dashed line disperse red light to the observer at a 42° angle; hence, the light forms a bow.

13.20. To see violet, you look 40° from the conical axis (so the thickness of glass in the cone of the previous paragraph is tapered—very thin at the tip and thicker with increased distance from the tip).

Your cone of vision that intersects the cloud of drops that creates your rainbow is different from that of a person next to you. So when a friend says, "Look at the pretty rainbow," you can reply, "Okay, move aside so I can see it too." Everybody sees his or her own personal rainbow.

Another fact about rainbows: A rainbow always faces you squarely, because of the lack of distance cues mentioned earlier. When you move, your rainbow moves with you. So you can never approach the side of a rainbow, or see it end-on as in the exaggerated view of Figure 13.19. You *can't* get to its end. Hence the expression

Figure 13.21 Under the rainbow, the sky is bright where thousands of overlapping rainbows combine into a whitish glow.

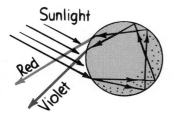

Figure 13.22 Double reflection in a drop produces a secondary bow.

"looking for the pot of gold at the end of the rainbow" means pursuing something you can never reach.

Often a larger, secondary bow with colors reversed can be seen arching at a greater angle around the primary bow. We won't treat this secondary bow except to say that it is formed by similar circumstances and is a result of double reflection within the raindrops (Figure 13.22). Because of this extra reflection (and extra refraction loss), the secondary bow is much dimmer and reversed.

QUESTIONS

1. If you point to a wall with your arm extended to make about a 42° angle to the wall, then rotate your arm in a full circle while keeping the 42° angle to the wall, what shape does your arm describe? What shape on the wall does your finger sweep out?

2. If light traveled at the same speed in raindrops as it does in air, would we still have rainbows?

13.4 TOTAL INTERNAL REFLECTION

Some Saturday night when you're taking your bath, fill the tub extra deep and bring a waterproof flashlight into the tub with you. Put the bathroom light out. Shine the submerged light straight up and then slowly tip it. Note how the intensity of the emerging beam diminishes and how more light is reflected from the water surface to the bottom of the tub. At a certain angle, called the **critical angle**, you'll notice that the beam no longer emerges into the air above the surface. The intensity of the emerging beam reduces to zero where it tends to graze the surface. When the flash-

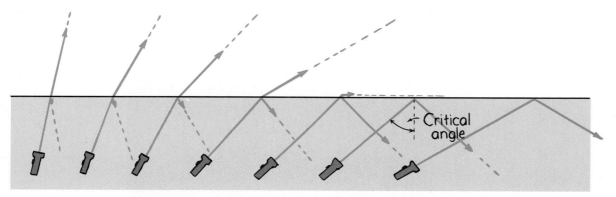

Figure 13.23 Light emitted in the water is partly refracted and partly reflected at the surface. The length of the arrows indicates the proportions refracted and reflected. Beyond the critical angle the beam is totally internally reflected.

ANSWERS

1. Your arm describes a cone, and your finger sweeps out a circle. Likewise with rainbows.

2. No.

light is tipped beyond the critical angle (48° from the normal for water), you'll notice that all the light is reflected back into the tub. This is *total internal reflection*. The light striking the air-water surface obeys the law of reflection: The angle of incidence is equal to the angle of reflection. The only light emerging from the water surface is that which is diffusely reflected from the bottom of the bathtub. This procedure is shown in Figure 13.23. The proportion of light refracted and light internally reflected is indicated by the relative lengths of the arrows.

Your pet goldfish in the bathtub looks up to see a compressed view of the outside world (Figure 13.24). The 180° view from horizon to opposite horizon is seen through an angle of 96°—twice the critical angle. A lens that similarly compresses a wide view is called a *fisheye lens*, used for special-effect photographs.

Figure 13.24 An observer underwater sees a circle of light at the still surface. Beyond a cone of 96° (twice the critical angle) an observer sees a reflection of the water interior or bottom.

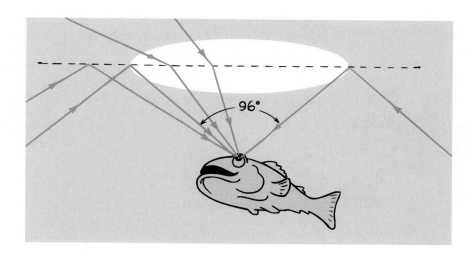

96°

The critical angle for glass is about 43°, depending on the type of glass. This means that within glass, light that is incident at angles greater than 43° will be totally internally reflected. No light will escape beyond this angle; instead, all of it will be reflected back into the glass. Whereas a silvered or aluminized mirror reflects only about 90 percent of incident light, glass prisms as shown in Figure 13.25 are more

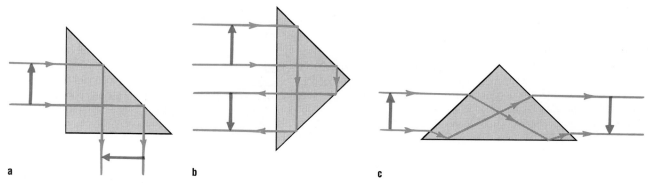

a   b   c

Figure 13.25   Total internal reflection in a prism. In (a) the prism changes the direction of the light beam by 90°, in (b) by 180°, and in (c) does not change the direction but instead turns the image upside down.

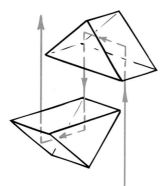

Figure 13.26 Total internal reflection in a pair of prisms.

Figure 13.27 Prism binoculars.

efficient. A little light is lost by reflection before it enters the prism, but once inside reflection on the 45°-slanted face is total—100 percent. Moreover, this light is not marred by any dirt or dust on the outside surface, which is the principal reason for the use of prisms instead of mirrors in many optical instruments.

A pair of prisms each reflecting light through 180° is shown in Figure 13.26. Binoculars use pairs of prisms to lengthen the light path between lenses and thus eliminate the need for long barrels. So a compact set of binoculars is as effective as a longer telescope (Figure 13.27). Another advantage of prisms is that whereas the image of a straight telescope is upside down, reflection by the prisms in binoculars re-inverts the image, so things are seen right-side up.

The critical angle for a diamond is about 24.5°, smaller than for any other known substance. The critical angle varies slightly for different colors, just as speed varies slightly for different colors. Once light enters a diamond gemstone, most is incident on the sloped backsides at angles greater than 24.5° and is totally internally reflected (Figure 13.28). Because of the great slowdown in speed as light enters a diamond, refraction is pronounced and there is great dispersion. Further dispersion occurs as the light exits through the many facets at its face. Hence we see unexpected flashes of a wide array of colors. Interestingly enough, when these flashes are narrow enough to be seen by only one eye at a time, the diamond "sparkles."

Total internal reflection also underlies the operation of optical fibers, or light pipes (Figure 13.29). An optical fiber "pipes" light from one place to another by a series of total internal reflections, much as a bullet ricochets down a steel pipe. Light rays bounce along the inner walls, following the twists and turns of the fiber. Optical fibers are used in decorative table lamps and to illuminate instrument displays on automobile dashboards from a single bulb. Dentists use them with flashlights to get light where they want it. Bundles of thin flexible glass or plastic fibers are used to see what is going on in inaccessible places, such as the interior of a motor or a patient's stomach. They can be made small enough to snake through blood vessels or through tubes such as the urethra. Light shines down some of the fibers to illuminate the scene and is reflected back along others.

Optical fibers are important in communications because they offer a practical alternative to copper wires and cables. In metropolitan areas, thin glass fibers now replace thick, bulky, and expensive copper cables to carry thousands of simultaneous telephone messages among the major switching centers. In many aircraft, control signals are fed from the pilot to the control surfaces by means of optical fibers. Signals are carried in the modulations of laser light. Unlike electricity, light is indifferent to

Figure 13.28 Paths of light in a diamond. Rays that strike the inner surface of a diamond at angles greater than the critical angle, about 24.5° depending on color, are internally reflected and exit via refraction at the top surface.

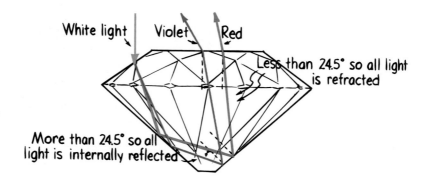

Figure 13.29 The light is "piped" from below by a succession of total internal reflections until it emerges at the top ends.

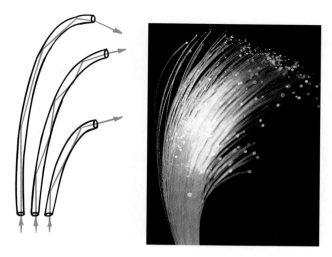

temperature and fluctuations in surrounding magnetic fields, so the signal is clearer. Also, it is much less likely to be tapped by eavesdroppers.

Optical fibers in nature are found on the polar bear—the hairs of its fur! The hairs of a polar bear are actually transparent optical fibers. Its fur appears white because visible light is reflected from the rough inner surface of each hollow hair. However, the hairs trap ultraviolet light. Like light within an optical fiber, the radiant energy is conducted through the hairs to the bear's skin. The skin is very efficient at absorbing all the solar energy it can get, and is black.

13.5 LENSES

A very practical case of refraction occurs in lenses. We can understand a lens by analyzing equal-time paths as we did earlier, or we can assume that it consists of a set of several matched prisms and blocks of glass arranged in the order shown in

Figure 13.30 The hairs of the polar bear are transparent light pipes that direct ultraviolet light to its skin—which is guess what color? Black! So what's white and black and warm all under? A polar bear under the Arctic sun.

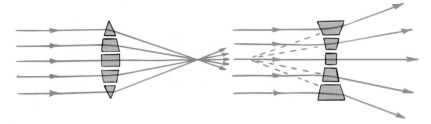

Figure 13.31   A lens may be thought of as a set of prisms.

Figure 13.31. The prisms and blocks refract incoming parallel light rays so they converge to (or diverge from) a point. The arrangement shown in Figure 13.31a converges the light, and we call such a lens a **converging lens**. Note that it is thicker in the middle.

The arrangement in 13.31b is different. The middle is thinner than the edges, and it diverges the light; such a lens is called a **diverging lens**. Note that the prisms diverge the incident rays in a way that makes them appear to come from a single point in front of the lens. In both lenses the greatest deviation of rays occurs at the outermost prisms, for they have the greatest angle between the two refracting surfaces. No deviation occurs exactly in the middle, for in that region the glass faces are parallel to each other. Real lenses are not made of prisms, of course, as is indicated in Figure 13.31; they are made of a solid piece of glass with surfaces ground usually to a circular curve. In Figure 13.32 we see how smooth lenses refract waves.

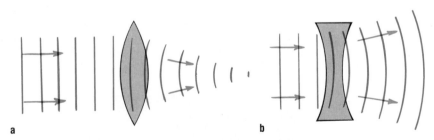

Figure 13.32   Wave fronts travel more slowly in glass than in air. In (a), the waves are retarded more through the center of the lens, and convergence results. In (b), the waves are retarded more at the edges, and divergence results.

Some key features in lens description are shown for a converging lens in Figure 13.33. The *principal axis* of a lens is the line joining the centers of curvatures of its surfaces. The *focal point* is the point at which a beam of parallel light, parallel to the principal axis, converges. Incident parallel beams that are not parallel to the principal axis focus at points above or below the focal point. All such possible points make up a *focal plane*. Since a lens has two surfaces, it has two focal points and two focal planes. When the lens of a camera is set for distant objects, the film is in the focal plane behind the lens in the camera.

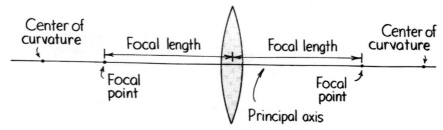

Figure 13.33   Key features of a converging lens.

In the diverging lens, an incident beam of light parallel to the principal axis is not converged to a point, but is diverged—so the light appears to come from a point in front of the lens. The *focal length* of a lens, whether converging or diverging, is the distance between the center of the lens and its focal point. For a thin lens, the focal lengths on either side are equal, even when the curvatures on the two sides are different.

Image Formation by a Lens

At this moment, light is reflecting from your face onto the page of this book. Light that reflects from your forehead, for example, strikes every part of the page. Likewise for the light that reflects from your chin. Every part of the page is illuminated with reflected light from your forehead, your nose, your chin, and every part of your face. You don't see an image of your face on the page because there is too much overlapping of light. But put a barrier with a pinhole in it between your face and the page, and the light that reaches the page from your forehead will not overlap the light from your chin. Likewise for the rest of your face. Without this overlapping, there will be an image of your face on the page. It will be very dim, for very little light reflected from your face gets through the pinhole. To be seen, you'd have to shield the page from other light sources.

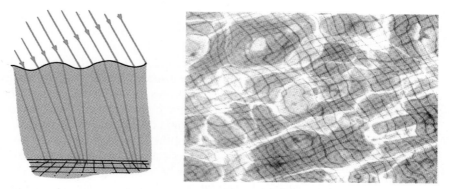

Figure 13.34   The moving patterns of bright and dark areas at the bottom of the pool result from the uneven surface of water, which behaves like a blanket of undulating lenses. A fish looking upward at the sun would see the sun shimmering in intensity. Because of similar irregularities in the atmosphere, we see the stars twinkle.

Your Eye

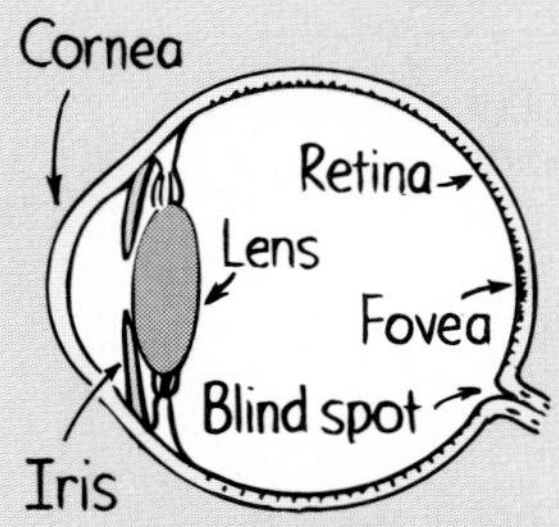

Figure 13.35 The human eye.

Light is the only thing we see with the most remarkable optical instrument known—the eye (Figure 13.35). Light enters through the *cornea,* which does about 70 percent of the necessary bending of the light before it passes through the *pupil* (the aperture in the iris). Light then passes through the *lens,* which provides the extra bending power needed to focus images of nearby objects on the extremely sensitive *retina* (only quite recently have artificial detectors attained greater sensitivity to light than the human eye). An image of the visual field outside the eye is spread over the retina. The retina is not uniform. There is a spot in the center of our field of view called the *fovea,* or region of most distinct vision. Greater detail is seen here than at the side parts of the eye. There is also a spot in the retina where the nerves carrying all the information exit; this is the *blind spot.*

You can demonstrate that you have a blind spot in each eye if you hold this book at arm's length, close your left eye, and look at the circle and X below with your right eye only. You can see both the circle and the X at this distance. Now move the book slowly toward your face, with your right eye fixed upon the circle, and you'll reach a position about 20–25 centimeters from your eye where the X disappears; when both eyes are open, one eye "fills in" the part to which the other eye is blind. Be glad if you have two eyes.

The light receptors in the retina do not connect directly to the optic nerve, but are interconnected to many other cells. Through these interconnections a certain amount of information is combined and "digested" in the retina. In this way the light signal is "thought about" before it goes to the optic nerve and then to the main body of the brain. So some brain functioning occurs in the eye itself. Amazingly, the eye does some of our "thinking."

⬤   ✖

The first cameras had no lenses and admitted light through a small pinhole. You can see why the image is upside down by the sample rays in Figure 13.36b. Long exposure times were required because of the small amount of light admitted by the pinhole. A somewhat larger hole would admit more light, but overlapping rays would produce a blurry image. Too large a hole would allow too much overlapping and no image would be discernible. That's where a converging lens comes in (Figure 13.36c). The lens converges light onto the screen without the unwanted overlapping of rays. Whereas the first pinhole cameras were useful only for still objects because of the long exposure time required, moving objects can be photographed with the lens camera because of the short exposure time. Now you know why photographs taken with lens cameras came to be called *snapshots.*

The simplest use of a converging lens is a magnifying glass. Magnification occurs when an image is observed through a wider angle with the use of a lens than without the lens. With unaided vision, an object far away is seen through a relatively small angle of view, while the same object when closer is seen through a larger angle of view. This wider angle permits the perception of greater detail. A magnifying glass is simply a converging lens that increases the angle of view.

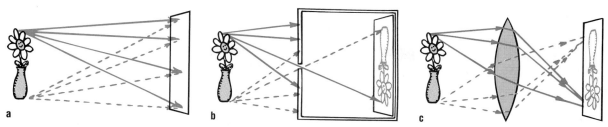

Figure 13.36 Image formation. (a) No image appears on the wall because rays from all parts of the object overlap all parts of the wall. (b) A single small opening in a barrier prevents overlapping rays from reaching the wall; a dim, upside-down image is formed. (c) A larger opening admits more light and a lens converges the rays upon the wall without overlapping; more light makes a brighter image.

Figure 13.37

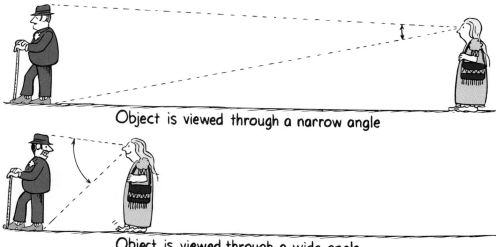

Object is viewed through a narrow angle

Object is viewed through a wide angle

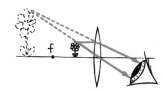

Figure 13.38 When an object is near a converging lens (inside its focal point *f*), the lens acts as a magnifying glass to produce a virtual image. The image appears larger and farther from the lens than the object.

When we use a magnifying glass, we hold it close to the object we wish to see magnified. This is because a converging lens will magnify only when the object is inside the focal point. The magnified image will be farther from the lens than the object, and it will be right-side up. If a screen is placed at the image distance, no image will appear on it. This is because no light is actually directed to the image position. The rays that reach our eye, however, behave virtually as if they came from the image position, so we call this a **virtual image**.

When the object is far away enough to be outside the focal point of a converging lens, instead of a virtual image a **real image** is formed. Figure 13.39 shows a case in which a converging lens forms a real image on a screen. A real image is inverted. A similar arrangement is used for projecting slides and motion pictures on a screen and for projecting a real image on the film of a camera. Real images with a single lens will always be inverted.

A diverging lens, when used alone, produces only a diminished virtual image. It makes no difference how far or how near the object is. When a diverging lens is used alone the image is always virtual, erect, and smaller than the object. A diverging lens is often used as a "finder" on a camera. When you look at the object to be photographed through such a lens, you see a virtual image that approximates the same proportions as the photograph.

Figure 13.39 When an object is far from a converging lens (beyond its focal point), a real upside-down image is formed.

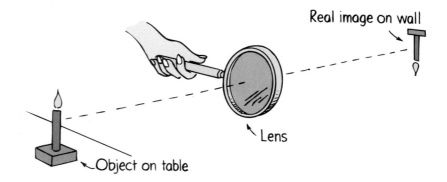

Real image on wall

Lens

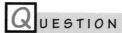

Object on table

QUESTION

Why is the greater part of the photograph in Figure 13.40 out of focus?

Figure 13.40 A diverging lens forms a virtual, right-side-up image of Jamie and his cat.

ANSWER

Both Jamie and his cat and the virtual image of Jamie and his cat are "objects" for the lens of the camera that took this photograph. Since the objects are at different distances from the camera lens, their respective images are at different distances with respect to the film in the camera. So only one can be brought into focus. The same is true of your eyes. You cannot focus on near and far objects at the same time.

Lens Defects

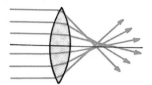

Figure 13.41 Spherical aberration.

No lens provides a perfect image. A distortion in an image is called an **aberration**. By combining lenses in certain ways, aberrations can be minimized. For this reason, most optical instruments use compound lenses, each consisting of several simple lenses, instead of single lenses.

Spherical aberration results from light passing through the edges of a lens and focusing at a slightly different place from where light passing through the center of the lens focuses (Figure 13.41). This can be remedied by covering the edges of a lens, as with a diaphragm in a camera. Spherical aberration is corrected in good optical instruments by a combination of lenses.

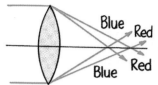

Figure 13.42 Chromatic aberration.

Chromatic aberration is the result of the different speeds of various colors and hence the different refractions they undergo (Figure 13.42). In a simple lens (as in a prism), red light and blue light do not come to focus in the same place. *Achromatic lenses,* which combine simple lenses of different kinds of glass, correct this defect.

The pupil of the eye regulates the amount of light that enters the eye by changing its size. Vision is sharpest when the pupil is smallest because light then passes through only the center of the eye's lens, where spherical and chromatic aberrations are minimal. Also, light bends the least through the center of a lens, so minimum focusing is required for a sharp image. An image that is formed by straight lines of light can appear in focus anywhere. You see better in bright light because your pupils are smaller.*

Astigmatism of the eye is a defect that results when the cornea is curved more in one direction than the other, somewhat like the side of a barrel. Because of this defect, the eye does not form sharp images. The remedy is eyeglasses with cylindrical lenses that have more curvature in one direction than in another.

Q UESTIONS

1. If light traveled at the same speed in glass as in air, would glass lenses still alter the direction of light rays?
2. Why is there chromatic aberration in light that passes through a lens, but no chromatic aberration in light that reflects from a mirror?

An option for those with poor sight in the last 500 years has been wearing spectacles or, in more recent times, contact lenses. It is interesting to note that at the

A NSWERS

1. No.

2. Different frequencies travel at different speeds in a transparent medium and therefore refract at different angles, which produces chromatic aberration. But the angles at which light reflects has nothing to do with the frequency of light. One color reflects the same as any other reflected color. Mirrors are therefore preferable to lenses in telescopes because with reflection there is no chromatic aberration.

---

*If you wear glasses and ever misplace them, or if you find it difficult to read small print as in a telephone book, hold a pinhole (in a piece of paper or whatever) in front of your eye, close to the page. You'll see the print clearly, and because you're close, it is magnified. Try it and see!

Lateral Inhibition

The human eye can do what no camera film can do: perceive degrees of brightness that range from about 500 million to 1. The difference in brightness between the sun and moon, for example, is about 1 million to 1. But, because of an effect called *lateral inhibition,* we don't perceive the actual differences in brightness. The brightest places in our visual field are prevented from outshining the rest, for whenever a receptor cell on our retina sends a strong brightness signal to our brain, it also signals neighboring cells to dim their responses. In this way, we even

out our visual field, which allows us to discern detail in very bright areas and in dark areas as well.

Lateral inhibition exaggerates the difference in brightness at the edges of places in our visual field. Edges, by definition, separate one thing from another. So we accentuate differences rather than similarities. The gray rectangle in Figure 13.43 appears dimmer than the gray rectangle on the right when the edge that separates them is in our view. But cover the edge with your pencil or your finger, and they look equally bright (try it now)! That's

Figure 13.43 Both rectangles are equally bright. Cover the boundary between them with your pencil and see.

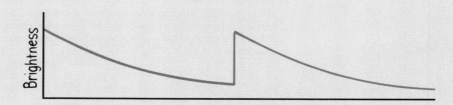

Figure 13.44 Graph of brightness levels for the rectangles in Figure 13.43.

because both rectangles *are* equally bright; each rectangle is shaded lighter to darker, moving from left to right. Our eye concentrates on the boundary where the dark edge of the left rectangle joins the light edge of the right rectangle, and our eye-brain system assumes that the rest of the rectangle is the same. We pay attention to the boundary and ignore the rest.

Questions to ponder: Is the way the eye picks out edges and makes assumptions about what lies beyond similar to the way we sometimes make judgments about other cultures and other people? Don't we in the same way tend to exaggerate the differences on the surface while ignoring the similarities and subtle differences within?

present time there is an alternative to both spectacles and contacts for people with poor eyesight. Experimental and controversial techniques today allow eye surgeons to reshape the cornea of the eye for normal vision. In tomorrow's world, the wearing of eyeglasses and contact lenses may be a thing of the past. We really do live in a rapidly changing world. And that can be nice.

13.6 WAVE-PARTICLE DUALITY

We have described light as a wave. The earliest ideas of light, however, were that light was composed of tiny particles. In ancient times Plato and other Greek philosophers held a particle view of light, as did Isaac Newton in the early 1700s, who first became famous for his experiments with light. A hundred years later the wave nature of light was demonstrated by Thomas Young in the double-slit experiment. The wave view was reinforced in 1862 by Maxwell's finding that light was energy carried in oscillating electric and magnetic fields of electromagnetic waves. The wave view of light was confirmed experimentally by Hertz, 25 years later.

Then in 1905, Albert Einstein published a Nobel Prize–winning paper that challenged the wave theory of light. Einstein stated that light in its interactions with matter was confined not in continuous waves, as Maxwell and others had envisioned, but in tiny particles of energy called *photons*. Einstein's particle model of light explained a perplexing phenomenon of that time—the *photoelectric effect*.

The Photoelectric Effect

When light shines on certain metal surfaces, electrons are ejected from the surfaces. This is the **photoelectric effect**, used in electric eyes, in the photographer's light meter, and in the sound track of motion pictures. What perplexed investigators at the turn of this century was that ultraviolet and violet light imparted sufficient energy to knock electrons from these surfaces, while lower-frequency light did not—even when the low-frequency light was very bright. Ejection of electrons depended only on the frequency of light, and the higher the frequency of light used, the greater the kinetic energy of the ejected electrons. Very dim high-frequency light ejected fewer electrons, but each with the same kinetic energy of electrons ejected in brighter light of the same frequency.

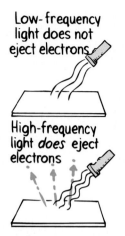

Low-frequency light does not eject electrons

High-frequency light *does* eject electrons

Figure 13.45 The photoelectric effect depends on frequency.

Einstein's explanation was that the electrons in the metal were being bombarded by "particles of light"—by **photons**. Einstein stated that the energy of each photon was proportional to its frequency: That is,*

$$E \sim f$$

So Einstein viewed light as a hail of photons, each carrying energy proportional to its frequency. One photon is completely absorbed by each electron ejected from the metal.

All attempts to explain the photoelectric effect by waves failed. A light wave has a broad front, and its energy is spread out along this front. For the light wave to eject a single electron from a metal surface, all its energy would somehow have to be concentrated on that one electron. But this is as improbable as an ocean wave hitting a beach and knocking only one single seashell far inland with an energy equal to the energy of the whole wave. Therefore, instead of thinking of light encountering a surface as a continuous train of waves, the photoelectric effect suggests we conceive of light encountering a surface or any detector as a succession of particle-like photons. The energy of each photon is proportional to the frequency of light, and that energy is given completely to a single electron in the metal surface. The number of ejected electrons had to do with the number of photons—the brightness of the light.

Experimental verification of Einstein's explanation of the photoelectric effect was made 11 years later by the American physicist Robert Millikan. Every aspect of Einstein's interpretation was confirmed. The photoelectric effect proves conclusively that light has particle properties. A wave model of light is inconsistent with the photoelectric effect. On the other hand, interference demonstrates convincingly that light has wave properties. A particle model of light is inconsistent with interference.

Recall Thomas Young's double-slit interference experiment, which we earlier discussed in terms of waves. When monochromatic light passes through a pair of closely spaced thin slits, an interference pattern is produced on photographic film (Figure 13.46). Now let's consider the experiment in terms of photons. Suppose we dim our light source so that in effect only one photon at a time reaches the thin slits. If the film behind the slits is exposed to the light for a very short time, the film becomes exposed as simulated in Figure 13.47a. Each spot represents the place where the film has been exposed to a photon. If the light is allowed to expose the film for a longer time, a pattern of fringes begins to emerge as in Figure 13.47b and c. This is quite amazing. Spots on the film are seen to progress photon by photon to form the same interference pattern characterized by waves!

---

*When the energy of a photon is divided by its frequency, the single number that results is the proportionality constant, called **Planck's constant**, h (6.6×10^{-34} J · s). Planck's constant is a fundamental constant of nature that serves to set a lower limit on the smallness of things. We can insert this constant in the above proportion and express it as an exact equation:

$$E = hf$$

This equation gives the smallest amount of energy that can be converted to light with frequency f. The radiation of light is not emitted continuously but is emitted as a stream of photons, which each throb at a frequency f and carry an energy hf.

Figure 13.46
(a) Arrangement for double-slit experiment.
(b) Photograph of interference pattern. (c) Graphic representation of pattern.

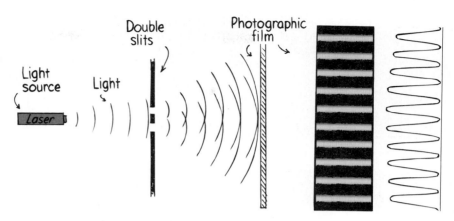

Evidently, light has both a wave and particle nature—a wave-particle duality.* This duality is evident in the formation of optical images. We understand the photographic image produced by a camera in terms of light waves, which spread from each point of the object, refract as they pass through the lens system, and converge to focus on the photographic film. The path of light from the object through the lens system and to the focal plane can be calculated using methods developed from the wave theory of light.

But now consider carefully the way in which the photographic image is formed. The photographic film consists of an emulsion that contains grains of silver halide crystal, each grain containing about 10^{10} silver atoms. Each photon that is absorbed gives up its energy to a single grain in the emulsion. This energy activates surrounding crystals in the entire grain and is used in development to complete the photochemical process. Many photons activating many grains produce the usual photographic exposure. When a photograph is taken with exceedingly feeble light, we find that the image is built up by individual photons that arrive independently and are seemingly random in their distribution. We see this strikingly illustrated in Figure 13.48, which shows how an exposure progresses photon by photon.

a   b   c

Figure 13.47 Stages of two-slit interference pattern. The pattern of individually exposed grains progresses from (a) 28 photons to (b) 1000 photons to (c) 10 000 photons. As more photons hit the screen, a pattern of interference fringes appears.

---

*From a Newtonian point of view, this wave-particle duality is mysterious. This leads some people to believe that photons and other "quanta" have some sort of consciousness, with each photon or quanta having "a mind of its own." (A quantum is the smallest "particle" of something, such as light, electricity, or energy itself.) The mystery, however, is like beauty. It is in the mind of the beholder rather than in nature itself. We conjure models to understand nature, and when inconsistencies arise, we sharpen or change our models. The wave-particle duality of light doesn't fit the model devised by Newton. An alternate model is that quanta have minds of their own. Another model is quantum physics. We subscribe to the latter.

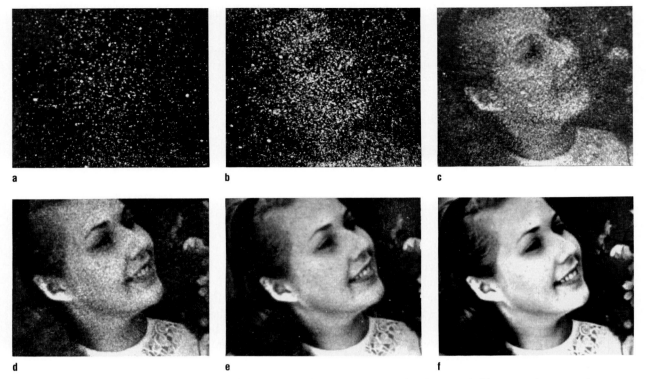

Figure 13.48 Stages of exposure revealing the photon-by-photon production of a photograph. The approximate numbers of photons at each stage were (a) 3×10^3, (b) 1.2×10^4, (c) 9.3×10^4, (d) 7.6×10^5, (e) 3.6×10^6, and (f) 2.8×10^7.

What all this means is that light has both wave and particle properties. Simply put, *light behaves as a stream of photons when it interacts with the photographic film or other detectors and behaves as a wave in traveling from a source to the place where it is detected.* Light travels as a wave and hits as a stream of photons. In interference experiments, photons strike the film at places where we would expect to see constructive interference of waves.

The fact that light exhibits both wave and particle behavior is one of the most interesting surprises that physicists discovered in this century. The finding that light comes in tiny bunches, tiny *quanta* as they are called, led to a whole new way of looking at nature—wave mechanics, or *quantum mechanics*. An outcome of this new mechanics is that just as light has particle properties, so do particles have wave properties. First electrons were found to have wave properties; a beam of electrons passing through slits exhibits the same type of diffraction pattern as light. Then other particles—even baseballs and orbiting planets, could be described by the new mechanics of waves. Quantum mechanics and Newtonian physics overlap in the macroworld, and both are seen as "correct." But only quantum mechanics, with its emphasis on waves, is wholly accurate in the microworld of the atom. More about this in Part 5. Onward!

SUMMARY OF TERMS

Reflection The return of light rays from a surface in such a way that the angle at which a given ray is returned is equal to the angle at which it strikes the sur-face. When the reflecting surface is irregular, light is returned in irregular directions; this is *diffuse reflection*.

Refraction The bending of an oblique ray of light when it passes from one transparent medium to another. This is caused by a difference in the speed of light in the transparent media. When the change in medium is abrupt (say, from air to water), the bending is abrupt; when the change in medium is gradual (say, from cool air to warm air), the bending is gradual, which accounts for mirages.

Law of reflection The angle of incidence equals the angle of reflection. The incident and reflected rays lie in a plane that is normal to the reflecting surface.

Critical angle The minimum angle of incidence at which a light ray is totally reflected within a medium.

Total internal reflection The total reflection of light traveling in a medium when it strikes on the surface of a less dense medium at an angle greater than the critical angle.

Converging lens A lens that is thicker in the middle than at the edges and refracts parallel rays passing through it to a focus.

Diverging lens A lens that is thinner in the middle than at the edges, causing parallel rays passing through it to diverge.

Virtual image An image formed by light rays that do not converge at the location of the image. A virtual image is that reflected by a mirror; it cannot be displayed on a screen.

Real image An image formed by light rays that converge at the location of the image. A real image can be displayed on a screen.

Aberration A limitation on perfect image formation inherent, to some degree, in all optical systems.

Photoelectric effect The emission of electrons from a metal surface when light shines on it.

• •

REVIEW QUESTIONS

Reflection

1. Distinguish between *reflection* and *refraction*.

2. What does incident light do to the electron clouds that surround the atomic nucleus?

3. What do electron clouds do when they are made to oscillate?

Law of Reflection

4. What is the law of reflection?

5. Compared to the distance of an object in front of a plane mirror, how far behind the mirror is the image?

6. Does the law of reflection hold for curved mirrors? Explain.

Diffuse Reflection

7. Does the law of reflection hold for diffuse reflection? Explain.

8. How can a surface be polished for some waves and not others?

Refraction

9. How does the angle at which light strikes a pane of window glass compare to the angle at which it passes out the other side?

10. How does the angle at which a ray of light strikes a prism compare to the angle at which it passes out the other side?

11. Does light travel faster in thin air or in dense air? What does this have to do with the duration of daylight?

12. What is a mirage?

13. Why do stars twinkle?

14. When a cart wheel rolls from a smooth sidewalk onto a plot of grass, the interaction of the wheel with the blades of grass slows the wheel. What slows light when it passes from air into glass or water?

15. When light passes from one material into another, it bends toward the normal to the surface. Or it may bend away from the normal. When does it do which?

16. Does refraction make the bottom of a swimming pool seem deeper or shallower?

Dispersion

17. What happens to light of a certain frequency when it is incident upon a material that has the same natural frequency?

18. Which travels more slowly in glass, red light or violet light?

19. If light of different frequencies has different speeds in a material, does it also refract at different angles in the same material? Explain.

Rainbows

20. What is it that prevents all rainbows from being complete circles?

21. Does a single raindrop illuminated by sunlight disperse a single color or a spectrum of colors?

22. Does a viewer see a single color or a spectrum of colors dispersed from a single faraway drop?

23. Why is a secondary rainbow dimmer than a primary bow?

Total Internal Reflection

24. What is meant by *critical angle*?

25. When is light totally reflected in a glass prism?

26. When is light totally reflected in a diamond?

27. Light normally travels in straight lines, but it "bends" in an optical fiber. Explain.

Lenses

28. Distinguish between a *converging lens* and a *diverging lens*.

29. What is the *focal length* of a lens?

Image Formation from a Lens

30. Distinguish between a *virtual image* and a *real image*.

31. What kind of lens can be used to produce a real image? A virtual image?

Lens Defects

32. Distinguish between *spherical aberration* and *chromatic aberration*.

33. What is astigmatism?

Wave-Particle Duality

34. What evidence can you cite for the wave nature of light? For the particle nature of light?

Photoelectric Effect

35. Which are more successful in dislodging electrons from a metal surface, photons of violet light or photons of red light? Why?

36. Does the brightness of a beam of light primarily depend on the frequency of photons or on the number of photons?

37. Does light behave primarily as a wave or as a particle when it interacts with the crystals of matter in photographic film?

38. Does light travel from one place to another in a wavelike way or a particlelike way?

39. Does light interact with a detector in a wavelike way or a particlelike way?

40. When does light behave as a wave? When does it behave as a particle?

• •

HOME PROJECTS

1. Make a pinhole camera, as illustrated below. Cut out one end of a small cardboard box, and cover the end with tissue or wax paper. Make a clean-cut pinhole at the other end. (If the cardboard is thick, make it through a piece of tinfoil placed over an opening in the cardboard.) Aim the camera at a bright object in a darkened room, and you will see an upside-down image on the tissue paper. If in a dark room you replace the tissue paper with unexposed photographic film, cover the back so it is light tight, and cover the pinhole with a removable flap, you are ready to take a picture. Exposure times differ depending principally on the kind of film and amount of light. Try different exposure times, starting with about 3 seconds. Also try boxes of various lengths. You'll find everything in focus in your photographs, but the pictures will not have clear-cut, sharp outlines. The principal difference between your pinhole camera and a commercial one is the glass lens, which is larger than the pinhole and therefore admits more light in less time—hence the name *snapshots*.

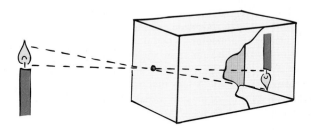

2. If you don't have a prism, you can produce a spectrum by placing a tray of water in bright sunlight. Lean a pocket mirror against the inside edge of the pan and adjust it until a spectrum appears on the wall or ceiling.

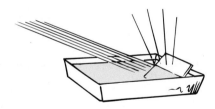

3. Stand a pair of mirrors on edge with the faces parallel to each other. Place an object such as a coin between the mirrors and look at the reflections in each mirror. Neat?

4. Set up two pocket mirrors at right angles and place a coin between them. You'll see four coins. Change the angle of the mirrors and see how many images of the coin you can see. With the mirrors at right angles, look at your face. Then wink. What do you see? You now see yourself as others see you. Hold a printed page up to the double mirrors and contrast its appearance with the reflection of a single mirror.

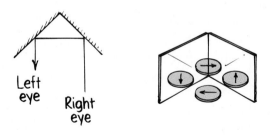

5. Determine the magnification power of a lens by focusing on the lines of a ruled piece of paper. Count the spaces between the lines that fit into one magnified space, and you have the magnification power of the lens. You can do the same with binoculars and a distant brick wall. Hold the binoculars so that only one eye looks at the bricks through the eyepiece while the other eye looks directly at the bricks. The number of bricks seen with the un-aided eye that will fit into one magnified brick gives the magnification of the instrument.

6. Look at the reflections of overhead lights from the two surfaces of eyeglasses, and you will see two fascinat-ingly different images. Why are they different?

. . . .

EXERCISES

1. Her eye at point P looks into the mirror. Which of the numbered cards can she see reflected in the mirror?

2. Cowboy Joe wishes to shoot his assailant by ricochet-ing a bullet off a mirrored metal plate. To do so, should he simply aim at the mirrored image of his assailant? Explain.

3. What must be the minimum length of a plane mirror in order for you to see a full view of yourself?

4. What effect does your distance from the plane mirror have in the above answer? (Try it and see!)

5. Hold a pocket mirror at almost arm's length from your face and note the amount of your face you can see. To see more of your face, should you hold the mirror closer or farther, or would you have to have a larger mirror? (Try it and see!)

6. The diagram shows a person and her twin at equal dis-tances on opposite sides of a thin wall. Suppose a win-dow is to be cut in the wall so each twin can see a com-plete view of the other. Show the size and location of the smallest window that can be cut in the wall to do the job. (Hint: Draw rays from the top of each twin's head to the other twin's eyes. Do the same from the feet of each to the eyes of the other.)

7. Suppose that you walk 1 m/s toward a mirror. How fast do you and your image approach each other?

8. What is wrong with the cartoon of the man looking at himself in the mir-ror? (Have a friend face a mirror as shown, and you'll see.)

9. Why is the lettering on the front of some vehicles "backward"?

ƎƆᴎA⅃UᙠMA

10. A person in a dark room looking through a window can clearly see a person outside in the daylight, whereas the person outside cannot see the person inside. Explain.

11. Why is it difficult to see the roadway in front of you when driving on a rainy night?

12. Does the reflection of a scene in calm water look ex-actly the same as the scene itself only upside down? (Hint: Can you ever see the reflection of the top of a stone that extends above water?)

13. Why does reflected light from the sun or moon appear as a column in the body of water as shown? How would it ap-pear if the water surface were per-fectly smooth?

14. Show with a simple diagram that when a mirror with a fixed beam incident on it is rotated through a certain angle, the reflected beam is rotated through an angle twice as large. (This doubling of displacement makes irregularities in ordinary window glass more evident.)

15. When you look at yourself in the mirror and wave your right hand, your beautiful image waves the left hand. Then why don't the feet of your image wiggle when you shake your head?

16. A pair of toy cart wheels are rolled obliquely from a smooth surface onto two plots of grass, a rectangular

plot as shown in (a) and a triangular plot as shown in (b). The ground is on a slight incline so that after slowing down in the grass, the wheels will speed up again when emerging on the smooth surface. Finish each sketch by showing some positions of the wheels inside the plots and on the other sides, thereby indicating the direction of travel.

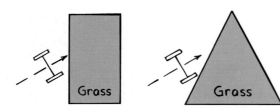

17. If light of all frequencies traveled at the same speed in glass, how would white light appear after passing through a prism?

18. Which is more likely to sparkle: a nearby diamond or a diamond farther away? Explain.

19. A beam of light bends as shown in (a), while the edges of the immersed square bend as shown in (b). Do these pictures contradict each other? Explain.

20. If you were spearing a fish, would you aim above, below, or directly at the observed fish to make a direct hit? If you instead used light from a laser as your "spear," would you aim above, below, or directly at the observed fish? Defend your answers.

21. If the fish in the previous exercise were small and blue, and your laser light were red, what corrections should you make? Explain.

22. When a fish looks upward at an angle of 45°, does it see the sky or the reflection of the bottom beneath? Defend your answer.

23. If you were to send a beam of laser light to a space station above the atmosphere and just above the horizon, would you aim the laser above, below, or at the visible space station? Defend your answer.

24. What accounts for the large shadows cast by the ends of the thin legs of the water strider? What accounts for the ring of bright light around the shadows?

25. Figure 13.17 shows the rainbow as an ellipse rather than as a circle to indicate it is being viewed from the side. Can one view a rainbow from the side so that it appears as the segment of an ellipse rather than the segment of a circle? Defend your answer.

26. Two observers standing apart from one another do not see the "same" rainbow. Explain.

27. A rainbow viewed from an airplane may form a complete circle. Where will the shadow of the airplane appear? Explain.

28. How is a rainbow similar to the halo sometimes seen around the moon on a night when ice crystals are in the upper atmosphere?

29. What is responsible for the rainbow-colored fringe commonly seen at the edges of a spot of white light from the beam of a lantern or slide projector?

30. Transparent plastic swimming-pool covers called *solar heat sheets* have thousands of small lenses made up of air-filled bubbles. The lenses in these sheets are advertised to focus heat from the sun into the water and raise its temperature. Do you think the lenses of such sheets direct more solar energy into the water? Defend your answer.

31. Would the average intensity of sunlight measured by a light meter at the bottom of the pool in Figure 13.34 be different if the water were still?

32. What would be the effect of a pinhole camera (Figure 27.47b and Home Project 1) that has two pinholes instead of one? Multiple holes?

33. What condition must exist for a converging lens to produce a virtual image? Can a diverging lens ever produce a real image?

34. Can you take a photograph of your image in a plane mirror and focus the camera on both your image and the mirror frame? Explain.

35. In terms of focal length, how far behind the camera lens is the film located when very distant objects are being photographed?

36. Maps of the moon are actually upside down. Why is this so?

37. In taking a photograph, what would happen to the image if you cover up the bottom half of the lens?

38. Why do older people who do not wear glasses read books farther away from their eyes than do younger people?

39. Why is chromatic abberation not a problem with mirrors?

40. Rays of light parallel to the principal axis of a converging lens pass through the lens and come to a focus a certain distance from the lens—its focal point. When the lens is under water, will the focal point be longer, shorter, or the same distance as for air?

41. No glass is perfectly transparent. Because of reflections, about 92 percent of light energy is transmitted by an average sheet of clear, dust-free windowpane. The 8-percent loss is not noticeable through a single sheet, but through several sheets it is apparent. How much light is transmitted by two sheets, each of which transmits 92 percent?

42. We speak of photons of red light and photons of green light. Can we speak of photons of white light? Why or why not?

43. A beam of red light and a beam of blue light have exactly the same energy. Which beam contains the greater number of photons?

44. Silver bromide (AgBr) is a light-sensitive substance used in some types of photographic film. To cause exposure of the film, it must be illuminated with light having sufficient energy to break apart the molecules. Why do you suppose this film may be handled without exposure in a darkroom illuminated with red light? How about blue light? How about very bright red light as compared to very dim blue light?

45. Suntanning produces cell damage in the skin. Why is ultraviolet radiation capable of producing this damage, while visible radiation is not?

46. In the photoelectric effect, does brightness or frequency determine the number of electrons ejected per second?

47. Explain how the photoelectric effect is used to open automatic doors when someone approaches.

48. Explain briefly how the photoelectric effect is used in the operation of at least two of the following: an electric eye, a photographer's light meter, the sound track of a motion picture.

49. Does the photoelectric effect *prove* that light is made of particles? Do interference experiments *prove* that light is composed of waves? (Is there a distinction between what something *is* and how it *behaves?*)

50. The camera that took the photograph of the woman's face (Figure 13.45) used ordinary lenses that are well known to refract waves. Yet the step-by-step formation of the image is evidence of photons. How can this be? What is your explanation?

PART 5

THE ATOM

Like everyone, I'm made of atoms. I get lots of them whenever I take a breath of air. Some I exhale right away but many stay for a while to become part of me -- then I may exhale them later. Whenever you take a breath, some of my atoms are in it and become part of you (and like-wise, yours become part of me). Atoms cycle and recycle through all of us. Their numbers and smallness are staggering -- there are many more atoms in a breath of air than the entire human population since time zero. So every time we take a breath we inhale atoms that were once a part of every person who ever lived. And atoms that are now part of us will be part of everyone else later on. In this sense, we're all one!

14

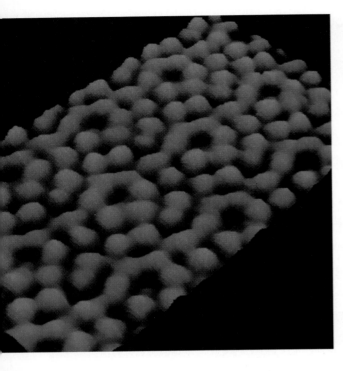

ATOMIC
STRUCTURE

All manner of things—shoes, and ships, and sealing wax; cabbages and kings—anything we can think of is composed of atoms. You might think that an enormous number of different kinds of atoms must exist to account for the rich variety of substances in the universe. But the number of different atoms is surprisingly small. The great variety of substances results not from a great variety of atoms, but from the many ways a few types of atoms can be combined. Just as three colors can be combined to form any color in a color print, a few kinds of atoms combine in different ways to produce all substances. To date (1993) we know of 109 distinct atoms. We call these atoms the chemical *elements*. Only 88 elements are found naturally; the others are made in the laboratory with high-energy atomic accelerators and nuclear reactors. These heavy synthetic elements are too unstable (radioactive) to occur naturally in appreciable amounts.

14.1 THE ELEMENTS

Hydrogen, which makes up over 90 percent of the atoms in the universe, was the original element. Elements heavier than hydrogen are manufactured in the deep interiors of stars. There, enormous temperatures and pressures cause hydrogen atoms

Figure 14.1 Co-author Leslie, shown here at the age of 16, is made of stardust—in the sense that the carbon, oxygen, nitrogen, and other atoms that make up her body originated in the deep interiors of ancient stars that have long since exploded.

to fuse into more complex elements. With the exception of some of the hydrogen and trace amounts of other light elements, all the elements in our surroundings are remnants of stars that exploded long before the solar system came into being.

These star remnants are the building blocks of all matter. All matter, however complex, living or nonliving, is some combination of the elements. From a pantry having 109 bins, each containing a different element, we have all the materials needed to make up any substance occurring in the universe.

The majority of elements are not found in great abundance, and some elements are exceedingly rare. Only a dozen or so elements compose the things we see every day.* Living things, for example, consist primarily of four elements: carbon [C], hydrogen [H], oxygen [O], and nitrogen [N]. The letters in brackets represent the chemical symbols for these elements.

It is difficult to imagine how small atoms are. Atoms are so small that they can't be seen with visible light—because they are smaller than the wavelengths of visible light. We could stack microscope on top of microscope and never "see" an atom because of the diffraction of visible light (Chapter 13). Photos of atoms, such as Figure 14.2, are obtained with a scanning tunneling microscope, a non-light imaging

Figure 14.2 An image of graphite obtained with a scanning tunneling microscope.

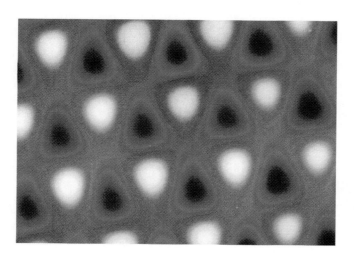

*Most common substances are formed out of combinations of two or more of these most common elements: hydrogen [H], carbon [C], nitrogen [N], oxygen [O], sodium [Na], magnesium [Mg], aluminum [Al], silicon [Si], phosphorus [P], sulfur [S], chlorine [Cl], potassium [K], calcium [Ca], and iron [Fe].

device that bypasses light and optics altogether. Even better detail can be seen with newer types of imaging devices that are presently revolutionizing microscopy.

The first fairly direct evidence for the existence of atoms was unknowingly discovered in 1827 by a Scottish botanist, Robert Brown, while he was studying the spores of pollen under a microscope. He noticed that the spores were in a constant state of agitation, always moving, always jumping about. At first he thought that the spores were some sort of moving life forms. But later he found that inanimate dust particles and grains of soot also showed this motion. Brown didn't realize it, but this perpetual jiggling of particles—called **Brownian motion**—results from bombardment by neighboring particles and atoms too small to be seen.

Today we know the atom is made up of a central nucleus that contains most of an atom's mass. Surrounding the nucleus are electrons. These are the electrons that make up the electric currents we studied in Part 4. The electrons also dictate the chemical properties of the element; we will treat this in Part 6. Let's now study the electron and its role in the atom.

A Breath of Air

Atoms are so small that there are about as many atoms of air in your lungs at any moment as there are breaths of air in the earth's atmosphere. That's because there are about 10^{22} atoms in a liter of air at atmospheric pressure and about 10^{22} liters of air in the atmosphere. Here's what that means: Exhale a deep breath so the number of molecules exhaled approximately equals the number of breathfuls of air in the earth's atmosphere. It will take about six years for your breath to become uniformly mixed in the atmosphere. Then anyone, anywhere on earth, who inhales a breath of air takes in, on the average, one of the molecules in that exhaled breath of yours. But you exhale many many breaths, so other people breathe in

many many molecules that were once in your lungs—that were once a part of you. And of course, vice versa. You breathe molecules that were once a part of everyone who ever lived, with each breath you take in. Considering that exhaled atoms are part of our bodies, it can be truly said that we are literally breathing in one another.

The origin of most atoms goes back to the origin of the universe, and nearly all atoms are older than the sun and earth. There are atoms in your body that have existed since the beginning of time, recycling throughout the universe among innumerable forms, both non-living and living. So you don't "own" the atoms that make up your body—you're simply the present caretaker. There will be many to follow.

14.2 THE ELECTRON

The name *electron* comes from the Greek word for *amber,* the material discovered by early Greeks to exhibit the effects of electrical charge. In the mid 1700s Ben Franklin further experimented with electrical charge. He discovered that lightning was a flow of electrical charge. Hence he realized that electricity is not restricted to solid objects and can travel through gases, such as the atmosphere.

Franklin's experiments inspired other scientists to send electric currents through various gases in sealed glass tubes. Glowing rays were produced when a voltage was applied across electrodes in these gas-filled tubes. It was observed

Atomic Imagery

Atoms, the building blocks of all matter, are smaller than the wavelength of light. Hence you can never see an atom—at least not with optical microscopes or any other instrument that depends on visible light. So during the first half of this century it was thought that human beings would never be able to see atoms. Then in the mid 1980s researchers at IBM first developed the scanning tunneling microscope (STM), an amazing tool that permits us to look at images of actual atoms. The underlying principle of this microscope is rather simple. To produce atomic images, an ultra-thin needle is dragged back and forth over the surface of a sample. Bumps the size of atoms on the surface cause the needle to move up and down. The vertical motion of the needle is detected and translated by a computer into a topographical image of bumps, which correspond to atoms. The images produced by scanning tunneling microscopes are truly fantastic (Figure 14.3). Micro imagery is presently an area of intense research.

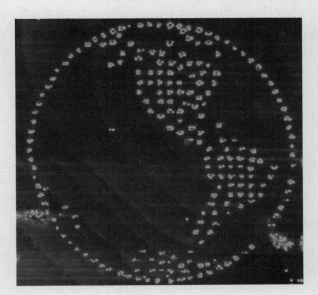

Figure 14.3 Each dot in the world's tiniest map consists of a few thousand atoms of gold. The atoms were moved into place by a scanning tunneling microscope.

that such rays were blocked from reaching the positive electrode (anode) when an object was placed in their path. It was thus reasoned that the ray emerged from the negative electrode (cathode). The apparatus was named the **cathode ray tube** (Figure 14.5).

Experiments showed that the intensity of the cathode ray did not diminish when the cathode ray tube was partially evacuated. Instead, the ray simply became less visible. This suggested that the ray could be generated with or without the

Figure 14.4 Franklin's kite-flying experiment.

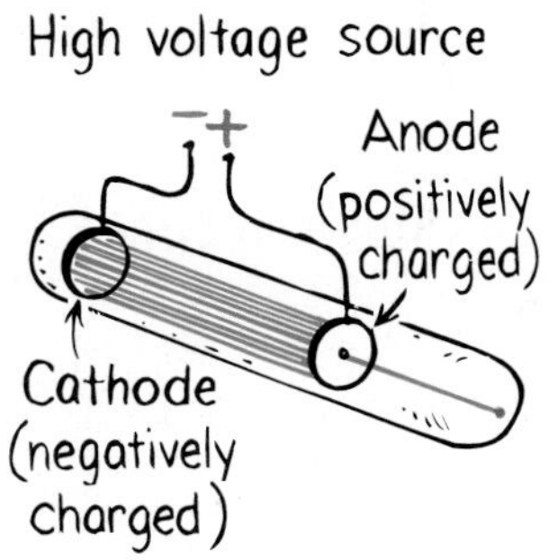

High voltage source

Anode (positively charged)

Cathode (negatively charged)

Figure 14.5 A simple cathode ray tube. The small hole in the anode permits the passage of a narrow beam, which strikes the end of the tube producing a glowing dot.

Figure 14.6 Cathode rays are deflected by magnetic fields.

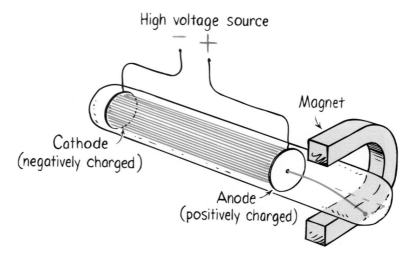

High voltage source

Magnet

Cathode (negatively charged)

Anode (positively charged)

presence of a gas and was thus an entity in itself. Further experiments showed that the ray was deflected by a magnetic field. This suggested that the ray consisted of rapidly moving charged particles. Also, when electric charges were brought nearby, the ray was deflected toward the positive charge and away from the negative charge. So it appeared that these particles were negatively charged.

In 1887, J. J. Thomson measured the deflection angles of cathode ray particles in a magnetic field. He reasoned that the deflection of particles depended on two things—mass and electrical charge. The greater each particle's mass, the greater the inertia and the less the deflection. The greater each particle's charge, the greater the deflection. From his measurements he was able to calculate the charge-to-mass ratio of the cathode ray particles (Figure 14.6). With the ratio alone, Thomson could not calculate the mass or the charge of the particles however. In order to calculate the mass he needed to know the charge, but in order to calculate the charge, he needed to know the mass.

Robert Millikan addressed this question. He calculated the numerical value of a single unit of electric charge on the basis of an experiment he carried out in 1911. Millikan sprayed tiny oil droplets into an electric field. When the field was strong, some of the tiny droplets moved upward, indicating they had a very slight negative charge. Millikan adjusted the field so droplets would hover motionlessly. He knew that the downward force of gravity on the motionless droplets was exactly balanced by the upward electrical force. Investigation showed that the electrical charges on the droplets were always some multiple of a single and very small value. Millikan proposed this smallest value to be a fundamental unit of all electrical charge. Using this value and the charge-mass ratio discovered by Thomson, Millikan calculated the mass of a cathode ray particle. He found it to be considerably less than that of

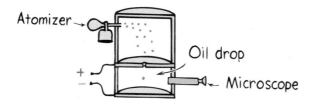

Atomizer

Oil drop

Microscope

Figure 14.7 Millikan's oil-drop experiment for determining the charge on the electron. The pull of gravity on the drops is balanced by the upward electrical force.

the smallest known atom, hydrogen. To many people this was very surprising for it meant that the atom was no longer the smallest known particle of matter.

The cathode ray particle is known today as the **electron**. The electron is a fundamental component of all atoms. An electron in one atom is identical to any electron in or out of any other atom. Electrons are negative charges. They therefore repel other electrons and are attracted to positively charged atomic nuclei. Electrons on the outer edges of atoms dictate whether and how atoms bond to become molecules, the melting and freezing temperatures of materials, the electrical conductivity of metals, as well as the taste, texture, appearance, and color of substances. The cathode ray—a stream of electrons—has found a great number of applications, most notably in producing images on your television set.

The discovery of the electron posed questions about the structure of the atom. For if atoms contained negatively charged particles, some balancing positively charged matter must also exist to account for the electrical neutrality of atoms. To account for the neutrality of atoms, Thomson put forth what he called a "plum pudding" model of the atom. In this model electrons were studded like plums in a sea of positively charged pudding. Further experimentation, however, soon proved the plum pudding model to be wrong.

14.3 THE ATOMIC NUCLEUS

A more accurate picture of the atom came to the British physicist, Ernest Rutherford, when in 1909 he oversaw the now-famous gold-foil experiment. This significant experiment showed that the atom is mostly empty space, with most of its mass concentrated in the central region—the *atomic nucleus*.

In Rutherford's experiment a beam of positively charged particles (alpha particles) from a radioactive source was directed through a very thin gold foil. Since alpha particles are thousands of times more massive than electrons, it was expected that the alpha particle stream would not be impeded as it passed through the "atomic pudding." This was indeed observed to be the case—for the most part. Nearly all alpha particles passed through the gold foil undeflected and produced spots of light when they hit a fluorescent screen around the gold leaf. But some particles were deflected from their straight-line path as they emerged (Figure 14.8). A few alpha particles were widely deflected, and a small number were even scattered backwards! These alpha particles must have hit something relatively massive, but what? Rutherford reasoned that the undeflected particles traveled through regions of the gold foil that

Figure 14.8 Rutherford's gold foil experiment. Deflection of alpha particles showed the atom to be mostly empty space with a concentration of mass at its center—the atomic nucleus.

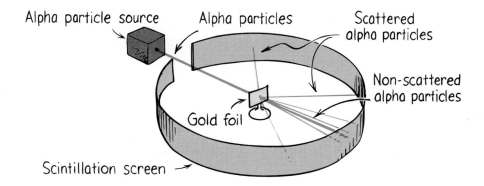

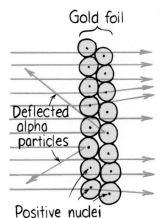

Gold foil

Deflected
alpha
particles

Positive nuclei

Figure 14.9 Most alpha particles pass through the gold foil undeflected. Those few that approach a positive nucleus are repelled and scattered through various angles.

were empty space, while the small number of deflected particles were repelled from extremely dense positively charged centers. Each atom, he concluded, must contain one of these centers, which he named the **atomic nucleus**.

Like the solar system, the atom is mostly empty space. The nucleus and surrounding electrons occupy only a tiny fraction of the atomic volume. If it weren't for the electrical forces of repulsion between the electrons of neighboring atoms, solid matter would be much more dense than it is. We and the solid floor are mostly empty space, because the atoms making up all materials are themselves mostly empty space. These same electrical repulsions prevent us from falling through the solid floor. Electrical forces keep atoms from caving in on each other under pressure. Atoms too close will repel (if they don't combine to form molecules). But when atoms are several atomic diameters apart, the electrical forces they exert on each other are negligible.

The nucleus occupies only a few quadrillionths of the volume of an atom. It is therefore extremely dense. If bare atomic nuclei could be packed against each other into a lump 1 centimeter in diameter (about the size of a large pea), the lump would weigh 133,000,000 tons! Huge electrical forces of repulsion prevent such close packing of atomic nuclei since each nucleus is electrically charged and repels other nuclei. Only under special circumstances are the nuclei of two or more atoms squashed into contact. When this happens, a violent reaction takes place. This reaction happens in the centers of stars and is what makes them shine. This will be further discussed in Chapter 16.

14.4 PROTONS AND NEUTRONS

The positive charge of the nucleus was found to be equal in magnitude to the combined negative charge of the electrons in an atom. It was thus reasoned, and then experimentally demonstrated, that positively charged particles, **protons**, make up the nucleus. Although the proton was found to be nearly 2000 times more massive than the electron, its charge is equal and opposite. The number of protons in the nucleus is electrically balanced by an equal number of electrons whirling about the nucleus. For example, a neutral oxygen atom has 8 electrons and 8 protons. All neutral atoms have a zero net charge. (Recall from Chapter 9 that when electric charge is not neutralized, the atom is called an *ion;* more about ions in Part 6.)

Atoms are classified by their **atomic number**, which is equal to the number of protons in the nucleus. Hydrogen has atomic number 1, helium 2, and so on in sequence to naturally occurring uranium with atomic number 92. The numbers continue through the artificially produced transuranic elements, at this writing up to 109. The arrangement of elements by their atomic numbers makes up the **periodic table**. See the front inside cover of this book.

If we compare the mass-to-charge ratios of different atoms we see that the atomic nucleus must be made up of more than protons. Nuclear charge and nuclear mass do not always go hand in hand. Helium, for example, has twice the charge of hydrogen but four times the mass. The added mass is due to another particle found in the nucleus, the **neutron**. The neutron has about the same mass as the proton. It, however, is without charge. Atoms of an element may or may not contain the same number of neutrons. For example, most hydrogen atoms have no neutrons. A small percentage, however, have one neutron and a smaller percentage two neutrons. Similarly, most iron nuclei have 30 neutrons. A smaller percentage, however, have 29 neutrons. Atoms of the same element that contain different numbers of neutrons are **isotopes**.

Physical and Conceptual Models

There are *physical models* and there are *conceptual models*. A physical model replicates an object on some different scale. A toy airplane, for example, is a physical model of a real airplane. The more accurate a physical model, the more it looks like the real thing (Figure 14.10).

A conceptual model, on the other hand, describes a system, not an object. A thunderstorm, for example, is best described using a conceptual model. Such a model shows how the various components of the system are related to one another. The components of a thunderstorm components include the humidity, atmospheric pressure, temperature, electric charge, and motion of large masses of air. The more accurate a conceptual model, the better we can predict the behavior of the system (Figure 14.11).

Like the thunderstorm, the atom is a complex and dynamic system, and it is best described using a conceptual model. We should be careful, therefore, not to misinterpret the atomic model as a re-creation of the actual atom. For practical purposes, the atomic model serves as a symbolic representation of atomic behavior.

Imaging devices such as scanning tunneling microscopes provide an exterior view of the atom, but no present-day imaging devices can provide actual views of the interior of atoms. This is where a model comes in. We can gain an understanding of an atom's interior by piecing together a model of the atom—shaped by information gained by many experiments. An atomic model is not valued as a means of knowing what the insides of an atom "look like," but rather to help explain why atoms behave as they do. One practical outcome of a good atomic model is extending the list of human-created modern materials such as metal alloys, plastics, pharmaceuticals, and semiconductors.

Figure 14.10 Physical models are physical representations.

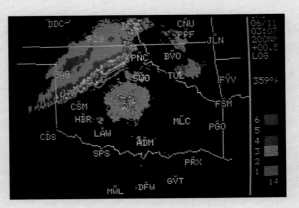

Figure 14.11 Conceptual models are conceptual representations.

Atoms interact with one another electrically. The behavior of any atom among other atoms, therefore, is largely determined by the charged particles it contains. Isotopes of an element only differ by mass, not by electrical charge. For this reason, isotopes of an element share many characteristics—in fact, as chemicals, they cannot be distinguished. For example, chemicals such as sugars, fats, and proteins containing carbon atoms with 7 neutrons are digested no differently than sugars, fats, or proteins containing carbon atoms with the more common 6 neutrons. About 1 percent of the carbon we eat contains 7 neutrons. The number of neutrons in an element does not affect how it interacts with other elements, even those elements of our bodies. We will return to isotopes in the next chapter.

So if neutrons have no chemical effect, what *do* they do? The presence of neutrons plays a significant role in the stability of the nucleus. Consider the fact that all protons have like charges. Since particles of like charge repel, how is it possible for protons to bunch together in the atomic nucleus? The answer is that although protons electrically repel one another, they are attracted to nearby protons and neutrons by the strong nuclear force (next chapter). Neutrons, which only attract in the atomic nucleus, therefore act to hold the nucleus together. The more protons there are in a nucleus, the more neutrons are needed. For light elements, it is sufficient to have about as many neutrons as protons. For example, most carbon nuclei have 6 protons and 6 neutrons. For heavier elements, however, extra neutrons are essential. Most lead nuclei have 82 protons and 126 neutrons, or about one and a half times as many neutrons as protons.

The total mass of an atom is its **atomic mass**. This is the sum of the masses of all the atom's components (electrons, protons, and neutrons). Electrons are light compared to the protons and neutrons, so their contribution to the total mass of an atom is negligible. Most elements have a variety of isotopes, each with its own atomic mass. For this reason the atomic mass for each element listed in the periodic table is not a whole number. Rather, each listed mass is actually the average of the masses of isotopes based upon their relative abundance.

Protons and neutrons are made of even more fundamental particles, *quarks*. In the same way that Rutherford was able to deduce the internal structure of the atom by bombarding it with alpha particles, evidence for the existence of quarks has been obtained by bombarding nuclei with highly energetic electrons. We will discuss quarks in the next chapter.

14.5 BOHR'S PLANETARY MODEL OF THE ATOM

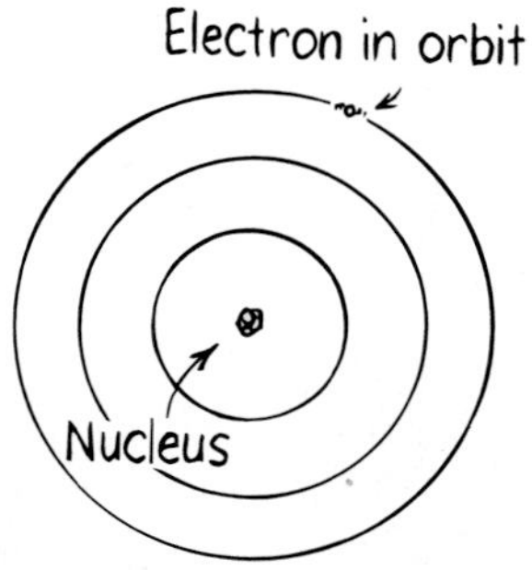

Electron in orbit

Figure 14.12 The Bohr model of the atom.

An early conceptual model of the atom is the classic planetary model in which electrons whirl around the small but dense nucleus like planets orbiting the sun (Figure 14.12). This model, modernized by the Danish physicist Niels Bohr in 1911, explains why atoms emit or absorb only particular frequencies of light.

Recall from the previous chapter the photon model of light, which explained the photoelectric effect. A photon is thought of as a vibrating corpuscle of light. Photons that vibrate at a high frequency are more energetic than photons that vibrate at lower frequencies. A photon of ultraviolet light, for example, is more energetic than a photon of visible light. Hence ultraviolet light produces sunburns while red, green, or blue light does not. In the visible spectrum, blue and violet photons have more energy than red and green photons. The energy of a photon is directly proportional to its frequency.

Figure 14.13 When an electron is boosted to a higher orbit the atom is excited. When it returns the atom de-excites and gives off light.

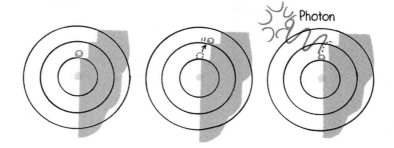

$$E \sim f$$

Where does the photon get this energy? Bohr's planetary model of the atom provides an explanation. Bohr reasoned that light is emitted when an electron jumps from a higher energy outer orbit to a lower energy inner orbit. Similarly, an electron is boosted from a lower energy inner orbit to a higher energy outer orbit when light is absorbed (Figure 14.13). An atom in this higher energy state is said to be *excited*. When a photon is emitted, the atom de-excites to a lower energy state.

In the Bohr model, an electron can only have particular (discrete) amounts of energy. This means that the energy difference between any two orbits has a particular value. Thus, an electron making a transition from one orbit to another emits or absorbs only light of energy equal to this energy difference. For example, if an electron jumps from a high-energy state to a low-energy state, the photon of light emitted has a relatively high energy associated with it. A smaller energy transition produces a photon of lower energy. So the energy of a photon is the energy difference in the atomic states of the atom from which it is emitted.

Atomic Spectra

Long before the Bohr model, clues to atomic structure were evident in the light emitted by atoms—**atomic spectra**. White light passed through a prism or a diffraction grating is dispersed into a spectrum of colors. When the yellow light from a sodium lamp is passed though a prism or grating, it is dispersed into two colors of yellow. If the light is first passed through a thin vertical slit, separation of the colors is evident in two images of the slit. These are spectral lines. All elements when excited emit their own frequencies of light, which produce a spectrum characteristic of the element. The atomic spectra of several elements is shown in Figure 14.14. Atomic

Figure 14.14 Typical spectral patterns of some elements.

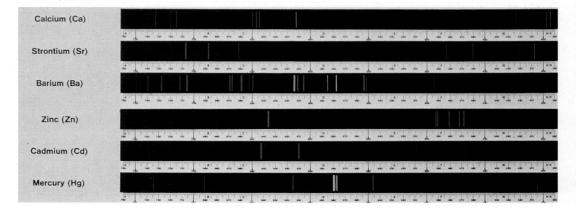

Figure 14.15 A simple spectroscope. Images of the illuminated slit are cast on a screen and make up a pattern of lines. The spectral pattern is characteristic of the light used to illuminate the slit.

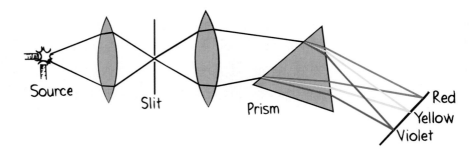

spectra are the fingerprints of the elements. Every element produces its own characteristic spectrum.

The arrangement of light source, thin slit, and prism or grating is called a *spectroscope* (Figure 14.15). Chemists used the spectroscope for chemical analysis, while physicists were busy trying to find order in the confusing arrays of spectral lines. It had long been noted that the lightest element, hydrogen, had a far more orderly spectrum than the other elements. Figure 14.16 shows a portion of the hydrogen spectrum. Note that spacing between successive lines decreases in a regular way. A Swiss schoolteacher, J. J. Balmer, expressed these line positions by a mathematical formula. The formula worked for hydrogen, and even predicted the positions of lines not yet measured.

Figure 14.16 A portion of the hydrogen spectrum.

Another regularity in atomic spectra was found by J. Rydberg. He noticed that the sum of the frequencies of two lines in the spectrum of hydrogen sometimes equals the frequency of a third line. Rydberg and other investigators could offer no explanation for this—not until the Bohr model. The Bohr model of the atom explains why the sum of two frequencies of light emitted by an atom often equals a third frequency of light emitted by the same atom. If an electron is raised to the third energy level—that is, the third highest orbit—it can return to the first orbit by two routes. It can return by a single jump from the third to the first orbit, or by a double jump from the third to the second, and then from the second to the first orbit (Figure 14.17). These two paths produce three spectral lines. Note that the energy jump for A plus B is equal to the energy jump for C. Since frequency is proportional to energy, the frequency for path A plus the frequency for path B equals the frequency for path C.

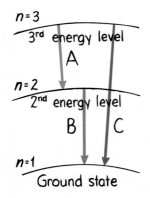

Figure 14.17 The frequency of light emitted or absorbed by an atom is proportional to the energy difference between electron orbits. As the energy differences between orbits are discrete, so are the frequencies of light that are either emitted or absorbed.

QUESTION

Suppose the frequency of light emitted along path A in Figure 14.17 is 5×10^9 Hz, and along path B is 7×10^9 Hz. What frequency of light is emitted when an electron makes a transition along path C?

ANSWER

Add the two frequencies from paths A and B together to get the frequency of path C: $(5 \times 10^9$ Hz$) + (7 \times 10^9$ Hz$) = (12 \times 10^9$ Hz$)$ or 1.2×10^{10} Hz.

The Bohr model accounts for X rays and shows that they are emitted when electrons jump from outermost to innermost orbits. Bohr predicted X-ray frequencies that were later confirmed by experiment. By measuring the frequencies of light emitted by atoms, energy levels can be mapped for the orbits of each element. Also, Bohr accounted for the general chemical properties of an unknown element, hafnium, leading to its discovery. So we see that the planetary model of the atom was a remarkable guide to explaining and predicting atomic behavior.

Despite its successes, Bohr knew his model was only a beginning. It was limited for it did not explain why an electron was restricted to certain energy levels. There were other theoretical difficulties. An important premise of his model was that an electron did not fall into the positively charged nucleus. Today, however, we find that electrons do occasionally fall into the nucleus and then bounce back out! Electrons, therefore, do not orbit the nucleus like planets orbit the sun. The Bohr model and its revisions are now seen as useful stepping stones to newer conceptual models that help explain most atomic behavior. The newer models involve the wave nature of electrons and particles in general.

14.6 PARTICLES AS WAVES

In 1924 the French physicist Louis de Broglie discovered an important clue to our modern day concept of the atom. Recall that the photoelectric effect shows that light waves behave as a hail of bullets with enough momentum to knock electrons off metal surfaces. Since waves have particle properties, de Broglie proposed that particles also have wave properties. According to de Broglie, all moving matter seems to possess wavelike properties and under certain conditions produce an interference or diffraction pattern. All material things—electrons, protons, atoms, bowling balls, planets, stars, and you—have a wavelength that is related to momentum by the equation:*

$$\text{Wavelength} = \frac{h}{\text{momentum}}$$

where h is Planck's constant, an extremely small number.

When the wavelength of something of large mass and ordinary speed is calculated by this equation, it is found to be so small that any interferences or diffractions are negligible. For example, a bullet of mass 0.02 kg traveling at the speed of sound (330 m/s) has a wavelength of 10^{-34} meters, or rather a million million million millionth the size of a hydrogen atom. The wavelength that you possess while running down the street is even less! An electron traveling at 2 percent the speed of light, on the other hand, has a wavelength of 10^{-10} meters, which is equal to the diameter of the hydrogen atom. According to de Broglie, traveling electrons should exhibit detectable interferences or diffraction patterns, as does light.

The wave properties of electrons were experimentally confirmed and found to conform to de Broglie's equation. Today, the wave nature of electrons and other small but fast moving particles are routinely detected (Figure 14.18). A practical

---

*Recall that the wavelength of a wave is the distance between successive parts of the wave, and is related to frequency and speed by $v = f\lambda$; recall also that momentum = mass × velocity.

Figure 14.18 Fringes produced by the diffraction (left) of an electron beam and (right) of light.

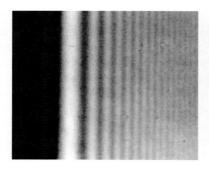
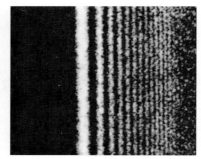

application is the electron microscope, which focuses electron waves. Because electron waves are much shorter than light waves, electron microscopes are able to show unusually great detail (Figure 14.19).

QUESTION

What wavelength does matter have when standing still?

Figure 14.19 (a) An electron microscope makes practical use of the wave nature of electrons. The wavelength of electron beams is typically thousands of times shorter than the wavelength of visible light, so the electron microscope is able to distinguish detail not visible with optical microscopes. (b) Detail of a female mosquito head as seen with a scanning electron microscope at a "low" magnification of 200 times.

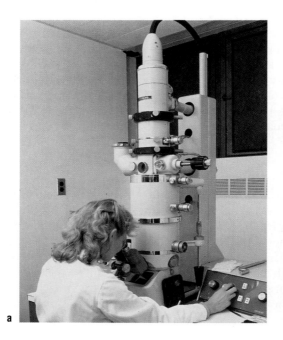

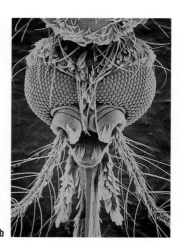

a   b

ANSWER

Standing still means zero velocity and zero momentum. When h is divided by zero, wavelength is infinitely long—or rather, nonexistent. But interestingly enough, all motion is relative. The reason an electron has a wavelength that is detectable to us is because it is moving very fast relative to us. From an electron's point of view, it is not moving at all, and we're the ones with a wavelength! We'll treat relativity in Chapter 29. Interesting stuff!

Electron Waves

A satellite can orbit the sun at any distance—there are no forbidden orbits. Similarly, according to the Bohr model, it seems like an electron should be able to orbit at any radial distance around the nucleus. But this doesn't happen. It can't. Why the electron occupies only discrete levels is understood by considering the electron to be not a particle, but a wave. Permitted orbits are a natural consequence of the electron wave closing on itself in phase. In this way, the wave is reinforced in each cycle, similar to the way a standing wave on a drum head or a spherical bell is reinforced by its successive reflections. Such reinforced waves give the appearance of standing still. So we can view an electron as a 3-dimensional standing wave surrounding the atomic nucleus. The wavelength of this electron equals the circumference of its orbit around the nucleus as shown in Figure 14.20.

Figure 14.20 (a) Orbiting electrons form standing waves only when the circumference of the orbit is equal to a whole-number multiple of wavelengths. (b) When the wave does not close in on itself in phase, it undergoes destructive interference.

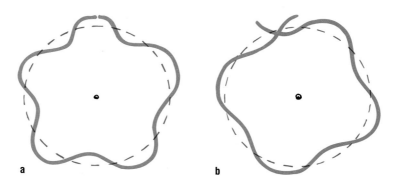

a

b

The circumference of the innermost orbit is equal to one wavelength of the electron wave. The second orbit has a circumference of two electron wavelengths, the third has three, and so forth (Figure 14.21). This is similar to a "chain necklace"

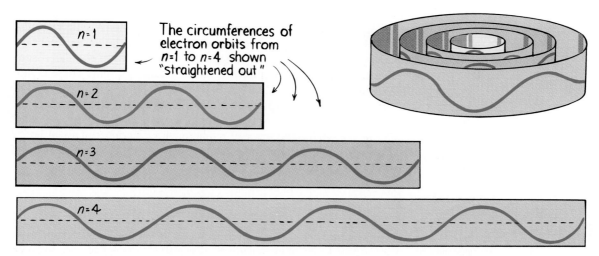

Figure 14.21 The orbits of an electron in an atom have discrete radii because the circumferences of the orbits are whole-number multiples of the electron wavelengths. This results in a discrete energy state for each orbit.

made of equal-size paper clips.* No matter what size necklace is made, its circumference is equal to some multiple of the length of a single paper clip. This shows how electrons have only certain quantities of energy. Since only certain sizes of the electron wave are permitted, only certain energy levels for the electron are permitted.

So we see how negative electrons don't spiral closer and closer to the positive nucleus that attracts them. By viewing each electron orbit as a standing wave, the circumference of the smallest orbit can be no smaller than a single wavelength—no fraction of a wavelength is possible in a constructive standing wave. As long as an electron carries the momentum necessary for wave-like behavior, atoms don't shrink in on themselves.

14.7 ELECTRON WAVE-CLOUD MODEL OF THE ATOM

It is easy to visualize the motion of a planet about the sun—macrophysics, but it's very different to visualize the motion of an electron about the atomic nucleus—microphysics. Electrons have wave properties; they also exhibit particle properties. So we are still led to ask how does an electron, behaving as a particle, actually orbit the nucleus? What path does it trace?

One way to answer this question would be to pinpoint the orbiting electron's location over time and plot its path. But here we find a big difference in the methods of macrophysics and microphysics. In noting the motion of planets we can passively observe the planets' positions in the sky. Observing them has no effect on them. Observing an electron, however, is a different story. To observe an electron requires probing it somehow—perhaps with electromagnetic radiation. This type of probing, however, moves the electron unpredictably away from its original location. All attempts of investigators to measure specific electron positions in the early 1920s were futile. Futility gave way to enlightenment when atomic processes were viewed in terms of *probabilities* rather than certainties.

In the late 1920s, an Austrian-German physicist, Erwin Schrodinger, and a German physicist, Werner Heisenburg, formulated equations for calculating the *probability* of finding an electron at any given location in an atom. Using their equations, Schrodinger and Heisenburg showed that there were certain regions of space centered about the nucleus where an electron of a given energy state was most likely to be found. One way to picture such a region is to plot the various possible locations of an electron as dots. Because some locations are more likely than others, the resulting pattern resembles a cloud having variably dense regions. These clouds are **probability clouds**.

As was suspected, the dimensions of probability clouds corresponded to the dimensions of different electron waves. For example, a single electron wave centered about the nucleus of an atom has the shape of a sphere with a diameter of about 10^{-10} meters. This corresponds to the shape and size of the probability cloud of an electron that is in its lowest energy state.

*For each orbit the electron has a unique speed, which determines its wavelength. Electron speeds are less and wavelengths longer for orbits of increasing radii; so to make our analogy accurate, we would have to use not only more paper clips to make increasingly longer necklaces but larger paper clips for each longer necklace as well.

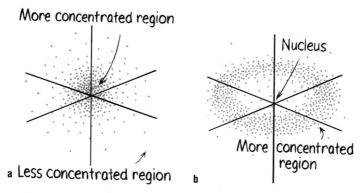

Figure 14.22 Probability clouds showing probable locations of the electron (a) relatively close to the nucleus and (b) relatively far from the nucleus.

We see this electron wave-cloud model of the atom as an extension of the planetary model. In the planetary model, the energy of an electron is dependent upon its distance from the nucleus. This is also true for the wave-cloud model of the atom. An electron in some low energy state will show a wave or probability cloud concentrated close to the nucleus (Figure 14.22a). An electron in a higher energy state, on the other hand, will show a wave or probability cloud concentrated farther away from the nucleus (Figure 14.22b).

Heisenburg took the idea of the probability cloud one step further and showed that a probability cloud was a special case of a more general concept, the **uncertainty principle**. He showed that the more specifically you know an electron's momentum, the less specifically you can know where it is. Likewise, the more you know about an electron's energy, the less you can know the time at which it has that energy. The more certain we are of one value, the more uncertain we are of the other. So, we view the distinct levels of energy for an electron as the electron's *probable locations* rather than exact locations. The probability-cloud model and the uncertainty principle go hand in hand.

14.8 THE QUANTUM MODEL: ATOMIC ORBITALS

Because of its wave-like properties, an electron in an atom can only possess particular quantities of energy. This concept is an important part of a branch of modern physics and chemistry called **quantum mechanics**, which embraces a conceptual model called the *quantum model*. Here an electron's wave and its probability cloud are recognized as two aspects of the same idea. Rather than viewing an electron as occupying a single point about the nucleus at any given moment, the electron is seen to occupy an entire region.

Rather than seeking specific electron locations in the atom, the equations of quantum mechanics specify regions of space where an electron of a certain energy level is most likely to be found. This region of space is an **atomic orbital**. For every different energy level of an electron, there will be a different atomic orbital in which it may be found. By convention, atomic orbitals are drawn to show where the electron is located 90 percent of the time. This gives the atomic orbital an apparent border (Figure 14.23). This border, however, is arbitrary for the electron may exist on either side of it.

Electron is out here somewhere 10% of the time

Electron is in here 90% of the time

Figure 14.23 A spherical atomic orbital.

Electron is 10% of the time out here somewhere

Electron is 90% of the time in here

Figure 14.24 An atomic orbital the shape of an hourglass.

There are a number of different types of atomic orbitals and they differ by shape and/or size. The simplest shape of an atomic orbital is a sphere, as shown in Figure 14.23. Another type of atomic orbital has the shape of an hourglass. The region of space defined by the hourglass-shaped orbital is where an electron may be located 90 percent of the time (Figure 14.24).

The hourglass-shaped orbital illustrates the significance of the wave nature of the electron. Unlike a real hourglass, the two lobes of this orbital are not open to each other. Yet the electron can occupy either lobe because of its wave properties. A guitar player can gently tap a guitar string at its mid-point (the 12th fret) and pluck it elsewhere at the same time to produce a high pitched tone called a *harmonic*. A close inspection of this string will reveal that it vibrates everywhere along the string except at the point directly above the 12th fret. This point is a **node** and although it has no motion, waves nonetheless travel through it. Thus the guitar string vibrates on both sides of the node when only one side is plucked. Similarly, the point between the two lobes of the hourglass-shaped orbital is a node through which the electron's waves pass.

An orbit is a distinct path described by a body in its revolution around another. An atomic orbital is different—it is a region of space where an electron of a given energy state will most likely be found about the nucleus. There are as many different atomic orbitals for an atom as there are different energy states for its electrons.

The planetary model of the atom explains the emission or absorption of light by an atom. Simply put, when an electron jumps from an outer orbit to an inner orbit, the atom emits light; and when an electron is boosted from an inner orbit to an outer one, the atom absorbs light. The emission or absorption of light can be explained in a similar fashion using the quantum model. When an electron jumps from an orbital of higher energy to one of lower energy light is emitted. Likewise, when light is absorbed, an electron is boosted from an orbital of lower energy to one of higher energy. The two models say much the same thing. The difference is that the quantum model incorporates the wave nature of the electron.

The more atomic behavior we observe, the more we are able to refine our model of the atom. A more refined model of the atom doesn't give us a better picture of what atoms "look like," but rather a better idea of how atoms will behave under a variety of conditions. For example, being familiar with a good atomic model helps us to understand how atoms can combine to form molecules, and how molecules can combine to form various substances. A good model enables prediction.

. . . .

SUMMARY OF TERMS

Brownian motion Random movement of very small particles suspended in a fluid that results from collisions with molecules in the fluid.

Cathode ray tube A device that emits a cathode beam, which is a beam of electrons. Oscilloscopes and television tubes are common examples.

Electron The negative particle in the shell of an atom.

Atomic nucleus The core of an atom, consisting of two basic nucleons—protons and neutrons.

Atomic number A number designating an atom, which is the same as the number of protons in the nucleus or the same as the number of electrons in a neutral atom.

Periodic table A highly ordered listing of the elements (as shown on the front inside cover of this book).

Proton The positively charged nucleon in an atomic nucleus.

Neutron The electrically neutral nucleon in an atomic nucleus.

Isotopes Atoms whose nuclei have the same number of protons but different number of neutrons.

Atomic mass The total mass of an atom.

Atomic spectra The range of frequencies of electromagnetic radiation emitted by atoms.

Uncertainty principle The ultimate accuracy of measurement is given by the magnitude of Planck's constant, *h*. Further, it is not possible to measure exactly both the position and the momentum of a particle at the same time, nor the energy and the time associated with a particle simultaneously.

Probability cloud The pattern of electron positions plotted over a period of time that show the likelihood of an electron's position at a given time.

Quantum mechanics The branch of science that deals with finding the probability clouds of atoms, and the rules for the behavior of particles at the atomic and subatomic level.

Atomic orbital The region of space where electrons of a given energy are likely to be located.

Node Point of zero amplitude in a standing wave.

. . . .

REVIEW QUESTIONS

1. Is hydrogen an atom or an element?

2. How many atoms are found naturally in the environment?

The Elements

3. What is the origin of hydrogen?

4. What is the origin of the heavier elements?

5. What are the major four elements that compose living things?

6. What is Brownian motion?

7. What part of the atom dictates its chemical properties?

The Electron

8. What is a cathode ray?

9. Why is a cathode ray deflected by a nearby electric charge and a nearby magnet?

10. What did J. J. Thomson discover about the electron?

11. What did R. Millikan discover about the electron?

The Atomic Nucleus

12. What did E. Rutherford discover about the atom?

13. What was the fate of the vast majority of alpha particles that were directed at gold foil in Rutherford's laboratory?

14. What was the fate of a tiny fraction of alpha particles that surprised Rutherford?

15. What kind of force prevents atoms from meshing into one another?

Protons and Neutrons

16. How massive is a proton compared to an electron?

17. How much charge does a proton have compared to an electron?

18. Exactly what is meant by the *atomic number* of an element?

19. Exactly what is a *periodic table?*

20. What role does atomic number play in the periodic table?

21. What is the principle role of the neutron in the atomic nucleus?

22. What role do isotopes play in the atomic mass of an element?

23. What role do quarks play in the protons and neutrons of an element?

Bohr's Planetary Model of the Atom

24. What does light have to do with atomic structure?

25. Why does ultraviolet light produce sunburns while visible light does not?

26. From where does light get its energy?

27. Which color of light comes from the higher energy transition, red or blue?

28. How does the energy of a photon compare to the energy difference in the atom that emits the photon?

Atomic Spectra

29. What is the relationship between the light emitted by an atom and its electron energy levels?

30. Why do we say the atomic spectra of elements are like atomic fingerprints?

31. What is a spectroscope?

32. What is the connection between the spectrum of an element and its atomic structure?

Particles as Waves

33. What was L. de Broglie's hypothesis about particles and waves?

34. What is the relationship between the wavelength of a particle and its momentum?

Electron Waves

35. Why are electron orbits at only certain distances from the atomic nucleus?

36. How many electron wavelengths make up the circumference of an electron in its second orbit?

Electron Wave-Cloud Model of the Atom

37. We can passively observe a satellite in earth orbit, but we cannot passively observe an electron in an atomic orbital. Why?

38. What did W. Heisenburg discover about probing subatomic particles?

39. How do Bohr's orbits compare to electron probability clouds?

40. What is uncertain in the uncertainty principle?

The Quantum Model: Atomic Orbitals

41. How do atomic orbitals relate to probability clouds?

42. How do the shapes of atomic orbitals relate to wave patterns?

43. How does the quantum model of light emission/absorption differ from the planetary model?

44. What is the function of an atomic model?

• •

EXERCISES

1. A cat strolls across your backyard. An hour later a dog with his nose to the ground follows the trail of the cat. Explain this occurrence from a molecular point of view.

2. If all the molecules of a body remained intact, would the body have any odor?

3. Which are older, the atoms in the body of an elderly person or those in the body of a baby?

4. From where do the atoms that make up the body of a newborn baby originate?

5. Why are atoms visible with electron microscopes yet invisible with even ideal optical microscopes?

6. A teaspoonful of a vegetable oil dropped on the surface of a quiet pond spreads out to cover almost an acre. The oil film has a thickness equal to the size of a molecule. If you know the volume of the oil and measure the film area, how can you calculate the thickness or the size of a vegetable oil molecule?

7. In what sense can you truthfully say that you are a part of every person around you?

8. What are the chances that at least one of the atoms exhaled in your first breath will be in your last breath?

9. If the particles of a cathode ray were more massive, would the ray be bent more or less in a magnetic field? How about if the particles had greater charges?

10. Why did Rutherford assume that the atomic nucleus was positively charged?

11. How does Rutherford's model of the atom account for the back-scattering of alpha particles directed at the gold leaf?

12. Since the atom is mostly empty space, why don't atoms simply ooze into one another under pressure?

13. If two protons and two neutrons are removed from the nucleus of an oxygen atom, what nucleus remains?

14. You could swallow a capsule of germanium without ill effects. But if a proton were added to each of the germanium nuclei, you would not want to swallow the capsule. Why? (Consult the periodic table of the elements.)

15. If an atom has 43 electrons, 56 neutrons, and 43 protons, what is its approximate atomic mass? What is the name of this element?

16. The nucleus of a neutral iron atom contains 26 protons. How many electrons does a neutral iron atom have?

17. Would you use a physical model or a conceptual model to describe the following: the brain; the mind; the solar system; the beginning of the universe; a stranger; your best friend; a gold coin; a dollar bill; a car engine; the greenhouse effect; a virus; the spread of sexually transmitted diseases?

18. Why is the light emitted by elements characteristic of the emitting elements?

19. How can a hydrogen atom, which has only one electron, have so many spectral lines?

20. Suppose that a certain atom possesses four distinct energy levels. Assuming that all transitions between levels are possible, how many spectral lines will this atom exhibit? Which transitions correspond to the highest-energy light emitted? To the lowest-energy light?

21. An electron de-excites from the fourth quantum level to the third and then directly to the ground state. Two photons are emitted. How do their combined energies compare to the energy of the single photon that would be emitted by de-excitation from the fourth level directly to the ground state?

22. In a process called *fluorescence,* ultraviolet light falls on certain dyes and visible light is emitted. When infrared light falls on these materials, visible light is not emitted. Why not?

23. How does the wave model of electrons orbiting the nucleus account for discrete energy values rather than arbitrary energy values?

24. Why does no stable electron orbit exist in an atom for a circumference of 2.5 de Broglie wavelengths?

25. Distinguish between *classical mechanics* and *quantum mechanics.*

PROBLEMS

1. There are approximately 10^{23} H_2O molecules in a thimbleful of water and 10^{46} H_2O molecules in the earth's oceans. Suppose that Columbus had thrown a thimbleful of water into the ocean and that the water molecules have by now mixed uniformly with all the water molecules in the oceans. Can you show that if you dip a sample thimbleful of water from anywhere in the ocean, you'll have at least one of the molecules that was in Columbus's thimble? (Hint: The ratio of the number of molecules in a thimble to the number of molecules in the ocean will equal the ratio of the number of molecules in question to the number of molecules the thimble will hold.)

2. There are approximately 10^{22} molecules in a single medium-sized breath of air and approximately 10^{44} molecules in the atmosphere of the whole world. Now the number 10^{22} squared is equal to 10^{44}. So how many breaths of air are there in the world's atmosphere? How does this number compare with the number of molecules in a single breath? If all the molecules from Julius Caesar's last dying breath are now thoroughly mixed in the atmosphere, how many of these on the average do we inhale with each single breath?

3. Assume that the present world population of about 4×10^9 people is about one-thirtieth the number of people who ever lived on earth. Then about how many more molecules are there in a single breath compared to the total number of people who have ever lived? Can you show that the number of molecules in a single breath compared to the total number of people who ever lived is about the same as the number of people who ever lived compared to one person?

15

THE NUCLEUS
AND RADIOACTIVITY

The alchemists of old tried in vain for over 2000 years to change the elements. They expended enormous effort and performed elaborate rituals in their quest to change lead into gold. They never succeeded. Interestingly enough, lead in fact *can* be changed to gold, but certainly not by the chemical means the alchemists employed.

Chemical reactions involve rearrangements of the outermost electrons of atoms and molecules. To change the identity of an element, one must go deep within its electron cloud to the central nucleus. The nucleus, however, is immune to even the most violent chemical reactions. To change lead to gold, three positive charges must be extracted from its nucleus. This can never happen in a chemical reaction.

Ironically, elements were constantly changing into other elements all around the alchemists, as they are around us today. The cause is *radioactive decay*. Radioactive decay of minerals in rocks has been occurring since the minerals' formation. But this was unknown to the alchemists. Even if they had discovered these radiations, the atoms so changed would most likely have escaped their notice. Why? Because they had no model of the atom that would make sense of such changes.

Certain atomic nuclei tend to break down—decay. In doing so, they release a relatively large amount of energy by the ejection of particles and by the emission of high-frequency electromagnetic radiation. Elements that contain unstable nuclei, and this includes all isotopes of elements with atomic numbers greater than 83 and some isotopes of smaller atoms, are said to be *radioactive*.

15.1 ALPHA, BETA, AND GAMMA RAYS

The radioactive elements emit three distinct types of radiation, called by the first three letters of the Greek alphabet, α, β, γ—*alpha*, *beta*, and *gamma*, respectively. **Alpha rays** are streams of particles that have a positive electrical charge, **beta rays** are streams of particles that have a negative charge, and **gamma rays** have no charge at all. The three rays can be separated by putting a magnetic field across their paths (Figure 15.1).

An **alpha particle** is the nucleus of a helium atom—a combination of 2 protons and 2 neutrons. Alpha particles are relatively easy to stop because of their relatively large size, and their double positive charge. They do not normally penetrate through light materials such as paper or clothing. Even when traveling through only a few centimeters of air, alpha particles pick up electrons and become nothing more than

Figure 15.1 The bending of a narrow beam of rays in a magnetic field. The beam comes from a radioactive source placed at the bottom of a hole drilled in a lead block.

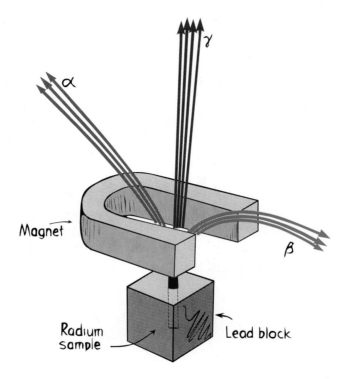

harmless helium. Interestingly enough, that's where the helium in a child's balloon comes from. Practically all the earth's helium atoms were once fast-moving alpha particles.

A **beta particle** is an electron formed and instantly ejected from a nucleus. A beta particle is normally faster than an alpha particle and carries only a single negative charge. They are not as easy to stop as alpha particles so they are able to penetrate light materials such as paper or clothing. They are not able to penetrate deeply into denser materials, however, such as water or aluminum. Beta particles, once stopped, simply become a part of the material they are in, like any other electron.

Gamma rays are the high frequency electromagnetic radiation emitted by radioactive elements. Like visible light, a gamma ray is pure energy. The amount of energy in a gamma ray, however, is much greater than visible light, ultraviolet light, or even X rays. Because gamma rays have no mass or electric charge and because of their high energies, gamma rays are able to penetrate through most materials, except for unusually dense materials such as lead. Molecules such as those inside our delicate cells that get zapped by gamma rays suffer structural damage. Hence, gamma rays typically have more harmful effects on our bodies than do alpha or beta particles (unless the alphas or betas are ingested).

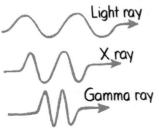

Figure 15.2 A gamma ray is simply electromagnetic radiation, much higher in frequency and energy than light and X rays.

Q UESTION

Pretend you are given three radioactive cookies—one alpha emitter, one beta emitter, and one gamma emitter. You must eat one, hold one in your hand, and put the other in your pocket. What can you do to minimize your exposure to radiation?

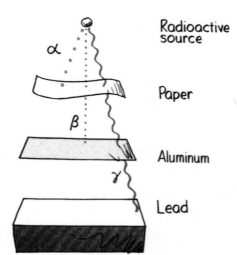

Figure 15.3 Alpha particles are the least penetrating form of radioactivity and can be stopped by a few sheets of paper. Beta particles will readily pass through paper, but not through a sheet of aluminum. Gamma rays penetrate several centimeters into solid lead.

A NSWER

Ideally get as far from the cookies as possible. But if you must eat one, hold one, and put one in your pocket, hold the alpha; the skin on your hand will shield you. Put the beta in your pocket; your clothing will likely shield you. Eat the gamma; it will penetrate your body in any of these cases anyway.

Marie and Pierre Curie

Marya Sklodovska was the youngest child born to two Polish teachers in 1867. As a child she was an avid reader and buried herself in books. She loved science and graduated from high school with honors. Since the University of Warsaw would not accept women, and neither she nor her father had enough money for her to study elsewhere, she resigned not to continue her formal education. But then she and her older sister Bronya agreed to a plan—Marya would work to put Bronya through medical school, and then Bronya would do the same for Marya. So Marya worked for about five years as a governess to sponsor Bronya through medical school. When Bronya became a physician, she helped Marya. In 1891 Marya left Poland with a fourth-class ticket to Paris and forty rubles (twenty dollars) in her pocket. She was accepted into the Sorbonne, where she became known as Marie. She dedicated herself to three years of study, solitude, and near famine. To the amazement of many who believed women were incapable of doing science, she placed first in her final examinations. She earned a master's degree in mathematics and physics. At the Sorbonne, Marie was introduced to Pierre Curie.

Pierre Curie, born in 1859, was the second son of a French physician. As a child he was considered to be slow mentally, and his father did not force him to go through the regular French schools. At the age of 14 a tutor taught him mathematics, whereupon his intellect developed very rapidly. By the age of 16 he held a Bachelor of Science, and at 24 he was chief of the laboratory at the School of Physics and Chemistry of Paris, where he remained for 22 years. Before meeting and falling in love with Marie, he discovered piezoelectricity and had achieved a first-rate scientific reputation.

Pierre courted Marie and a year later they were married. Throughout their life together their main relaxation consisted of exploring nature on bicycle trips. At the time of their marriage, Marie had already passed her qualifying exam. After giving birth to a daughter Irene, who was later to follow in her parents' scientific footsteps, Marie decided to study radioactivity for her doctor's thesis. She used an electrometer designed by Pierre to measure radioactive decay. To Marie, the measurements of radioactive decay from uranium and thorium ore suggested the presence of other elements in the ore. Other investigators disagreed with Marie and cited uranium and thorium as the sources of the radioactivity. With Pierre's help she bought eight tons of waste ore from a uranium mine. Using chemical techniques they divided the ore into fractions with different properties, keeping those fractions of high radioactivity and discarding the rest. They worked enormously hard under very poor conditions testing the eight tons of ore. But ultimately they succeeded in proving the existence of two elements unknown at the time. She named the first element *polonium*, after her native country. She named the second element *radium*, after the element's intense radioactivity. For their work on radioactivity the Curies shared the 1903 Nobel Prize with Henri Becquerel, the discoverer of radioactivity. Eight years later Marie Curie was awarded a second Nobel Prize for discovering polonium and radium. She is one of only three scientists to receive two Nobel Prizes.

When fame reached the Curies, they were both in poor health and exhausted. They lived in great seclusion, enjoying the countryside as much as possible. They received medals and honors, but not the better laboratory they dearly wanted. They did not patent their refining techniques, choosing to place no restrictions on others pursuing science. Sadly, Marie lacked the money to buy ore for further experiments. In 1906, at the age of 48, while crossing a street in Paris, Pierre was killed by a run-away horse and carriage. This terrible tragedy left Marie alone and overburdened with responsibilities and work, and a great fame that she disliked. Marie, known as Madame Curie, remained an incessant worker until leukemia, likely caused by the radium she discovered, ended her life at the age of 67.

15.2 THE NUCLEUS

As described in the previous chapter, the atomic nucleus occupies only a few quadrillionths of the volume of the atom. Thus, the atom is mostly empty space. The nucleus is composed of **nucleons**, which when electrically charged are protons and when electrically neutral are neutrons. The neutron's mass is slightly greater than the proton's mass. Nucleons have nearly 2000 times the mass of electrons, so the mass of an atom is practically equal to the mass of its nucleus. The positive charge of the proton is the same in magnitude as the negative charge of the electron.

Nuclear radii range from about 10^{-15} meters for hydrogen to about six times larger for uranium. The shapes of nuclei are generally spherical but deviate from that shape sometimes in the "football" way and sometimes in the "doorknob" way. Protons and neutrons tend to cluster into alpha particles, making a generally lumpy nucleus. A "skin" of neutrons is thought to make up the outer portion of the heavier nuclei.

Just as there are energy levels for the orbital electrons of an atom, there are energy levels within the nucleus. Whereas electrons making transitions to lower orbits emit photons of light, similar changes of energy states within the nucleus result in the emission of gamma-ray photons. The emission of alpha particles from the nucleus can also be understood from the viewpoint of quantum mechanics. Just as orbital electrons form a probability cloud around the nucleus, inside the radioactive nucleus there is a similar probability cloud for alpha particles (and other particles as well). Probability is extremely high that alpha particles will be inside the nucleus. But if the probability wave extends outside, there is a finite chance that the alpha particle will be outside. Once outside, it is hurled violently away by electrical repulsion.

In addition to alpha, beta, and gamma rays, more than 200 various other particles have been detected coming from the nucleus when it is clobbered by energetic particles. We do not think of these so-called elementary particles as being buried within the nucleus and then popping out, just as we do not think of a spark as being buried in a match before it is struck. These particles come into being when the nucleus is disrupted. Regularities in the masses of these particles as well as the particular characteristics of their creation are explained by the existence of a fairly compact family of six subnuclear particles—the *quarks*.

Quarks are the fundamental building blocks of all nucleons. An unusual property of quarks is that they carry fractional electrical charges. One kind, the *up* quark, carries $+\frac{2}{3}$ the electron charge; another kind, the *down* quark, has $-\frac{1}{3}$ the electron charge. (The name *quark,* inspired by a quotation from *Finnegan's Wake* by James Joyce, was chosen in 1963 by Murray Gell-Mann, who first proposed their existence.) Each quark has an antiquark, which is a quark with opposite electric charge. The proton consists of the combination *up up down,* and the neutron of *up down down.* As with magnetic monopoles, no quarks have been isolated and experimentally observed. Most investigators think quarks by nature cannot be isolated. We will not delve further into the nature of quarks though you may want to follow up on your own.

Lighter particles like muons and electrons, and still lighter particles called *neutrinos,* make up a class of six particles called *leptons.* Leptons are not composed of quarks. At the present time, the six quarks and six leptons are thought to be the truly *elementary particles,* particles not composed of more elementary constituents. Elementary particles are the smallest components of matter. Investigation of elementary particles is at the frontier of our present knowledge and the area of much current excitement and research.

15.3 ISOTOPES

The nucleus of a hydrogen atom consists of a single proton. Helium has two protons, lithium has three, and so forth. Every succeeding element in the list of elements has one more proton than the preceding element. In neutral atoms, there are as many protons in the nucleus as there are electrons outside the nucleus. The number of protons in the nucleus is the same as the atomic number.

The number of neutrons in the nucleus of a given element may vary somewhat. For example, every electrically neutral atom of chlorine has 17 protons and hence 17 orbital electrons, but the number of accompanying neutrons varies. We discussed this briefly in the previous chapter. Atoms that have like numbers of protons but unlike numbers of neutrons are called **isotopes.** The two chief chlorine isotopes have 35 and 37 times the mass of a single nucleon, respectively. We denote these by $^{35}_{17}Cl$ and $^{37}_{17}Cl$, where the numbers refer to the *atomic number,* which is the number of protons, and the *atomic mass number,* which is the total number of protons and neutrons (the atomic mass number is approximately but not exactly equal to the *atomic mass*).

The mass number corresponds to the total number of nucleons in the nucleus. Since for both isotopes of chlorine, 17 of these nucleons are protons, the lighter isotope has $35 - 17 = 18$ neutrons, and the heavier isotope has $37 - 17 = 20$ neutrons. In nature the isotopes of chlorine are found mixed, with three times as many $^{35}_{17}Cl$ atoms as there are $^{37}_{17}Cl$ atoms. So the average atomic mass of naturally occurring chlorine is about 35.5. Chlorine is only one of many elements found in nature that consist of two or more stable isotopes. Even hydrogen, the lightest element, has three known isotopes (Figure 15.4). The double-weight hydrogen isotope $^{2}_{1}H$ is called *deuterium.* "Heavy water" is the name usually given to H_2O in which one or both of the H's are deuterium atoms. In water, hydrogen gas, and all other hydrogen compounds in nature there is 1 deuterium atom for every 6000 or so atoms of "ordinary" hydrogen. The triple-weight hydrogen isotope $^{3}_{1}H$, which is not stable but lives long enough to be a known constituent of atmospheric water, is called *tritium.* Tritium is present only in extremely minute amounts—less than 1 per 10^{17} atoms of ordinary hydrogen.

There are three isotopes of uranium that naturally occur in the earth's crust; the most common is $^{238}_{92}U$. In briefer notation we can drop the atomic number and simply say U-238. (We will see in the next chapter that it is the isotope U-235 that undergoes nuclear fission.) Of the 83 elements present on earth in significant amounts, 20 have a single stable isotope. The others have from 2 to 10 stable isotopes. Taking radioactive isotopes into account, over 2000 distinct isotopes are known.

Figure 15.4   Three isotopes of hydrogen. Each nucleus has a single proton, which holds a single orbital electron, which in turn determines the chemical properties of the atom. The different number of neutrons changes the mass of the atom but not its chemical properties.

QUESTION

State the numbers of protons and neutrons in each of the following nuclei: ^1_1H, $^{14}_6\text{C}$, $^{235}_{92}\text{U}$.

15.4 THE STRONG NUCLEAR FORCE

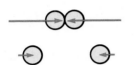

Figure 15.5 The nuclear strong interaction is a very short-range force. For nucleons very close or in contact, the strong force is very strong. But a few nucleon diameters away it is nearly zero.

We know that electrical charges of like sign repel one another. So how is it possible that all the positively charged protons of the nucleus can stay clumped together? This question led to the discovery of an interaction, called the **strong nuclear force**. It acts between nucleons and is very strong over extremely short distances (about 10^{-15} meters, the diameter of a typical atomic nucleus). Electrical interactions, on the other hand, are relatively long-ranged. For protons that are close together, as in small nuclei, the nuclear force of attraction easily overcomes the electrical force of repulsion. But for distant protons, like those on opposite edges of a large nucleus, the attractive nuclear force may be small in comparison to the repulsive electrical force.* Hence, a larger nucleus is not as stable as a smaller nucleus.

Recall from the previous chapter the role of the stability of neutrons in the nucleus. Neutrons act as a sort of nuclear cement. Although protons both attract and repel one another (due to the presence of the strong nuclear and electrical forces), neutrons only attract—they attract both protons and other neutrons (since they interact only through the strong nuclear force). Hence, the presence of neutrons adds to the net attraction and helps to hold the nucleus together. The more protons there are in a nucleus, the more neutrons are needed. For light elements, it is sufficient to have about as many neutrons as protons. The common isotope of carbon has equal numbers of each—6 protons and 6 neutrons. Extra neutrons are essential in heavy elements. Lead, for example, has about one and a half times as many neutrons as protons.

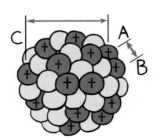

Figure 15.6 Proton A both attracts (nuclear force) and repels (electrical force) proton B, but mainly repels proton C (because the nuclear attraction is weaker at greater distances). The greater the distance between A and C, the more unstable the nucleus because of this repulsion. (The shaded particles are neutrons.)

ANSWER

One proton and no neutron in ^1_1H; 6 protons and 8 neutrons in $^{14}_6\text{C}$; 92 protons and 143 neutrons in $^{235}_{92}\text{U}$.

---

*Fundamental to the strong interaction is the *color force* (which has nothing to do with visible color). This color force interacts between quarks and holds them together by the exchange of "gluons." Read more about this in H. R. Pagels, *The Cosmic Code: Quantum Physics as the Language of Nature* (New York: Simon & Schuster, 1982).

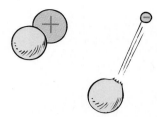

Figure 15.7 A neutron with a proton is stable, but a neutron by itself is unstable and turns into a proton by emitting an electron and an even smaller subatomic particle, the anti-neutrino (not shown).

The stabilizing effect of neutrons, however, can only go so far. There are two reasons for this. First, there is the fact that neutrons are not stable by themselves. A lone neutron will spontaneously eject a beta particle and turn into a proton (Figure 15.7). A neutron seems to need protons around to keep this from happening. After the size of a nucleus reaches a certain point, however, the neutrons so outnumber the protons that there are not enough protons in the mix to prevent the neutrons from changing. As neutrons within a nucleus change into protons, the stability of the nucleus decreases. The second reason for the limitation of the stabilizing effect of neutrons is that protons are attracted only to adjacent protons but are repelled by *all* protons in the nucleus. As more and more protons are squeezed into the nucleus, the electrical forces of repulsion increase substantially. For example, each of the two protons in a helium nucleus feels the repulsive effect of the other. Each proton in a nucleus containing 84 protons, however, feels the repulsive effects of 83 protons! The size of the atomic nucleus, therefore, is limited. For example, we find that all nuclei having more than 82 protons are unstable. As a consequence, this limits the number of elements in the periodic table.

A nuclear force distinct from the strong interaction is responsible for beta emission. This weaker nuclear force is called the *weak interaction*. It principally affects lighter particles like electrons and still lighter particles called *neutrinos*. A detailed account of the weak nuclear force is beyond the scope of this book.

15.5 HALF-LIFE

The rate of decay for a radioactive element is measured in terms of a characteristic time, the **half-life**. This is the time it takes for half of an original quantity of the element to decay. For example, radium-226 has a half-life of 1620 years. This means that half of a sample of radium will decay by the end of 1620 years. In the next 1620 years, half of the remaining radium will decay, leaving only one fourth of the original amount of radium. After 20 half-lives, the initial quantity of radioactive radium will be diminished by a factor of about one million.

Figure 15.8 Every 1620 years the amount of radium decreases by half.

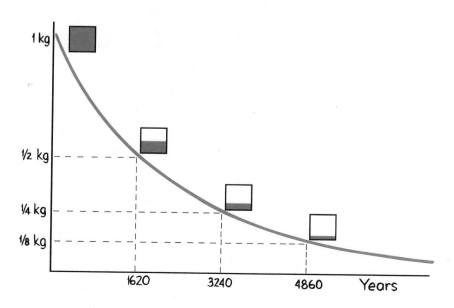

Figure 15.9 Some radiation detectors. (a) A Geiger counter detects incoming radiation by its ionizing effect on enclosed gas in the tube. (b) A scintillation counter detects incoming radiation by flashes of light that are produced when charged particles or gamma rays pass through it.

a   b

The rate of decay for any radioactive element is remarkably constant and is not affected by external conditions. There are a wide variety of half-lives demonstrated by different radioactive elements. Some have half-lives that are less than a millionth of a second, while others have half-lives of more than a billion years. Uranium-238 has a half-life of 4.5 billion years. All uranium eventually decays in a series of steps to lead. In 4.5 billion years, half the uranium presently in the earth will be lead.

It is not necessary to wait through the duration of a radioactive half-life in order to measure it. The half-life of an element can be calculated at any given moment by measuring the rate of decay of a known quantity. This is easily done using a radiation detector (Figure 15.9). In general, the shorter the half-life of a substance, the faster it disintegrates, and the more radioactivity per amount is detected.

QUESTIONS

1. If a sample of radioactive isotopes has a half-life of 1 day, how much of the original sample will be left at the end of the second day? The third day?

2. Which will give a higher counting rate on a radiation detector, radioactive material that has a short half-life or radioactive material that has a long half-life?

15.6 NATURAL TRANSMUTATION OF ELEMENTS

When a nucleus emits an alpha particle, a different element is formed. The changing of one chemical element to another is called **transmutation**. Consider uranium-238, which has 92 protons and 146 neutrons. When an alpha particle is ejected, the nucleus is reduced by two protons and two neutrons (since an alpha particle is a helium nucleus consisting of two protons and two neutrons). The 90 protons and 144 neutrons left behind are then the nucleus of a different element. This element is *thorium*. This reaction is expressed as

ANSWERS

1. One-fourth of the original sample will be left—the three-fourths that underwent decay is now a different element altogether. At the end of 3 days, one-eighth of the original sample will remain.

2. The material with the shorter half-life is more active and will give a higher counting rate on a radiation detector.

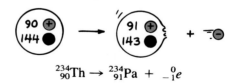

$$^{238}_{92}U \rightarrow {}^{234}_{90}Th + {}^{4}_{2}He$$

An arrow is used here to show that the $^{238}_{92}U$ changes into the other elements. When this happens, energy is released, partly in the form of gamma radiation, partly in the kinetic energy of the alpha particle ($^{4}_{2}He$), and partly in the kinetic energy of the thorium atom. In this and all such equations, the mass numbers at the top balance $(238 = 234 + 4)$ and the atomic numbers at the bottom also balance $(92 = 90 + 2)$.

Thorium-234, the product of this reaction, is also radioactive. When it decays, it emits a beta particle. Recall that a beta particle is an electron—not from an atomic orbital, but from the nucleus. A beta particle is emitted by a neutron in the nucleus. You may find it useful to think of a neutron as a combined proton and electron (although it's not really the case) because when the neutron emits an electron, it becomes a proton. A neutron is ordinarily stable when it is locked in the nucleus of an atom, but a free neutron is radioactive and has a half-life of 12 minutes. It decays into a proton by beta emission.* So in the case of thorium, which has 90 protons, beta emission leaves the nucleus with one less neutron and one more proton. The new nucleus then has 91 protons and is no longer thorium, but the element *protactinium*. Although the atomic number has increased by 1 in this process, the mass number (protons + neutrons) remains the same. The nuclear equation is

$$^{234}_{90}Th \rightarrow {}^{234}_{91}Pa + {}^{0}_{-1}e$$

We write an electron as $^{0}_{-1}e$. The 0 indicates that its mass is insignificant when compared to that of the protons and neutrons that alone contribute to the mass number. The -1 is the charge of the electron. Remember that this electron is a beta particle from the nucleus and not an electron from the electron cloud that surrounds the nucleus.

---

*Beta emission is always accompanied by the emission of a neutrino, a neutral particle with about zero mass that travels at about the speed of light. The neutrino ("little neutral one") was predicted from theoretical calculations by Wolfgang Pauli in 1930 and detected in 1956. Neutrinos are hard to detect because they interact very weakly with matter. To capture a neutrino is extremely difficult. Whereas a piece of solid lead a few centimeters thick will stop most gamma rays from a radium source, it would take a piece of lead about 8 light-years thick to stop half the neutrinos produced in typical nuclear decays. Thousands of neutrinos are flying through you every second of every day, because the universe is filled with them. Only occasionally, one or two times a year or so, does a neutrino or two interact with the matter of your body.

At this writing, whether or not the neutrino has mass is not known. Present speculation is that if neutrinos do have any mass, they are so numerous that they may make up about 90% of the mass of the universe—enough to halt the present expansion and ultimately close the cycle from the Big Bang to the Big Crunch. Neutrinos may be the "glue" that holds the universe together.

So we see that when an element ejects an alpha particle from its nucleus, the mass number of the resulting atom decreases by 4, and its atomic number decreases by 2. The resulting atom belongs to an element two spaces back in the periodic table (see the insert in Chapter 18). When an element ejects a beta particle (electron) from its nucleus, the mass of the atom is practically unaffected so there is no change in mass number, but its atomic number *increases* by 1. The resulting atom belongs to an element one place *forward* in the periodic table. Thus the emission of an alpha or beta particle by an atom produces a different atom in the periodic table—alpha emission lowers the atomic number and beta emission increases it. Gamma emission results in no change in either the mass number or the atomic number. So we see that radioactive elements can decay backward or forward in the periodic table.*

Figure 15.10 U-238 decays to Pb-206 through a series of alpha and beta decays.

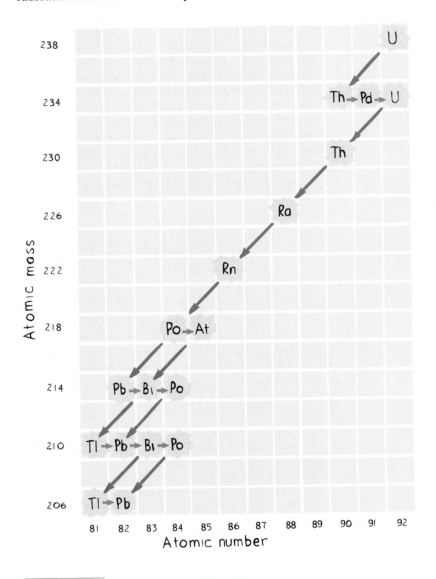

*Sometimes a nucleus emits a positron, which is the "antiparticle" of an electron. In this case, a proton becomes a neutron, and the atomic number is decreased.

The successions of radioactive decays of $^{238}_{92}U$ to $^{206}_{82}Pb$, an isotope of lead, are shown in Figure 15.10. The steps in the decay process are shown in the diagram, where each nucleus that plays a part in the series is shown by a burst. The vertical column containing the burst shows its atomic number, and the horizontal column shows its mass number. Each arrow that slants downward toward the left shows an alpha decay, and each arrow that points to the right shows a beta decay. Notice that some of the nuclei in the series can decay in both ways. This is one of several similar radioactive series that occur in nature.

QUESTIONS

1. Complete the following nuclear reactions.
   (a) $^{226}_{88}Ra \rightarrow {}^{?}_{?}? + {}^{0}_{-1}e$
   (b) $^{209}_{84}Po \rightarrow {}^{205}_{82}Pb + {}^{?}_{?}?$

2. What finally becomes of all the uranium that undergoes radioactive decay?

15.7 ARTIFICIAL TRANSMUTATION OF ELEMENTS

Ernest Rutherford, in 1919, was the first of many investigators to succeed in transmuting a chemical element. He used a piece of radioactive ore as a source of alpha particles and bombarded nitrogen gas. The impact of an alpha particle on a nitrogen nucleus transmutes nitrogen into oxygen:

$$^{14}_{7}N + {}^{4}_{2}He \rightarrow {}^{17}_{8}O + {}^{1}_{1}H$$

Rutherford used a *cloud chamber* to record this event (Figure 15.11). In a cloud chamber, moving charged particles show a trail of ions along their path in a way similar to the ice crystals that mark the trail of jet planes high in the sky. From a quarter-of-a-million cloud-chamber tracks photographed on movie film, Rutherford showed seven examples of atomic transmutation. Analysis of tracks bent by a strong external magnetic field showed that when an alpha particle collided with a nitrogen atom, a proton bounced out and the heavy atom recoiled a short distance. The alpha particle disappeared. The alpha particle was absorbed in the process, transforming nitrogen to oxygen.

ANSWERS

1. (a) $^{226}_{88}Ra \rightarrow {}^{226}_{89}Ac + {}^{0}_{-1}e$
   (b) $^{209}_{84}Po \rightarrow {}^{205}_{82}Pb + {}^{4}_{2}He$

2. All uranium will ultimately become lead. On the way to becoming lead, it will exist as an element in a series, as indicated in Figure 15.10.

Figure 15.11 A cloud chamber. Charged particles moving through supersaturated vapor leave trails. When the chamber is in a strong electric or magnetic field, bending of the tracks provides information about the charge, mass, and momentum of the particles.

Since Rutherford's announcement in 1919, we have created many such nuclear reactions, first with natural bombarding projectiles from radioactive ores and then with still more energetic projectiles, protons and electrons hurled by giant particle accelerators. Artificial transmutation has been used to produce the hitherto unknown synthetic elements from atomic numbers 93 to 109. All these artificially made elements have short half-lives. If they ever existed naturally when the earth was formed, they have long since decayed.

Figure 15.12 Tracks of elementary particles in a bubble chamber, a similar yet more complicated device than a cloud chamber. Two particles have been destroyed at the points where the spirals emanate, and four others created in the collision.

15.8 ISOTOPIC DATING

The earth's atmosphere is continuously bombarded by cosmic rays—energetic particles and high-energy gamma rays. This results in the transmutation of many atoms in the upper atmosphere to produce a scattering of many protons and neutrons. Most of the protons quickly capture electrons and become hydrogen atoms in the upper atmosphere. The neutrons, however, keep going for longer distances because they have no charge and do not interact electrically with matter. Eventually, many of them col-

Radioactive Tracers

Radioactive isotopes of all the elements have been made by bombardment with neutrons and other particles. These isotopes are extremely useful in scientific research and industry.

In order to check the action of a fertilizer in plants, researchers combine a small amount of a radioactive isotope with the fertilizer, which is then applied to a few selected plants. The amount of fertilizer taken up by the plants can be easily measured with radiation detectors. From such measurements, scientists can inform farmers of the proper amount of fertilizer to use. When used in this way, radioactive isotopes are called **tracers.**

Tracers are used in medicine to study the process of digestion and the way in which chemicals move about the body. Food that contains small amounts of a radioactive isotope is fed to a patient and traced through the body with a radiation detector. Also, chemicals containing small amounts of radioactive isotopes can be injected into a patient to study circulation of the blood.

Engineers can study engine wear by making the cylinder walls in the engine radioactive. While the engine is running, the piston rings rub against the cylinder walls. The tiny particles of radioactive metal that were worn away fall into the lubricating oil, where they can be measured with a radiation detector. This test is repeated with different oils. In this way the engineer can determine which oil gives the least wear and the longest life to the engine.

There are many other applications for radioactive elements. In the following chapter we shall see how they can be used for the production of significant amounts of energy.

Figure 15.13 Radioisotopes are used to check the action of fertilizers in plants and the progress of food in digestion.

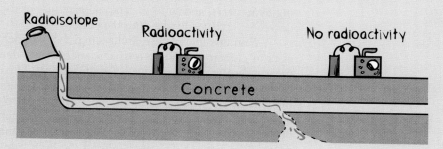

Figure 15.14 Tracking pipe leaks with radioactive isotopes.

---

lide with the nuclei in the denser lower atmosphere. When captured by the nuclei of nitrogen atoms, the following reaction takes place:

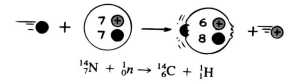

$$^{14}_{7}N + ^{1}_{0}n \rightarrow ^{14}_{6}C + ^{1}_{1}H$$

Nitrogen captures a neutron and becomes an isotope of carbon by emitting a proton. This carbon isotope is radioactive carbon-14. It has 8 neutrons, unlike the most common isotope, carbon-12, which has 6 neutrons. Less than one-millionth of 1 percent of the carbon in the atmosphere is carbon-14. Both carbon-12 and carbon-14 combine with oxygen to become carbon dioxide, which is taken in by plants. This means that all plants have a tiny bit of radioactive carbon-14. All animals eat plants (or at least plant-eating animals), and therefore have a little carbon-14 in them. All living things on earth contain some carbon-14.

Carbon-14 is a beta emitter and decays back into nitrogen by the decay of one of its neutrons:

$$^{14}_{6}C \rightarrow \, ^{14}_{7}N \, + \, ^{0}_{-1}e$$

Because living plants continue to take in carbon dioxide, this decay is accompanied by a replenishment of carbon-14. As carbon-14 atoms decay, new ones take their place. In this way, a radioactive equilibrium is reached where there is a ratio of about one carbon-14 atom to every 100 billion carbon-12 atoms. When a plant or animal dies, replenishment stops. Then the percentage of carbon-14 decreases at a constant rate given by its radioactive half-life.* The longer an organism is dead, the less carbon-14 remains.

The half-life of carbon-14 is about 5760 years. This means that half of the carbon-14 atoms that are now present in a body, plant, or tree will decay in the next 5760 years. Half of the remaining carbon-14 atoms will then decay in the following 5760 years, and so forth. The radioactivity of living things, therefore, gradually decreases at a steady rate after they die.

With this knowledge, archaeologists and other investigators can calculate the ages of carbon-containing artifacts, such as wooden tools or skeletons, simply by measuring their current level of radioactivity. This process, known as **carbon-14 dating**, enables us to probe as much as 50,000 years into the past. Because of fluctuations in the production of carbon-14 through the centuries, this technique gives an uncertainty of about 15 percent. This means, for example, that the straw of an old adobe brick that is dated to be 500 years old may really be only 425 years old on the low side, or 575 years old on the high side. For many purposes this is an acceptable level of uncertainty. If greater accuracy is desired, other techniques are employed.

Carbon dating would be an extremely simple and accurate dating method if the amount of radioactive carbon in the atmosphere had been constant over the ages.

20 930 BC   15 200 BC   9470 BC   3740 BC   1990 AD

Figure 15.15   The radioactive carbon isotopes in the skeleton diminish by one half every 5730 years.

---

*A 1-g sample of contemporary carbon contains about 5×10^{22} atoms, 6.5×10^{10} of which are C-14 atoms, and has a beta disintegration rate of about 13.5 decays per minute.

But it hasn't been. Fluctuations in the sun's magnetic field as well as changes in the strength of the earth's magnetic field affect cosmic-ray intensities in the earth's atmosphere, which in turn produce fluctuations in the production of C-14. In addition, changes in the earth's climate affect the amount of carbon dioxide in the atmosphere. The oceans are great reservoirs of carbon dioxide. When the oceans are cold, they release less carbon dioxide into the atmosphere than when they are warm. These complications require complex corrective procedures for the accurate dating of ancient organic objects. Nevertheless, recalibration and refined techniques make carbon dating the most versatile chronometric method we have. There are also other techniques for assigning dates to relics of the past.

UESTION

Suppose an archaeologist extracts a gram of carbon from an ancient axe handle and finds it one-fourth as radioactive as a gram of carbon extracted from a freshly cut tree branch. About how old is the axe handle?

Uranium Dating

The dating of older, but nonliving, things is accomplished with radioactive minerals, such as uranium. The naturally occurring isotopes U-238 and U-235 decay very slowly and ultimately become isotopes of lead—but not the common lead isotope Pb-208. For example, U-238 decays through several stages to finally become Pb-206, whereas U-235 finally becomes the isotope Pb-207. Lead isotopes 206 and 207 that now exist were at one time uranium. The older the uranium-bearing rock, the higher the percentage of these remnant isotopes.

From the half-lives of uranium isotopes, and the percentage of lead isotopes in uranium-bearing rock, it is possible to calculate the date at which the rock was formed. Rocks dated in this way have been found to be as much as 3.7 billion years old. Samples from the moon, where there has been less obliteration of the early rocks than occurs on earth, have been dated at 4.2 billion years, which begins to agree closely upon the estimated 4.6-billion-year age of the earth and the solar system.

We'll return to isotopic dating when we get into geology in Part 7.

15.9 EFFECTS OF RADIATION ON HUMANS

A common misconception is that radioactivity is something new in the environment. But radioactivity has been around far longer than the human race. It is as much a part of our environment as the sun and the rain. It is what warms the interior of the earth and makes it molten. In fact, radioactive decay inside the earth is what heats the

ANSWER

Assuming the ratio of C-14/C-12 was the same when the axe was made, the axe handle is two half-lives of C-14—about 11,460 years old.

Figure 15.16 Origins of radiation exposure for an average individual in the United States.

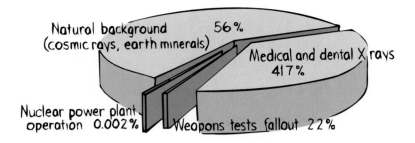

water that spurts from a geyser or that wells up from a natural hot spring. Even the helium in a child's balloon comes from the alpha particles that were produced by radioactive decay.

As Figure 15.16 shows, most of the radiation we encounter originates in the natural surroundings. It is in the ground we stand on and in the bricks and stones of surrounding buildings. This natural background radiation was present before humans emerged in the world. If our bodies couldn't tolerate it, we wouldn't be here. Even the cleanest air we breathe is somewhat radioactive due to cosmic ray bombardment. At sea level the protective blanket of the atmosphere reduces it, while at higher altitudes radiation is more intense. In Denver, the "mile-high city," a person receives more than twice as much radiation from cosmic rays as at sea level. A couple of round-trip flights between places as distant as New York and San Francisco exposes us to as much radiation as we receive in a normal chest X ray. The air time of airline personnel is limited because of this extra radiation.

Exposure to radiation greater than normal background should be avoided because of the damage it can do. The cells of living tissue are composed of intricately structured molecules in a watery, ion-rich brine. When X radiation or nuclear radiation encounters this highly ordered soup, it produces chaos on the atomic scale. A beta particle, for example, passing through living matter collides with a small percentage of the molecules and leaves a randomly dotted trail of altered or broken molecules along with newly formed, chemically reactive ions and free radicals that may cause further damage. Gamma radiation produces a similar effect. As a high-energy gamma-ray photon moves through matter, it may rebound from an electron and give it a high kinetic energy. The electron then may career through the tissue, creating havoc. All types of high-energy radiation break or alter the structure of some molecules and create conditions in which other molecules will be formed that may be harmful to life processes.

The cells are able to repair most kinds of molecular damage if the radiation is not too intense. A cell could survive an otherwise lethal dose of radiation if the dose were spread over a long period of time to allow intervals for healing. Also when radiation *is* sufficient to kill a cell, the dead cell can be replaced by a new one. Important exceptions to this are most nerve cells, which are irreplaceable. Sometimes a cell survives with a damaged DNA molecule. Defective genetic information will be transmitted to its daughter cell when it reproduces, and a cell mutation will occur. This will sometimes be insignificant. But if significant, it will probably result in cells that do not function as well as the original one. In rare cases the mutation will be an improvement. A genetic change of this type could also be part of the cause of a cancer that will develop in the tissue at a much later time.

The concentration of disorder produced along the trajectory of a particle depends upon its energy, charge, and mass. Gamma-ray photons and very energetic beta particles cause the lowest intensity of damage. A particle of their type penetrates deeply with widely separated interactions, like a very fast BB fired through a hailstorm. Slow, massive, highly charged particles such as low-energy alpha particles are most disruptive. They have collisions that are close together, more like a bull charging through a flock of sleepy sheep. They do not penetrate deeply because their energy is absorbed by many closely spaced collisions. Especially damaging particles of this type are the assorted nuclei (called *heavy primaries*) flung outward by the sun in solar flares. These include all the elements found on earth. Particles that approach the earth several minutes after leaving the sun are partly captured in the earth's magnetic field. The others are absorbed by collisions in the atmosphere, so practically none reach the earth. We are partly shielded from these dangerous particles by the very property that makes them a threat—their tendency to have many interactions close together.

Astronauts do not have this protection, and they absorb large doses of radiation during the time they spend in space. Every few decades there is an exceptionally powerful solar flare that would almost certainly kill any conventionally protected astronaut far from the earth who is unprotected by its atmosphere and magnetic field.

Radiation dosage is generally measured in *rads* (short for radiation), a unit of absorbed energy of ionizing radiation. The number of rads indicates the amount of radiation energy absorbed per kilogram of exposed material. When concerned with the potential ability of radiation to affect human beings, we measure the radiation in *rems* (*r*oentgen *e*quivalent *m*an). In calculating the dosage in rems, we multiply the number of rads by a factor that allows for the different health effects of different types of radiation. For example, 1 rad of slow alpha particles has the same biological effect as 10 rads of fast electrons. We call both of these dosages 10 rems.

The average person in the United States is exposed to about 0.2 rem a year. This comes from within the body itself, from the ground, buildings, cosmic rays, diagnostic X rays, television, and so on. It varies widely from place to place on the planet, but it is strongest near the poles where the earth's magnetic field does not act as a shield. It is also stronger at higher altitudes where the atmosphere provides less protection.

The lethal dose of radiation begins at 500 rems; and a person has about a 50-percent chance of surviving a dose of this magnitude if it is received over a short period of time. Under radiotherapy—the use of radiation to kill cancer cells—a patient may receive localized doses in excess of 200 rems each day for a period of weeks. A typical diagnostic chest X ray exposes a person to 5 to 30 millirems, less than

Figure 15.17 The internationally used symbol to indicate an area where radioactive material is being handled or produced.

one ten-thousandth of the lethal dose. However, even small doses of radiation can produce long-term effects due to mutations within the body's tissues. And, because a small fraction of any X ray dose reaches the gonads, some mutations occasionally occur that are passed on to the next generation. Medical X rays for diagnosis and therapy have a far larger effect on the human genetic heritage than any other artificial source of radiation. It is important to keep in mind that we normally receive significantly more radiation from natural minerals in the earth than from all artificial sources of radiation combined.

Taking all causes into account, most of us will receive a lifetime exposure of less than 20 rems, distributed over several decades. This makes us a little more susceptible to cancer and other disorders. But more significant is the fact that all living beings have always absorbed natural radiation and that the radiation received in the reproductive cells has produced genetic changes in all species for generation after generation. Small mutations selected by nature for their contributions to survival over billions of years can gradually come up with some interesting organisms—*us,* for example!

SUMMARY OF TERMS

Alpha ray A stream of helium nuclei ejected by certain radioactive nuclei.

Beta ray A stream of beta particles ejected by certain radioactive nuclei.

Gamma ray High-frequency electromagnetic radiation emitted by the nuclei of radioactive atoms.

Alpha particle The nucleus of a helium atom, which consists of two neutrons and two protons, ejected by certain radioactive elements.

Beta particle An electron (or positron) emitted during the radioactive decay of certain nuclei.

Nucleon A nuclear proton or neutron; the collective name for either or both.

Quarks The elementary constituent particles of building blocks of nuclear matter.

Isotopes Atoms whose nuclei have the same number of protons but different numbers of neutrons.

Atomic number The number associated with an atom, which is equal to the number of protons in the nucleus or, equivalently, to the number of electrons in the electric cloud of a neutral atom.

Atomic mass number The number associated with an atom, which is equal to the number of nucleons in the nucleus.

Half-life The time required for half the atoms of a radioactive element to decay.

Transmutation The conversion of an atomic nucleus of one element into an atomic nucleus of another element through a loss or gain in the number of protons.

REVIEW QUESTIONS

Alpha, Beta, and Gamma Rays

1. How do the electric charges of alpha, beta, and gamma rays differ?

2. Which of the three rays has the greater penetrating power?

The Nucleus

3. Give two examples of a nucleon.

4. How does the mass of a nucleon compare to the mass of an electron?

5. When beta emission occurs, what change takes place in an atomic nucleus?

6. In what way is the emission of gamma rays from a nucleus similar to the emission of light from an atom?

7. Why does an alpha particle leave at high speed once it gets outside an atomic nucleus?

8. What are *quarks?*

Isotopes

9. Distinguish between an *isotope* and an *ion.*

10. Distinguish between *atomic number* and *atomic mass number.*

11. Distinguish between *deuterium* and *tritium.*

The Strong Nuclear Force

12. Why does the repulsive electric force of protons in the atomic nucleus not cause the protons to fly apart?

13. Why do protons in a very large nucleus have a greater chance of flying apart by electrical repulsion?

14. Why is a smaller nucleus generally more stable than a larger nucleus?

15. What is the fate of a neutron away from other nucleons?

16. What nuclear force is responsible for beta emission?

Half-Life

17. What is meant by *radioactive half-life*?

18. What is the half-life of radium-226?

19. How does the decay rate of an isotope compare to its half-life?

Natural Transmutation of Elements

20. What exactly is a *transmutation?*

21. When thorium, atomic number 90, decays by emitting an alpha particle, what is the atomic number of the resulting nucleus?

22. When thorium decays by emitting a beta particle, what is the atomic number of the resulting nucleus?

23. How does the atomic mass change for each of the above two reactions?

24. What is the effect on the makeup of a nucleus when it emits an alpha particle? A beta particle? A gamma ray?

25. What is the long-range fate of all the uranium that exists in the world?

Artificial Transmutation of Elements

26. The alchemists of old believed that elements could be changed to other elements. Were they correct? Were they effective? Why or why not?

27. When did the first successful intentional transmutation of an element occur?

28. Why are the elements beyond uranium not common in the earth's crust?

Isotopic Dating

29. What do cosmic rays have to do with transmutation?

30. How is carbon-14 produced in the atmosphere?

31. Which is radioactive, carbon-12 or carbon-14?

32. Why is there more carbon-14 in new bones than in old bones of the same mass?

33. Why would the carbon dating method be useless for dating old coins but not old pieces of cloth?

Uranium Dating

34. Why is there lead mixed in with all deposits of uranium ores?

35. What does the proportion of lead and uranium in rock tell us about the age of the rock?

Effects of Radiation on Humans

36. From where does most of the radiation you encounter originate?

37. Which is worse—having cells in your body *damaged* by radiation or *killed* by radiation?

38. Is radioactivity in the world something relatively new? Defend your answer.

• •
HOME PROJECT

Some watches and clocks have luminous hands that continuously glow. These have traces of radium bromide mixed with zinc sulphide (safer clock faces use light rather than radioactive disintegration as a means of excitation and become progressively dimmer in the dark). If you have a luminous watch or clock available, take it into a completely dark room and, after your eyes have become adjusted to the dark, examine the luminous hands with a very strong magnifying glass or the eyepiece of a microscope or telescope. You should be able to see individual tiny flashes, which together seem to be a steady source of light to the unaided eye. Each flash occurs when an alpha particle ejected by a radium nucleus strikes a molecule of zinc sulphide.

• •
EXERCISES

1. What is the evidence that radioactive decay is the changing of one element to another?

2. Why is a sample of radium always a little warmer than its surroundings?

3. Some people say that all things are possible. Is it at all possible for a hydrogen nucleus to emit an alpha particle? Defend your answer.

4. Why are alpha and beta rays deflected in opposite directions in a magnetic field? Why are gamma rays undeflected?

5. The alpha particle has twice the electric charge of the beta particle but deflects less than the beta in a magnetic field. Why is this so?

6. How would the paths of alpha, beta, and gamma radiations compare in an electric field?

7. Which type of radiation—alpha, beta, or gamma—results in the greatest change in mass number? Atomic number?

8. Which type of radiation—alpha, beta, or gamma—results in the least change in mass number? Atomic number?

9. In bombarding atomic nuclei with proton "bullets," why must the protons be accelerated to high energies to make contact with the target nuclei?

10. Why would you expect alpha particles to be less able to penetrate in materials than beta particles of the same energy?

11. Within the atomic nucleus, which interaction tends to hold it together and which interaction tends to push it apart?

12. What evidence supports the contention that the strong nuclear interaction is stronger than the electrical interaction at short internuclear distances?

13. If a sample of radioactive isotope has a half-life of 1 year, how much of the original sample will be left at the end of the second year? Third year? Fourth year?

14. A sample of a particular radioisotope is placed near a Geiger counter, which is observed to register 160 counts per minute. Eight hours later the detector counts at a rate of 10 counts per minute. What is the half-life of the material?

15. Radiation from a point source obeys the inverse-square law. If a Geiger counter 1 m from a small sample reads 360 counts per minute, what will be its counting rate 2 m from the source? 3 m?

16. When the isotope bismuth-213 emits an alpha particle, what new element results? What new element results if it instead emits a beta particle?

17. When $^{226}_{88}$Ra decays by emitting an alpha particle, what is the atomic number of the resulting nucleus? What is the resulting atomic mass?

18. When $^{218}_{84}$Po emits a beta particle, it transforms into a new element. What are the atomic number and atomic mass of this new element? What are they if the polonium instead emits an alpha particle?

19. State the number of neutrons and protons in each of the following nuclei: $^{2}_{1}$H, $^{12}_{6}$C, $^{56}_{26}$Fe, $^{197}_{79}$Au, $^{90}_{38}$Sr, and $^{238}_{92}$U.

20. How is it possible for an element to decay "forward in the periodic table"—that is, to decay to an element of higher atomic number?

21. When radioactive phosphorus (P) decays, it emits a positron. Will the resulting nucleus be another isotope of phosphorus? If not, what?

22. How could a physicist test the following statement? "Strontium-90 is a pure beta source."

23. Elements above uranium in the periodic table do not exist in any appreciable amounts in nature because they have short half-lives. Yet there are several elements below uranium in atomic number with equally short half-lives that do exist in appreciable amounts in nature. How can you account for this?

24. You and your friend journey to the mountain foothills to get closer to nature and escape such things as radioactivity. While bathing in the warmth of a natural hot spring she wonders aloud how the spring gets its heat. What do you tell her?

25. Coal contains minute quantities of radioactive materials, yet there is more environmental radiation surrounding a coal-fired power plant than a fission power plant. What does this indicate about the shielding that typically surrounds these power plants?

26. When we speak of dangerous radiation exposure, are we generally speaking of alpha radiation, beta radiation, or gamma radiation? Discuss.

27. People who work around radioactivity wear film badges to monitor the amount of radiation that reaches their bodies. These badges consist of small pieces of photographic film enclosed in a light-proof wrapper. What kind of radiation do these devices monitor and how can they determine the amount of radiation the people receive?

28. A friend produces a Geiger counter to check the local background radiation. It ticks. Another friend, who normally fears most that which is understood least, makes an effort to keep away from the region of the Geiger counter and looks to you for advice. What do you say?

29. Why is the carbon-dating technique currently not accurate for estimating the ages of materials older than 50,000 years?

30. The age of the Dead Sea Scrolls was found by carbon dating. Could this technique have worked if they were carved in stone tablets? Explain.

16

NUCLEAR FISSION AND FUSION

Nuclear power is nothing new. It was around long before the solar system was formed, and will be around long after the solar system has run its life cycle. The source of nuclear power is the energy locked in atomic nuclei—energy that is released in nuclear processes. Nuclear processes make the stars shine, create sunlight, and produce volcanoes, geysers, and natural hot springs. One form of nuclear power—radioactivity—was discussed in the previous chapter. In this chapter we will treat the more energetic side of nuclear power—*nuclear fission* and *nuclear fusion.*

16.1 NUCLEAR FISSION

Biology students know that living tissue grows by the division of cells. The splitting in half of living cells is called fission. In a similar way, the splitting of atomic nuclei is called **nuclear fission.**

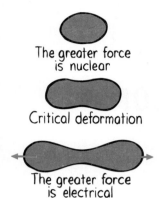

The greater force is nuclear

Critical deformation

The greater force is electrical

Figure 16.1 Nuclear deformation may result in repulsive electrical forces exceeding attractive nuclear forces, in which case fission occurs.

Nuclear fission involves the delicate balance between attracting strong nuclear forces and repelling electrical forces within the nucleus. In nearly all nuclei the strong nuclear forces dominate. In uranium, however, this domination is tenuous. If the uranium nucleus is stretched into an elongated shape (Figure 16.1), the electrical forces may push it into an even more elongated shape. If the elongation passes a certain point, electrical forces overwhelm strong nuclear forces, and the nucleus splits. This is nuclear fission.

The absorption of a neutron by a nucleus such as uranium-235 is apparently enough to cause such an elongation. The resulting fission process may produce any of several combinations of smaller nuclei. A typical example is

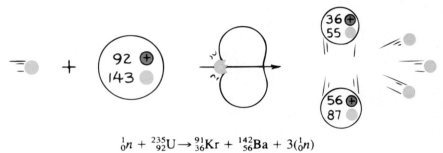

$$^1_0n + ^{235}_{92}U \rightarrow ^{91}_{36}Kr + ^{142}_{56}Ba + 3(^1_0n)$$

The energy that is released by the fission of one U-235 atom is enormous—about seven million times the energy released by the explosion of one TNT molecule. This energy is mainly in the form of kinetic energy of the fission fragments that fly apart from one another, with some energy given to ejected neutrons and the rest to gamma radiation.

Note in the above reaction that one neutron starts the fission of the uranium nucleus, and three more neutrons are produced when the uranium fissions. Two or three neutrons are produced in typical nuclear fission reactions. These new neutrons can, in turn, cause the fissioning of two or three other atoms, releasing from four to nine more neutrons. If each of these neutrons succeeds in splitting an atom, the next step in the reaction will produce between 8 and 27 neutrons, and so on. This makes a chain reaction (Figure 16.2).

Why do chain reactions not occur in naturally occurring uranium ore deposits? They would if all uranium atoms fissioned so easily. Fission occurs mainly for the rare isotope U-235, which makes up only 0.7 percent of the uranium in pure uranium metal. When the prevalent isotope U-238 absorbs neutrons from fission, it typically does not undergo fission. So any chain reaction is snuffed out by the neutron-absorbing U-238.

If a chain reaction occurred in a baseball-size chunk of pure U-235, an enormous explosion would result. If the chain reaction were started in a smaller chunk of pure U-235, however, no explosion would occur. This is because of geometry. A small piece has relatively more surface area compared to its volume than a large piece has. (Similarly, there is more skin on a kilogram of small potatoes than on a single, large, one-kilogram potato.) In a small piece of uranium, neutrons leak through the surface before an explosion can occur. In a bigger piece, the chain reaction builds up to enormous energies before the neutrons get to the surface and escape (Figure 16.4). For masses greater than a certain amount, called the **critical mass,** an explosion of enormous magnitude may take place.

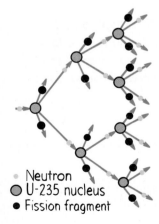

• Neutron
◯ U-235 nucleus
● Fission fragment

Figure 16.2 A chain reaction.

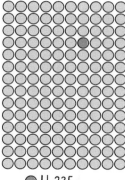

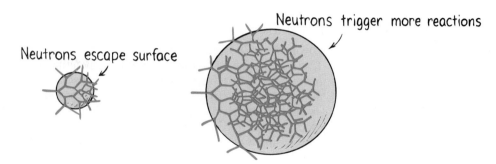

Figure 16.4 The exaggerated view shows that a chain reaction in a small piece of pure U-235 runs its course, because neutrons leak from the surface too soon. The small piece has a lot of surface compared to mass. In a larger piece, more uranium and less surface is presented to the neutrons.

○ U-235
○ U-238

Figure 16.3 Only 1 part in 140 of naturally occurring uranium is U-235.

3 units + 3 units = 6 units of area

Only 4 units of area

Figure 16.5 Each piece is subcritical. The amount of surface area is relatively large compared to the mass of uranium. When combined, the total surface area decreases, and fewer neutrons escape.

Consider a large quantity of U-235 divided into two units, each smaller than that of critical mass (Figure 16.5). The units are subcritical. Neutrons readily reach a surface and escape before a sizable chain reaction builds up. But if the pieces are suddenly driven together, the total surface area decreases. If the timing is right and the combined mass is greater than critical, a violent explosion takes place. Such a device is a nuclear fission bomb.

Constructing a fission bomb is a formidable task. The difficulty is separating enough U-235 from the more abundant U-238. Project scientists took more than two years to extract enough U-235 from uranium ore to make the bomb that was detonated on Hiroshima in 1945. To this day uranium isotope separation remains a difficult process.

QUESTION

A kilogram of U-235 broken up into small chunks is not critical, but if the chunks are slammed together in a ball shape, it is critical. Why?

Figure 16.6 Simplified diagram of a simple fission bomb.

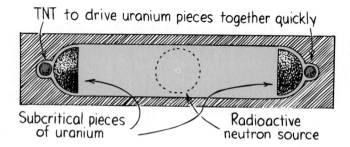

TNT to drive uranium pieces together quickly

Subcritical pieces of uranium

Radioactive neutron source

ANSWER

A kilogram of U-235 in small chunks will not support a sustained reaction because of the relatively greater surface area of the chunks (like the greater combined surface area of gravel compared to the surface area of a boulder of the same mass). Neutrons escape via the surface before a sustained chain reaction can build up. But when the pieces are slammed together, there is more uranium compared to the surface, and a chain reaction ensues.

16.2 NUCLEAR FISSION REACTORS

Power by way of nuclear fission was introduced to the world by nuclear bombs. The violent image still impacts our thinking about nuclear power, making it difficult for many people to recognize its potential usefulness. Since the second world war, however, nuclear fission has been used for power production. Currently, about 20 percent of electric energy in the United States is generated by nuclear fission reactors. These reactors are simply nuclear furnaces. They, like fossil fuel furnaces, do nothing more elegant than boil water to produce steam for a turbine (Figure 16.7). The greatest practical difference is the amount of fuel involved. One kilogram of uranium fuel, less than the size of a baseball, yields more energy than 30 freightcar loads of coal.

A fission reactor contains three components: the nuclear fuel, the control rods, and liquid (usually water) to transfer heat from the reactor to the turbine and generator. The nuclear fuel is primarily U-238 with about 3 percent U-235. Because the U-235 isotopes are so highly diluted with U-238, an explosion like that of a nuclear bomb is not possible. The reaction rate, which depends on the number of neutrons that initiate the fission of other U-235 nuclei, is controlled by rods inserted into the reactor. The control rods are made of a neutron-absorbing material, usually the metal cadmium or boron. Heated water around the nuclear fuel is kept under high pressure to keep it at a high temperature without boiling. It transfers heat to a second lower pressure water system, which operates the turbine and electric generator in a conventional fashion. In this design two separate water systems are used so that no radioactivity reaches the turbine.

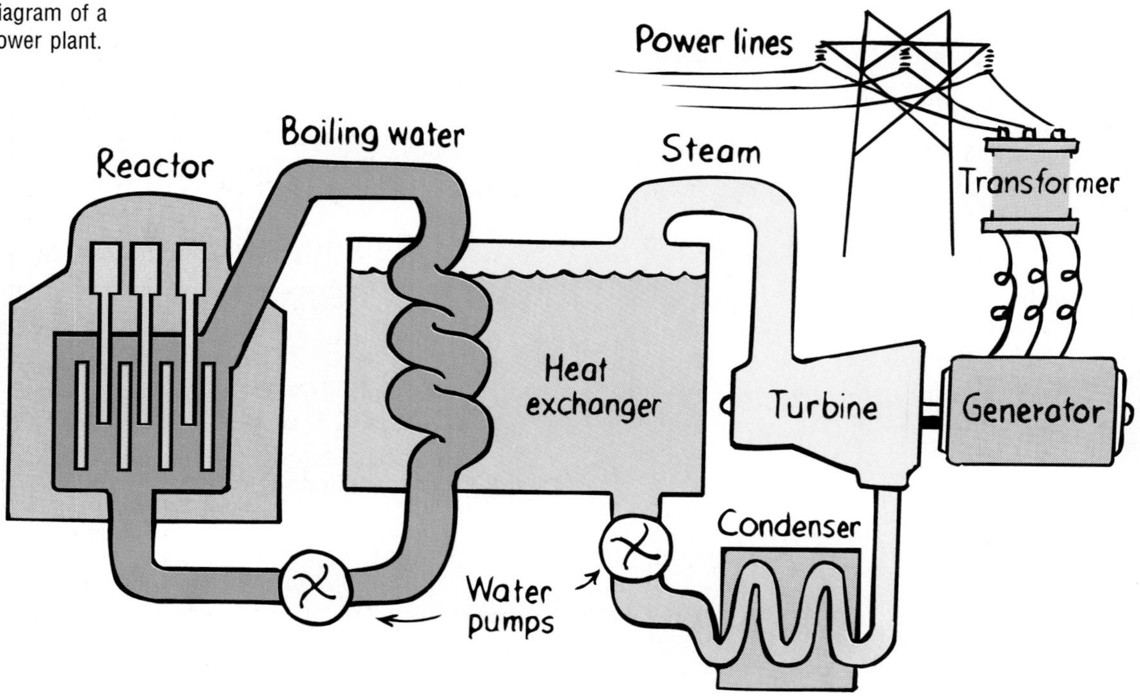

Figure 16.7 Diagram of a nuclear fission power plant.

Figure 16.8 A typical nuclear fission power plant.

One disadvantage of fission power is the generation of radioactive waste products. Recall from the previous chapter that light atomic nuclei are most stable when composed of equal numbers of protons and neutrons, and that heavy nuclei need more neutrons than protons for stability. There are more neutrons than protons in uranium—143 neutrons compared to 92 protons in U-235, for example. When uranium fissions into two medium-weight elements, the extra neutrons in their nuclei make them unstable. They are radioactive, most with very short half-lives, but some with half-lives of thousands of years. Safely disposing of these waste products as well as materials made radioactive in the production of nuclear fuels requires special storage casks and procedures. Although fission power goes back nearly a half century, the technology of radioactive waste disposal is still in the developmental stage.

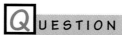

QUESTION

What is the function of the control rods in a nuclear reactor?

The Breeder Reactor

One of the fascinating features of fission power is the breeding of fission fuel from nonfissionable U-238. This breeding occurs when small amounts of fissionable isotopes are mixed with U-238 in a reactor. Fissioning liberates neutrons that convert the relatively abundant nonfissionable U-238 into U-239, which beta decays to become Np-239, which in turn beta decays to become the fissionable Pu-239. So in addition to the abundant energy produced, fission fuel is bred from the relatively abundant U-238 in the process. This breeding process occurs to some extent in all fission reactors. A reactor designed to breed more fissionable fuel than is put into it is called a **breeder reactor.** Using a breeder reactor is like filling a gas tank in a car with water, adding some gasoline, then driving the car and having more gasoline after the trip than at the beginning, at the expense of common water! The basic principle of the breeder reactor is very attractive, for after a few years of operation a breeder-reactor power utility could produce vasts amounts of power while at the same time breeding twice as much fuel as its original fuel. The downside is the enormous complexity of successful and safe operation. The United States gave up on breeders more than a decade ago, and only France and Germany have operational breeder power plants.

The benefits of fission power are plentiful electricity, conservation of the many billions of tons of coal, oil, and natural gas that every year are literally turned to heat and smoke (while in the long run they may be far more precious as sources of organic molecules than as sources of heat), and the elimination of the megatons of sulfur oxides and other poisons that are put into the air each year by the burning of these fossil fuels.

ANSWER

Control rods absorb neutrons and thereby control the amount of neutrons that participate in a chain reaction.

The drawbacks include the problems of storing radioactive wastes, the production of bomb-grade materials, low-level release of radioactive materials into the air and ground water, the risk of an accidental release of large amounts of radioactivity as well as thermal pollution, which is an inevitable thermodynamic consequence of all forms of power production.

Plutonium

Early in the nineteenth century, the most distant known planet in the solar system was Uranus. The first planet to be discovered beyond Uranus was named Neptune. In 1930 a planet beyond Neptune was discovered, and was named Pluto. During this time the heaviest element known was uranium. Appropriately, the first transuranic element to be discovered was named *neptunium,* and the second transuranic element was named *plutonium.*

Neptunium is produced when a neutron is absorbed by a U-238 nucleus. Rather than undergoing fission, the nucleus emits a beta particle and becomes neptunium, the first synthetic element beyond uranium. The half-life of neptunium is only 2.3 days, so it isn't around very long. Neptunium is a beta emitter, and very soon becomes plutonium. The half life of plutonium is about 24,000 years, so plutonium lasts a considerable time. The isotope plutonium-239, like U-235, will undergo fission when it captures a neutron. Whereas the separation of fissionable U-235 from uranium metal is a very difficult process (because U-235 and U-238 have the same chemistry), the separation of plutonium from uranium metal is relatively easy. This is because pluto-

nium is an element distinct from uranium, with its own chemical properties. The element plutonium is chemically poisonous in the same sense as lead and arsenic. It attacks the nervous system and can cause paralysis. Death can follow. Fortunately, plutonium does not remain in its elemental form for long because it rapidly combines with oxygen to form three compounds: PuO, PuO_2, and Pu_2O_3, all of which chemically are relatively benign. They will not dissolve in water or in biological systems. These plutonium compounds do not attack the nervous system and have been found to be biologically harmless.

Plutonium in any form, however, is radioactively toxic. It is more toxic than uranium, although less toxic than radium. Plutonium emits high-energy alpha particles, which kill cells rather than simply disrupting them and leading to mutations. Interestingly enough, damaged cells rather than dead cells contribute to cancer so plutonium ranks relatively low as a cancer-producing substance. The greatest danger that plutonium presents to humans is its potential for use in nuclear fission bombs. Its usefulness is in fission reactors—particularly breeder reactors.

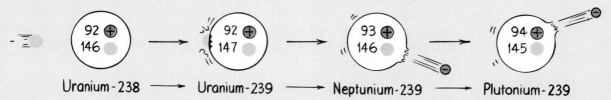

Uranium-238 ⟶ Uranium-239 ⟶ Neptunium-239 ⟶ Plutonium-239

Figure 16.9   After U-238 absorbs a neutron, it emits a beta particle, which means that a neutron in the nucleus becomes a proton. The atom is no longer uranium, but neptunium. After the neptunium atom emits a beta particle, it becomes plutonium.

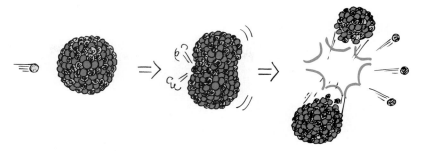

Figure 16.10   Pu-239, like U-235, undergoes fission when it captures a neutron.

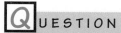UESTION

In a breeder reactor, what is bred from what?

Reasoned judgment is not made by considering only the benefits or the draw-backs of fission power. You must also compare its benefits and its drawbacks to those of alternate power sources. All power sources have drawbacks. The benefits versus the costs of fission power is a subject of much debate.

16.3   MASS-ENERGY RELATIONSHIP

Figure 16.11   Work is required to pull a nucleon from an atomic nucleus. This work increases the energy and hence the mass of the nucleon outside the nucleus.

Early in this century Albert Einstein discovered that mass is actually congealed energy. Mass and energy are two sides of the same coin, as stated in the celebrated equation $E = mc^2$. In this equation E stands for the energy that mass has when at rest, m stands for mass, and c is the speed of light. The quantity c^2 is the proportionality constant of energy and mass. We will discuss $E = mc^2$ later when we treat the special theory of relativity in Chapter 29. Here we see that this relationship of energy and mass is the key to understanding why energy is released in nuclear reactions.

The more energy associated with a particle, the greater is the mass of the particle. Is the mass of a nucleon inside a nucleus the same as that of the same nucleon outside a nucleus? This question can be answered by considering the work that would be required to separate all the nucleons from a nucleus. Recall that work, which is expended energy, is equal to the product of force and distance. Then think of the amount of force that would be required to pull nucleons apart through a sufficient distance to overcome the attractive nuclear strong force. Enormous work would be required. The work you would have to put into such a task would be manifest in the energy of the protons and neutrons that are pulled out. They would have more

ANSWER

Fissionable Pu-239 is bred from nonfissionable U-238. (Of the average 2.5 neutrons per fission of plutonium, about 1 neutron keeps the chain reaction going, and approximately 1.5 neutrons breed new fuel.)

Figure 16.12   The mass spectrometer. Ions are directed into the semicircular "drum," where they are swept into semicircular paths by a strong magnetic field. Because of inertia, heavier ions are swept into curves of large radii and lighter ions are swept into curves of smaller radii. The radius of the curve is directly proportional to the mass of the ion.

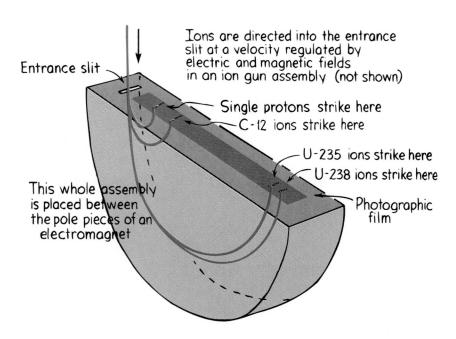

Entrance slit

Ions are directed into the entrance slit at a velocity regulated by electric and magnetic fields in an ion gun assembly (not shown)

Single protons strike here

C-12 ions strike here

U-235 ions strike here

U-238 ions strike here

This whole assembly is placed between the pole pieces of an electromagnet

Photographic film

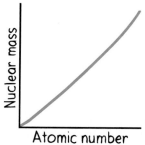

Nuclear mass

Atomic number

Figure 16.13   The plot shows how nuclear mass increases with increasing atomic number.

energy outside the nucleus—an amount equal to the energy or work required to separate them. This energy, in turn, would be manifest in the nucleons' mass. Therefore the mass of nucleons outside a nucleus is greater than the mass of the same nucleons when locked inside a nucleus. For example, in the units used to measure masses of atoms and atomic particles, a carbon-12 atom has a mass of exactly 12.00000 units.* However, outside the nucleus a proton has a mass of 1.00728 units, and a neutron has a mass of 1.00866 units. The combined mass of six free protons and six free neutrons is greater than the mass of one carbon-12 atom with a nucleus of six protons and six neutrons.

The masses of the isotopes of various elements can be very accurately measured with a mass spectrometer (Figure 16.12). This important device uses a magnetic field to deflect ions of these isotopes into circular arcs. The greater the inertia (mass) of the ion, the more it resists deflection, and the greater the radius of its curved path. The magnetic force sweeps lighter ions into shorter arcs and heavier ions into larger arcs.

A graph of the nuclear masses for the elements from hydrogen through uranium is shown in Figure 16.13. The graph slopes upward with increasing atomic number as expected: Elements are more massive as atomic number increases. The slope curves because there are proportionally more neutrons in the more massive atoms.

A more important graph results from the plot of nuclear mass *per nucleon* from hydrogen through uranium (Figure 16.14). This is perhaps the most important graph in this book, for it is the key to understanding the energy associated with nuclear processes like those that cause our sun to burn. To obtain the nuclear mass per nucleon, the nuclear mass is simply divided by the number of nucleons in the particular nucleus. (If you divided the mass of a roomful of people by the number of people in the room, you would get the average mass per person.) Note that the

---

*These units are called atomic mass units and are the units for atomic mass used in chemistry (Section 20.6).

Figure 16.14 The graph shows that the average mass of a nucleon depends on which nucleus it is in. Individual nucleons have the most mass in the lightest nuclei, the least mass in iron, and intermediate mass in the heaviest nuclei.

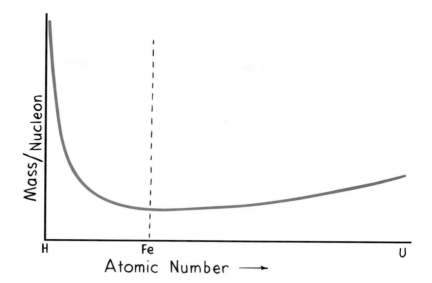

masses of the nucleons are different when combined in different nuclei. A proton has the greatest mass when it is the nucleus of a hydrogen atom, and has progressively less and less mass as it occurs in atoms of increasing atomic number. The proton has the least mass when it is in the nucleus of the iron atom. Beyond iron, the process reverses itself as protons (and neutrons) have progressively more and more mass in atoms of increasing atomic number. This continues all the way to uranium and the transuranic elements.

From the graph we can see why energy is released when a uranium nucleus is split into nuclei of lower atomic number. When a uranium nucleus splits in two, the masses of the fission fragments lie about halfway between uranium and hydrogen on the horizontal scale of the graph. Most importantly, note that the masses of nucleons in these fission fragments are *less than* the masses of the same nucleons when combined in the uranium nucleus. The protons and neutrons decrease in mass. When this decrease in mass is multiplied by the speed of light squared, it is equal to the energy yielded by each uranium nucleus that undergoes fission.

QUESTION

Correct the following incorrect statement: When a heavy element such as uranium undergoes fission, there are fewer nucleons after the reaction than before.

We can think of the mass-per-nucleon graph as an energy hill that starts at hydrogen (the highest point) and drops steeply to the lowest point (iron), then rises gradually to uranium. Iron is at the bottom of the energy hill and is the most

ANSWER

When a heavy element such as uranium undergoes fission, the nucleons after the reaction have less mass than the same nucleons before the reaction. Here we distinguish between the mass of the nucleons and the number of nucleons.

stable nucleus. It is also the most tightly bound nucleus; more energy per nucleon is required to separate nucleons from its nucleus than other nuclei.

So the decrease in mass is detectable in the form of energy—much energy—when heavy nuclei undergo fission. A drawback to this process involves the fission fragments. They are radioactive isotopes because of their greater-than-normal number of neutrons for their atomic numbers. A more promising source of energy is to be found on the left side of the energy hill.

16.4 NUCLEAR FUSION

Note from the graph of Figure 16.14 that the steepest part of the energy hill goes from hydrogen to iron. Energy is gained as light nuclei fuse, or combine, rather than split apart. This process is **nuclear fusion.** It's the opposite of nuclear fission. Not only is energy released when heavy nuclei split apart in the fission process, energy is also released when light nuclei fuse. We see from the graph that the nucleons of the fused product have less mass than they had before fusion (Figure 16.15).

Figure 16.15 The mass of a nucleus is not equal to the sum of the mass of its parts. (a) The fission fragments of a heavy nucleus like uranium are less massive than the uranium nucleus. (b) Two protons and two neutrons are more massive in their free states than when combined to form a helium nucleus.

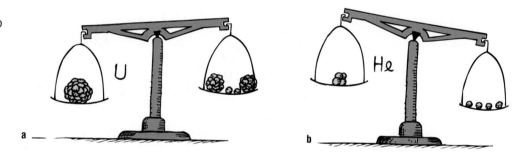

Although the fusion energy per reaction of individual hydrogen atoms is less than the energy given up by the fissioning of individual uranium atoms, gram for gram, fusion is several times more energy-producing than fission. This is because there are more hydrogen atoms in a gram of hydrogen than there are heavier uranium atoms in a gram of uranium.

For a fusion reaction to occur, the nuclei must collide at a very high speed in order to overcome their mutual electric repulsion. The required speeds correspond to the extremely high temperatures found in the sun and stars. Fusion brought about by high temperatures is called **thermonuclear fusion**—that is, the fusing together of atomic nuclei by high temperature. In the high temperatures of the sun, approximately 657 million tons of hydrogen are converted into 653 million tons of helium each second. The missing 4 million tons of mass are discharged as radiant energy. Such reactions are, quite literally, nuclear burning.

Thermonuclear fusion is analogous to ordinary chemical combustion. In both chemical and nuclear burning, a high temperature starts the reaction; the release of energy by the reaction maintains a high enough temperature to spread the fire. The net result of the chemical reaction is a combination of atoms into more tightly bound molecules. In nuclear reactions, the net result is more tightly bound nuclei. In both cases mass decreases as energy is given off.

$\mathbb{Q}$UESTIONS

1. First it was stated that nuclear energy is released when atoms split apart. Now it is stated that nuclear energy is released when atoms combine. Is this a contradiction? How can energy be released by opposite processes?

2. To get energy from the element iron, should iron be fissioned or fused?

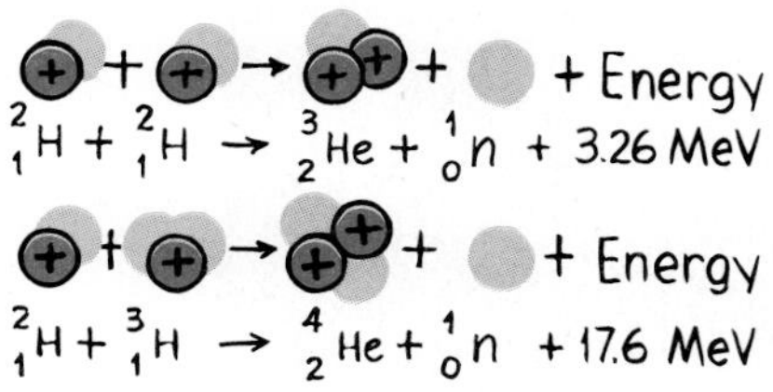

$$\text{(+)} + \text{(+)} \rightarrow \text{(+)(+)} + \text{(●)} + \text{Energy}$$

$$^2_1H + ^2_1H \rightarrow ^3_2He + ^1_0n + 3.26 \text{ MeV}$$

$$\text{(+)} + \text{(+)} \rightarrow \text{(+)(+)} + \text{(●)} + \text{Energy}$$

$$^2_1H + ^3_1H \rightarrow ^4_2He + ^1_0n + 17.6 \text{ MeV}$$

Figure 16.16   Fusion reactions of hydrogen isotopes. Most of the energy released is carried by the neutrons, which are ejected at high speeds.

16.5   CONTROLLING FUSION

Carrying out fusion reactions under controlled conditions requires temperatures of millions of degrees. Producing and sustaining such high temperatures is the goal of much current research. There are a variety of techniques for attaining high temperatures. No matter how the temperature is produced, a problem is that all materials melt and vaporize at the temperatures required for fusion. The solution to this problem is to confine the reaction in a nonmaterial container.

A nonmaterial container is a magnetic field, which can exist at any temperature and can exert powerful forces on charged particles in motion. "Magnetic walls" of sufficient strength provide a kind of magnetic straightjacket for hot gases called plasmas. Magnetic compression further heats the plasma to fusion temperatures. At about a million degrees, some nuclei are moving fast enough to overcome electrical repulsion and slam together and fuse. The energy output, however, is small compared to the energy used to heat the plasma. Even at 100 million degrees, more energy must be put into the plasma than is given off by fusion. At about 350 million degrees, the fusion reactions will produce enough energy to be self-sustaining. At this ignition temperature, nuclear burning yields a sustained power output without further input of energy. A steady feeding of nuclei is all that is needed to produce continuous power.

$\mathbb{A}$NSWERS

1. Energy is released only in a nuclear reaction in which the mass of the nucleons decreases. Light nuclei, such as hydrogen, have less mass after they combine (fuse) to form heavier nuclei. They release energy in this reaction. Heavy nuclei, such as uranium, have less mass after they split to become lighter nuclei. For energy release, "Decrease Mass" is the name of the game—any game, chemical or nuclear.

2. Iron will release no energy at all, because it is at the very bottom of the energy hill. If fused, it "climbs the right side of the hill" and gains mass. If fissioned, it climbs the left side of the hill and gains mass. In gaining mass, it absorbs energy instead of releasing energy.

Figure 16.17 The International Thermonuclear Experimental Reactor (ITER), the largest fusion reactor in the world, has a 12-m diameter plasma chamber and powerful super-conducting magnets to confine the plasma. A neutron-shielding blanket protects the magnets from overheating and shutting down.

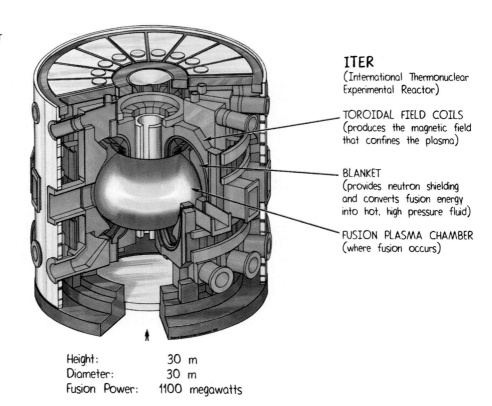

ITER
(International Thermonuclear Experimental Reactor)

TOROIDAL FIELD COILS
(produces the magnetic field that confines the plasma)

BLANKET
(provides neutron shielding and converts fusion energy into hot, high pressure fluid)

FUSION PLASMA CHAMBER
(where fusion occurs)

Height:	30 m
Diameter:	30 m
Fusion Power:	1100 megawatts

Fusion has already been achieved in several devices, but instabilities in the plasma current have thus far prevented a sustained reaction. A big problem has been devising a field system that will hold the plasma in a stable and sustained position while an ample number of nuclei fuse. A variety of magnetic confinement devices are the subject of much present-day research. The latest plasma device, the ITER reactor shown in Figure 16.17, is promising.

Another promising approach bypasses magnetic confinement altogether with high-energy lasers. One proposed technique is to align an array of laser beams at a common point and drop solid pellets composed of hydrogen isotopes through the synchronous cross fire (Figure 16.18). The energy of the multiple beams should crush pellets to densities 20 times that of lead. Such a fusion "burn" could produce several hundred times more energy than is delivered by the laser beams that compress and ignite the pellets. Like the succession of small fuel/air explosions in an automobile engine's cylinders that convert into a smooth flow of mechanical power, the successive ignition of dropping pellets in a fusion power plant may similarly produce a steady stream of electric power.* The success of this technique requires precise tim-

---

*The rate of pellet fusion is 5 per second on the projected Cascade power plant, on the drawing boards at Lawrence Livermore Laboratory. For comparison, about 20 explosions per second occur in each automobile engine cylinder in a car that travels at highway speed. Such a plant could produce 1000 million W of electric power, enough to supply a city of about 600,000 people. Five fusion burns per second will provide about the same power as 60 L of fuel oil or 70 kg of coal per second from conventional power plants.

Figure 16.18 Fusion with multiple laser beams. Pellets of deuterium are rhythmically dropped into synchronized laser crossfire. The resulting heat is carried off by molten lithium to produce steam.

Figure 16.19 Pellet chamber at Lawrence Livermore Laboratory. The laser source is Nova, the most powerful laser in the world, which directs 10 beams into the target region.

ing, for the necessary compression must occur before a shock wave causes the pellet to disperse. Success also requires the development of more efficient lasers so the electricity generated will be greater than the amount required to operate the lasers.

Other fusion schemes involve the bombardment of fuel pellets not by laser light but by beams of electrons, light ions, and heavy ions.

QUESTIONS

1. Fission and fusion are opposite processes, yet each releases energy. Isn't this contradictory?
2. Would you expect the temperature of the core of a star to increase or decrease as a result of the fusion of intermediate elements to manufacture heavy elements?

ANSWERS

1. No, no, no! This is contradictory only if the same element is said to release energy by both the processes of fission and fusion. Only the fusion of light elements and the fission of heavy elements result in a decrease in nucleon mass and a release of energy.
2. Energy is absorbed and the star core tends to cool at this late stage of its evolution. This, however, allows the star to collapse, which produces an even greater temperature.

Cold Nuclear Fusion

There is a way of bypassing the high temperatures needed for thermonuclear fusion altogether. Non-thermonuclear fusion can be accomplished by effectively neutralizing the positive charge of the proton by a very nearby negative particle—but not an electron. In a hydrogen atom, the electron orbits the nucleus at a relatively large distance that is dictated by the relatively long-wavelength matter wave of the tiny electron. When the electron is replaced by a *muon,* an elementary particle that has the same negative charge as the electron but a mass more than 200 times greater, the relatively short-wavelength matter wave puts it 200 times closer to the proton. The tight orbit of the muon about the nucleus results in a smaller version of the hydrogen atom, a "muonic atom." In the muonic atom the muon orbits so close to the nucleus that the muon and nucleus appear as a single neutral particle to distant protons. Thus there is no electrical repulsion between a muonic atom and a hydrogen nucleus or any charged particle. Voila! Ordinary thermal motion is all that is needed for these muonic atoms and hydrogen nuclei to bump into one another, whereupon they form a short-lived molecule, then fuse. This is *cold nuclear fusion,* known as *muon-catalyzed fusion.**

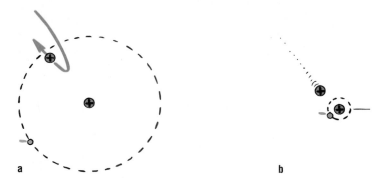

a   b

Figure 16.20 (a) In an ordinary hydrogen atom, a stray proton that gets inside the electron's orbit is electrically repelled by the positive charge of the nucleus. High temperatures are required to slam the protons together. (b) When a muon takes the place of an electron in a hydrogen atom, its orbit is so close to the nucleus that the combination is like a particle with neutral charge. If a stray proton gets inside the muon's orbit, it is in the clutches of the strong nuclear force. Fusion rather than electrical repulsion occurs.

---

*This cold nuclear fusion is very different from the controversial 1989 "cold-fusion-in-a-jar" experiments conducted in Utah by Fleischmann and Pons. Muon-catalyzed fusion was suggested on theoretical grounds in the late 1940s by F. C. Frank and Andrei D. Sakharov. A decade later Luis W. Alvarez and his colleagues found evidence of muon-catalyzed fusion in bubble chamber tracks at the University of California at Berkeley. This discovery generated much excitement. Enthusiasm waned, however, when calculations showed that most muons catalyzed only a single fusion before decay, producing too little energy and too few muons to catalyze later reactions. The Alvarez group had studied reactions involving only ordinary hydrogen and deuterium. More current findings involving a mixture of deuterium and tritium show much more promising results. See the article *Cold Nuclear Fusion* by Johann Rafelski and Steven E. Jones in the July 1987 issue of *Scientific American.*

Fusion Torch and Recycling

A fascinating application for the abundant energy that fusion of whatever kind may provide is the *fusion torch,* a star-hot flame or high-temperature plasma into which all waste materials—whether liquid sewage or solid industrial refuse—could be dumped. In the high-temperature region the materials would be reduced to their constituent atoms and separated by a mass-spectrometer-type device into various bins ranging from hydrogen to uranium. In this way, a single fusion plant could, in principle, not only dispose of thousands of tons of solid wastes per day but also provide a continuous supply of fresh raw material—thereby closing the cycle from use to reuse.

The fusion torch would bring a major turning point in materials economy (Figure 16.21). Our present concern for recycling materials will reach a grand fruition with this or a comparable achievement, for it would be recycling with a capital *R!* Rather than gut our planet further for raw materials, we'd be able to recycle our existing stock over and over again, adding new material only to replace the relatively small amounts that are lost. Fusion power can produce abundant electrical power, desalinate water, help to cleanse our world of pollution and wastes, recycle our materials, and in so doing can provide the setting for a better world—not in the far-off future, but perhaps in the first part of the coming century. If and when fusion power plants become a reality, they are likely to have an even more profound impact upon almost every aspect of human society than did the harnessing of electromagnetic energy in the previous century.

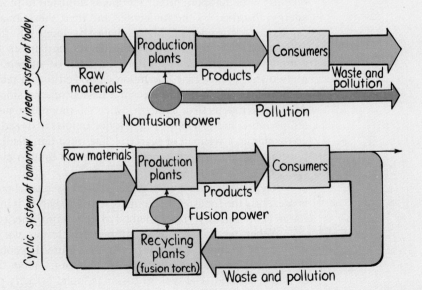

Figure 16.21 A closed materials economy could be achieved with the aid of the fusion torch. In contrast to present systems (a) which are based on inherently wasteful linear material economies, a stationary-state system (b) would be able to recycle the limited supply of material resources, thus alleviating most of the environmental pollution associated with present methods of energy utilization. (Redrawn from "The Prospects of Fusion Power" by William C. Gough and Bernard J. Eastlund. Copyright ©February 1971 by Scientific American, Inc. All rights reserved.)

Most of the energy of muon-catalyzed fusion is in the form of kinetic energy—mainly of the neutrons that are ejected in the reaction. The neatest thing about this type of fusion is that most reactions eject the muons as well, which go on to catalyze other fusion reactions. Surrounding atoms are essentially unaffected, except for the increased temperature from the high kinetic energies of the neutrons. There are several ways in which the kinetic energy of the neutrons can be harnessed, the simplest of which is the conversion to heat to drive turbines for generating electricity.

The not-so-neat thing about this is that muons are short-lived and hard to come by. Muons are found in nature in cosmic rays. Muons can be created by colliding high-energy ions from a particle accelerator with ordinary matter such as carbon. The collisions produce particles called *pions,* which quickly decay to make both positive and negative muons. Muons themselves are unstable and have a half-life of about 2 microseconds. If muon-catalyzed fusion is to produce commercial power, the muon must catalyze enough reactions in its short lifetime to do more than just power the accelerator that generates the muons to begin with.

Recent experiments are somewhat encouraging. In such experiments, negatively charged muons are fired into a gaseous mixture of deuterium and tritium molecules. The muons collide with and replace orbital electrons about the nuclei of these molecules. When the muon enters its tight orbit about one of these isotopes, the molecule breaks apart, leaving a slow-moving muonic atom that easily invades the nucleus of any deuterium or tritium molecule it encounters. Fusion occurs, and the muon is released to initiate the process again. To increase the number of reactions, the temperature of the gas is adjusted to produce resonance between the energy absorbed by the molecules and their vibrational states. Some investigators report the occurrence of this resonance at about 900°C, which produces well over 100 fusions per muon. Higher temperatures dampen the reactions, which means a muon-catalyzed fusion reactor would not be susceptible to runaway reactions or meltdowns. In any case, the muon-catalyzed fusion reaction is not a chain reaction such as that which takes place in a fission reactor since the reaction does not produce more muons than come in. Once the muon is gone the reaction stops. The low temperature and low power density of muon-catalyzed fusion also makes it a poor contender as a weapon of warfare. The most optimistic schemes for muon production fall far short of enough reactions at one time to produce bomblike amounts of energy.

Cold nuclear fusion, whether caused by muons or particles other than muons, is presently the focus of much interest and research. It may turn out to be a contender for an economically viable method of generating energy.

When we think of our continuing evolution, we can see that the universe is well suited to those who will live in the future. If people are one day to dart about the universe in the same way we are able to jet about the world today, their supply of fuel is assured. The fuel for fusion is found in every part of the universe, not only in the stars but also in the space between them. About 91 percent of the atoms in the universe are estimated to be hydrogen. For people of the future, the supply of raw materials is also assured; all the elements result from fusing more and more hydrogen nuclei together. Simply put, if you fuse 8 deuterium nuclei, you have oxygen; 26, you have iron; and so forth. Future humans might synthesize their own elements and produce energy in the process, just as the stars do.

Humans may one day travel to the stars in ships fueled by the same energy that makes the stars shine.

SUMMARY OF TERMS

Nuclear fission The splitting of the nucleus of a heavy atom, such as uranium-235, into two main parts, accompanied by the release of much energy.

Chain reaction A self-sustaining reaction that, once started, steadily provides the energy and matter necessary to continue the reaction.

Critical mass The minimum mass of fissionable material in a reactor or nuclear bomb that will sustain a chain reaction.

Breeder reactor A fission reactor designed for the production of fissionable plutonium-239 from nonfissionable uranium-238.

Nuclear fusion The combination of the nuclei of light atoms to form heavier nuclei, with the release of much energy.

Thermonuclear fusion Nuclear fusion produced by high temperature.

REVIEW QUESTIONS

1. Is nuclear power something new, or has it been around for a long time? About how long a time?

Nuclear Fission

2. What is the role of electrical forces in nuclear fission?

3. What is the role of a neutron in nuclear fission?

4. Of what use are the neutrons that are produced when a nucleus undergoes fission?

5. Why does a chain reaction not occur in uranium mines?

6. Which isotope of uranium is most common?

7. Which isotope of uranium characterizes nuclear fission?

8. What does surface area have to do with neutron leakage?

9. Which has more total surface area—an apple or the same apple cut into two equal pieces?

10. Which has more total surface area—two separate pieces of uranium or the same pieces stuck together?

11. Which will leak more neutrons—two separate pieces of uranium or the same pieces stuck together?

12. Will an energetic chain reaction be more likely in two separate pieces of U-235 or in the same pieces stuck together?

Nuclear Fission Reactors

13. What controls the chain reaction in a nuclear reactor?

14. Are the fission fragments from a nuclear reactor light, medium, or heavy elements?

15. Why are the fission-fragment elements radioactive?

16. What happens when U-238 absorbs a neutron?

17. How can plutonium be created?

18. Is plutonium an isotope of uranium or is it a completely different element?

The Breeder Reactor

19. What is the effect of putting a little Pu-239 with a lot of U-238?

20. What are the three steps in the conversion of U-238 to Pu-239?

21. Does the process of breeding plutonium from uranium occur only in breeder reactors?

Mass-Energy Relationship

22. Does a nucleon have more mass, or less mass, when outside of an atomic nucleus?

23. What device can be used to measure the relative masses of ions of isotopes?

24. What is the primary difference in the graphs shown in Figures 16.13 and 16.14?

25. What becomes of the loss in mass of nucleons when heavy atoms split?

Nuclear Fusion

26. How does nuclear fusion primarily differ from nuclear fission?

27. Why does helium not yield energy if fissioned?

28. Why does uranium not yield energy if fused?

29. Why does iron not yield energy if fissioned or fused?

30. What becomes of the loss in mass of nucleons when light atoms fuse to become heavier ones?

Controlling Fusion

31. What kind of containers are used to contain million-degree plasmas?

32. How is fusion accomplished with lasers?

33. How do the product particles of fusion compare with the product particles of fission?

34. Why are fusion reactors not yet a present-day reality like fission reactors?

Cold Nuclear Fusion

35. Which is smaller in size, an atom consisting of a proton and an electron, or an atom consisting of a proton and a muon?

36. What is the net charge of an atom consisting of a proton and a muon?

37. Why can a stray proton get closer to the nucleus of a muonic atom than a regular hydrogen atom?

38. What happens when a stray proton gets very close to the nucleus of a muonic atom?

39. Why is not cold fusion harnessed for power production?

40. What is the fuel for nuclear fusion? Which element makes up most of the universe? Which element is the primary building block for all elements? Why will there be no shortage of nuclear fusion fuel and material in the "hydrogen age"?

• •

EXERCISES

1. Why is it that uranium ore doesn't spontaneously explode?

2. Why will nuclear fission probably not be used for powering automobiles?

3. How is chemical burning similar to a nuclear chain reaction?

4. Why does a neutron make a better nuclear bullet than a proton or an electron?

5. Why will the escape of neutrons be proportionally less in a large piece of fissionable material than in a smaller piece?

6. Which shape is likely to produce a larger critical mass, a cube or a sphere? Explain.

7. Does the surface area as a whole increase or decrease when pieces of fissionable material are assembled into one piece? Does this assembly increase or decrease the probability of an explosion?

8. U-235 releases an average of 2.5 neutrons per fission, while Pu-239 releases an average of 2.7 neutrons per fission. Which of these elements might you therefore expect to have the smaller critical mass?

9. Which will provide the faster chain reaction, U-235 or Pu-239?

10. Why does plutonium not occur in appreciable amounts in natural ore deposits?

11. Discuss and make a comparison of pollution by conventional fossil-fuel power plants and nuclear-fission power plants.

12. The water that passes through a reactor core does not pass into the turbine. Instead, heat is transferred to a separate water cycle that is entirely outside the reactor. Why is this done?

13. Is the mass of an atomic nucleus greater or less than the sum of the masses of the nucleons composing it?

14. The energy release of nuclear fission is tied to the fact that the heaviest nuclei weigh about 0.1 percent more per nucleon than nuclei near the middle of the periodic table of elements. What would be the effect on energy release if the 0.1 percent figure were instead 1 percent?

15. To predict the approximate energy release of either a fission or a fusion reaction, explain how a physicist makes use of the curve of Figure 16.14 or a table of nuclear masses and the equation $E = mc^2$.

16. Which process would release energy from gold, fission or fusion? From carbon? From iron?

17. If uranium were to split into three segments of equal size instead of two, would more energy or less energy be released? Defend your answer in terms of Figure 16.14.

18. Suppose the curve of Figure 16.14 for mass per nucleon versus atomic number took the shape of the curve shown in Figure 16.12. Then would nuclear fission reactions produce energy? Would nuclear fusion reactions produce energy? Defend your answers.

19. The "hydrogen magnets" in Figure 16.15 weigh more when apart than when combined. What would be the basic difference if the fictitious example instead consisted of "uranium magnets"?

20. Which produces more energy, the fissioning of a single uranium atom or the fusing of a pair of deuterium atoms? The fissioning of a gram of uranium or the fusing of a gram of deuterium? (Why are your answers different?)

21. Why is there, unlike fission fuel, no limit to the amount of fusion fuel that can be safely stored in one locality?

22. If a fusion reaction produces no appreciable radioactive isotopes, why does a hydrogen bomb produce significant radioactive fallout?

23. List at least two major advantages to power production by fusion rather than by fission.

24. Nuclear fusion is a present hope for abundant energy in the near future. Yet the energy that has always sustained us has been the energy of nuclear fusion. Explain.

25. What effect on the mining industry can you foresee in the disposal of urban waste by a fusion torch coupled with a mass spectrometer?

26. The world has never been the same since the discovery of electromagnetic induction and its applications to electric motors and generators. Speculate and list some of the worldwide changes that may likely follow the advent of successful fusion reactors.

PART 6

CHEMISTRY

Atoms of the same kind can be combined in different ways to produce totally different materials. For example, carbon atoms arranged in flat planes like playing cards that slide over one another make up graphite, used as a lubricant and as the "lead" in pencils. Arranged in a three-dimensional structure, carbon atoms form diamonds, the hardest natural substance known. Carbon atoms may also bond in this soccer-ball shape --a "buckyball," which when crystallized can conduct electricity with zero resistance. All these and more from only carbon! Onward to chemistry!

17

BASIC CONCEPTS
OF CHEMISTRY

Chemistry is often called the central science, for it bridges physics and biology. Chemistry springs from the principles of physics as applied to the atom, and lays the chemical foundation for the most complex science of all—the science of living things—biology. Chemistry is the science of materials, so it also underlies the earth sciences—geology and its branches. The applications of chemistry are far reaching—from the manufacture of useful materials to understanding the mechanisms of life.

17.1 A BRIEF HISTORY OF MODERN CHEMISTRY

We humans have long been curious about the materials around us. It almost seems that it was our destiny to tinker with these materials and use them to our advantage. Stone would be carved into arrowheads. Flint would be used to create fire. Rare naturally occurring copper, silver, and gold would be used for jewelry. Colorful minerals would be ground up and used as pigments for paintings and cosmetics. Dyes, ointments, and perfumes would be extracted from plants.

The Impact of Materials

Standards of living throughout history and across cultures have been closely tied to the materials available for use. Stone-Age people learned how to design tools from stone, which made hunting, food preparation and, consequently, life in general easier. The advent of new materials in the Bronze Age and Iron Age further raised standards of living. Later, the introduction of brick, paper, glass, and gunpowder also enhanced human capability. In the present time the list of new materials is being extended daily.

Unlike previous eras, our present time is not an age of any one type of material. It is the rapidly growing diversity of materials that makes our time unique. Fabrication of new materials is possible because we have learned how to manipulate atoms. Thus we can say we are living in an atomic age. Evidence of the atomic age is all around us. Steel and other alloys are used for buildings, cars, and appliances. We walk on carpets of synthetic fibers.

We cook on non-stick Teflon surfaces. Food is kept sanitized by plastic wrapping. Semiconductor materials enable computers. Superconductors levitate trains. Modern ceramics boost efficiency in automobile engines. Our ability to manipulate atoms for the production of nuclear energy is also a hallmark of our atomic age.

The chemist's command of materials goes far beyond providing more "material things." The effects on life itself are direct. Nitrogen combined with hydrogen makes ammonia for fertilizers. Formulations of herbicides, pesticides, and fungicides aid in the large-scale production of foodstuffs. Formulations of other materials produce substances we use to treat ills ranging from headaches to cancer. Expanded knowledge of DNA may lead to a cure for genetic diseases such as muscular dystrophy, cystic fibrosis, and sickle-cell anemia.

The science of materials is an important one, indeed.

The Origins of Chemistry

With the advent of fire, an explosion of new substances became available. Moldable wet clay, for example, was found to harden into ceramic when heated by fire. Pottery was thus invented. Baked clay ceramic figurines have been found that date back to 18,000 BC. By 5000 BC, pottery fire pits gave rise to furnaces hot enough to melt copper. Eventually, even higher temperatures were achieved in furnaces, and it was found that copper could be produced by firing the mineral ore malachite (a form of copper carbonate). Other mineral ores were similarly treated and new metals discovered. One such metal, tin, was melted and mixed with copper to give the useful alloy bronze. This discovery ushered in the bronze age, a time when many metals, excluding iron, were commonly used for tools and weapons. By 1200 BC, furnaces were hot enough to convert iron ores into iron, a metal most notable for its strength. This significantly improved the durability of metal tools and weapons, and promoted the achievements of high Chinese, Egyptian, and Greek civilizations.

To ancient people, the creation of new materials forged out of fire seemed like the magic of gods or demons. Rituals were developed to maintain the favor of these deities. Ancient people had no concept of natural law. Things that could not be explained were supernatural and could not be questioned. Under these circumstances, most discoveries were made by chance.

This unquestioning climate began to change after the Greek philosopher Thales of Miletus accurately predicted the solar eclipse of May 25, 583 BC. To future generations of Greek philosophers the prediction meant there was an underlying order to the universe—a set of natural laws that even the heavens had to obey. The Greek philosophers reasoned that all things had both cause and effect. Soon they were trying to explain natural phenomena in terms of natural laws.

The very influential Greek philosopher Aristotle advanced an explanation for the composition of matter. To Aristotle, all substances were just different displays of what he called *prime matter,* which could take on the four qualities of hot, cold, moist, and dry. The combination of various qualities gave rise to the four basic elements: hot and dry gave fire; moist and cold gave water; hot and moist gave air; and dry and cold gave earth. These relationships are depicted by the following diagram.

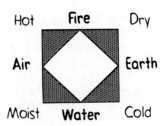

Various substances were considered to be made of various proportions of the four qualitites. For example, an unusually hard substance such as rock, had an excess of the dry quality. A soft substance like clay, on the other hand, had an excess of the moist quality.

Although we know Aristotle's model to be wrong, it was nonetheless a remarkable achievement because it was based upon observed facts rather than imagination. People using the model, therefore, found much common sense. Wet clay, for example, was converted to earth (ceramic) because the heat of the fire drove out the moist quality and replaced it with dry. In fact, Aristotle's views on the composition of matter made so much sense to people that other less obvious views were difficult to accept. One such alternative was the forerunner to our present-day model: that matter is composed of a finite number of discrete particles. This model was advanced by several Greek philosophers including Democritus, who coined the term "atom." So compelling were the ideas of Aristotle, however, that the atomic model would not reappear for some 2000 years.

According to Aristotle, it was theoretically possible to transform any substance into another simply by altering the relative proportion of qualities. This meant that under the proper conditions common metals such as lead could be transformed into gold. Equipped with Aristotelian concepts, profit-seekers of the Middle Ages, also known as alchemists (first mentioned in Chapter 16), tried in vain to convert various metals into gold. Despite the futility of their efforts, much was learned about the behavior of many different chemicals. Significantly, it was found that chemicals such as distilled alcohol, ether, opium, and salts of lead, arsenic, and mercury acted as curatives. (Usually these chemicals acted as disinfectants. Sometimes they merely gave the body time to heal before medieval physicians implemented destructive treatments such as bleeding.) Such discoveries were the basis of *medical alchemy,* which by the 1400s had gained a fair amount of respect due to its successes. Other aspects of working with chemicals had also gained respect. In particular, chemical work that led to the production of useful materials such as steel, glass, paper, and gunpowder had real value. Traditional alchemy, on the other hand, failed to live up to its promise of the cheap production of gold. Eventually it became a swindler's profession and was outlawed by many governments.

Democritus

In the 1400s the printing press was introduced to Europe and there followed an explosion of information. This in turn enhanced the quality of living for many. Craftsmen, for example, could improve their techniques by reading about the successes and failures of others. Similarly, knowledge of different chemicals and chemical processes became organized. Eventually, an accumulation of evidence showed that Aristotle's model was wrong. One of the most convincing arguments came from a well-known English experimentalist, Robert Boyle. He proposed in 1661 that a substance was not an element if it could be decomposed into simpler substances. Yet Boyle did not completely divorce himself from Aristotelian concepts. For example, he believed in prime matter. Perhaps this prevented him from proposing the existence of elements as we know them today.

The Development of Modern Chemistry

Antoine Lavoisier

About a century after Boyle, huge steps toward our present understanding of the elements were taken by the French chemist, Antoine Lavoisier. He, without question, was the first leading figure in the development of modern chemistry. Free from Aristotelian bias, Lavoisier completed Boyle's statement about elements by recognizing a chemical element as any material that cannot be broken down into simpler substances by chemical analysis. A few examples include materials like copper, tin, carbon, and oxygen. Elements, in turn, can combine to form a wide variety of *chemical compounds* that have unique properties of their own. A few examples include bronze (from copper and tin), copper carbonate (from copper, carbon, and oxygen), and carbon dioxide (from carbon and oxygen). Under the proper conditions chemical compounds can be separated back into elements. These definitions were supported by much experimental evidence. They led, however, to some very important and fundamental questions. Why can't elements be broken down into simpler substances? What makes one element different from another? How exactly do elements combine to make chemical compounds? The answers to these questions came only after further experimentation.

By 1774 Lavoisier had developed a balance that could measure 1/100 the mass of a drop of water. In a most important experiment, he carefully measured the mass of a sealed glass vessel that contained tin, known to be a reactive element. Upon heating the vessel he noted that the tin reacted by changing into a white powder. Lavoisier again measured the vessel's mass and found it had not changed. From this and similar experiments performed by himself and other investigators Lavoisier proposed the principle of the **conservation of mass,** which states:

There is no detectable change in the total mass during a chemical process.

This fundamental principle remains as one of the most important principles in chemistry today.* It is easy to see, however, why the conservation of mass eluded so many

---

*We have seen in Chapter 16 that mass is a form of energy called *rest energy*, E, as shown by the equation $E = mc^2$. Total energy, including rest energy, is always conserved in any process. In nuclear reactions, as well as chemical reactions, kinetic and potential energies are produced or absorbed with a corresponding loss or gain in rest energy (mass), such that the total energy does not change. In a nuclear reaction the energy released or absorbed is so great that the change in rest energy (mass) is measurable, about 1 part in 10^3. In a chemical reaction, the amount of energy released or absorbed is so relatively small that the change in rest energy (mass), about 1 part in 10^9, is generally not detected. Hence, from a chemistry point of view, to a close approximation, the conservation of mass principle holds true.

earlier investigators. After all, when wood burns, the remaining ashes always have less mass; hence they weigh less. Also, it was known that some substances, such as quicklime, have the tendency to gain weight. Early investigators, however, failed to recognize the role that gases play in many reactions. When wood burns, carbon dioxide and water vapor are released, and so after the fire, ashes weigh less than wood. Also, quicklime gains weight as it absorbs atmospheric carbon dioxide. Lavoisier was exceedingly careful in attending to details, he recognized the role that gases might play, and he knew the importance of sealing his apparatus before performing the chemical reaction.

Lavoisier continued his experiments with tin beyond the conservation of mass. When he opened the sealed vessel in which the white powder formed, he observed that air rushed in. He hypothesized that as the tin formed the white powder, it absorbed the air inside, or perhaps something in the air. To find out what percentage of the air had reacted with the tin he performed the same experiment using a different set-up (Figure 17.1). Lavoisier placed a piece of tin on a wooden block floating in water, which he then covered with a glass jar. Then he heated the tin by focusing sunlight on it with a magnifying glass. As the tin reacted, the water level in the jar rose. This continued until the tin no longer reacted and the volume of air in the jar had been reduced by about 20 percent. This suggested to Lavoisier that the air was made of at least two components. One component, making up about 20 percent of the air, disappeared from its gaseous phase by combining with the metal. The second component, making up the other 80 percent, remained in the gaseous phase because it did not combine with the metal.

Joseph Priestley

Figure 17.1 Lavoisier added sunlight to tin floating on a block of wood underneath a glass jar. As the tin reacted to form a powder, the water level in the jar rose.

Soon after Lavoisier had completed these experiments he learned that an English chemist, Joseph Priestley, had recently prepared a new gas with remarkable properties (Figure 17.2). This new gas caused candles to burn brighter, glowing charcoal to burn hotter, and a mouse kept in a closed jar with the gas to live longer than one kept in a closed jar of air. Lavoisier investigated this gas. Since the gas was also found to produce acidic solutions, he gave it the name *oxygen*, which means "acid former."* He found that oxygen was the reacting gaseous substance in his tin experiments. Interestingly enough, Lavoisier also inferred from his previous observations that air itself is composed of 20% oxygen. In this he was remarkably accurate.†

---

*The nature of acids is discussed in Chapter 20.

†Sadly, Lavoisier was beheaded during the French Revolution. Two years after his execution, however, the French government was erecting statues in honor of him and his important contributions.

Phlogiston

During the time of Lavoisier many investigators thought that matter contained a substance they called *phlogiston*. They believed that when a substance burns or otherwise reacts with air, phlogiston is released from the reacting substance and absorbed by the air. They reasoned that matter did not react or burn in the absence of air because there was nothing to absorb the phlogiston. According to this model, all matter weighs less after reacting. Observations of wood burning supported the model as it was commonly observed that ashes weighed less than the wood before burning. This phlogistan model, developed in the 1600s by two Germans, Johann Becher and his student Georg Stahl, was an extension of Aristotle's teachings.

Lavoisier showed by careful use of his balance, however, that the phlogiston model could not be right. He showed that after a metal such as tin reacts it weighs *more*. How could a metal gain weight while releasing phlogiston? Lavoisier's experiments with tin caused many phlogiston supporters to abandon their model. Others, including Priestley, however, were more conservative and defended their well-guarded views of phlogiston. They concluded that phlogiston could, in some instances, have a negative mass. By the early 1800s, however, so much evidence had mounted against the phlogiston model that it was abandoned for good.

In 1766, the English physicist and chemist Henry Cavendish isolated a new gas that could be ignited in air to produce water along with heat. He named this new gas *hydrogen*, which means "water former." After the discovery of oxygen Cavendish and others showed that hydrogen and oxygen combine to form water. In the late 1790s the French chemist J. L. Proust was one of the first to recognize that elements combine with one another in a definite ratio by mass. Proust noted, for example, that 8 grams of oxygen always combines with 1 gram of hydrogen (no more and no less) to produce 9 grams of water. Equivalently, 32 grams of oxygen always combine with 4 grams of hydrogen to produce 36 grams of water. In all cases, the ratio of oxygen to hydrogen by mass is 8:1.

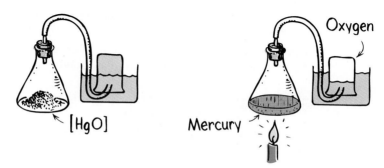

Figure 17.2　Priestley discovered oxygen by heating a metallic compound, known today as mercuric oxide, HgO. When heated, mercuric oxide decomposes to liquid mercury and oxygen gas. Priestley collected the oxygen using an apparatus similar to the one illustrated here. As oxygen is formed in this apparatus, it displaces water in the submerged inverted glass.

QUESTIONS

How many grams of water will be produced by the combination of

1. 8 grams of oxygen and 8 grams of hydrogen?

2. 800 grams of oxygen and 8 grams of hydrogen?

John Dalton

The observations of Proust and others led John Dalton, a self-educated English school teacher, to reintroduce the atomic ideas of Democritus. In 1803 Dalton published a paper announcing the following postulates:

1. Each element consists of indivisible, minute particles called atoms.
2. All the atoms of a given element are identical.
3. The atoms of different elements have different masses.
4. Atoms chemically combine in definite whole-number ratios to form chemical compounds.
5. Atoms can neither be created nor destroyed in chemical reactions.

These postulates, having accounted for all the experimental evidence that had been collected, were readily accepted by the scientific community.

Dalton's postulates were elegantly straightforward and simple. Through them, the formation of water, for example, was seen to be merely the combination of oxygen and hydrogen atoms. Today, we know Dalton was correct. Dalton, however, went on to assume that the most fundamental unit of water, a water molecule, consists of oxygen and hydrogen atoms in a 1:1 ratio. (As we shall see in Section 17.4, a *molecule* is a group of tightly bonded atoms.) Dalton's formula for water, therefore, was HO. He reasoned that since 8 grams of oxygen always combine with 1 gram of water, the oxygen atom must be 8 times as massive as the hydrogen atom. In spite of the correctness of his postulates, Dalton clung to the erroneous notion that the majority of compounds were the combination of two single atoms.

In 1809, Joseph Gay-Lussac showed that two volumes of hydrogen gas were required to combine with one volume of oxygen gas to form two volumes of water vapor (all gases being at the same temperature and pressure). Based on these results and similar ones for other gas phase reactions, Gay-Lussac established the **principle of combining volumes**:

> **Volumes of gases, at the same temperature and pressure, combine chemically with one another in the ratio of small whole numbers.**

Such an idea did not agree with Dalton's formula for water. If, as Dalton assumed, the combining gases consisted of single H and O atoms, then a volume ratio of 2:1

ANSWERS

1. Of the 8 grams of hydrogen, only 1 gram of hydrogen will react with the oxygen. So 9 grams of water will form, with 7 grams of hydrogen left over.

2. Seventy-two grams of water will form, with 736 grams of oxygen left over. Since oxygen and hydrogen react in an 8:1 ratio, 800 grams of oxygen need 100 grams of hydrogen for a complete reaction. But only 8 grams of hydrogen are available. So 8 times as much oxygen will react. This is 64 grams (64:8 is the same as 8:1). This leaves 800 − 64 = 736 grams of oxygen left over. For the amount of water formed we add the amount of oxygen and hydrogen that reacted: 64 grams oxygen + 8 grams hydrogen = 72 grams of water.

suggested that an O atom split in half. This would be a clear violation of Dalton's notion of indivisible atoms. Or perhaps two hydrogen atoms combined with a single oxygen atom. Yet this notion violated Dalton's way of thinking as well.

Dalton did not accept Gay-Lussac's results, but the Italian physicist and lawyer, Amadeo Avogadro did. In 1811, Avogadro extended Gay-Lussac's principle by correctly suggesting that hydrogen and oxygen atoms were paired in the gas phase, so the fundamental units were H_2 and O_2, rather than single atoms. Thus, two volumes of H_2 must have twice the number of molecules as one volume of O_2. Since the formation of water involves twice as many hydrogen atoms as oxygen atoms, the formula for water must be H_2O.

$$2\,H_2 + 1\,O_2 \rightarrow 2\,H_2O$$

Avogadro's explanation, however, was not widely accepted, largely because it was believed that atoms of the same kind were unable to bond. It took almost half a century before Avogadro's important contributions were recognized—in 1860, four years after his death.

Today's model of the atomic nature of matter has not changed much since these times, though it has undergone some refinement. For example, we find that the atom is indeed divisible and is made of even more fundamental units—the electron, proton, and neutron (Chapter 14).

In the history of the initial chemical discoveries, we see how reluctant investigators were to change their points of view—even in the face of evidence. Priestley, in spite of evidence to the contrary, clung to the phlogiston model. Dalton refused to believe Gay-Lussac's experimental results. This human trait has its benefits. In science, skepticism, whatever its cause, forces people to support their ideas with well-documented experiments that can be duplicated by other scientists with the same results. Lavoisier's critical arguments against phlogiston, for example, would not have been initially accepted without his insistence on careful and accurate mass measurements. In science, ideas are usually accepted only after much supporting evidence. Science has always been and still is a field of restless human activity (Table 17.1).

Amadeo Avogadro

TABLE 17.1   Players in the Game

Democritus (460–370 BC)	Proposed an atomic model for matter.
Aristotle (384–323 BC)	Proposed a convincing but erroneous model on the nature of matter based upon common observations.
Boyle (1627–1691)	A great experimentalist who showed how Aristotle's concepts of matter were wrong.
Becher (1635–1682) and Stahl (1660–1734)	Developed phlogiston model.
Cavendish (1731–1810)	1766 Discovered hydrogen.
Lavoisier (1743–1794)	1774 Developed the conservation of mass principle by experimentation. Recognized importance of oxygen. An important founder of modern chemistry.
Priestley (1733–1804)	1774 Discovered oxygen.
Proust (1755–1826)	1790 Showed that elements combine in definite ratios by mass.
Dalton (1766–1844)	1803 Developed the atomic theory by way of a set of postulates.
Gay-Lussac (1778–1850)	1808 Mixed known volumes of reactive gases.
Avogadro (1776–1856)	1811 Explained Gay-Lussac's observations. Discovered that equal volumes contain equal numbers of molecules.

17.2 PHYSICAL AND CHEMICAL PROPERTIES

Substances can be described by their **physical properties**. Physical properties are characteristics such as color, density, hardness, electrical and thermal conductivity, heat capacity, and (for a particular temperature) phase (Figure 17.3). Changes in a substance's physical properties—for example, freezing, melting, and condensation—are called **physical changes** (Figure 17.4). The identity of a substance is not lost in a physical change. Water remains water even after it freezes or boils. Ice and steam are still water. Similarly, iron is still iron, even though it may have melted, expanded, or contracted due to changes in temperature.

Figure 17.3 The phases of all the elements at 30°C (86°F). At this temperature, most elements are solids. Five elements—gallium [Ga], bromine [Br], cesium [Cs], mercury [Hg], and francium [Fr]—are liquids. Elements shown in light blue, such as hydrogen [H], and helium [He], are gases.

Figure 17.4 The phase of an element changes at certain temperatures. (a) When the temperature of sulfur, S_8, is brought above 150°C, its phase changes from a solid to a liquid. (b) When the temperature of nitrogen, N_2, is brought down to −210°C, its phase changes from a liquid to a solid.

a

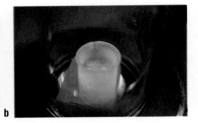
b

A substance can also be characterized by its **chemical properties**. Chemical properties relate to the tendency of a substance to change into a different substance. For example, hydrogen and oxygen share the chemical property of combining to form water. Iron, oxygen, and water have the chemical property of combining to form rust. Likewise, it is a chemical property of helium gas to *resist* combining with any other element, and a chemical property of gold to *resist* combining with oxygen.

A substance transforms into a different substance by a rearrangement of its atoms. New atomic connections are made and/or old ones are broken as substance transformation takes place. This type of change is called a **chemical change**.

Examples include the rusting of iron, the fermenting of grape juice into wine, and the burning of gasoline. In general, the identifying feature of a chemical change is that a new material has been formed (Figure 17.5).

Figure 17.5 The turning colors of a tree and the growth of a human being both involve the formation of new materials—chemical changes.

17.3 ELEMENTS AND COMPOUNDS

Certain substances contain only one kind of atom. Such substances are the **elements** as Dalton defined them (Section 17.1). Hydrogen is an element since it is made of only a single kind of atom. All hydrogen atoms are made of a single proton and a single electron. Similarly, gold is an element because it is made of only a single kind of atom, each having 79 protons in its nucleus. Presently, we know of 109 elements. These elements are listed in the **periodic table** (see insert in Chapter 18), discussed in Chapter 18.

Each element is designated by its **atomic symbol**. The atomic symbol is derived from the first letters of the element's English or Latin name. For example, the atomic symbol for carbon is C, and the atomic symbol for chlorine is Cl. Gold has the atomic symbol Au, after its Latin name "aurum." Sodium has the atomic symbol Na, after the Latin name "natrium." Elements with symbols derived from Latin names are typically those discovered earliest.

When atoms of the same element combine they make an **elemental compound**. Many elemental compounds occur naturally. Examples include the precious metals gold, platinum, and palladium, the mineral sulfur, and the atmospheric gases nitrogen and oxygen. The way in which atoms combine in an elemental compound may vary. In gold, for example, atoms are pretty much all grouped together in a single mass. In sulfur, however, atoms can be connected to one another in the form of rings, each ring containing 8 sulfur atoms. The nitrogen and oxygen that we breathe contain paired atoms. We represent elemental compounds with an **elemental formula**. The elemental formula consists of the atomic symbol along with a numerical subscript to indicate the number of atoms that are grouped together. Since gold atoms do not group in any particular fashion, gold is not usually given an elemental formula and is simply represented using the atomic symbol, Au. Sulfur, on the other hand, is given the elemental formula S_8, and nitrogen and oxygen are given the elemental formulas N_2 and O_2, respectively.

QUESTION

The oxygen we breathe, O_2, is converted to ozone, O_3, in the presence of electrical sparks. Is this an example of a physical or chemical change?

When atoms of different elements combine, they make a **chemical compound**. Most elements in nature are found as chemical compounds. Sodium and chlorine, for example, are found combined as sodium chloride (table salt), and iron and oxygen are found combined as iron oxide (rust). A chemical compound is represented by the **chemical formula** in which the symbols for the different elements of the compound are written together. The chemical formula for sodium chloride and iron oxide, for example, are NaCl and Fe_2O_3, respectively. Numerical subscripts indicate the ratio in which atoms combine to make the chemical compound. By convention, subscripts of 1 are understood and omitted. So, we see that for NaCl there is 1 sodium atom for every 1 chlorine atom and for Fe_2O_3 there are 2 iron atoms for every 3 oxygen atoms.

Chemical compounds have physical and chemical properties that are uniquely different from their elemental components. For example, sodium chloride, NaCl, is very different from either sodium or chlorine. Sodium is a soft silvery metal that can be cut with a knife; sodium's melting point is 97.5°C, and it reacts violently with water. Chlorine is very toxic and works to kill bacteria in swimming pools; chlorine's boiling point is −34°C and as a gas has a yellow-green color. Sodium chloride, however, is a translucent, brittle, colorless crystal with a melting point of 800°C. Sodium chloride is very stable in water, and is an essential chemical compound for living organisms (Figure 17.6). Sodium chloride is not sodium, nor is it chlorine; it is uniquely sodium chloride.

Figure 17.6 The physical and chemical properties of sodium chloride (a) are very different from the physical and chemical properties of sodium (b) and chlorine (c).

ANSWER

When atoms are regrouped the result is an entirely new substance. For example, while the oxygen we breathe, O_2, is odorless and life-giving, ozone, O_3, is toxic and has the pungent smell commonly associated with electric motors. The conversion of O_2 to O_3, therefore, is an example of a chemical change.

17.4 IONIC, COVALENT, AND METALLIC COMPOUNDS

In our study of electricity in Chapter 9 we learned about Coulomb's law, which states that electric charges attract or repel depending on their sign of charge, and that the force of attraction or repulsion depends on the square of the distance between the charges. If the charges are of like signs, like two electrons, they repel each other; if the signs are opposite, like an electron and a proton, they attract. Charged things interact electrically with other charged things. Now we learn that all types of bonding that hold atoms together to form chemical compounds are electrical in nature. Bonds are Coulomb's law at the atomic level. In this section we will see how electrical interactions produce three different types of bonds and therefore three different types of chemical compounds. These are the *ionic, covalent,* and *metallic* compounds.

Ionic Compounds

An **ionic compound** is a chemical compound that contains **ions**. Recall from Chapter 9 that an ion is an atom that has either lost or gained one or more electrons and therefore has a net electrical charge. A sodium atom, for example, that loses one electron becomes a positively charged sodium ion, Na^+. Similarly, a chlorine atom that gains one electron becomes a negatively charged chlorine ion, Cl^-. Ions of opposite charge in an ionic compound are held together by the attractive electrical force, commonly referred to as the **ionic bond**. We find that elements located toward the left of the periodic table tend to lose electrons to become positively charged ions, and elements towards the right tend to gain electrons to become negatively charged ions. For this reason, ionic compounds typically consist of elements found on opposite sides of the periodic table. Sodium chloride, NaCl, is an example of an ionic compound. Other examples are potassium iodide, KI, which is added in minute quantities to commercial salt to prevent a medical condition known as goiter, and calcium fluoride, CaF_2, which is a natural source of fluorine in many people's drinking water.

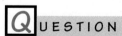QUESTION

What is the electrical charge of a calcium atom that loses two electrons?

Ions of ionic compounds are frequently found in regular patterns in which ions of opposite charges surround one another. For sodium chloride, each sodium ion is surrounded by six chlorine ions. In turn, each chlorine ion is surrounded by six sodium ions (Figure 17.7). As a whole, there is one sodium ion for each chlorine ion, but there are no identifiable sodium-chlorine pairs. When the atoms of an ionic compound are arranged in this type of array they form an **ionic crystal**.

ANSWER

Any atom that loses two electrons is a +2 ion. A neutral calcium atom, which has 20 protons and 20 electrons, has a net charge of zero. After losing two electrons, calcium's 20 protons are surrounded by only 18 electrons, which gives it a net charge of +2, written as Ca^{2+}. Since the calcium ion has a +2 charge it is able to bond with two fluorine ions, which each carry a −1 charge, to give the ionic compound calcium fluoride, CaF_2.

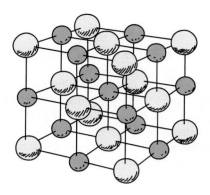

Figure 17.7 Sodium chloride, as well as other ionic compounds, form ionic crystals in which every internal ion is surrounded by others of opposite charge. (For simplicity only a 3 by 3 ion lattice is shown here. A typical NaCl crystal involves millions and millions of ions.)

Covalent Compounds

Two children can be held together by their mutual attraction for the toys they share. In a similar fashion, a **covalent compound** is a compound in which atoms are held together by their mutual attraction for electrons they share. (*Co* signifies sharing, and *valent* refers to chemical combining power, so *covalent* means a sharing of chemical combining power). The elemental compound hydrogen, H_2, is a covalent compound. In this compound each of the two hydrogen atoms shares an electron with the other so that they are both attracted to the same electrons, hence, they are held together (Figure 17.8a). This type of electrical attraction in which two atoms share two electrons between them is called the **covalent bond**. By convention, two atoms that are covalently bonded are represented by drawing a straight line between their atomic symbols (Figure 17.8b).

Figure 17.8 (a) The covalent compound hydrogen, H_2, consists of two hydrogen nuclei that are mutually attracted to two shared electrons. (b) The conventional representation of two hydrogen atoms held together by the covalent bond.

Elements that commonly form covalent bonds include hydrogen and elements in the upper right-hand side of the periodic table, such as fluorine, oxygen, nitrogen, and carbon. While each hydrogen atom forms only one covalent bond, other atoms form more. Oxygen atoms, for example, form two covalent bonds (as in water, H_2O). Nitrogen atoms form three (as in ammonia, NH_3, which is used as a cleanser). And carbon atoms form four covalent bonds (as in methane, CH_4, which is the primary component of natural gas) (Figure 17.9). More than one covalent bond may occur between two atoms. For example, the covalent compound oxygen, O_2, consists of

Figure 17.9 Certain atoms are able to form more than one covalent bond. Hence, they may be connected to more than one atom.

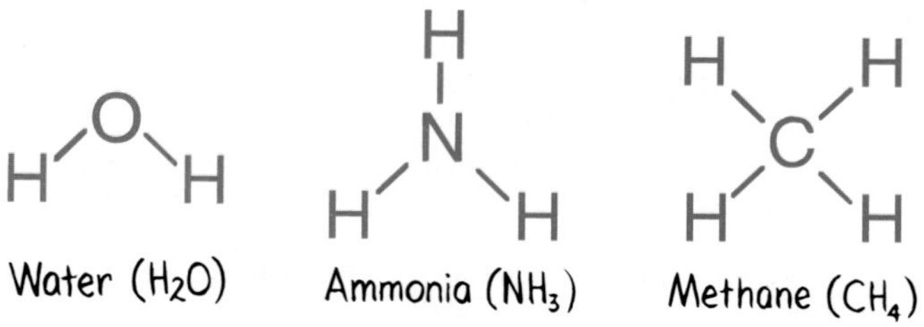

Water (H_2O)   Ammonia (NH_3)   Methane (CH_4)

two oxygen atoms that are connected by two covalent bonds, called a *double bond* (Figure 17.10). Within a double bond there are a total of four electrons being shared, two for each covalent bond. Similarly, the covalent compound nitrogen, N_2, which comprises about 79 percent of our atmosphere, consists of two nitrogen atoms that are connected by three covalent bonds (6 electrons), called a *triple bond*. The covalent compound carbon dioxide, CO_2, which we exhale, consists of two double bonds that connect two oxygen atoms to a central carbon atom.

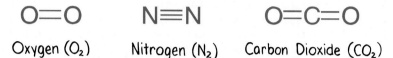

Figure 17.10 Multiple covalent bonds may occur between atoms.

Since atoms within a covalent compound remain bonded only to each other, they can be identified as a single unit—a **molecule**. A molecule of hydrogen, for example, is H_2, and a molecule of water is H_2O. Generally speaking, a molecule is any group of atoms held together by covalent bonds. Most molecules are far too small to be seen with powerful microscopes. Some atoms, however, are able to bond repeatedly to produce molecules large enough to be seen with the naked eye! An example is a diamond, which is a large molecule made of many carbon atoms connected by single covalent bonds (Figure 17.11).

Figure 17.11 A diamond is a large single molecule consisting of many covalently bonded carbon atoms.

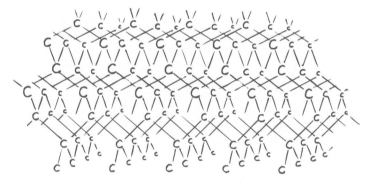

Molecules that arrange themselves in an orderly and periodic fashion form what is called a **covalent crystal**. An example of a covalent crystal is ice, which consists of water molecules that have been frozen into a solid 3-dimensional array (Figure 17.12). Diamond is also considered to be an example of a covalent crystal even though it is only a single molecule.

In a covalent bond, two atoms share two electrons. If the atoms are identical, then their nuclei have the same positive charge and the electrons are shared evenly. The electrons of a hydrogen-hydrogen covalent bond, for example, experience identical nuclear charges from both sides of the bond. So, the electrons are attracted to both sides equally and they do not tend to accumulate to one side or the other. We can represent these electrons as being centrally located, as shown in Figure 17.13a. If, on the other hand, a covalent bond is between a pair of non-identical atoms, then the nuclear charges will be different and the bonding electrons will be shared unevenly. This occurs for a hydrogen-chlorine bond where electrons are more attracted to chlorine's greater nuclear charge, and are drawn to the chlorine side of the covalent bond, as shown in Figure 17.13b.

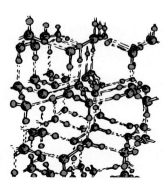

Figure 17.12 Water molecules pack together in an orderly fashion to produce ice—a covalent crystal.

Figure 17.13 (a) Hydrogen nuclei attract electrons with equal forces. The bonding electrons are therefore not pulled to one side or the other. (b) The chlorine nucleus attracts the bonding electrons with a greater force because of its greater nuclear charge; the bonding electrons are therefore closer to the chlorine nucleus.

a H : H   b H : Cl

We find the electrons in the hydrogen-chlorine bond spend more time around chlorine, much like a bunch of bees spend more time around a sweeter flower. For this reason we find the chlorine side of the bond to be slightly negative, and since electrons have been drawn away from the hydrogen atom, the hydrogen side of the bond is slightly positive. This separation of charge, commonly called a **dipole**, may be represented using the characters "$\delta-$" and "$\delta+$", which, respectively, read "slightly negative" and "slightly positive," Figure 17.14a. In Chapter 19 we will see how molecular dipoles account for such oddities as why water sticks to glass and how fish are able to breathe. A second way to represent a dipole is to draw a crossed arrow that points to the negatively charged side of the bond, shown in Figure 17.14b.

Figure 17.14 (a) In a molecule of hydrogen chloride, HCl, the hydrogen atom is slightly positive, $\delta+$, while the chlorine atom is slightly negative, $\delta-$. (b) A crossed arrow that points towards the negatively charged side of the bond may also be used to represent a dipole.

a $\overset{\delta+}{H} — \overset{\delta-}{Cl}$ b $H — Cl$

A covalent bond with a dipole is said to be **polar**. Recall from Chapter 9 how the electrical clouds of atoms can be distorted by the presence of a neighboring charge. Charge polarization in atoms and molecules plays a very important role in chemical processes. As we will soon discuss, the polarity of bonds dictates whether a substance is a gas, liquid, or solid at particular temperatures. Also, the polarity of bonds dictates the freezing and boiling points of substances.

Some covalent bonds are more polar than others. The magnitude of polarity can be deduced from the position of the elements in the periodic table. We will see in the following chapter that the farther apart two elements are from each other in the periodic table, the greater the difference in their nuclear charges. A covalent bond involving two elements far apart in the periodic table, therefore, tends to be more polar than a covalent bond involving elements that are closer together. For example, we find that the hydrogen-fluorine, H—F, bond is more polar than the hydrogen-oxygen, H—O, bond, which is more polar than the hydrogen-nitrogen, H—N, bond (Figure 17.15). We will see that the degree of polarity plays a strong role in the chemical properties of substances. When an element bonds with itself, as in molecular hydrogen, H_2, oxygen, O_2, or nitrogen, N_2, no dipole is formed and the bond is said to be **nonpolar**.

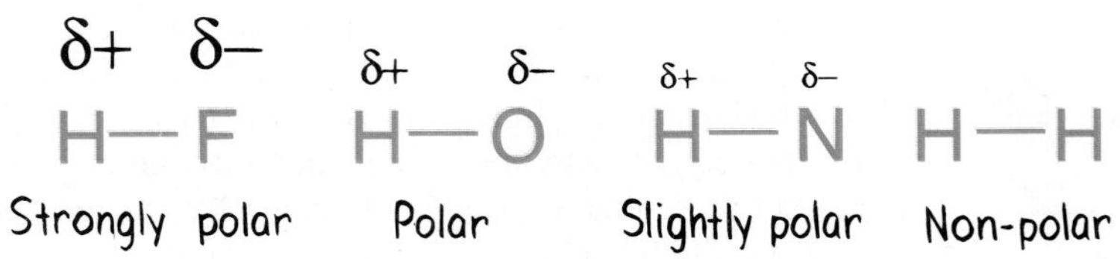

Figure 17.15 The farther apart two elements are positioned from each other in the periodic table, the more polar the covalent bond between them. A covalent bond involving the same element is nonpolar.

$\overset{\delta+}{H}—\overset{\delta-}{F}$ $\overset{\delta+}{H}—\overset{\delta-}{O}$ $\overset{\delta+}{H}—\overset{\delta-}{N}$ $H—H$

Strongly polar   Polar   Slightly polar   Non-polar

A molecule may be polar or nonpolar depending upon the type of covalent bonds within it and also upon the relative orientation of these bonds. If all the bonds

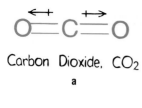

Carbon Dioxide, CO₂

a

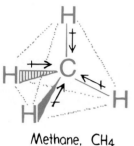

Methane, CH₄

b

Figure 17.16 There is no net dipole for either the carbon dioxide, CO_2, or the methane, CH_4, molecules. (The shape of a methane molecule is that of a *tetrahedron*, which is outlined here by light broken lines. The solid triangle represents a single bond that comes out of the plane of the page, while the dashed triangle represents a single bond that extends behind the plane of the page.)

in a molecule are nonpolar then the molecule is also nonpolar—like H_2, O_2, and N_2, as previously mentioned. We find, however, that dipoles are *vector quantities* in that they have *direction* in addition to magnitude. As was seen with velocity vectors in Chapter 1 and momentum vectors in Chapter 3, vector quantities can cancel or diminish one another when pointing in opposite directions. We find that oppositely-facing dipoles in a molecule may cancel to give the entire molecule a net dipole of zero. This occurs in the carbon dioxide molecule, CO_2, where the dipoles of each carbon-oxygen bond are equal but opposite to each other (Figure 17.16a). Similarly, the methane molecule, CH_4, has no net dipole because all the dipoles of the carbon-hydrogen bonds in its 3-dimensional structure are aligned so that they cancel one another (Figure 17.16b). Since carbon dioxide, CO_2, and methane, CH_4, have no net dipole they are nonpolar molecules.

Electrically neutral nonpolar molecules have relatively little attraction for one another. They are easily separated. This is why liquids composed of nonpolar molecules tend to boil at relatively low temperatures. When a liquid boils, molecules separate from one another into a gaseous phase. The weaker the attractions among molecules, the less energy is required to separate them and the lower the boiling temperature. The boiling temperatures of hydrogen, H_2, oxygen, O_2, nitrogen, N_2, carbon dioxide, CO_2, and methane, CH_4, for example, are well below room temperature. This of course is why we commonly find these substances as gases.

If the dipoles in a molecule are not exactly equal and opposite, then the molecule will have some degree of polarity. For example, the shape of a water molecule, H_2O, is bent so that the dipoles diminish, but do not cancel each other. With a water molecule, therefore, there is a net dipole. Oxygen is slightly negative and the two hydrogens are slightly positive (Figure 17.17a). Similarly, the pyramidal shape of an ammonia molecule, NH_3, results in a net dipole with the nitrogen slightly negative and the hydrogens slightly positive (Figure 17.17b). Based upon its 3-dimensional structure, we find that the organic solvent dichloromethane, CH_2Cl_2, also carries a net dipole (Figure 17.17c).

Because of their electric orientation, polar molecules attract one another much like a bunch of tiny magnets. They are relatively difficult to separate. So we find substances composed of polar molecules have typically higher boiling temperatures than those composed of nonpolar molecules (Table 17.2). Water, H_2O, for example, boils at 100°C, whereas methane, CH_4, boils at −164°C. This 264°C difference is quite astounding, for water and methane differ only by one type of atom and their masses are nearly identical.

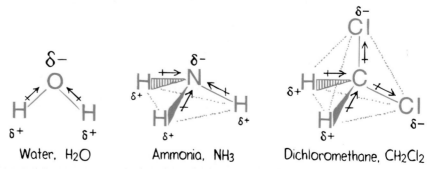

Water, H₂O   Ammonia, NH₃   Dichloromethane, CH₂Cl₂

Figure 17.17 The dipoles of these molecules add up to some net value. These molecules are polar.

TABLE 17.2   Boiling Temperatures of Some Polar and Nonpolar Substances

Polar		Nonpolar	
Hydrogen fluoride, HF	20°C	Hydrogen, H_2	−253°C
Water, H_2O	100°C	Methane, CH_4	−164°C
Ammonia, NH_3	−33°C	Nitrogen, N_2	−196°C
Dichloromethane, CH_2Cl_2	40°C	Oxygen, O_2	−183°C
		Carbon dioxide, CO_2	−79°C

UESTION

Liquids made of polar molecules tend to have boiling temperatures higher than liquids made of nonpolar molecules. How come?

Interestingly enough, we find that the attractions between polar molecules and nonpolar molecules are rather weak. As a consequence, polar and nonpolar substances, such as water and oil, are difficult to mix. The reason, however, is *not* that water and oil molecules repel each other. Instead, polar water molecules are attracted mostly to themselves and this causes them to come together into a single mass. Nonpolar oil molecules are excluded and left to themselves. In Chapter 19 we explore many other examples of how the polarity of molecules affects the general behavior of substances.

Metallic Compounds

About 85 percent of the known elements are metals. Metals are generally shiny, opaque to light, and can conduct electricity and heat. We find all metallic elements grouped to the lower left of a zig-zag diagonal line as shown in Figure 17.18.

Figure 17.18   An outline of the periodic table. Metallic elements are grouped together toward the lower left-hand side.

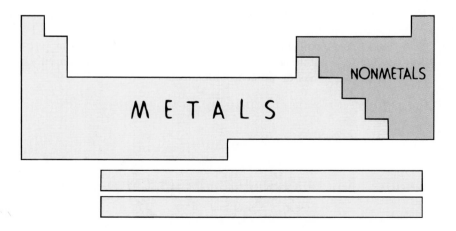

NSWER

Polar molecules stick to one another making it difficult to separate them into a gaseous phase. Nonpolar molecules, on the other hand, do not stick together so well, and separation into a gaseous phase is easier.

Naming Compounds

The naming of chemical compounds, for the most part, is quite straightforward and systematic. For some of the more complex compounds, it becomes complicated. We won't go through the task of learning all the rules for naming compounds at this point. But we will acquaint you with how the system works for many simple compounds.

Naming compounds of only two elements is easy. The name of the element farthest to the left in the periodic table is followed by the name of the element farthest to the right with the suffix −*ide*. For example:

NaCl	Sodium chloride
LiH	Lithium hydride
CaF_2	Calcium fluoride
HCl	Hydrogen chloride
MgO	Magnesium oxide
$Sr_3 P_2$	Strontium phosphide

Carrying this further, some molecules contain atoms that are ionized. These ion-containing molecules are called **polyatomic ions**. Polyatomic ions for the most part behave as single units in a chemical reaction and so are named as single units as listed below. Polyatomic ions form ionic compounds, as do monoatomic ions such as sodium or chlorine. The rules for naming polyatomic ionic compounds are the same as for monoatomic ionic compounds, except that the name of the polyatomic ion is never modified with a suffix. For example:

Polyatomic ions	
$(NH_4)^+$	Ammonium
$(OH)^-$	Hydroxide
$(CO_3)^{2-}$	Carbonate
$(NO_3)^-$	Nitrate
$(CN)^-$	Cyanide
$(PO_4)^{3-}$	Phosphate
$(SO_4)^{2-}$	Sulfate

Polyatomic ionic compounds	
NH_4Cl	Ammonium chloride
NaOH	Sodium hydroxide
$CaCO_3$	Calcium carbonate
NH_4NO_3	Ammonium nitrate
KCN	Potassium cyanide
Na_3PO_4	Sodium phosphate
$MgSO_4$	Magnesium sulfate

Prefixes are added to remove ambiguity between compounds that have different numbers of the same elements. The prefixes designate the number of atoms of the element that occur in the molecule by the following: mono- ("one"), di- ("two"), tri- ("three"), tetra- ("four").

CO	Carbon monoxide*
CO_2	Carbon dioxide
NO_2	Nitrogen dioxide
N_2O_4	Dinitrogen tetraoxide
SO_2	Sulfur dioxide
SO_3	Sulfur trioxide

Common names are used to designate many chemical compounds when the systematic names are tedious to use repeatedly, or when they're difficult to pronounce. Common names are also used because of tradition. Many compounds were named long before today's systematic approach was developed. Examples of common names include "water," "ammonia," and "methane." The systematic names for these compounds are dihydrogen oxide [H_2O], trihydrogen nitride [NH_3], and tetrahydrogen carbide [CH_4], respectively. Care should be taken in using common names, however; they are useful only to those who are already familiar with the composition and structure or identity of the compound in question. A mixture of common and systematic naming is often quite useful. For example, the name "deoxyribonucleic acid" (DNA) does not specifically tell us how all the atoms in this molecule are arranged or even what they are, but it does tell us that this molecule contains a ribose sugar ("ribo"), which is lacking an oxygen atom ("deoxy") and is connected to a type of acid ("nucleic acid").

Interestingly enough, it was Lavoisier (Section 17.1) who pioneered the systematic naming of chemical compounds. Prior to Lavoisier most chemicals were given common names that had nothing to do with chemical composition. For chemistry students the systematic method was a blessing, for it meant they no longer had to memorize the common names of the many known compounds. Instead, a compound's name could be derived systematically.

* The prefix *mono-* is typically omitted from the beginning of the name.

The outermost electrons of all metal atoms are weakly coupled with the atoms to which they belong. Consequently, they are easily dislodged, leaving behind positively charged metal ions. Dislodged electrons make up a sort of electron fluid that can move freely through the assembly of metal ions. The fluid of electrons serves to hold the oppositely charged ions together, as depicted in Figure 17.19. This action constitutes the **metallic bond,** a bond that is quite unlike the ionic or covalent bonds.

Figure 17.19 Metal ions are held together by freely flowing electrons that have been dislodged. These loose electrons form a kind of "electronic fluid" that can flow though the lattice of positively charged ions.

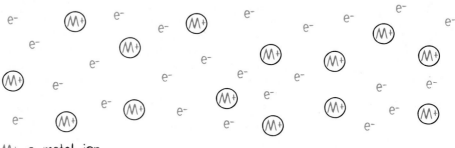

M^+ = metal ion
e^- = electron

The mobility of electrons in metals accounts for their electrical and thermal conductivities. Metals are opaque and shiny because free electrons easily vibrate to the oscillations of light when it falls upon them, effectively reflecting most of the incident light energy. Since metal atoms are not bonded in specific orientations, they are malleable (can be hammered into different shapes or bent) and ductile (can be drawn into wires).

Metal atoms mix together to form **metallic compounds**. Gold atoms, for example, may mix with palladium atoms, to form a metallic compound known as *white gold*. Because atoms in a metal are not linked by specific bonds, atoms of different metals can be combined in almost any proportion. These are **alloys**—metallic compounds composed of more than one element in any desired ratio. By playing around with proportions, the properties of the alloys can be modified. For example, sterling silver is an alloy containing 92.5 percent silver, and 7.5 percent copper. Change the proportion of silver and copper to 70 percent and 10 percent, respectively, throw in 18 percent tin, and 2 percent mercury and you have dental amalgam, which has a coefficient of thermal expansion the same as human teeth!

17.5 CHEMICAL MIXTURES

Most materials are **mixtures** of elements, chemical compounds, or both. Stainless steel, which is also an example of an alloy, is a mixture of the elements iron, chromium, nickel, and carbon. Sparkling mineral water is an example of a mixture of chemical compounds—water, mineral salts, and carbon dioxide gas. Our atmosphere is a mixture of elements nitrogen (78 percent), oxygen (21 percent), and argon (<1 percent) and small amounts of chemical compounds such as carbon dioxide, water vapor, and various pollutants.

What's in a Glass of Water?

The answer to this question is more than meets the eye, because tap water is anything but pure water. Tap water contains a variety of chemical compounds such as calcium carbonate, magnesium carbonate, calcium fluoride, chlorine disinfectants, the ions of metals such as iron and potassium, trace amounts of heavy metals such as lead, mercury, and cadmium, and trace amounts of organic compounds as well as dissolved gaseous materials such as oxygen, nitrogen, and carbon dioxide.

But there's no need to panic and go thirsty. It's unnecessary and undesirable to remove all substances from the water we drink. Some of the dissolved gases and minerals give water a pleasing taste. Many dissolved substances promote human health: The fluorine ion has been shown to protect teeth, a small amount of chlorine destroys harmful bacteria, and as much as 10 percent of our daily requirements for iron, potassium, calcium, and magnesium is obtained from ordinary drinking water. Bottoms up!

Elements can be separated from mixtures by physical means. Consider air, for example, which is a colorless liquid at −200°C. The boiling points of its major components, nitrogen and oxygen, however, differ by 13°C (−196°C and −183°C, respectively). So, if we raise the temperature of liquid air to −196°C, the nitrogen boils away, leaving the oxygen. Or, we can cool air to −183°C and watch the oxygen condense while the nitrogen remains a gas.

Chemical compounds can also be separated from mixtures by physical means. Consider sea water, which is a mixture of water and a variety of salts. The boiling point of water is 100°C, which is much less than the boiling or even the melting points of sea salts. One way to isolate water from sea water is to heat sea water to the boiling point of water. Water is vaporized and recollected elsewhere into its liquid form by condensation. This process of recollecting a vaporized substance is **distillation** (Figure 17.20). After all the water has been removed from sea water, all that remains are the dry sea salts. Sea salts, also a mixture of chemical compounds, contain a variety of valuable materials such as sodium chloride, potassium bromide, and small amounts of various metals such as gold! Further separation of the components of sea salts is of significant commercial interest.

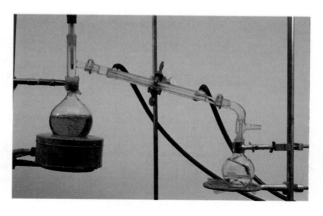

Figure 17.20 A simple distillation set-up that may be used to separate water from sea water.

Q|UESTION

Multiple Choice: Impure water can be purified by (a) removing the impure water molecules. (b) removing everything that is not water. (c) breaking down the water to its simplest components. (d) adding some disinfectant such as chlorine.

17.6 CLASSIFICATION OF MATTER

Matter can be classified as pure or impure (Figure 17.21). If the material is **pure** it consists of only a single element or chemical compound. In pure gold, for example, there is nothing but the element gold. In pure salt there is nothing but the chemical compound sodium chloride. If the material is **impure**, on the other hand, it is a mixture of more than one element or chemical compound.

Mixtures may be **heterogeneous** or **homogeneous**. In a heterogeneous mixture the different components can be seen as individual substances, like sand in water, or a mixture of oil and water. The different components are obvious. In a homoge-

Figure 17.21   The chemical classification of matter.

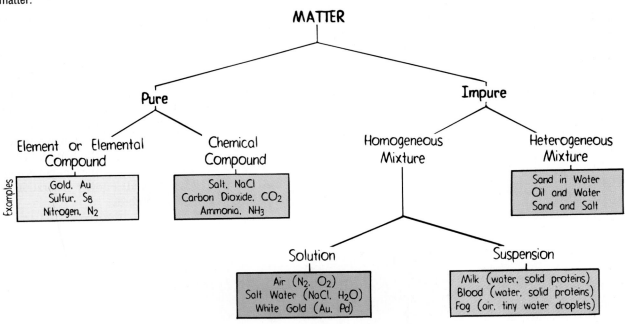

A|NSWER

(b) Water, H_2O, is a chemical compound made of the elements hydrogen and oxygen in a 2 to 1 ratio. Every H_2O molecule is exactly the same and there's no such thing as an impure H_2O molecule. Just about anything, including you, beach balls, rubber ducks, dust particles, bacteria, and even other chemical compounds can be found in water. When something other than water is found in water we say that the water is not pure. It is important to see that the impurities are *in* the water and not part *of* the water, which means that it is possible to remove them by a variety of physical means, like filtration or distillation.

Figure 17.22 A common laboratory centrifuge.

neous mixture, however, substances are mixed together so finely that they cannot be distinguished. Homogeneous mixtures have the same composition throughout—one portion of the mixture has the same ratio of substances as does any other portion. A homogeneous mixture may be either a *solution* or a *suspension*.

In a **solution,** all components are of the same phase. The atmosphere we breathe is an example of a gaseous solution that consists of the elemental compounds nitrogen and oxygen, as well as other gaseous materials. Salt water is an example of a liquid solution because both sodium chloride and water are found in a single liquid phase. An example of a solid solution is white gold, which is a homogeneous mixture of the elements gold and palladium. We will be discussing solutions in more detail in Section 19.5.

A **suspension** is a homogeneous mixture in which the different components are of different phases, like solids in liquids, or liquids in gases. In a suspension these different phases are finely mixed together. Milk is a suspension, for it is a homogeneous mixture of solid proteins finely dispersed throughout water. Blood is also a suspension, composed of finely dispersed solid proteins in water. Another example of a suspension is fog, or clouds, which are homogeneous mixtures of tiny water droplets in air. The easiest way to distinguish a suspension from a solution is to apply centrifugal forces, as with a centrifuge (Figure 17.22). A centrifuge separates suspensions, but not solutions. The common solution saltwater, for example, cannot be separated into salt and water in a centrifuge. Nor can air be separated into nitrogen and oxygen. But put blood or any suspension in a centrifuge and the different phases are separated based upon differing densities. When blood is centrifuged, the blood proteins, including red and white blood cells, are nudged outward away from the blood plasma. The plasma in turn is a solution of water and various water soluble substances, such as salt.

SUMMARY OF TERMS

Element Any material that cannot be broken down into simpler substances by chemical analysis. We find that each element consists of only one type of atom.

Conservation of mass A principle stating that matter is neither created nor destroyed in a chemical reaction, as far as we are able to detect.

Phlogiston A long-since disproven hypothetical substance that is released by a material as it burns.

Combining volumes A principle which states that volumes of gases, at the same temperature and pressure, combine chemically with one another in the ratio of small whole numbers.

Physical property The physical characteristics of a substance, such as color, density, and hardness.

Physical change A change in which a substance changes its physical properties without changing its chemical identity. The vaporization of water is an example of a physical change.

Chemical property The tendency of a substance to change chemical identity. For example, it is a chemical property of iron to change into rust.

Chemical change A change in which a substance changes its chemical identity. During a chemical change atoms are rearranged to give a new substance.

Periodic table A tabular listing of all the known elements.

Atomic symbol The first letter or letters of the name of an element used to denote that element.

Elemental compound A material in which only atoms of the same element are chemically bonded to one another.

Elemental formula A notation that uses the atomic symbol and a numerical subscript to denote the composition of an elemental compound.

Chemical compound A material in which atoms of different elements are chemically bonded to one another.

Chemical formula A notation used to denote the composition of a chemical compound. In a chemical formula the atomic symbols for the different elements of the compound are written together along with numerical subscripts that indicate their proportions.

Ion An atom that has an excess or deficiency of electrons, compared to protons. An ion is electrically charged positive if it lacks electrons, and negative if it has excess electrons.

Ionic compound Any chemical compound containing ions.

Ionic bond The electrical force of attraction that holds ions of opposite charge together.

Ionic crystal A group of many ions held together in an orderly and periodic 3-dimensional array.

Covalent bond A chemical bond in which atoms are held together by their mutual attraction for two electrons that they share.

Covalent compound An elemental or chemical compound in which atoms are held together by the covalent bond.

Double bond Two covalent bonds between the same two atoms.

Triple bond Three covalent bonds between the same two atoms.

Molecule Any group of atoms held together by covalent bonds.

Covalent crystal A group of molecules arranged in an orderly and periodic fashion.

Dipole A separation of charge.

Polar The state of a chemical bond or molecule having a dipole.

Nonpolar The state of a chemical bond or molecule having no dipole.

Metallic bond A nonspecific chemical bond in which metal atoms are held together by their attraction to a common pool of electrons.

Metallic compound An elemental or chemical compound containing metal atoms that are held together by the metallic bond.

Polyatomic ion An ionized molecule.

Chemical mixture The combination of different elements, elemental or chemical compounds.

Distillation The process of recollecting a vaporized substance.

Pure The state of a material that consists of only a single element or chemical compound.

Impure The state of a material that consists of more than one element, elemental or chemical compound. A chemical mixture is impure.

Heterogeneous A chemical mixture in which the different components can be seen as individual substances.

Homogeneous A chemical mixture composed of components so finely mixed that the composition throughout is the same.

Solution A homogeneous mixture in which all components are of the same phase. Salt water is an example of a solution.

Suspension A homogeneous mixture in which different components are of a different phase. Milk is an example of a suspension.

REVIEW QUESTIONS

A Brief History of Chemistry

1. In what ways was Aristotle's erroneous model for the composition of matter a remarkable achievement?

2. How did Antoine Lavoisier define an element?

3. How did John Dalton define an element?

4. What evidence supported Lavoisier's principle of the conservation of mass?

5. How did Lavoisier's data refute the phlogiston model?

6. What led Dalton to propose his postulates on the atomic theory of matter?

7. Why were people so reluctant to accept Avogadro's conclusion that equal volumes of gases under the same conditions have equal numbers of molecules?

8. What did Proust contribute to the atomic theory of matter?

Physical and Chemical Properties

9. How is a physical change different from a chemical change?

10. How is a chemical property much like one's personality?

Elements and Compounds

11. What distinguishes one element from another?

12. What is so unique about an elemental compound?

13. Which elements are some of the older known elements? What is your evidence?

14. What is the difference between an elemental and a chemical compound?

15. What is the number of atoms of each element and the total number of atoms per molecule in H_3PO_4?

16. Eat sodium and you die. Inhale chlorine and you die. But mix them together and you sprinkle them on your tomatoes for better taste. What is going on?

Ionic, Covalent, and Metallic Compounds

17. What elements of the periodic table tend to form ionic compounds?

18. Suppose an oxygen atom gains two electrons to become an oxygen ion. What is the electrical charge of this ion?

19. What is an ionic crystal?

20. By what means are the atoms of a covalent bond held together?

21. Which is the true statement: Atoms are composed of molecules; or, molecules are composed of atoms?

22. How many electrons are shared per covalent bond?

23. How many electrons are shared in a triple covalent bond?

24. Cite an example of a molecule easily seen by the naked eye.

25. How is frozen water different from liquid water?

26. What is a dipole?

27. Does a dipole have a net electric charge? Does it have electric properties?

28. Which is more polar, a carbon-oxygen bond or a carbon-nitrogen bond?

29. An individual carbon-oxygen bond is polar. Yet carbon dioxide, CO_2, which has two carbon-oxygen bonds, is nonpolar. Why is this so?

30. Why do nonpolar substances tend to boil at relatively low temperatures?

31. How do the outer electrons in metal atoms differ from the outer electrons of nonmetal atoms?

32. How are metal atoms held together in the metallic bond?

33. Your friend says all chemical bonds are electrical. Do you agree or disagree?

34. What is an alloy?

Naming Compounds

35. What is the name for the chemical compound of the formula KF? How about TiO_2?

36. Why are common names often used instead of systematic names?

37. What is the advantage of using a systematic name over a common name?

Chemical Mixtures

38. What is the difference between a chemical mixture and a covalent compound?

39. How is it possible to separate the various components of a mixture?

40. How might you separate a mixture of hydrogen, H_2, and methane, CH_4? Use data from Table 17.2.

Classification of Matter

41. Classify the following as A. Homogeneous mixture, B. Heterogeneous mixture, C. Elemental Compound, or D. Chemical Compound.

Milk: _____   Steel: _____

Ocean Water: _____   Blood: _____

Sodium: _____   Planet Earth: _____

42. By what means can a solution and a suspension be distinguished?

·······································

HOME PROJECTS

1. Tap water contains a variety of chemical compounds other than water, such as calcium carbonate, $CaCO_3$, and calcium fluoride, CaF_2. Add water from a full glass to a clean cooking pot and boil the water to dryness. Be sure to turn off the burner when the water is almost gone. Examine the resulting residue by scraping it with a knife. This is what you ingest with every glass of tap water you drink. As the box on page 417 suggests, however, this is not necessarily bad. We consume about 10 percent of our mineral requirements from the water we drink.

2. Stuff a nonsoapy steel wool cleansing pad into the bottom of a narrow jar, such as an olive jar. A soapy pad can be rinsed soap-free, and a pad that doesn't fit through the jar opening may be cut into strips with a pair of scissors. Invert the narrow jar into a wider jar that is a quarter filled with water. Note the water level in the inverted narrow jar. Leave the entire set-up alone for a couple of hours or until the wool, which is primarily iron, has gained some rust. What has happened to the water level in the inverted jar? How come? Hint: Iron is simply an element. Rust, however, is a chemical compound consisting of iron and oxygen atoms in a 2:3 ratio, Fe_2O_3.

3. Compare the chemical reactivity of various metals. Place a piece of sponge soaked in salty water at the bottom of a wide-mouth jar. On the sponge place a shiny copper penny (pre-1982), a shiny copper/zinc penny (1982 and later), a silver dime (pre-1965), a silver/copper dime (1964 and later), a nickel, an iron nail, a small piece of aluminum foil, and, if you're daring, a piece of gold, or gold-plated jewelry. Avoid contact among the metals and use more than one jar if necessary. Close the jar and observe the metal pieces daily for a couple of days and note the extent of any chemical changes. Which metals are more chemically reactive? Which are less?

4. Use toothpicks and different color jelly beans or gum drops to build models of the molecules shown in Figures 17.16 and 17.17. Assign a certain color to represent each type of atom. As a general rule, nonpolar molecules are always more symmetric than polar molecules. Your models should show this in that the atoms of nonpolar molecules appear more evenly distributed. Polar molecules, on the other hand, appear lopsided. If ammonia, NH_3, were flat, rather than puckered, might you expect it to be a polar or nonpolar molecule?

• •

EXERCISES

1. Briefly explain how chemistry and alchemy were different. In what sense did chemistry not arise from alchemy?

2. Match the following scientists with their description:

   Aristotle _____ (a) Said Dalton's version of water was wrong

   Cavendish _____ (b) Kept mice alive longer in jars

   Lavoisier _____ (c) An English Physicist

   Priestley _____ (d) Clued Dalton into thinking about the atom

   Proust _____ (e) Influenced 20 centuries of intellectual thought

   Gay-Lussac _____ (f) Often considered to be the "Father of Modern Chemistry"

3. Describe how Lavoisier employed the scientific approach with one of his accomplishments (i.e., observation, hypothesis, experiment, theory).

4. Classify the following changes as physical or chemical.

   (a) grape juice turns into wine . . . . _____
   (b) wood burns to ashes . . . . _____
   (c) photographic film is exposed to light . . . . _____
   (d) water begins to boil . . . . _____
   (e) a broken leg mends itself . . . . _____
   (f) grass grows . . . . _____
   (g) an infant gains 10 pounds . . . . _____

5. Why is it impossible for us to breathe water when by mass 88.88 percent of water is oxygen?

6. Do chemical compounds have the same physical and chemical properties as the elements from which they are made? Give an example to support your answer.

7. What elements tend to form ionic bonds?

8. If magnesium ions have a +2 charge and chlorine ions have a −1 charge, what would be the chemical formula for the ionic compound magnesium chloride?

9. How are ionic and covalent crystals different?

10. Cite four sample elements that tend to form covalent bonds.

11. Classify the following molecules as polar or nonpolar:

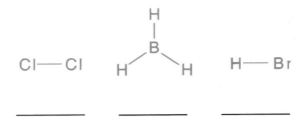

_____   _____   _____

12. Which is the most polar bond in the following molecules:

   (a) H—N (b) N—C (c) C—O
   (d) C—C (e) O—H (f) C—H

$$H—O—\underset{\underset{H}{|}}{\overset{\overset{H}{|}}{C}}—\underset{\underset{H}{|}}{\overset{\overset{H}{|}}{C}}—N—H$$

13. List the following in order of increasing boiling temperatures (1 = lowest and 3 = highest). Briefly explain your reasoning.

Compound A

Compound B

Compound C

14. Water, H_2O, and methane, CH_4, have about the same mass and differ by only one type of atom. Why are their boiling temperatures so different?

15. Classify the following bonds as *ionic, covalent,* or *metallic* (O, atomic #8; F, atomic #9; Na, atomic #11; Cl, atomic #17; U, atomic #92).

O with F	Ca with Cl
Na with Na	U with Cl

16. What is the name for the chemical compound of the formula NH_4CN? How about $(NH_4)_2SO_4$?

17. Suggest how you might separate a heterogeneous mixture of sand and salt. How about iron and sand?

. . . .

PROBLEMS

1. How many grams of water will form from the combination of 25 grams of hydrogen and 225 grams of oxygen? How much of which element will be left over?

2. Nitrogen and hydrogen react in a 14 to 3 ratio by mass to form ammonia. How many grams of ammonia will be formed from the combination of 7 grams of nitrogen and 6 grams of hydrogen? How many grams of which element will be left over?

18

THE PERIODIC TABLE

You go to the pantry to get ingredients for a cake. You get sugar, flour, eggs, milk, baking soda plus an assortment of other ingredients, all of which you will measure and mix in some particular fashion. But if you change your measurements you can create any of a large number of different products—pancakes, muffins, biscuits, or who knows what—instead of a cake.

Nature is a bit like a baker. The ingredients in nature's pantry are the elements. Different combinations of the elements give rise to the diversity of materials in our environment. Change the arrangement of elements and you have a varied assortment of substances. Cotton is made primarily of carbon, hydrogen, and oxygen. The same elements combined in a different way make sugar. Arranged still differently, we have a potato.

This chapter is all about nature's elements. Similarities and differences in their characters place them into a grand organization—the **periodic table** (see insert). The periodic table of the elements is to a chemist what a dictionary is to a writer. Both are useful as references and should always be available. Neither, of course, should be memorized. The focus of this chapter is a tour of the periodic table, which can be your roadmap to understanding chemical processes—the subject of the following three chapters. We begin by organizing the elements.

18.1 ORGANIZING THE ELEMENTS

A cook organizes spices in a spice rack. We'll do the same sort of thing with the elements. The simplest way to organize the elements is to list them in a vertical column according to increasing atomic number. The list begins with hydrogen (Table 18.1).

If various physical or chemical properties of the elements are listed alongside their atomic number, some interesting trends become apparent. First, consider the atomic sizes of the elements as measured by *atomic radius*. We find that the atomic radii gradually get smaller with increasing atomic number. This is understandable, for the greater nuclear charge pulls electrons closer to the nucleus. At certain intervals, however, the atomic radius of an element is dramatically *greater* than that of the previous element. This occurs for the elements lithium [Li], sodium [Na], potassium [K], rubidum [Rb], cesium [Cs], and francium [Fr]. We see these jumpy intervals when atomic radius is plotted against atomic number (Figure 18.1a).

A similar pattern arises when we look at the *ionization energy* of an element. Ionization energy is the amount of energy required to pull an electron away from an atom. We find that the ionization energy gradually increases with increasing atomic

TABLE 18.1   Elements Listed by Atomic Number

Element	Atomic Number	Atomic Radius (picometers)	Ionization Energy (kJ/mole)
[H]	1	37	1300
[He]	2	50	2380
[Li]	3	152	520
[Be]	4	111	900
[B]	5	88	800
[C]	6	77	1100
[N]	7	70	1400
[O]	8	66	1310
[F]	9	64	1700
[Ne]	10	70	2050
[Na]	11	186	480
[Mg]	12	160	700
[Al]	13	143	600
[Si]	14	117	800
[P]	15	110	1010
[S]	16	104	1000
[Cl]	17	99	1280
[Ar]	18	94	1540
[K]	19	231	400
⋮	⋮	⋮	⋮

Figure 18.1 (a) The atomic radii of elements gradually decrease with increasing atomic number, but occasionally increase sharply. (b) The ionization energy of elements gradually increases with increasing atomic number, but occasionally decreases sharply. Note that the discontinuities occur in the same places for both plots.

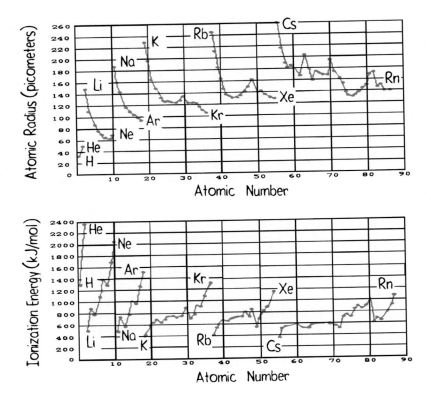

number. But at the same break points where atomic radii increase there is a dramatic *decrease* in ionization energy. A plot of ionization energy versus atomic number results in the same set of intervals (Figure 18.1b).

Other properties of the elements follow the same pattern. Therefore it is natural to group the elements according to these intervals in our effort to organize them, as we have done in Table 18.1. Similarly, we can list the elements in a series of horizontal rows such that elements of the same interval appear in the same row. Each row is commonly referred to as an **atomic period**. When we place the seven known atomic periods on top of one another we get a list that looks like the following:

1st Period [H] [He]
2nd Period [Li] [Be] [B] [C] [N] [O] [F] [Ne]
3rd Period [Na] [Mg] [Al] [Si] [P] [S] [Cl] [Ar]
4th Period [K] [Ca] [Sc] [Ti] [V] [Cr] [Mn] [Fe] [Co] [Ni] [Cu] [Zn] [Ga] [Ge] [As] [Se] [Br] [Kr]
5th Period [Rb] [Sr] [Y] [Zr] [Nb] [Mo] [Tc] [Ru] [Rh] [Pd] [Ag] [Cd] [In] [Sn] [Sb] [Te] [I] [Xe]
6th Period [Cs] [Ba] [La*] [Hf] [Ta] [W] [Re] [Os] [Ir] [Pt] [Au] [Hg] [Tl] [Pb] [Bi] [Po] [At] [Rn]
7th Period [Fr] [Ra] [Ac*] [Unq] [Unp] [Unh] [Uns] [Uno] [Une]

Note that different periods include different numbers of elements. We see, for example, the first period has only two elements, the second and third periods each have eight elements.*

---

*The latter elements in the 7th period, Unq–Une, have been named systematically, rather than after some place or person. Elements heavier than these have not been discovered.

This horizontal listing of the elements leads to further organization. We find similar physical and chemical properties for all the first elements of each period. For instance, hydrogen [H], lithium [Li], sodium [Na], potassium [K], rubidum [Rb], cesium [Cs], and francium [Fr] share many of the same chemical and physical properties.

With some slight modifications we are able to generate even more organization. The second element of the first period, helium [He], has nothing in common with the second elements of the other periods, which all have very similar properties. Helium, however, is an inert gas as are the elements neon [Ne], and argon [Ar]. For these reasons, we slide the position of helium toward the right.

1st Period [H] ————————slide right————————→ [He]
2nd Period [Li] [Be] [B] [C] [N] [O] [F] [Ne]
3rd Period [Na] [Mg] [Al] [Si] [P] [S] [Cl] [Ar]
4th Period [K] [Ca] [Sc] [Ti] [V] [Cr] [Mn] [Fe] [Co] [Ni] [Cu] [Zn] [Ga] [Ge] [As] [Se] [Br] [Kr]
5th Period [Rb] [Sr] [Y] [Zr] [Nb] [Mo] [Tc] [Ru] [Rh] [Pd] [Ag] [Cd] [In] [Sn] [Sb] [Te] [I] [Xe]
6th Period [Cs] [Ba] [La*] [Hf] [Ta] [W] [Re] [Os] [Ir] [Pt] [Au] [Hg] [Tl] [Pb] [Bi] [Po] [At] [Rn]
7th Period [Fr] [Ra] [Ac*] [Unq] [Unp] [Unh] [Uns] [Uno] [Une]

Also, for the purpose of grouping elements with similar properties above or below one another, we slide helium along with the second period elements boron [B], through neon [Ne], and the third period elements aluminum [Al], through argon [Ar], to the far right.

1st Period [H] [He]
2nd Period [Li] [Be] ————————slide right————————→ [B] [C] [N] [O] [F] [Ne]
3rd Period [Na] [Mg] ————————slide right————————→ [Al] [Si] [P] [S] [Cl] [Ar]
4th Period [K] [Ca] [Sc] [Ti] [V] [Cr] [Mn] [Fe] [Co] [Ni] [Cu] [Zn] [Ga] [Ge] [As] [Se] [Br] [Kr]
5th Period [Rb] [Sr] [Y] [Zr] [Nb] [Mo] [Tc] [Ru] [Rh] [Pd] [Ag] [Cd] [In] [Sn] [Sb] [Te] [I] [Xe]
6th Period [Cs] [Ba] [La*] [Hf] [Ta] [W] [Re] [Os] [Ir] [Pt] [Au] [Hg] [Tl] [Pb] [Bi] [Po] [At] [Rn]
7th Period [Fr] [Ra] [Ac*] [Unq] [Unp] [Unh] [Uns] [Uno] [Une]

The resulting configuration produces the *periodic table*. The organization of elements in the periodic table is astonishing for we find that not just a few, but *all* elements listed directly above or below one another share similar physical and chemical properties. For example, all elements of the last column, helium [He] through radon [Rn], are inert gases; all elements of the second-to-last column, fluorine [F] through astatine [At], readily form ionic compounds with the elements of the first two columns—hydrogen [H] through francium [Fr], and beryllium [Be] through radium [Ra].

A cook finds it useful to group foods together based on their similar properties. Spices, for example, may be placed together near the stove, fresh vegetables are put into the bottom drawer of the refrigerator, and dry ingredients are stored in a cool pantry. Likewise, it is useful to group elements together based upon their similar physical or chemical properties. The periodic table is naturally organized for this purpose; elements with similar properties are found in the same vertical columns.

Dmitri Mendeleev—Father of the Periodic Table

Science is the search for patterns and regularities in nature. More than a century ago many scientists working independently had noted how the physical and chemical properties of elements tended to repeat themselves with increasing atomic mass. Many charts were produced in which elements of similar properties were listed close to one another. The chart upon which our modern periodic table is based was first created in 1869 by a Russian chemistry professor, Dmitri Mendeleev. Mendeleev's chart was unique in that it resembled a calendar with all the Sundays in one column, all the Mondays in the next, and so on, with the days of the week in rows. Down each column Mendeleev placed elements of similar properties and across each row he placed all the elements that appeared within one period of repeating properties. What was particularly interesting about Mendeleev's table was that there were obvious gaps—blank spaces that could not be filled by any known element. Instead of looking upon these blank spaces as defects, Mendeleev boldly predicted the existence of elements that had not been discovered. Furthermore, he predicted the properties of some of those missing elements, which ultimately led to their discoveries!

18.2 ATOMIC GROUPS

Each of the 18 vertical columns of elements in the periodic table is called an **atomic group**. Since the elements within an atomic group have similar properties, we can get an overview of the properties of all the elements by learning the general characteristics of each atomic group.

Atomic groups are numbered from left to right across the periodic table. We'll consider the groups in sequence. We'll see that many atomic groups have traditional names that describe the properties of the elements within them.

Group 1: Alkali Metals

Figure 18.2 Ashes and water make for a slippery alkaline solution once used to clean hands.

Early in human history people discovered that ashes mixed with water produced a slippery solution that was useful for removing grease. By the Middle Ages such mixtures were described as *alkaline,* which is derived from the Arab word for ashes, *al-qali.* Alkaline mixtures found many uses, particularly in the original preparation of soaps. We now know that alkaline ashes contain compounds of group 1 elements, most notably the compound potassium carbonate (potash). Since group 1 elements are metallic, we now call them the **alkali metals**.

The element hydrogen [H], is commonly found on earth in the gaseous phase and is not normally considered a metal. But at high enough pressures, hydrogen takes on the general properties of a metal—electrical conductivity, luster, and opaqueness to light. Hydrogen has recently been pressed to the liquid-metal phase in the laboratory. Further it is believed that hydrogen exists naturally as a liquid metal deep beneath the surfaces of giant planets, such as Jupiter and Saturn (Figure 18.4). These

The Periodic Table

Atomic masses are averaged by isotopic abundance in the earth's surface, expressed in atomic mass units (amu). Atomic masses for radioactive elements shown in parentheses are the whole number nearest the most stable isotope of that element.

The Periodic Table

Keyboard to the Elements

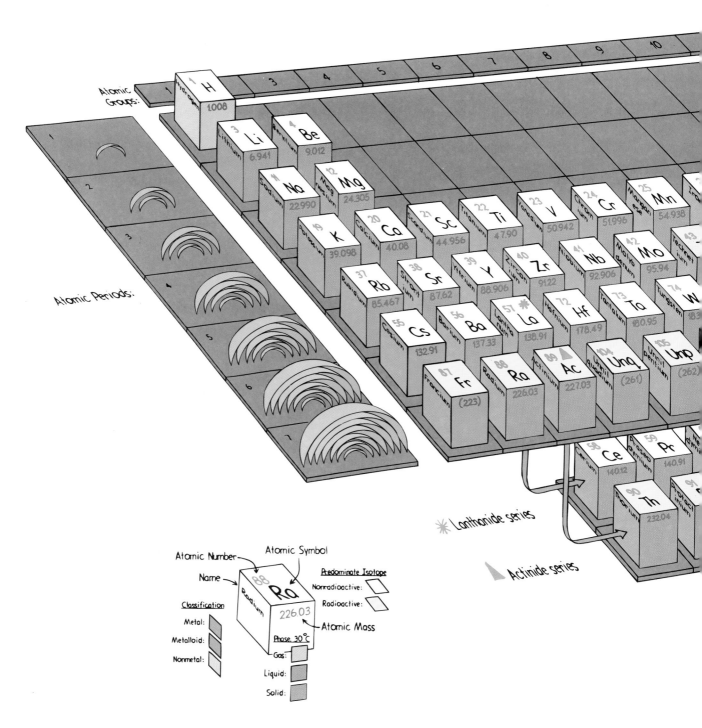

Atomic Groups:

Atomic Periods:

* Lanthanide series

◣ Actinide series

Atomic Number

Atomic Symbol

Name

Classification

Predominate Isotope
Nonradioactive:
Radioactive:

88
Ra
Radium
226.03

Atomic Mass

Phase, 30°C
Gas:
Liquid:
Solid:

Metal:
Metalloid:
Nonmetal:

The Periodic Table

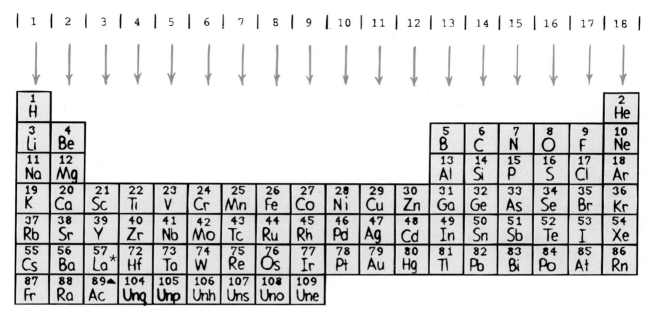

Figure 18.3 Each vertical column is an *atomic group,* having similar chemical and physical properties.

planets are composed mostly of hydrogen. Inside them, internal pressures exceed 3 million times earth's atmospheric pressure. At this pressure hydrogen would be pressed to a liquid-metal phase. Scientists think that abundant electric currents in the liquid-metal hydrogen produce the enormous magnetic fields monitored by space probes about these planets. Back here on earth, hydrogen is normally a gas.

All alkali metals tend to lose one electron to form the +1 ion. Many of these ions play important roles in our health. For example, strong concentrations of the hydrogen ion, H^+, are found in our stomachs where they help digest food. Trace amounts of the lithium ion, Li^+, are important for mental health; people suffering from psychological depression are sometimes treated with lithium compounds. The sodium ion, Na^+, is important in many ways. It affects water retention and plays a role in the transmission of nerve signals. The potassium ion, K^+, does the same. We ingest alkali ions in the form of salts such as sodium chloride, $NaCl$, or potassium chloride, KCl. Potassium chloride and sodium chloride have such similar properties that potassium chloride is often used as a salt substitute. Anyone who watches the news knows that too much or too little of these salts in our diet can be detrimental to our health so it is important that our intake be well balanced.

Figure 18.4 Jupiter's interior is thought to be made of liquid metallic hydrogen.

Group 2: Alkali-Earth Metals

Medieval alchemists noted that certain minerals (which we now know are made up of group 2 elements) would not melt nor change when put in fire. These fire-resistant substances were known to the alchemists as "earth." As a lay-over from these ancient times, we call this group of elements the **alkali-earth metals**. The group 2 elements are similar to the group 1 elements. Both groups are metals and form compounds that are alkaline in water. The oceans of the earth, for example, are slightly alkaline primarily due to the presence of calcium carbonate, $CaCO_3$, which contains the group 2 element calcium. Group 2 elements form a number of common minerals such as calcite, $CaCO_3$, magnesite, $MgCO_3$, and gypsum, $CaSO_4 \cdot 2\,H_2O$.

Figure 18.5 Brick mortar, which is a mixture of lime, CaO, sand, and water, has been used for thousands of years. The mortar hardens as it reacts with atmospheric carbon dioxide to make limestone, $CaCO_3$, reinforced by sand particles.

Calcium is the fifth most abundant element in the earth's crust. It is found in vast deposits of limestone, which is a form of calcium carbonate, $CaCO_3$ (more about this in Chapter 22). Calcium carbonate is also the principal substance in sea shells, corals, and chalk. When limestone is heated to high temperatures it is transformed into lime. Lime is the compound calcium oxide, CaO. Each year about 35 billion pounds of lime are produced in the United States. Almost half this amount is consumed in the manufacture of steels. The other half finds its use in the "softening" of hard water, a process that involves the removal of excess calcium carbonate. Lime is also the key ingredient of brick mortar (Figure 18.5).

All alkali-earth metals tend to lose two electrons to form the +2 ion, and like the group 1 elements, they are important to our health. Most notably, the calcium ion, Ca^{2+}, is the principal constituent of our bones. It's also involved in muscle contraction. The magnesium ion, Mg^{2+}, is the central component of chlorophyll, the molecule in plants that makes it possible for sunlight to transform carbon dioxide and water into sugar (Figure 18.6).

Groups 3–12: The Transition Metals

The elements of groups 3 through 12 are all metals that do not form alkaline solutions with water. These metals tend to be harder than the alkali metals and less reactive with water, hence they are used for structural purposes. Collectively they are known as the **transition metals,** a name that denotes their central position in the periodic table.

The transition metals include some of the most familiar and important elements, such as iron, copper, nickel, chromium, silver, and gold. They also include many lesser-known elements that are nonetheless important in modern technology. Persons with hip implants appreciate transition metals titanium, molybdenum, and manganese for these metals are used to make alloys for the implant devices. The properties of transition metals are diverse, although those in the same atomic group have quite similar properties. We find, for example, that the "coinage" metals copper [Cu], silver [Ag], and gold [Au], are all within group 11.

Our standard of living and economy are completely dependent upon the production of transition metals. We get them either from natural sources or through recycling. Approximately 500 million tons of iron, 8 million tons of copper, and 750,000 tons of nickel are produced each year worldwide for the manufacture of coins, cars, appliances, bridges, buildings, and other metal-requiring commodities. Production involves the mining of naturally occurring ores and subsequent chemical processing. Many transition metal compounds are highly colored. This makes them useful as pigments in paints and dyes. The white of this page, for example, is produced by the presence of titanium dioxide, TiO_2. Even the colors of rubies, sapphires, and other gemstones are due to trace quantities of transition metals such as chromium and iron.

Transition metals tend to lose one or more electrons to form a variety of different ions. Iron, for example, may lose two electrons to form the Fe^{2+} ion, or three electrons to form the Fe^{3+} ion. These different states of the same element have remarkably different properties. Fe^{2+}, for example, is used in blood to carry oxygen. Fe^{3+}, on the other hand, is a component of rust. Many transition elements, in their ionized state, are essential for bodily functions (Table 18.2). We require less of some than others, and as with anything we ingest, their quantities need to be well balanced. Too much of these elements in our diet may do more harm than good.

Figure 18.6 Green leafy vegetables are a good dietetic source of magnesium.

TABLE 18.2   Transition Metal Ions in Our Diet

Transition Metal (RDA)	Function	Some Sources
Chromium, Cr^{3+} (< 0.2 mg)	Sugar tolerance	Whole wheat, honey
Cobalt, Co^{3+} (< 1.8 mg)	In vitamin B_{12}	Meats, dairy products
Copper, Cu^{2+} (2–3 mg)	Metabolism	Seafood, copper pipes
Iron, Fe^{2+} (12 mg)	In hemoglobin	Dried fruit, meats, eggs
Manganese, Mn^{2+} (2–5 mg)	Skin pigment	Nuts, spinach
Molybdenum, Mo^{3+} (< 0.5 mg)	Protein synthesis	Vegetables, liver
Vanadium, V^{3+} (< 0.5 mg)	Metabolism	Vegetables
Zinc, Zn^{2+} (15 mg)	Metabolism	Yeast, oysters, nuts

Figure 18.7   Iron is by far the most widely used transition metal.

After lanthanum [La] (atomic number 57) in the sixth period we find a subset of 14 metallic elements (atomic numbers 58–71) that are quite unlike any of the other transition elements. A similar subset (atomic numbers 90–103) is found after actinium [Ac] (atomic number 89) in the seventh period. These two subsets are the **inner transition metals**. Since they are so unique, they are shown separately from the main body of the periodic table (Figure 18.8).

The sixth period inner transition metals that fall after lanthanum are the **lanthanides.** The physical and chemical properties of all lanthanides are very similar. As a consequence, they tend to occur mixed together in the same geologic zones. Also, because of their similarities, lanthanides are unusually difficult to purify.

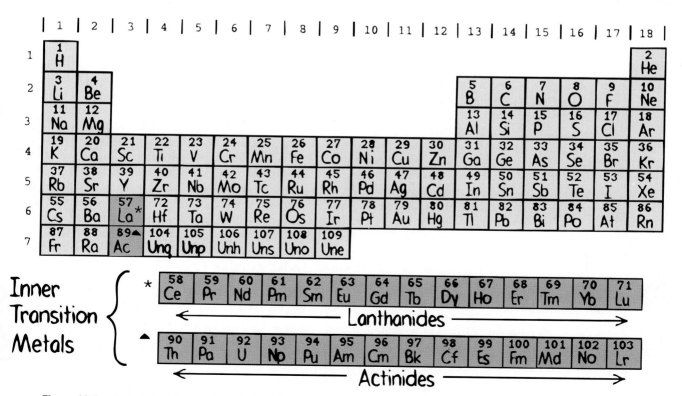

Figure 18.8   The sixth and seventh periods of the periodic table each include a subset of elements, the lanthanides and the actinides. They are not included in the main body of the periodic table because of their unique properties.

However, recently they have been used in an increasing number of applications. For example, several lanthanide elements are used in the fabrication of the flat TV screens that will soon be hung on walls like paintings.

The seventh period inner transition metals that fall after actinium are the **actinides.** They too are a group of elements with similar properties and, hence, are not easily purified. The nuclear power industry faces this obstacle because it requires purified samples of two of the most publicized actinide elements, uranium [U] and plutonium [Pu]. Actinides heavier than uranium are not found in nature but are synthesized in nuclear reactions.

Groups 13 and 14

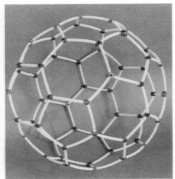

Figure 18.9 A model of a buckyball molecule made of 60 interconnected carbon atoms.

Groups 13 and 14 include aluminum and silicon. These are the third and second most abundant elements of the earth's crust respectively. With the exception of carbon, none of the elements in groups 13 and 14 occur naturally in an elemental state, that is, in a form where they are not combined with other elements. Nearly all the earth's silicon is combined with oxygen as silicon dioxide, SiO_2. In its purest form SiO_2 is known as quartz. Likewise, most aluminum is combined with oxygen as aluminum oxide, which further combines with water to form the mineral bauxite, $Al_2O_3 \cdot H_2O$. Carbon combines with a large variety of other elements, such as hydrogen, oxygen, nitrogen, sulfur, and chlorine, to form a class of compounds that compose living organisms. These are the *organic* compounds and their chemistry is discussed in Chapter 21. Carbon in nature can be found in its elemental state either as diamond or as graphite. There is also a third newly discovered state of carbon in which carbon atoms bond together to form molecules that resemble soccer balls. These are the so-called *buckyballs,* which are the subject of much research due to their unusual and potentially useful properties (Figure 18.9).

The elements of groups 13 and 14 have many applications. Elemental silicon has its greatest application in the manufacture of computer chips, as has its fellow group 14 element germanium. Metallic aluminum has many uses as a structural material. Interestingly enough, pure aluminum is soft and weak, so it is rarely used. To strengthen aluminum it is alloyed with small amounts of other metals such as copper, magnesium, and manganese. Microcapsules of aluminum and silicon oxides surrounded by water make up clay. Clay can be shaped easily because the water serves as a lubricant allowing the microcapsules to slip over one another. When dry, the microcapsules become locked in position and the clay holds its shape. Heating the dried clay to high temperatures causes the microcapsules to bond to one another. At this point, the clay is transformed into a type of hard and water-resistant ceramic.

In general, a ceramic is a solid that has been hardened by heating to high temperatures. Most ceramics contain nonmetallic elements such as oxygen, carbon, or silicon, but they may also include metallic elements, such as aluminum or a transition metal. Ceramics tend not to be metallic in character. For example, unlike metals they cannot be pounded into thin sheets or drawn into wires. They tend to fracture as anyone who has dropped a ceramic dinner plate knows. But ceramics can be hard. An example of a ceramic that is almost as hard as diamond is silicon carbide, SiC, also known as carborundum. This material is able to withstand very high temperatures and is now used to make parts for ceramic automobile engines. Recall from Chapter 8 that the high temperatures of ceramic engines increase efficiency providing significant fuel savings. Ceramics are presently an area of intense research, and are often used in place of metals. For example, at many hardware stores you can now buy ceramic knives and scissors with improved resistance to fracturing.

Figure 18.10 Wine kept open under a nitrogen atmosphere will not degrade.

Figure 18.11 All these products contain phosphates.

Figure 18.12 Stockpiles of mined sulfur.

Group 15: The Nitrogen Group

The air we breathe is composed of 78 percent nitrogen, the lightest element of group 15. Atmospheric nitrogen occurs as a diatomic molecule, N_2. Since this form of nitrogen is relatively inert, it is used quite extensively as the gas under which foods are packaged and wine is bottled (Figure 18.10). Nitrogen in the form of ammonia, NH_3, is an essential nutrient for all plants. Most plants, however, are unable to produce ammonia and they rely on external sources. The only significant natural source of ammonia is from a process called *nitrogen fixation*. This is the biological transformation of atmospheric nitrogen into ammonia. Only blue-green algae and some field crops such as alfalfa and soybeans are capable of nitrogen fixation. Many investigators recognized in the early 20th century that nitrogen fixation alone might not be sufficient for growing crops large enough to feed the growing human population. Then in 1914 two German chemists, Fritz Haber and Carl Bosch, developed a means for preparing large amounts of ammonia. Today, in the United States alone, over 34 billion pounds of ammonia are produced annually for fertilizer and other useful items.

Of all the elements in group 15, phosphorus is the most abundant in the earth's crust, ranking 11th among all the elements. It occurs primarily bonded to oxygen in the form of phosphates, which make useful fertilizers. Phosphates are also added to soft drinks to add bite to their flavor, to hot cereals and rice to make them cook within minutes, and to detergents to help soften water (Figure 18.11). Washing with phosphate-containing detergents, however, is to be discouraged. Since phosphates are plant nutrients, they cause an overgrowth of vegetation in rivers and streams that receive them. This overgrowth decreases dissolved oxygen in the water, which causes the gradual elimination of marine life.

One example of the similarity among group 15 elements is that they all react with hydrogen to produce compounds in which there are three hydrogen atoms per molecule. Nitrogen, for example, forms ammonia, NH_3. Similarly, phosphorus forms PH_3, arsenic forms AsH_3, antimony forms SbH_3, and bismuth forms BiH_3.

Group 16: The Oxygen Group

The air we breathe is composed of 21 percent oxygen, the lightest element of group 16. Atmospheric oxygen occurs as a diatomic molecule, O_2. Unlike atmospheric nitrogen, diatomic oxygen is rather reactive. It causes wood, oil, and gasoline to burn when ignited. It makes our cells function, and our skin wrinkle. Oxygen is the most abundant element in the earth's crust where it occurs bonded to other elements such as silicon, aluminum, and iron. Interestingly enough, the earth's crust is about 46 percent oxygen by mass. The other major reservoir of oxygen on this planet is in the oceans. There it is bonded to hydrogen in the form of water.

The second most common element in group 16 is sulfur [S]. Sulfur, like oxygen, forms compounds with many other elements. Familiar sulfur minerals include pyrite (fool's gold), FeS_2, cinnabar, HgS, and galena, PbS. Sulfur combines with phosphorus to produce the compound P_4S_3, which is the essential ingredient in "strike anywhere" matches. Sulfur also combines with arsenic to produce the compound As_2S_3, the form of arsenic used by ancient physicians and assassins. Vast amounts of sulfur are found near the Gulf of Mexico (Figure 18.12). About 88 percent of all sulfur is used to make sulfuric acid. Sulfuric acid is produced in far larger quantities than any other product of the chemical industry—about 80 billion pounds annually. Most of this acid goes toward the manufacture of metals and various other chemical products.

Figure 18.13 The element selenium gives a traffic light its bright red color.

Sulfur-containing mineral deposits commonly contain minor quantities of other group 16 elements, such as selenium [Se] and tellurium [Te]. They are found together in nature because of their chemical similarities. Selenium has found several technological applications. For example, a calcium selenium sulfide compound, CaSSe, is mixed into glass to give it a bright red color. So it's used for traffic lights (Figure 18.13). Selenium has a useful property for xerography: Elemental selenium that is electrically charged will discharge when exposed to light. When a film of selenium coating an aluminum drum is charged and then exposed to light, the exposed areas are discharged. A black toner then sticks only to the areas that remain charged. A copy is made when the toner is transferred to a sheet of plain paper. Interestingly enough, selenium is also the active ingredient of many anti-dandruff shampoos.

Group 17: The Halogens

All the elements of group 17 are able to combine with elements of groups 1 or 2 to form ionic compounds, commonly known as salts. These elements have thus come to be known as the **halogens,** which is derived from a Greek word meaning "salt-forming." Chlorine, for example, combines with sodium to produce sodium chloride. Iodine combines with potassium to give potassium iodide, KI, which is added to table salt to help prevent a thyroid disorder called goiter. Also, fluorine combines with calcium to produce calcium fluoride, CaF_2, a natural source of fluorine.

In the elemental state, halogens exist as diatomic molecules. Fluorine is found as F_2, chlorine as Cl_2, bromine as Br_2, and iodine as I_2. We find that hydrogen bears many similarities to the halogens. For example, hydrogen exists as a diatomic molecule in its elemental state, H_2, and it forms single covalent bonds as do the halogens. For this reason, some periodic tables list hydrogen in both group 1 and group 17.

Group 18: The Noble Gases

All the elements in the last group of the periodic table are colorless monoatomic gases at room temperature. They also share a lack of chemical reactivity. Hence, they are

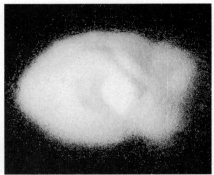

Figure 18.14 The yellow vapors of bromine, which is a halogen, react with metallic sodium to produce the salt sodium bromide, NaBr.

Figure 18.15 Noble gases in glass tubes glow their characteristic colors when energized with electricity.

called the **noble gases**, a term that refers to their inertness.* The lightest noble gas, helium, was first discovered in the sun by analysis of the sun's spectrum, and was thus named after the Greek word for sun, *helios*. An interesting quality of the noble gases is that they glow bright colors when an electrical current runs through them. The element that glows the brightest for a given voltage is neon, named after the Greek word for new, *neos*. It is this element that causes the red/orange glow seen in "neon" lights. When a "neon" light glows blue, however, it does not contain neon. A blue "neon" light probably contains the noble gas argon. Argon is named after the Greek word for inactive, *argos*. Or the blue "neon" light may contain xenon, which is named after the Greek word for stranger, *xenon*.

Chlorine in Water

One of the major uses of chlorine is to disinfect water. Chlorination of public water supplies has prevented much illness and increased the quality of living for many. There are several ways to add chlorine to water. Chlorine gas, Cl_2, may be bubbled through the water, or chemical compounds containing chlorine may be dissolved in the water. One such compound is calcium hypochlorite, $Ca(OCl)_2$, the powdery solid commonly added to swimming pools. Another compound is sodium hypochlorite, NaOCl, the active ingredient of laundry bleach. Chlorine gas, calcium hypochlorite, and sodium hypochlorite each mix with water to form the chemical compound hypochlorous acid, HOCl. This compound kills disease-producing microorganisms. Monitors watch the levels of chlorine in water carefully. Too little chlorine does not control waterborne diseases. On the other hand, too much chlorine can give the water a bad taste and even cause sickness.

---

*It was once thought that none of the elements of group 18 could react to form chemical compounds. Since 1962, however, many compounds of the heavier noble gases, krypton [Kr], xenon [Xe], and radon [Rn], have been reported.

Figure 18.16 Elements as *metal*, *metalloid*, or *nonmetal*.

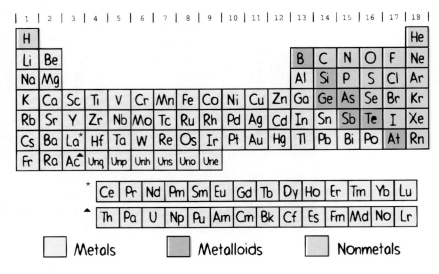

The heaviest noble gas, radon, is a decay product of uranium. Like uranium, it is radioactive. You might remember from Chapter 15 that radon seeps naturally from soils and is a hazard, for it tends to accumulate in enclosed and unventilated areas such as basements. Good sense dictates that basement areas be well ventilated. One out of fourteen lung cancer deaths is estimated to be caused by radon exposure.

Metals, Nonmetals, and Metalloids

The organization of the periodic table tells us much more than the identity of elements in the atomic groups. A glance at the table shows that metallic elements appear toward the lower left and nonmetallic elements appear toward the upper right. In between we find a third class of elements, the **metalloids,** which have both metallic and nonmetallic properties (Figure 18.16).

Metalloids such as silicon and germanium are weak conductors of electricity. Elements with this property can be used as **semiconductors**. Semiconductors have their greatest application in electronics—particularly in the integrated circuits that computers contain. We see in Figure 18.16 that germanium [Ge] is closer to the metals than the nonmetals. Germanium has more metallic properties than silicon and is a slightly better conductor of electricity. So we find that integrated circuits fabricated with germanium operate faster. Because silicon is much more abundant and less expensive to obtain, silicon computer chips remain quite common (Figure 18.17).

Figure 18.17 Purified silicon cylinders are sliced into wafers for the manufacture of silicon-based integrated circuits.

QUESTION

Do elements necessarily have to be directly above or below one another in order to have similar properties?

ANSWER

No. We find there are many different ways in which elements are grouped in the periodic table—by vertical columns is one; by metal, nonmetal, or metalloid is another.

The natural organization of the periodic table played a large role in leading early 20th century atomic researchers toward the models of the atom discussed in Chapter 14. The general scheme of successful models, whether the simple planetary model or the more complex quantum mechanical model, has to do with how electrons arrange themselves about the nucleus. This can get quite detailed, too much for the level of this book. A good understanding can be gained, however, from the brief overview in the remaining sections of this chapter.

18.3 ATOMIC SHELLS

Figure 18.18   A cut-away view of the atomic shell model. The numbers indicate the maximum number of electrons each shell may hold.

The quality of a song depends upon the arrangement of its musical notes. In a similar fashion, the chemical and physical properties of all substances depend upon the arrangements of electrons in its atoms. In this section we explore possible arrangements by way of the atomic shell model. First, we should note that we have abbreviated the atomic shell model, which as a whole is beyond the scope of this text. The simplified model presented here, therefore, has its limitations, which become apparent after studying chemistry in more detail. For our purposes, however, even an abbreviated version of the atomic shell model helps us to understand many things about atoms.*

According to the atomic shell model, electrons behave as though they are arranged about the atomic nucleus in concentric shells. There are as many as seven shells and each shell can hold only a limited number of electrons. The inner shell can hold two; the second and third shells, eight; the fourth and fifth shells, 18; and the sixth and seventh shells, 32 (Figure 18.18).

Since the electrons of atoms are attracted to the positive charge of the nucleus, they tend to cluster as close to the nucleus as possible. For this reason we find that electrons tend to fill inner shells before outer shells.

Shells and the Periodic Table

The organization of the periodic table is consistent with atomic shells. The seven atomic shells account for the seven periods in the periodic table. Elements of the first period, hydrogen and helium, have electrons only in the first atomic shell (Figure 18.19). Elements of the second period, lithium through neon, have electrons that occupy the first and second atomic shells. Elements in the third period, sodium through argon, have electrons occupying the first, second, and third atomic shells. This pattern continues through to the seventh period, which has electrons in all seven shells.

The number of elements in each period is restricted to the capacity of the shells. The first atomic shell, for example, only has a capacity for two electrons, so only two elements, hydrogen and helium, are found in the first period (Figure 18.19). Likewise, the second and third atomic shells each have capacities for eight electrons, so as many as eight elements are found in both the second and third periods. These elements are lithium through neon for the second, and sodium through argon for the

---

*A note to science types: The atomic shells discussed here are the noble gas shells, which chemists use to explain the organization of the periodic table, not the electron shells that follow principal quantum numbers, which quantum mechanists use to explain the structure of the atom.

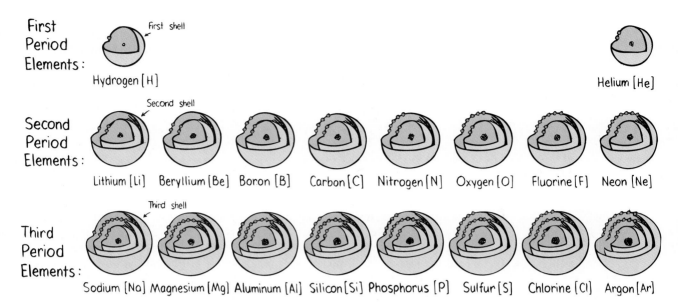

First Period Elements:
First shell
Hydrogen [H]
Helium [He]

Second Period Elements:
Second shell
Lithium [Li]   Beryllium [Be]   Boron [B]   Carbon [C]   Nitrogen [N]   Oxygen [O]   Fluorine [F]   Neon [Ne]

Third Period Elements:
Third shell
Sodium [Na]   Magnesium [Mg]   Aluminum [Al]   Silicon [Si]   Phosphorus [P]   Sulfur [S]   Chlorine [Cl]   Argon [Ar]

Figure 18.19   The first three periods of the periodic table according to the atomic shell model. Elements in the same period have electrons in the same shells. Elements differ from one another by the number of electrons in the outermost shell.

third. Similarly, because of atomic shell capacities, the fourth and fifth periods have 18 elements each (potassium [K] through krypton [Kr], and rubidium [Rb] through xenon [Xe]), and the sixth and seventh periods have 32 elements each (cesium [Cs] through radon [Rn], and francium [Fr] through to an element yet to be discovered).*

Valence Electrons

The outside surface of the atom is composed of outermost-shell electrons. These outer-shell electrons are the **valence electrons** we very briefly discussed in the previous chapter. Since valence electrons are directly exposed to the external environment, they are the first to interact with other atoms. For this reason, we find that elements with the same number of valence electrons often behave quite similarly.

Valence electrons interact not only with the world outside the atom, but interact among themselves. For example, since all electrons repel one another, valence electrons tend to be as far apart as possible. We find, however, that when the number of valence electrons is more than half the capacity of the shell, the valence electrons begin to pair up.† A good analogy for this is a bunch of strangers filling double seats on a bus. The strangers represent electrons and the bus represents an atomic shell. These particular strangers prefer to occupy the double seats alone. Only when all the seats are occupied will they begin to pair up. Atomic shells are similar. When the shell is filled more than halfway, electrons are forced to pair up.

Figure 18.20   Electrons pair up in atomic shells in a way similar to single passengers who pair up on the double seats in a bus.

*To date, only 23 out of the 32 possible elements of the seventh period have been observed.

†All electrons behave like tiny magnets and electrons are able to pair up when their north and south poles align themselves in opposite directions.

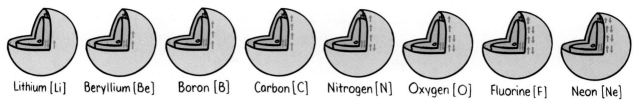

Figure 18.21   Valence electrons begin to pair once the atomic shell is half filled. (Electrons are represented here with arrows. Paired electrons are coupled and pointing in opposite directions.)

Consider the configurations of the electrons in elements of the second period (Figure 18.21). The lithium atom has three electrons, two paired together in the first shell and one unpaired valence electron in the second shell. The beryllium atom has four electrons, two paired in the first shell and two unpaired in the second. The boron atom, we find, has three unpaired valence electrons, and the carbon atom has four. With the nitrogen atom, however, two valence electrons are forced to pair, which leaves only three unpaired. For similar reasons, oxygen has only two unpaired valence electrons, fluorine has one, and neon, which has enough electrons to fill both the first and second shells, has none.

The periodic table is organized so that elements within the same atomic group have the same number of unpaired valence electrons. All alkali metals (group 1), for example, have one unpaired valence electron, and all alkali-earth metals (group 2) have two unpaired valence electrons. Similarly, we find that all halogens (group 17) have one unpaired valence electron and all noble gases (group 18) have no unpaired valence electrons.

QUESTIONS

How many unpaired valence electrons are there in a phosphorus atom [P] that has 15 electrons? How many in an arsenic atom [As] that has 33 electrons? (Use shell models to deduce your answers.)

The number of *unpaired valence electrons* greatly affects the chemistry of an element—very much so. This is because the strongest interactions between atoms involve the *unpaired* valence electrons. Again for emphasis: *the strongest interactions between atoms involve the unpaired valence electrons.* Alkali metals (group 1), for example, all tend to form +1 ions because they lose their single

ANSWERS

Three for both phosphorus and arsenic. Two of phosphorus' electrons pair up in the first shell, and eight pair up in the second shell. The remaining 5 electrons occupy the third shell. Because the third shell has a capacity for 8 electrons, two are forced to pair up, leaving three unpaired. Arsenic has enough electrons to fill the first 3 atomic shells. This leaves 15 to fill the 4th shell, which has a capacity for 18 electrons. Filling this outer shell like double seats on a bus results in 3 unpaired valence electrons. So, we see another touch of grandeur in the periodic table—elements in the same atomic group have the same number of unpaired valence electrons.

TABLE 18.3 *Second-Period Elements Bonding with Hydrogen*

Element	Number of Unpaired Valence Electrons	Hydrogen Compound Formed (common name)
Li	1	LiH (lithium hydride)
Be	2	BeH_2 (beryllium hydride)
B	3	BH_3 (borane)
C	4	CH_4 (methane)
N	3	NH_3 (ammonia)
O	2	H_2O (water)
F	1	HF (hydrogen fluoride)
Ne	0	*no compound formed*

unpaired valence electrons during chemical reactions. Similarly, alkali-earth metals (group 2) all tend to form $+2$ ions as they have two unpaired valence electrons to lose. The number of times an atom bonds also depends upon its number of unpaired valence electrons. For example, we find that the number of times that second period elements bond with hydrogen is the same as their number of unpaired valence electrons (Table 18.3). The noble gases also show how important unpaired valence electrons are to chemical reactivity. Noble gases are unreactive simply because they have no unpaired valence electrons.

Inner Shell Shielding

What is it that holds the negatively charged electron in a hydrogen atom? The answer, of course, is its attraction to the positively charged nucleus. Atoms heavier than hydrogen, however, are complex in that they contain many electrons, which in addition to being attracted to the nucleus, repel one another because of their like charges. What is the effect of these repulsive forces on an electron's attraction to the nucleus? To answer this question we turn to the shell model.

All electrons within a single atomic shell have nearly the same access to the nucleus—they all have about the same "line of sight," and do not get in one another's way. Imagine you are one of two electrons in the innermost shell of a helium atom. You share this shell with one other electron, but it does not stand in your way of the nucleus. You and your neighboring electron sense a nucleus of two protons and you each are attracted to it as such. This is very much like two honeybees buzzing closely around a single flower. Both bees can be equally attracted to the flower despite the other bee's presence.

The situation is different for atoms beyond helium, which have more than one shell of electrons. In this case, inner shell electrons weaken the attraction between outer shell electrons and the nucleus. Imagine, for example, that you are an electron in the second atomic shell of a lithium atom looking toward the nucleus (Figure 18.22). What do you sense? Not only the nucleus, for electrons in the innermost shell are in your way. These two electrons, with their repelling charge, have the effect of weakening your electrical attraction to the nucleus. This is **inner shell shielding**. Returning to the bee analogy, the sight of the flower for an incoming bee is obscured by the first wave of bees already hovering close to the flower. As a bee within a second wave of bees, you are partially shielded from the flower and consequently your attraction to the flower is reduced.

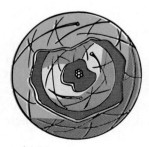

Lithium –
3 electrons in 2 shells

Figure 18.22 Lithium's outer shell electron, the valence electron, is partially "shielded" from the nucleus by two electrons in the inner shell.

Inner shell electrons always diminish the attraction of outer shell electrons to the nucleus. In effect, the nuclear charge sensed by an outer shell electron is always less than the actual charge of the nucleus. The diminished nuclear charge experienced by an outer shell electron is called the **effective nuclear charge**. The valence electron for lithium shown in Figure 18.22, for example, does not "sense" the full effect of lithium's +3 nuclear charge (there are three protons in the nucleus of lithium). Instead, the total charge of all inner shell electrons, −2, subtracts from the charge of the nucleus, +3, to give an effective nuclear charge of about +1. We say *about* +1 and not *exactly* +1 because the distances between all charges is a factor. Recall Coulomb's law in Chapter 9, where the electrical force between charges decreases as the square of the distance. The effective charge in this case is actually less than +1.

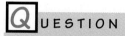

UESTION

Approximately, what is the effective nuclear charge for a valence electron in beryllium [Be]?

18.4 PERIODIC TRENDS

Atomic Size

Reading from left to right through any row of the periodic table, the atoms of elements get *smaller* in size. If the number of electrons were the only consideration, this might at first seem illogical because in this direction, atomic shells contain a greater number of electrons. But atomic nuclei contain the same greater number of protons, which pull on the electrons and tighten their shells.

Let's look at this from the point of view of the effective nuclear charge. Consider lithium's valence electron experiencing an effective nuclear charge of about +1. Then look at the valence electrons for neon, which experience an effective nuclear charge of about +8! Since the neon electrons experience a greater net attraction to the nucleus, they are pulled in closer to the nucleus. So neon, which is nearly three times as massive as lithium, is considerably smaller in size! In general, across any atomic period from left to right the atoms of elements become smaller.

In reading down an atomic group, the atoms of elements get larger because of an increasing number of filled inner shells. Looking at the periodic chart as a whole, elements towards the upper right corner are smaller, while elements toward the lower left corner are larger. This is evident in Figure 18.23a.

NSWER

A beryllium atom has 4 protons, hence, a nuclear charge of +4. Two inner shell electrons, however, partially shield beryllium's valence electrons from this charge causing each of them to experience an effective nuclear charge of about +2.

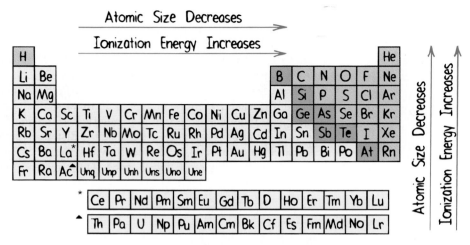

Figure 18.23 (a) The atoms of elements toward the upper right corner of the periodic chart are smaller while those toward the lower left corner are larger. (b) Ionization energies are typically larger for elements toward the lower right of the periodic table and smaller for elements toward the lower left.

QUESTION

What would be the effect on the atomic sizes of atoms of increasing atomic number if the first atomic shell had an unlimited capacity for electrons? (Could life as we know it exist?)

Figure 18.24 Relatively small fluoride ions penetrate the tooth enamel and make stronger teeth.

Some real-life applications can be related to this periodic trend. Have you ever wondered why the fluorine ion is added to toothpaste and drinking water? We can deduce from fluorine's position in the periodic table that fluorine ions are relatively small. The ionic charge of the fluorine ion, therefore, is concentrated into a small volume. This, in turn, allows for stronger ionic bonds. When fluorine ions are introduced to tooth enamel, they replace larger ions that are not held so tightly. With strongly bonded fluorine ions in place, the tooth enamel is harder and more resistant to decay.

Ionization Energies

We stated in Section 18.1 that the energy required to remove an outer electron from an atom is the *ionization energy*. We find that the larger the atom, the less the ionization energy. This is mainly because outer-shell electrons in larger atoms are farther

ANSWER

Atoms become larger because of a greater number of filled shells. If the first shell had an unlimited capacity, electrons would not occupy outer shells. And with no inner shells to act as sheilds, the effective nuclear charge would be the same as the number of protons in the nucleus, and would increase with increasing atomic number. The undiminished positive nucleus would pull the lone shell tighter and tighter, and smaller and smaller. Nobody knows, by the way, why shells have limited capacities. At this time we simply say it's a basic fact. And depressingly enough, if atomic shells had unlimited capacities, the nature of matter would be entirely different and life as we know it would not exist.

from the nucleus. Because electric forces decrease over greater distances, outer electrons are less tightly held. Since elements toward the lower left corner of the periodic table are larger in size, they have smaller ionization energies. How about elements toward the upper right corner that are smaller in size? You guessed it. They have higher ionization energies. This is indicated in Figure 18.23b.

UESTION

From which atom would an electron be more easily removed, a sodium atom (atomic number 11) or an argon atom (atomic number 18)?

18.5 CHEMICAL BONDING AND THE PERIODIC TABLE

Through the periodic table we gain a deeper understanding of how atoms bond to one another, either covalently or ionically. Furthermore, the periodic table is a useful guide for telling us just how polar a covalent bond may or may not be. There are two closely related concepts that lead us to this understanding; they are "electron affinity" and "electronegativity."

Electron Affinity

The ability of an atom to attract one or more additional electrons is its **electron affinity**. A neutral hydrogen atom, for example, has one positive proton balanced by a single negative electron. Hydrogen's single shell has room for an additional electron, and given the opportunity, will pick up a second electron to become a negatively charged ion (Figure 18.25). Interestingly enough, a negative hydrogen ion is a more likely configuration for single hydrogen than a neutral atom.

Figure 18.25 A neutral hydrogen atom will attract an electron to become a negatively charged hydrogen ion.

Hydrogen atom, H

Negatively charged hydrogen ion, [H]⁻

How a perfectly neutral atom attracts an additional electron can be explained using our honeybee analogy. We know that bees do not exclude one another from the smell of a sweet flower. Instead, many bees can hover around the same flower, provided they do not get in one another's way. Likewise, as long as there is sufficient room, electrons do not exclude one another from the charge of the nucleus. So, if an electron were to cross paths with a hydrogen atom, it would be attracted to the nucleus

NSWER

Since sodium is farther to the left in the periodic table, it is likely larger than argon. The larger the atom, the lower its ionization energy. So, it should be easier to pull an electron away from sodium. In fact, the sodium atom is about twice as large as the argon atom and its ionization energy about three times less.

Figure 18.26 More than one bee can be attracted to a single flower. Likewise, more than one electron can be attracted to a single nucleus.

Figure 18.27 The hydrogen molecule, H_2, consists of two hydrogen atoms that are attracted to each other's electron.

of that atom and have the tendency to join in with the electron that is already there (Figure 18.26).

Would a third electron be similarly attracted to the hydrogen nucleus? The answer would be yes if space were available around the nucleus. The first atomic shell has room for only two electrons. Any third electron added to the hydrogen atom would be forced to reside in the second atomic shell, which is not only farther away from the nucleus but also shielded from the nucleus by electrons in the first atomic shell. The hydrogen atom, therefore, has an affinity for only two electrons— its original electron, plus one additional.

It is the affinity that hydrogen atoms have for an additional electron that causes hydrogen to form covalent bonds. For instance, when that additional electron is attached to another hydrogen nucleus the result is a hydrogen molecule, H_2 (Figure 18.27). In the hydrogen molecule, the hydrogen atoms share their electrons with each other so that each atom has access to two electrons—its own plus the electron of its neighbor. As discussed in Section 17.4 in the previous chapter, atoms held together by this sharing of electrons are covalently bonded. Strictly speaking, atoms within a covalent bond are not held together simply because they "share electrons." Rather, they are held together by the affinity they have for the electrons that they share.

Consider oxygen, which has eight protons in the nucleus, two electrons in the first atomic shell, and six electrons in the second atomic shell. How many additional electrons might the oxygen atom attract? Because of inner shell shielding, electrons in the vicinity of an oxygen atom would sense an effective nuclear charge of about +6, which from the point of view of an electron is quite attractive. Furthermore, because the capacity of the second atomic shell is eight electrons, two "open spaces" are available. An oxygen atom, therefore, has an affinity for not one, but two additional electrons.

The oxygen atom finds two such electrons when it encounters two hydrogen atoms and reacts to form water, H_2O. In water, not only does the oxygen atom have access to two additional electrons by covalently bonding to two hydrogen atoms, but each hydrogen atom has access to an additional electron by bonding to the oxygen atom (Figure 18.28). Everybody's happy!

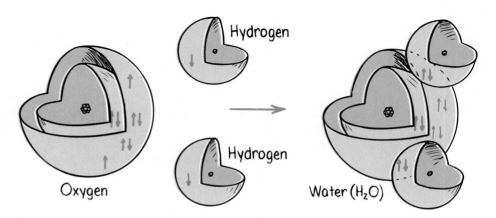

Figure 18.28 Two unpaired valence electrons of oxygen get together with the unpaired valence electrons of two hydrogen atoms to produce water, H_2O. Note how the shells overlap where the covalent bonds are formed.

Electron affinity plays a strong role in an atom's ability to form chemical bonds. Most elements have some degree of electron affinity, and therefore, the ability to bond with other atoms. The noble gases are an exception, for they have no space for additional electrons in their filled shells. With no electron affinity, they tend to be non-reactive.

QUESTION

Hydrogen tends to form one covalent bond. Is this because it has one unpaired valence electron or because there is room in its shell for one additional electron?

Electronegativity

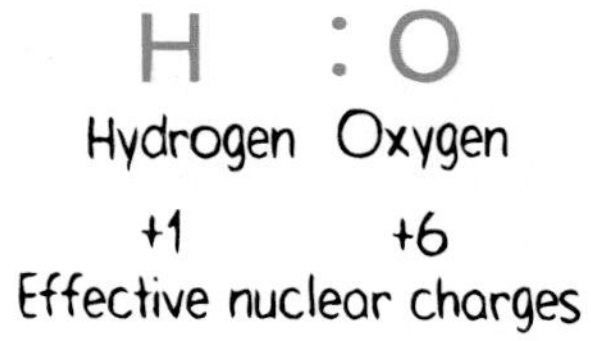

H : O

Hydrogen Oxygen

+1   +6

Effective nuclear charges

Figure 18.29 The effective nuclear charge for the outer shell of oxygen is greater, so shared electrons on the average are closer to oxygen than to hydrogen.

The ability of an atom to attract electrons to itself when already bonded to another atom has to do with its **electronegativity**. When atoms combine to form molecules, such as H_2O, they share electrons in a covalent bond. If one of the bonded atoms has a greater effective nuclear charge, the shared electrons will spend most of their time closer to that atom (Figure 18.29).

The bonded atom with the greater effective nuclear charge is said to be more *electronegative*. For example, oxygen is more electronegative than hydrogen. Typically, elements toward the upper right corner of the periodic table are more electronegative than elements toward the lower left (Figure 18.30). This is because elements towards the upper right normally have greater effective nuclear charges. Hydrogen is an exception to this trend since its electronegativity is more similar to that of boron [B] and carbon [C], and not lithium [Li]. For this reason, the periodic table shown in Figure 18.30 is modified to show hydrogen situated above boron and carbon, rather than in group 1 above lithium.

Atoms with the same electronegativity form covalent bonds where the shared electrons are distributed evenly. When bonded atoms have the same electronegativities, they pull equally on the electrons they share. So the electrons don't accumulate on one side or the other of the bonded atoms. As discussed in Section 17.4, a covalent bond like this is said to be *nonpolar*. We find examples of nonpolar covalent bonds in molecular hydrogen, H_2, nitrogen, N_2, and oxygen, O_2.

Atoms of different electronegativities form bonds in which the shared electrons tend to accumulate on one side—the side of the more electronegative atom. This results in a bond polarity with one side negatively charged and the other side positively charged.

In general, the greater the difference in the electronegativities of two bonded atoms the greater the polarity of the bond. We can easily estimate the electronegativity difference of two elements by noting their relative positions in the periodic

ANSWER

Both! Hydrogen bonds only once because it has only one unpaired valence electron to share. Likewise, however, since hydrogen's valence electron is unpaired, there is one vacant space to which an additional electron may be attracted. The number of unpaired valence electrons in an atom is always the same as the number of additional electrons it is able to attract. Is your cup half full or half empty? It just depends on how you look at it.

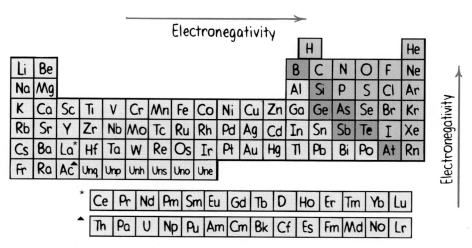

Figure 18.30 Elements toward the upper right corner of the periodic table typically have greater electronegativities while elements toward the lower left corner have smaller electronegativities. In this version, hydrogen [H] is situated above boron [B] and carbon [C] because the electronegativity of hydrogen is between that of these two elements.

table. If the elements are close to each other, their electronegativities are similar. This, in turn, indicates only a slighty polar covalent bond between these two elements. But for elements farther apart, there is a greater difference in their electronegativities, and a bond formed between them will be more polar. So a chemist doesn't need to memorize which bonds are more polar than others; to find out he or she needs only to look at the relative positions of the bonded elements in the periodic table.

QUESTION

Which is more polar, a bond between fluorine [F] and oxygen [O], or a bond between fluorine [F] and boron [B]?

When elements at opposite ends of the periodic table combine, the element toward the right side of the table has an electronegativity that is so much greater that it literally strips electrons away from the element toward the left. This results in two oppositely charged ions, which are held to each other by simple ionic attraction. These ions combine to form an ionic compound, such as sodium chloride, Na^+Cl^- (Section 17.4). We find, therefore, that all ionic compounds are composed of elements at opposite ends of the periodic table. Furthermore, the element toward the right bears the negative charge, like Cl^-, while the element toward the left bears the positive charge, like Na^+.

ANSWER

Fluorine and boron are farther apart in the periodic table, so a covalent bond between them will be more polar.

The science of chemistry, like other basic sciences, has many intricacies that can be difficult to comprehend—especially for the beginning student. The periodic table, however, reveals much basic chemistry in one convenient picture. It is a roadmap not only to the nature of the elements, but also to the hows and whys of elements interacting with one another. Much information about an element's chemical or physical properties can be inferred simply from its position in the periodic table. Knowing the properties of elements, we can deduce molecular properties and then, at a higher level, the properties of materials around us. Periodic tables also list factual information about elements, such as atomic masses, atomic numbers, and most common isotopes. All this makes the periodic table an excellent reference for anyone concerned with chemistry and the nature of the fundamental ingredients of everything.

● ●

SUMMARY OF TERMS

Periodic table A highly organized chart listing all the known elements.

Atomic radius A measure of the size of an atom.

Ionization energy The amount of energy required to pull an electron away from an atom.

Atomic period A horizontal row in the periodic table.

Atomic group A vertical column in the periodic table.

Alkali metals Group 1 elements.

Alkali-earth metals Group 2 elements.

Transition metals The elements of groups 3 through 12.

Inner transition metals Two subgroups of metals within the transition metals.

Lanthanides The inner transition metals within the sixth period.

Actinides The inner transition metals within the seventh period.

Ceramic Water- and heat-resistant materials usually made of elements from groups 13 or 14.

Nitrogen fixation A biological process whereby ammonia, NH_3, is produced from atmospheric nitrogen, N_2.

Halogens Group 17 elements.

Noble gases Group 18 elements.

Metalloid Elements such as silicon and germanium that exhibit both the properties of a metal and a nonmetal.

Semiconductor Any material that partially conducts electricity, usually fabricated from metalloid elements.

Atomic shell A spherical region of space about the atomic nucleus where electrons may reside.

Valence electron Any electron in the outermost shell of an atom.

Inner shell shielding The tendency of inner shell electrons to partially shield outer shell electrons from the nuclear charge.

Effective nuclear charge The nuclear charge "sensed" when one or more electrons partially shield the nucleus.

Electron affinity The ability of an atom to attract one or more additional electrons.

Electronegativity The ability of an atom to attract electrons to itself when bonded to another atom.

● ●

REVIEW QUESTIONS

Organizing the Elements

1. What about the properties of elements becomes apparent as we list them in a single vertical column?

2. What is represented by each horizontal row of the periodic table?

3. Helium [He] and beryllium [Be] are both the second elements in their respective intervals. Why are these two elements not placed directly above and below each other in the periodic table?

4. Why is the position of helium [He] shifted so that it appears directly above neon [Ne]?

Atomic Groups

5. How many atomic groups are there in the periodic table?

6. Why are the group 1 elements called the *alkali metals?*

7. Why does hydrogen take on the properties of a metal in the planet Jupiter, but not on the planet Earth?

8. Why are the group 2 elements called the *alkali-earth metals?*

9. What group 2 element is common in sea shells?

10. What is the primary use of transition metals?

11. In what ways are transition metals important for our health?

12. Why are the inner transition metals not listed in the main body of the periodic table?

13. Why is it difficult to purify an inner transition metal?

14. In what form are the elements aluminum and silicon most often found in nature?

15. Why is pure aluminum not used in the fabrication of road signs?

16. What is a ceramic and how is it made?

17. What is the most abundant element in our atmosphere?

18. What is the most abundant element in the earth's crust?

19. Why are the group 17 elements called the halogens?

20. Where was the noble gas helium first discovered?

21. Are most elements in the periodic table metallic or non-metallic?

Atomic Shells

22. What is the relationship between the number of atomic shells and the number of periods in the periodic table?

23. What is the relationship between the electron capacities of atomic shells and the number of elements in each period of the periodic table?

24. What is a *valence electron?*

25. When do valence electrons start to pair up?

26. Place the proper number of electrons in each shell for sodium [Na] (atomic number 11), rubidium [Rb] (atomic number 37), krypton [Kr] (atomic number 36), and chlorine [Cl] (atomic number 17). Use arrows to represent electrons.

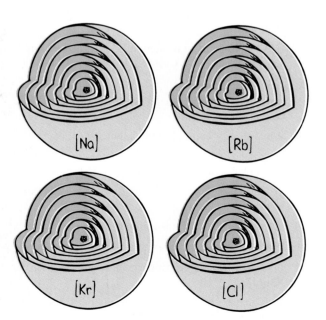

27. How many atomic shells are occupied by electrons in a gold atom, [Au] (atomic number 79)?

28. How many electrons are there in the outermost shell of a carbon atom?

29. The nucleus of a carbon atom has a charge of +6, but this is not the charge experienced in carbon's valence electrons. How come?

30. What is meant by the *effective nuclear charge?*

31. What is the approximate effective nuclear charge for a valence electron in krypton [Kr] (atomic number 36)?

Periodic Trends

32. How is it possible that as atoms get more massive they become smaller in size?

33. How is it possible that as atoms get more massive they become larger in size?

34. Smaller atoms tend to have greater ionization energies. How come?

Chemical Bonding and the Periodic Table

35. How is it possible for a neutral hydrogen atom to attract an additional electron?

36. How many additional electrons is an oxygen atom able to attract?

37. Distinguish between *electron affinity* and *electronegativity.*

38. Which should be more electronegative, a chlorine atom or a phosphorus atom?

39. Which should be more polar, a fluorine-sulfur (F—S) bond or an oxygen-aluminum (O—Al) bond? How come?

40. Why are salts made of elements found on opposite ends of the periodic table?

EXERCISES

1. A radioactive isotope of strontium [Sr] is especially dangerous to humans because it tends to accumulate in calcium-dependent bone marrow tissues. Suggest how this fact relates to what you know about the organization of the periodic table.

2. Germanium computer chips operate faster than silicon computer chips. So, how might a gallium [Ga] computer chip compare to a germanium computer chip? (We find that gallium [Ga] melts at only 30°C, so to be useful as a semiconductor it is alloyed with the metalloid, arsenic [As].)

3. If an element of atomic number 118 is ever synthesized, what properties might you expect it to have?

4. Another interesting periodic trend is the density of the elements. We find that osmium [Os] (atomic number 76) has the greatest density of all elements, and, with some exceptions, the closer an element is positioned to osmium, the greater its density. Based upon this trend, list the following elements in order of increasing density: copper [Cu], gold [Au], platinum [Pt], and silver [Ag].

5. How many valence electrons are there in magnesium [Mg] (atomic number 12)? How many of these are unpaired?

6. How many valence electrons are there in fluorine [F] (atomic number 9)? How many of these are unpaired?

7. Which experiences a greater effective nuclear charge, an electron in the outer shell of neon [Ne], or an electron in the outer shell of sodium [Na]? How come?

8. An electron in the outermost shell of which of these elements experiences the greatest effective nuclear charge?

   (a) sodium [Na]
   (b) potassium [K]
   (c) rhubidium [Rb]
   (d) cesium [Cs]
   (e) all the same

9. List the following atoms in order of *increasing* atomic size: Thallium [Tl]; Germanium [Ge]; Tin [Sn]; Phosphorus [P].

   (smallest) _____ < _____ < _____ < _____ < (largest)

10. Arrange the following elements in order of increasing ionization energy: Tin [Sn]; Lead [Pb]; Phosphorus [P]; Arsenic [As].

    (weakest) _____ < _____ < _____ < _____ < (strongest)

11. How are electron affinity and electronegativity different?

12. Fluorine has an affinity for how many additional electrons?

13. How many electrons are there in the third shell of the sodium ion, Na^+? How about in the third shell of the chlorine ion, Cl^-?

14. How are the arrangements of electrons in the neon atom, Ne, and the calcium ion, Ca^{2+}, similar? How are they different?

15. Based upon the atomic shell model, which would you expect to be larger: a neutral sodium atom, or a positively charged sodium ion, Na^+?

16. Based upon the atomic shell model, which would you expect to be larger: a neutral fluorine atom, or a negatively charged fluorine ion, F^-?

17. Based upon the atomic shell model, explain why the alkali metals (group 1) tend to form +1 ions while the alkali-earth metals (group 2) tend to form +2 ions.

18. It is relatively easy to pull an electron away from a sodium atom, but very difficult to remove a second one. Use the atomic shell model and the idea of effective nuclear charge to explain why this is so.

19. Oxygen [O] has an affinity for two additional electrons. Its affinity for the second electron, however, is somewhat less than its affinity for the first electron. Explain.

20. The effective nuclear charge in the outer shell of a neon atom is relatively strong (about +8). Why is it that the neon atom has no affinity for an additional electron?

21. Oxygen tends to form two covalent bonds. Is this because it has two unpaired valence electrons or because there is room in its valence shell for two additional electrons? Briefly explain.

22. In each of the following diatomic molecules, indicate which atom is more positively charged:

    hydrogen chloride, HCl   _____
    bromine monofluoride, BrF   _____
    carbon monoxide, CO   _____
    molecular bromine, Br_2 _____

23. List the following bonds in order of increasing polarity:

    (a) N—N   (b) N—F   (c) N—O   (d) H—F

    _____ < _____ < _____ < _____

24. Which should be more polar, a sulfur-bromine (S—Br) bond or a selenium-chlorine (Se—Cl) bond?

19

CHEMICAL INTERACTIONS

All materials are composed of molecules or other small chemical units, such as ions. The activities and interactions of these units produce most of the physical properties of substances. For example, molecules of graphite (a form of pure carbon) are flat, so they can pile on top of one another like playing cards. Rub graphite between your fingers and you'll find it has a slippery feel. The individual graphite molecules are gliding over one another like cards in a stack, producing that slippery feel. Graphite makes a great lubricant, because of its molecular behavior. Graphite is also used to make pencils because it glides so easily onto paper. (Pencil "lead" is really graphite; lead is too toxic for this purpose.)

The phase of any material (solid, liquid, or gas) also depends on the interactions between its chemical units. For example, water is a solid at temperatures below 0°C because its molecules stick together through strong interactions, which are electrical in character. At room temperature these interactions perpetually break and reform, allowing water molecules to tumble over one another. This gives us water in a liquid phase. At higher temperatures these interactions are overcome altogether, which gives us water in a gaseous phase.

Chemical interactions between different compounds in a mixture account for properties of the mixture. For example, we can explain what happens when sugar is added to water, and when soap, water, and grease are mixed, and when oxygen dissolves in blood on the basis of what happens between different chemical units. All of this makes up the study of chemical interactions.

19.1 TYPES OF CHEMICAL INTERACTIONS

Four important types of chemical interactions are shown in Table 19.1. Like the covalent and ionic bonds within chemical units, these interactions between chemical units are electrical. Therefore they are described by Coulomb's law: Opposite charges attract and like charges repel. Any one of these interactions, however, is over 100 times weaker than covalent or ionic bonds. Chemical interactions between molecules or other chemical units differ from one another by their relative strengths, and by the nature of their formation.

TABLE 19.1 Relative Strengths of Chemical Interactions

Type of interaction	Relative Strength
ion-dipole	Strongest
dipole-dipole	
dipole-induced dipole	↑
induced dipole-induced dipole	Weakest

Ion-Dipole

In many molecules, electrons are distributed closer to one side of the molecule than the other. The result is that one side of the molecule is slightly negative in charge, $\delta-$, and the other slightly positive, $\delta+$. The water molecule is an example. Its oxygen side is slightly negative and its hydrogen sides are slightly positive. The water molecule is called a *dipole* for this reason (Section 17.4).

What happens when water molecules approach an ionic compound such as sodium chloride? The positive sodium ion is attracted to the negative side of water, while the negative chlorine ion is attracted to the positive side of water. Such an interaction between an ion and a dipole is called an **ion-dipole** interaction (Figure 19.1).

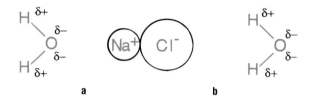

Figure 19.1 (a) An ion-dipole interaction occurs between the negative side (oxygen side) of a water molecule and the positively-charged sodium ion. (b) Another ion-dipole interaction occurs between the positive side (hydrogen side) of a water molecule and the negatively-charged chlorine ion.

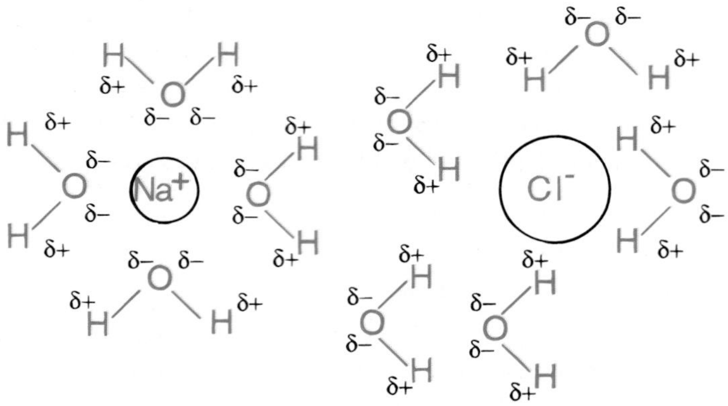

Figure 19.2 In salt water, tightly bound sodium and chlorine ions are separated from one another by the collective attractions of many water molecules.

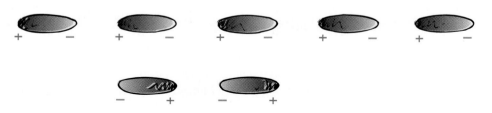

Ion-dipole interactions are much weaker than ionic bonds. A large number of them, however, can act collectively to disrupt an ionic bond. This occurs when salt is mixed with water. The sodium and chlorine ions of salt are strongly held together, but a multitude of ion-dipole interactions with water can serve to pull them apart. The result is a solution of salt water (Figure 19.2).

Dipole-Dipole

When dipoles interact among themselves, we have a **dipole-dipole** interaction. If we represent dipole molecules as cigar-shaped objects with negative and positive ends, we can see that their charges align in an orderly and mutually attractive fashion as shown in Figure 19.3. Such dipoles are very much attracted to one another.

Figure 19.3 Dipoles align with one another like compass needles.

An example of an unusually strong dipole-dipole interaction is the **hydrogen bond.** This interaction occurs between molecules that have hydrogen atoms covalently bonded to highly electronegative atoms, such as nitrogen, oxygen, and fluorine. As we saw in Section 18.5, highly electronegative atoms have strong nuclear charges that readily attract bonding electrons. In a water molecule a highly electronegative oxygen atom pulls electrons away from two hydrogen atoms. This makes the oxygen side of the water molecule negatively charged. Since electrons are pulled away from the hydrogen atoms, their positively charged protons become exposed. This makes the hydrogen sides of a water molecule positively charged. A hydrogen bond occurs when the hydrogen side of one molecule is attracted to the negatively charged side of a neighboring molecule. The strength of the hydrogen bond lies in the magnitude of the dipoles involved. Shake some water onto a dry surface and you see the water form beads. Water molecules are mutually attracting one another by hydrogen bonding (Figure 19.4).

The term "hydrogen bond" is somewhat of a misnomer, and shouldn't be confused with a real chemical bond. Recall from Section 17.4 that a chemical bond is

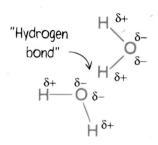

"Hydrogen bond"

Figure 19.4
A dipole-dipole interaction that involves a hydrogen atom is called a hydrogen bond.

either covalent, ionic, or metallic. A hydrogen bond, however, is none of these for it is simply a dipole-dipole interaction. As we shall see in future sections of this chapter, the hydrogen bond is responsible for many of the unusual properties of water and other similar compounds. The hydrogen bond is of great importance. Although its strength is greater than other dipole-dipole interactions, it is much weaker than any chemical bonds. Perhaps, elevation to the status of a bond owes to the importance of the hydrogen bond rather than its strength.

Dipole-Induced Dipole

Electrons in many molecules *are* distributed evenly, so there is no dipole. As briefly discussed in Section 17.4, oxygen, O_2, has no dipole (Figure 19.5a). But the oxygen molecule can be induced into becoming a temporary dipole when it is brought close to a water molecule or any other permanent dipole molecule (Figure 19.5b). The negative side of a water molecule, for example, will push the oxygen's electrons to the opposite side of the molecule. This gives it an uneven distribution of electrons and a charge polarity for the oxygen molecule. The resulting attraction between the permanent dipole of water and the induced dipole of oxygen is an example of the **dipole-induced dipole** interaction.

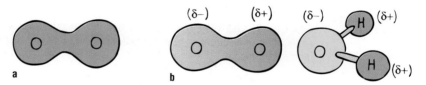

Figure 19.5 (a) An isolated oxygen molecule, O_2, has no dipole—its electrons are distributed evenly. (b) An adjacent water molecule induces a redistribution of charge in the oxygen molecule.

QUESTION

How would the charge distribution in an oxygen molecule differ if the hydrogen end of a water molecule were nearby?

Induced dipoles are only temporary. If the water molecule were knocked away by thermal motion, the oxygen molecule would return to its normal *nonpolar* state. As a consequence, dipole-induced dipole interactions are weak compared to dipole-dipole interactions (Table 19.1). But they're strong enough to hold relatively small quantities of oxygen dissolved in water. This is vital for fish and other forms of marine life. Dipole-induced dipole interactions also occur between molecules of non-polar carbon dioxide and water. This helps to keep carbonated beverages, which are prepared by mixing carbon dioxide in water, from losing their fizz too quickly after they've

ANSWER

Since the hydrogen side of a water molecule is slightly positive, the electrons in oxygen would be pulled toward it so that a dipole of opposite polarity would be induced.

Figure 19.6 Polar induction of the normally nonpolar molecules in plastic wrap makes it stick to the highly polar molecules of glass. Plastic wraps don't stick well to wet surfaces because water disrupts this interaction.

been opened. Dipole-induced dipole interactions are responsible for holding plastic wraps, such as polyethylene, to glass (Figure 19.6). These wraps are made of very long nonpolar molecules, which are induced to have dipoles when placed in contact with the highly polar glass molecules.

Distinguish between a dipole-dipole interaction and a dipole-induced dipole interaction.

Induced Dipole-Induced Dipole

Nonpolar atoms or molecules, on the average, have a fairly even distribution of electrons. Due to the randomness of electron motion, however, at any given moment the electrons may be bunched to one side. The consequence is a momentary dipole (Figure 19.7).

Figure 19.7 (a) The electron distribution in a nonpolar atom is normally even. (b) The distribution of electrons at any moment, however, may be less than even. This results in a momentary dipole.

So the permanent dipole of a polar molecule can induce a temporary dipole in a nonpolar molecule. A momentary dipole can do the same thing. This gives rise to the relatively weak **induced dipole-induced dipole** interaction (Figure 19.8).

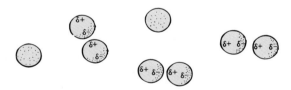

Figure 19.8 Nonpolar argon [Ar] atoms can be attracted to one another by induced dipole-induced dipole interactions.

Momentary dipoles are more significant for larger atoms. This is because electrons in larger atoms have more space available for random motion, and a higher likelihood of bunching together on one side of the nucleus. The electrons in smaller atoms are less able to bunch to one side for they are confined to smaller space where greater electrical repulsion tends to keep them spread out. So it is the larger atoms, and molecules made of larger atoms, that have the stronger induced dipole-induced dipole interactions.

ANSWER

The dipole-dipole interaction is stronger and it involves the attraction between two permanent dipoles. The dipole-induced dipole interaction is weaker and it involves the attraction between a permanent dipole and a temporary one.

TABLE 19.2   Boiling Points of Halogens

Halogen	Atomic Radius	Boiling Point	Phase at 25°C
Iodine, I_2	1.3 Å	184°C	solid
Bromine, Br_2	1.1 Å	59°C	liquid
Chlorine, Cl_2	0.98 Å	−35°C	gas
Fluorine, F_2	0.68 Å	−188°C	gas

This is illustrated by the different boiling points of different sized molecules. Compare the boiling points of the halogens iodine, I_2, bromine, Br_2, chlorine, Cl_2, and fluorine, F_2 (Table 19.2). Iodine is a large molecule. This allows more frequent momentary dipoles and many induced dipole-induced dipole interactions among iodine molecules. Because of these combined interactions, iodine molecules are relatively hard to pull apart. Hence iodine has a high boiling point of 184°C. At room temperature, iodine's induced dipole-induced dipole interactions are strong enough to hold it together as a solid. Fluorine is the smallest of the halogen molecules. Momentary dipoles for fluorine are infrequent, hence, induced dipole-induced dipole interactions are weak. As a consequence, fluorine has a low boiling point of −188°C and, at room temperature, fluorine molecules remain separated from one another in the gaseous phase.

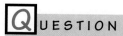UESTION

Distinguish between a dipole-induced dipole interaction and an induced dipole-induced dipole interaction.

Because the fluorine atom is very small, momentary dipoles are not so easily formed, nor are they easily induced. A substance made with nothing but fluorine, therefore, only experiences very weak chemical interactions—so weak that it becomes difficult for anything to stick. This is the principle behind the Teflon nonstick surface. The chemical structure of Teflon consists of a long chain of carbon atoms that are chemically bonded to fluorine atoms (Figure 19.9).

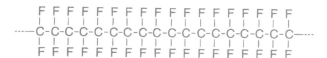

Figure 19.9   Few things stick to Teflon because of the high proportion of fluorine atoms it contains. The structure depicted here is only a portion of the full length of the molecule.

ANSWER

The dipole-induced dipole interaction is stronger and it involves the attraction between a permanent dipole and a temporary one. The induced dipole-induced dipole interaction is weaker and it involves the attraction between two temporary dipoles.

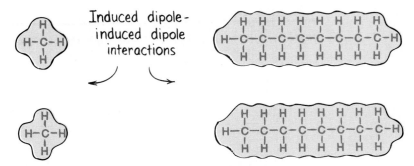

Figure 19.10 (a) Methane molecules, CH_4, are attracted to one another by induced dipole-induced dipole interactions. The number of these interactions, however, is relatively low. (b) Molecules found in gasoline, such as isooctane, C_8H_{18}, are similar to methane but larger in size. The number of induced dipole-induced dipole interactions among these larger nonpolar molecules is greater.

Induced dipole-induced dipole interactions explain why natural gas is a gas and gasoline is a liquid. The major component of natural gas is methane, CH_4. The chemical composition of methane is similar to that of molecules found in gasoline, such as isooctane, C_8H_{18}. We see in Figure 19.10 that the number of induced dipole-induced dipole interactions that can take place between two methane molecules is appreciably less than can occur between two isooctane molecules. Isooctane molecules are larger and therefore more induced dipole-induced dipole interactions are possible. Have you noticed that two small pieces of Velcro are easier to pull apart than two long pieces? Like short pieces of Velcro, methane molecules can be pulled apart with little energy. That's why methane has a low boiling point, $-161°C$, and is a gas at room temperature. But isooctane molecules, like long strips of Velcro, are relatively hard to pull apart because they take part in a larger number of induced dipole-induced dipole interactions. The boiling point of isooctane, $98°C$, is therefore much higher than that of methane. So we see why isooctane is a liquid at room temperature.

QUESTIONS

1. Gasoline is a vapor at the temperatures of an operating car engine whereas motor oil, which is used for lubricating the engine, is not. Both gasoline and motor oil are made of nonpolar molecules that are held together by induced dipole-induced dipole interactions. Suggest how the sizes of gasoline molecules and motor oil molecules may differ.

2. Methanol, CH_3OH, which can also be used as a fuel, is not much larger than methane, yet it is a liquid at room temperature. Suggest why.

ANSWERS

1. Motor oil molecules are longer, and like long strips of velcro, are harder to separate.

2. The polar oxygen-hydrogen covalent bonds in methanol lead to strong dipole-dipole interactions among methanol molecules. These strong interactions hold methanol molecules together as a liquid at room temperature.

How Soap Works

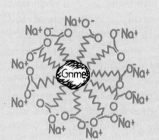

Why are dirt and grime so difficult to remove with water alone? Most dirt and grime is made of nonpolar molecules that do not mix well with polar water molecules. But they mix quite readily with nonpolar organic solvents such as turpentine. This is because organic solvent and dirt and grime molecules are all nonpolar and able to interact strongly by way of induced dipole-induced dipole interactions.

Rather than washing our dirty hands and clothes with turpentine, however, we have a more pleasant alternative—soap and water. Soap works because soap molecules have both nonpolar and polar properties. A typical soap molecule has a long nonpolar tail consisting of carbon and hydrogen atoms. At one end of this tail is a group of atoms containing at least one ionic bond. This region is known as the polar head. A typical soap molecule is shown below.

Since most of a soap molecule is nonpolar, it interacts quite well with dirt and grime. In fact, dirt or grime quickly finds itself surrounded in three dimensions by the nonpolar tails of soap molecules. This interaction is usually enough to lift the dirt or grime away from the surface that is to be cleaned.

With the nonpolar tails of soap molecules faced inward toward the grime, the polar heads are all directed outward where they are attracted to water by relatively strong ion-dipole interactions. If the water is flowing, the whole enterprise of grime and soap molecules flows with it, away from your hands or clothes and down the sink.

Nonpolar tail   Polar head

19.2 PHASES OF MATTER

There are three common phases of matter: *solid, liquid,* and *gas.* A solid has definite volume and shape and is not readily deformed. A liquid has definite volume and takes the shape of its container. Matter in the gaseous phase is diffuse, having neither definite volume nor shape. These principal distinguishing features of solids, liquids, and gases can be understood by considering chemical interactions.

Matter is a solid when chemical interactions between atomic or molecular particles are strong enough to hold them together in some fixed three-dimensional arrangement. When the particles have an ordered periodic structure, the solid is a **crystal**. For example, in sodium chloride (table salt) the sodium and chlorine

Figure 19.11 When molten glass doped with heavy metal compounds like lead oxide cools slowly, the metal compounds take on a fixed and orderly crystalline array. The result is crystalware.

atoms occupy alternate corners of a cubic structure (see Section 17.4). A solid with particles that are more randomly oriented is **amorphous**. Glass is an example of an amorphous solid.

As previously discussed in Chapter 8, adding heat to a solid will increase particle motion and weaken or break chemical interactions until particles can slide past one another. The solid becomes a liquid—it melts. Conversely, removing heat will decrease particle motion until permanent interactions form and the liquid becomes a solid—it freezes.

A special class of liquids is **liquid crystals**. These materials consist of rod-shaped molecules that align with one another in an orderly array. This array is unlike the solid crystalline phase only in that the particles are not held in fixed orientations—they can still tumble around one another like wooden matches in a matchbox. Liquid crystals have many uses, most notably in the displays of wrist-watches and computers.

Matter is in the gaseous phase when atomic or molecular particles are widely separated (more than 10 particle diameters is typical) and there is little interaction. The gaseous phase can occur at low pressures where the number of particles is relatively few, or at high temperatures where particles move at high speeds and bounce rather than stick when they collide.* You can simulate this effect by throwing a small magnet against a refrigerator door. Throw the magnet gently and it will stick; throw it hard and it will bounce back.

Since particles of a gas are separated by relatively large average distances, matter in the gaseous phase occupies much more volume than it does in the solid or liquid phase. Although the particles of a gas move at high speeds, they do not go far because the particles are continually hitting one another. You can find evidence for the drift speed of gas particles in your home when someone opens an oven door after baking: a shot of aromatic gas molecules escape. In the next room you don't smell the aroma until the particles drift from the oven to your nose.

Figure 19.12 The computer screen is composed of liquid crystals, an array of crystals that exhibit the properties of both solids and liquids.

19.3 SPECIAL PROPERTIES OF WATER

Compared with other substances, water is most unusual. Its melting and boiling temperatures are remarkably high. Also, as was discussed in Section 6.4, water has one of the highest heat capacities, which means that relatively large amounts of energy are needed to change the temperature of a body of water to any considerable extent. Ice floats on water, lakes freeze from the top down and not from the bottom up, and snow flakes form with an intriguing six-sided geometry. We can understand why water has these and other unusual physical properties by examining it at the molecular level.

Water's unique features are the consequence of the ability of water molecules to cling tenaciously to one another by hydrogen bonding. We saw in Section 19.1 that the hydrogen bond is an example of an unusually strong dipole-dipole interaction. Each water molecule forms up to four of these interactions with neighboring water molecules. This gives rise to some unusually strong packing, and explains why water has such a high melting point. In order to melt, the packing must be broken apart. This only happens at a relatively high temperature of 0°C (Figure 19.13).

---

*Gaseous air molecules at room temperature travel in excess of 1500 kilometers per hour!

Figure 19.13 Water molecules pack together quite strongly on account of multiple hydrogen bonding. The result is an open crystalline structure through which there are many tunnels of unoccupied space.

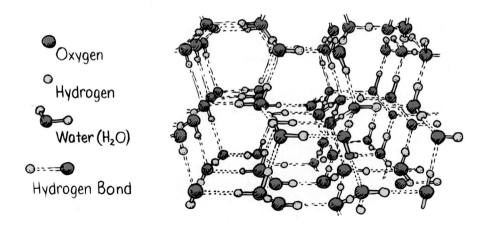

Oxygen

Hydrogen

Water (H₂O)

Hydrogen Bond

When water molecules pack together in a solid they form an open crystal structure. Upon melting this open structure collapses. Therefore, the water molecules take up less space when melted than when frozen in their crystal form. As was discussed in Section 6.5, this means that solid water (ice) is less dense than liquid water. The density difference is small, but it is all it takes to make ice float on top of water and to cause ponds and lakes to freeze from the top down.

A close examination of the open crystalline structure of frozen water reveals a network of 6-sided rings. Most snowflakes share a similar "6-sidedness." In fact, snowflake patterns are the macroscopic consequence of this molecular geometry.

Melting and Freezing

The molecules in ice crystals are held together by hydrogen bonds. Once there is sufficient heat to break these hydrogen bonds, the ice starts to melt. That is, the crystal structure starts to break apart and the material becomes fluid. Breaking a hydrogen bond, however, requires energy, just as breaking any interaction requires energy. For example, try to pull two magnets apart and you will find it requires the *input* of energy. Similarly, to melt ice we must supply energy to it.

Recall from Section 6.4 that 80 calories (335 joules) of energy are needed to melt 1 gram of ice at its melting temperature, 0°C. This energy is known as water's **heat of fusion** (which is small compared to water's *heat of vaporization*, the 540 calories [2260 joules] per gram needed to pull water molecules completely apart from one another at 100°C to form steam). Adding heat to melt ice doesn't change its temperature, for the added energy goes into breaking the hydrogen bonds. Only when all the ice has melted will temperature rise with the addition of heat. Each calorie of heat supplied will then raise the temperature 1°C for each gram of water.

The reverse also holds true. If we wish to cool water we must take away 1 calorie per gram to reduce the temperature by 1°C. Then to freeze water we must reduce its temperature to 0°C and then extract 80 calories per gram.

Molecular Differences between Solid and Liquid Water

Hydrogen bonds hold water molecules together in the solid phase (as in ice), but they also serve to hold water molecules together in the liquid phase. Compared to the solid phase, there are fewer hydrogen bonds in the liquid phase. Just after ice melts, for example, approximately 85 percent of the hydrogen bonds remain. As the

Figure 19.14 Add heat to liquid water and the temperature increases. Add heat to melting ice, however, and there is no change in temperature.

Figure 19.15 Thermal energy is readily stored in the breaking of hydrogen bonds. But hydrogen bonds are not broken as heat is applied to ice (providing it doesn't melt) or water vapor. Consequently, the heat capacities of ice and water vapor are about half that of liquid water.

temperature of liquid water increases, the percentage of hydrogen bonds decreases even further. From a molecular point of view, however, it is the orientation of the hydrogen bonds, not the number of them, that is most significant. As a solid, hydrogen bonds are fixed *permanently* within the crystalline structure. Since water molecules cannot move relative to one another, the water is solid in character. In the liquid phase it is quite different as hydrogen bonds are *impermanent,* continually breaking and reforming. Since water molecules are constantly changing partners, the water is fluid in character.

QUESTION

How can you add heat to ice without it melting?

Specific Heat Capacity of Water

We discussed water's most important thermal property in Chapter 6—its relatively high specific heat capacity—its ability to store vast amounts of thermal energy for correspondingly small increases in temperature. We return to this topic here, in terms of hydrogen bonding. Recall that heat applied to a pot of water increases its temperature only slightly, while the same heat applied to an empty metal pot increases its temperature a lot. Water, compared to other materials, has a great capacity for storing thermal energy. Water acts like a sort of "heat sponge."

ANSWER

A common misconception is that ice cannot have a temperature less than 0°C. In fact, ice can have any temperature below 0°C, on down to absolute zero, −273°C. An effect of adding heat to ice is to raise its temperature—say from −200°C to −100°C—as long as its temperature stays below 0°C the ice will not melt.

Why water has a high specific heat capacity involves guess what? Once again, hydrogen bonds! When heat is applied to water, much of it is consumed in breaking hydrogen bonds. As we saw with water's heat of fusion, broken hydrogen bonds are a form of potential energy (just as two magnets pulled apart are a form of potential energy). Much of the heat added to water, therefore, is stored as this potential energy. Consequently, less heat is available to increase the kinetic energy of the water molecules. Since temperature is a measure of kinetic energy, we find that as water is heated, its temperature rises slowly. By the same token, when water is cooled, its temperature drops slowly, primarily because as the kinetic energy decreases, molecules slow down and more hydrogen bonds are able to reform. This in turn releases heat that helps to maintain the temperature.

So because of water's high specific heat capacity, the temperature of water is difficult to change. As was discussed in Section 6.4, this means that people who live around oceans and lakes have milder summers and winters than people who do not. Actually, the overall temperature of our entire planet is moderated by the fact that over three-fourths of the surface is covered by water. Without such moderate temperatures, the earth would be much less hospitable to life.

19.4 SURFACE TENSION AND CAPILLARY ACTION

Take the spring out of a ball point pen and drop it into water. Since the weight of the spring is more than the buoyant force acting on it, the spring sinks. Now, dry the spring thoroughly and very gently lay it on top of the water. If done properly, the spring floats! Why?

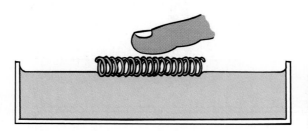

Figure 19.16 A spring lies on the surface of the water, pushing the water down slightly, but not sinking.

Molecules on the surface are attracted to neighboring surface molecules, and it is this sideways attraction that allows the spring to stay afloat. Consider a circle of people holding hands. To go into the circle you must break through the holding hands. Similarly, if the spring is to enter the water it must break through the grip of cohesive water molecules (Figure 19.17). The surface behaves as if it were a tightened elastic film upon which various light objects are able to rest, provided they do not pierce through it. Dry steel needles on their sides or razor blades lying flat can be made to

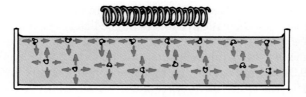

Figure 19.17 The spring encounters the surface molecules, which are clinging to one another so as to form a somewhat impenetrable barrier.

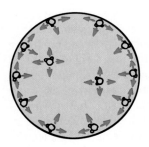

Figure 19.18 The molecules on the surface have a net pull inward toward the center. This causes the material to reduce its surface area as much as possible. When the liquid is not contained, it forms a sphere, which has a minimum surface area per volume.

rest on the water in this fashion. Also, many insects that "skate" on top of streams and ponds use this means of support.

The **surface tension** of a liquid is defined as the energy required to break through the surface. Liquids with strong chemical interactions, such as water, typically have high surface tensions. Liquid mercury has one of the highest (about six times greater than that of water). Soap molecules disrupt the dipole-dipole interactions among water molecules, so we find that soap interferes with the surface tension of water. Get the ball point pen spring floating and then carefully touch the water a few centimeters away with the corner of a bar of wet soap. You will be amazed by how quickly the surface tension is destroyed.

Surface tension accounts for the spherical shape of liquid drops. The surface molecules of a liquid are pulled sideways, but attractions from underneath also serve to pull them downward into the liquid. This pulling of surface molecules into the liquid causes the surface to contract and become as small as possible. Guess what geometrical shape has the least surface for a given volume? That's right—a sphere. So we see why raindrops, drops of oil, and falling drops of molten metal are all spherical (Figure 19.18).

Turn the ball point pen spring on-end and touch it to the surface, then pull the spring upward. You'll find that for a short distance, the water is brought up with the metal (Figure 19.19). What force is it that allows the water to be pulled up against gravity?

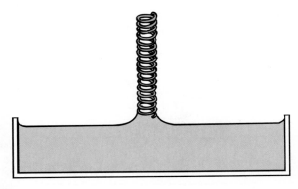

Figure 19.19 Water molecules are attracted to the metal spring.

Figure 19.20 Adhesive forces account for the observation that when you empty a glass of water, some water drops invariably cling behind. Cohesive forces account for the beading together of water drops.

Water molecules are polar and the metal is full of loose electrons. As discussed in Section 19.1, this makes for relatively strong dipole-induced dipole interactions. This is how the water and the metal stick to each other. Chemical interactions that arise between two different substances are called **adhesive forces**. (When between like substances they are called *cohesive forces*.) Glass is made of polar molecules, hence, there are adhesive forces between glass and water. This causes water to creep up the sides of glass containers, as is depicted in Figure 19.21. We call the curving of water (or any liquid) at the interface of a container a **meniscus**.

When a small diameter glass tube is placed in water, adhesive forces initially cause a relatively steep meniscus (Figure 19.21a). The attractive forces among water molecules, the **cohesive forces**, immediately respond to minimize the surface area of the meniscus (Figure 19.21b). The result is a rise of the water level within the tube. Adhesive forces again cause the formation of another steep meniscus (Figure 19.21c). This is followed by the action of cohesive forces, which cause the steep meniscus to be "filled in" (Figure 19.21d). This cycle is repeated until the upward

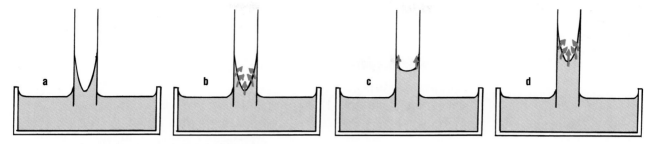

Figure 19.21 Water is spontaneously drawn up a narrow glass tube.

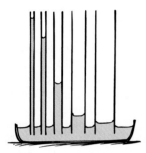

Figure 19.22 Capillary tubes.

adhesive force equals the weight of raised water. This rise in the fluid due to adhesive and cohesive forces is called **capillary action**. In a tube with a bore of about 0.5 millimeter in diameter the water will rise slightly higher than 5 centimeters. With a still smaller bore, the water will rise much higher (Figure 19.22).

We see capillary action at work in many phenomena. If a paintbrush is dipped into water, the water will rise up into the narrow spaces between the bristles by capillary action. Hang your hair in the bathtub, and water will seep up to your scalp in the same way. This is how oil soaks upward in a lamp wick and water into a bath towel when one end hangs in water. Dip one end of a lump of sugar in coffee, and the entire lump is quickly wet. The capillary action occurring between soil particles is important in bringing water to the roots of plants.

19.5 SOLUTIONS

Why does sugar disappear as it is stirred into water? Is the sugar destroyed? We know this is not so because the sugar sweetens the water. Does the sugar disappear because it somehow ceases to occupy space, or because it fits within the nooks and crannies of the water? These hypotheses too are incorrect because the presence of sugar also affects the water's volume. This may not be noticeable at first, but continue to add sugar and eventually you'll see that the volume increases just as it would if you were adding sand.

What happens to the sugar is that it loses its crystalline form. Each crystal of sugar consists of billions upon billions of sugar molecules packed tightly together. When the sugar crystals are exposed to water, they are pulled apart by an even greater number of water molecules (Figure 19.23). With a little stirring we soon find the sugar molecules dissolve throughout the water. In place of sugar crystals and water

Figure 19.23 Water molecules serve to break sugar molecules away from one another.

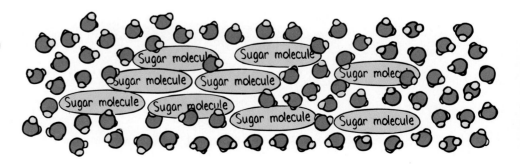

Figure 19.24 When copper sulfate is added to water the result is a blue homogeneous mixture.

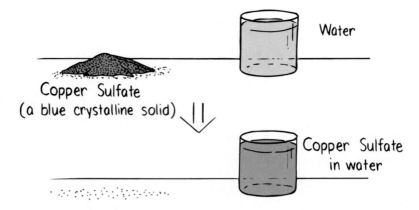

Water

Copper Sulfate
(a blue crystalline solid)

Copper Sulfate
in water

we have a *homogeneous mixture* of sugar and water. * We will discuss the details of sugar dissolving in water shortly, when we discuss the concept of solubility.

When a homogeneous mixture consists of only a single phase it is called a **solution.** Sugar in water is a solution where the phase is liquid. Copper sulfate in water is another example (Figure 19.24). The phase of a solution, however, is not restricted to liquids. Solutions can also be solid or gaseous. Gem stones are examples of solid solutions. A ruby, for example, is a solid solution of trace quantities of red chromium ions, Cr^{3+}, in transparent aluminum oxide. A blue sapphire is a solid solution of trace quantities of light green iron ions, Fe^{2+}, and blue titanium ions, Ti^{4+}, in aluminum oxide. We find that a gem's quality of color depends entirely upon the relative proportions of the trace metal ions it contains. Other important examples of solid solutions include metal alloys. Brass, for example, is a solid solution of copper and zinc, and stainless steel is a solid solution of metallic iron, chromium, nickel, and carbon. The air we breathe is a gaseous solution of 78 percent nitrogen, 21 percent oxygen, and 1 percent other gaseous materials, which include water vapor and carbon dioxide. The air we *exhale,* is a gaseous solution of 75 percent nitrogen, 14 percent oxygen, 5 percent carbon dioxide, and 6 percent water vapor.

There are many terms that are commonly used to describe solutions. Typically, these terms refer to the relative proportions of the components. It is usual to think of the component present in the largest amount as the **solvent** and the other component(s) as the **solute(s)**. For example, when a teaspoon of sugar is mixed with a liter of water, we identify the sugar as the solute and the water as the solvent. The process of mixing a solute in a solvent is called **dissolving**. In order to make a solution, a solute must *dissolve* in a solvent, that is, the solute and solvent must form a homogeneous mixture that consists of only one phase.

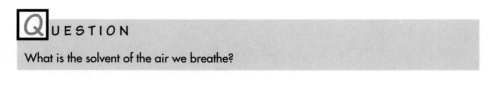

QUESTION

What is the solvent of the air we breathe?

ANSWER

Nitrogen is the solvent because it is the component of air that is in greatest quantity.

---

*As was discussed in Section 17.4, homogeneous means that a sample from one part is the same as a sample from any other part. For example, the first sip of the sugar water is the same as the last.

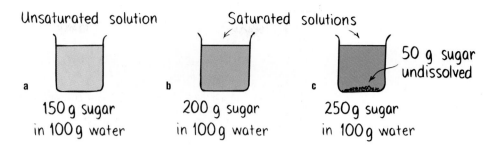

Unsaturated solution

Saturated solutions

a 150 g sugar in 100 g water

b 200 g sugar in 100 g water

c 250 g sugar in 100 g water

50 g sugar undissolved

Figure 19.25   A maximum of about 200 grams of sugar will dissolve in 100 grams of water at 20°C. (a) 150 g of sugar in 100 g of water produces an unsaturated solution. (b) 200 g of sugar in 100 g of water produces a saturated solution. (c) If 250 g of sugar is mixed into 100 g of water, 50 g will remain undissolved.

Solutions are often described by the amount of solute they contain. When there is a relatively large amount of dissolved solute, the solution is **concentrated**. Strong coffee is an example of a concentrated solution, for it contains large amounts of solutes, such as caffeine. When there is a relatively small amount of dissolved solute, the solution is **dilute**. Weak coffee is an example of a dilute solution.

In many instances, the amount of solute that can dissolve in the solvent is limited. When you add sugar to a glass of water, for example, the sugar rapidly dissolves. But as you continue to add sugar, there comes a point when the solid no longer dissolves and it simply collects at the bottom of the glass, even after stirring. At this point we find that the solution contains a maximum amount of solute—the solution is **saturated**. A solution that has *not* reached the limit of solute that will dissolve in it is said to be **unsaturated** (Figure 19.25).

The quantity of solute dissolved in a solution is described in mathematical terms by the solution's **concentration**—the amount of solute dissolved per volume of solution.

$$\text{Concentration} = \frac{\text{Amount of Solute}}{\text{Volume of Solution}}$$

For example, sugar water may have a concentration of 1 gram of sugar for every liter of solution.* This can be compared with concentrations of other solutions. A sugar solution containing 2 grams of sugar per liter of solution, for example, would be more *concentrated*. A sugar solution containing only 0.5 grams of sugar per liter of solution would be less concentrated, or more *dilute*.

We can calculate the amount of solute in a solution if we know the concentration of the solution and its volume. That is,

$$\text{Amount of Solute} = \text{Concentration} \times \text{Volume}$$

For example, if the concentration of sugar water is 1 gram per liter and the volume of solution is 1 liter, then we find the solution contains 1 gram. In using this relation, notice that the units of volume must be the same:

$$\text{Amount of Solute} = \left(\frac{1 \text{ gram}}{1 \text{ liter}}\right) \times (1 \text{ liter}) = 1 \text{ gram}$$

*Note carefully that the definition of concentration used here refers to liters of *solution* and not the solvent.

1. How many grams of sugar are there in 5 liters of sugar water that has a concentration of 0.5 grams per liter of solution?

2. A saturated solution of salt water has a concentration of about 300 grams per liter of solution. How many grams of sodium chloride are required to make 3 liters of a saturated solution?

Chemists are often more interested in the number of molecules of solute in a solution rather than in the number of grams. Molecules, however, are so very small that the number of them in any observable sample is incredibly large. To get around cumbersome numbers a new unit called **the mole** was defined. For reasons that are explained in Section 20.7, one mole is an incredibly large number—6.02×10^{23}. A mole of marbles, for example, would be enough to cover the entire land area of the United States to a depth greater than 4 meters! But molecules are incredibly small. There is a mole of sugar molecules in 330 grams. Therefore, a solution of sugar water that has a concentration of 330 grams per liter also has a concentration of 6.02×10^{23} molecules per liter, or, by definition, a concentration of 1 mole per liter. We'll return to the mole in the next chapter.

The most common unit of concentration used by chemists is the **molarity,** which is the solution's concentration expressed in moles per liter.

$$\text{Molarity} = \frac{\text{Number of Moles}}{\text{One Liter of Solution}}$$

A solution that contains 1 mole of solute per liter of solution is said to have a concentration of 1 *molar,* which is often abbreviated 1 M. Similarly, a more concentrated 2 molar solution would be abbreviated 2 M.

The importance of referring to the number of molecules of a solute rather than the number of grams will become more apparent in the next chapter on chemical reactions. For now, consider the following question: The saturated solution of sugar water in Figure 19.25b contains 200 g of sugar and 100 g of water. Which is the solvent—the sugar or the water? We find there are 3.5×10^{23} molecules of sugar in 200 g of sugar, but there are almost ten times as many molecules of water in 100 g of water—3.3×10^{24}. As defined here, the solvent is the component present in the largest amount. What do we mean by *amount?* If amount is taken to be the number of molecules, then the water is the solvent. If amount is taken to be mass, then the sugar is the solvent. So the answer depends on how you look at it. From a chemist's point of view it is typically by the number of molecules, so water would be the solvent in this case.

1. Multiply the concentration by the volume: (0.5 g/liter)(5 liters) = 2.5 grams.

2. Multiply the concentration by the volume to obtain the amount of solute required: (300 g/liter)(3 liters) = 900 grams.

QUESTIONS

1. How many moles of sugar are there in 0.5 liters of a 4 M solution? How many molecules of sugar is this?
2. Does 1 liter of a 1 M solution of sugar water contain one liter of water, less than one liter of water, or more than one liter of water?

Solubility

The **solubility** of a solute refers to its *ability* to dissolve in a solvent. As we might expect, this ability depends in part on the chemical interactions that take place between the solute and the solvent. With strong chemical interactions, a solute may be very **soluble**—a lot can be dissolved before the solution is saturated. With weaker chemical interactions, a solute may be less soluble, and the solution may become saturated after only a very little is dissolved. To illustrate the various degrees of solubility, we consider several examples.

In some instances, chemical interactions are so significant that there is no limit to the amount of solute that can be dissolved—there is no point of saturation. In these cases, the two materials are said to be *infinitely soluble*. This is the situation for ethanol and water, which are attracted to each other by the hydrogen bond. We can even add ethanol to water until the ethanol rather than the water may be considered the solvent. In fact, ethanol and water stick so well to each other that even after distillation, the purest ethanol we can get is 95 percent (Figure 19.26).

As discussed in Section 19.1, sodium chloride has excellent solubility in water because of relatively strong ion-dipole interactions. Sugar also has excellent solubility in water, not because of ion-dipoles, but because of dipole-dipole interactions. A sugar molecule (chemical name *sucrose*) has many hydrogen-oxygen bonds. Each of these bonds is polar with the hydrogen side slightly positive, $\delta+$, and the oxygen side slightly negative, $\delta-$. This attracts many water molecules, which are also polar. The result is that each sugar molecule becomes surrounded by a multitude of water molecules and the sugar dissolves (Figure 19.27).

Considering the many dipole-dipole interactions that can take place between sugar and water, we may wonder why sugar is not infinitely soluble in water. Solubility has not only to do with interactions between the solute and the solvent, but also with interactions among solute molecules and interactions among solvent molecules. How about interactions among sugar molecules themselves? These interactions,

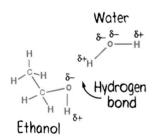

Figure 19.26 Ethanol and water attract each other by an unusually strong dipole-dipole interaction, the "hydrogen bond." The amounts of ethanol and water that can be mixed is unlimited.

ANSWERS

1. First you need to understand that 4 M is equal to 4 moles per liter. Then multiply the concentration by the volume to obtain the amount of solute: (4 moles/liter)(0.5 liters) = 2 moles. Since 1 mole equals 6.02×10^{23} molecules, 2 moles equals twice this much, or 12.04×10^{23} molecules.

2. The definition of molarity refers to the number of liters of solution, not liters of solvent. When sugar is added to water, the volume increases. So, if 1 mole of sugar were added to 1 liter of water, the result would be *more* than 1 liter of solution. One liter of a 1 M solution, therefore, requires less than one liter of water. If you wish to calculate the concentration, it is best to add the solvent to the solute, rather than the other way around. Discuss with a classmate or your instructor why this is so.

Figure 19.27 Many water molecules are attracted to a sugar molecule by way of dipole-dipole interactions.

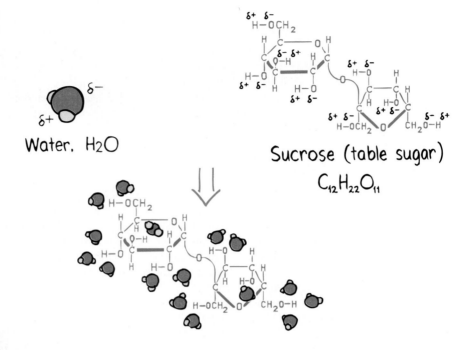

Water, H_2O

Sucrose (table sugar) $C_{12}H_{22}O_{11}$

Figure 19.28 Glass is frosted by dissolving in hydrofluoric acid.

Figure 19.29 Styrofoam cups dissolve in acetone.

after all, are strong enough that sugar is a solid at room temperature and doesn't melt until a temperature of 185°C is reached (which is 185° greater than the melting point of water). The fact that water molecules must work to pull sugar molecules away from one another serves to diminish, somewhat, the solubility of sugar in water. So although sugar is very soluble in water, we see why it is not infinitely soluble in water.

An example of a solute that has low solubility in water is oxygen, O_2. In contrast to sugar, only 0.004 grams of oxygen can dissolve in 100 grams of water. We can account for oxygen's low solubility in water by noting that only weak dipole-induced dipole interactions occur. More importantly, however, the stronger attraction of water molecules to water molecules effectively excludes oxygen molecules.

A material that does not dissolve in a solvent to any appreciable extent is said to be **insoluble**. There are many substances that we consider to be completely insoluble in water, including sand, glass, and Styrofoam. With sand and glass, the chemical interactions with water are not sufficient to overcome the chemical forces within these substances. With Styrofoam, it is again a case of not being able to compete with water's attraction for itself. But just because a material is not soluble in one solvent does not mean it won't dissolve in another. Sand and glass, for example, are soluble in hydrofluoric acid, which is used to give glass a decorative frosted look. Styrofoam is soluble in acetone, which is a solvent used in finger nail polish remover. Pour a little acetone into a Styrofoam cup and you will see the acetone dissolve right through the bottom.

Solubility and Temperature

We know from experience that water soluble solids, such as sugar, dissolve better in hot water than in cold water. A molecular explanation is that hot water molecules have greater kinetic energy, and are able to "attack" the solid solute more

Perfluorocarbons

Oxygen is not very soluble in water because polar water molecules are so attracted to themselves that they exclude nonpolar oxygen molecules. We find, however, that oxygen is very soluble in a new class of nonpolar compounds known as perfluorocarbons, which like Teflon, consist of carbon atoms bonded to many fluorine atoms. The structure of a typical perfluorocarbon molecule is shown at left.

Interestingly enough, a saturated solution of oxygen in liquid perfluorocarbons is about 20 percent more concentrated in oxygen than the atmosphere we breathe. Astoundingly, when this solution is inhaled, the lungs are able to absorb the oxygen in much the same way they absorb it from air, and since perfluorocarbons themselves are as inert as Teflon, negative side-effects of having such a fluid in the lungs are minimal (Figure 19.30).

Much research is presently being conducted on perfluorocarbons and their potential applications. For example, it is very difficult and nearly impossible for babies born before 7 months gestation to breathe air. This is because their lungs lack an inner lining that prevents the lungs from collapsing due to the cohesiveness of water. Researchers at the University of Pittsburgh have found that premature infants can breathe quite effectively when inhaling oxygenated perfluorocarbons. Adults may also benefit from inhaling perfluorocarbons. When the liquid is drained from the lungs it carries with it much foreign matter that has accumulated over time. Have you had your lungs cleaned lately?

Another exciting application that has already been demonstrated is that perfluorocarbons may be used as a blood substitute in humans. Among the many advantages of such an "artificial blood" would be low cost, long-term storage capability, and the elimination of the transmission of diseases such as hepatitis and AIDS through blood transfusions.

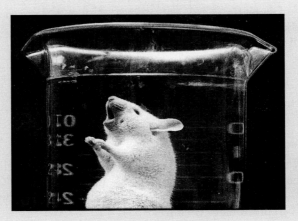

Figure 19.30 A mouse alive and well breathing liquid perfluorocarbon saturated with oxygen gas.

vigorously. But while the solubilities of some solid solutes are greatly affected by temperature changes, the solubilities of other solid solutes are only mildly affected (Figure 19.32). Why this is so has to do with a number of factors, including the strength of chemical forces within the solid, and the way the molecules of the solid are packed together.

When a solution saturated at a high temperature is allowed to cool, some of the solute usually comes out of solution. When this happens, the solute is said to have **precipitated**. For example, at 100°C the solubility of sodium nitrate, $NaNO_3$,

Oxygen in Our Blood

Figure 19.31 A hemoglobin molecule.

If oxygen is not very soluble in water, how is it possible for blood, which is mostly water, to deliver oxygen efficiently to cells? Blood is more than water. It is actually both a solution of ionic compounds such as sodium chloride and a suspension of blood cells. Red blood cells contain the large oxygen capturing molecule hemoglobin (Figure 19.31). At the center of each hemoglobin molecule there is an iron atom in its +2 ionized state, Fe^{2+}. We find that this form of iron readily combines with oxygen to become Fe^{3+}. Interestingly enough, this is the same chemistry that causes an iron nail to rust. So, when red blood cells are passed through the blood/lung interface, the Fe^{2+} ions of hemoglobin attract available oxygen molecules. These oxygen molecules, however, never make actual contact with the iron because neighboring proteins in the red blood cell get in the way and hold the oxygen molecule just out of reach. Oxygen is normally a nonpolar molecule, but in the presence of such a strong ion, it is induced to become polar. The result is a significant ion-induced dipole interaction. This interaction is strong enough to hold the oxygen molecule onto the hemoglobin as it gets transported to other parts of the body. But it is also weak enough so that once the hemoglobin gets to an oxygen-requiring cell, the oxygen molecule is readily released.

• •

Figure 19.32 The solubility of many water soluble solids changes with temperature.

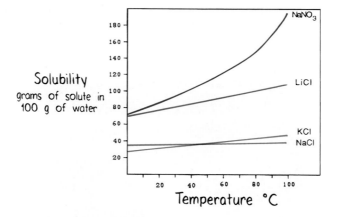

is 180 g per 100 g. As we cool this solution, the solubility of $NaNO_3$ decreases, and this causes some of the $NaNO_3$ to precipitate. At 20°C, the solubility of $NaNO_3$ is 87 g per 100 g. So, if we were to cool to 20°C, we would find that 93 g (180 g − 87 g) precipitates (Figure 19.33).

Figure 19.33 The solubility of sodium nitrate, $NaNO_3$, is 180 g per 100 g of water at 100°C but only 87 g/100 g at 20°C. Cooling a 100°C saturated solution of $NaNO_3$ to 20°C causes 93 g of the solute to precipitate.

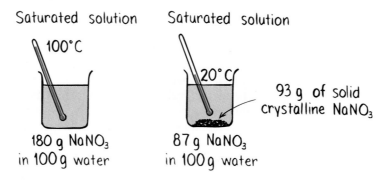

Figure 19.34 Rock candy is grown out of a supersaturated solution of sugar water.

If a hot saturated solution is allowed to cool slowly and without disturbance, it is possible that the solute may stay in solution. The result is a **supersaturated** solution. Supersaturated solutions of sugar water are fairly easy to make. Just dissolve as much sugar as possible in some boiling water, then allow the solution to cool. You will find that this solution is somewhat unstable, for if you disturb it, by stirring for example, massive quantities of sugar will suddenly precipitate. It is possible to grow some rather large sugar crystals, also known as "rock candy," out of a supersaturated solution. One good way is to tie some string to a weight, such as a nut or bolt, and lower the string into the solution *before* it has a chance to cool. Support the string with a pencil such that the weight does not touch the bottom. Leave the mixture undisturbed for about a week, but check it periodically. The longer you wait, the larger the crystals (Figure 19.34).

In contrast to solids, the solubilities of gases in liquids *decrease* with increasing temperature. With an increase in temperature the solvent has more kinetic energy. This makes it more difficult for the gaseous material to stay in solution because it is literally being kicked out by the solvent. We have all noticed that warm carbonated beverages go flat faster than cold ones. Greater warmth causes the carbon dioxide molecules of carbonated beverages to leave at a higher rate.

The concepts of chemistry enable one to think critically. An experienced chemist, for example, need not memorize some long list of what solutes are soluble in which solvents. Instead, fairly accurate predictions about solubility can be made based upon an understanding of chemical interactions. Similarly, recall from Chapter 18 that the properties of any element or compound can be inferred using the periodic table as a guide. The concepts of chemistry also tell us much about how the world and all its materials are put together, and even provide a framework for understanding and appreciating living material—life itself. Chemistry, however, is a laboratory science—very much so. Using only chemistry concepts, there is no way to predict with absolute certainty the outcome of every chemical experiment. For example, the predicted properties of a substance may be somewhat different from its actual properties. So, knowledge in chemistry is twofold: a blend of conceptual understanding and actual experience. In fact, much of the flavor of chemistry comes from hands-on experience, which you will hopefully be able to gain through a laboratory that accompanies your physical science course.

Figure 19.35 Oxygen is less soluble in warm tropical waters. This is one of the reasons why tropical waters are not as fertile as colder polar waters.

So having some familiarity with the basic concepts of chemistry in Chapter 17, and seeing the rhyme and reason that underlies the periodic table in Chapter 18, and now learning about the nature of chemical interactions in this chapter, we move on to the heart of chemistry—chemical reactions, as shown in Chapter 20. Onward!

SUMMARY OF TERMS

Chemical interaction The electrical force of attraction or repulsion between two or more molecules or other chemical units.

Dipole A chemical unit wherein electric charge is separated. For example, a molecule negatively charged on one side and positively charged on the other is a dipole.

Polar The state of having a dipole. A molecule may be described as *polar* when it has a dipole.

Nonpolar The state of having no dipole. A molecule may be described as *nonpolar* when it has no dipole.

Ion-dipole The chemical interaction involving an ion and a dipole.

Dipole-dipole The chemical interaction involving dipoles.

Hydrogen bond A strong dipole-dipole interaction that involves a hydrogen atom chemically bonded to a strongly electronegative element, such as nitrogen, oxygen, or fluorine.

Induced dipole A dipole temporarily created in an otherwise nonpolar molecule. It is *induced* by a neighboring charge or dipole.

Dipole-induced dipole The chemical interaction involving a dipole and an induced dipole.

Induced dipole-induced dipole The chemical interaction involving only induced dipoles. This is a relatively weak chemical interaction.

Solid A phase of matter characterized by definite volume and shape.

Crystal A material, usually a solid, in which atomic or molecular particles are arranged in an ordered periodic fashion.

Amorphous A term used to describe a solid in which atomic or molecular particles are randomly oriented.

Liquid A phase of matter characterized by definite volume but indefinite shape.

Liquid crystal A liquid in which atomic or molecular particles have a resemblance of order and periodicity.

Gas A phase of matter characterized by indefinite volume and shape.

Freeze To change from a liquid to a solid phase. This process involves the output of energy.

Melt To change from a solid phase to a liquid phase. This process involves the input of energy.

Heat of fusion The heat released as a substance freezes or the heat absorbed as it melts.

Heat capacity The ability of a substance to retain thermal energy. Water has a large heat capacity because much of the heat added to it is stored in the breaking of hydrogen bonds.

Surface tension The energy required to break through the surface of a liquid.

Adhesive forces Chemical interactions that arise between two different substances.

Cohesive forces The attractive forces within a substance.

Meniscus The curving of a liquid at the interface of its container.

Capillary action The rising of liquid into a small vertical space due to adhesion between the liquid and the sides of the container and to cohesive forces within the liquid.

Solution A homogeneous mixture consisting of only a single phase. Salt water is an example of a solution.

Suspension A homogeneous mixture consisting of more than a single phase. Examples include milk, blood, and clouds.

Solvent The component in a solution present in the largest amount.

Solute Any component in a solution that is not the solvent.

Dissolving The process of mixing a solute in a solvent.

Concentrated A solution containing a relatively large amount of solute.

Dilute A solution containing a relatively small amount of solute.

Saturated A solution containing as much solute as will dissolve.

Unsaturated A solution that will dissolve additional solute if added.

Concentration A quantitative measure of the amount of solute in a solution.

Mole A very large number equal to 6.02×10^{23}. This number is a unit commonly used when describing a number of molecules.

Molarity A common unit of concentration measured by the number of moles in one liter of solution.

Solubility The ability of a solute to dissolve, which depends not only on chemical interactions between the solute and the solvent, but upon the interactions among both solute molecules and among solvent molecules.

Soluble Capable of dissolving in a solvent.

Insoluble Not capable of dissolving to any appreciable extent in a solvent.

Precipitate A solute that has come out of solution.

Supersaturated A solution that contains more solute than it normally contains.

REVIEW QUESTIONS

Types of Chemical Interactions

1. What distinguishes a chemical interaction from a chemical bond?

2. Which is stronger, the ion-dipole interaction or the induced dipole-induced dipole interaction?

3. Why are water molecules attracted to sodium chloride?

4. How are ion-dipole interactions able to break apart the ionic bond, which is relatively strong?

5. Are electrons distributed evenly or unevenly in a molecular dipole? Why?

6. What exactly is a hydrogen bond?

7. Two molecules are attracted to each other by the dipole-dipole interaction. Is it possible for a third molecule to join in the attraction? If this third molecule had no dipole?

8. How are oxygen molecules attracted to water molecules?

9. Are induced dipoles normally permanent?

10. How can nonpolar argon atoms induce dipoles in other nonpolar argon atoms?

11. According to Table 19.2, which is the larger molecule, iodine, I_2, or fluorine, F_2?

12. Why is it difficult to induce a dipole in a fluorine atom?

13. Why is the boiling point of isooctane, C_8H_{18}, so much higher than the boiling point of methane, CH_4?

Phases of Matter

14. How are the molecules of a solid arranged differently from those of a liquid?

15. What distinguishes a crystalline solid from an amorphous solid?

16. How does the arrangement of molecules in a gas differ from the arrangements in liquids and solids?

17. Which occupies the most volume: 1 gram of ice, 1 gram of liquid water, or 1 gram of water vapor?

18. Gas particles travel at speeds of up to 1,500 km/hr. Why, then, does it take so long for aromatic gas molecules to travel the length of a room?

Special Properties of Water

19. Why is water sometimes described as a sticky substance?

20. What kind of dipole-dipole interaction takes place between water molecules?

21. What accounts for ice being less dense than water?

22. Why do lakes and ponds freeze from the top down rather than from the bottom up?

23. When water freezes, is heat released to the surroundings, or absorbed from the surroundings?

24. Are there hydrogen bonds between water molecules in the liquid phase?

25. Why does water have such a large heat capacity?

26. Why doesn't the temperature of melting ice rise as it is heated?

27. How many joules of heat are needed to melt one gram of ice?

28. How many calories of heat are needed to melt one gram of ice?

29. Is it relatively easy or difficult to change the temperature of a substance with a low heat capacity?

30. Which of the processes in the figure below involve the input of energy and which involve the output of energy? (Recall from Chapter 6 that sublimation occurs as a substance changes between solid and gaseous phases.)

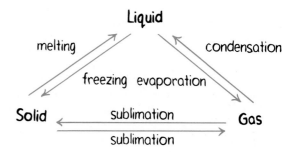

Surface Tension and Capillary Action

31. What is the cause of surface tension?

32. Why do liquids with stronger chemical interactions have greater surface tension?

33. What kind of chemical interactions take place between water and metal?

34. In which does water rise higher: a narrow tube or a wide tube?

35. What determines the height that water will rise by capillarity?

Solutions

36. What happens to the volume of water as sugar is dissolved in it?

37. Why is a ruby gemstone considered to be a solution?

38. Distinguish between a solute and a solvent.

39. What does it mean to say that a solution is concentrated?

40. Distinguish between a saturated and an unsaturated solution.

41. How is the amount of solute in a solution calculated? What information about the solution is needed?

42. Is a *mole* a very large or very small number?

43. By what means are ethanol and water molecules attracted to each other?

44. Why is sugar not infinitely soluble in water?

45. Why does oxygen have such a low solubility in water?

46. What effect does temperature have on the solubility of many solutes?

47. How are supersaturated solutions made?

. . . .

HOME PROJECTS

1. Black ink is made by combining many different-colored inks, such as blue, red, and yellow. Together, these inks serve to absorb all the frequencies of light. Since no light is reflected, the ink appears black. Use chemical interactions to separate the components of black ink through a special technique called "paper chromatography." Place a concentrated dot at the center of a piece of porous paper, such as a paper towel, napkin, or coffee filter. Next, carefully place a drop of solvent, such as water, acetone (fingernail polish remover), rubbing alcohol, or white vinegar on top of the dot and watch the ink spread radially with the solvent. Since the various components have differing affinities for the solvent, they travel with the solvent at differing rates. Just after your drop of solvent is completely absorbed, add a second drop, then a third, and so on until the components have separated to your satisfaction. How the components separate depends on several factors, including your choice of solvent and your technique. Black felt-tip pens tend to work best. It's also interesting to watch the leading edge of the moving

ink under a strong magnifying glass or microscope. Check for capillary action!

2. Add a few drops of food coloring to a shallow dish of water, stir, and then immerse the cut end of a stalk of celery that still has its leaves. Support the celery stalk upright and let it stand overnight. Slice open the celery stalk the next day to see how the water has been drawn up into the stalk. Explain your observations based upon capillarity and also upon the fact that water tends to evaporate from the leaves.

3. Just because a solid dissolves in a liquid doesn't mean that it no longer occupies space. Fill a tall glass to its brim with warm water. Carefully pour all the water into a larger container. Now add three or four heaping tablespoons of sugar or salt to the empty glass. Return half of the warm water to the glass and stir so as to dissolve all of the solid. Return the remaining water, and as you get close to the top ask your friend to predict whether the water level will be less than before, about the same as before, or whether the water will spill over the edge. If your friend doesn't understand the result, ask him or her what would happen if you had added a single large crystal of sugar or salt directly to the full glass of water.

4. Be sure to make those sugar crystals from a supersaturated solution of sugar water as described at the end of Section 19.5. Supersaturated solutions of salt and sodium bicarbonate (baking soda) also make interesting crystals. Crystal shape directly relates to how the molecules of the substance pack together. In fact, as we will see in Chapter 22, substances are often characterized by the shape of the crystals they form. Note how different solutes give rise to differently shaped crystals.

. . . .

EXERCISES

1. Why are ion-dipole interactions stronger than dipole-dipole interactions?

2. Dipole-induced dipole forces exist between water and gasoline, yet these two substances do not mix. Explain.

3. Fish don't breathe water, they breathe the oxygen, O_2, that is dissolved in the water. Since water is a polar substance and oxygen is a nonpolar substance, how can there be any oxygen, O_2, in the water?

4. Chlorine, Cl_2, is a gas at room temperature, yet bromine, Br_2, is a liquid at room temperature. Explain.

5. Dipole-induced dipole forces of attraction exist between water and gasoline, yet these two substances do not mix because water has such a strong attraction for itself. Which of the following compounds might best help to make these two substances mix into a single liquid phase?

6. Estimate the amount of heat that is released when a bathtub of water freezes.

7. About 10 million calories of heat are released when a particular sample of water freezes. Why doesn't this heat simply remelt the ice?

8. Like water, hydrogen fluoride, HF, and ammonia, H_3N, have somewhat high boiling points. Explain.

9. Why must a razor blade be lying flat in order for it to be held up on the surface of water by surface tension?

10. Capillary action causes water to climb up the centers of narrow glass tubes. Why does the water not climb so high when the glass tube is wider?

11. Mercury forms a convex meniscus with glass. What does this tell you about the cohesive forces within mercury versus the adhesive forces between mercury and glass? Which are stronger?

12. Does it make sense to talk about the surface tension of a solid? If so, might the surface tension of a solid be greater or weaker than that of a liquid? Defend your answer.

13. Would you expect the surface tension of water to increase or decrease with temperature? Defend your answer.

14. Consider the boiling points of the following compounds and their solubilities in water. Then briefly explain how the solubilities in water go down as the boiling points of these alcohols go up.

	CH_3OH	$CH_3CH_2CH_2CH_2-OH$	$CH_3CH_2CH_2CH_2CH_2-OH$
Boiling point:	65°C	117°C	138°C
Solubility:	infinite	8g/100g	2.3g/100g

15. The boiling point of 1,4-butanediol is 230°C. Would you expect this compound to be *soluble* or *insoluble* in water? Support your answer with a brief statement.

1, 4-butanediol

16. Are noble gases infinitely soluble in noble gases? Defend your answer.

17. Give two ways to tell whether a sugar solution is saturated or not.

18. At 10°C, which is more concentrated—a saturated solution of sodium nitrate, $NaNO_3$, or a saturated solution of sodium chloride? (See Figure 19.33.)

19. Oil has a very small solubility in water. What are the benefits and drawbacks of this fact when it comes to massive oil spills in the oceans?

20. Suggest why salt and gasoline are insoluble. Consider the chemical interactions that are involved within and between these substances.

21. When 50.0 mL of water are combined with 50.0 mL of ethanol the result is 98.0 mL of solution. Why is this so?

PROBLEMS

1. How many grams of salt are needed to make 15 liters of a solution that has a concentration of 3.0 grams per liter of solution?

2. If 1 mole of salt is put into water to make a total of one liter of solution, what is the molarity? What about for 2 moles in one half liter?

3. There is 1 mole of oxygen molecules, O_2, in every 32 grams of oxygen gas. If 0.04 grams of O_2 are able to dissolve in 1 liter of water, how many moles is this? How many molecules is this?

20

CHEMICAL REACTIONS

How do we tell substances apart? We identify them by their physical and chemical properties. Water, for example, is a colorless, odorless liquid at room temperature and atmospheric pressure. Water, we find, reacts with sodium metal to produce hydrogen gas and sodium hydroxide. A liquid that doesn't behave like this chemically isn't water. Physical and chemical properties characterize substances much as fingerprints or personalities characterize people.

Under proper conditions, substances change their identities. Grape juice treated with yeast changes to wine. Iron exposed to oxygen and water changes to rust. When one substance changes into another substance, there is a *chemical change*—one chemical becomes another by an alteration in atomic composition (Section 17.2). The process through which chemical change occurs is a **chemical reaction.**

Chemical reactions go on around us and within us. Around us, wood that catches fire reacts with oxygen in the atmosphere to form carbon dioxide and water plus large amounts of heat. The water vapor and carbon dioxide in the atmosphere, in turn, react to form a different substance called carbonic acid. Rain

Chemical reactions are at the heart of chemistry – a science geared to producing new and useful materials.

is naturally acidic because of this carbonic acid.* Inside us, the chemical reactions are abundant. Living bodies are like fast-paced cities where activity goes on in every corner day and night. But in living bodies chemical reactions make up the relentless activity. Chemical reactions link different life forms to each other as well. Plants take carbon dioxide molecules out of the atmosphere and combine them to assemble the biomaterials for a new leaf. Animals eat plants to form structural materials such as skin, bone, muscle, and fat. Predators then eat plant-eating animals to form their own versions of skin, bone, muscle, and fat. A web of complex chemical reactions continually creates new substances out of old ones in this realm of life.

20.1 THE CHEMICAL EQUATION

Figure 20.1 Ammonia and hydrogen chloride gases react to form a white cloud of solid ammonium chloride.

On paper, a chemical reaction is represented by a chemical equation. The **reactants** (materials present before the reaction) are shown before an arrow that points to the **products** (new materials formed by the chemical reaction). Typically, the reactants and products are represented by their atomic or molecular formulas. Phases are also often shown: (*s*) solid, (*l*) liquid, or (*g*) gas. Materials dissolved in water are designated as (*aq*) aqueous. A chemical reaction, for example, occurs when gaseous ammonia, NH_3, and gaseous hydrogen chloride, HCl, are combined. The result is solid ammonium chloride, NH_4Cl (Figure 20.1). This is represented by the following chemical equation:

$$NH_3\ (g)\quad +\quad HCl\ (g)\quad \rightarrow\quad NH_4Cl\ (s)$$
$$\text{reactants} \qquad\qquad\qquad \text{product}$$

In this example, ammonia, NH_3 (*g*), and hydrogen chloride, HCl (*g*), are the reactants and ammonium chloride, NH_4Cl (*s*), is the product.

Numbers are placed in front of the chemicals to show the molecular ratio in which reactants combine and products form. These numbers are called **coefficients**. For example, we find that NH_3 molecules react with HCl molecules to produce NH_4Cl in a 1:1:1 ratio. Hence, the equation can be written as:

$$1\ NH_3\ (g) + 1\ HCl\ (g) \rightarrow 1\ NH_4Cl\ (s)$$

Likewise, hydrogen, H_2, and oxygen, O_2, react to form water, H_2O, in a 2:1:2 ratio. This is indicated by using coefficients:

$$2\ H_2\ (g) + 1\ O_2\ (g) \rightarrow 2\ H_2O\ (l)$$

By convention, a coefficient of 1 is omitted so that the above two chemical equations are typically shown as:

$$NH_3\ (g) + HCl\ (g) \rightarrow NH_4Cl\ (s)$$
$$2\ H_2\ (g) + O_2\ (g) \rightarrow 2\ H_2O\ (l)$$

According to the principle of the conservation of mass, matter is neither created nor destroyed in a chemical reaction (to the limits of our detection, Section 17.1). The chemical equation, therefore, must be *balanced;* that is, the number of

*See the box entitled "Acid Rain and Basic Oceans."

times each element appears on both sides of the arrow must be the same. For example, the above equation for the formation of NH_4Cl is balanced as each side shows one nitrogen, four hydrogen, and one chlorine atom. Similarly, the equation for the formation of water is also balanced. Note that a coefficient in front of a chemical tells you the number of times that chemical must be counted. For example, $2\ H_2O$ indicates two units of H_2O, which altogether contain 4 hydrogen and 2 oxygen atoms.

Practicing chemists have developed a knack for balancing equations. This knack can involve much creative energy. We leave the art of balancing equations to introductory chemistry courses where it is treated more fully. Here, you may want to try your hand at balancing a few relatively simple examples. One rule to keep in mind when balancing a chemical equation is that the coefficients can be changed however necessary in order to come up with correct proportions, but the subscripts cannot and should not be altered. Changing H_2O to H_2O_2, for example, means you no longer have water—you now have the formula for hydrogen peroxide! Also, never insert a coefficient between the atoms of a chemical. Coefficients must appear before a chemical compound, not within.

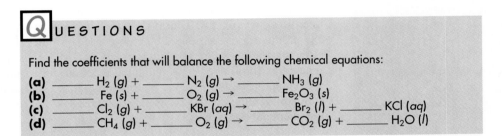

QUESTIONS

Find the coefficients that will balance the following chemical equations:

(a) _____ $H_2\ (g)$ + _____ $N_2\ (g)$ → _____ $NH_3\ (g)$
(b) _____ $Fe\ (s)$ + _____ $O_2\ (g)$ → _____ $Fe_2O_3\ (s)$
(c) _____ $Cl_2\ (g)$ + _____ $KBr\ (aq)$ → _____ $Br_2\ (l)$ + _____ $KCl\ (aq)$
(d) _____ $CH_4\ (g)$ + _____ $O_2\ (g)$ → _____ $CO_2\ (g)$ + _____ $H_2O\ (l)$

20.2 ENERGY AND CHEMICAL REACTIONS

Where do rockets get the energy to lift the space shuttle into orbit? Where does a campfire get the energy to glow red hot? We eat food to gain energy, but from where does this energy ultimately come? The answer to all these questions is: *chemical reactions*. We use the energy that comes from chemical reactions for many daily purposes, such as generating electricity, running cars, and cooking food. In this section we study the intimate relationship between energy and chemical reactions.

During a chemical reaction, chemical bonds are broken, atoms rearrange themselves, and then chemical bonds are reformed. Breaking and forming chemical bonds, however, is not something that can be done without the input or output of energy. Consider two magnets that are held together. If you want to pull them apart, it requires some "muscle energy." Similarly, we find that when magnets collide to become attached, energy is released in the form of heat (as they strike each other). The same principle applies to atoms. To pull bonded atoms apart, there must

Chemical reactions involve breaking or forming chemical bonds, which means changes in energy.

ANSWERS

(a) 3, 1, 2.
(b) 4, 3, 2.
(c) 1, 2, 1, 2.
(d) 1, 2, 1, 2. (By convention, the 1's are not typically shown in the final balanced equation.)

be energy input to overcome the electrical force of attraction between the atoms. Likewise, when bonding atoms accelerate toward each other to form a chemical bond, energy is released. Energy output is usually in the form of heat or light. It is important to note that chemical reactions work in accord with the conservation of energy principle: The amount of energy required to pull two bonded atoms apart is the same amount of energy released when they are brought together into a chemical bond.

The role of energy in a chemical reaction is nicely illustrated by the reaction of hydrogen with oxygen to form water:

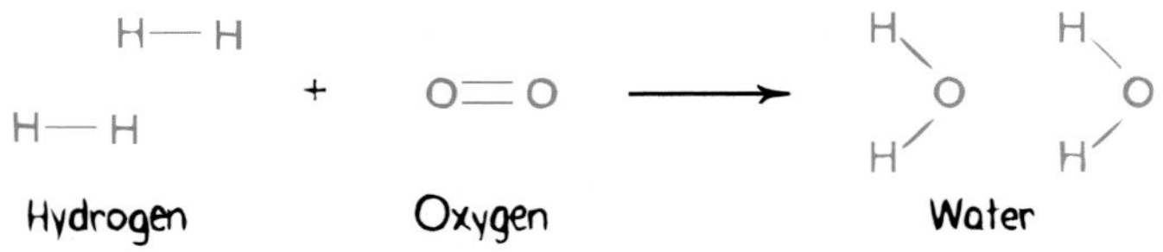

With the reactants we see that hydrogen atoms are bonded to hydrogen atoms and oxygen atoms are bonded to oxygen atoms. In the products, however, we find that hydrogen atoms are bonded to oxygen atoms. So in order for the reaction to proceed, hydrogen-hydrogen and oxygen-oxygen bonds must first be broken. Only then can the atoms rearrange themselves to form hydrogen-oxygen bonds.

Breaking chemical bonds requires energy. How much energy can be calculated with experimental data. For example, we find that 436 kJ (kilojoules) of energy are required to break the hydrogen-hydrogen bond.* Since there are two hydrogen-hydrogen bonds that need to be broken, according to the balanced chemical equation, twice this amount is required, that is, 872 kJ. We find that 498 kJ are required to break the oxygen-oxygen double bond, so the total amount of energy required to break all the bonds in the reactants is 872 kJ + 498 kJ = 1370 kJ (Table 20.1).

TABLE 20.1   Bond Energies in the Formation of Water

Amount of energy **required** to break chemical bonds in reactants

Type of Bond	Number of Bonds	Energy per Bond	Total
H—H	2	436 kJ	872 kJ
O=O	1	498 kJ	498 kJ
		Total energy required:	1370 kJ

Amount of energy **released** upon bond formation in products

Type of Bond	Number of Bonds	Energy per Bond	Total
H—O	4	464 kJ	1856 kJ
		Total energy required:	1856 kJ

Net change in energy: 1370 kJ required + 1856 kJ released = 486 kJ released

*This is a false simplification. Actually, 436 kJ is the amount of energy required to break 602 billion trillion (6.02×10^{23}) hydrogen-hydrogen bonds (1 mole). Atoms are so small that they are commonly measured in bulk, not individually. Why we use 6.02×10^{23} is discussed in Section 20.7.

Figure 20.2 A balloon filled with a mixture of hydrogen and oxygen is ignited to produce water and lots of energy.

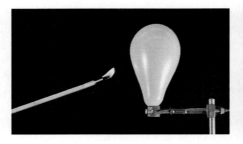 

Although energy is required to break the bonds in the reactants, energy is released upon the *formation* of bonds in the products. For example, when each hydrogen-oxygen bond forms, 464 kJ of energy are released. Since there are a total of 4 hydrogen-oxygen bonds that form (2 per molecule), the total amount of energy released is 464 kJ × 4 = 1856 kJ.

When we compare the amount of energy used to break the bonds of the reactants with the amount of energy that is released upon the formation of the bonds in the products we find that there is a net release of energy—a total of 486 kJ. This is summarized in Table 20.1. When hydrogen reacts with oxygen there is a release of energy.

$$2\,H_2 + O_2 \rightarrow 2\,H_2O + \text{Energy}$$

Figure 20.3 The space shuttle uses chemical reactions to lift off from the earth's surface.

The amount of energy released depends upon the amount of hydrogen and oxygen that react. The above example corresponds to 4 grams of hydrogen reacting with 32 grams of oxygen. With greater quantities of these reactants, much more energy is released. In fact, it is the reaction of large amounts of hydrogen and oxygen that lifts the space shuttle into orbit. There are two vast compartments in the large central tank upon which the orbiter is attached. One is filled with liquid hydrogen and the other is filled with liquid oxygen. Upon ignition, the hydrogen and oxygen are piped into the orbiter where they mix and react chemically to produce water and energy (the energy is in the form of accelerating water and much heat). These products are directed out through the cones on the back of the orbiter to produce thrust. Also included in the space shuttle lift-off assembly are two solid-fuel rocket boosters, each containing a mixture of the chemicals ammonium perchlorate, NH_4ClO_4, and powdered aluminum, Al. Upon ignition, these chemicals also react to produce chemical products that are energetically expelled out the back of the rocket, thereby producing additional thrust. This reaction is

$$3\,NH_4ClO_4 + 3\,Al \rightarrow Al_2O_3 + AlCl_3 + 3\,NO + 6\,H_2O + \text{Energy}$$

A chemical reaction that produces energy is **exothermic.** The reactions that lift the space shuttle into orbit are exothermic reactions. For exothermic reactions, an initial amount of energy is required to break apart the first reactant molecules. When hydrogen and oxygen are mixed, for example, they do not react until they are ignited with a spark. Once the reaction is going, however, enough energy is generated to make an exothermic reaction self-sustaining.

A chemical reaction that has the net result of absorbing energy is said to be **endothermic.** Endothermic reactions require the continual input of energy from an external source in order to proceed. The conversion of sodium chloride into its elemental components, sodium metal and chlorine gas, is an endothermic reaction. This reaction occurs upon the input of electrical energy.

$$\text{Electricity} + 2\,NaCl\ (aq) \rightarrow 2\,Na\ (metal) + Cl_2\ (g)$$

Figure 20.4 Punch the seal of this cold pack and it will quickly become cold to your touch—and your bruises if need be.

Some endothermic reactions are able to sustain themselves by absorbing heat from their surroundings. An example of such a reaction occurs when solid ammonium nitrate, NH_4NO_3 (s), dissolves in water and separates into ammonium, NH_4^+ (aq) and nitrate, $NO_3^-(aq)$, ions:

$$\text{Heat} + NH_4NO_3\ (s) \rightarrow NH_4^+\ (aq) + NO_3^-(aq)$$

As this endothermic reaction proceeds, heat is removed from the solvent, the container, and any material in contact with the reaction vessel. We can sense this action by a resulting drop in temperature. The "cold packs" used to diminish the swelling of bruises use this type of endothermic reaction. To activate the pack, it must be punched. This breaks an inner seal and allows the ammonium nitrate to mix with water. As the ammonium nitrate dissolves, the endothermic reaction shown above takes place and the temperature of anything in contact with the pack—including sprained ankles—decreases.

QUESTION

Is the following an example of an exothermic or an endothermic reaction? Should heat be written as a reactant or product?

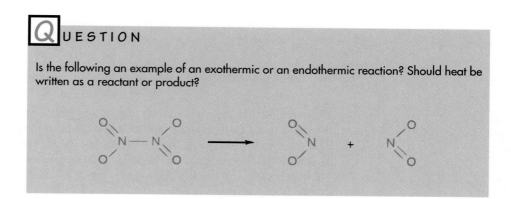

Catalysis

A **catalyst** is any substance that serves to increase the rate of a chemical reaction. Iron rusts faster, for example, in the presence of salt; food molecules are more efficiently digested in the presence of vitamins; and gasoline is more thoroughly combusted in the presence of platinum, the active component of most automotive catalytic converters. In these examples, salt, vitamins, and platinum all serve as catalysts.

Interestingly enough, rather than being consumed during a chemical reaction, a catalyst is regenerated. This means that a single catalyst molecule can be recycled to facilitate the formation of many thousands of product molecules. A catalyst, therefore, need only be present in tiny amounts. This is clearly an advantage, especially since many catalysts like platinum are expensive. After the reaction is complete, a catalyst can be recovered and used for future reactions.

Figure 20.5 The interior of a catalytic converter contains catalysts that help to degrade noxious exhaust fumes.

ANSWER

We see that the nitrogen-nitrogen bond is broken during this reaction. Since energy is *required* to break a chemical bond, this reaction is **endothermic.** Heat should be written as a reactant: Heat + N_2O_4 → NO_2+ NO_2. (If the reaction were to run backwards, such that NO_2 and NO_2 combined to form N_2O_4, we would see the formation of a chemical bond, hence, the *release* of energy and an exothermic reaction: NO_2+ NO_2 → N_2O_4+ Heat.)

The chemical industry is very dependent upon catalysts because they lower the costs of manufacturing. Without catalysts the price of gasoline would be much higher, as would the price of rubber, carpets, plastics, pharmaceuticals, automobile parts, clothing, all food grown with chemical fertilizers, and most other chemical products. If it were not for catalysts many chemical reactions would not even be feasible. Carbon dioxide, water and sunlight, for example, do not normally react to form glucose, which is an important food source. But this reaction proceeds quite efficiently in green plants due to a special biochemical catalyst known as chlorophyll. The action of catalysts can be very nice.

Not so nice is the role of catalysis in the layer of ozone high in the earth's atmosphere. Ozone, which absorbs most of the ultraviolet radiation that impinges on earth, is quickly destroyed by the presence of chlorine. Chlorine acts as a catalyst to turn ozone, O_3, into diatomic oxygen, O_2. Initially, chloromonoxide, ClO, and molecular oxygen, O_2, are formed by the following reaction:

Cl	+	O_3	$\rightarrow$	ClO	+	O_2
atomic chlorine		ozone		chloromonoxide		molecular oxygen

ClO	+	O	$\rightarrow$	Cl	+	O_2
chloromonoxide		atomic oxygen		atomic chlorine		molecular oxygen

Chloromonoxide then reacts with atomic oxygen, which is abundant in the upper atmosphere, to form another molecule of oxygen, and to reform the chlorine atom. The net result is the destruction of ozone and the regeneration of the chlorine atom, which then catalyzes more ozone destruction (see box on next page). Chlorine is not the only catalyst in the destruction of ozone: Sulfur dioxide, SO_2 (much of which comes from volcanoes), and nitric oxide, NO (the product of lightning strikes) to a lesser extent also catalyze the destruction of ozone. In short, there appear to be many causes (natural and human). Reformation of ozone occurs, but due to catalytic actions, the present rate of depletion has been seen to be greater than its rate of reformation.

20.3 CHEMICAL EQUILIBRIUM

The arrow of a chemical equation indicates the direction of the chemical reaction. In the equation for the formation of water, for example, the arrow indicates that oxygen and hydrogen combine to form water. The reverse reaction is also possible; that is, oxygen and hydrogen can be formed from water:

$$2\ H_2\ (g) + O_2\ (g) \leftarrow 2\ H_2O\ (l)$$

We find that most chemical reactions are *reversible* and that the extent of the reverse reaction depends very much on the conditions. Under usual conditions, water does not convert back into oxygen and hydrogen. When electricity is passed through the water, however, the reverse endothermic reaction dominates.

Figure 20.6 In principle, apply a strong electric current to the ocean for an incredibly long time and all the water will be converted into gaseous oxygen and hydrogen. The first strike of a match, however, would initiate a massive explosion as oxygen and hydrogen exothermically convert back into water.

Atmospheric Ozone and CFCs

Ozone is a gas formed of three oxygen atoms, O_3. Some ozone is formed in auto emissions then expelled as a pollutant in city air. Ozone is also formed naturally high in the atmosphere at an altitude of over 25 kilometers. At this altitude high energy ultraviolet (UV) radiation breaks apart molecular oxygen, O_2, into atomic oxygen, O, which then reacts with additional O_2 to form ozone, O_3.

$$
\begin{array}{llll}
& O_2 + UV & \rightarrow & 2\,O \\
& O + O_2 & \rightarrow & O_3 \\
& O + O_2 & \rightarrow & O_3 \\
\hline
\text{Net} & & & \\
\text{reaction:} & 3\,O_2 + UV & \rightarrow & 2\,O_3
\end{array}
$$

This synthesis of ozone is of great benefit for it involves the absorption of high energy UV radiation, which, if it were to reach the earth's surface, would cause immediate harm to living tissues and significant changes in weather patterns. Once formed, ozone itself absorbs UV radiation, which causes the ozone to fragment into molecular and atomic oxygen. These fragments eventually come back together to reform ozone. Since chemical bonds are formed through ozone's reformation, energy is released, mostly in the form of heat.

$$
\begin{array}{llll}
& O_3 + UV & \rightarrow & O_2 + O \\
& O_2 + O & \rightarrow & O_3 + \text{heat} \\
\hline
\text{Net} & & & \\
\text{reaction:} & O_3 + UV & \rightarrow & O_3 + \text{heat}
\end{array}
$$

The net reaction of ozone is to transform harmful UV radiation into not so harmful heat. We see from the above equation that ozone is not lost by this transformation so it can continue to shield us from UV for an indefinite amount of time.

The concentration of ozone in the upper atmosphere is quite small—if it were brought to the surface where atmospheric pressure is greater, it would comprise a layer of gas only 3 millimeters thick. Nevertheless, this ozone layer absorbs over 95% of the high-frequency ultraviolet radiation that comes to our planet from the sun. It is the safety blanket of life on earth.

Ozone is presently being depleted, mainly by the catalytic action of chlorine (Section 20.2 on catalysts). Where do the harmful chlorine atoms originate from? Chlorine in the upper atmosphere comes mainly from chlorofluorocarbons (CFCs) made here at the earth's surface. CFCs are inert gases commonly used in air conditioners and aerosols. Estimates are that CFCs are so stable they remain in the atmosphere from 80 to 120 years from the time of release. Some are carried by winds to regions above the ozone layer. Here they are broken apart by strong ultraviolet rays and chlorine atoms are let loose. These chlorine atoms fall back into the ozone layer where they catalyze the destruction of ozone. Estimates are that one chlorine atom causes the destruction of at least 100,000 ozone molecules before a year or two passes and it is finally carried away. We'll return to a discussion of ozone depletion, especially over the earth's poles, in Chapter 26.

How much harm will be done over the next century by the CFCs we have already released? Are we ready for alternatives to CFCs? Can there be any real solution to ozone depletion that does not involve a global effort to restrict or prohibit the manufacture of these chemicals?

Nitrogen dioxide, NO_2, a brown gas and one of the toxic components of air pollution, provides a good example of a reversible chemical reaction. At room temperature nitrogen dioxide molecules pair off to become colorless dinitrogen tetroxide molecules, N_2O_4, as is shown by the following chemical equation:

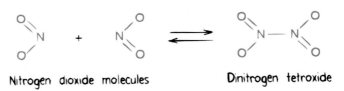

Nitrogen dioxide molecules   Dinitrogen tetroxide

Once dinitrogen tetroxide molecules are formed, however, they break apart to reform nitrogen dioxide molecules.

Dinitrogen tetroxide   Nitrogen dioxide molecules

Of course, once the nitrogen dioxide molecules are reformed, they can get together to reform the dinitrogen tetroxide. The net result is two competing reactions, which we depict using double arrows.

Nitrogen dioxide molecules   Dinitrogen tetroxide

Figure 20.7 The NO_2 and N_2O_4 in this flask are in equilibrium. Although no activity is noticed, individual molecules are continually reacting, with a balance between forward and reverse reactions.

Initially, we may start with pure nitrogen dioxide. In time, however, the percentage of nitrogen dioxide decreases, while the percentage of dinitrogen tetroxide increases. This continues to a point where the two reactions balance each other and the percentages of NO_2 and N_2O_4 settle to constant values (at room temperature this is about 16 percent and 84 percent, respectively). At this point, the reactions have achieved what is called **chemical equilibrium,** where the frequency of the products being formed is equal to the frequency of their being converted back into reactants (Figure 20.7).

Chemical equilibrium is like a department store during steady business. With steady business the number of customers entering through the doors is equal to the number of customers exiting through the doors. Chemical equilibrium does *not* mean that the amounts of reactants and products are equal. Like the department store, there may be 5 billion people outside during steady business and only a few hundred on the inside. What is special is that the rates they come in and go out are the same.

Chemical equilibrium is a very dynamic state. Although the concentrations of products and reactants remain constant, individual molecules are continuously changing back and forth. Again, this is like the department store during steady business where there is no change in the number of customers versus outsiders. Individual people, however, find that at any moment they may be one or the other.

If the department store suddenly holds a big sale, the equilibrium of customers is affected. For a while, the rate people enter the store will be greater than the rate

When reacting chemicals combine, they push toward equilibrium. But even at equilibrium, chemicals continue reacting – there's simply a balance between forward and reverse reactions. ⁻ Sigh ⁻

they leave. Hence the number of people inside the store increases. Likewise, we find that chemical equilibria are affected by various conditions. For example, when dinitrogen tetroxide is exposed to energy, it tends to break apart into molecules of nitrogen dioxide. Increasing the temperature of the nitrogen dioxide/dinitrogen tetroxide equilibrium, therefore, favors the rate of nitrogen dioxide formation. At high temperatures nitrogen dioxide predominates, so the overall color is dark brown. At lower temperatures dinitrogen tetroxide predominates, so the overall color is light brown. This is one of the reasons why air pollution becomes particularly noticeable on warm days; the higher the temperature, the greater the concentration of brown nitrogen dioxide.

Figure 20.8 Higher temperatures favor the rate of nitrogen dioxide formation. Lower temperatures favor the rate of dinitrogen tetroxide formation.

QUESTION

A mixture of nitrogen dioxide and dinitrogen tetroxide is brought to 100°C and kept at that temperature for a few minutes. Which, then, is greater: the rate of nitrogen dioxide formation or the rate of dinitrogen tetroxide formation?

Another factor that affects chemical equilibrium is the availability of reactants or products. For example, cobalt chloride, $CoCl_2$, is a blue solid that reacts with moisture to form cobalt chloride hexahydrate, $CoCl_2(H_2O)_6$, a red solid, according to the following chemical equilibrium:

$$CoCl_2 \ + \ 6\,H_2O \ \rightleftarrows \ CoCl_2(H_2O)_6$$
$$\text{Blue} \qquad\qquad\qquad\qquad \text{Red}$$

Under dry conditions when there are few water molecules available, cobalt chloride predominates, and the sample is blue. Under wet conditions, however, when there is

ANSWER

Neither! After a few minutes a new equilibrium is achieved. It's like the department store holding a sale. Initially, the equilibrium is disturbed as more customers pile in. Eventually, however, the store becomes so packed that the rate people come in slows down and the rate people leave speeds up. After a given period of adjustment a new equilibrium is achieved where, although there are now more people in the store, the rate they come in is again equal to the rate they go out.

an abundance of water molecules, cobalt chloride hexahydrate predominates, and the sample is red. Many rain-predicting devices once incorporated small but observable quantities of cobalt chloride. When rain was imminent, and it was humid outside, and the "predictor" would turn a shade of red; $CoCl_2(H_2O)_6$ predominated. With fair weather, however, the predictor would turn back toward a shade of blue, in which case $CoCl_2$ predominated. Interestingly enough, these devices are no longer in use because of the toxicity of cobalt chloride.

Figure 20.9 Cobalt chloride hexahydrate (reddish solid on left) loses water when heated to become cobalt chloride (bluish solid on right). Conversely, adding water to cobalt chloride changes it back into cobalt chloride hexahydrate.

Equilibrium in Nature

Equilibrium is the balance between opposing forces or actions, and certainly not restricted to chemical reactions and department stores. Processes of equilibrium are at work all around us. The action of the rain falling on the earth is in equilibrium with the action of the evaporation of water from oceans, lakes, and rivers. As we live, our bodies are in equilibrium with the environment, for actions such as UV radiation are causing our bodies to decay. To a finite extent these actions are balanced by our bodies' ability to adapt and rebuild. Within any ecological system there are many equilibria. Reindeer, for example, have a tendency to overpopulate. During the winter months, the scarcity of food threatens their survival. Wolves, on the other hand, kill and eat reindeer, usually weaker reindeer. Here, as with any ecological system, the survival of a species is dependent upon a state of equilibrium. Remove the wolves and the reindeer are threatened; remove the reindeer, and the wolves are threatened. Planet Earth is actually one large system of related equilibria where all forces and actions affect one another.

Events have occurred throughout earth's history that have altered its many states of equilibrium. More often than not, the resulting environmental changes have been too sudden for species to adapt—extinction has claimed over 99 percent of all species that have ever lived. Today, primarily due to our deforestation activities, an estimated one species becomes extinct every 15 minutes. As we emit chlorofluorocarbons in the northern hemisphere, we lose ozone over Antarctica. As we use DDT to protect plants and ourselves from insects, birds and other species are forced to extinction. As rainforests are devastated for lumber and cattleland, a multitude of irreparable consequences are incurred, including an alteration of weather patterns.

To understand the changes we humans have introduced to the global environment we must understand and appreciate the concept of equilibrium. Since we now dominate the planet, we have a responsibility to maintain favorable living conditions for its inhabitants. We must not cause the balance of forces to change too hastily. It's not Planet Earth we must protect, for it will be here whatever our actions. It's the dynamic equilibria of the biosphere that requires our utmost attention.

20.4 ACIDS AND BASES

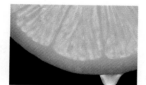

Figure 20.10 Citrus fruits contain many types of acids including ascorbic acid, which is vitamin C.

A compound may be an acid or a base, or even a little bit of both. Acids are sour and bases are bitter. The term *acid* comes from the Latin word *acidus,* which means "sour." The sour taste of vinegar and citrus fruits is due to the presence of acids. Food is digested in the stomach with the help of acids. Acids are essential in the chemical industry. Over 85 billion pounds of sulfuric acid, for example, are produced annually in the United States. Sulfuric acid is used in the manufacture of fertilizers, detergents, paint dyes, plastics, pharmaceuticals, storage batteries, iron, and steel. Sulfuric acid is so important in the manufacturing of goods that its production is considered a standard measure of a nation's industrial strength.

Bases are characterized by their bitter taste and slippery feel. Interestingly enough, bases themselves are not slippery. Rather, they react with skin oils to produce slippery solutions of soap. Most commercial preparations for unclogging drains are composed of sodium hydroxide, NaOH, (also known as lye), which is extremely basic and hazardous because it dissolves flesh. Bases are also heavily used in industry. Each year in the United States about 25 billion pounds of sodium hydroxide are manufactured for use in the production of various chemicals such as soaps, and for use in the pulp and paper industry. Solutions containing bases are often called *alkaline.* As was discussed in Section 18.2, this term is derived from the Arabic word for ashes (al-qali), which are slippery when wet due to the base potassium carbonate.

The first person to recognize the essential nature of acids and bases was Svante Arrhenius. Before the turn of the 20th century, Arrhenius postulated that **acids** produce hydrogen ions, H^+, when dissolved in water, whereas **bases** produce hydroxide ions, HO^-.* For example, when hydrogen chloride, HCl, dissolves in water it breaks apart into hydrogen, H^+, and chloride, Cl^-, ions.

$$HCl\ (g) \rightarrow H^+(aq) + Cl^-(aq)$$

Since hydrogen ions are produced by this process, hydrogen chloride is considered an acid. Sodium hydroxide, NaOH, on the other hand, dissolves in water to produce hydroxide, HO^-, ions.

Figure 20.11 All these supermarket products contain bases.

ANSWER

It would not be a good predictor of sunshine. This is because while it is raining, there are too many water molecules in the air to allow the formation of noticeable amounts of cobalt chloride. The device will only turn back to the blue color *after* all the moisture in the air has dried up; that is, *after* the sun is already shining.

---

*The hydroxide ion consists of a hydrogen atom covalently bonded to an oxygen atom: $H\!-\!O^-$. The single negative charge of this "polyatomic" ion resides on the oxygen atom, which has a total of 9 electrons surrounding its 8 proton-containing nucleus.

$$NaOH \ (s) \rightarrow Na^+ (aq) + HO^- (aq)$$

Then according to Arrhenius, sodium hydroxide is a base.

The Arrhenius definition of an acid and a base is somewhat limited. For example, there are many substances like ammonia, NH_3, that are bases, and do not have hydroxide ions as part of their formula. In 1923 a more general definition of acids and bases was suggested by the Danish chemist Johannes Brønsted and the English chemist Thomas Lowry. In the Brønsted-Lowry definition, an acid is a hydrogen ion, H^+, donor, and a base is a H^+ acceptor.* For example, consider what happens when hydrogen chloride is introduced to water:

Water   Hydrogen chloride   Hydronium   Chloride
(base)   (acid)   ion   ion

Hydrogen chloride donates a hydrogen ion to an accepting water molecule. In this case, hydrogen chloride behaves as an *acid,* and water behaves as a *base.* The result is the chloride ion, Cl^-, and the hydronium ion, H_3O^+.†

The Brønsted-Lowry definition accounts for the basic properties of ammonia. Ammonia, NH_3, behaves as a base in that it accepts a hydrogen ion from water. This results in the ammonium ion and the hydroxide ion.

Ammonia   Water   Ammonium   Hydroxide
(base)   (acid)   ion   ion

The products of a Brønsted-Lowry acid/base reaction themselves can behave as acids and bases. The ammonium ion, for example, may donate the hydrogen ion back to the hydroxide ion to reform ammonia and water:

Ammonium   Hydroxide   Ammonia   Water
ion   ion
(acid)   (base)

---

*The acronym BAAD is useful for remembering this definition. A **B**ase **A**ccepts a hydrogen ion and an **A**cid **D**onates.

†The hydronium ion consists of three hydrogen atoms covalently bonded to an oxygen atom. The single positive charge of the hydronium ion resides on the oxygen atom, which has a total of 7 electrons surrounding its 8 proton-containing nucleus.

Forward and reverse acid/base reactions eventually reach a state of chemical equilibrium. The above two reactions, therefore, can be represented as:

$$NH_4^+ \; + \; HO^- \; \rightleftharpoons \; NH_3 \; + \; H_2O$$

$$\text{acid} \qquad\quad \text{base} \qquad\qquad \text{base} \qquad \text{acid}$$

When the equation is viewed from left to right, NH_4^+ behaves as an acid for it donates a hydrogen ion to the hydroxide ion, HO^-, which therefore acts as a base. In the reverse direction, we see that water donates a hydrogen ion to ammonia, NH_3. In this case water behaves as an acid and ammonia behaves as a base.

Q UESTIONS

Identify each of the following chemicals as acting as an acid or a base.

1. $H_3O^+ \; + \; Cl^- \; \rightleftharpoons \; H_2O \; + \; HCl$

_____   _____   _____   _____

2. $NH_2^- \; + \; H_3O^+ \; \rightleftharpoons \; NH_3 \; + \; H_2O$

_____   _____   _____   _____

Acid Strength

The stronger an acid, the greater its ability to donate the hydrogen ion. Hydrogen chloride, HCl, is an example of a strong acid for it forcefully donates the hydrogen ion to water. This results in both chloride and hydronium ions. Since HCl is such a strong acid, nearly all of it is converted into these ions. At equilibrium, only a very few unionized hydrogen chloride molecules remain.

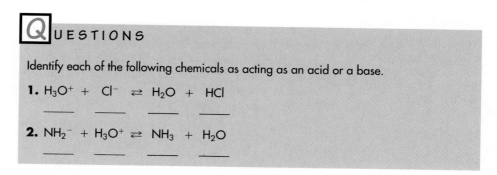

$$HCl \; + \; H_2O \; \rightleftharpoons \; Cl^- \; + \; H_3O^+$$

Hydrogen chloride Water Chloride ion Hydronium ion

An example of a weak acid is acetic acid, $C_2H_3O_2H$, which donates the hydrogen ion to water with much less force. When dissolved in water, only a small portion of acetic acid is converted into ions. The majority of acetic acid molecules remain intact in their original unionized form.

$$C_2H_3O_2H \; + \; H_2O \; \rightleftharpoons \; C_2H_3O_2^- \; + \; H_3O^+$$

Acetic Acid   Water   Acetate ion   Hydronium ion

A NSWERS

1. acid, base, base, acid.

2. base, acid, acid, base.

Figure 20.12 Pure water is unable to conduct an electric current. The light bulb in the above circuit, therefore, remains unlit.

What is actually going on in solution can be determined by measuring the solution's ability to conduct an electric current. Electricity does not pass very well through pure water because there are practically no ions to conduct the current from one electrode to the other (Figure 20.12). A strong acid dissolved in water, however, generates a lot of ions that are able to conduct electric current (Figure 20.13a). A weak acid dissolved in water, on the other hand, releases a relatively small number of ions so that only a small, but significant, electric current is conducted (Figure 20.13b).

Water as an Acid and a Base

A substance that behaves as either an acid or a base is said to be **amphoteric.** Water is a good example because of its ability to accept and donate hydrogen ions. Through its amphoteric nature water even has the ability to react with itself. In behaving as an acid, a water molecule donates a hydrogen ion to a neighboring water molecule, which in accepting the hydrogen ion is behaving as a base. This produces a hydroxide ion, HO^-, and a hydronium ion, H_3O^+.

$$H_2O \quad + \quad H_2O \quad \rightleftharpoons \quad HO^- \quad + \quad H_3O^+$$
$$\text{acid} \qquad\quad \text{base}$$

Water, however, is a very weak acid just as it is a very weak base. This means that in pure water there are very few ions—so few that the light bulb in Figure 20.12 remains unlit.

Acidic, Basic, or Neutral

With pure water or any solution that contains water, we find an interesting rule that pertains to the relative concentrations of hydronium and hydroxide ions. At equilibrium the concentration of hydronium ions, $[H_3O^+]$, multiplied by the concentration of the hydroxide ions, $[HO^-]$, equals a constant, K_w, which is a very small number.

$$[H_3O^+][HO^-] = K_w = 0.00000000000001$$

Written in scientific notation,

$$[H_3O^+][HO^-] = K_w = 1.0 \times 10^{-14}$$

The significance of K_w is the fact that it is constant. *No matter what is dissolved in the water,* the product of the hydronium ion concentration $[H_3O^+]$, and the hydroxide ion concentration $[HO^-]$, always equals 1.0×10^{-14}. This means that if the concen-

Figure 20.13 Since HCl is a strong acid it breaks apart completely in water giving a high concentration of ions, which are able to conduct an electric current that lights the bulb. Acetic acid, AcOH, is a weak acid and it breaks into ions less frequently. Since fewer ions are generated, only a weak current is conducted and the bulb is dimmer.

tration of H_3O^+ goes up, the concentration of HO^- must come down—so that the product of the two is still 1.0×10^{-14}. For example, if HCl gas is dissolved in water, increasing the hydrogen ion concentration, $[H_3O^+]$, then the hydroxide ion concentration, $[HO^-]$, correspondingly decreases. The reasons for this are best explored in a follow-up chemistry course. Briefly, however, K_w remains constant because the equilibrium is able to shift to accommodate new conditions.

In measuring the concentrations of hydronium and hydroxide ions, we use a standard unit known as the *moles per liter,* often abbreviated as capital *M,* which was introduced in Section 19.5. Recall that one mole equals the very large number, 6.02×10^{23}. So if the concentration of hydronium ions in a solution is 1 mole per liter (1 *M*), that means there are 6.02×10^{23} hydronium ions in every one liter of solution.

QUESTIONS

1. What is the concentration of hydronium ions if the concentration of hydroxide ions equals 1.0×10^{-9} *M*?

2. What is the concentration of hydronium ions if the concentration of hydroxide ions equals 1.0×10^{-5} *M*?

3. In pure water the hydroxide ion concentration equals 1.0×10^{-7} *M*. What is the hydronium ion concentration in pure water?

A solution can be described as **acidic, basic,** or **neutral.** By definition, an acidic solution is one where the hydronium ion concentration is *greater* than the hydroxide ion concentration. An acidic solution is made by adding an acid, such as hydrogen chloride, to water. The effect of this is to *increase* the concentration of hydronium ions, which necessarily *decreases* the concentration of hydroxide ions. A basic solution, on the other hand, is one where the *hydroxide ion* concentration is greater than the hydronium ion concentration. A basic solution is made by adding a base, such as sodium hydroxide, to water. This effectively increases the concentration of hydroxide ions, while decreasing the concentration of hydronium ions. A neutral solution is one where the hydronium ion concentration *equals* the hydroxide ion concentration. Pure water is an example of a neutral solution.

ANSWERS

1. The hydronium ion concentration multiplied by the hydroxide ion concentration always equals 1.0×10^{-14}.

$$[H_3O^+][1.0 \times 10^{-9}] = 1.0 \times 10^{-14}$$

Algebraically, we solve for the hydronium ion concentration by dividing each side by 1.0×10^{-9}.

$$\frac{[H_3O^+]1.0 \times 10^{-9}}{1.0 \times 10^{-9}} = \frac{1.0 \times 10^{-14}}{1.0 \times 10^{-9}}$$
$$[H_3O^+] = 1.0 \times 10^{-5} \, M$$

2. 1.0×10^{-9} *M*.

3. 1.0×10^{-7} *M*.

A neutral solution is also obtained when equal quantities of acid and base are combined. In short:

1. In an **acidic** solution, $[H_3O^+] > [HO^-]$.
2. In a **basic** solution, $[HO^-] > [H_3O^+]$.
3. In a **neutral** solution, $[H_3O^+] = [HO^-]$.

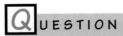UESTION

How does adding ammonia to water make a basic solution when there are no hydroxide ions within its formula?

The pH Scale

The pH scale is a method for expressing the acidity or basicity of a solution. Mathematically, the term is equal to the negative logarithm of the hydronium ion concentration.*

$$pH = -\log [H_3O^+]$$

When a solution is neutral, the hydrogen ion and hydroxide ion concentrations both equal 1.0×10^{-7} *M*. To find the pH of this solution, we first take the log of this value, which equals -7. We then change the sign, which gives a positive value. Hence, in a neutral solution where the hydrogen ion concentration equals 1.0×10^{-7} *M*, the pH = 7. Acid solutions, we find, have pH values that are less than 7; the

Logarithms

The logarithm of a number can be conveniently found on any scientific calculator. Briefly, when finding the logarithm of a number, often abbreviated "log," you are finding the exponent to which the number 10 must be raised to equal this number. The log of 100, for example, equals 2 because 10 raised to the second power, 10^2, equals 100. Similarly, the log of 1000 equals 3 because 10 raised to the third power, 10^3, equals 1000.

Quiz What is the log of 10,000?

Answer The number 10,000 is 10^4, the log of which is 4.

Any positive number, including a very small one, has a log value. The log of 0.0001, which is 10^{-4} in scientific notation, for example, is simply -4 (the power to which 10 is raised to equal this number).

ANSWER

Ammonia indirectly increases the hydroxide ion concentration by reacting with water: $NH_3 + H_2O \rightarrow NH_4^+ + HO^-$.

---

*The hydrogen ion, H^+, and the hydronium ion, H_3O^+, are often used interchangeably so that pH may also be defined as the negative log of the hydrogen ion concentration: $pH = -\log[H^+]$.

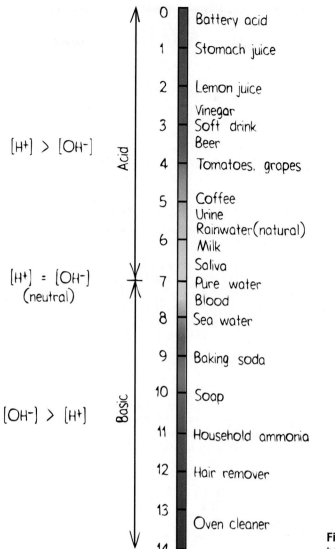

Figure 20.14   Some typical pH values.

more strongly acidic they are, the lower the pH. Basic solutions, also referred to as *alkaline,* have pH values of more than 7; the more strongly basic they are, the higher the pH. Figure 20.14 shows typical pH values of some familiar solutions.

20.5   OXIDATION-REDUCTION REACTIONS

One of the most common and important class of reactions is *oxidation-reduction.* The common feature of all oxidation-reduction reactions is the transfer of electrons between reacting atoms. **Oxidation** is the process whereby a reactant loses one or more electrons. **Reduction** is the opposite process whereby a reactant gains one or more electrons. Oxidation and reduction are complementary processes. They always

Acid Rain and Basic Oceans

All rain, not just "acid rain," is acidic. One source of this acidity is carbon dioxide, CO_2, the same gas that gives fizz to soda drinks. There are some 2.5 trillion tons of CO_2 in our atmosphere, most of it from natural sources such as volcanoes and decaying organic matter, and a growing amount from human activities. Water in the atmosphere reacts with carbon dioxide to form *carbonic acid:*

$$CO_2 + H_2O \rightarrow H_2CO_3$$
(carbon dioxide) (water) (carbonic acid)

Carbonic acid, as its name implies, lowers the pH of water. The concentration of CO_2 in the atmosphere brings the pH of rain water to about 5.6—noticeably below the neutral pH value of 7! Due to local fluctuations the normal pH of rain water varies between 5 and 7.

By definition, *acid rain* is rain with a pH *less* than 5. Lower values result when airborne pollutants like sulfur dioxide, SO_2, are absorbed by water in the atmosphere. Sulfur dioxide is readily oxidized to sulfur trioxide, which reacts with water to form *sulfuric acid:*

$$SO_2 \xrightarrow{\frac{1}{2} O_2} SO_3$$
(sulfur dioxide) (sulfur trioxide)

$$SO_3 + H_2O \rightarrow H_2SO_4$$
(sulfur trioxide) (water) (sulfuric acid)

Each year about 30 million tons of SO_2 are released into the atmosphere by the combustion of sulfur-containing coal and oil. Sulfuric acid is a much stronger acid than carbonic acid. Rain laced with sulfuric acid eventually corrodes metals, paints, and other exposed substances. Each year the damages cost billions of dollars. More importantly, rivers and lakes receiving acid rain become less capable of sustaining life. Much vegetation that receives acid rain doesn't survive. This is particularly evident in regions around heavy industry.

One solution is preventing most of the generated sulfur dioxide and other pollutants from entering the atmosphere. Because this is costly even with present-day technology, many industries and individuals are reluctant to comply. Recent innovations, however, hold the promise of lowering these costs. A second solution is to reduce energy consumption, which also reduces the amount of fossil fuels burned in the first place. A third solution is to utilize alternate, clean-air sources of energy, such as nuclear energy, solar energy, and hydroelectric power.

The amount of carbon dioxide put into the atmosphere by human activities is growing at an ever increasing rate. Studies show, however, that atmospheric concentrations of CO_2 are not increasing proportionately. One probable explanation has to do with the oceans. When CO_2 dissolves in ocean water it forms carbonic acid, just as it does in fresh water. We find, however, that the oceans are alkaline (pH = 8.2) because of dissolved alkaline compounds such as calcium carbonate, $CaCO_3$. In the ocean, carbonic acid is quickly neutralized to form a water soluble salt, such as calcium bicarbonate, $Ca(HCO_3)_2$:

$$H_2CO_3 + CaCO_3 \rightarrow Ca(HCO_3)_2$$
(carbonic acid) (calcium carbonate) (calcium bicarbonate)

This neutralization prevents CO_2 from being released back into the atmosphere, as it is when dissolved in fresh water. The ocean, therefore, is a genuine "carbon dioxide sink." Pushing more CO_2 into our atmosphere means pushing more of it into the vast oceans. So this is another of the many ways the ocean moderates our environment.

Nevertheless, concentrations of atmospheric CO_2 *are* increasing, even though not as much as would happen without ocean absorption! Carbon dioxide is being produced faster than the ocean can absorb it. As is discussed in Chapter 26, increased atmospheric levels of CO_2 may cause global warming, a phenomenon that has much negative potential for the environment.

So we find the pH of rain dependent, in great part, on the concentration of atmospheric CO_2, which in turn is dependent on the pH of the oceans. These equilibria, in turn, are in equilibrium with global temperatures, which in turn naturally connect to the countless equilibria of all living systems on earth.

Figure 20.15 Sodium metal is oxidized by chlorine gas, which is reduced by the sodium metal, in the exothermic formation of sodium chloride.

occur together; you cannot have one without the other. Electrons that are lost by one chemical don't simply disappear; they must be gained by another.

Different materials have different oxidation and reduction tendencies—some lose electrons more readily, while others gain electrons more readily. When a reactant loses an electron we say it has been *oxidized*. When a reactant gains an electron we say it has been *reduced*. We can look at a material that tends to oxidize and see that it has the effect of causing other materials to be reduced. Because of this tendency, a material that loses its electrons (becomes oxidized) is referred to as a *reducing agent*. Similarly, a material that accepts electrons (becomes reduced) is referred to as an *oxidizing agent*.

QUESTIONS

True or false:

1. Reducing agents themselves are oxidized in a reaction.
2. Oxidizing agents themselves are reduced in a reaction.

A straightforward example of an oxidation-reduction reaction occurs when elemental sodium [Na] and chlorine, Cl_2, react to form the ionic compound sodium chloride, Na^+Cl^-.

$$2 \, Na \, (s) + Cl_2 \, (g) \rightarrow 2 \, Na^+Cl^- \, (s)$$

Through this reaction, the sodium atom changes from an electrically neutral state into a positively charged ion. Each sodium atom loses an electron. Therefore, the sodium is oxidized.

$$Na \rightarrow Na^+ + e^-$$

Similarly, the chlorine atoms change from an electrically neutral state into negatively charged ions. The chlorine, therefore, is reduced.

$$Cl + e^- \rightarrow Cl^-$$

ANSWERS

Both statements are true, in accord with one of the great conservation principles in physics and chemistry—the *conservation of charge*. The electrical charge lost by one substance is the charge gained by another. Charge is simply transferred in reactions.

Figure 20.16 Zinc has a greater tendency to oxidize than does iron. For this reason, many iron articles, such as nails, are "galvanized" by coating them with a thin layer of zinc. When the zinc oxidizes, it turns into zinc oxide, which is an inert and insoluble substance that serves to protect the inner iron from rusting!

When sodium and chlorine are combined, the net result is sodium atoms release their electrons to chlorine atoms. Since sodium causes the chlorine to be reduced, we refer to the sodium as a reducing agent. Since the chlorine causes sodium to be oxidized, the chlorine is referred to as an oxidizing agent.

Electrochemistry

Electrochemistry concerns the relationship between electrical energy and chemical change. Electrochemistry typically involves either the production of an electric current from an oxidation-reduction reaction, or the use of an electric current to produce a chemical change.

To understand how an oxidation-reduction reaction can be used to generate an electric current, consider what happens when an oxidizing substance is placed in direct contact with a reducing substance. Electrons flow between them. This flow of electrons is an electric current that can be harnessed.

Iron [Fe] is a better reducing agent than the copper ion, Cu^{2+}. So when iron and copper ions are placed in contact, electrons tend to flow from iron to copper ions. The result is the oxidation of iron and the reduction of copper ions (Figure 20.17).

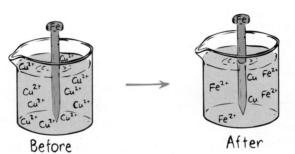

Before   After

Figure 20.17 An iron nail, Fe, placed in a solution of copper ions, Cu^{2+}, tends to oxidize into Fe^{2+} ions, which dissolve in the water. Copper ions, meanwhile, are reduced to metallic copper, Cu, which tends to coat the nail. For simplicity, the negatively charged ions that balance these positively charged ions in solution are not depicted.

To produce electric current from an oxidation-reduction reaction, one might separate the iron nail from the copper ions and connect the two by a conducting wire (Figure 20.18). Then if electrons were to flow from the iron to the copper ions, the resulting electric current in the wire could be attached to some useful device (such as a light bulb).

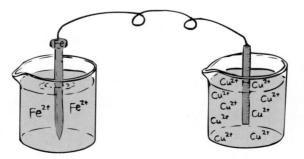

Figure 20.18 An iron nail, Fe, is placed in water and connected to a solution of copper ions, Cu^{2+}, by way of a conducting wire. Nothing happens because this arrangement does not make a complete circuit.

We find, however, that an electric current is not sustained by this arrangement because it is an incomplete circuit and it would result in an impossible build-up of charge in each beaker. The circuit can be completed, however, by allowing ions to migrate between the solutions. This can be done using a salt bridge, which is a

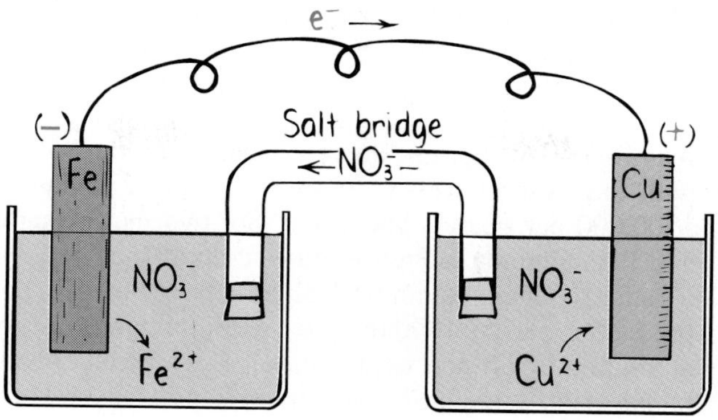

Figure 20.19 The salt bridge completes the circuit. As iron ions, Fe^{2+}, enter the water in the left vessel, oppositely charged ions, NO_3^-, from the right vessel migrate through the salt bridge to balance the charge.

U-shaped tube filled with salt and enclosed by semi-porous plugs. A salt bridge allows ions to pass freely between the two solutions. This permits the flow of electrons through the conducting wire (Figure 20.19).

So we see that with the proper set-up it is possible to harness electrical energy from a chemical reaction. The apparatus shown in Figure 20.19 is one example. Such devices are called **batteries** (also *galvanic cells*). Although there are many different kinds of batteries, they all function by the same principle. Two materials that oxidize and reduce each other are connected by a conducting wire.

A battery lasts for only so long because electron-producing chemicals are consumed. In a car battery, for example, lead oxide, PbO_2, and lead, Pb, are consumed to form lead sulfate, $PbSO_4$.

$$PbO_2 + Pb + 2\,H_2SO_4 \rightarrow 2\,PbSO_4 + 2\,H_2O + \text{Electrical energy}$$

A useful characteristic of the car battery, however, is that it can be recharged by forcing an electric current in the opposite direction. As a result, $PbSO_4$ is consumed and PbO_2 and Pb are replenished.

$$\text{Electrical energy} + 2\,PbSO_4 + 2\,H_2O \rightarrow PbO_2 + Pb + 2\,H_2SO_4$$

This is the task of your car's alternator, which is powered by the engine.

Electrolysis is the use of electrical energy to produce chemical change. The recharging of a car battery is an example of electrolysis. Another important example is the process of passing an electric current through water, which causes the water to break down into its elemental components, hydrogen, H_2, and oxygen, O_2.

$$\text{Electrical energy} + 2\,H_2O \rightarrow 2\,H_2\,(g) + O_2\,(g)$$

Since car batteries also contain water, they cannot be recharged without some electrolysis of water also occurring. The result is a potentially explosive mixture of hydrogen, H_2, and oxygen, O_2. That's why it is so important not to produce a spark near a car battery while it is charging.

Electrolysis is used to produce metals from their ores. The metal produced in the greatest quantity by electrolysis is aluminum, which, as we shall see in Chapter 22, is the third most abundant element in the earth's crust. Aluminum occurs naturally bonded to oxygen in an ore called *bauxite*.* Aluminum metal wasn't known until 1827, when it was first prepared by reacting bauxite with hydrochloric acid.

Figure 20.20 The electrolysis of water produces hydrogen and oxygen in a 2:1 ratio by volume. Why must ions be dissolved in the water in order for this electrolysis to work?

---

*Bauxite is named after the place of its discovery, Les Baux, France.

Figure 20.21 For a nerve-wracking experience, bite a piece of aluminum foil with a tooth filled with dental amalgam. Aluminum is a better reducing agent than dental amalgam, and gives electrons to the amalgam. The slight current that results produces a jolt of pain.

This gave the aluminum ion, Al^{3+}, which was reduced to aluminum metal using sodium.

$$Al^{3+} + 3\,Na \rightarrow Al + 3\,Na^{+}$$

Aluminum thus became an expensive rarity. In 1855 pieces were exhibited in Paris with the crown jewels of France. The price of aluminum at that time was about $100,000 per pound. Then, in 1886, two men, Charles Hall in the United States and Paul Heroult in France, almost simultaneously discovered an electrolytic process for producing aluminum directly from bauxite. This greatly facilitated its mass production and by 1890 the price of aluminum dropped to about $2 per pound. Pure aluminum is soft and weak. So, to be structurally useful, aluminum is alloyed with metals such as zinc or manganese. Today, about 16,000 kilowatt-hours of energy are required to produce each ton of aluminum from its ore. Considering that the world-wide production of aluminum is about 16 million tons annually, this is a lot of energy. The process of recycling aluminum, on the other hand, consumes only about 700 kilowatt-hours for every ton. Recycling aluminum not only reduces litter in the environment, but helps to reduce the load on electric companies, which in turn reduces air pollution.

20.6 ATOMIC AND FORMULA MASSES

When a baker is baking cookies, ingredients must be mixed in proper proportions. Similarly, when a chemist prepares a chemical substance by way of a chemical reaction, atoms or molecules must be mixed in proper proportions. In baking, we need not be extremely precise in our measurements. The size of the egg, for example, can vary and it doesn't really matter if we add a few extra drops of milk. In chemistry, however, much precision is required. A chemical recipe for ammonium chloride, for example, requires that we measure and combine the same number of ammonia molecules as hydrogen chloride molecules:

| 1 molecule of ammonia | + | 1 molecule of hydrogen chloride | = | 1 ammonium chloride unit |

But these molecules are incredibly small. How, then, is it possible to measure equal numbers? Instead of counting, we use a scale to measure equal numbers according to mass.

Measuring equal masses of different substances, however, doesn't produce equal numbers of atoms or molecules. One kilogram of golf balls contains fewer balls than one kilogram of ping pong balls. Likewise, because different atoms and molecules have different masses, there are different numbers of atoms or molecules in 1-gram samples of different substances. There are more lighter ammonia molecules in a gram of ammonia than there are heavier hydrogen chloride molecules in a gram of hydrogen chloride. Weighing equal masses of these different molecules won't give us the same number of these molecules.

By knowing the *relative masses* of the different substances we can measure equal numbers. For example, golf balls are about 40 times as massive as ping pong balls. Hence, for equal numbers, the mass of the golf balls should be 40 times the mass of the ping pong balls. In other words, the number of golf balls in 40 kilograms of them equals the number of ping pong balls in 1 kilogram of them. Similarly, if we know the relative masses of ammonia and hydrogen chloride, we can weigh out an equal number of them. Since hydrogen chloride is about twice as massive as ammonia, we weigh out twice as much hydrogen chloride.* Then we have an equal number of hydrogen chloride and ammonia molecules.

One molecule of ammonia combines with one molecule of hydrogen chloride to produce one unit of ammonium chloride.† From this we can deduce that 2 molecules of ammonia will combine with 2 molecules of hydrogen chloride to produce 2 units of ammonium chloride. Likewise, a billion molecules of ammonia will combine with a billion molecules of hydrogen chloride to produce a billion units of ammonium chloride.

Although the number of units of ammonium chloride are the same as the number of units of ammonia and hydrogen chloride, we find that the mass of the resulting ammonium chloride is the sum of the masses of ammonia and hydrogen chloride that reacted. We distinguish between the number of units and the masses of the units. For example, when 1 gram of ammonia (3.5×10^{22} molecules) combines with 2 grams of hydrogen chloride (3.5×10^{22} molecules), it forms 3 grams of ammonium chloride (3.5×10^{22} units). So we see that a single ammonium chloride unit is 3 times as massive as a single ammonia molecule and 1.5 times as massive as a single hydrogen chloride molecule.

QUESTION

Assuming that ammonia is exactly one half as massive as hydrogen chloride, how much ammonium chloride (in grams) will a chemist make if she combines 2 grams of ammonia with 2 grams of hydrogen chloride?

With modern techniques the absolute masses of individual atoms are measured to a high degree of accuracy. The hydrogen atom has a mass of 1.674×10^{-24} grams and the oxygen atom has a mass of 2.657×10^{-23} grams.‡ These numbers are unwieldy and can be simplified by a change of units. Just as a grocer calls 12 eggs a dozen, and a printer calls 500 sheets of paper a ream, a physical scientist handles the masses of atoms in **atomic mass units**—amu's. For several reasons that include its stability, the carbon-12 atom is chosen as a standard. The amu is defined as exactly $\frac{1}{12}$ the mass of the carbon-12 atom.

ANSWER

Three grams of ammonium chloride will again be produced. One gram of ammonia will remain left over, unreacted. As was shown by Proust, Dalton, and others, molecules react in a definite proportion by mass, which in this case is 1 gram of ammonia for every 2 grams of hydrogen chloride. Ammonia is less massive than hydrogen chloride, and if equal weights were combined you'd be combining more ammonia molecules than hydrogen chloride molecules. It is important that the chemist know the relative masses of atoms and compounds.

---

*Hydrogen chloride is actually 2.14 times as massive as ammonia.

†Ammonium chloride is an ionic salt, like sodium chloride; hence, we use the term *unit* rather than *molecule.*

‡These are the average values based on the different isotopes of these elements and their abundances.

$$\text{Carbon-12} = 1.993 \times 10^{-23} \text{ grams}$$
$$\downarrow$$
$$\text{divide by 12}$$
$$\text{or}$$
$$\downarrow$$
$$1 \text{ amu} = 1.661 \times 10^{-24} \text{ grams}$$

In these units, the mass of the hydrogen atom is 1.01 amu (rather than 1.674×10^{-24} grams), and the mass of the oxygen atom is 16.00 amu (rather than 2.657×10^{-23} grams).

The mass of an element given in amu is its **atomic mass.** The atomic mass of each element is written in the periodic table directly below the atomic symbol. With these values, the relative masses of any two elements can be calculated. For example, by dividing the atomic mass of oxygen (16.00 amu) by the atomic mass of hydrogen (1.01 amu) we find that oxygen is 15.84 times more massive than hydrogen.

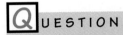 UESTION

What is the mass of the carbon-12 isotope in atomic mass units?

The **formula mass** of a substance is the sum of the atomic masses of elements in its chemical formula. For example, the formula mass of ammonia, NH_3, is the atomic mass of nitrogen plus three times the atomic mass of hydrogen, or 17.04 amu:

$$
\begin{array}{lll}
\text{Nitrogen (14.01 amu)} \times 1 & = & \text{(14.01 amu)} \\
\text{Hydrogen (1.01 amu)} \times 3 & = & \underline{\text{(3.03 amu)}} \\
& & \text{17.04 amu } NH_3
\end{array}
$$

Similarly, the formula mass of hydrogen chloride, HCl, is the atomic mass of hydrogen plus the atomic mass of chlorine, or 36.46 amu:

$$
\begin{array}{lll}
\text{Hydrogen (1.01 amu)} \times 1 & = & \text{(1.01 amu)} \\
\text{Chlorine (35.45 amu)} \times 1 & = & \underline{\text{(35.45 amu)}} \\
& & \text{36.46 amu HCl}
\end{array}
$$

By knowing the formula masses of different compounds we can determine their relative masses. From the above data, for example, we find that ammonia is $(36.46)/(17.04) = 2.14$ times as massive as hydrogen chloride (not exactly 2 as we assumed in the beginning of this section). Atomic masses give us a handle on how *atoms* should be weighed so that they are in correct proportions. Formula masses, on

ANSWER

12.0000 amu. You may notice in the periodic table that the atomic mass of carbon is 12.011. This represents the isotopic abundance of carbon in the earth's surface. Only 98.8 percent of naturally occurring carbon is the carbon-12 isotope. A small but significant percentage includes the carbon-13 isotope (1.1 percent) and the carbon-14 isotope (< 0.1 percent). Hence the average value is slightly greater than 12.0000. Similarly, the atomic mass of each element is somewhat modified by the abundance of various isotopes.

the other hand, are useful because we can determine what proportion, by mass, two or more *compounds* should be mixed so that the number of molecules of each are in a desired ratio.

UESTION

Which has the greater number of molecules: 17.04 grams of ammonia or 36.46 grams of hydrogen chloride?

So we see that the rules of cooking and chemistry are similar in that they both require the measuring of ingredients. Just as a cook looks to a recipe to find the necessary quantities measured by the cup or the tablespoon, a chemist looks to the periodic table to find the necessary quantities, which are measured by the atom and the molecule.

20.7 AVOGADRO'S NUMBER AND THE MOLE

It's time to learn more about that big number we've been talking about (Sections 19.5, 20.2, and 20.4). If we were given 1.993×10^{-23} grams of carbon-12, how many atoms would we have? Look back to the previous section and you'll find this happens to be the mass of a single carbon-12 atom. Therefore, the answer is one. Knowing the mass of a single carbon-12 atom permits us to calculate the number of carbon-12 atoms in any amount of carbon-12. In 12 grams, for example, we find that there are 6.02×10^{23}.* Take a moment to appreciate the enormity of this number.

$$6.02 \times 10^{23} = 602,000,000,000,000,000,000,000$$

This many grains of wheat would occupy 20 million cubic kilometers (about the volume of the Arctic ocean). Atoms are so small, however, that this many atoms of carbon-12 fit within a 12-gram sample (about the size of 5 sugar cubes if the carbon is in the form of graphite).

By knowing the mass of a single atom of an element, we can calculate the number of atoms in any given sample of that element. Interestingly enough, for any element we find the same number of atoms (6.02×10^{23}) when the mass of the sample in grams equals the value of the atomic mass. For example, in a 16.00-gram sample of atomic oxygen (atomic mass = 16.00 amu) there are 6.02×10^{23} atoms. This same number of atoms are in a 55.85-gram sample of iron (atomic mass = 55.85 amu). Likewise, when the mass of a compound in grams equals the

NSWER

There are the same number of molecules in 17.04 grams of ammonia as there are in 36.46 grams of hydrogen chloride. Do you see why?

---

*Take 12 grams and divide by the mass of a single carbon-12 atom, 1.993×10^{-23} grams, to derive this answer.

formula mass, we get the same number. So, there are 6.02×10^{23} molecules of ammonia in 17.04 grams of ammonia (formula mass = 17.04 amu), just as there are 6.02×10^{23} molecules of hydrogen chloride in 36.46 grams of hydrogen chloride (formula mass = 36.46 amu). These are hardly coincidences for we find this large number is simply the ratio of *relative mass* (expressed not in amu but in grams) to *actual mass* (the actual mass of an atom or molecule is very small). A chemical with a larger relative mass has a proportionately larger actual mass, and the ratio of the two will always be the same—6.02×10^{23}.

The existence of this special number was recognized in the late 1800s. Its value, however, could not be determined until the early 1900s, when techniques for measuring the mass of a single atom were developed (see Box). Because of Amadeo Avogadro's important contributions to the development of the atomic theory (see Section 17.1), this number bears his name—**Avogadro's number.***

As a matter of convenience, we refer to 6.02×10^{23} atoms or molecules of a substance as 1 **mole** of that substance. Accordingly, the atomic or formula mass of a substance given in grams is equal to 1 mole of that substance. So we say that 12 grams of carbon-12 (12 amu) is 1 mole of carbon-12, which equals 6.02×10^{23} atoms of carbon-12. And 17.04 grams of ammonia (17.04 amu) is 1 mole of ammonia, which equals 6.02×10^{23} molecules of ammonia.

One mole of any substance always contains the same number of particles, 6.02×10^{23}. The mole is therefore an ideal unit when dealing with chemical reactions. For example, 1 mole of ammonia (17.04 grams) reacts with 1 mole of hydrogen chloride (36.46 grams) to give 1 mole of ammonium chloride (53.50 grams). In many instances, the ratio in which chemicals react is not one to one. This is most easily depicted using the mole unit. For example, 1 mole of molecular oxygen, O_2 (32.00 grams) reacts with 2 moles of molecular hydrogen, H_2 (4.04 grams) to give 2 moles of water, H_2O (36.04 grams) (Figure 20.22).

Figure 20.22 One mole of molecular oxygen, O_2, reacts with 2 moles of molecular hydrogen, H_2, to give 2 moles of water, H_2O. This is the same as saying 32 grams of O_2 react with 4 grams of H_2 to give 36 grams of H_2O, or equivalently, that 6.02×10^{23} molecules of O_2 react with 12.04×10^{23} molecules of H_2 to give 12.04×10^{23} molecules of H_2O.

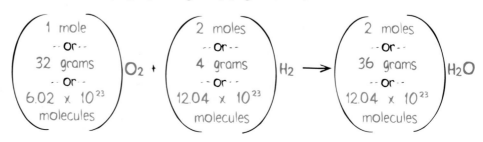

*Avogadro died in 1856, some 50 years before this honor was bestowed upon him.

QUESTION

Which has a greater mass, 12.04×10^{23} molecules of molecular hydrogen, H_2, or 12.04×10^{23} molecules of water, H_2O?

ANSWER

The water has more mass, just as a bunch of golf balls has more mass than the same number of ping pong balls. The water in fact has about 9 times as much mass because each water molecule is about 9 times (18 amu/2 amu) as massive. The big numbers don't change anything: 12.04×10^{23} molecules of water have a greater mass than 12.04×10^{23} molecules of molecular hydrogen.

SUMMARY OF TERMS

Reactant The starting material for a chemical reaction. It appears before the arrow in the chemical equation.

Product The new material formed by a chemical reaction. It appears after the arrow in the chemical equation.

Chemical equation A representation of a chemical reaction showing the relative numbers of reactants and products.

Coefficients Numbers used in chemical equations to show the ratio in which reactants combine and products form. In balancing a chemical equation, coefficients are used to make the number of times an atom appears before and after the arrow the same.

Chemical reaction The energetic process whereby atoms are rearranged to give rise to new substances; a chemical change.

Chemical change The change of a substance into another substance or substances through a reorganization of atoms.

Chemical bond The electrical force that holds atoms together. Energy is required to break a chemical bond and it is released when a chemical bond is formed.

Exothermic reaction A chemical reaction that has the net effect of producing energy.

Endothermic reaction A chemical reaction that has the net effect of absorbing energy.

Chemical equilibrium A dynamic state in which the rate of the forward chemical reaction is equal to the rate of the reverse chemical reaction. At chemical equilibrium the concentrations of reactants and products remain constant.

Acid A substance that produces or donates hydrogen ions in solution.

Base A substance that produces hydroxide ions in solution or accepts hydrogen ions.

Hydronium Ion A water molecule after accepting a hydrogen ion, H_3O^+.

Hydroxide Ion A water molecule after donating a hydrogen ion, HO^-.

Amphoteric A substance that can behave as either an acid or a base.

Acidic A solution in which the hydronium ion concentration is greater than the hydroxide ion concentration.

Basic A solution in which the hydroxide ion concentration is greater than the hydronium ion concentration.

Neutral A solution in which the hydronium ion concentration is equal to the hydroxide ion concentration.

pH A measure of the acidity of a solution. The pH is equal to the negative logarithm of the hydronium ion concentration. The lower the pH, the greater the acidity.

Oxidation The process whereby a reactant loses one or more electrons.

Reduction The process whereby a reactant gains one or more electrons.

Electrochemistry A branch of chemistry concerned with the relationship between electrical energy and chemical change.

Battery A device in which the chemical energy from oxidation and reduction reactions is transformed into electrical energy.

Electrolysis The use of electrical energy to produce chemical change.

Atomic mass unit A very small unit of mass used for atoms and molecules. One atomic mass unit (amu) is equal to $\frac{1}{12}$ the mass of the carbon-12 atom, or 1.661×10^{-24} grams.

Atomic mass The mass of an element given in amu.

Formula mass The mass of a chemical compound given in amu.

Avogadro's number A very large number: 6.02×10^{23}. This is the number of atoms in exactly 12 grams of carbon-12.

Mole One mole represents 6.02×10^{23} units (Avogadro's number). When the atomic mass of an element is expressed in grams, this is the number of atoms that are present. For example, the atomic mass of nitrogen is 14 amu. There is 1 mole of nitrogen atoms in 14 grams. Similarly, there is 1 mole of oxygen atoms in 16 grams (O = 16 amu).

· ·

REVIEW QUESTIONS

The Chemical Equation

1. What is the purpose of coefficients in chemical equations?

2. What is meant by saying a chemical equation is balanced?

3. How many chromium atoms are indicated on the right hand side of the following balanced chemical equation? How many oxygen atoms?

$$4 \, Cr \, (s) + 3 \, O_2 \rightarrow 2 \, Cr_2O_3$$

Energy and Chemical Reactions

4. Why is energy required to break apart a chemical bond? Use the analogy of trying to pull apart two magnets that are stuck together.

5. What's the difference between an exothermic and endothermic reaction?

6. In what sense does an exothermic reaction consume energy?

Chemical Equilibrium

7. What does it mean to say that chemical reactions are *reversible?*

8. Are all chemical reactions, in principle, reversible?

9. What does it mean to say that the condition of chemical equilibrium is a dynamic situation? Although the reaction overall may appear to have stopped, what is still going on in the system?

Acids and Bases

10. What role do acids and bases play in our economy?

11. When an acid is dissolved in water, what ion does the water form?

12. What does it mean to say that an acid is strong in aqueous solution?

13. What does the value of K_w say about the extent of the ionization of pure water?

14. How do we characterize solutions as acidic, basic, or neutral, in terms of the relative concentrations of hydronium and hydroxide ions?

15. As the hydronium ion concentration of a solution *increases,* does the pH of the solution increase or decrease?

Oxidation-Reduction Reactions

16. Write an equation showing a potassium atom (atomic symbol = K) being oxidized. Write one showing a bromine atom (atomic symbol = Br) being reduced.

17. What is the purpose of the salt bridge depicted in Figure 20.19?

18. What chemical reaction is forced to occur while a car battery is being recharged?

19. Why should sparks not be produced around a car battery that is being recharged?

20. Why is electrolysis important to the national economy?

Atomic and Formula Masses

21. You are given two chemicals, 10 grams each. Why are the number of atoms in each of these samples not the same?

22. What is the atomic mass unit?

23. What is the mass of a sodium atom in atomic mass units?

24. What is the relationship between the amu and the gram?

Avogadro's Number and the Mole

25. Would Avogadro's number of golf balls occupy a large volume or a relatively small volume—compared to the size of a classroom, for example?

26. If you had 1 mole of marbles, how many marbles would you have?

27. If you had 2 moles of pennies, how many pennies would you have?

28. How many molecules of water are there in 18 grams?

· ·

HOME PROJECTS

1. Examples of chemical reactions are all around us, especially in kitchens. A great home project is to combine small amounts of baking soda (a base) with vinegar (an acid) to produce carbon dioxide, which is a gas heavier than air. Add a teaspoon of baking soda to the bottom of a tall drinking glass. Follow with 3 or 4 capfuls of vinegar. Add the capfuls slowly so that the bubbles don't overflow the glass. After the bubbles subside light a match and hold it adjacent to the lip of the glass. Extinguish the match by pouring the *gaseous* contents of the glass over the match.

2. An example of an endothermic reaction is as near as the salt in your cupboard. Add lukewarm water to two plastic cups. Transfer the liquid back and forth between containers to assure equal temperatures. Dissolve a lot of table salt in one of the cups (stirring helps). What happens to its temperature relative to the untreated lukewarm water? (Hold the cups up to your cheeks to tell.) Heat energy can be used to break chemical bonds. What type of chemical bonds are being broken here?

3. The pH of a solution can be approximated with a pH indicator—any chemical whose color changes with pH. Many pH indicators are found in plants; a good example is red cabbage. Shred about a quarter of a head of red cabbage. Boil the shredded cabbage in 2 cups of water for about 5 minutes. Strain the cabbage while collecting the broth, which contains the pH indicator. Red cabbage indicator is red at low pHs (pH = 1-4), a light purple at neutral pHs (pH = 7), green at moderately alkaline pHs (pH = 8–11), and yellow at very alkaline pHs (pH = 13). Add small amounts of cabbage broth to various solutions, such as white vinegar, rain water, ammonia, baking soda, or bleach, to estimate their pHs.

4. Perform the electrolysis of water by immersing the top end of a 9 V battery into a solution of salt water. The bubbles that form contain oxygen, which results from the decomposition of water. Why does this activity work better with salt water than with pure water? Why does this activity quickly ruin your battery? (We recommend you dry off and then properly dispose of your battery after performing this home project.)

5. Silver tarnish is an outer coating of silver sulfide, Ag_2S, formed when silver reacts with trace quantities of airborne hydrogen sulfide, a smelly gas produced naturally by the digestion of food in living beings. Silver atoms in silver sulfide have lost electrons to sulfur atoms. You can remove this tarnish with home electrochemistry. Convert silver atoms back to their elemental state by restoring their electrons. Sulfur atoms won't relinquish electrons to silver, but with the proper connection, aluminum atoms will. Add about a liter of water with several heaping tablespoons of baking soda to an aluminum pan that has been scoured clean. If you don't have an aluminum pan, use aluminum foil at the bottom of another pan. Bring the water to boiling and remove it from the heat source. Slowly immerse a tarnished piece of silver; you'll see an immediate effect as the silver and aluminum make contact. (Add more baking soda if you don't.) Also, as silver is brought back to its elemental state, hydrogen sulfide is released back into the air. You'll smell it! The baking soda serves to remove a thin and transparent coating of aluminum oxide, which normally coats aluminum and thus prevents direct contact. The baking soda also serves as a conductive ionic so-

lution to permit the passage of electrons from the aluminum to the silver. What is the advantage of this approach over polishing the silver with an abrasive?

• •

EXERCISES

1. Find the coefficients that will balance the following chemical equations:

   (a) ____ Fe (s) + ____ S (s) → ____ Fe_2S_3 (s)

   (b) ____ P_4 (s) + ____ H_2 (g) → ____ PH_3 (g)

   (c) ____ NO (g) + ____ Cl_2 (g) → ____ NOCl (g)

   (d) ____ $SiCl_4$ (l) + ____ Mg (s)

   → ____ Si (s) + ____ $MgCl_2$ (s)

2. Some chemical equations can be somewhat difficult to balance (unless your instructor provides you with a successful methodology). For a challenge, try this one:

   _____ Na_2SO_3 + _____ S_8 → _____ $Na_2S_2O_3$

3. Consider both the chemical reaction

   $$3 H_2 + N_2 → 2 NH_3$$

   and the following bond energies (the energy required to break the bond, or, conversely, the energy released upon bond formation):

Bond	Bond Energy (kJ/mole)
H—H	436
N—H	389
N≡N	946

   Then follow the example given in Table 20.1 and calculate whether the formation of ammonia, NH_3, from hydrogen, H_2, and nitrogen, N_2, is exothermic or endothermic. How much energy is released or absorbed by this reaction?

4. Many people hear about atmospheric ozone depletion and wonder why we don't simply replace that which has been destroyed. Knowing about CFCs and catalysis, explain how this would not be a lasting solution.

5. Increasing temperature favors the formation of nitrogen dioxide, NO_2, from dinitrogen tetroxide, N_2O_4. Why is it true that at higher temperatures the quantity of N_2O_4 never completely disappears?

6. Equilibrium is a balance of opposing processes. Give an example of an "equilibrium" encountered in everyday life, showing how the processes involved oppose each other.

7. Suggest why people once washed their hands with ashes.

8. The Arrhenius definition of a base is that which increases the concentration of hydroxide ions. How does ammonia, NH_3, which is a base, satisfy the Arrhenius definition?

9. Which should conduct electricity better, a solution saturated with a strong acid, or one saturated with a weak acid? Defend your answer.

10. A strong acid gives rise to a weak base, whereas a weak acid gives rise to a relatively strong base. Briefly explain.

11. When a hydronium ion concentration equals 1×10^{-10} moles per liter, what is the pH of the solution? Is this acidic or basic?

12. When a hydronium ion concentration equals 1×10^{-4} moles per liter, what is the pH of the solution? Is this acidic or basic?

13. When a hydrogen ion concentration equals 1 mole per liter, what is the pH of the solution? Is this acidic or basic?

14. When a hydrogen ion concentration equals 2 moles per liter, what is the pH of the solution? Is this acidic or basic?

15. Is sodium metal oxidized or reduced when used in the production of aluminum?

16. Does an oxidizing agent donate or accept electrons? Answer the same question for a reducing agent.

17. Consider the oxidation-reduction reaction

$$Mg \ (s) + \ Cu^{2+} \ (aq) \rightarrow Mg^{2+} \ (aq) + \ Cu \ (s)$$

Sketch a galvanic cell that uses this reaction. Which metal ion is reduced? Which metal is oxidized?

18. Jewelry is often manufactured by plating an expensive metal such as gold over a cheaper metal. How might such a process be set up as an electrolysis reaction?

19. What's the mass of a single oxygen atom, O, in atomic mass units?

20. What's the mass of a single oxygen atom, O, in grams?

21. What's the mass of a single molecule of water, H_2O, in atomic mass units?

22. What's the mass of a single molecule of water, H_2O, in grams?

23. Which has more atoms: 17.03 grams of ammonia, NH_3, or 72.92 grams of hydrogen chloride, HCl?

24. Which has the greatest number of molecules?

   (a) 32 grams of nitrogen, N_2
   (b) 32 grams of oxygen, O_2
   (c) 32 grams of methane, CH_4
   (d) 38 grams of fluorine, F_2

25. What are the formula masses (in amu) of the following compounds: water, H_2O; propene, C_3H_6; 2-propanol, C_3H_8O.

PROBLEMS

1. How many grams of water, H_2O, and grams of propene, C_3H_6, can be formed from the reaction of 6.0 grams of 2-propanol, C_3H_8O? (*Hint:* Find their formula masses.)

2. How many molecules of aspirin are there in a 0.250 gram sample (aspirin = 180 amu)?

3. How many grams of water, H_2O, can be formed from the combination of 10 grams of hydrogen, H_2, and 10 grams of oxygen, O_2?

$$2 \ H_2 + \ O_2 \rightarrow 2 \ H_2O$$

4. When small samples of oxygen gas are needed in the laboratory, the gas may be generated by any number of simple chemical reactions, such as

$$2 \ KClO_3 \ (s) \rightarrow 2 \ KCl \ (s) + 3 \ O_2 \ (g)$$

What mass of oxygen (in grams) should be produced when 122.55 grams of $KClO_3$ (formula mass = 122.55 amu) has reacted?

5. How many grams of carbon dioxide, CO_2, can be formed from the combination of 16 grams of methane, CH_4, and 32 grams of oxygen, O_2?

$$CH_4 + 2 \ O_2 \rightarrow CO_2 + 2 \ H_2O$$

21

ORGANIC CHEMISTRY

Organic chemistry is the study of carbon-containing compounds. Over 90 percent of all compounds contain carbon. This is why the study of carbon makes up a separate branch of chemistry. There are more than 3 million known organic compounds. Over 100,000 new organic compounds are added to the list each year. By contrast, there are only about 200,000 to 300,000 known inorganic compounds. The distinction between organic compounds and inorganic compounds is a simple one—organic compounds contain carbon; inorganic compounds do not.

Why does carbon form so many compounds? Carbon's unique ability to bond with itself to form rings or long chains and its talent for bonding with other elements make it the champion compound maker. Numerous organic chemicals are introduced in this chapter. Don't worry about memorizing their names or their features, but focus instead on gaining an appreciation for their diversity. Because so many organic chemicals are being introduced to our environment so quickly, it's important to have a basic understanding of them—what they can do and can't do, and what their implications in medicine and the biological sciences are. We begin with the simplest organic molecules—those that consist only of carbon and hydrogen.

The First Artificial Urine

Until about 1850, scientists generally believed that organic compounds could only be formed by living organisms. Also, scientists thought organic compounds contained a "vital force" associated with the life process. This vital force theory was gradually abandoned after Friedrich Wohler discovered that the organic compound urea (present in human urine) could be synthesized from the inorganic salt ammonium cyanate.

$$NH_4OCN + Heat \rightarrow NH_2CONH_2$$

ammonium cyanate · · · · · · · · · · · · urea

Soon many other "natural" organic compounds such as methyl alcohol, ethyl alcohol, and acetic acid were synthesized by other chemists. Organic chemicals not found in nature were synthesized too. The synthesis of new organic compounds has since continued at an ever-increasing pace.

21.1 HYDROCARBONS

Compounds that contain just carbon and hydrogen are **hydrocarbons**. Hydrocarbons have simple structures. Nevertheless, they exist in a tremendous variety. Hydrocarbons include natural gas, gasoline, motor and heating oil, and many plastics.

Hydrocarbons differ in the number of carbon and hydrogen atoms per molecule (Table 21.1). Natural gas, for example, is a mixture of hydrocarbons that contains mostly methane. Methane is a hydrocarbon that has only 1 carbon atom per molecule. Gasoline, on the other hand, is a mixture of hydrocarbons that contains between 5 and 10 carbons per molecule. Motor and heating oil are mixtures of hydrocarbons that contain about two times more carbon atoms per molecule.

TABLE 21.1   Properties of Hydrocarbons

Molecular Formula	Name	Melting Point (°C)	Boiling Point (°C)	Phase at Room Temp.
CH_4	Methane	−184	−161	
C_2H_6	Ethane	−183	−88	
C_3H_8	Propane	−188	−42	Gas
C_4H_{10}	n-Butane	−138	−0.5	
C_5H_{12}	n-Pentane	−130	36	
C_6H_{14}	n-Hexane	−94	69	
C_7H_{16}	n-Heptane	−91	98	Liquid
C_8H_{18}	n-Octane	−57	126	
$C_{16}H_{34}$	n-Hexadecane	18	288	
$C_{17}H_{36}$	n-Heptadecane	23	303	
$C_{18}H_{38}$	n-Octadecane	28	317	Solid
C_nH_{2n+2}*	Polyethylene	136	dec.**	

* For the formula C_nH_{2n+2}, n equals an unlimited number.
** decomposes before boiling

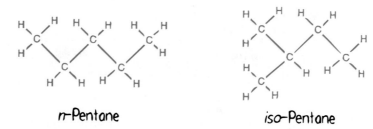

Figure 21.1 *n*-pentane and isopentane are two different hydrocarbons that have the same molecular formula: C_5H_{12}.

Polyethylene, a solid at room temperature, is a hydrocarbon formed by joining small hydrocarbon molecules. The result is a molecule that contains a very large number of carbon atoms.

Hydrocarbons also differ by the way the carbon atoms are connected. Consider the two hydrocarbons *n*-pentane and isopentane (Figure 21.1). These hydrocarbons have the same molecular formula, C_5H_{12}, yet they are chemically different because their atoms are bonded differently. The carbon framework of isopentane branches but the carbon framework of *n*-pentane does not. Molecules that have the same molecular formula but differ in their chemical structures are **structural isomers**.* Structural isomers have different chemical and physical properties. For example, *n*-pentane has a boiling point of 36°C, whereas isopentane's boiling point is 30°C.

Figure 21.2 The carbon framework of isopentane is branched, whereas that of *n*-pentane is not.

n-Pentane   iso-Pentane

We can see more clearly that *n*-pentane and isopentane are structural isomers by ignoring the hydrogen atoms (Figure 21.2). Structures shown in Figure 21.2 can be abbreviated by drawing only a series of connected sticks, and assuming that carbon atoms exist where sticks intersect and where sticks remain open-ended (Figure 21.3).

Figure 21.3 Stick structures for structural isomers *n*-pentane and isopentane.

n-Pentane   iso-Pentane

*The "iso" prefix in front of many of the chemical names in this chapter is derived from the term "isomer." It is used to indicate a branched molecular structure. The "*n*-" prefix indicates a "normal" unbranched molecular structure.

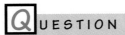UESTION

Two structural isomers of pentane are shown in Figure 21.3. What would a third structural isomer of pentane look like? (Make a sketch.)

For more complex hydrocarbons, the number of structural isomers increases rapidly as the number of carbon atoms increases. For example, there are 5 structural isomers of hexane, C_6H_{14}, 9 of heptane, C_7H_{16}, 18 of octane, C_8H_{18}, 75 of decane, $C_{10}H_{22}$, and a whopping 366,319 structural isomers of $C_{20}H_{42}$!

Hydrocarbons are obtained primarily from coal or petroleum. Most of the coal and petroleum that presently exists was formed between 280 and 395 million years ago—about 55 million years before the first dinosaurs appeared. The formation of coal begins with the slow bacterial decay of plant cellulose under water, in the absence of air. In the decay process oxygen is gradually removed from the cellulose. A residue that is largely hydrogen and carbon remains.

Beds of coal generally are the remains of ancient swamps. Swamps have produced coal ever since the appearance of plant life. Seldom, however, were conditions as favorable as occurred over 280 million years ago. At that time, there were extensive swamps lying close to sea level. These swamps periodically became submerged. The partially decayed vegetation there was trapped underneath layers of marine sediments.

The origin of petroleum (also called crude oil) is more obscure. Fossils are not preserved in fluids, so they cannot be found in petroleum. Also oil often migrates long distances after formation. Most geologists believe that petroleum is formed in much the same way that coal is formed—through the anaerobic decay of organic matter. Both plant and animal matter probably contribute to the formation of petroleum. However, proteins, fats, and waxes rather than cellulose are the compounds involved.

A NSWER

When molecules are structural isomers, their carbon frameworks are different. Bonds can rotate to give a single molecule different orientations just as your elbow can flex to give different orientations to your arm. None of the following stick structures, therefore, represents a third structural isomer of pentane. They are merely different orientations of the same molecule, *n*-pentane. The framework of *n*-pentane is 5 carbon atoms connected sequentially:

Similarly, the following stick structures represent different orientations for *iso*-pentane:

A third structural isomer for pentane is a central carbon bonded to four different carbon atoms:

This is *neo*-pentane, boiling point 10°C. What is its molecular formula?

To Burn or To Build

A great number of commercial items are made from petroleum based materials. This includes all items fabricated from plastics (auto parts, utensils, and furniture), synthetic fibers (clothing, carpets, and draperies), and many pharmaceuticals (even aspirin is produced from oil). Other items derived from petroleum include cosmetics, perfumes, chewing gum, wax, artificial sweeteners, flavorings, dyes, asphalt, paints, ceramics, and explosives.

Dmitri Mendeleev, the 19th century Russian chemist who proposed the modern periodic table, was one of the first to recognize the value of petroleum as a raw material for industry. He cautioned, however, that burning petroleum as a fuel would be like burning money. Today, his warning holds special truth for we find that the world's supply of petroleum is quickly being exhausted. Many petroleum experts suspect that world oil production will have peaked by the mid-1990s. After that, they predict a sharp decline in petroleum production as the world's finite reserves dwindle. Within a century, oil production may fall back to the meager level of the early 1900s.

Burning petroleum for its energy content may be worse than burning money. Money only represents wealth. Petroleum, however, *is* wealth. It is rich with usable energy, and it is the most versatile of all building materials. There are alternate sources of energy, such as solar, hydroelectric, and nuclear, that may be developed to substitute for petroleum. But there are no alternatives to petroleum or recycled petroleum products for the fabrication of new petroleum-based products. The continued burning of petroleum means that the price of all petroleum-based products will gradually increase. Eventually these now pervasive valuables will become less affordable, and the growing human population will have to turn to some other resource for its building needs.

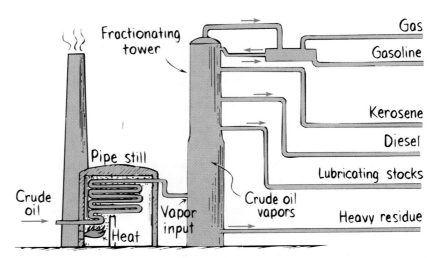

Figure 21.4 A schematic for the fractional distillation of crude oil (oil pumped directly from the ground). Crude oil is volatilized in the pipe still and then sent to a fractionating tower, which is warmer toward the bottom and cooler toward the top. Hydrocarbons with low boiling points such as natural gas and gasoline can travel to the top of the tower before condensing into a liquid. Hydrocarbons with higher boiling points, such as lubricating stocks, condense into liquids at lower heights. Pipes then drain the various hydrocarbon fractions from the tower.

About 17 million barrels of petroleum are consumed each day in the United States. Approximately 8 million of these are converted to gasoline, and another 8 million barrels are converted to other fuels such as heating oil, diesel fuel, jet engine fuel, and oil for electrical power plants. The remaining one million barrels used daily provide the raw material for production of organic chemicals and polymers. Thus only one seventeenth of the hydrocarbons consumed daily go into useful materials while the rest end up as heat and smoke.

Petroleum is a complex mixture of various hydrocarbons and other compounds. To obtain useful materials at a reasonable cost, the components must be separated efficiently. Since most of the materials are volatile, *fractional distillation* is used. Through this method, the components of crude petroleum are separated into fractions according to their boiling temperatures (Figure 21.4).

Special refinery processes are used to increase the production of high-demand hydrocarbons like gasoline. For example, large-molecule hydrocarbon oils are broken down into smaller less viscous gasoline hydrocarbons through the **catalytic cracking** process. In catalytic cracking hydrocarbon vapor is passed over a silica and aluminum oxide (SiO_2-Al_2O_3) catalyst at 450°C to 550°C (Figure 21.5).

Figure 21.5 Larger hydrocarbons can be broken down into smaller hydrocarbons by catalytic cracking.

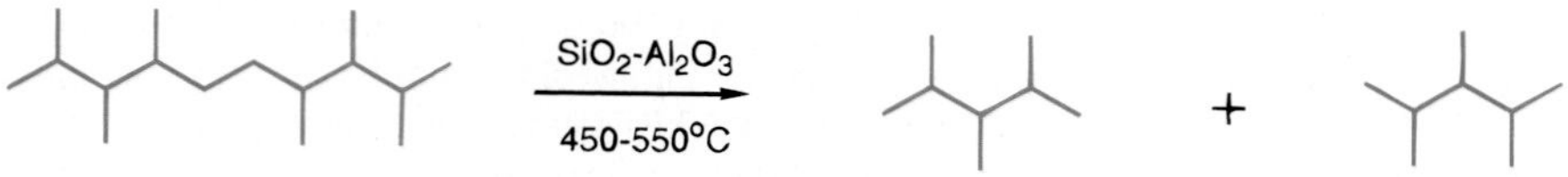

$$C_{16}H_{34} \xrightarrow[\text{450-550°C}]{SiO_2\text{-}Al_2O_3} C_8H_{18} + C_8H_{18}$$

A second procedure to increase the yield of gasoline from crude oil is combining lightweight hydrocarbons to form heavier-weight gasoline hydrocarbons. To do this, lightweight hydrocarbons are treated with hydrofluoric acid, HF, sulfuric acid, H_2SO_4, or other acids (Figure 21.6).

Figure 21.6 By using an acid catalyst, lightweight hydrocarbons can be combined to form gasoline hydrocarbons.

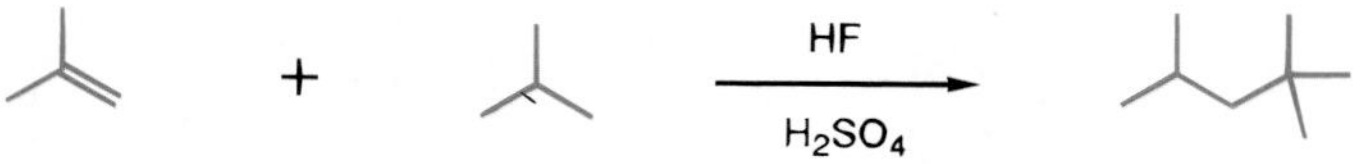

$$C_4H_8 + C_4H_{10} \xrightarrow[H_2SO_4]{HF} C_8H_{18}$$

Some hydrocarbons burn more efficiently than others in a car engine. In general, more branching in a gasoline hydrocarbon provides better efficiency. Isooctane, for example, combusts quite well; *n*-hexane, which has no branching, causes an engine to fire irregularly. This is commonly called engine "knock" (Figure 21.7).

iso-Octane

n-Hexane

Figure 21.7 Isooctane has more branching in its structure and burns smoothly, while *n*-hexane with no branching burns irregularly to produce engine "knock" in car engines.

Figure 21.8 The higher the octane number, the greater the degree of hydrocarbon branching.

These two compounds are used as standards in assigning octane ratings to gasoline. Isooctane is arbitrarily assigned an octane number of 100, and *n*-hexane is assigned 0. The antiknock performance of a particular gasoline is compared with that of various mixtures of isooctane and *n*-hexane, and an octane number is assigned.

QUESTION

Hydrocarbons burn and release a lot of energy when ignited. Where does this energy come from?

21.2 UNSATURATED HYDROCARBONS

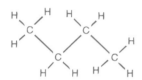

n-Butane

Figure 21.9 An example of a saturated hydrocarbon is *n*-butane.

Two classes of hydrocarbons are **saturated** and **unsaturated**. In a saturated hydrocarbon, all the carbon atoms are bonded to four atoms. For example, consider the saturated hydrocarbon *n*-butane (Figure 21.9). The two terminal carbons in *n*-butane are each bonded to three hydrogens and one carbon—a total of four atoms. Similarly, the two internal carbons are each bonded to two hydrogens and two carbons—also a total of four atoms.

In an unsaturated hydrocarbon, carbons bond to only three or two atoms. This occurs when multiple covalent bonds are present. For example, consider the unsaturated hydrocarbons 2-butene* and acetylene (Figure 21.10). For 2-butene, each of the internal carbons bonds to two carbons and one hydrogen—a total of three atoms. For acetylene, each of the carbons bonds to only one carbon and one hydrogen—a total of two atoms. So the term "unsaturated" means that multiple bonds are present and therefore a lesser number of hydrogen atoms are found in these molecules.

Figure 21.11 The unsaturated hydrocarbon acetylene burned in this torch produces a flame hot enough to melt iron and steel.

2-Butene   Acetylene

Figure 21.10 Examples of unsaturated hydrocarbons are 2-butene and acetylene.

ANSWER

Since most hydrocarbons were once living matter, the original source of their energy is the sun. (In Chapter 28 we'll see that solar energy in turn has a source.) So the energy in gasoline is solar energy that has been transformed into chemical bonds through the process of photosynthesis, and stored for millions of years!

---

*The number in front of this chemical name indicates the location of the multiple bond. In this case, it is the *second* carbon.

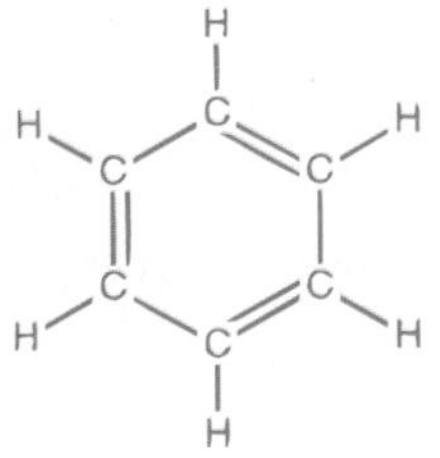

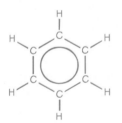

Benzene

Figure 21.12 Benzene is a special case of an unsaturated hydrocarbon.

Figure 21.14 A circle represents the three migratory double bonds of the benzene ring.

QUESTION

Draw and compare stick structures for 2-butene and acetylene.

A special case of an unsaturated hydrocarbon is benzene, C_6H_6. Because of its arrangement of three double bonds, the benzene molecule forms a flat hexagonal ring (Figure 21.12). The flat structure of benzene is partly responsible for the fact that benzene is *carcinogenic* (cancer-causing). Because it is flat, the benzene molecule can literally slice into the DNA double helix and disrupt the cell's ability to regulate itself.

The benzene structure shown in Figure 21.12 is simplified, for the electrons in benzene double bonds migrate throughout the ring (Figure 21.13). This being the case, we show the electrons of the double bonds distributed about the whole ring rather than at distinct locations. This electron distribution is indicated by the circle within the benzene ring (Figure 21.14).

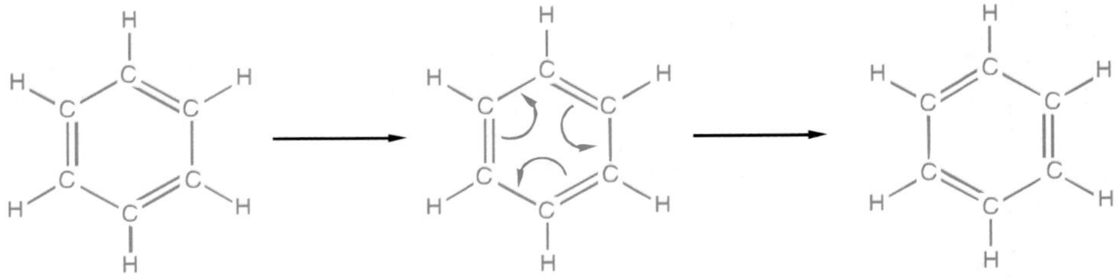

Figure 21.13   The location of the double bonds of benzene shifts because of the migration of electrons.

Many organic compounds contain benzene rings within their structure. Many of these compounds are also fragrant. So by convention, any organic molecule containing the benzene ring is classified as an *aromatic* compound.* Toluene, a common solvent and paint thinner, is such an aromatic compound. Toluene gives airplane glue its distinctive odor (Figure 21.15a). Some aromatic compounds such as naphthalene contain two or more benzene rings fused together. Naphthalene gives rise to the smell of moth balls (Figure 21.15b).

ANSWER

Stick structures ignore the presence of hydrogens and focus on the configuration of the carbon framework. The stick structure for 2-butene can be represented as

The stick structure for acetylene can be represented as

$$\equiv$$

*Even if it is not especially fragrant.

Toluene

Naphthalene

Figure 21.15 The stick structures for two odoriferous benzene ring-containing organic compounds: toluene and naphthalene.

21.3 NON-CARBON ATOMS IN ORGANIC MOLECULES

There is a large number of different hydrocarbons because carbon atoms can bond to themselves and hydrogen in so many ways. Carbon can bond to other common elements, such as oxygen, nitrogen, sulfur, and chlorine, as well. This ability further increases the number of different possible organic molecules. Adding even a single non-carbon or non-hydrogen atom to a hydrocarbon may change its physical and chemical properties remarkably. For example, consider a change that occurs when an oxygen atom is added to the hydrocarbon ethane, C_2H_6. Ethane has a melting point of $-183°C$, a boiling point of $-88°C$, is a gas at room temperature, and does not dissolve in water. By adding an oxygen atom to the ethane structure we come up with a new molecule, ethanol, C_2H_6O (Figure 21.16). Ethanol has a melting point of $-130°C$, a boiling point of $78°C$, is a liquid at room temperature, and is infinitely soluble in water. Organic chemist types call a non-carbon or non-hydrogen atom in an organic molecule a *heteroatom*.* Adding a heteroatom such as oxygen, nitrogen or sulfur to the hydrocarbon structure can have an enormous effect.

Figure 21.16 The oxygen atom accounts for the remarkably different physical and chemical properties of ethane and ethanol.

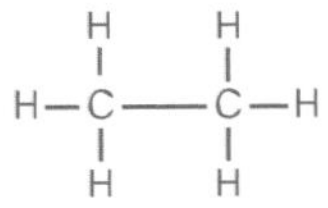

Ethane

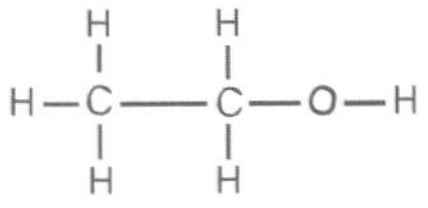

Ethanol

Hydrocarbons are relatively inert. The fact they persist for millions of years suggests this. Their inert structure serves as a three-dimensional carbon framework, upon which various heteroatoms can be added. We have seen that adding an oxygen atom to an ethane molecule produces ethanol, a substance many people use (or misuse) for its effects on the nervous system. Adding a nitrogen atom there instead produces a substance called *ethylamine*, a corrosive and potentially lethal gas that smells like ammonia (Figure 21.17).

Figure 21.17 Heteroatoms give character to organic molecules. Ethanol is drinkable, but ethylamine is not.

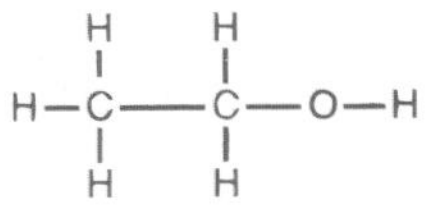

Ethanol

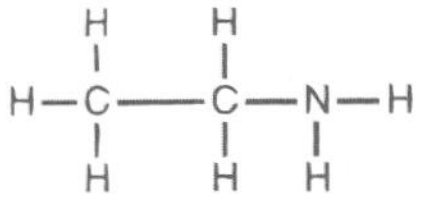

Ethylamine

---

*"Hetero" means different. In organic chemistry a "heteroatom" is different from carbon or hydrogen.

Because the properties of organic molecules are largely determined by heteroatoms, organic molecules are classified by which ones they contain and how they are attached to the carbon framework. Examples of organic compounds so classified are the alcohols, ethers, amines, ketones, aldehydes, amides, and carboxylic acids.

21.4 ALCOHOLS, ETHERS, AND AMINES

Alcohols are a class of organic molecules that contains the *hydroxyl group,* an oxygen atom bonded to a hydrogen, OH. Because of the polarity associated with the oxygen-hydrogen bond, smaller alcohols are often soluble in water. Some common alcohols are listed in Table 21.2.

More than 1 billion gallons of methanol, CH_3OH, are produced annually in the United States. Most of it is used for making formaldehyde and acetic acid, important components of various polymers. In addition, methanol is used as a solvent and as an octane booster and anti-icing agent in gasoline. Methanol is sometimes called wood alcohol, for it can be obtained from the distillation of wood. Methanol (wood alcohol) should never be ingested. Once in the body it is metabolized into formaldehyde. Formaldehyde accumulates in the eyes and leads to blindness. Formaldehyde is familiar to biology students for it is used to preserve dead bodies as lab specimens.

Ethanol is probably the oldest and one of the most important synthetic chemicals. It is the "alcohol" of alcoholic beverages. Ethanol is prepared for drink by fermenting sugar from various plants. Industrial ethanol is used as a solvent. For many years it too was made by fermentation. However, in the last several decades, it became cheaper to make ethanol from petroleum byproducts (Figure 21.18).

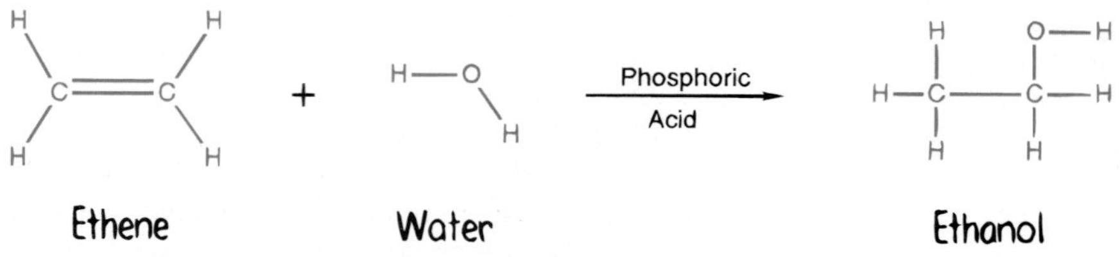

Figure 21.18 Ethanol can be synthesized from the hydrocarbon ethene with the use of phosphoric acid as a catalyst.

Ethene   +   Water   →(Phosphoric Acid)→   Ethanol

A third well-known alcohol is isopropyl or "rubbing" alcohol, also called 2-propanol. Although 2-propanol has a relatively high boiling point it readily evaporates at lower temperatures. Also, 2-propanol readily dissolves grime and has some antiseptic properties. For all these reasons rubbing alcohol is used to swab skin prior to needle injection.

TABLE 21.2   Properties of Some Simple Alcohols

Structure	Scientific Name	Common Name	Melting Point (°C)	Boiling Point (°C)
CH_3—OH	Methanol	Methyl alcohol	−97	65
CH_3—CH_2—OH	Ethanol	Ethyl alcohol	−115	78

CH_3—CH—CH_3
    |
    OH
	2–Propanol	Isopropyl alcohol	−126	97

Ethers are a class of organic compounds structurally related to alcohols. The oxygen atom in ethers, however, is not situated in a hydroxyl group. Instead, it is bonded to two carbon atoms (Figure 21.19). Although ethanol and dimethyl ether have the same chemical formula, C_2H_6O, their physical properties are vastly different. While ethanol mixes with water and boils at 78°C, dimethyl ether does not mix with water and boils at −25°C. These differences are due to the fact that ethers lack the polar hydroxyl group of alcohols. Ethers do not mix with water because, without the hydroxyl group, they are unable to form strong dipole-dipole interactions with water. Furthermore, without the polar hydroxyl group, the chemical interactions among ether molecules are relatively weak. Therefore ether molecules are easy to pull apart. This is why ethers have relatively low boiling points.

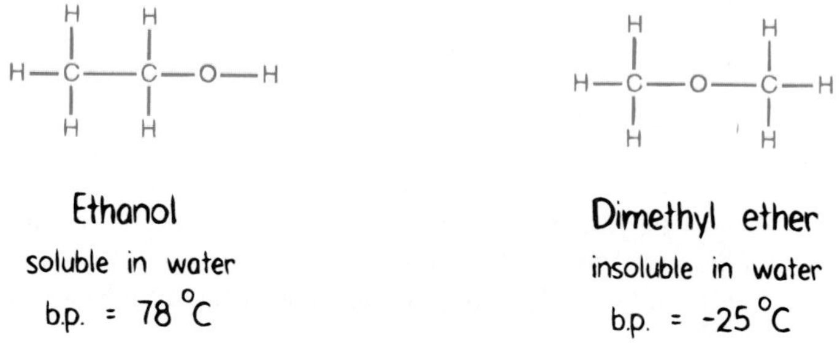

Figure 21.19 The oxygen of alcohols is bonded to one carbon atom and one hydrogen atom. The oxygen of ethers, however, is bonded to two carbon atoms. Alcohols and ethers of similar molecular mass have vastly different physical properties.

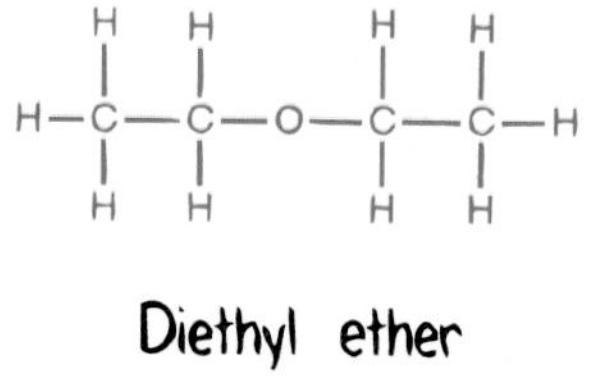

Ethanol

soluble in water

b.p. = 78 °C

Dimethyl ether

insoluble in water

b.p. = -25 °C

Diethyl ether

Figure 21.20 Diethyl ether is the technical name for the "ether" that historically was used as an anesthetic.

Diethyl ether, with a boiling point of 35°C, was one of the first anesthetics (Figure 21.20). The anesthetic properties of diethyl ether were discovered in the early 1800s. The discovery revolutionized the practice of surgery. Because of its high volatility at normal room temperatures diethyl ether can be rapidly administered to the blood stream by way of inhalation. But since it has low solubility in water, it quickly leaves the blood stream once introduced. Due to these physical properties a surgical patient can be brought in and out of anesthesia at will simply by regulating the gases that he breathes. Modern-day gaseous anesthetics, which have fewer side effects such as nausea and headaches, work by the same principle.

Amines are a class of organic compounds that contain nitrogen. The compounds shown in Table 21.3 are examples. The nitrogen of an amine may be connected to one, two, or three carbons. Since the polarity of the nitrogen-hydrogen or the nitrogen-carbon bond is not as great as the polarity of the oxygen-hydrogen bond, amines are typically less soluble in water than alcohols are. Also their boiling points are typically somewhat less than those of alcohols of similar molecular mass.

One of the most notable physical properties of many lightweight amines is their offensive odor. Two appropriately named amines, putrescine and cadaverine, for example, are responsible for the odor of decaying fish.

TABLE 21.3 *Properties of Some Simple Amines*

Structure	Scientific Name	Melting Point (°C)	Boiling Point (°C)
CH_3CH_2—NH_2	Ethylamine	−81	17
CH_3CH_2—NH—CH_2CH_3	Diethylamine	−50	55
CH_3CH—N—CH_2CH_3 $\quad\quad\;\;$| $\quad\quad CH_2CH_3$	Triethylamine	−7	89

$$H_2N - CH_2CH_2CH_2CH_2 - NH_2$$

Putrescine

(1,4-butanediamine)

$$H_2N - CH_2CH_2CH_2CH_2CH_2 - NH_2$$

Cadaverine

(1,5-pentanediamine)

Organic amines typically are alkaline. That is, they can raise the pH of an aqueous solution above the neutral value of 7. The alkalinity arises because the nitrogen in an amine readily accepts a hydrogen ion. An amine put into water, therefore, tends to increase the hydroxide ion concentration by the reaction shown in Figure 21.21.

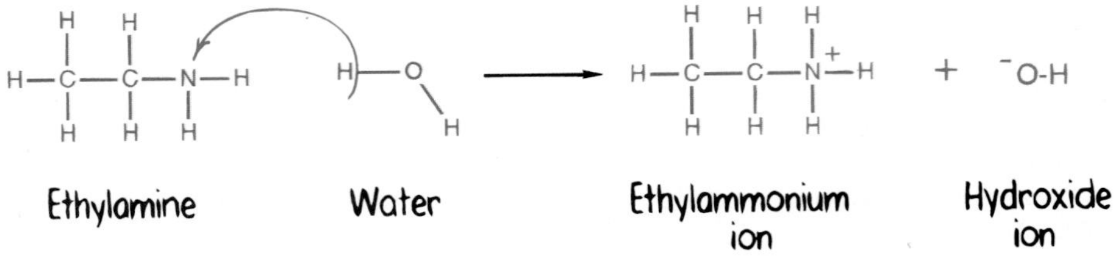

Figure 21.21 Ethylamine acts as a base and accepts a hydrogen ion from water to become the ethylammonium ion. This generates the hydroxide ion, which then increases the pH of the solution.

Ethylamine   Water   Ethylammonium ion   Hydroxide ion

In general, any organic molecule containing a nitrogen tends to be slightly alkaline. Molecules found in nature that are alkaline because of nitrogens are often called **alkaloids**. Alkaloids may also have hydroxyl groups, ether, or other groups (Figure 21.22).

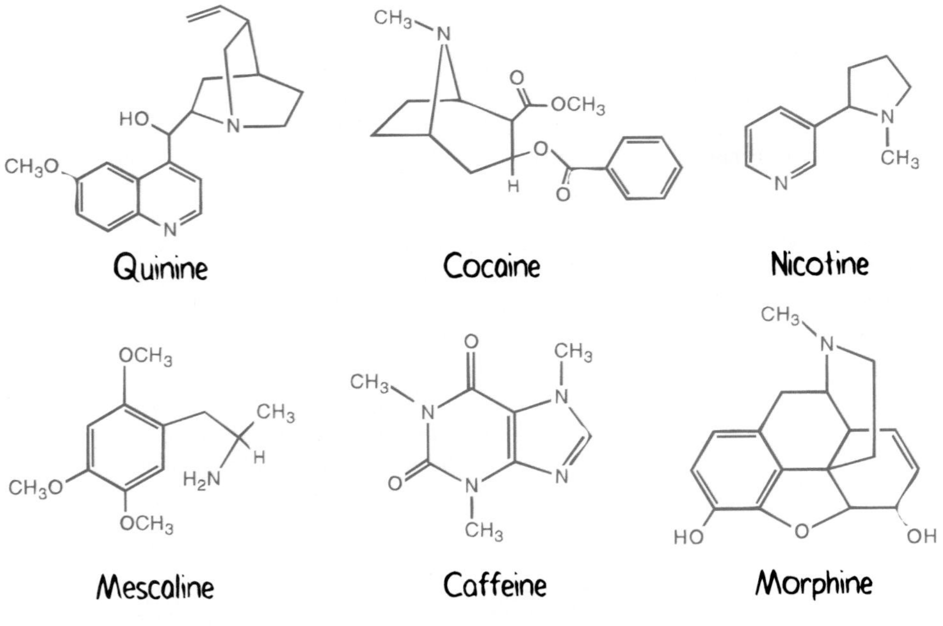

Figure 21.22 Alkaloids are a class of naturally occurring compounds that are alkaline due to the presence of nitrogens. Alkaloids may also have other types of atoms in them. Note the great variety of the carbon frameworks.*

Quinine   Cocaine   Nicotine

Mescaline   Caffeine   Morphine

*The stick structures shown in Figure 21.22 are meant to represent the 3-dimensional shapes of these molecules as much as possible. In the structures of quinine, cocaine, and morphine there are bonds that appear in front of other bonds. The bonds that lie behind, therefore, are obscured at the point of overlap. For this reason, they appear discontinuous, while in fact, they are not. Because organic molecules have intricate 3-dimensional structures, drawing them on paper is often severely limited. Model kits provide a better representation. Also, more realistic shapes of organic molecules can be represented 3-dimensionally with computer molecular modeling programs.

Figure 21.23 Cocaine, an alkaloid, reacts with hydrochloric acid, HCl, to form the hydrogen chloride salt of cocaine, a substance that is soluble in water. (Any alkaloid found in a nonsalt form is often referred to as a free base.)

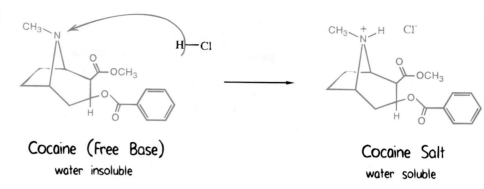

Cocaine (Free Base)
water insoluble

Cocaine Salt
water soluble

Figure 21.24 Tannins are responsible for the brown stains left behind in coffee mugs or on a coffee drinker's teeth. Since tannins are acidic they can be readily removed using alkaline cleanser. For the coffee mug, use a little bleach. For your teeth, brush them with baking soda.

Many alkaloids have medicinal value. Hence, it is of great interest to isolate these compounds from the materials where they occur (typically of plant or marine origin). This can be conveniently done with some fairly simple acid-base chemistry. Since alkaloids are basic, they react with acids to form neutral salts (Figure 21.23). Salts are usually quite soluble in water. An efficient way to isolate alkaloids, therefore, is to expose alkaloid-containing material to a solution of acid in water. By this method, all the alkaloids that are present convert to a water-soluble salt form. Alkaloid salts are then carried away by water. Once the alkaloids are isolated, other chemical separation techniques can be used to purify them.

Nature is usually a step ahead of us in forming the salts of alkaloids. Frequently, natural salts are made using certain organic acids called tannic acids. Tannic acid salts are usually soluble only in hot water. This is why we use hot water when we brew coffee beans or tea leaves for their caffeine content. What we are actually doing is extracting the tannic acid salt of caffeine.

Humanity has developed a real taste for the alkaloid caffeine. But caffeine by itself is not very soluble in water. The beverage industry overcomes this by making beverages acidic. Read the label on your next can of caffeinated cola soft drink. You'll find that it contains phosphoric acid. The phosphoric acid guarantees that the caffeine stays in its water soluble salt form. The phosphoric acid also enhances the flavor of the soft drink!

Many people also have a taste for spicy foods. What makes a hot sauce "hot" are the alkaloids it contains. The reason that the "hot" stays in your mouth for so long is that these alkaloids are in their free base (nonsalt) form and they are not very soluble in water—or your saliva. So, once these alkaloids have come in contact with the tissue of your mouth, they don't go down your throat no matter how hard you swallow.

Q UESTION

If, by chance, your mouth is on fire after taking in a bit too much hot sauce, which would be best to drink: a cola soft drink, a tall glass of ice water, or an Alka-Seltzer?

Figure 21.25 If the alkaloids in hot chili are too much for you, the acid in a cola drink will help.

A NSWER

What we need to do here is convert the alkaloids into their water soluble salt forms. Only the cola drink has the acid content that can do this. Dairy products are also somewhat acidic, and they too can be used to alleviate a burning mouth. So order a cola or a glass of milk with that next hot taco!

21.5 CARBONYL-CONTAINING ORGANIC MOLECULES

Four additional major classes of organic compounds are the ketones, aldehydes, amides, and carboxylic acids. All these compounds have a special group called the *carbonyl* in common (Figure 21.26a). The carbonyl consists of a carbon atom double bonded to an oxygen atom. A **ketone** is a type of organic molecule containing a carbonyl group in which the carbon of the carbonyl is bonded to two carbon atoms. A good example of a ketone is *acetone,* often used in nail polish remover (Figure 21.26b). An **aldehyde** is another class of organic molecules. Aldehydes contain a carbonyl group in which the carbon of the carbonyl is bonded to both one carbon atom and one hydrogen atom (Figure 21.26c).

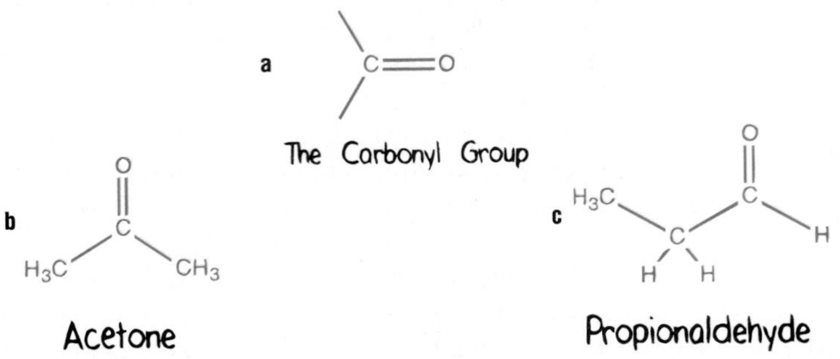

Figure 21.26 (a) The carbonyl group. (b) When the carbon of the carbonyl group is bonded to two carbon atoms, the result is ketone. An example is acetone. (c) When the carbon of the carbonyl group is bonded to at least one hydrogen atom, the result is an aldehyde called propionaldehyde.

The Carbonyl Group

Acetone

Propionaldehyde

Structurally, ketones and aldehydes are quite similar. With aldehydes, however, the carbonyl group must always come at the end of a carbon chain (if it were in the middle, the carbonyl would be surrounded by two carbons and the molecule would be classified as a ketone).

Many aldehydes are particularly fragrant. A number of flowers, for example, owe their pleasant odors to the presence of simple aldehydes such as *propionaldehyde.* The smell of lemons is due to the aldehyde *citral.* The smells of cinnamon, vanilla, and almond are due to the aldehydes *cinnamonaldehyde, vanillin,* and *benzaldehyde,* respectively (Figure 21.27).

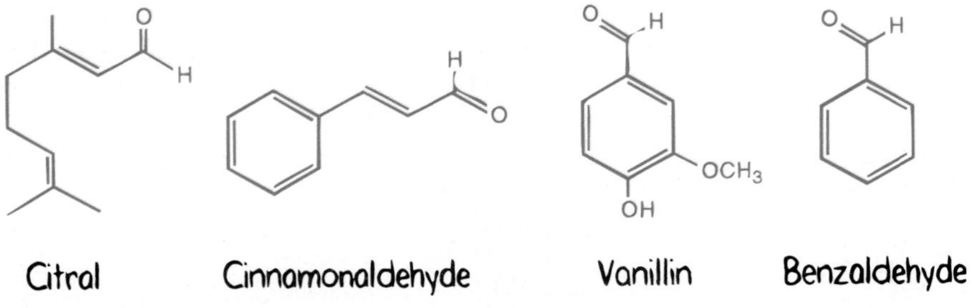

Figure 21.27 Aldehydes are responsible for many familiar fragrances.

Citral   Cinnamonaldehyde   Vanillin   Benzaldehyde

When the carbon of the carbonyl group is bonded to a nitrogen, a group called an **amide** is formed. Molecules containing the amide group are classified as *amides* (Figure 21.28). Campers and hikers find one particular amide especially useful— *N,N-diethyl-m-toluamide* (also known as DEET). This is the active ingredient in most effective insect repellents. Diethyltoluamide is actually not an insecticide. Rather, it causes certain insects, such as mosquitoes, to lose their sense of direction. This effectively protects DEET wearers from being bitten.

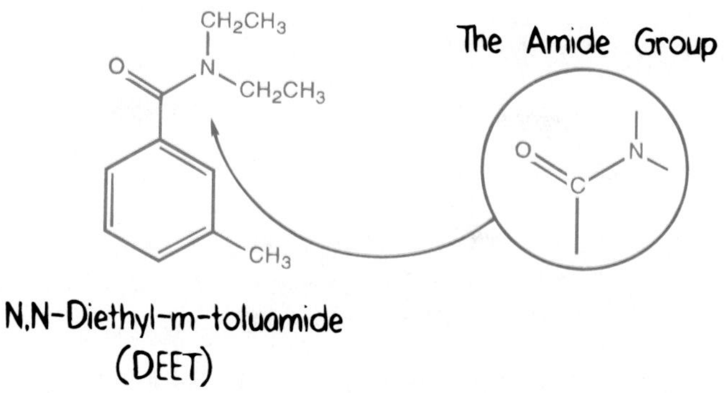

Figure 21.28 Amides contain the amide group in which a nitrogen is bonded to the carbon of the carbonyl.

The Amide Group

N,N-Diethyl-m-toluamide
(DEET)

The amide group is important in many polymers, both human-made and natural. Nylon is a polymer of many carbon chains held together by the amide group. Protein is a natural polymer that contains amide linkages (Figure 21.29).

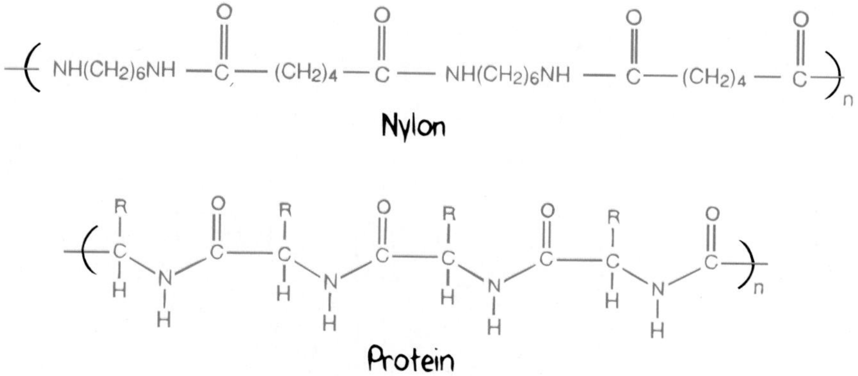

Figure 21.29 Polymers are exceedingly long molecules that consist of repeating units. The repeating units of nylon and protein polymers are held together by amide groups.

Nylon

Protein

A *carboxylic acid* group is formed when a hydroxyl group, —OH, bonds to the carbon of the carbonyl. Molecules containing a carboxylic acid group are classified as **carboxylic acids** (Figure 21.30). An example is *salicylic acid,* found in the bark of the willow tree. Once brewed for its antipyretic (fever reducing) effect, salicylic acid became an important analgesic (painkiller). But salicylic acid also causes nausea and severe stomach upset. In 1899, Friederich Bayer and Company, in Germany, introduced a chemically modified version of this compound that has fewer side effects. Acetylsalicylic acid—known as aspirin—was discovered (Figure 21.31).

Carbon frameworks form in almost any imaginable 3-dimensional shape. With heteroatoms added to these frameworks, a wide spectrum of physical and

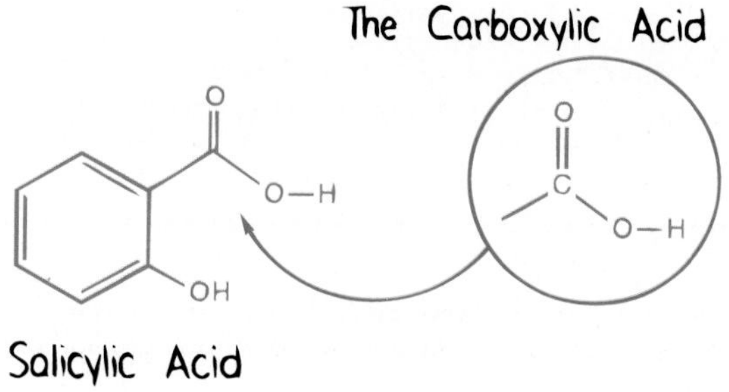

Figure 21.30 The carboxylic acid group consists of a hydroxyl group bonded to a carbonyl group.

The Carboxylic Acid Group

Salicylic Acid

Aspirin
(Acetylsalicylic Acid)

Figure 21.31 Aspirin is a chemical that was originally synthesized from the naturally occurring salicylic acid found in willow trees.

chemical properties is then obtained. We find these compounds everywhere in our environment. The vast oceans and rainforests undoubtedly encompass a great number of undiscovered compounds. But because the potential for different organic compounds is endless, an even greater number may yet be created.

QUESTION

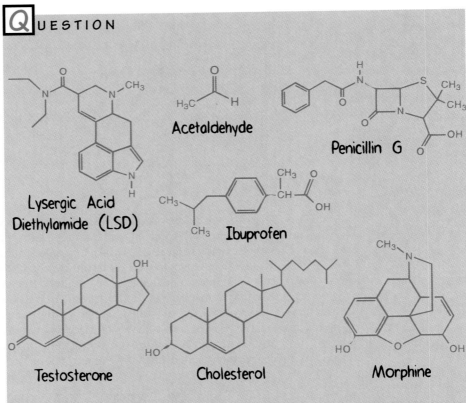

Classify each of the above organic molecules. They may be any one or more than one of the following: alcohol, ether, amine, ketone, aldehyde, amide, carboxylic acid.

ANSWER

LSD: amide, amine; *acetaldehyde:* aldehyde; *ibuprofin:* carboxylic acid; *penicillin G:* amide, carboxylic acid (the sulfur group was not discussed); *testosterone:* alcohol, ketone; *cholesterol:* alcohol; *morphine:* alcohol, ether, amine.

Drugs

A drug may be loosely defined as any chemical that elicits a biological effect. Most drugs used today are organic chemicals, and they come from different sources (Table 21.4). Many drugs come directly from terrestrial or marine plants or animals. Other drugs are natural products that have been chemically modified to increase potency or decrease harmful side effects. There is also a growing number of drugs that are purely synthetic.

Aspirin can cure a headache. But when you pop an aspirin pill, how does the aspirin know to go to your head rather than your big toe? The answer is it doesn't! After aspirin dissolves in your stomach, it gets absorbed into your blood stream, which distributes the drug over your whole body—from head to toe. Consequently, aspirin is good for headaches, muscle aches, backaches, *and* toe aches.

One effect of aspirin is to alleviate pain. But, aspirin can also cause ringing in your ears, and can even inhibit your blood from clotting. Like most drugs, aspirin has more than one effect on the body. Much effort in the pharmaceutical industry is directed toward modifying the chemical structure of drugs so that their effects are more specifically aimed at treating ailments.

An interesting example of the above approach is cancer chemotherapy. Many anticancer drugs work by killing cells that are in the process of dividing. This selectively kills cancer cells because, unlike normal cells, cancer cells are nearly always dividing. This treatment has had much success; however, it's not perfect. One problem is that even normal cells divide occasionally. Some cells, such as those in your intestines and hair follicles, divide quite frequently. These anticancer drugs, therefore, take their toll on normal cells too. But, by focusing on other differences between cancer cells and normal cells, it is hoped that more specific anticancer drugs may be found. For example, it now appears that many cancer cells have unusual outer surfaces. Anticancer

TABLE 21.4   Some Commonly Known Drugs

Drug	Biological Effect	Origin
Caffeine	Nerve stimulant	
Reserpine	Reduces hypertension	
Vincristrine	Anticancer agent	Natural product
Penicillin	Antibiotic	
Morphine	Analgesic	
Prednisone	Antirheumatic	
Ampicillin	Antibiotic	
LSD	Hallucinogen	Chemical derivative
Chloroquinine	Antimalarial	of a natural product
Ethynodiol diacetate	Contraceptive	
Valium	Antidepressant	
Benadryl	Antihistamine	
Allobarbital	Sedative-hypnotic	Synthetic
Phencyclidine	Veterinary anesthetic	
Methadone	Analgesic	

drugs that are able to selectively stick to this outer surface might help to increase the success of cancer chemotherapy many fold.

Many drugs work like a lock and key. There are different types of receptor sites in the body, each acting like a lock in a door. When a drug molecule fits into one of these sites, like a key fits into a lock, a particular biological effect, such as a nerve impulse or change in cellular morphology, is triggered. In order for a molecule to fit into a particular re-

ceptor site, however, it must have the proper shape, like a key must have properly shaped notches in order to fit the lock. According to this model, the problem with nonspecific drugs is that they fit into many different receptor sites. Hence, like a skeleton key, they unlock a wide variety of biological effects. Knowing the precise shape of a target receptor site, however, such as one on the surface of a cancer cell, allows chemists to design molecules that have an optimum fit, and a specific biological effect.

UESTION

Why are organic chemicals so suitable for making drugs?

21.6 BIOMOLECULES

At one time the physical and biological worlds were thought to be two separate realms. They interacted with each other, but nevertheless were thought to be distinct. Some intangible "life force" was believed present in living things but absent elsewhere. Evidence now indicates that there is no "life force" that distinguishes the animate from the inanimate. Rather, there appears to be a continuous chain of development from simple chemical compounds to more elaborate organic chemical compounds that make up viruses, primitive one-celled organisms, and even complex plants and animals. Three important classes of organic compounds found in living matter are the *carbohydrates, fats,* and *proteins.*

Carbohydrates

Carbohydrate means "hydrate of carbon." All carbohydrates have the general molecular formula $C_x(H_2O)_y$ where x and y are integers. Although carbohydrates account for only 1 percent of total body mass, they are centrally important in human biochemistry, especially as a source of energy. Carbohydrates are formed in plants by the process of photosynthesis from carbon dioxide, water, and energy from the sun.

$$x\,CO_2 \;+\; y\,H_2O \;+\; energy \;\rightarrow\; \underset{\text{carbohydrate}}{C_x(H_2O)_y} \;+\; x\,O_2$$

ANSWER

It is the vast diversity of organic chemicals that permits the manufacture of the many different types of medicines needed to match the many different types of illnesses.

Not only do they serve as the ultimate food source for animals, but the carbohydrate cellulose is the main structural component of plants. Furthermore, carbohydrates are major components of wood, cotton, paper, and many other important materials derived from natural sources.

Carbohydrates are divided into three classes, depending on molecular size. *Monosaccharides* are the simplest, consisting of only a single saccharide unit, which can be identified by a single ring of five or six atoms. Monosaccharides all have one unit of water per carbon [$C_x(H_2O)_y$, where $x = y$] (Figure 21.32).

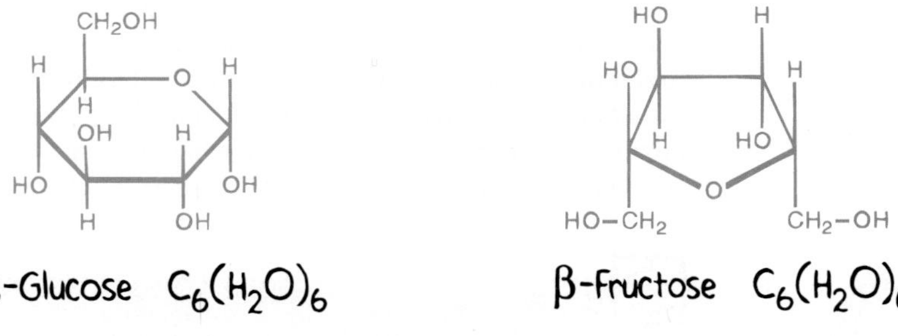

Figure 21.32 α-Glucose and β-fructose are examples of monosaccharides.

α-Glucose $C_6(H_2O)_6$

β-Fructose $C_6(H_2O)_6$

Monosaccharides are the building blocks of *disaccharides,* which contain two monosaccharide units. Common table sugar, sucrose, is an example of a disaccharide (Figure 21.33).

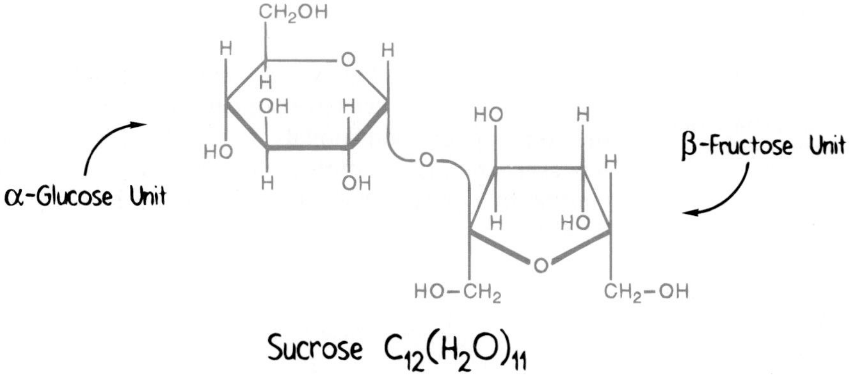

Figure 21.33 Disaccharides consist of two chemically bonded monosaccharide units.

α-Glucose Unit

β-Fructose Unit

Sucrose $C_{12}(H_2O)_{11}$

Honey, a mixture of the monosaccharides glucose and fructose, has been used for centuries as a natural sweetener for foods. In contrast, sucrose, derived from sugar cane or sugar beets, is a disaccharide. Honey is a popular sweetener because sugar is not as sweet as a mixture of pure glucose and fructose. To convert cane sugar into glucose and fructose requires treatment with acid or with a natural enzyme or catalyst called "invertase." The sugar industry goes to considerable expense to convert sugar chemically into "dextrose" or "levulose" (common names for glucose and fructose).

Monosaccharides are also the building blocks of *polysaccharides,* which contain a larger number of monosaccharide units. Two important polysaccharides are **cellulose** and **starch**. Cellulose is the most abundant organic compound on earth, and its purest form is cotton. This polysaccharide, which typically consists of 300 to 3000 glucose units, is also found as the woody part of trees and the supporting material in plants and leaves. Since cellulose is so abundant, it would be advantageous if humans could use it for food. Unfortunately, we cannot digest cellulose because the glucose units are bonded in such a way that humans cannot digest (metabolize) them. Microorganisms that live inside the digestive tracts of termites, cows, sheep, goats, and other ruminants, however, possess the necessary enzyme to break away the tightly held glucose units. Strictly speaking, these insects and animals cannot digest cellulose. They can, however, digest the by-products of the microorganisms that live inside them.

Because we cannot digest cellulose, we rely partly on starch as a source of glucose. Starch, like cellulose, is a polysaccharide consisting of many glucose units. The glucose units in starch, however, are connected differently, and we do possess an enzyme that is able to break them away. This enzyme, "amylase," is found in our pancreatic juices and in our saliva. You know this because if you hold a piece of bread in your mouth for a few minutes, it begins to taste sweet. This is a signal that sweet tasting glucose is being produced. When the bread is swallowed, the acid and enzymes in your stomach continue what your saliva began, and the freed glucose can be absorbed into the bloodstream and carried off to be used for its energy content.

QUESTION

Dieticians often refer to mono and disaccharides as "simple" carbohydrates and polysaccharides as "complex." Which do you suppose is more useful for quick energy, and why?

Figure 21.34 A candy bar is good for a quick energy fix, but chow down on a spaghetti feast the night before a very strenuous workout for long-run energy.

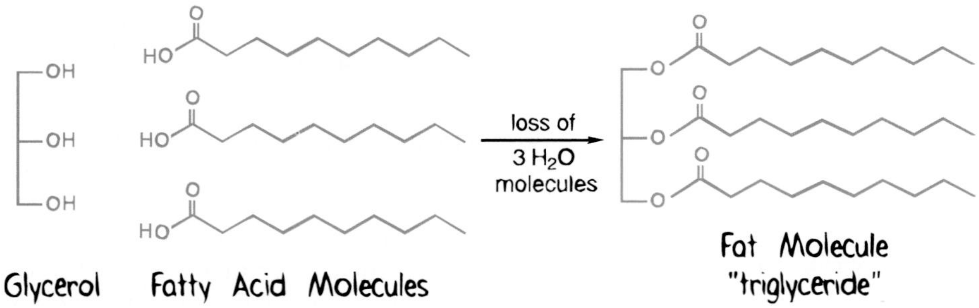

Figure 21.35 One typical fat molecule is the combination of one glycerol molecule and up to three fatty acid molecules. This type of fat molecule is often called a "triglyceride."

Fats

Like carbohydrates, fats contain only the elements carbon, hydrogen, and oxygen. Fats and carbohydrates have similar composition because fats themselves are synthesized in plants and animals from carbohydrates. The chemical structures of fats, however, are vastly different than those of carbohydrates. A typical fat molecule consists of a glycerol molecule with three *fatty acid* molecules attached to it (Figure 21.35).

Glycerol   Fatty Acid Molecules   loss of 3 H_2O molecules   Fat Molecule "triglyceride"

ANSWER

Simple carbohydrates give energy quicker. To get energy from complex carbohydrates, they must first be broken down. This can be a relatively long process because of the size of these molecules. The result is a delay in the release of their energy content. Marathon runners take advantage of this by eating lots of complex carbohydrates (bread and spaghetti) the evening before a race. Simple carbohydrates, like those found in a candy bar, are much easier to process and they can be absorbed into your bloodstream and used for energy the moment they're in your mouth.

Fats are used for energy storage and other purposes, such as insulation against cold. The digestion of a fat molecule involves breaking the bonds between the fatty acids and glycerol portions. The subsequent oxidation of fat is accompanied by the release of considerably more energy than is produced by an equivalent amount of carbohydrate. There are 9 kilocalories of energy in every gram of fat and 4 kilocalories of energy per gram of carbohydrate.

Proteins

Proteins, the principal constituents of living cells, are compounds of carbon, hydrogen, oxygen, nitrogen, and sometimes sulfur. The basic chemical units of protein molecules are amino acids (Figure 21.36). Only about 20 amino acids are primary to life, and they differ from one another only by the chemical identity of a side group. Typical protein molecules consist of several hundred amino acids joined together in chains, and their structures are quite complex. The molecular formula of one of the proteins found in milk is $C_{1864}H_{3012}O_{576}N_{468}S_{21}$, which gives an idea of the size of some protein molecules.

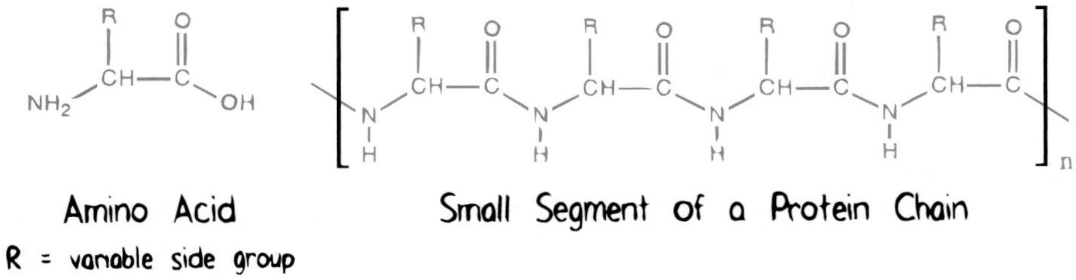

Figure 21.36 Amino acids are the fundamental building blocks of proteins.

Amino Acid

Small Segment of a Protein Chain

R = variable side group

Plant and animal tissues contain proteins both in solution and in insoluble form. Dissolved proteins reside in intracellular fluid and in other fluids such as blood. Insoluble proteins form skin, muscles, hair, nails, and horns. The human body contains about 100,000 different proteins. They all are made from the amino acids and other starting materials obtained from the digestion of food. A great success of modern chemistry is the discovery of how living cells build the complex arrangements of amino acids found in proteins. We leave the story of how this happens to biochemistry.

SUMMARY OF TERMS

Organic chemistry The study of carbon compounds.

Hydrocarbon A chemical compound containing only carbon and hydrogen atoms.

Structural isomer Molecules that have the same molecular formula yet differ by their chemical structures.

Stick structure A shorthand notation of the structure of an organic molecule where the carbon framework is represented as a series of connected sticks.

Fractional distillation A method whereby the components of crude petroleum are separated into fractions according to their boiling temperatures.

Catalytic cracking A process whereby larger hydrocarbons are broken down into smaller less viscous hydrocarbons.

Saturated hydrocarbon A hydrocarbon containing no multiple covalent bonds. In such a hydrocarbon, carbon atoms are "saturated" with hydrogen atoms.

Unsaturated hydrocarbon A hydrocarbon containing at least one multiple covalent bond.

Aromatic compound Any organic molecule containing a benzene ring.

Heteroatom Any non-carbon or non-hydrogen atom appearing in an organic molecule.

Alcohols A class of organic molecules that contain the hydroxyl group.

Hydroxyl group An oxygen atom bonded to a hydrogen atom, —OH. The hydroxyl group is found in alcohols.

Ethers A class of organic molecules in which two carbon atoms are bonded to a single oxygen atom.

Amines A class of organic molecules that contain the element nitrogen.

Alkaloids Molecules found in nature that are alkaline because of the nitrogen atoms they contain. Many alkaloids have biological effects on humans and other organisms.

Carbonyl A carbon atom double-bonded to an oxygen atom, C═O. The carbonyl group is found in ketones, aldehydes, amides, and carboxylic acids.

Ketone A class of organic molecules containing a carbonyl group in which the carbon of the carbonyl is bonded to two carbon atoms.

Aldehyde A class of organic molecules containing a carbonyl group in which the carbon of the carbonyl is bonded to one carbon atom and one hydrogen atom.

Amide A class of organic molecules containing a carbonyl group in which the carbon of the carbonyl is bonded to one carbon atom and one nitrogen atom.

Carboxylic acid A class of organic molecules containing a carbonyl group in which the carbon of the carbonyl is bonded to one carbon atom and one hydroxyl group.

Biomolecule A chemical important for life.

Carbohydrates Biomolecules made of hydrated carbon. Carbohydrates are produced in plants by photosynthesis and used primarily as a source of energy in humans.

Monosaccharide The simplest carbohydrate consisting of a single saccharide unit, which can be identified by a single ring of five or six atoms. Examples are α-glucose and β-fructose.

Disaccharide A carbohydrate consisting of two monosaccharides. An example is sucrose, commonly known as table sugar.

Polysaccharide A complex carbohydrate consisting of many monosaccharides.

Cellulose The most abundant polysaccharide on earth. Its purest form is cotton. Cellulose is found as the woody part of trees and the supporting material in plants and leaves.

Starch A digestable polysaccharide for humans and most other species.

Fat A biomolecule that packs a lot of energy per gram. A typical fat molecule consists of a glycerol molecule with three fatty acid molecules attached to it (see Figure 21.35).

Protein A structural biomolecule consisting of many amino acid units linked together.

Amino acid The fundamental building block of proteins. In an amino acid, amine and carboxylic acid groups are bonded to a central carbon atom. About 20 different types of amino acids are primary to life.

· ·

REVIEW QUESTIONS

Hydrocarbons

1. What are some examples of hydrocarbons?

2. What are some uses of hydrocarbons?

3. How do two structural isomers differ from each other?

4. How are two structural isomers similar to each other?

5. How is the formation of coal different from the formation of petroleum?

6. What physical property of hydrocarbons is used for fractional distillation?

7. How is it possible to increase the yield of light hydrocarbons from crude petroleum?

8. What types of hydrocarbons are more abundant in higher octane gasoline?

Unsaturated Hydrocarbons

9. What is the difference between a saturated and an unsaturated hydrocarbon?

10. What does the circle in the following stick structure of benzene signify?

Benzene

11. What property of benzene makes it a cancer-causing agent?

12. Aromatic compounds contain what kind of ring system?

Non-Carbon Atoms in Organic Molecules

13. What is a heteroatom?

14. Why do heteroatoms make such a difference in the physical and chemical properties of an organic molecule?

15. Which molecule should have the higher boiling point and how come?

$CH_3CH_2CH_2CH_3$ $CH_3CH_2CH_2CH_2 - OH$

16. How are heteroatoms on a hydrocarbon molecule like ornaments on a tree?

Alcohols, Ethers, and Amines

17. Why are many alcohols soluble in water?

18. Is ingesting methanol directly or indirectly harmful to one's eyes? Briefly explain.

19. What distinguishes an alcohol from an ether?

20. Why do ethers typically have lower boiling points than alcohols?

21. What heteroatom is characteristic of an amine?

22. Do amines tend to be acidic (pH<7), neutral (pH = 7) or basic (pH > 7)?

23. Where might one find an alkaloid?

24. What are some examples of alkaloids?

Carbonyl-Containing Organic Molecules

25. What elements make up the carbonyl group?

26. How are ketones and aldehydes related? How are they different?

27. What is one commercially useful feature of aldehydes?

28. How are amides and carboxylic acids related? How are they different?

29. From what naturally occurring compound is aspirin prepared?

30. Identify each of the following molecules as a hydrocarbon, alcohol, or carboxylic acid:

$$CH_3-C_6H_4-CH(CH_3)-C(O)-OH$$

$$CH_3CH_2CH_2CH_3 \qquad CH_3CH_2CH_2CH_2-OH$$

Biomolecules

31. What does the word carbohydrate mean?

32. When we eat carbohydrates we gain energy. Where does this energy ultimately come from?

33. What is another biological use for carbohydrates besides energy?

34. Table sugar is an example of what kind of saccharide?

35. In what ways are cellulose and starch similar? How are they different?

36. Why does starch begin to taste sweet after it has been in your mouth for a few minutes?

37. What two types of molecules come together to make a fat molecule?

38. Which yields more energy, a gram of fat or a gram of carbohydrate?

39. What is the building block of a protein molecule?

40. How do various amino acids differ from one another?

41. Why does carbon give rise to so many different kinds of molecules?

HOME PROJECTS

1. In your saliva there are chemicals called enzymes that start digesting food as soon as it is in your mouth. Polysaccharides like starch are broken down into monosaccharides by the enzyme amylase. The action of amylase is demonstrated by this home project. Prepare a solution of starch by boiling several potato slices in a cup of water. Add about a teaspoon of the broth to the bottom of a glass along with a few drops of a solution containing iodine. Look for this in your medicine cabinet, as it is a disinfectant. Note the deep blue color that appears. This color is diagnostic for the presence of starch. Next, collect a good wad of saliva in your mouth and then spit into the glass. Swirl the solution around. What happens to the color of the solution? How does this indicate that the starch is slowly being transformed into monosaccharides? Enzymes such as amylase are destroyed by heat. How could you confirm this experimentally? Many instant Cream of Wheat cereals contain a related enzyme, papain. Can it be said that these instant cereals start being digested *before* reaching your mouth?

2. Make your own white glue! White glue is often made from a protein in milk called *casein,* which coagulates as curd upon the addition of an acid and heat. Add a couple of capfuls of white vinegar (acetic acid) to a half cup of skim milk in a sauce pan. Heat gently and stir until lumps begin to form. Remove from heat and continue to stir until all of the milk has curdled. Filter the curds from the liquid (whey) using a coffee filter and a funnel. Squeeze out any excess liquid from the curds and place them in a glass containing a couple of tablespoons of water. To neutralize any remaining acetic acid, sprinkle in some baking soda while stirring. Continue adding baking soda until bubbles stop forming. The resulting material is glue. Test its adhesive properties by gluing together two coffee filters, two light pieces of wood, or other porous objects.

EXERCISES

1. Which has more hydrogen atoms, a saturated hydrocarbon molecule with five carbons or an unsaturated hydrocarbon molecule with five carbons?

2. Draw all the structural isomers for a hydrocarbon with the molecular formula C_4H_{10}. Use stick formulas.

3. Draw all the structural isomers for a hydrocarbon with the molecular formula C_6H_{14}. Use stick formulas.

4. The temperatures within a fractionating tower at an oil refinery are important, but so are the pressures. Where might the pressure within the fractionating tower be greatest, at the bottom or at the top? Defend your answer.

5. Identify the following functionalities in the complex organic molecule shown below: amide, ketone, ether, alcohol, aldehyde, amine.

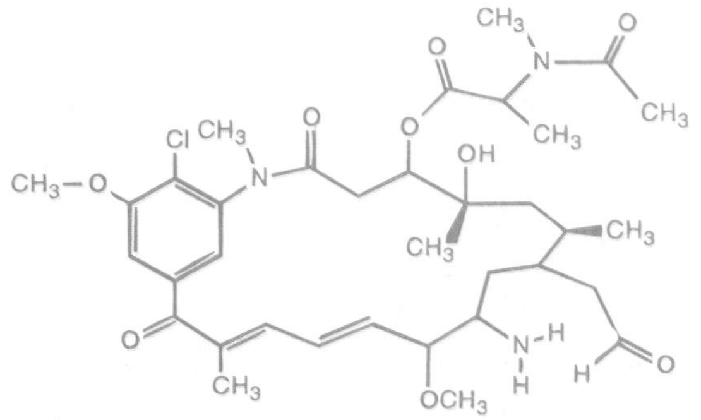

6. Briefly describe how to remove the caffeine from a cola drink. (*Hint:* the free base of caffeine is soluble in the organic solvent diethyl ether. Also, we find that diethyl ether and water are immiscible.)

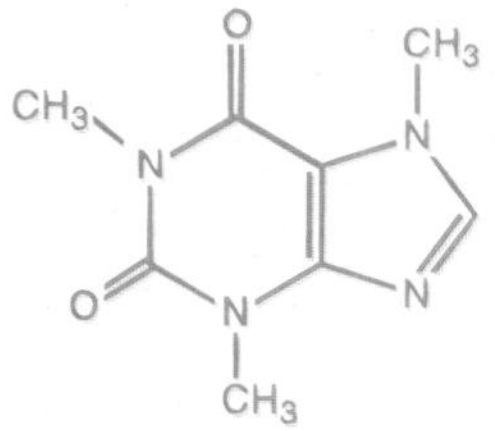

Caffeine

7. Explain why caprylic acid, $CH_3(CH_2)_6COOH$, is soluble in 5 percent aqueous NaOH but caprylaldehyde, $CH_3(CH_2)_6CHO$, is not.

8. If you saw this label on a decongestant: phenylephrine·HCl, would you worry that consuming it exposes you to the strong acid HCl? Explain.

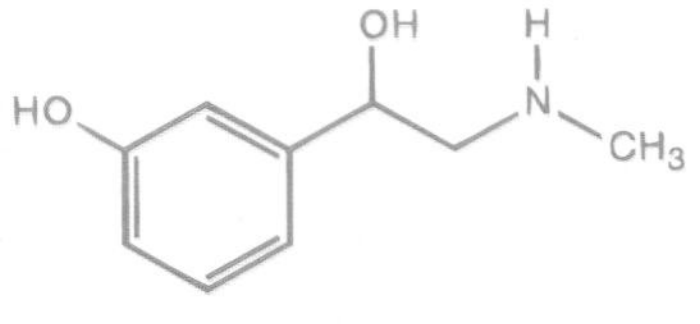

Phenylephrine

9. In water, the following molecule tends to act as:

   (a) an acid
   (b) a base
   (c) neither, for it is neutral
   (d) an acid *and* a base

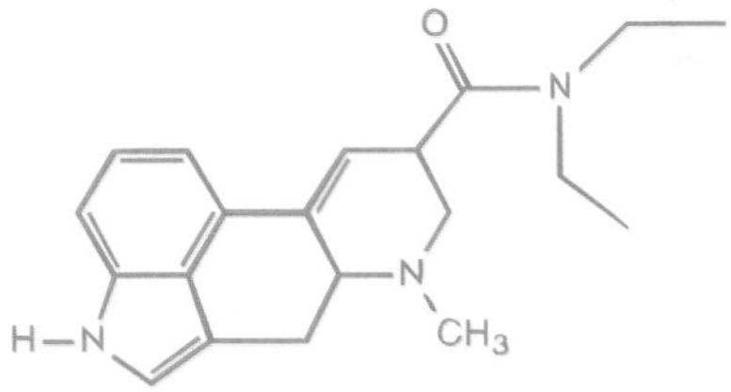

Lysergic Acid Diethylamide

10. Examine the compounds in Table 21.1. Apply some of the concepts you learned from Chapter 19 and suggest why the melting and boiling points of these hydrocarbons increase as the molecular formulas get larger.

PART 7

EARTH SCIENCE

I find it amazing that the rock formations here at Bryce Canyon are composed of dissolved seashells -- limestone -- which means this land was once an ocean bottom. After uplift by compressive forces, eroding waters carved the beautiful rock formations we now see. Iron and other leached minerals produce its stunning rusty orange color. Planet Earth is continually changing -- just as these rock formations weren't here a billion years ago, they won't be here a billion years from now. Continued erosion will flatten them completely. I love learning about our planet's history and its changes. Onward to geology!

22

ROCKS AND MINERALS

Figure 22.1 The erosion that cut through these layers of the Grand Canyon continues, and one day it will appear as another stretch of the Great Plains that now dominate central North America.

When we apply physics and chemistry to Planet Earth, we have the science of geology. We'll begin our study of geology with an investigation of what is directly beneath our feet—the earth's crust. We'll take a close look at the rocks and minerals that make up the earth's crust. Most of the crust is covered with water; we'll examine earth's waters in Chapter 23. Then in Chapter 24 we'll profile the continents. We'll see that today's continents are not permanent land masses. Like ice flows drifting downstream, continents are amalgams of lands repeatedly broken up, dispersed, then crunched together into new shapes. We'll see that the drifting of continents is driven by processes deep in the earth's interior. In Chapter 25 we'll probe geologic time and emphasize an underlying theme in all geology chapters—that change is ever-present and ongoing. Just as an insect with a few-hour life span has no concept of the growth rate of the plant on which it lives, it is difficult for us to imagine this planet's dynamic nature. An exploration of the geology of the Grand Canyon will provide a glimmer of the change earth has undergone (Figure 22.1). We'll conclude our study of geology with a study of the atmosphere and its processes—meteorology.

So we begin our study of geology with the outer skin of our planet—the earth's crust. The crust is very thin in comparison to the earth's radius—it is only from 5 to 40 kilometers (3 to 25 miles) thick. The thinnest parts are the ocean floors and the thickest parts support the mountains. The earth's crust is less than 1 per cent of the earth's radius—about the thickness of tomato skin relative to a tomato. Yet this very thin layer is a storehouse of the materials found on the earth's top surface—the minerals and the aggregates of minerals that we know as rocks.

22.1 ROCKS

Rocks are so familiar to us; we walk on them, we throw them, and sometimes we collect them. The rocks of the earth's crust are classified into three types according to origin. The first two types are generated by processes that occur below the crust's surface. The third type is generated by processes at the surface (Figure 22.2). The three rock types are:

1. **Igneous rocks**—formed by the cooling and consolidation of hot, molten rock material called *magma*. Igneous rocks make up about 95 percent of the earth's crust. *Igneous* means "formed by fire." Basalt and granite are common igneous rocks.

Figure 22.2   The many different rock types. (a) Basalt and granite are igneous rocks. (b) Marble and slate are metamorphic rocks. (c) Sandstone, shale, and limestone are sedimentary rocks.

2. **Metamorphic rocks**—formed from pre-existing rocks that have been changed or transformed by high temperature, high pressure, or both—without melting. The word *metamorphic* means "changed in form." Marble and slate are common metamorphic rocks.

3. **Sedimentary rocks**—formed from weathered material (sediments) carried by water, wind, or ice. Sedimentary rocks are the most common rocks in the uppermost part of the earth's crust. They cover over two thirds of the earth's surface. Sandstone, shale, and limestone are common sedimentary rocks.

Although formed by different processes, the three rock types are related. All three types of rock are composed of one or more minerals. We will now focus on minerals and their properties. With a firm grasp of the nature of minerals, we'll then return to the formation of igneous, metamorphic, and sedimentary rocks later in this chapter.

22.2 MINERALS

A **mineral** is a naturally formed inorganic solid composed of an ordered array of atoms. The atoms in different minerals are arranged in their own characteristic ways. The characteristic arrangement of atoms in a mineral is known as its crystalline structure.

A few minerals, such as gold, copper, and iron, are composed of single elements. Most minerals, however, are compounds of different chemical elements. The manner in which its elements are combined and the size and electric charge of the atoms determine the properties of each mineral. Physical properties, such as crystal form, luster, color, streak, hardness, cleavage, and specific gravity are often used to identify minerals. Let's consider these physical properties in turn.

Crystal Form

The orderly internal arrangement of atoms in a crystal is expressed in its shape. Each crystal face corresponds to a plane of atoms in the crystal structure. The angles between crystal faces are identical for all crystals of the same mineral. Measurements of angles and symmetry allow the classification of six major crystal systems (Figure 22.3). In order of increasing symmetry these are *triclinic, monoclinic, orthorhombic, tetragonal, hexagonal,* and *isometric.**

Every mineral has its own characteristic crystal form. Some minerals possess such a unique crystal form that a measure of crystal face angles is not necessary. The mineral pyrite, for example, commonly forms as intergrown cubes, while quartz commonly forms as six-sided prisms that terminate in a point (Figure 22.4). Unfortunately, well shaped crystals are rare in nature because minerals typically grow in cramped spaces. Even if the crystal form is imperfect, most minerals can still be identified by their most-often-seen crystal growth patterns. The mineral hematite

---

*These names are not presented for you to memorize (for in memorizing them you'd learn nothing of value anyway). Of importance is that these six systems account for the thousands of different crystals that exist. If you develop an interest in crystals, their names will become familiar as a matter of course.

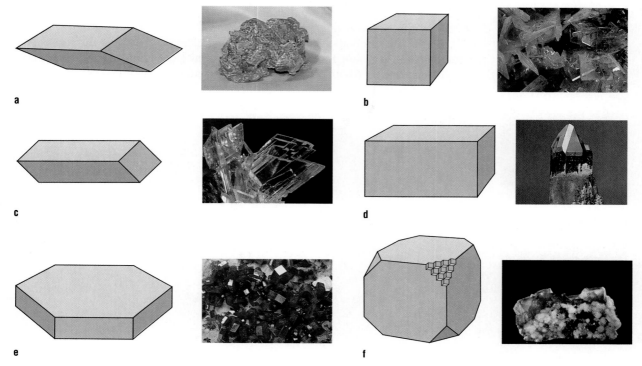

Figure 22.3   Six major crystal systems—distinguished by their internal symmetrical arrangement of atoms.

(a) The *triclinic system* forms an asymmetrical crystal in which no intersecting edges form right angles. An example is the mineral *chalcanthite*.

(b) The *tetragonal system* forms crystals with right-angled corners. An example is the square-faced mineral *wulfenite*.

(c) The *monoclinic system* forms a skewed crystal in which two of three intersecting edges form at a right angle. An example is the mineral *gypsum*.

(d) The *orthorhombic system* forms crystals in which all pairs of intersecting edges form at a right angle. An example is the mineral *topaz*.

(e) The *hexagonal system* forms crystals in which six identical sides meet at equal angles. An example is the mineral *vanadinite*.

(f) The *isometric system*—the most symmetrical of all the crystal systems, based on a repeating cube-shaped arrangement of atoms, in which equally proportioned crystal faces meet at right angles. The mineral *fluorite* is an example of the isometric system.

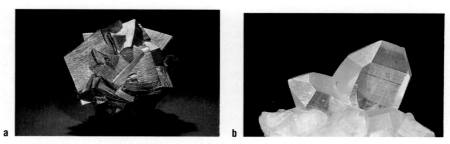

Figure 22.4   The crystal form of a mineral is unique—it is the external expression of a mineral's internal arrangement of atoms. (a) The mineral pyrite commonly forms as intergrown cubes with striated faces. (b) The mineral quartz commonly forms as six-sided prisms that terminate in a point.

a

b

Figure 22.5   Well-shaped crystal forms do not develop when growing occurs in a confined space. Nevertheless, the distinctive growth patterns of many minerals are apparent. (a) The asbestos group minerals often resemble narrow threadlike fibers. (b) The mineral hematite often forms a botryoidal shape that resembles a bunch of grapes.

often assumes a globular form that resembles a bunch of grapes. Asbestos minerals often resemble narrow threadlike fibers (Figure 22.5).*

Different minerals that contain the same elements but have different crystal structures are **polymorphs** (many forms). Graphite and diamond are examples of polymorphs for they both consist only of carbon atoms—yet they exhibit vastly different properties (Figure 22.6). Since the formation of polymorphs depends on particular temperatures and pressures, they are good indicators of the geological conditions at their sites of formation.

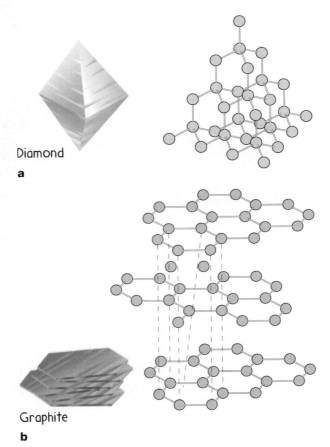

Diamond

a

Graphite

b

Figure 22.6   Polymorphs graphite and diamond are pure carbon. (a) Diamond, the hardest substance known, has a tightly packed symmetric structure. (b) Graphite has an open and layered structure. When rubbed between fingers, individual graphite molecules glide over one another like cards in a stack, giving it a slippery feel—hence its use as a lubricant. It also glides easily when stroked onto paper leaving an opaque tracing—hence its use in pencils, where it is also much less toxic than lead.

---

*The fibrous flexible nature, incombustability, and low conductivity of asbestos has prompted its use as insulation material against heat and electricity. Asbestos comes in two principal forms—*chrysotile,* which is white and accounts for 95 percent of asbestos production, and *crocidolite,* which is blue and accounts for the remaining 5 percent. Over the years, asbestos has been given much attention due to its potential health hazards. The blue form, crocidolite, has been found to be much more hazardous than chrysotile.

Crystallization of Minerals

Crystalline minerals form when molten rock cools. Like any substance, when the temperature of molten rock decreases, its molecules move slower. At sufficient slowness, electrical forces pull molecules to one another and bond them into orderly crystalline arrangements. Just as water turns to solid ice when the temperature drops to 0°C, atoms or molecules in a molten material likewise solidify at their respective freezing points as cooling occurs. Since the earth's molten material is made up of a variety of elements, all of the molten material does not solidify at the same time. The rate of cooling affects crystal size. Very slow cooling allows elements to migrate over greater distances so relatively large crystals form. Rapid cooling, on the other hand, produces the formation of a larger number of smaller crystals. When molten material is cooled so quickly that there is not time for atoms to arrange their respective pattern, the solid formed is *glass*. Its atoms are as unordered as those in ordinary window glass.

Crystallization also occurs from a vapor, and has the same underlying principles. The most familiar is the formation of snowflakes from air rich with water vapor. The systematic study of crystal form is called *crystallography*. Although originally developed as a branch of mineralogy, it is now considered a separate science that concerns all crystalline matter.

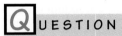

QUESTION

To what crystal system do the minerals pyrite and quartz belong?

Luster

The luster of a mineral is the appearance of its surface as it reflects light. Luster is independent of color; minerals of the same color may have different lusters and minerals of the same luster may have different colors. Mineral lusters are classified as either *metallic* or *nonmetallic*. The metallic luster of minerals that resemble metals may be polished or dull. Nonmetallic lusters are listed in Table 22.1.

TABLE 22.1   Non-Metallic Mineral Lusters

Mineral Lusters	Appearance
Vitreous	Glassy
Resinous	Resins, waxy
Greasy	Oily glass, also may feel greasy
Pearly	Pearly iridescence
Silky	Sheen of silk
Adamantine	Diamond

ANSWER

Pyrite commonly forms as intergrown cubes and belongs to the isometric system. Quartz forms as a six-sided prism that terminates in a point and belongs to the hexagonal system.

Figure 22.7 The mineral corundum (Al_2O_3) comes in a variety of colors as a result of chemical impurities. The addition of small amounts of chromium in place of aluminum produces the gemstone *ruby,* and with the addition of small amounts of iron and titanium, the result is the gemstone *sapphire.*

Color and Streak

Although color is an obvious feature of a mineral, it is not a very reliable means of identification. Minerals such as copper or turquoise have a distinctive greenish-blue color, but the majority of minerals may occur in a variety of colors or be colorless. Impurities in the chemical composition of a mineral affect color. For example, the common mineral quartz, SiO_2, displays a variety of colors depending on slight impurities. Quartz can be clear and colorless, milky white from minute fluid inclusions, rose colored from small amounts of titanium, violet from small amounts of iron, or smokey gray to black from a radiation-damaged crystal lattice. The color of the mineral corundum, Al_2O_3, is commonly white or grayish, but replacing small amounts of aluminum with chromium gives us the precious gem, the ruby. When small amounts of iron and titanium are present, the corundum is deep blue and we have the prized gemstone, the sapphire (Figure 22.7).

Another test for color is the streak test. When a mineral is rubbed across a nonglazed porcelain plate it leaves a thin layer of powder—a streak. Although minerals often vary in color, the color of the streak is constant and is thus more useful in mineral identification (Figure 22.8). Minerals with a metallic luster generally leave a dark streak that may be different from the color of the mineral. For example, the mineral hematite is normally reddish-brown to black but always streaks red. Magnetite is normally iron-black but streaks black. Limonite is normally yellowish-brown to dark brown and always streaks yellowish-brown. Minerals with a nonmetallic luster either leave a light streak or no streak at all.

Hardness

Just as a diamond will scratch glass, so will corundum and quartz. But the mineral fluorite will not scratch glass—instead glass will scratch fluorite. The resistance of a mineral to being scratched or its ability to scratch is a measure of the mineral's hardness. The varying degrees of hardness are represented by Mohs scale of hardness (Table 22.2).

Figure 22.8 The streak of a mineral is used for identification.

QUESTION

When pieces of calcite and fluorite are scraped together, which scratches which?

ANSWER

By looking at Mohs scale of hardness we see that fluorite is harder than calcite. So fluorite will scratch calcite.

TABLE 22.2   Mohs Scale of Hardness

Mineral	Scale Number	Hardness of Common Objects
Diamond	10	Jewelry
Corundum	9	Machine tools
Topaz	8	Jewelry
Quartz	7	Steel file
Orthoclase	6	Window glass
Apatite	5	Pocket knife
Fluorite	4	
Calcite	3	Copper wire or coin
Gypsum	2	Fingernail
Talc	1	

Cleavage and Fracture

Cleavage is the tendency of a mineral to break along its planes of weakness. Planes of weakness are a function of crystal structure and symmetry. Some minerals have very distinct cleavage. Mica, for example, has perfect cleavage in one direction and breaks apart to form thin, flat sheets (Figure 22.9). The hexagonal mineral, calcite, has perfect cleavage in three directions and breaks to produce rhombohedral faces that intersect at 75-degree angles. On the other hand, garnet, an isometric mineral, has no cleavage.

Figure 22.9   A mineral's cleavage is very useful in its identification. Mica has perfect cleavage in one direction and breaks into very thin sheets.

A break other than along cleavage planes is a *fracture*. When a mineral fractures with a smooth curved surface resembling broken glass, the fracture is *conchoidal*. Quartz and olivine display conchoidal fractures when broken (Figure 22.10). Some minerals such as hematite and serpentine break into splinters or fibers, but most minerals fracture irregularly. The degree and type of cleavage or fracture are useful guides for the identification of minerals.

Figure 22.10   Quartz does not exhibit cleavage. When it breaks it instead develops a conchoidal fracture—a curved smooth surface that resembles broken glass.

Specific Gravity

An obvious physical property of a mineral is its density. In practical terms, density is how heavy a mineral feels for its size. The standard measure of density is *specific gravity*—the ratio of the weight of a substance to the weight of an equal volume of water. For example, if a mineral weighs three times as much as an equal volume of water, its specific gravity is 3. The specific gravities of some minerals are shown in Table 22.3.

Gold's particularly high specific gravity of 19.3 is neatly taken advantage of by miners panning for gold. Fine gold pieces hidden in a mixture of sediments settle to the bottom of the pan when the mixture is swirled in water. Water and less dense materials spill out upon swirling. After a succession of dousings and swirls, only the substance with the highest specific gravity remains—gold!

TABLE 22.3 Specific Gravity of Various Minerals

Borax	1.7
Quartz	2.65
Talc	2.8
Mica	3.0
Olivine	3.6
Chromite	4.6
Pyrite	5.0
Hematite	5.26
Silver	10.5
Gold	19.3

QUESTION

Why are there no units for specific gravity?

Chemical Properties of Minerals

A mineral's physical properties are one thing; its chemical properties are another. Two simple chemical tests for identifying minerals are the taste test and the "fizz" test. The taste test is commonly used to identify the mineral halite, NaCl (common table salt) due to its distinctive salty taste. The fizz test on carbonate minerals is also common. Carbonate minerals effervesce (fizz) in dilute hydrochloric acid, HCl. The fizz is carbon dioxide, CO_2, produced by a chemical reaction between carbonate minerals and HCl. Some carbonate minerals react more readily with HCl than others.

22.3 BUILDING BLOCKS OF ROCK-FORMING MINERALS

Of the 109 known elements, 88 occur naturally in the earth's crust. These elements combine to make up the more than 3400 different types of minerals. Of these, only about two dozen are abundant (Table 22.4). The abundant minerals are composed predominantly of eight elements. These eight elements represent about 88 percent of the mass of the earth's crust. Almost half of this mass is the element oxygen, which is found in common minerals such as the *silicates, oxides,* and *carbonates.* With few exceptions, all rock-forming minerals are members of these groups.

The Silicates

After oxygen, the second most abundant element in the earth's crust is silicon. Oxygen and silicon combine to form the most common mineral group, the **silicates**.

ANSWER

Specific gravity is a ratio of densities. Density units divided by density units cancel out. For example, the density of the mineral hematite, Fe_2O_3, is 5.26 g/cm^3. So compared to the density of water its specific gravity is $\frac{5.26 \text{ g/cm}^3}{1.0 \text{ g/cm}^3} = 5.26$

TABLE 22.4   Most Common Chemical Elements in the Crust

Element	Symbol	Percent by Mass	Percent by Volume
Oxygen	O	46.60	93.8
Silicon	Si	27.72	0.9
Aluminum	Al	8.13	0.5
Iron	Fe	5.00	0.4
Calcium	Ca	3.63	1.0
Sodium	Na	2.83	1.3
Potassium	K	2.59	1.8
Magnesium	Mg	2.09	0.3
TOTAL		98.59	100.0

From *Principles of Geochemistry* by Brian Mason and Carleton B. Moore.
Copyright 1982 by John Wiley & Sons Inc.

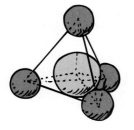

Figure 22.11 The silicon-oxygen tetrahedron is constructed of four oxygen atoms with a silicon atom at the center.

Quartz, the second most common mineral in the earth's crust, is composed only of oxygen and silicon. Silicates may also contain elements other than oxygen and silicon. Feldspars, the most common and abundant minerals, for example, also contain aluminum, sodium, potassium, and/or calcium. All silicates have the same fundamental structure, the silicon-oxygen tetrahedron (Figure 22.11). Stability is achieved when tetrahedra electrically link to other tetrahedra. This is *polymerization,* the linking of tetrahedra to form chains, sheets, and various network patterns (Figure 22.12).

Mineral		Idealized Formula	Cleavage	Silicate Structure
Olivine		$(Mg,Fe)_2SiO_4$	None	Single tetrahedron
Pyroxene		$(Mg,Fe)SiO_3$	Two planes at right angles	Chains
Amphibole		$(Ca_2Mg_5)Si_8O_{22}(OH)_2$	Two planes at 60° and 120°	Double chains
Micas	Muscovite	$KAl_3Si_3O_{10}(OH)_2$	One plane	Sheets
	Biotite	$K(Mg,Fe)_3Si_3O_{10}(OH)_2$		
Feldspars	Orthoclase	$KAlSi_3O_8$	Two planes at 90°	Three-dimensional networks
	Plagioclase	$(Ca,Na)AlSi_3O_8$		
Quartz		SiO_2	None	

Figure 22.12 As silicate tetrahedra link to other tetrahedra, they polymerize to form chains, sheets, and various network patterns. The complexity of the silicate structure increases down the chart.

The Oxides

The oxide minerals are chemical compounds in which oxygen is combined with one or more metals. Some are of great economic importance. These include iron (hematite and magnetite), chromium (chromite), manganese (pyrolusite), tin (cassiterite), and uranium (uraninite).

The Carbonates

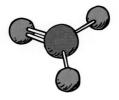

Figure 22.13 The fundamental carbonate ion structure CO_3^{2-}, is composed of a carbon atom centrally bonded to three oxygen atoms.

The carbonate minerals are much simpler in structure than the silicate minerals. The structure is triangular in shape, with a carbon atom centrally bonded to three oxygen atoms, CO_3^{2-} (Figure 22.13). Two common carbonate minerals are calcite, which is the chemical compound calcium carbonate, $CaCO_3$, and dolomite, which is a mixture of calcium carbonate and magnesium carbonate, $MgCO_3$. Carbonate minerals are predominantly formed in sedimentary environments. For example, the sedimentary rock chalk is composed of calcium carbonate ($CaCO_3$) formed from the shells of tiny organisms.

22.4 IGNEOUS ROCKS

Most of the earth's crust—about 95 percent—is igneous rock. On the continents the most common igneous rocks are granite and andesite. On the ocean floor, basalt is predominant. Igneous rock originated as magma—molten rock from the earth's interior.

Magma and the Evolution of Igneous Rocks

Just as there is water and ice, there is magma and rock. **Magma** that cools and solidifies becomes rock, and rock that is heated melts to become magma. We will see in Chapter 24 when we study plate tectonics that part of the earth's crust is continually forming as magma rises through fissures and solidifies. Meanwhile part of the earth's crust is being destroyed as it is pushed downward below the crust where it returns to the mantle, the high temperature rocky layer beneath the crust. Just as ice melts at the same temperature that water freezes, the temperature at which a solid mineral melts is the same temperature at which the same mineral in molten form solidifies. So when we discuss the melting temperature of a mineral, we imply the same temperature for solidification.

For rock to melt, the temperature needs to be very high—usually above 750°C for granitic rocks, and above 1000°C for basaltic rocks. The earth's temperature increases with depth—about 30°C for each kilometer (Figure 22.14). Pressure also increases with depth due to the increased load of rock above. In general as pressure on rock increases, the melting point of the rock increases. (This is similar to increased atmospheric pressure that causes an increase in the boiling point of water.) The water content of a rock also affects its melting point. Rocks with a high water content have a lower melting point because water dissolves in the magma. Rocks with a low water content have a higher melting point and therefore require higher temperatures to melt.

Keep in mind that rock is composed of various minerals. As rock is heated, the first minerals to melt are those with the lowest melting point. Thus the melting of rock into magma occurs over a broad range of temperatures. Solid minerals can be

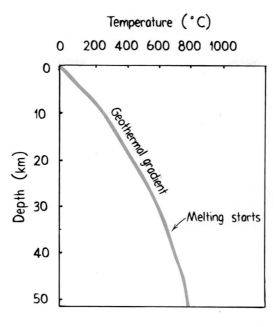

Figure 22.14 The earth's temperature increases about 30°C for each kilometer of depth in the upper crust. This increase of temperature with depth is known as the *geothermal gradient.* At depths below the earth's crust, temperatures range from 750° to 1200°—hot enough to melt rock.

found in virtually all magmas. When conditions are such that all minerals within a rock can completely melt, the composition of the resulting magma is the same as the composition of the original rock. Most often, however, melting is not complete; we have **partial melting**. This is analogous to the partial distillation of crude oil (Chapter 21), where the various component hydrocarbons in the oil depend upon its temperature. Similarly, the magma resulting from partial melting is composed only of those minerals that have melted, the ones with the lowest melting points. All this results in magmas of many different compositions and hence, the many different igneous rocks.

Because silica has a relatively low melting point, partial melting produces magmas with a higher silica content than the parent rock. There are three major types of magma, *basaltic, andesitic,* and *granitic*. **Basaltic magma** is about 50 percent SiO_2. Basaltic magma that has solidified is the dark rock—basalt—that makes up the Hawaiian Islands. **Andesitic magma,** composed of 60 percent SiO_2, is produced from partial melting of basaltic oceanic crust. The rock andesite, produced from andesitic magma, gets its name from the Andes Mountains in South America where it is very common. When water-rich andesitic rocks undergo partial melting, **granitic magma,** composed of about 70 percent SiO_2, is produced. This magma, when solidified, forms the rock granite and other granitic type rocks. Of all igneous rocks in the crust, oceanic and continental combined, approximately 80 percent form from basaltic magma, 10 percent from andesitic magma, and 10 percent from granitic magma.

Igneous Rock Crystallization

All magma that rises upward through the mantle and crust cools and crystallizes to form igneous rock. We find that the process of crystallization is very similar to the process of partial melting but in the reverse order. Minerals with the highest melting points crystallize first, followed by minerals with lower melting points. As solidification proceeds, the composition of the remaining liquid changes continuously as it becomes depleted in minerals that have already crystallized and becomes enriched in minerals yet to crystallize. This is the process of **fractional crystallization**.* If the crystallization process is very slow and is uninterrupted, the resulting solid rock will have the same bulk composition as the original magma. If the crystals are prevented from reacting with the melt, either by settling to the bottom of a magma chamber or by the magma moving out of a chamber, the composition of the final solid rock will be different from the original bulk composition of the magma. The process of fractional crystallization allows a single magma to generate several different igneous rocks.

QUESTION

How are partial melting and fractional crystallization similar?

Igneous Rock Formation beneath the Earth's Surface

Igneous rocks may form either at or below the earth's surface. Igneous bodies formed at the surface can be readily observed at active volcanic sites; but igneous bodies formed below the surface can only be studied after the processes of uplift and erosion have exposed them. Rocks formed by the intrusion of magma below the earth's surface are called **intrusive** rocks. Large intrusive bodies are called **plutons**. Plutons occur in a great variety of shapes and sizes, ranging from slablike or tabular to quite massive, nondescript blobs (Figure 22.16).

A common pluton is a **dike**—formed by the intrusion of magma into fractures that cut across the bedding of existing rock. When a dike is more resistant to erosion than the rock it intrudes, the dike forms wall-like ridges that range in thickness from less than a centimeter to hundreds of meters. If the dike is more prone to erosion than the surrounding rock, it leaves a trench or ditch at the surface. Dikes, the ancient channel ways for rising magma, are closely associated with volcanic vents. A spectacular example of this association are the radiating dikes around the eroded volcanic neck at Shiprock, New Mexico (Figure 22.17).

Another pluton is a **sill**—formed by the intrusion of magma into fractures that are parallel to the bedding of existing rock. Most sills are formed by the intrusion of low-viscosity basaltic magma at a shallow depth. Because sills form at shallow depths, they often resemble buried lava flows. A variation of a sill is a **laccolith**.

ANSWER

Both produce temperature-dependent materials, one liquid, the other solid, but by opposite processes. Partial melting produces *magmas* of various compositions that depend on the melting temperatures of minerals. Fractional crystallization produces *crystals* of various compositions that depend on the temperatures at which minerals crystallize (which is the same as the melting temperature). In both the processes of partial melting and fractional crystallization, minerals separate from one another as a function of temperature, much like crude oil is fractionally distilled into its various components as discussed in Section 21.1.

---

*If you continue your study of geology, you'll learn about *Bowen's Reaction Series,* the arrangement of minerals by fractional crystallization.

Igneous Textures

The texture of a rock reveals a great deal about the environment in which it formed. The texture of igneous rocks is related to the cooling rate of the magma. Magma that solidifies below the earth's surface cools slowly, forming large, visible interlocking crystals that can be identified with the unaided eye. This coarse-grained texture is described as *phaneritic.* By contrast, magma that reaches the surface tends to cool rapidly forming very fine-grained or even glassy rocks. This fine-grained texture is described as *aphanitic,* in which the individual crystals are too small to be identified with the naked eye or even with a small hand lens. Many aphanitic rocks contain cavities left by gases escaping from the rapidly cooling magma. These gas bubbles tend to solidify in the upper portion of a lava where cooling is fastest. These "frozen-bubble" cavities are called *vesicles,* and the rocks that contain them are said to have a *vesicular* texture. In a volcanic eruption, gases escape with enough force to cause rock fragments to be torn from the sides of the volcanic vent. These rock fragments and molten magma splatter and shatter into small hot fragments called volcanic ash and cinders. Such igneous rocks are said to be *pyroclastic* and have a pyroclastic, or fragmental, texture. Because pyroclastic rocks are composed of individual rock fragments rather than interlocking crystals, their overall textures are often more similar to sedimentary rocks than to igneous rocks.

A change in the cooling rate during crystallization may produce a texture in which some crystals are conspicuously larger than others. Larger crystals, called *phenocrysts,* represent crystallization during a period of slower cooling than the remainder of the rock. Sometimes the rate of cooling increases to produce a fine-grained or glassy texture between the earlier-formed larger crystals, producing a texture said to be *porphyritic.* So we find that all the minerals in an igneous rock do not form simultaneously—certain minerals begin crystallization before others. These various igneous textures are displayed in Figure 22.15.

Figure 22.15 Different cooling rates of magma produce igneous rocks of different textures: (a) Aphanitic texture, (b) Phaneritic texture, (c) Porphyritic texture, (d) Glassy texture, (e) Vesicular texture.

a

b

c

d

e

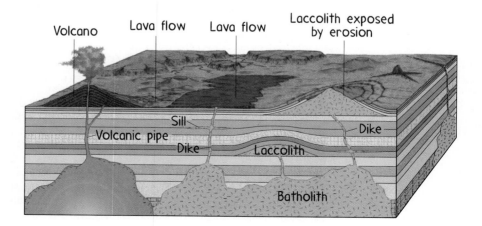

Figure 22.16 Intrusive igneous structures in cross-sectional view.

Figure 22.17 Shiprock, New Mexico. Radiating dikes surround the eroded remains of a volcanic vent.

Laccoliths are created when viscous magma rising upward in the earth's crust encounters a more resistant layer that forces the magma to spread, forming a mushroom shape. Unlike sills, laccoliths push the overlying layers upward in domelike fashion.

Batholiths are the largest of the plutons, and are defined as having more than 100 square kilometers of surface exposure. A batholith is usually not generated by a single intrusion. Instead, numerous intrusive events over millions of years account for the development of a massive batholith. Batholiths form the cores of many major mountain systems of the world. Many existing mountains are the cores of the batholiths of larger mountains that have long since eroded. Some of the largest batholiths in North America include the Coast Range Batholith and the Sierra Nevada Batholith. These batholiths continue to push upwards, increasing the height of the mountain range (Figure 22.18).

Figure 22.18 Some of the largest batholiths in North America include the Coast Range Batholith and the Sierra Nevada Batholith.

Igneous Rock Formation at the Earth's Surface

Igneous rocks that form at the earth's surface are called **extrusive** rocks. Magma that penetrates or extrudes onto the earth's crust is called **lava**. Lava may be extruded at the earth's surface through cracks and fractures, or through a central vent— a **volcano**. Although eruptions from a volcano are the most familiar, the outpourings of basaltic magmas through fissures are much more common.

Fissure Eruptions Lava outpourings commonly known as *flood basalts* have flooded extensive areas to make lava plains or have piled up to form lava plateaus. The Columbia Plateau in the Pacific northwest is an extensive flood basalt; so is the Deccan Plateau in India.

Most fissure eruptions occur on the ocean floor. Because hot magma is quickly quenched by cold sea water, flood basalts do not form. Instead the rapidly cooling lava takes forms that depend on the proximity of the source. When cooling occurs close to the fissure, sheeted flows of thin layers with glassy surfaces are produced. When cooling takes place farther from the fissure, a pillow basalt results (Figure 22.19). Pillow basalts are the most common of the submarine lava flows.

Figure 22.19 Fissure eruptions along the mountainous ridges of the ocean floors have produced enormous amounts of basalt— enough to build the crust of the earth's entire present sea floor in the past 200 million years. The most common of these submarine lava flows are *pillow basalts*.

The Formation of Batholiths

How batholiths form is not altogether certain. The magmatic origin hypothesis suggests that batholiths are formed when buoyant magma migrates upward from great depths. But if this is so, what happens to the large volume of rocks displaced by the batholith? One viewpoint suggests that the batholith makes room for itself by pushing aside pre-existing rocks. It may also be that when hot liquid magma works its way upward, large blocks of the roof of the magma chamber break off and fall into the melted material. These dislodged blocks may sink and melt into the hot magma and thus become incorporated into the batholith. This process is called *assimilation.*

An opposing viewpoint suggests that batholiths are created more or less in place by the metamorphism of pre-existing rocks. This is the process of *granitization*—the alteration of earlier-formed sedimentary or metamorphic rocks into granite without melting. Evidence for this theory is supported in the metamorphic rocks that surround most batholiths. But granitization generates very small quantities of the abundant granite found at these sites. Although some granitization occurs, most evidence for the formation of batholiths points to a magmatic origin. How batholiths form is still controversial. Like other scientific controversies, we are reminded that the sciences are not simply bodies of knowledge, but are human activities that serve to increase our knowledge and understanding of the world.

QUESTION

Why is it incorrect to say that igneous rocks may form from the intrusion of lava?

Volcanic Eruptions Eruptions from a central vent give rise to a variety of volcanic forms that depend on the fluidity, or viscosity,* of lava. Flows of basaltic lava often form a smooth skin, sometimes wrinkling as the flow of lava advances. These flows are known by the Hawaiian name **pahoehoe** (pronounced pa-ho′e-ho′e) and resemble the twisting braids in ropes (Figure 22.20a). Another common type of basaltic lava has a rough jagged surface with dangerously sharp edges and spiny projections. The Hawaiian name **aa** (pronounced ah-ah) is given to this type of flow. Because aa flows are generally cooler and more viscous than pahoehoe flows, they have the appearance of an advancing mass of lava rubble (Figure 22.20b). The movement downslope results in cooling and the escape of gas which increases the viscosity and converts the pahoehoe to aa. So smooth ropelike pahoehoe commonly grades into aa as the lava flow progresses downslope.

ANSWER

Terminology in the statement is wrong. To begin with, the term intrusion refers to solidification that occurs in the earth's interior and therefore has nothing to do with lava. Secondly, lava is not a synonym for magma, but is the term for magma that has been extruded at the earth's surface in molten form. Furthermore, lava is depleted of most of the gaseous components in magma. There is no lava beneath the earth's surface.

---

*Viscosity, a measure of internal resistance to flow, is directly related to a magma's silica content. Basaltic magmas with their low silica contents have a low viscosity and tend to be quite fluid, while granitic magmas have a high viscosity and flow so slowly that movement is often difficult to detect. A magma's behavior therefore depends very much on silica content.

Figure 22.20 (a) Smooth flowing *pahoehoe* lava is characterized by a twisting, rope-like appearance. (b) Jagged *aa* lava moves as an advancing mass of rubble.

UESTION

Is it correct to say that igneous rocks may form from the extrusion of lava?

Volcano Formation Volcanoes come in a variety of shapes and sizes. Volcanoes built by a steady supply of fluid basaltic lava produce a broad gently sloping cone that resembles a shield. These are *shield volcanoes*, built from the accumulation of successive flows that pour out in all directions to cool as thin gently dipping sheets. Some of the largest volcanoes in the world are shield volcanoes. The enormous size of Mauna Loa in Hawaii is the result of the accumulation of individual lava flows each only a few meters thick. Mauna Loa is the largest volcano on earth, projecting 4145 meters (13,599 ft.) above sea level and more than 9750 meters (31,988 ft.) above the deep ocean floor (Figure 22.21).

Cinder cones are common in many areas of active volcanoes. They are very steep and rarely rise more than 300 meters or so above ground level. They are formed from the piling up of ash, cinders, and rocks that have been explosively erupted from a single vent. As debris showers down, the larger fragments pile up near the summit to form a symmetrical, steep-sided cone around the vent. The finer particles fall farther from the vent to form gentle slopes at the base. Two well known examples of cinder cones are Sunset Crater in Arizona and Parícutin in Mexico.

The story of Parícutin presents an account of the birth of a volcano. Parícutin burst forth from a Mexican cornfield on February 20, 1943. For two weeks before its birth, numerous earth tremors jolted the countryside. Once the eruptions began, the earth tremors ceased. Then, beginning as a small hole in a cornfield, the volcano grew with astonishing speed! On the first day of activity it grew 50 meters and by the end of the first week it had grown 150 meters high! Within one year Parícutin topped off at 360 meters in elevation. Spewing forth over a billion tons of lava and vast quantities of ash, the fury of Parícutin buried two villages and covered the countryside for 35 km around. Abruptly, nine years and twelve days following its birth Parícutin fell silent.

Ⓐ NSWER •

Yes, once magma is extruded from the earth it is called lava, which when solidified becomes igneous rock.

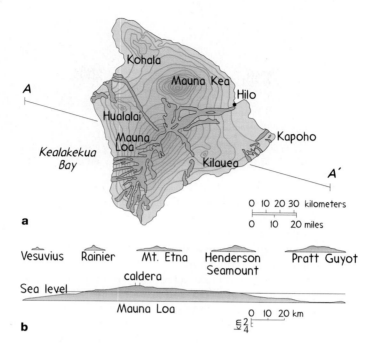

a

b

Figure 22.21 Mauna Loa, a shield volcano on the Island of Hawaii, is the largest volcano on earth. When compared to other large volcanoes, its immense size and volume is dramatic.

When a volcano erupts both lava and ash, a *composite cone* of alternating lava flows, ash, and mud flow debris is produced. The layers build up to form steep-sided summits and rather gently sloping flanks. Mt. Fujiyama is a classic example of a majestic composite cone (Figure 22.22). Composite cones tend to erupt explosively. In a volcanic blast, the pressure and temperature of the magma chamber can become so great that the confining pressure is exceeded and the whole mass of sticky magma and overlying rock explode into a mass of dust and rubble, which when mixed with volcanic ash, expands and engulfs everything in its path. Examples of such volcanic activity are Mt. Vesuvius in 79 AD, Mt. Pelee in 1902, Mount St. Helens in 1980, and Mt. Pinatubo in 1991.

Figure 22.22
Mt. Fujiyama—a composite volcano in Japan—is one of the earth's most picturesque volcanoes.

Figure 22.23 Crater Lake in Oregon is a remnant of the eruption of Mt. Mazama 7000 years ago.

Craters and Calderas Craters are commonly formed above the central vent of erupting volcanoes. During eruption, the upwelling lava overflows the crater walls, and then sinks back into the vent. The walls of a crater often collapse after an eruption, enlarging the area of the central vent. The size of a crater can grow up to more than a kilometer in diameter. A very large crater is referred to as a *caldera*. Calderas range from 5 to 30 kilometers or more in diameter. Most calderas are formed when the central part of a cone collapses into the partially emptied magma chamber below, but a few have been formed by explosive eruptions in which the top of a volcano was blown out. The volcanic eruption 7000 years ago of Mount Mazama in Oregon is one of these few catastrophic events. The eruption blasted ash throughout the northwestern United States. After the eruption most of the cone collapsed into the emptied magma chamber. The caldera, filled with rainwater, resulted in Oregon's famous Crater Lake, which is 9 kilometers wide and 590 meters deep (Figure 22.23).

Yellowstone National Park, located in one of the most seismically active regions of the Rocky Mountains, is a "hotspot" in the earth's crust. Situated in a caldera 70 kilometers long and 45 kilometers wide, Yellowstone Park is the remnant of an ancient volcano that violently erupted about 600,000 years ago. Most of the hot springs, bubbling muds, steaming pools, and spouting geysers for which the park is famous lie within the caldera. Heat from the enormous reservoir of molten rock that produced the massive eruption still remains not too far below the earth's surface, sustaining the present thermal activity. Although no eruption has occurred in historical time, the molten rocks remain so close to the earth's surface that the possibility of an eruption in the future cannot be disregarded. Yellowstone is an example of geology in action, where the eruption of scalding water may be but a precursor to more violent activity in the near future.

22.5 SEDIMENTARY ROCKS

While the process of volcanism continually generates new material at the earth's surface, the opposing force of **weathering** acts to break down and decompose the earth's surface. All rocks at the earth's surface are in contact with the atmosphere and are thus susceptible to weathering. The weathering process occurs by mechanical and chemical means. Mechanical weathering disintegrates and fragments rocks into smaller and smaller pieces. Chemical weathering alters and decomposes rock into substances more in balance with the surrounding environment. As rock is weathered, it erodes. **Erosion** is the process by which particles are transported away by water, wind, or ice.

Figure 22.24 Erosion and sedimentation follow a downhill path in response to gravity.

Sedimentation

Sediments produced by mechanical means are called *clastic* sediments, whereas sediments produced by chemical means are called *nonclastic*. By whatever means, weathering produces sediment. Sedimentation starts where erosion stops. As the wind dies down, dust settles; as water currents subside, sand settles; as glaciers melt, rock remains. Erosion and sedimentation follow a downhill path in response to gravity (Figure 22.24). The larger the particle, the stronger the current must be to carry it. Thus, particle size and shape provide clues to the method, time, and distance of transport, and the depositional environment. Larger particles are usually the first to settle out and become deposited, while smaller particles are able to stay with the flow. During transportation sediment is continually sorted and abraded, so the degree of rounding or the shape of the particles depends on the length of time and the distance of transport. Poorly sorted or angular grains of various shapes imply a short transportation distance, whereas well sorted and well rounded grains imply a greater transportation distance. Sediment is found almost everywhere.

As deposited sediment accumulates, the process of **lithification** turns sediment into sedimentary rock. Lithification occurs when sediments accumulate and press down upon deeper layers, squeezing out water that often has compounds such as silica, calcite, and hematite in solution. These solutions act as cementing agents as they percolate between the particles of sediment and fill the spaces with mineral matter. Thus calcite, silica, and iron oxide make up the most common cementing agents. Silica cement, the most durable cementing agent, produces some of the hardest and most resistant sedimentary rocks. When iron oxide acts as a cementing agent it produces the red or orange stain of many sedimentary rocks. The colors of Bryce Canyon (cover of this text) provide a picturesque example of iron oxide stain.

Sedimentary rocks are the most common rocks in the uppermost portion of the crust. They cover two thirds of the earth's surface, and form a thin and extensive blanket over igneous and metamorphic rocks below. Because sedimentary rocks are the weathered and eroded remains of the pre-existing rocks that they cover, they provide information about geological events that occurred at the earth's surface.

Clastic Sediments

Clastic sedimentary rocks are classified by particle size. The most abundant clastic sedimentary rocks are *shale,* composed of fine sized particles, *sandstone,* composed of medium sized particles, and *conglomerate,* made up of a wide variety of particle sizes (Figure 22.25).

Shale, formed by the compaction of silt and clay-sized particles, is finely laminated and exhibits *fissility,* the ability to split into thin layers or flakes parallel to bedding planes. The extremely fine grain size suggests that particle deposition occurred in relatively quiet waters, such as deep ocean basins, flood plains, deltas, lakes, or lagoons. The color of shale ranges from gray to black, and red to brown to green, indicating its environment of formation. Gray to black shale indicates buried organic matter, which can be preserved only in an oxygen deficient swampy environment. Black shale is commercially important, for it is the principle source rock for petroleum. Red to brown shale indicates ferric oxide (red), or hydroxide (brown). The absence of these shows the characteristic green color of shale.

Sandstone is the second most abundant clastic sedimentary rock and can be classified into three types. When quartz is the primary mineral, the rock is simply called *quartz sandstone.* Quartz sandstone is composed of well sorted and well rounded quartz grains. Sandstone which has considerable amounts of the mineral feldspar is called *arkose.* The grains in arkose tend to be poorly rounded and not as well sorted as quartz sandstone. Sandstone composed of quartz, feldspar, and angular rock fragments is called a *graywacke.* Sandstones form in a variety of environments such as desert dunes, beaches, marine sand bars, river channels, and as we shall see, in alluvial and submarine fans.

Conglomerates are composed of gravels and rock fragments. The rock fragments are usually large enough for easy identification, which provides useful information about the source areas of sediments. Larger rock fragments were transported by currents strong enough to carry them, likely a steep stream gradient with strong turbulence to quickly abrade and round them. So the roundness of their edges and corners are good guides to the distance they have traveled. Conglomerates are often found in old river channels, alluvial fans, and in rapidly eroding coastlines.

Clastic Sedimentary Environments

The most dominant feature of sedimentary rocks is the way the particles of sediment are laid down, layer upon layer. These layers are referred to as *beds.* Ranging in thickness and size, each bed is a representation of individual episodes of deposition. The deposition of clastic sediments occurs in many different environments—alluvial and desert environments, or in delta and shoreline environments.

The term *alluvial* refers to unconsolidated gravels, sand, and clay that have been deposited by streams. When a fast flowing downhill stream leaves its narrow confinement and abruptly emerges onto a broad relatively flat valley or plain, the velocity of the flow slows and the stream dumps its load of sediment. These deposits are generally fan-shaped *alluvial fans* (Figure 22.26). Steep upper slopes of the fan are dominated by boulders, cobbles, and gravels, while the base of the fan and the alluvial plain are made up of sand, silt, and mud. With each episode of deposition the alluvial fan grows upward and outward (Figure 22.27).

Stream channels continue their downhill path as they meander back and forth across a river valley, depositing sediments as they go. The meandering movement creates a wide belt of almost flat plain—a floodplain. As the name implies, it is this

a

b

c

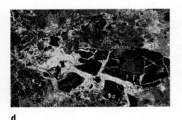

d

Figure 22.25
Sedimentary rocks:
(a) *shale,* composed of fine particles (this sample contains a fossil brachiopod), (b) *sandstone,* composed of medium-sized particles, (c) *conglomerate,* made up of a wide variety of particle sizes, (d) *breccia,* a conglomerate composed of angular grains and particles.

Figure 22.26 An alluvial fan in Death Valley, California.

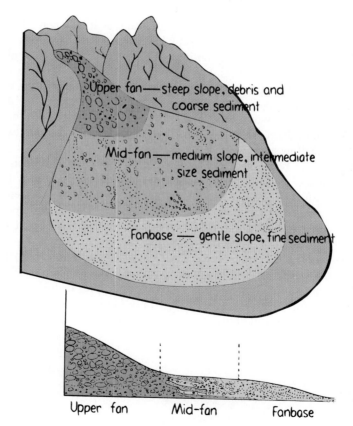

Upper fan — steep slope, debris and coarse sediment

Mid-fan — medium slope, intermediate size sediment

Fanbase — gentle slope, fine sediment

Upper fan   Mid-fan   Fanbase

Figure 22.27 An alluvial fan grows upward and outward with each episode of deposition. Coarser material is deposited at the apex and finer particles are deposited down the fan.

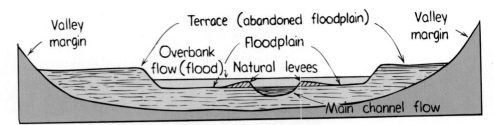

Figure 22.28 Cross-section of an alluvial valley. A floodplain is created when a river overflows its banks. Sands and gravel settle out first and act as natural levees to confine the river. Because the finer silt and clay particles are able to flow as a suspended load, they move beyond the levees and settle on the floodplain.

Figure 22.29 Desert erosion on a cliff face. A desert has many extremes—the scorching heat of the day, the chilling night air, and the strong blowing winds. Mechanical weathering physically breaks down the rocks to smaller and smaller pieces.

section of the river valley that becomes flooded with water when a river overflows its banks. When flooding occurs, sands and gravels are deposited on the river banks creating natural levees that confine the river, while silt and clay particles spread out over the floodplain (Figure 22.28).

The arid desert environment is characterized by angular hills, sheer canyon walls, and sand dunes. Lacking moisture, mechanical weathering predominates (Figure 22.29). Although the desert lacks moisture, water is the main cause of erosion and transportation of sediments. Rare as it is, when a heavy rain falls in the desert, it does not have time to soak in and causes "flash floods." These "flash floods" carry and deposit great quantities of debris and sediment as alluvial fans at the bases of mountain slopes and as alluvium on the floors of wide valleys and basins. The alternation of wet and dry conditions can be seen in the unique bedding features of mudcracks found in the desert basins (Figure 22.30). Mudcracks indicate that the sediment was alternately wet and dry at the time of formation. When exposed to air, mud dries out and shrinks, producing cracks. Mudcracks are also associated with shallow lakes and tidal flats.

Wind also acts to transport and deposit sediments. If you've ever been in a wind storm or at the beach on a windy day, you may have felt the sand-blasting effect of the wind. Once in the air, particles of sediment can be carried great distances by the wind. Red dust from the the Sahara desert is found on glaciers in the Alps, and fine quartz from central Asia has been detected on the Hawaiian Islands! In the desert, winds move over surfaces of dry sand picking up the small, more easily transported particles, leaving the large, harder-to-move particles behind. The small particles bounce across the desert floor knocking more particles into the air. Through this process alternating ripples of fine and coarse sand develop on the desert floor (Figure 22.31). (These ripple marks are similarly formed by the movement of sand grains in water currents, seen in shallow streams or under the waves at beaches.)

Figure 22.30 The alternation of wet and dry conditions can be seen in the unique bedding features of mudcracks in the desert basins.

Figure 22.31 Generated by blowing winds, ripples of alternating fine and coarse sand develop on the desert floor.

Sand dunes are formed when air flow becomes obstructed by an obstacle, such as a rock or clump of vegetation. As the wind sweeps over and around the obstacle, sand grains settle in the wind shadow. As more sand settles, mounds form and further impede the flow of air. With more sand and more wind the mound becomes a dune. As a dune grows, the whole mound starts moving downwind as sand grains on the windward slope move up and over the crest of the dune to fall on the leeward slope (Figure 22.32). The action of this continual process moves the entire dune. A unique feature of this action can be seen in the *cross-bedding* found on the leeward side of the dune. The direction of cross-bedding indicates the direction of the wind (or water current) that deposited the sediments (Figure 22.33). Cross-bedding is also a common feature in river deltas and certain stream channel deposits.

Figure 22.32 Formation of a sand dune. When air flow is obstructed, sand grains settle in the wind shadow. With more wind, more sand settles, and a dune is formed. As a dune grows, sand grains on the windward slope move up and over the crest of the dune to fall on the leeward slope, which results in motion of the whole mound downwind.

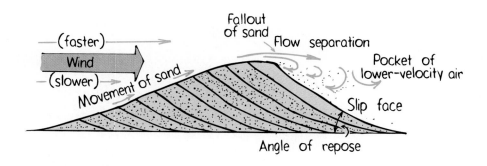

A delta environment resembles an alluvial fan in two respects. Both share the same fan-like shape and both are formed by a stream's inability to transport sediment indefinitely. A delta is the end of the line for rivers as they enter the sea, a bay, or a lake. A river that encounters a standing body of water dumps its load of sediment as its speed decreases. Coarser material settles first, then medium, and finer

Figure 22.33
Cross-bedding in an ancient sand dune. Can you tell the direction of windflow?

material farther out. With incoming sediment, the stream channel becomes choked and distributaries form off the main river as water seeks shorter paths (Figure 22.34). Some of the world's greatest rivers have massive deltas built at their mouths. Millions of years ago the mouth of the Mississippi River was where Cairo, Illinois is today. Since that time the delta has extended 1600 kilometers south to the city of New Orleans. Less than 5000 years ago the site of New Orleans was underwater in the Gulf of Mexico!

Shoreline environments are dominated by beaches and barrier islands. Winds blowing across the ocean surface generate waves; as the waves approach the shallow waters near land the waves become higher and steeper until they finally collapse, or break. This is the surf zone, where wave activity moves sediment back and forth, shoreward and seaward. Sandy beaches are the result of the turbulent motion of the surf zone. Barrier islands, low offshore islands of sand that parallel the coast, build up along the world's lowland coasts. Ridges of dune sand are built up on the island from successive wave action. During large storms, surf washes over the lowlands, making inlets into the lagoon area between the barrier island and the shore. The lagoon area is a more quiet environment with finer grained silts and muds that feature crossbedding and oscillation ripples. On shore, smooth stones, rounded pebbles, and/or sand make up the beaches.

Figure 22.34 Remote imagery of the Mississippi Delta. Note how the distributaries form from the main river as water seeks a shorter path to its destination in the Gulf.

Chemical Sediments

The most abundant chemical sedimentary rocks are limestone, dolomite, gypsum, anhydrite, and halite (Figure 22.35). The chemical sediments that compose these rocks are formed from the precipitation of minerals in a solution, usually water. The process can occur directly, as a result of inorganic processes, or indirectly, as a result of a biochemical reaction. Carbonates are the best example of rocks formed by biochemical reactions, whereas evaporites are good examples of rocks formed by inorganic processes.

Figure 22.35 Chemical sedimentary rocks: (a) limestone, (b) dolomite (pink rhombohedral crystals), (c) gypsum, (d) halite.

a

b

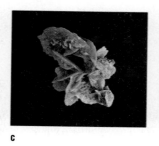
c

d

Figure 22.36 Calcium carbonate precipitating from dripping water forms icicle-shaped stalactites and cone-shaped stalagmites in caves.

Carbonate rocks are composed mostly of calcite, $CaCO_3$, or dolomite, $CaMg(CO_3)_2$. They make up about 10 percent of the total volume of sedimentary rocks. Most carbonate minerals are precipitated directly from seawater or as a result of organic precipitation. Cave dripstones such as stalactites and stalagmites provide an interesting example of calcium carbonate precipitating from dripping water (Figure 22.36).

The common rock **limestone,** the most abundant carbonate rock, is formed predominantly by organic precipitation. Many organisms living in the sea extract calcium carbonate from water to build hard protective shells. When these organisms die, their shells accumulate on the sea floor. Due to compaction and the high solubility of calcium carbonate, the original textures and structures are often obliterated. Closely related to limestone, **dolomite** results from the replacement of calcium by magnesium.

Evaporites are minerals precipitated by the evaporation of a restricted body of seawater or the waters of saline lakes. Upon evaporation the least soluble minerals such as gypsum (used for the making of plaster of Paris) precipitate first, followed by the more soluble minerals, such as common table salt. Although carbonates make up the bulk of chemical sediments, evaporites make up a small but significant portion.

Chemical Sedimentary Environments

Warm climates favor carbonate deposition because carbonates are less soluble in warm water than cold. Carbonate depositional environments include coral reefs and carbonate platforms.

Coral reefs are made up of actively growing coral, an organism that grows as colonies of individual coral joined together. Secreting calcium carbonate as they grow, the coral colonies cement themselves to the dead coral below, and the reef grows outward. Reef building corals require clear shallow water that allows the infiltration of light. (Corals feed on algae that need light to live. The coral and algae live in mutual support; the coral protects the algae, and the algae feed the coral.) Reefs therefore, are built at or close to sea level. A coral reef can be divided into

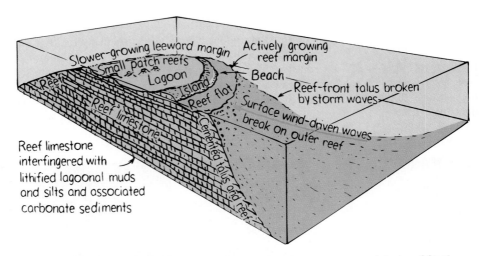

Reef limestone interfingered with lithified lagoonal muds and silts and associated carbonate sediments

Figure 22.37 A coral reef is composed of several sections—a wave-resistant reef front, a flat reef platform, a shallow protected reef lagoon, and the coral reef island.

several sections: a wave resistant reef front, a flat reef platform, a shallow protected reef lagoon, and the coral reef island (Figure 22.37). Each section of the reef has its own characteristics. Because the coral reef is partially destroyed by wave action as it grows, the carbonate particles range in size from blocks several meters across to fine mud. Ancient coral reefs, composed of alternating layers of porous material and impermeable muds, have the potential to act as traps for oil and gas. So coral reefs are economically important. An example of ancient coral reefs can be found in the Guadalupe Mountains of western Texas (Figure 22.38).

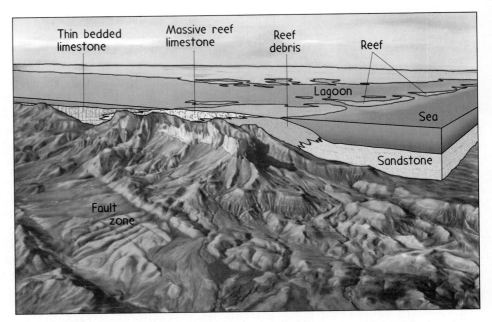

Figure 22.38 Representation of ancient coral reefs in the Guadalupe Mountains of western Texas.

Carbonate platforms are similar to coral reefs but are much larger. Carbonate platforms are the graveyards of calcium-secreting organisms, and are formed in the shallow waters in close proximity to or attached to continents. They account for the largest portion of carbonate sediment produced in the ocean.

Sand-sized fragments from coral reefs and carbonate platforms comprise the white sand beaches in many island areas such as Hawaii (Figure 22.39). Look carefully at the sand in such tropical beaches, and you'll see it is predominantly composed of shell fragments. In contrast, the sand on the beaches of the continents is predominantly the sediments of silicate minerals. So whereas sand in Hawaii is organic in origin, beach sand along the American west coast is largely inorganic—but not all inorganic. Vast carbonate deposits on continental land are evidence of the great seas that periodically covered the land surfaces in the past.

Evaporite deposits require a dry, arid climate conducive to the evaporation of lake or sea water. As the body of water dries out, evaporite minerals precipitate. Modern day as well as ancient evaporites are found in desert basins, tidal flats, and restricted sea basins. The ancient evaporite deposits give further testimony to the great seas that periodically covered the surface of the earth.

Figure 22.39 The white sand on Hawaiian beaches is fundamentally different from the white sand on California beaches: (a) The white sand beaches of California are composed of silicate minerals and can be classified as inorganic. (b) The white beaches of Hawaii are composed of carbonate minerals—the sediment remains of tiny shells—and can be classified as organic.

a

b

Fossil Fuels

Most of the organic matter of animals and plants of the past was quickly decomposed by bacteria and converted to nutrients that became available for use by new life. Sediments that escaped bacterial decay were stored as diffuse organic matter or converted to petroleum or coal. Although coal, oil, gas, and oil shale are all fossils in the sense that they are the remains of past organisms, these remains have been so changed after burial that the form and even the composition of the accumulated organisms are beyond recognition. The source of oil and gas is fossil organic matter integrated throughout buried sediments. When buried organic-rich sediment is heated over a sufficient period of time, chemical changes take place that create oil. Under pressure of the overlying sediments the minute droplets are squeezed out of the source rocks and into overlying porous rocks that become reservoirs. Oil shale is formed when algae produce an ooze rich in waxy hydrocarbons on the bottom of a lake or seaway. Just like in the metamorphism of rocks, deeper burial results in higher temperatures that generate gas rather than oil. Coal is formed from plants that do not completely decay and are so altered that the original structure is destroyed. Coal, petroleum, and natural gas are the primary fuels of our modern economy.

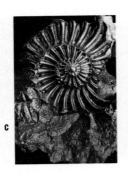

a b c d

Figure 22.40   Some of the many methods of fossilization are:

(a) *Permineralization* occurs when the porous remains of an organism become filled with water that is rich in dissolved minerals—like petrified wood.

(b) *Impression* is made by an organism buried quickly, before it could decompose, thereby preserving its impression.

(c) *Replacement* occurs when the remains of organisms are replaced by a mineral. Pyrite has replaced the original shell in this specimen.

(d) *Carbonization* occurs when a plant or an organism is preserved as a thin film of carbon.

Fossils

Because sedimentary rocks are formed at the earth's surface they often contain remains of pre-existing life forms—fossils—that provide important information for interpreting the earth's geologic past. As we shall see in Chapter 25, fossils play an important role as time indicators and in the correlation of rocks from different places of similar age. Some fossils consist of whole organisms, but most are just their parts. Other fossils are simply an impression, made in the rock before it hardened. Plants commonly leave their impression as a thin film of carbon. There are many methods of fossilization (Figure 22.40).

22.6   METAMORPHIC ROCKS

Igneous or sedimentary rocks may undergo change—**metamorphism**—if they are heated and/or compressed for long periods of time. Clay in the soil, for example, is soft and pliable. When heated it turns to a hard ceramic. Limestone subjected to enough heat and pressure becomes marble; granite is similarly metamorphosed to become gneiss. It is important to note that the minerals aren't melted. Change instead occurs in the distinctive processes of **mechanical deformation** of rock, and **recrystallization** of pre-existing minerals.

Mechanical deformation occurs when a rock is subjected to stress. Surface rocks that become deeply buried and undergo increased pressure may lose elasticity, and flow plastically, bending into intricate folds. Or applied pressure may deform and flatten the rock, or shear it, breaking it and grinding it into fragments.

Recrystallization often occurs when minerals in rocks subjected to high temperatures and pressures go through a change in mineral assemblage—usually by the loss of H_2O or CO_2. Although temperature and pressure are often high enough to cause a metamorphic reaction, fluids enclosed in the open pore spaces act as a catalyst to aid the reaction. When the rock is subjected to increased temperature and pressure the amount of pore space decreases, and the fluid is squeezed out. The fluid can then readily react with the surrounding rock. In general, the more fluid, the faster the reaction.

Types of Metamorphism

Different types of metamorphism are: *dynamic metamorphism, contact metamorphism,* and *regional metamorphism.* Each type is characterized by differences in mechanical deformation and chemical recrystallization.

Dynamic metamorphism primarily involves the process of mechanical deformation. For example, the shearing and grinding that take place in a fault zone crush, flatten, and elongate pre-existing crystals to produce broken and distorted textures (Figure 22.41).

Contact metamorphism is the alteration of rock by chemical recrystallization brought about by the intrusion or close proximity of magma (Figure 22.42). The high temperature of the magma produces a concentric zone of alteration (thermal metamorphism) that surrounds the intrusive rock at the contact. The width of the altered zone may range from a few centimeters to several hundred meters. In a small intrusive body, such as a dike, the altered zone is very narrow and resembles "baked" rock. But with larger intrusive bodies, such as a batholith, the altered zone may be a hundred meters thick or more. The degree of metamorphism changes with the distance from the actual contact. One of the most common changes is an increase in grain size due to recrystallization. The grain size is greatest at the igneous contact, and decreases with distance from the contact zone. The water content of rock also changes with distance. At the intrusion where temperature is high, water content is low. So we find dry, high-temperature minerals such as garnet and pyroxene at the contact border. Farther away, where rock has a higher water content, we find water-rich, low-temperature minerals such as mica and chlorite.

Regional metamorphism is the alteration of rock by both thermal and mechanical means. Regionally metamorphosed rocks are found in all of the major mountain belts of the world. During the process of mountain building, the earth's

Figure 22.41 Mechanical deformation. The elongated rock fragments of this metaconglomerate were at one time the rounded rock fragments of a conglomerate.

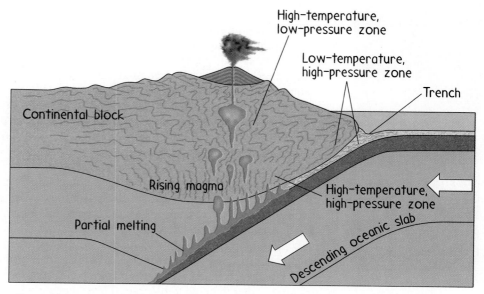

Figure 22.42 Contact metamorphism is the result of rising molten magma that intrudes rock.

Figure 22.43 Folded rock layers in the Rocky Mountains.

crust is severely compressed into a mass of highly deformed rock. This deformation can be seen in the folded and faulted rock layers in many mountain ranges (Figure 22.43). The effects of regional metamorphism are most pronounced in the cores of deformed mountains. Rocks develop characteristic textures, distinctly foliated and layered, and zonal sequences of minerals and textures. Because of the large-scale nature of regional metamorphism, these zones tend to be quite broad and extensive. Regions of structural deformation are the hunting grounds of gem prospectors, for the heat and pressure that accompany these changes produce beautiful minerals.

Metamorphic rocks are defined by their texture and their mineralogy. For classification and identification, metamorphic rocks can be divided into two groups: *foliated* and *nonfoliated*.

a

Foliated Metamorphic Rocks

Foliated metamorphic rocks have a directional texture and layered appearance. For example, sheet-structured minerals such as the micas start to grow and orient themselves so that the sheets are perpendicular to the direction of maximum formation pressure. We say the parallel flakes are *foliated*. The most common foliated metamorphic rocks are slate, schists, and gneiss (Figure 22.44).

Slate is the lowest grade foliated metamorphic rock, meaning that it is formed under relatively low temperatures and pressure. Metamorphosed shale, slate is a very fine-grained foliated rock composed of minute mica flakes. The most noteworthy characteristic of slate is its excellent rock cleavage. The best pool tables and chalk boards are made from slate quarried in metamorphic terrains where slaty cleavage is well developed. Slate is also used as roof and floor tile.

b

Schists are some of the most distinctive metamorphic rocks. A further increase in temperature and pressure causes the constituent minerals to grow large enough to be identified with the naked eye. Schists typically contain 50 percent platy minerals, most commonly muscovite and biotite. Schists are named according to the major minerals in the rock (biotite schist, staurolite-garnet schist, etc.).

Gneiss (pronounced "nice") is foliated metamorphic rock that contains mostly granular, rather than platy minerals. This change in texture is caused by still greater temperature and pressure conditions than those for schists. The most common minerals found in gneisses are quartz and feldspar. The foliation in this case is due

c

Figure 22.44 Common foliated metamorphic rocks: (a) slate, (b) schist, (c) gneiss.

a

b

Figure 22.45 Nonfoliated metamorphic rocks: (a) marble, (b) quartzite.

to the segregation of light and dark minerals rather than alignment of platy minerals. Gneisses have a composition very similar to granite and are often derived from granite.

Nonfoliated Metamorphic Rocks

Marble is a coarse crystalline metamorphosed limestone or dolomite. Pure marble is white and is made of virtually 100 percent calcite. Because of its color and relative softness (hardness 3) marble is a popular building stone. Often the limestone from which marble forms contains impurities that produce various colors. Thus marble can range from pink, gray, green or even black.

Quartzite is metamorphosed quartz sandstone, and therefore is very hard (hardness 7). The recrystallization of quartzite is so complete that the rock will split across the original quartz grains when broken, rather than between them. Although pure quartzite is white, like marble it commonly contains impurities and thus can be a variety of colors such as pink, green, and light gray. The nonfoliated metamorphic rocks, marble and quartzite, are shown in Figure 22.45.

22.7 THE ROCK CYCLE

We have seen that the igneous, sedimentary, and metamorphic rocks of the earth's crust have different origins. Although formed by different processes, the three rock types are related. This interrelationship is graphically portrayed in the model of the rock cycle (Figure 22.46). By following the different pathways in the model we can

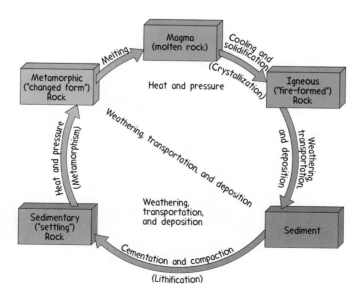

Figure 22.46 There are many possible paths in the rock cycle. For example, igneous rock subjected to heat and pressure far below the earth's surface may become metamorphic rock. Or metamorphic or sedimentary rocks at the earth's surface may decompose to become sediment that becomes new sedimentary rock. Whatever the variety of routes, molten rock rises from the depths of the earth, cools and solidifies to form a crust that over eons is reworked by shifting and erosion, only to eventually return to become magma in the earth's interior.

determine the origin of the three basic rock types and the various geologic processes that transform one rock type into another. The figure helps to summarize this chapter.

We have seen that igneous rock, the most abundant of all three rock types, makes up about 95 percent of the earth's crust. Igneous rock is formed when the molten magma beneath the earth's crust cools and crystallizes. Magma can crystallize into many different kinds of igneous rock. Although most of the earth's crust is igneous or derived from rock that was initially igneous, the rock we see at the surface is mainly sedimentary.

We have seen that sedimentary rock is the result of the decomposition and disintegration of igneous (or other) rocks by weathering and erosion. The effects of weathering slowly decompose the crust into loose noncemented rock particles, which are then transported by any of a number of erosional agents—gravity, running water, ice, wind, or waves. As the deposited sediment accumulates, it may compact from the weight of overlying layers or become cemented as percolating water fills the spaces between particles of sediment with mineral matter. By compaction or cementation, the process of lithification gradually transforms sediment into sedimentary rock.

When sedimentary rock is buried deep within the earth, or is involved in mountain building, we have seen that great pressures and heat have transformed it into the third rock type, metamorphic rock. When subjected to still greater heat and pressure, metamorphic rock melts and turns to magma, which eventually solidifies as igneous rock to complete the rock cycle.

The rock cycle varies in its routes. Igneous rock, for example, may be subjected to the heat and pressure far below the earth's surface to become metamorphic rock. Or metamorphic or sedimentary rocks at the earth's surface may decompose to become sediment that becomes new sedimentary rock. Cycles within cycles occur in the dynamics of the earth's crust. Whatever the routes, molten rock rises from the depths of the earth, cools and solidifies to form a crust that over eons is reworked by shifting and erosion, only to eventually return to the interior where it once again becomes magma (Figure 22.47).

Figure 22.47 Plate tectonic model of the rock cycle.

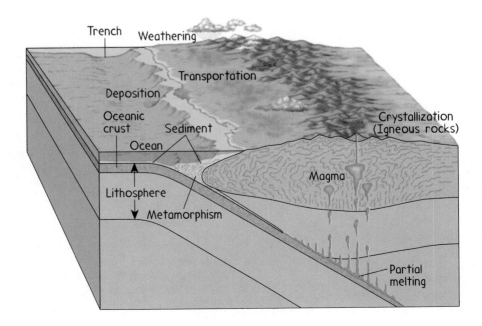

SUMMARY OF TERMS

Igneous rocks Rocks formed by the cooling and consolidation of hot, molten rock material called magma. Igneous rocks make up about 95 percent of the earth's crust. *Igneous* means "formed by fire."

Metamorphic rocks Rocks formed from pre-existing rocks that have been changed or transformed by high temperature, high pressure, or both. The word *metamorphic* means "changed in form."

Sedimentary rocks Rocks formed from weathered material (sediments) carried by water, wind, or ice. Sedimentary rocks are the most common rocks in the uppermost part of the earth's crust and cover over two-thirds of the earth's surface.

Mineral A naturally formed, inorganic solid composed of an ordered array of atoms chemically bonded to form a particular crystalline structure.

Polymorphs Minerals that have the same chemical composition but different crystal structures.

Silicates The most common mineral group. Silicates are constructed of four oxygen atoms (O^{2-}) with a single silicon atom (Si^{4+}) at the center in a structure called a tetrahedron.

Magma Rock heated to its melting point, molten rock.

Partial melting The incomplete melting of rocks resulting in magmas of different compositions.

Fractional crystallization The process and sequence by which minerals crystallize. Minerals with the highest melting points crystallize first, followed by minerals with lower melting points. The sequence of crystallization allows a single magma to generate several different igneous rocks.

Intrusive rocks Rocks that crystallize below the earth's surface.

Pluton A large intrusive body formed below the earth's surface.

Pahoehoe A basaltic lava flow characterized by a smooth wrinkly skin that resembles the twisting braids in a rope.

Aa A basaltic lava flow characterized by a rough jagged surface with dangerously sharp edges and spiny projections.

Extrusive rocks Igneous rocks that form at the earth's surface.

Lava The fluid rock that flows onto the earth's surface. Magma at the earth's surface.

Volcano The central vent through which lava, gases, and ash erupt and flow.

Weathering The breakdown and decomposition of the earth's surface. Weathering occurs by mechanical means that physically disintegrate and fragment rocks into smaller and smaller pieces, and by chemical alteration of rock into substances more in balance with the surrounding environment.

Erosion The process by which rock particles are transported away by water, wind, or ice.

Lithification The process by which sediment turns into sedimentary rock.

Mechanical deformation The process of metamorphism that results as a rock is subjected to stress, such as increased pressure.

Recrystallization The process of metamorphism that results when a rock is subjected to such high temperatures that the minerals go through a change in mineralogy, usually by the loss of H_2O or CO_2. Recrystallization generally results in larger grain size.

Rock cycle A sequence of events involving the formation, destruction, alteration, and reformation of rocks as a result of the generation and movement of magma, the weathering, erosion, transportation, and deposition of sediment, and the metamorphism of pre-existing rocks.

REVIEW QUESTIONS

Rocks

1. Name the three major types of rocks, and cite the conditions of their origin.

Minerals

2. A rock may be defined as an aggregate of one or more minerals. What, then, is a mineral?

3. What physical properties are used in the identification of minerals?

Crystal Form

4. All minerals are defined by an orderly internal arrangement of their constituent atoms—the crystal form. Yet, most mineral samples do not display their crystal form. Why?

5. What is a polymorph?

Luster

6. What are the two classifications for mineral luster?

Color and Streak

7. Although color is an obvious feature of a mineral, it is not a very reliable means of identification. Why?

Hardness

8. Will the mineral topaz scratch quartz, or will quartz scratch topaz? Why?

Cleavage and Fracture

9. The minerals calcite, halite, and gypsum are all nonmetallic, light, softer than glass, and have three

directions of cleavage. In what ways can they be distinguished from one another?

Specific Gravity

10. Silver has a density of 10.5 g/cm^3. What is its specific gravity?

11. What is the relationship of density to specific gravity?

Chemical Properties of Minerals

12. What are two common chemical tests for identifying some minerals?

Building Blocks of Rock-Forming Minerals

13. What is the most abundant element in the earth's crust? What is the second most abundant element?

14. What is the most abundant mineral in the earth's crust? What is the second most abundant mineral?

15. What is polymerization?

Igneous Rocks

16. What are the most common igneous rocks and where do they generally occur?

17. What percentage of the earth's surface is composed of igneous rocks?

Magma and the Evolution of Igneous Rocks

18. What is meant by *partial melting*?

19. What are the three main types of magma? Relate the different magmas to silica content.

20. What two reasons make silica an important element for the different magmas?

Igneous Rock Crystallization

21. What is meant by *fractional crystallization* and what are its consequences?

22. How are partial melting and fractional crystallization similar?

23. How do the different textures of igneous rocks relate to their environment of formation?

Igneous Rock Formation Beneath the Earth's Surface

24. Because sills form at shallow depths they are often confused with buried lava flows. In what ways would a sill differ from a buried lava flow?

25. In what way does a sill differ from a dike?

Igneous Rock Formation at the Earth's Surface

26. Where are lava flows most common?

27. What are the three major types of volcanoes?

28. What is viscosity?

29. What type of volcano produces the most violent eruptions? What type produces the most quiet eruptions?

30. What does Yellowstone National Park have to do with volcanic activity?

Sedimentary Rocks

31. How does weathering produce sediment? Distinguish between weathering and erosion.

Sedimentation

32. What does the roundness of sediment grains tell us about a rock?

33. What can we say about a rock that is composed of various sized sediments in a disorganized pattern?

34. Relate particle size and shape to the transportation of sand grains.

35. What can we say about a rock that is composed of very angular sediments?

36. What is the process of lithification?

Clastic Sediments

37. What are the three most common clastic sedimentary rocks?

38. What is meant by a clastic sedimentary rock?

39. Give two examples of sedimentary rocks that provide information about past geologic events at the earth's surface.

Clastic Sedimentary Environments

40. What is the most dominant feature of a sedimentary rock environment?

41. What is alluvium? Where do we find alluvium?

42. Deserts are generally dry areas. Why is water still a major factor of erosion in the desert environment?

43. Name two environments where cross-bedding occurs. What information do cross-beds provide?

44. What is a delta?

45. Are all beaches sandy? Why or why not?

Chemical Sediments

46. What is a chemical sediment?

47. What are three common chemical sedimentary rocks?

48. When water evaporates from a body of water what type of sediment is left behind?

49. What is the most common chemical sediment?

Chemical Sedimentary Environments

50. Why do coral-building reefs require clear shallow water?

51. The deposition of carbonate sediments is more prevalent in warm water environments. What does this tell us about ancient carbonate deposition near Dallas, Texas?

Fossils

52. What is a fossil? How are they used in the study of geology?

Metamorphic Rock

53. What is metamorphism? What are the agents of metamorphism?

54. What are the two processes by which rock is changed?

Types of Metamorphism

55. What changes are characteristic of contact metamorphism?

56. What changes are characteristic of dynamic metamorphism?

57. What changes are characteristic of regional metamorphism?

58. Distinguish between foliated and nonfoliated metamorphic rocks.

59. How does gneiss differ from granite?

The Rock Cycle

60. Explain the different cycles of rock formation.

HOME PROJECTS

1. Look at some table salt crystals under a microscope or magnifying glass and observe their generally cubic shapes. There's no machine at the salt factory specifically designed to give these cubic shapes, as opposed to spherical or triangular. The cubic shapes of salt crystals occur naturally and are a reflection of how the atoms of salt are organized—cubically. Smash a few of these salt cubes then look carefully. What you'll see are smaller salt cubes! Explain these results based upon the cleavage properties of salt (Section 22.2).

2. A physical property of any material is its density—its mass per volume. We know that lead is more dense than metals like zinc and copper. Pennies fabricated after 1982 contain both copper and zinc. Since zinc is less dense than copper, post-1982 pennies are less dense—hence, less massive. Dig into your penny collection and find 20 pre-1982 and 20 post-1982 pennies. Measure their masses on a sensitive scale, such as a home postage scale. Alternatively, hold the pennies in opposite hands to see if you can feel the difference in their masses. How few pennies can you hold and still feel the difference? Try holding single pennies on your left and right index fingers. Can you tell the difference with your eyes closed? Try this with a friend.

EXERCISES

1. We speak of the physical and chemical properties of minerals. Clearly distinguish between physical and chemical properties, and give examples of each.

2. Yellowstone National Park has a volcanic origin. Name at least one other national park that was formed by volcanic or plutonic activity.

3. What type of rock is formed when magma rises slowly and solidifies before it reaches the earth's surface?

4. What difference in density would you expect in magma that rises slowly, compared to magma that rises quickly?

5. Are the rocks that surround a batholith older, younger, or equal in age to the batholith? Defend your answer.

6. Are the Hawaiian Islands primarily made up of igneous, sedimentary, or metamorphic rock? Explain.

7. Two volcanoes that have erupted in recent times are Mauna Loa in Hawaii and Mt. St. Helens in Washington. What differences would you expect in the compositions of their lavas?

8. Why do we find few craters on earth from the intense bombardment of meteorites that cratered the nearby moon billions of years ago?

9. Clastic sedimentary rocks are composed primarily of clay minerals and quartz. Clay minerals are the by-product of the chemical weathering of feldspar. Because feldspar is the most abundant silicate, clay minerals are the most abundant clastic sediment. The abundance of quartz is due to its durability and resistance to weathering. What are the most abundant clastic sedimentary rocks?

10. Conglomerates are composed of many different rock fragments. Pebbles of granite are very common in a conglomerate, whereas pebbles of marble are relatively uncommon. Why is this? What are the properties of the different minerals in these rocks?

11. In the formation of a river delta, why is it that coarser material is deposited first, followed by medium and finer material further out? What type of bedding and gradation of sediments results from this sequence? Defend your answer.

12. Dig a hole in an alluvial fan and the size of the rocks changes with depth. How do they change, and most important, why?

13. Which of the following rocks would be the first to weather in a humid climate? Which would be the last? Defend your answer.

(a) granite (b) sandstone (c) limestone
(d) quartz (e) halite

14. How do chemical sediments produce rock? Name two rock types that form by chemical sedimentation.

15. Which type of rock is most sought by petroleum prospectors: igneous, sedimentary, or metamorphic rock? Explain.

16. What properties of slate make it good roofing material?

17. In metamorphosed rocks there is a sequence of mineral assemblages that reflect pressure and temperature conditions during formation. Metamorphosed rocks also show a sequence of texture characteristics corresponding to the different conditions. What would you choose as the better property to determine metamorphic grade—texture or mineralogy? Why?

18. Each of the following statements describes one or more characteristics of a particular metamorphic rock. For each statement, name the metamorphic rock that is being described.

   (a) Foliated rock derived from granite.
   (b) Hard, nonfoliated, monomineral rock, formed under high to moderate metamorphism.
   (c) Foliated rock possessing excellent rock cleavage. Generally used in making blackboards.
   (d) Nonfoliated rock composed of carbonate minerals.
   (e) Foliated rock containing about 50 percent platy minerals; named according to the major minerals in the rock.

23

WATER AND SURFACE PROCESSES

A view of Earth from outer space shows it to be a vast expanse of water interrupted by island-like continents. About 70 percent of the earth's surface is covered with water. Water plays an important role in just about every natural process on Earth's surface.

Water is vital to life, and has been since the very first life forms evolved. Our bodies are composed of over 65 percent water by weight and require water for bodily functions. Because water is essential to our lives, people have generally settled near or close to bodies of water. With the growth of population in today's world the need for water has increased. Water is not only used for drinking, but for agriculture, industry, sanitation, and transportation. Where does the water we use come from, and more important, will it last?

Slightly more than 97 percent of the total water is in the world's oceans, and a little more than 2 percent is frozen in the polar icecaps and glaciers. All the remaining water, less than 1 percent, makes up the water vapor in the atmosphere, the water in the ground, and the water in rivers and lakes. This one percent is the water we are aware of and rely on in our daily life (Figure 23.2).

In this chapter we will focus on the earth's fresh water supply—the water frozen in our glaciers and icecaps and the water found in our rivers and lakes, and underground.

569

23.1 THE HYDROLOGIC CYCLE

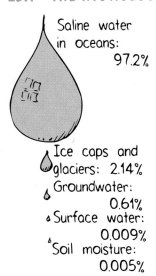

Saline water in oceans: 97.2%

Ice caps and glaciers: 2.14%

Groundwater: 0.61%

Surface water: 0.009%

Soil moisture: 0.005%

Figure 23.2 Distribution of the world's water supply.

The earth's water is constantly circulating, powered by the heat of the sun and the force of gravity. As the sun's energy evaporates ocean water, a cycle begins. Water molecules move from the earth's surface to become part of the atmosphere. The resulting moist air may be transported over great distances by winds. Some of the water molecules condense to form clouds, and then precipitate as rain or snow. Precipitation falling on the ocean completes the cycle from ocean back to ocean. Completion of a cycle is somewhat more complex when precipitation falls on land, for water may drain to streams, then to rivers, and then journey back into the ocean. Or it may percolate into the ground, or evaporate back into the atmosphere to begin another cycle. Also water falling on land may become part of a snow pack or glacier. Although snow or ice may lock water up for many years, it eventually melts or evaporates and gets back into the cycle. This natural circulation of water from the oceans to the air, to the ground, then to the oceans and then back to the atmosphere is called the **hydrologic cycle*** (Figure 23.3).

The total amount of water vapor in the atmosphere remains relatively constant. Therefore there must be a balance between evaporation and precipitation. Since most of the earth's surface area is ocean, it makes sense that evaporation and precipitation are greatest over the oceans. In fact, 85 percent of the atmosphere's water vapor

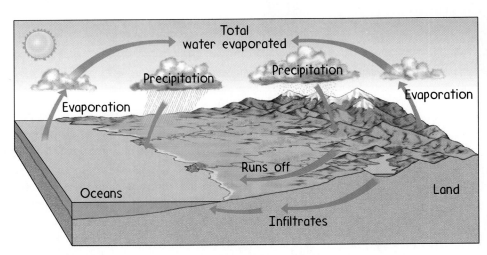

Figure 23.3 The hydrologic cycle. Water evaporated at the earth's surface enters and becomes part of the atmosphere as water vapor, where it condenses into clouds, then precipitates as rain or snow, and falls back to the surface. Movement into the atmosphere by evaporation is matched by precipitation.

---

*This key concept is another version of the conservation principle, which we saw in Chapter 3—the conservation of momentum and the conservation of energy. In Chapter 9 we saw it as the conservation of electric charge. In Chapter 16 we saw it as the conservation of nucleons, where in nuclear reactions the number of nucleons before and after a reaction are the same. Similarly, in Chapter 19, as the conservation of atoms, where the atoms before and after a chemical reaction are the same. So now we learn that the amount of water on earth is conserved. A lack of it in one place means an abundance someplace else, which in most instances, is the ocean.

is evaporated from the oceans and 75 percent of the atmosphere's water vapor is precipitated back to the oceans. On the continents, however, precipitation exceeds evaporation. Fifteen percent of the atmosphere's water vapor is evaporated from the continents and 25 percent of the atmosphere's water vapor is precipitated back to the land. So an overall balance is maintained between the amount of water taken up into the atmosphere and the amount of water precipitated out of the atmosphere. The rain or snowfall that reaches the continents is the earth's only natural supply of fresh water. There is no other natural source.

More than three quarters of the earth's fresh water is in the polar ice caps and glaciers. Of the fresh water not locked in glacial ices, most is not in lakes and rivers. Most water is beneath the surface—*groundwater.*

QUESTIONS

1. What fraction of the earth's water supply is fresh water?

2. The amount of water evaporated from over the oceans is 320,000 cubic kilometers while the amount evaporated from over the continents is 60,000 cubic kilometers per year. Thus, the total water evaporated is equal to 380,000 cubic kilometers. The amount of water precipitated over the oceans is 284,000 cubic kilometers while the amount of water precipitated over land is 96,000 cubic kilometers. Thus, the total water precipitated is equal to 380,000 cubic kilometers. If evaporation of water over the continents is 60,000 cubic kilometers and precipitation over the continents is 96,000 cubic kilometers, what happens to the excess 36,000 cubic kilometers that is precipitated?

23.2 GROUNDWATER

Except in polar regions, the water in lakes, ponds, rivers, and puddles is the only fresh water that meets our eye. Most of the earth's fresh water (other than ice caps and glaciers) resides in porous regions beneath the earth's surface. When we exclude the polar ice caps and glaciers, more than 90 percent of all the earth's fresh water is found beneath the earth's surface as groundwater!

Have you ever noticed how during a rainstorm sandy ground surfaces soak up rain like a sponge? The water literally disappears into the ground. The nature of the surface material influences the amount of water that penetrates the ground. Some soils, like sand, are very permeable and readily soak up water. Other soils, like clay, impede the infiltration of water and cause runoff. Rocky surfaces with little or no soil are the poorest absorbers of water, with penetration only through joints and cracks in the rock. Once water saturates the underground rock or soil, it is called **groundwater**.

ANSWERS

1. Less than 3 percent! About 77 percent of all fresh water is in the polar ice caps and glaciers; about 22 percent is in groundwater; and less than one percent of fresh water is in rivers and lakes.

2. Although precipitation exceeds evaporation over the continents by 36,000 cubic kilometers, the total overall balance between evaporation and precipitation is unaffected. The excess water eventually works its way back to the oceans. On its journey the water erodes the land surface.

The amount of water that can be contained underground depends on the **porosity** of the soil or rock material. Porosity is the volume of open space, or voids, in a sample compared to the total volume, solids plus voids. Porosity depends on the size distribution and shape of the soil or rock particles, and how tightly these particles are packed. For example, if a soil is composed of rounded particles that are all about the same size, it will have a higher porosity than a soil with rounded particles that has a wide distribution of sizes. This is because the smaller particles will fill up the pores formed by the larger particles, thereby reducing the overall porosity. In addition, angular particles can fill in pores created by other particles because of their irregular shape.

Groundwater moves through the small open pore spaces of the subsurface. The ability of a material to transmit fluid is its **permeability**. If the space between grain particles is extremely small, like in clay for example, water may not move at all. So although clay is very porous, it is practically impermeable. In contrast, sand and gravel, because of the large pore spaces, is highly porous and highly permeable. Porosity and permeability of surface and subsurface material is very important to the storage and movement of groundwater.

The Water Table

Water percolating into the subsurface fills the pore spaces in the soil until it reaches a point of saturation where every pore space is completely filled with water. The upper boundary of this saturated zone is called the **water table**. Beneath the water table, the ground is filled to capacity with water.

The depth of the water table beneath the earth's surface varies with precipitation and climate. It ranges from zero in marshes and swamps to thousands of meters in some parts of the deserts. The water table tends to rise and fall with the contours of the surface topography (Figure 23.4). We often find that lakes and streams form where the water table is above the surface and swamps form where the water table is at the surface. Above the water table is the unsaturated zone, the *zone of aeration*, where pore spaces are filled with air.

Aquifers and Springs Any water-bearing layer through which groundwater can flow is an *aquifer*. These underground reservoirs underlie the land everywhere and represent an enormous amount of water—approximately 35 times the total volume of water found in fresh water lakes, rivers, and streams. More than half the land area in the United States is underlain by aquifers. One such aquifer is the Ogallala aquifer, which stretches from South Dakota to Texas and from Colorado to Arkansas!

Figure 23.4 The water table roughly parallels the surface. In times of drought, the water table falls, reducing stream flow and drying up water wells. The water table also falls if the pumping of a well exceeds groundwater recharging.

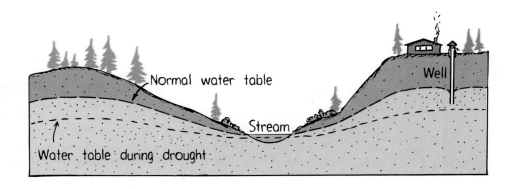

Figure 23.5 An artesian system is formed when groundwater, confined between layers of impermeable rock, rises upward to the surface through any opening that taps the aquifer. The water at the opening will rise to the height of the water table in the recharge area. Water flows freely if the water table's height in the recharge area is greater than the height of the opening. If the height of the opening (spring or well) is less than or equal to the height of the water table in the recharge area, the water will not flow.

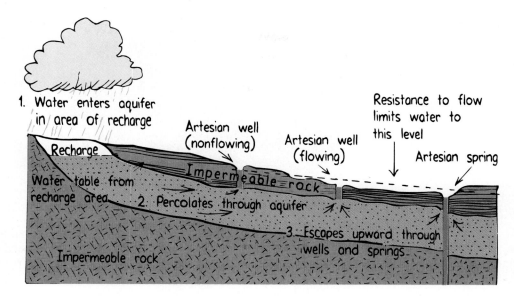

The flow of groundwater in an aquifer can be complicated by impermeable beds of rock or soil that hinder or prevent water movement.* Sometimes an aquifer becomes confined between two impermeable layers as if in a tunnel. If the confined portion of the aquifer is at a lower elevation than the unconfined part in the recharge area (the area where water enters the aquifer), the confined groundwater will be under pressure from the height of water above it and will flow out of the ground at any opening in the aquifer. This is an **artesian system** (Figure 23.5). If the opening is natural, it is an *artesian spring*. If the opening has been drilled, it is an *artesian well*. An impermeable layer can also act to intercept water above the main water table; when this happens a *perched* water table is created (Figure 23.6).

Figure 23.6 A perched water table is separated from the main water table by an impermeable layer.

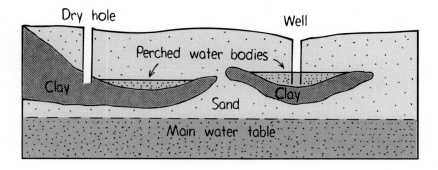

When the water table intersects the land surface, groundwater emerges from an aquifer as a spring (Figure 23.7). Springs can generally be found on the lower slopes or base of a hill or on the side of a valley or a coastal cliff. Because water tends to leak out of the ground through cracks and breaks in a rock, springs are often associated with faults. In fact, field geologists can often locate faults by looking for springs.

---

*Geologist types call these beds *aquicludes* or *aquitards*.

Figure 23.7 When the water table intersects the land surface, groundwater is released via a spring.

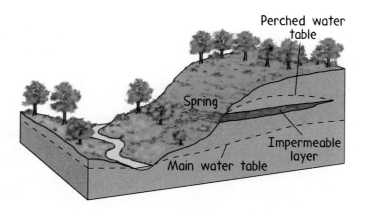

The rate of groundwater flow is directly proportional to the aquifer's permeability. Naturally, how fast water flows in an aquifer depends on the aquifer's ability to transmit water. But there's another factor that affects the rate of groundwater flow—*hydraulic gradient*. To understand how hydraulic gradient affects flow rate you will need to know what hydraulic head is. *Hydraulic head* is the height water will rise in a well. In an unconfined aquifer, that height equals the height of the water table. The hydraulic gradient is the difference in hydraulic head between two points divided by the horizontal distance between those points (Figure 23.8).*

Figure 23.8 The hydraulic gradient is the vertical difference in elevation between any two locations on an aquifer ($h_2 - h_1$), divided by the horizontal distance between them (L).

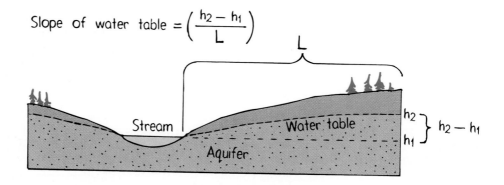

$$\text{Slope of water table} = \left(\frac{h_2 - h_1}{L}\right)$$

Groundwater Movement The flow of water underground depends on several geological conditions. Most water flows though pore spaces in rock. This movement is influenced by the force of gravity. Groundwater flows "downhill" underground, but the path it takes is dependent on hydraulic head and not topography. Hydraulic head

---

*The rate of groundwater flow in relation to hydraulic head is stated in the formula called *Darcy's Law*, developed in 1856 by the French engineer Henri Darcy:

$$V = K\frac{\Delta h}{\Delta l}$$

where V represents the velocity, or flow rate, Δh is the change in head, Δl is the change in horizontal distance, and K is the measure of permeability, or hydraulic conductivity.

is higher where the water table is high, such as beneath a hill, and lower where the water table is low, such as beneath a stream valley. So, responding to the force of gravity, water moves from a high water table to a low water table (Figure 23.9).

Figure 23.9 Groundwater flows from a high hydraulic head area, such as beneath a hill, to a low hydraulic head area, such as beneath a stream valley. The curved arrows indicate flow, which shows that the stream is fed from below.

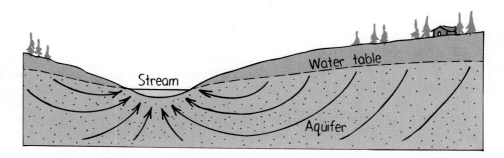

The speed of groundwater movement is generally very slow. The more permeable the aquifer, the faster the flow; and the greater the hydraulic gradient, the faster the flow. The speed and route of groundwater flow can be measured experimentally by introducing dye into a well and noting the time it takes to travel to the next well. In most aquifers, groundwater speed is only a few centimeters per day, enough to keep underground reservoirs full.

Mining Groundwater

The most common way to remove groundwater is by drilling an opening into an aquifer—a well. In an unconfined aquifer a well fills with water up to the water table level. As water is withdrawn, the water table immediately surrounding the well is lowered creating a depression in the water table. The size of the depression is dependent on the amount of water withdrawn from the well. Generally, domestic wells don't cause problems, but if the well is used for irrigation or industrial purposes, the depression can be quite steep and wide (Figure 23.10).

Figure 23.10 The water table depression around a pumping well takes the shape of a cone. This cone is almost imperceptible for low-volume domestic type wells, but can be very steep and wide for wells used for irrigation or industrial purposes. Excessive pumping may lower the water table and cause shallow wells to dry up.

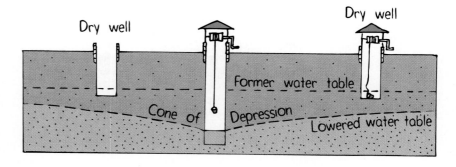

Although the reservoir of groundwater is great, when pumping exceeds water recharge there can be a problem. Recall that the supply of fresh water depends on rainfall. As populations grow, the demand for water grows. In areas such as the Pacific Northwest the climate is wet and extraction of water is balanced by recharge. But areas that are dry, such as Southern California and the High Plains, have very slow recharge. These areas are very dependent on the water reserves in the ground. The Ogallala aquifer located under the High Plains has supplied water to this thirsty agricultural region for more than a hundred years. Withdrawal has so greatly exceeded recharge that it will take thousands of years for the water table to recover to its original level. In certain parts of Texas the water table has dropped so low that water is now literally "mined." Likewise for many of the world's drier areas.

Recharge of an aquifer is dependent on the amount of rainfall and the nature of the subsurface material. It's also related to the length of time water spends in one place. Table 23.1 provides an illustration of **residence time,** the *average* time that a water molecule spends in a region. As residence time varies, the time required to complete the hydrologic cycle varies. Water that resides in polar ice and glaciers has both a long residence time and a long time before completing the hydrologic cycle. The residence time of thousands of years for deep groundwater means, for all practical purposes, that this water is nonrenewable. Much of it was accumulated thousands of years ago, perhaps under conditions of wetter climates. Just as coal and petroleum are called "fossil" fuels, such deep groundwater is called "fossil" water. Water is being recognized more and more as the most precious of our mineral resources.

TABLE 23.1 *Water Resource Residence Times*

Location	Average Residence Time
Atmosphere	1–2 weeks
Ocean	
Shallow depths	100–150 years
Deep depths	30,000–40,000 years
Continents	
Rivers	2–3 weeks
Lakes	10–100 years
Shallow groundwater	up to hundreds of years
Deep groundwater	up to thousands of years
Glaciers	10,000–20,000 years

Ground Subsidence In areas where the withdrawal of groundwater has been extreme, the ground surface is lowered—it *subsides.* Ground subsidence is most pronounced in areas underlain by sand, gravel, or other unconsolidated sediments.

ANSWERS

1. An aquifer is simply a body of rock or sediment through which groundwater easily moves.

2. An artesian system forms when an aquifer confined between impermeable beds and under sufficient pressure is tapped into, either naturally or by a human-made well, and water begins to rise above the top of the aquifer.

QUESTION

Why is ground subsidence most evident in regions where the underlying ground is composed of unconsolidated sediment?

As water is removed from the ground, and thus from the pore spaces, the sediments compact, and the ground subsides. Probably the most well known example of ground subsidence is the Leaning Tower of Pisa in Italy, built on the unconsolidated flood-plain sediments of the Arno River. As groundwater over the years has been withdrawn from underground aquifers to supply the growing city, the tilt of the tower has increased (Figure 23.11). Another region where ground subsidence is evident is Mexico City, built in the middle of an ancient shallow lake. The withdrawal of groundwater beneath this city now finds many "street level" buildings at basement level. Some areas have subsided by as much as 6 to 7 meters (19 to 22 ft). In the United States, extensive groundwater withdrawal for irrigation in the San Joaquin Valley of California has caused the water table to drop 75 meters (246 ft) in 20 years, lowering the ground surface by as much as 9 meters (29 ft). Corrective steps have been taken and water for irrigation is now provided by canals, and the aquifer is slowly recharging.

Figure 23.11   The Leaning Tower of Pisa. Construction began about 1173 and was suspended when builders realized the slightly more than 2-meter-deep foundation was inadequate. But work was later resumed, and the 60-m tower was completed 22 years later. Deviation from the vertical is about 4.6 meters. Its foundation has been recently stabilized by groundwater withdrawal management, so the tower should remain stable for years to come.

23.3   GROUNDWATER AND TOPOGRAPHY

The vast carbonate deposits that underlie millions of square kilometers of the earth's surface provide storage areas for groundwater. The effect of groundwater on limestone and other carbonate rocks is very unique with some interesting results. Rainwater naturally reacts with carbon dioxide in the air and the soil to produce carbonic acid. When this slightly acidic rainwater comes in contact with carbonate rocks, the carbonic acid partially dissolves the rocks into calcium and magnesium bicarbonate, which is then carried away in solution.* As groundwater steadily dissolves the limestone and other carbonate rocks, it creates unusual erosional features.

ANSWER

Unconsolidated sediment is loosely arranged. Before the withdrawal of groundwater the open pore spaces of the sediment are filled with water. With excessive groundwater withdrawal the pore spaces collapse and the ground subsides.

---

*Water from these source rocks is considered "hard" because it is rich in calcium and magnesium bicarbonates.

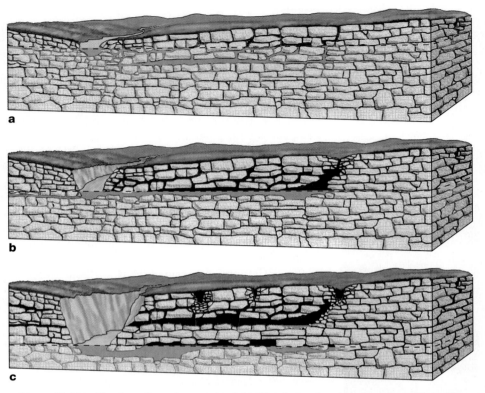

a

b

c

Figure 23.12 The formation of a cave begins with a layer of carbonate rock, mildly acidic groundwater, and an enormous span of time. In (a), horizontal movement of groundwater makes its way toward a stream. (b) As the stream valley deepens, the water table is lowered. The carbonate rock is eaten away as acidified water erodes and enlarges the existing fractures into small caves. (c) Further deepening of the stream valley causes the water table to drop even lower; water in the cave seeps downward, leaving an empty cave above a lowered groundwater level.

Caverns and Caves

The dissolving action of subterranean water has carved out magnificent underground caverns and caves. As groundwater seeps into the tiny fractures in limestone, the rock is dissolved. This enlarges the fractures. Working horizontally along the water table, the groundwater moves toward a natural outlet, such as a stream. This creates an underground channel which in time becomes a cave or cavern. As the stream valley deepens the water table drops, causing the water in the main cave channel to seep downward to begin a new level (Figure 23.12). Dripping water, rich in dissolved calcium carbonate, trickles down from the cave ceiling, creating icicle-shaped stalactites as water evaporates and carbonate precipitates. Some solution drips off the end of the stalactites to build corresponding cone-shaped stalagmites on the floor.

Carlsbad Caverns in southeastern New Mexico is one of the most impressive caverns in the United States. The cavern descends to a depth of 253 meters and covers over 11 kilometers (Figure 23.13). Other famous caves and caverns include the Mammoth Cave in Kentucky, the Adelsberg Cave in Austria, and the Good Luck Cave in Borneo.

Figure 23.13 Cave dripstone formations at Carlsbad Cavern.

Sinkholes

A sinkhole is a funnel-shaped cavity open to the sky. It is formed in much the same way as a cave: by the dissolution of carbonate rock by groundwater. Some sinkholes are caves whose roofs have collapsed. The Florida peninsula is well known for its sinkholes, which occasionally form overnight. In an area of about 25 square kilometers, more than 1000 sinkholes have formed in a matter of a few years (Figure 23.14). Some sinkholes are formed by drought conditions or excessive groundwater pumping.

When sinkholes, caves, and caverns define the land surface, the terrain is called **karst topography,** after the Karst region of the former Yugoslavia, where pronounced erosion of highly soluble rocks characterizes the landscape. The drainage pattern in this type of landscape is very irregular; streams and rivers disappear into the ground surface and reappear as springs. Some karst areas appear as soft rolling hills with large depressions that dot the landscape; the depressions are old sinkholes now covered with vegetation (Figure 23.15). In general, karst areas have sharp and rugged surfaces with thin to almost nonexistent soil due to high runoff and dissolution of surface material.

Figure 23.14 View of a large sinkhole that formed in Winter Park, Florida.

Figure 23.15 Karst topography covered by vegetation makes up the rolling hills in Bowling Green, Kentucky.

Karst regions can be found throughout the world: in the Mediterranean basin, in sections of the Alps and the Pyrenees, in southern China, and in Kentucky, Missouri, and Tennessee in the United States. The beauty of southern China's karst landscape is depicted in Figure 23.16.

Figure 23.16 The karst landscape of China has been an inspiration to classical Chinese brush artists for centuries.

23.4 THE QUALITY OF WATER

The quality of the water we drink as well as the water in lakes, streams, rivers, and oceans is a crucial factor in the quality of our lives. Rainwater is used as the standard of water purity. Most of our water supply is of good quality, as good as rainwater. There are important variations though. The quality of water depends on many factors and needs to be continually tested, and when needed, treated.

If rain falls through clean air and soaks into a bed of quartz sand, the water quality after filtering through the sand will be about the same as before filtration. On the other hand, groundwater in a karst area is "hard" from dissolved calcium and magnesium bicarbonate. This can affect the taste of the water as well as its utility. The quality of groundwater depends very much on the type of soil and rock it flows through.

The amount of dissolved substances in drinking water is very small. Good quality water averages 150 ppm (parts per million) for total dissolved substances with an upper limit of 500 to 1000 ppm. The taste of water depends on the type of dissolved substances. Water with about 1000 ppm of dissolved calcium tastes fine, but water with several hundred ppm of sodium chloride tastes salty. Several other dissolved substances, many introduced by human activities, have a strong effect on the quality of water; some are beneficial to health, while others can be quite dangerous. Added fluoride, for example, helps reduce tooth decay. Added zeolite minerals in water filters soften hard water (Figure 23.17). In contrast, lead and arsenic, two naturally occurring minerals, can make water unsafe to drink even if they are present in minute amounts. Bacteria from sewage is also a possible creator of contaminated water.

Figure 23.17 Zeolite is a hydrous silicate mineral that softens water by exchanging sodium ions for the calcium and magnesium ions that make water hard.

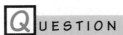UESTION

Why is soft water usually preferable to hard water? How is hard water softened?

Figure 23.18 When bathing in hard water, dissolved calcium prevents soap from developing a sudsy lather, and also from rinsing clean. Bathtub rings are more prevalent in areas with hard water.

Answer

Unlike hard water, soft water allows the formation of soap suds and can be rinsed clean (Figure 23.18). Hard water can be made soft by removal of calcium and magnesium ions. This is accomplished by passing it through a zeolite filter that absorbs calcium and magnesium ions while releasing an equivalent number of sodium ions.

Water Supply Contamination

The primary source of water supply contamination is human activity. As rivers and streams receive discharge from factories, sewage, and chemical spills, surface water is polluted. We can usually see or smell surface water contamination. Groundwater contaminants, however, are more difficult to detect. Surface waters are linked to groundwaters; what affects one affects the other. The groundwater supply can be adversely affected by a variety of sources, including septic tank disposal systems and sewage treatment facilities; agricultural fertilizers; municipal landfills and toxic- and hazardous-waste landfills; leaking underground storage tanks; and chemical and petroleum product spills. The contamination of groundwater is a very serious problem.

The most common source of groundwater contamination is sewage. Sewage includes drainage from septic tanks, inadequate or broken sewer lines, and barnyard wastes. Sewage water contains bacteria which, if untreated, can cause waterborne diseases such as typhoid, cholera, and infectious hepatitis. Sewage contamination can be treated naturally. If the contaminated water travels through very porous sediment or rock, such as gravel or cavernous limestone, it can travel long distances in short periods of time without much change. On the other hand, if the contaminated water travels through smaller pores, such as those in sand, the flow of water takes longer but the water can be purified within short distances. The sand acts to separate the bacteria from the water. Sand is a good filter for bacteria and viruses and is often used in sewage treatment plants to purify the water.

Agricultural areas where lots of nitrate fertilizers are used also contribute to groundwater contamination. Recall from Chapter 19 that nitrates are very soluble. Nitrate fertilizer spread over the land is used by plants, but some disperses through the air. The excess percolates down into the groundwater as a contaminant. Nitrate levels in groundwater have to be closely monitored due to their toxicity. Nitrates in amounts as small as 15 ppm are toxic to humans.

When a population grows, so does its garbage. The most common means of waste and refuse disposal is burial in a landfill. Even radioactive, toxic, and hazardous wastes are disposed of by burial. The placement of underground storage sites is tricky to decide. In order for a site to be considered safe it must be located where waste products and their containers cannot be affected chemically by water, physically by earth movements, or accidentally by people. Precipitation infiltrating the site may dissolve a variety of compounds from the solid waste. The resulting liquid, known as **leachate,** can move down from the landfill into the water table and cause groundwater contamination. When leachate mixes with groundwater it forms a plume that spreads in the direction of the flowing groundwater. Corrective steps can be taken. To reduce the chances of groundwater contamination, the landfill can be capped with layers of compacted clay soil or a synthetic membrane to prevent generation of leachate. It can also be lined with the same material, plus a collection system to catch any draining leachate and prevent its distribution.

Groundwater contamination also occurs as a result of spills and leaks of toxic and hazardous chemicals. These discharges can be sudden, as in a train or tanker truck accident, or as a result of slow leakage from a holding container. If the contaminant dissolves in the water, it will flow along with the groundwater. If less dense than water, it will float on the water table (Figure 23.19).

The process of cleaning up groundwater contamination can take many years and be very expensive. First of all the source must be cleaned up, removed, or isolated. Then begins the restoration of the aquifer, ideally to its precontaminated state.

Figure 23.19
A contaminant plume forms and spreads in the direction of flowing groundwater. In the case of a petroleum product, the specific gravity is less than water and the contaminant floats on the water table. The concentration of contaminant decreases from the source.

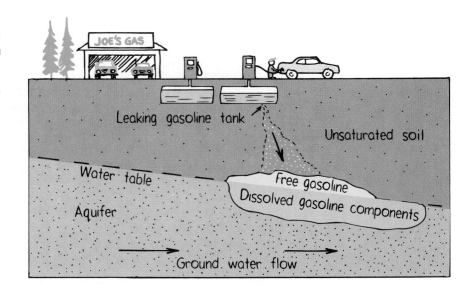

23.5 SURFACE WATER

The Grand Canyon is testimony to the mighty powers of the Colorado River. For millions of years the Colorado River has been carving out the canyon walls, cutting deeper and deeper into the rock as it makes its way to the ocean. Rivers and streams are dynamic systems that impact both the surface of the land and the people who live on that land. Rivers carve out and alter the landscape. Rivers also provide energy, irrigation, and a means of transportation for people. What better place to set up a home than a fertile river valley? The force of a river is indeed a powerful instrument of erosion as it leaves its mark on the land surface. Even where rivers are no longer flowing, their impact remains.

Stream Flow and Erosion

As rain falls on land, it begins a complex journey back to the oceans. Some water percolates into the ground, some evaporates back into the atmosphere, and some runs off into streams. Streams come in a variety of forms; they can be straight or bent, fast or slow-flowing. When water moves erratically downstream, stirring everything it comes in contact with, the flow is **turbulent**. When water flows steadily downstream with no mixing of sediment the flow is **laminar** (Figure 23.20). Stream flow depends on the nature of the stream bed and the water velocity.

Stream velocity depends on the *gradient* of the stream, and the dimensions and shape of the channel. Like the hydraulic gradient we discussed earlier, the **gradient** of a stream is the ratio of vertical drop to the horizontal distance for that drop. The greater the gradient, the greater the velocity. Most often, as a river moves downstream the gradient gradually decreases and the stream velocity slows. Friction between water and its contact with the channel also slows water flow. The greater the contact area, the greater the friction. Channel dimensions and the shape of a stream determine the contact area between water and the channel. If the stream channel is

Figure 23.20 Stream forms develop according to whether stream flow is laminar or turbulent. Laminar flow is slow and steady with no mixing of sediment in the channel. Turbulent flow is fast and jumbled, stirring up everything in the flow.

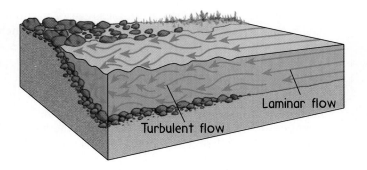

rounded and deep, as opposed to flat bottomed and relatively shallow, the stream velocity will be faster due to less channel contact. In a large river with a straight channel the maximum flow velocity is found mid-channel at the surface. In a river with bends and loops, the maximum flow velocity shifts toward the outside of each bend and is slightly below the surface (Figure 23.21). The amount of water that passes a given point in a channel for a given time is the *discharge.** The discharge of a stream depends on the cross sectional area and the average velocity.

Movement of water erodes stream channels in several different ways. Much like groundwater, river water contains many dissolved substances that chemically erode the rocks they encounter. In fact, the high content of dissolved substances in many streams is often from groundwater sources. (The high content of asbestos in much of California's drinking water, for example, originates in natural

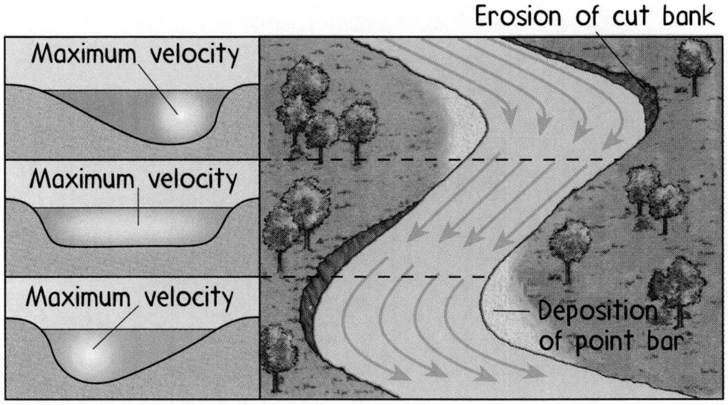

Figure 23.21 In a straight-channel stream, the maximum velocity of stream flow is mid-channel and close to or near the surface. In a stream that bends, maximum flow velocity is toward the outside of each bend and slightly below the surface. Erosion of the stream channel occurs where stream velocity is greatest; deposition occurs where stream flow slows.

*The discharge of a stream can be determined by multiplying the channel's dimensions by the stream's velocity:

$$Q = w \times d \times v$$

Discharge	width	depth	velocity
(m^3/s)	(m)	(m)	(m/s)

groundwater.)* Another form of erosion is caused by turbulent flow, where sediments and particles physically abrade and scour a channel, much like sandpaper on wood. When powered by turbulent eddies, rock particles rotate like drill bits as they carve out deep potholes (Figure 23.22). The stronger the current, the greater the turbulence, and the greater the erosion.

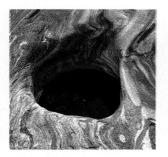

Figure 23.22 When powered by turbulent eddies, rock particles rotate like drill bits and carve out deep potholes.

Sediment Transport So streams carry more than just water—they transport great quantities of sediment. A turbulent current gathers and moves particles downstream mainly by lifting them into the flow or by rolling and sliding them along the channel bottom. The smaller, finer particles are easily lifted into the flow and remain suspended to make the water murky or muddy. Larger particle grains roll and slide along the bottom of the channel, occasionally bumping into the stream flow. We say the amount of sediment a stream can carry is its *capacity,* and the size of the sediment transported is its *competence.* Capacity and competence both depend on the nature of the particles and the stream discharge.

Streams are constantly changing shape. Upstream at the headwaters, the gradient is high, channels are narrow and shallow, and velocities are rapid. These sections are often called "rapids." As streams progress downslope, their channels widen as gradients and velocities decrease (Figure 23.23). Competence also decreases and only smaller particles are carried. By the time the flow reaches the sea, only the fine alluvium and sand particles are left. Thus form the alluvial fans and delta environments discussed in Chapter 22.

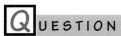UESTION

Which is more effective in transporting sediment, laminar or turbulent flow? Why?

Figure 23.23 At a stream's headwaters the gradient is high, the channels are narrow and shallow, and the stream flow is rapid. As the stream progresses downslope, the channel widens as gradients and velocities decrease.

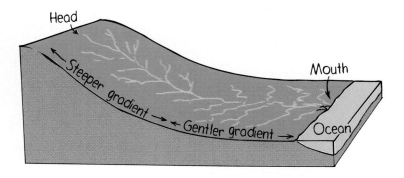

ANSWER

Turbulent flow is more effective in transporting sediment. This is because in turbulent flow the water motion is irregular and sediments have a greater tendency to remain in suspension. In laminar flow, on the other hand, water moves steadily in a straight line path with no mixing of sediment in the channel.

---

*Recall from Chapter 22 that not all asbestos minerals are hazardous to health. The asbestos found in the California water supply comes from the mineral chrysotile—a less harmful form of asbestos.

Stream Valleys

The power of erosion enables a stream to widen and deepen its channel, transport sediment, and, in time, create a stream valley. Other sources contribute to the creation of a stream valley. Rainfall loosens soil, and its pores become filled with water. With the impact of more rain, gullies form, and soil particles funnel into a stream. The erosive action of a stream cuts down into the underlying rock to ultimately form a V-shaped valley. Fast moving rapids and beautiful waterfalls are prominent early features of a stream valley (Figure 23.24).

When a stream has cut its channel to base level it can no longer cut down so it adjusts its profile and modifies its channel. Because of the reduced gradient, the stream becomes sinuous, winding side to side and widening the valley floor. Farther downstream, the stream becomes more sinuous as it meanders back and forth further widening the valley to form a floodplain (Figure 23.25).

Figure 23.24 Fast moving rapids and beautiful waterfalls are prominent features of an early stream valley.

Figure 23.25 The development of a flood plain.

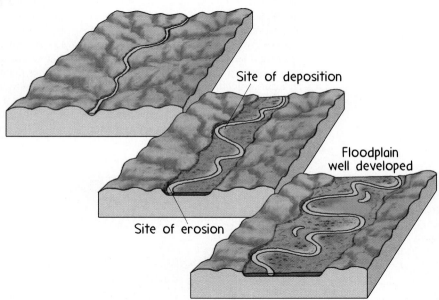

Site of deposition

Floodplain well developed

Site of erosion

Drainage Networks

A stream is a small segment of a much larger system—a drainage **basin,** the total area that contributes water to a stream, and a **divide,** the line that separates adjacent basins. A drainage basin can cover vast areas or be as small as one square kilometer. Since a divide separates basins, it can be very long as it separates two enormous drainage basins or it can be a ridge separating two small gullies. The **Continental Divide,** a continuous line running north to south down the length of North America, separates the Pacific basin on the west from the Atlantic basin on the east. Water west of the line eventually flows to the Pacific Ocean, and water east of the line flows to the Atlantic Ocean. Drainage networks are made up of interconnected streams, which as a unit, form different drainage patterns (Figure 23.27).

Streams and their drainage patterns can give a geologist useful information for interpreting former land conditions. Streams begin their journey from the high-

Stages of Valley Development

Stream erosion carves a landscape in stages. The evolution of a valley progresses from the high rugged mountains of youth, to the rounded hills of maturity, and finally to the worn down plains of old age (Figure 23.26).

A youthful valley is characterized by a downcutting stream in a rather straight course, fast-moving rapids, waterfalls, and a narrow V-shaped valley. The gradient of both the stream and valley walls are steep. Examples of youthful valleys are found in the Sierra Nevada Mountain Range.

A mature valley is characterized by a meandering stream that erodes both downward and laterally, deepening and widening its valley to its maximum extent as it forms a floodplain. The gradient of the stream is moderate, and rapids are inconspicuous. The valley walls have a moderate slope. Examples of mature valleys are found in the Appalachian Mountains.

An old-age valley is characteristically a very wide valley. The work of several floodplains increases the valley's width to the point where the stream is far from the valley walls. Channel cutting has run its course and erosion only reworks the floodplain deposits. The stream gradient is low and the valley walls are gentle slopes. An example of an old age valley is the Mississippi River Valley.

The stages of valley development depend on the ability of the stream to erode, the surface material through which the stream cuts, and the height above *base level,* the lowest elevation to which a stream can erode its channel. A stream that begins its journey close to base level reaches maturity faster than a stream that begins at a higher level. So we find that youth, maturity, and old age do not follow in a chronological sequence; rather, they follow the sequence of erosion.

Figure 23.26 The evolution of a valley progresses from (a) the high rugged mountains of youth, to (b) the rounded hills of maturity, and finally to (c) the worn-down plains of old age.

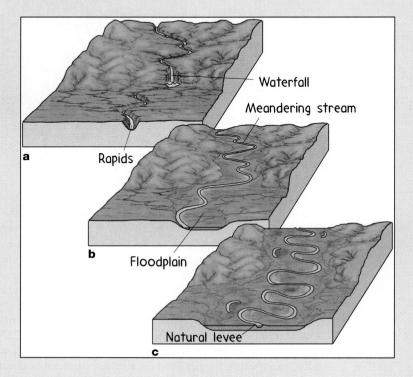

Figure 23.27 Different drainage patterns develop according to surface material and surface structure: (a) dentritic, (b) radial, (c) rectangular, and (d) trellis.

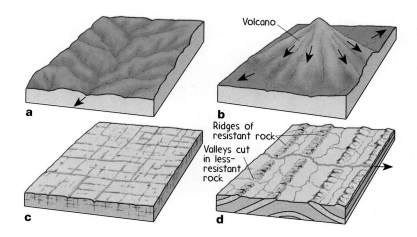

lands and work their way to the oceans. Each step of the journey brings changes that are recorded in the stream channel and its surroundings. Patterns change, channels change, and sediments change. From the patterns and changes, a stream's energy history is recorded.

QUESTION

What is the ultimate destination of all water flow and hence the eventual site of deposition of most sediments?

23.6 GLACIERS AND GLACIATION

Figure 23.28 Geologist Bob Abrams observes the grandeur of the Juneau Ice Field, Alaska.

The mightiest rivers on earth are frozen solid and normally flow a sluggish few centimeters per day. These great icy currents are **glaciers**, among the most spectacular and powerful agents of erosion. Glaciation has given us the beautiful landscapes of Tibet, Nepal, and Bhutan in Asia, the Alps of Switzerland, the fjords of Norway, Yosemite Valley, and the Great Lakes in North America. Glaciation is still at work in many regions of the world, as small alpine glaciers in mountainous areas, as large alpine ice fields, and as huge Arctic and Antarctic continental ice sheets.

Glacier Formation and Movement

The ice of a glacier is formed from recrystallized snow. After snowflakes fall, their accumulation slowly changes the individual flakes to rounded lumps of frozen icy material. As more snow falls, the pressure on the bottom layers of icy snow compacts and recrystallizes it into glacial ice. The ice does not become a glacier until it moves

ANSWER

Water flows eventually to the ocean, and sediments to the ocean floor.

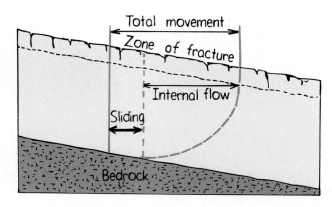

Figure 23.29 Cross section of a glacier. Glacial movement is slowest at the base due to frictional drag and greatest at the center due to pressure. Movement of ice in the center carries surface ice along in a piggyback fashion.

Figure 23.30 Geological features are best viewed from an airplane. Next time you are flying in an airplane request a window seat and enjoy the geology below.

under its own weight. This occurs when the glacier reaches a thickness of some 50 meters. Once this "critical thickness" is met the glacier begins its flow. Sliding is enhanced by melted ice—*meltwater*—at the base, which further contributes to the glacier's movement.* Pressure causes the glacier to slide over itself, without cracking or breaking.

A glacier moves very slowly. From top to bottom, the glacier moves differentially according to the amount of pressure. At the base, frictional drag slows movement; in the center, movement is greater. At the top, the surface of ice behaves as a rigid brittle mass that breaks and cracks as it is carried along "piggyback" style on the ice below (Figure 23.29). Huge gaping cracks called *crevasses* develop in the surface ice, which extend to great depths and can therefore be quite dangerous.

Glacier velocity is measured by placing a line of markers across the ice and recording their changes in position over a period of years. Ice is found to move fastest in the center and slower toward the edges due to frictional drag (Figure 23.31). The average velocities vary from glacier to glacier and can range from a few centimeters to a few meters per day. Sudden surges of rapid movement occur in the otherwise steady flow. These surges are likely caused by periodic melting of the base and sud-

Figure 23.31 Top view of a glacier. Movement is fastest at the center and gradually decreases along the edges due to valley-wall friction.

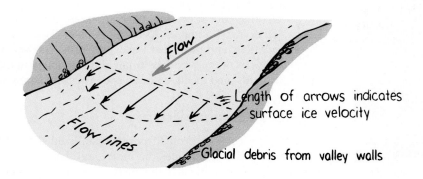

*The cause of the meltwater may result from the pressure of the overlying ice, the internal heat flow of the earth, or the generation of heat from frictional drag as the glacier moves. Whatever the reason, meltwater contributes to movement of the glacier.

a

b

Figure 23.32 Glacial flows: (a) normal and (b) surge flow.

den redistribution of mass. The flow rate in these relatively brief surges can be 100 times faster than the normal rate. Viewed from the air, flow bands of rock debris and ice normally have a parallel pattern, but during a surge the flow bands become intricately folded (Figure 23.32).

From season to season, and over long periods of time, the mass of a glacier changes. Glaciers grow as snow falls and accumulates on their surfaces. During warm periods, glaciers shrink as ice melts and evaporates. So the front of a glacier may advance down slope, or retreat, or when growing and shrinking processes balance each other, remain stationary. Interestingly enough, in all cases, the glacier is always flowing forward.

QUESTION

Under what conditions does a glacier remain stationary?

Glacial Erosion

Glaciers are powerful agents of erosion. They have dramatically carved out the land and created beautiful valleys—California's Yosemite Valley is considered one of the world's outstanding examples. As the glacier moves across the surface, it loosens and lifts up blocks of rock incorporating them into the ice. Every particle, no matter how big or small, is picked up and carried in the glacier's load. The large rock fragments carried at the bottom of a glacier's load scrape the underlying bedrock and leave long parallel scratches aligned in the direction of ice flow (Figure 23.33).* These are called *striations*. Millions of years ago, the Arctic glacier moved over the North American continent as far south as the Ohio River, and then, withdrawing, left huge boulders and visible striations as it retreated.

Figure 23.33 Striations mark the presence of a former glacier.

ANSWER

The front of a glacier remains stationary when the rate of growth equals the rate of shrinking, but interestingly enough, the glacier itself is always flowing downhill—like water, only much slower.

---

*Because glacial striations are aligned in the direction of flow, they are often used to reconstruct ancient glaciation periods. As we will learn in Chapter 24, they were instrumental in establishing the theory of Continental Drift.

Figure 23.34   The power of glaciation carved out this characteristic U-shaped valley.

Glacial Landforms

The two main types of glaciers, *alpine glaciers* and *continental glaciers,* have different erosional effects and different features. Alpine glaciers develop in previously formed stream valleys and are characterized by sharp, angular features. As the glacier carves out the existing features, it further accentuates them. A V-shaped valley is transformed by the glacier as it deepens, widens, and straightens the valley floor into a characteristic U-shape (Figure 23.34). The focal point of glacial erosion is most active near the head of the glacier, as it cuts deep into the rock, to form a steep-sided bowl-shaped depression.* After the glacier has melted away, meltwater often fills this depression and forms a small lake. When two such depressions work to carve out rock on opposite sides of a divide, the cutting action steepens the divide to form a jagged, knife-sharp, linear ridge. When the depressions are grouped together on a mountain top, they converge into one another to steepen the peak creating a *horn* (Figure 23.35). Down below, the main glacial valley contains

Figure 23.35   The Matterhorn—named for its characteristic "horn" feature.

---

*Geologists call the steep bowl-shaped depressions *cirques;* and when filled with meltwater, *tarns;* and the jagged knife-sharp ridges they make, *aretes.*

Figure 23.36 Hanging valleys are a spectacular feature in alpine glaciers as illustrated by Bridal Falls in Yosemite National Park.

a much greater amount of ice than does a tributary-stream valley; it is therefore eroded deeper than a tributary valley. The tributary valley stands suspended above the valley to which it enters and is thus called a *hanging valley.* Beautiful waterfalls are associated with hanging valleys (Figure 23.36). The erosional features of alpine glaciation are depicted in Figure 23.37.

Continental glaciers erode the land surface in a way quite similar to alpine glaciers, but with less dramatic effects. Continental glaciers are not confined to valleys, but spread over all the land surface, smoothing and rounding the underlying topography. The direction of ice flow can often be deciphered by small asymmetrical hills.* In the direction of ice flow the hill's slope is smooth and gentle; on the downflow side it is rough and steep (Figure 23.38). Continental glaciation, as well as alpine glaciation, creates great amounts of erosional debris. These deposits form their own type of landforms.

Glacial Deposits

As a glacier moves and advances across the land it acquires and transports great quantities of debris. When the glacier retreats or is melted, this debris is deposited. Because a glacier abrades and picks up everything in its path, glacial deposits are characteristically composed of unsorted mixtures of rock fragments in a variety of shapes and sizes.† Many fragments are striated and polished by the pulverizing action of the ice. Often huge boulders of unknown origin are mixed in with the debris. These *glacial erratics* differ from the underlying rock and provide proof of a glacier's ability to carry and transport heavy loads for great distances. When the origin of the erratic is found the distance of glacial flow can be estimated.

The most common landform created of glacial debris is the *moraine,* a ridge-shaped landform that marks the boundaries of ice flow. The farthest point of a glacier's advance is marked by deposition in a tongue-shaped *terminal* moraine. Another landform created by glacial debris is a *drumlin*—a streamlined hill that

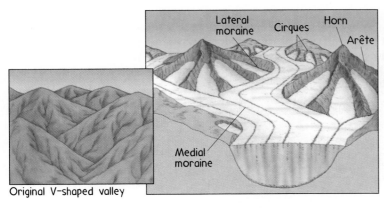
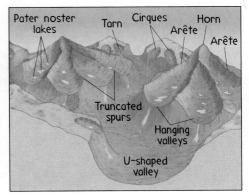

Figure 23.37 The many erosional features of alpine glaciation.

---

*The geologic term for this type of asymmetrical hill is *roches moutonnées.*

†Debris deposited directly by the glacier is called *till,* and debris deposited by glacial melt water *outwash.* The general term for both till and outwash is *drift.*

parallels the direction of ice movement. The steep, blunt side of the hill faces the direction of ice movement, while the longer, more gentle slope points in the direction of ice movement. Retreating glaciers leave a myriad of large dish-shaped hollows that become what are called *kettle lakes*. These make up the "10,000 lakes" of Minnesota, and the five Finger Lakes in upstate New York. These features of glacial deposits, and others, are illustrated in Figure 23.39.

So in summary, we see the cycle of water greatly affects the surface of our land. Water is precious; it helps to sustain the life of our planet and hence our lives. Water works its way into the ground where it carves out beautiful caves and caverns. It enters our streams and gives the land beautiful canyons and scenic stream valleys. The frozen rivers of ice, glaciers, carve out and resculpt the land, to further enhance the grandeur of former river valleys. Water is a powerful geologic force.

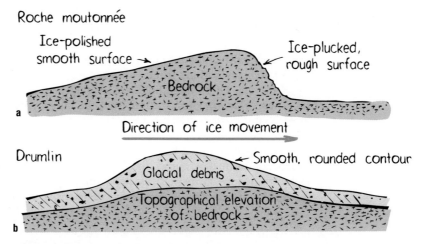

Figure 23.38 Continental glaciers leave in their path small asymmetrical hills that show the direction of glacial movement. (a) One type (*rouche moutonée*) has a smooth gentle slope in the direction of flow, and a steep opposite downslope. (b) Another type (*drumlin*) has a steep, blunt side facing the direction of ice movement, with a longer, more gentle slope along the direction of ice movement.

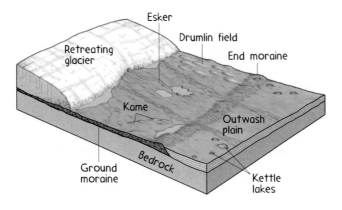

Figure 23.39 Some of the many features of glacial deposits.

SUMMARY OF TERMS

Hydrologic cycle The natural circulation of water from the oceans to the air, to the ground, then to the oceans and then back to the atmosphere.

Porosity The ratio of open space in a rock or in sediment to the total volume.

Permeability The ability of a material to transmit fluid.

Water table The upper boundary of the zone of saturation below which every pore space is completely filled with water.

Groundwater Subsurface water that is in the zone of saturation.

Aquifer A rock body (or sediment body) through which groundwater easily moves.

Artesian system A system in which groundwater is under great enough pressure to rise above the level of an aquifer. The confined groundwater will flow out of the ground at any opening that taps the aquifer. If the opening is natural, it is an artesian spring; if the opening has been drilled, it is an artesian well.

Residence time The *average* time that a water molecule spends in a region.

Subsidence Sinking or downward settling of the earth's surface due to compaction, the natural withdrawal of fluid, or human withdrawal of subsurface material by pumping of groundwater or oil.

Karst Topography formed by the dissolution of carbonate rock. Karst topography is characterized by sinkholes, caves, and underground drainage.

Leachate A solution formed by water that has percolated through soil containing soluble substances.

Laminar Steady flow of water in a straight line path with no mixing of sediment.

Turbulent Erratic flow of water that moves in a jumbled manner stirring up everything with which it comes in contact.

Gradient The ratio of vertical drop compared to horizontal distance.

Divide The line that separates adjacent basins.

Drainage basin The total area that contributes water to a stream.

Continental Divide The continuous line running north to south down the length of North America separating the Pacific basin on the west from the Atlantic basin on the east. Water west of the line flows to the Pacific Ocean, and water east of the line flows to the Atlantic Ocean.

Striations Long parallel scratches produced in a rock by geological forces such as glaciers, stream flow, or faults. In the case of glaciers, striations are aligned in the direction of ice flow.

Glacier A large mass of ice formed by the compaction and recrystallization of snow. The ice does not become a glacier until it is able to move downslope under its own weight.

REVIEW QUESTIONS

Groundwater

1. Distinguish between *porosity* and *permeability.*

2. Name two rock types that can have a high porosity but low permeability.

The Water Table

3. An ordinary table is usually flat. But the water table is generally not flat. Why?

4. Compare and contrast the zones of aeration and saturation.

5. What types of soils allow for the greatest filtration of rainfall?

6. How does an aquiclude differ from an aquifer?

7. Can an aquifer be composed of igneous rocks? Explain.

8. What is an artesian system and how is it formed?

9. What factors affect the rate of groundwater movement?

Mining Groundwater

10. What are the consequences of the practice of overpumping groundwater?

11. What two factors contribute to the ability of an aquifer to recharge?

12. What is meant by the statement "In parts of Texas the groundwater is now literally 'mined'"?

Groundwater and Topography

13. What are the effects of groundwater on carbonate rocks?

14. Carbonate rocks are formed in marine environments. Why do we find carbonate deposits on continental land?

15. What is karst topography, and where is it found in the landscape?

16. How does a stalactite form? How does a stalagmite form?

The Quality of Water

17. Which aquifer would be most effective in purifying contaminated groundwater: course gravel, sand, or cavernous limestone? Defend your answer.

18. How does rainwater become acidic? How does this affect groundwater?

Water Supply Contamination

19. Does groundwater affect surface water? If so, why? If not, how come?

20. List three ways our water supply is being contaminated.

21. High nitrate levels in groundwater can come from what source?

22. How does leachate form?

Surface Water

23. Rivers are important to human culture. Why?

Stream Flow and Erosion

24. What is meant by stream gradient and how does it affect stream velocity?

25. What is the greater transporter of sediment, a laminar flow or a turbulent flow? Why?

26. What are the consequences when (a) the discharge of a stream increases? (b) the velocity of a stream increases?

27. How does the shape of a stream channel affect flow?

28. Name three ways the movement of water erodes the stream channel. Which is responsible for creating potholes?

29. Distinguish between capacity and competence.

30. Relate competence to the deposition of an alluvial fan and a delta.

31. In what ways can a stream transport the debris of erosion?

Stream Valleys

32. What factors are responsible for the formation of a stream valley?

33. Under what conditions do sinuous, meandering rivers form along a floodplain?

34. What type of stream flow do we generally find in high mountainous regions?

Drainage Networks

35. What is a continental divide?

36. What is the significance of the Continental Divide in North America with respect to water flow to the Atlantic and Pacific Oceans?

Glaciers and Glaciation

37. Glaciers are one of the most powerful agents of erosion. What well known landscapes have been carved by these great icy currents?

Glacier Formation and Movement

38. What are the conditions for the formation of a glacier?

39. What distinguishes a huge block of ice from a glacier?

40. What are the main features of glacial flow?

41. Does all the ice in a glacier move at the same speed? Explain.

42. Why do crevasses form on the surface of glaciers?

43. Under what conditions does a glacier front advance?

44. Under what conditions does a glacier front retreat?

45. Under what conditions does a glacier front remain stationary?

46. What is a glacial surge?

Glacial Erosion

47. What are striations? What is their significance?

48. How do glacially deposited rocks differ from river deposited rocks?

49. What erosional features might you find in an area of alpine glaciation?

50. What features might you find as a glacial deposit?

• •

HOME PROJECT

Water is considered to be "hard" due to dissolved minerals such as calcium and magnesium bicarbonate. To test the hardness of the water in your area, collect four water samples from local sources—a nearby pond or well, a stream, your kitchen faucet, and bottled distilled water. Label your sample bottles accordingly. Add a drop of liquid soap to each water sample and shake. The bottle with the most suds should be the softest water. Record your observations.

• •

EXERCISES

1. The oceans are salt water, yet evaporation over the ocean surface produces clouds that precipitate fresh water. Why no salt?

2. In a confined aquifer the water in a well rises above the top of the aquifer. What is this system called? How high does the water level rise in an unconfined aquifer?

3. How does the practice of drawing drinking water from a river, then returning sewage to the same river affect the local hydrologic cycle?

4. Why is pollution of groundwater a greater hazard than pollution of surface waters?

5. Some heavy metals can be extremely dangerous to water supplies. Aluminum has been linked to Alzheimer's and Parkinson's diseases, cadmium is known to cause liver damage, and lead affects the circulatory, reproductive, nervous, and kidney systems. In what ways can these heavy metals get into our water supply?

6. When a water supply becomes overly rich in nitrogen and phosphorus, plant life thrives to the point of destruction. The overgrowth of algae causes unsightly scum, unpleasant odors, and robs the water supply of dissolved oxygen. What effect may this have on other aquatic life? What are the sources of this type of pollution?

7. What factors determine how long a well will produce water?

8. By what means could one predict the discharge of a stream after a rainstorm?

9. In the San Joaquin Valley of California the withdrawal of groundwater for irrigation purposes caused the water table to drop by 75 meters in a 20-year period. Since that time, the aquifer has been recharged and water for irrigation has been provided by canals that get water from the Sierra Nevada Mountains. Do you think the problem of water is now solved in this area? How about other areas? Defend your answers.

10. Removal of groundwater causes subsidence. If removal of groundwater is stopped, do you think the land will rise again to its original level? Defend your answer.

11. What is a sinkhole? What factors contribute to its formation?

12. As a population increases so does the amount of garbage produced by that population. In many areas the way to deal with increasing wastes is by burial in a landfill or underground storage facilities. What factors must be considered in the planning and building of such sites?

13. How can a youthful valley be older (in years) than a mature valley?

14. The Mississippi River is large and extensive. Discuss the river's capacity and related competence.

15. Recall from Chapter 22 that the Mississippi Delta has moved south from near Cairo, Illinois, to its present location in Louisiana. Other than the length of time the river has been in existence, why has the delta moved so far?

16. Which of the three agents of transportation—wind, water, or ice—is able to transport the largest boulders; which is most limited as to the largest size it can transport? Defend your answers.

17. In regard to residence time, how may the process of cleaning up groundwater contamination differ from cleaning up surface contamination in a lake?

18. What effect does a dam have on the water table in the vicinity of the dam?

19. Once a dam is constructed and filled, what effect does increased evaporation of water behind the dam have on the ratio of water flow into the dam and downstream from the dam? What effect does the dam have on erosional activity downstream?

20. What effect does the accumulation of sediments behind a dam have on its capacity for storing water?

21. Does groundwater flow into streams or does stream water flow into the ground? Explain.

22. In what way does a glaciated mountain valley differ from a non-glaciated mountain valley?

23. The earth's periods of glaciation have had a major impact on the surface features of our planet. What places, other than the examples cited in this chapter, owe their striking features to glaciers?

24. How does "frictional drag" play a role in the external movement of a glacier? How about the internal movement?

24

THE EARTH'S INTERNAL PROPERTIES

When you were a child, did you ever think about digging a hole to China? The idea is intriguing, for why go all the way around the earth if you could take a shortcut straight through? Digging such a hole, unfortunately, is not a very realistic possibility. But if it were possible to dig such a hole, it might be more valuable as a source of geological information than as means of travel—at least to a geologist type.

Much of our knowledge of the earth's interior comes from the study of waves that travel through the earth. These are *seismic* waves—the waves generated by earthquakes and underground nuclear explosions. Recall from Chapter 11 that a wave's velocity depends on the medium through which it travels. We know that the sound waves generated by two rocks clicking together will travel faster through water than through air, and even faster through a solid. Like sound waves, the speed of seismic waves depends on the material they're travelling through. So for clues about the composition of the earth we measure the speeds of seismic waves.

The energy generated in the earth's interior during an earthquake or an underground nuclear explosion radiates in all directions. The energy travels in the

Figure 24.1 Model of a seismograph. When the earth moves, the support unit attached to the ground also moves, but because of inertia, the mass at the end of the pendulum tends to stay in place. A pen on the mass marks the relative displacement on the slowly rotating drum beneath. In this way the magnitude of ground movement is recorded.

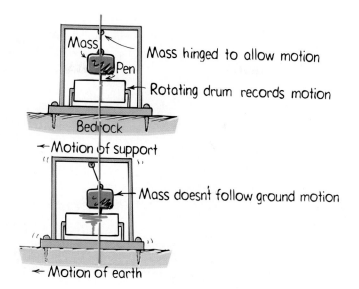

form of seismic waves to the earth's surface. Evidence of the waves is gathered on a seismograph (Figure 24.1). The seismograph output, when carefully analyzed, provides a map of the earth's interior.

24.1 SEISMIC WAVES

Before we examine the earth's composition and structure we should first understand seismic-wave propagation through the layers of the earth. There are two types of seismic waves: **body waves** that travel through the earth's interior, and **surface waves** that travel on the earth's surface (Figure 24.2). Body waves are further classed as either **primary waves** (P-waves) or **secondary waves** (S-waves).

Primary waves, like sound waves, are longitudinal—they compress and expand rock or other media as they move through it. Like vibrations in a bell, a primary wave moves out in all directions from its source. P-waves are the fastest of all seismic waves (6.4 kilometers per second, 4.0 miles per second) and so they are the

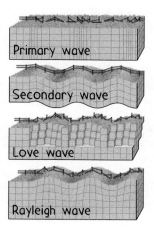

Figure 24.2 Block diagrams show the effects of the four different types of seismic waves. The yellow portion on the side of each diagram represents undisturbed crust. (a) Primary waves alternately compress and expand the earth's crust similar to the action of a spring. (b) Secondary waves cause the crust to oscillate up and down and side to side. (c) Love waves whip back and forth in horizontal motion. (d) Rayleigh waves operate much like secondary waves, but affect only the surface of the earth.

Figure 24.3 Cross section of the earth's internal layers.

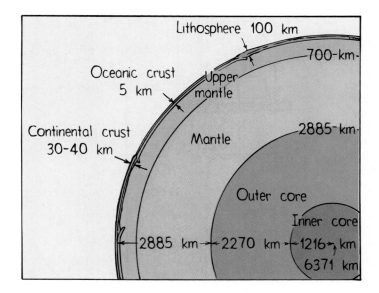

first to register on a seismograph. Since solids and fluids respond to compression, P-waves travel through any type of material—from solid granite to lava to water to air.

Secondary waves, like a vibrating violin string, are transverse—they vibrate the particles of their medium up and down and side to side. S-waves travel more slowly than P-waves, and are the second to register on a seismograph. Fluids can't support such motion, so they do not transmit S-waves. S-waves can only travel through solids.

When body waves reach the earth's surface, they transform into surface waves. Surface waves come in two varieties: **Rayleigh waves** that move in an up-and-down motion, and **Love waves** that move in a side-to-side whiplike motion. Both waves travel at slower velocities than P- and S-waves, and are the last to register on a seismograph.

Seismic waves are reflected by surfaces within the earth's interior and are refracted where wave velocity changes. Thus geologists study the reflections and refractions, as well as the velocities of the various seismic waves in their detective work. Research with seismic waves shows that the earth's interior consists of three pronounced boundaries separating four fundamental zones. The four fundamental zones of the earth's structure are the crust, the mantle, the outer core, and the inner core (Figure 24.3).

24.2 EARTH'S INTERNAL LAYERS

In 1909 the Croatian seismologist Andrija Mohorovičić presented the first convincing evidence that the earth's "innards" are layered. Studying the seismograms from a recent earthquake, he discovered that seismic waves suddenly picked up speed at a certain depth below the surface. Since the speeds of these waves depend on the density of the material they pass through, Mohorovičić concluded that the speed increase he observed was due to variation in the density of the earth. The data from the seismograms had literally drawn a map of the upper boundary of the earth's mantle, a layer of dense rock underlying the lighter crust. This boundary, known as the **Mohorovičić discontinuity** ("Moho"), separates the earth's crust from the rocks of different composition in the mantle below.

Two years after the discovery of the Moho another boundary was found. This time, the boundary between the mantle and the core was detected. Both P- and S-waves are strongly influenced by a pronounced boundary 2900 kilometers (1800 miles) deep. When P-waves reach that boundary, they are reflected and refracted so strongly that the boundary actually casts a P-wave shadow over part of the earth. This wave shadow develops 105° to 140° of arc from the origin of an earthquake, and allows no direct penetration of seismic waves. Because the boundary is so pronounced, we infer that it is the place where comparatively light silicate material of the mantle meets the dense metallic iron of the core. This boundary casts an even more pronounced S-wave shadow because transverse waves cannot travel through liquid. Therefore, the outer core is liquid (Figure 24.4).

In 1936, the discovery that seismic waves are reflected from a boundary within the core showed there was another layer. The waves passing through the inner portion of the core had greater velocity than those in the outer core. Therefore the inner core had to be solid. Question: Do you suppose these layers in the earth's interior influence the changes our planet experiences? The answer is *yes*, as you will now see.

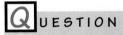

QUESTION

What evidence supports the concept that the inner core is solid and the outer core liquid?

The Core

The **core** has two regions—the inner core and the outer core. The regions differ in their phase. The inner and outer core are both composed mostly of iron and nickel.* In the inner core, the iron and nickel are solid. Although it is very hot, intense pressure from the weight of the rest of the earth prevents the material of the inner core from melting (much as a pressure cooker prevents high-temperature water from boiling).

Surrounding the solid inner core is the liquid outer core. It generally has the same chemical composition as the inner core. But since less weight is exerted on it,

ANSWER

The evidence is in the differences between P- and S-wave propagation through the earth's interior. As these waves encounter the boundary at 2900 km, a very pronounced wave shadow develops. P-waves are refracted at the boundary, and S-waves are reflected. S-waves cannot travel through liquids, implying a liquid outer core. As P-waves continue to propagate through the earth's interior the wave velocity increases. Since waves travel faster in solids, we infer a solid inner core.

---

*The density of rocks at the earth's surface is typically 2.7 to 3.0 g/cm^3, while the average density of the earth as a whole is 5.5 g/cm^3. Thus the rocks at the earth's surface are not representative of the planet's interior. To account for the earth's high average density, the density of the core must be at least 10 g/cm^3. This and other reasons point to iron and nickel, the most abundant of the heavier elements. Many meteorites, assumed to be representative of the material from which the earth originally accreted, are also composed of iron and nickel.

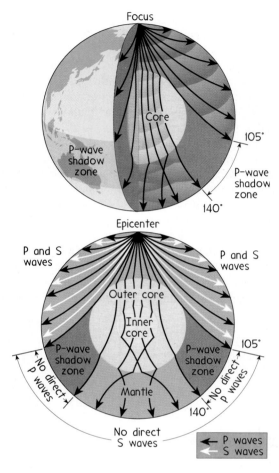

Figure 24.4 Cut-away and cross-sectional diagrams showing the change in wave paths in the major internal boundaries and the P-wave shadow. The P-wave shadow between 105° and 140° from an earthquake's epicenter is caused by the refraction of the P-waves at the core-mantle boundary. Note that any location that is more than 105° from an earthquake epicenter will not receive S-waves because the liquid outer core does not transmit S-waves.

the pressure on it is less and it is liquid. The outer molten core flows at the slow rate of several kilometers per year. This flow is evident far outside the earth's surface as it generates the electric current which powers the earth's magnetic field.* This magnetic field is not stable, but has changed throughout geologic time. Recall from Chapter 10 that there have been times when the earth's magnetic field has diminished to zero, only to build up again with the poles reversed. These magnetic pole reversals probably result from changes in the direction of outer core flow.†

---

*The ultimate energy that drives such an electric current is thermal energy. Several theories exist to explain the source of thermal energy and the mechanism that converts it to fluid motion. One theory suggests that the heat is from the radioactive decay of elements such as potassium, thorium, and uranium that reside in the earth's interior. This heat, in turn, causes thermal convection in the outer core. An alternate theory states that as dense iron-nickel crystals from the molten outer core sink toward the solid inner core they produce a convective motion in the fluid of the outer core. Regardless of how the convective flow is generated, the convection currents combined with the rotational effects of the earth produce the earth's magnetic field.

†When rocks exhibit the same magnetism (preserved by crystals from time of rock's origin) as the present magnetic field, they are said to possess normal polarity. Rocks exhibiting the opposite magnetism are said to have reverse polarity. A time table of normal and reverse polarity reveals that 171 reversals have occurred in the past 76 million years.

The Dynamic Mantle

Surrounding the core of the planet is the mantle. The **mantle** is a rocky layer, some 3000 kilometers thick. Composed of hot, iron-rich silicate rocks, the mantle behaves like plastic. We say the mantle is plasticlike because it acts in a semi-fluid manner. The upper mantle, called the **asthenosphere,** is especially plasticlike. Thermal convection currents in the asthenosphere contribute to its gradual flow. The constant flowing movements in the asthenosphere greatly affect the surface features of our planet. The **lithosphere,** above the asthenosphere, includes the entire crust and the uppermost portion of the mantle. This layer is relatively rigid and brittle, with a resistance to deformation. Hence the brittle lithosphere is broken into many individual pieces called *plates*. The lithosphere is, in a sense, floating on top of the asthenosphere.

The lithospheric plates are continually in motion. They float on the circulating asthenosphere. Although asthenosphere currents move at a leisurely pace—taking hundreds of millions of years to complete one loop—they are powerful enough to move continents and reshape many of our surface features. The movement of the lithospheric plates cause earthquakes, volcanic activity, and the deformation of large masses of rock which create mountains.

The Crustal Surface

The outermost layer of our planet, the layer on which we live, is the **crust.** Its density, composition, and thickness vary markedly from the deep ocean basins to the lofty continental plateaus. The crust of the ocean basins is compact. It's only about 10 kilometers (6 miles) thick and is composed of dense basaltic rocks. The part of the crust we know as continents is between 20 to 60 kilometers (12 to 40 miles) thick. Continental masses are composed of granitic rocks which are less dense than basaltic rock. The lower density of the continental crust makes it buoyant, and allows it to ride high on the lithosphere. We can compare the mountainous regions of the continental crust to a floating iceberg—both have most of their mass below the surface (Figure 24.5). The greater thickness of the less dense continental crust is compensated for by the greater density of the thinner oceanic crust (Figure 24.6). This state of equilibrium with the rock masses in balance is known as **isostasy.**

Figure 24.5   Because the density of ice is only slightly less than that of seawater, a floating iceberg has most of its mass below the water line.

Figure 24.6 Like a floating iceberg, continental crust is thicker beneath mountainous regions. Mountains are topographically high because the thick continental crust is light and buoyant, and ocean basins are topographically low because the thin ocean crust is very dense.

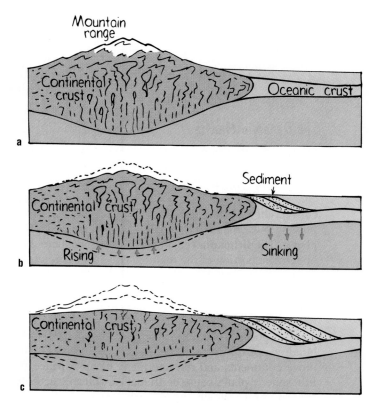

<table>
<tr><td></td></tr>
</table>

QUESTION

If you wished to drill the shortest hole to the mantle, would you drill in western Colorado, or in Florida?

24.3 STRUCTURAL GEOLOGY

Convection within the earth's mantle causes the overlying lithospheric plates to be in slow but constant motion. This motion creates stress within the lithosphere, and more importantly, within the earth's crust. Rocks subjected to stress begin to deform into intricate or broad folds, and faults, which can be small and virtually unnoticeable, or large and loaded with the potential to devastate.

ANSWER

Put the question another way: If you wanted to drill the shortest hole through ice to the water below, would you drill atop an iceberg, or through a slab of ice that hardly extends out of the water? Drill your hole in the slab, and likewise through the thinner crust of nonmountainous Florida. Mountainous western Colorado, on the other hand, like an iceberg, goes deeper than it is high. If you really want the shortest hole, drill through the ocean floor—exactly what scientists have done in Project MoHole, in the East Pacific Ocean.

Folds

When a rock is subjected to compressive stress it begins to buckle and fold. Suppose you had a throw rug on your floor, and your friend was standing on one end of it. If you push the rug toward your friend while keeping it on the floor, a series of ripples, or **folds,** develops in the rug. This is what happens to the earth's crust when it is subjected to compressive stress. We can compare the ripples created on the rug to folding of rocks on the earth's surface. Both are generated by compressive stresses.

We know that sediments settling from water in an ocean or in a bay are deposited in horizontal layers. The layer at the bottom is deposited first. It is therefore the oldest in the sequence of deposited layers. Each new layer is deposited on top of the previous layer. So in a sequence of sedimentary layers the oldest layer is at the bottom of the sequence and the youngest layer is at the top. As originally flat sedimentary rock layers are subjected to compressive stress they tilt and become folded. Each fold has an axis. When the tilted layers dip toward the fold axis, the fold is called a **syncline**. The rocks in the center, or core, are younger than those away from the core. If the tilted layers dip away from the axis, the fold is called an **anticline**. The rocks in the core of the fold of an anticline are older than the rocks away from the core. Anticlines and synclines almost always occur in pairs (Figure 24.7). Their geometry can be simple, or complex as a fold axis itself tilts and folds (Figure 24.8).

QUESTION

Why are rocks that make up the core of a syncline young, while rocks at the core of an anticline are relatively older?

Figure 24.7 Anticline and syncline folds. Layer 1 is the oldest rock and layer 6 is the youngest. The limbs of an anticline dip away from the axis of the fold and the rock layers are older in the core of the fold. The limbs of a syncline dip toward the axis of the fold and the rocks are younger in the core.

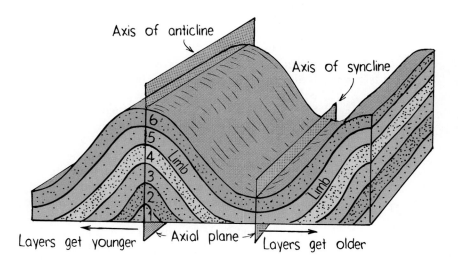

ANSWERS

Think of the rug example. Assume the top surface of the rug is young, and the lower surface old. When you push the rug it can fold two ways: (1) with the top surface folded onto itself, or (2) with the bottom surface folded onto itself. In the first case, the top surface makes up the core—a syncline. In the second case the bottom surface makes up the core—an anticline. Makes sense!

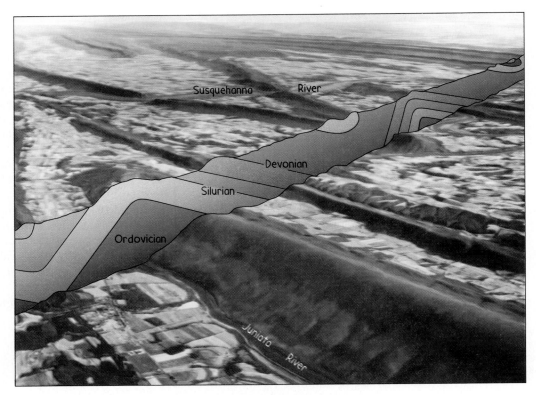

Figure 24.8 The erosional remnants of plunging anticline-syncline pairs in the Valley and Ridge belt 50 km NW of Harrisburg, Pennsylvania with the cross-sectional view superimposed.

Faults

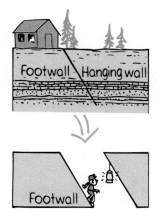

Figure 24.9 The terms *footwall* and *hanging wall* were coined by miners because one could hang a lamp on a hanging wall, and one could stand on a footwall.

When compressional stress overcomes the strength of rock, it will break and fracture. If displacement can be seen on both sides of the fracture or parallel to it, the fracture is called a **fault**. Look at Figure 24.9, and note the angle of the fault with respect to the horizontal ground surface. Imagine you could pull the block diagram apart at the fault. The half of the inclined plane where someone could stand is the **footwall** block. The other half is inclined to make standing impossible. This is the **hanging wall** block. These terms were coined by miners because one could hang a lamp on a hanging wall, and one could stand on a footwall.

Now we can discuss what happens in a zone of compressional faulting. Forces of compression cause rocks in the hanging wall to be pushed over rocks in the footwall. Depending on the angle between the fault plane and the horizontal, the fault is either a **thrust fault** (less than 45°) or a **reverse fault** (more than 45°). We see this in Figure 24.10. Older, lower rock is pushed over younger, higher rock. The Rocky Mountain foreland, the Canadian Rockies, and the Appalachian Mountains, to name a few, were formed in part by reverse and thrust faulting.

In addition to compression, stress can also occur by tension. Whereas compression pushes, tension pulls. Tension causes rocks in the hanging wall to drop down relative to the footwall, producing a **normal fault,** placing younger, higher rock over older, lower rock (Figure 24.11). Virtually the entire state of Nevada, and eastern

Figure 24.10 In a zone of compressional faulting, rocks in the hanging wall are pushed up over rocks in the footwall. (a) Reverse fault before erosion; (b) reverse fault after erosion.

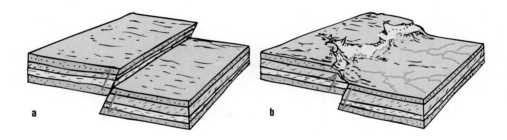

California, southern Oregon, southern Idaho, and western Utah are greatly affected by normal faulting.

The faults described so far have purely vertical or **dip-slip** motion. But some of the world's most famous faults, like the San Andreas fault in California, have virtually no vertical motion. These faults are called **strike-slip faults**. Their relative motion is horizontal (Figures 24.12 and 24.13). Devastating earthquakes are associated with horizontal faults like the San Andreas. The Great San Francisco Earthquake and Fire of 1906 registered near 8.3 on the Richter scale (a 10-point scale of earthquake severity). It caused 700 deaths and extensive fire damage. The Loma Prieta earthquake near Santa Cruz, California, in 1989, registered 7.1 on the Richter scale. It sadly caused 62 deaths and more than six billion dollars in damage. Still larger and more catastrophic earthquakes have occurred along reverse faults. The 1964 earthquake in Anchorage, Alaska, registered 8.5 on the Richter scale, and caused 131 deaths and 300 million dollars in damage.* One of the greatest tragedies of recent times was the Mexico City earthquake of 1985. It measured 8.5 on the Richter scale, and caused 7000 deaths!

Figure 24.11 In a zone of tensional faulting, rocks in the hanging wall drop down relative to the footwall. (a) Normal fault before erosion; (b) normal fault after erosion.

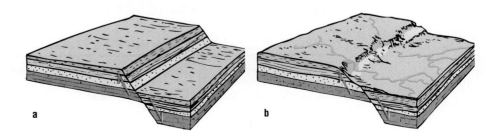

Understanding earthquakes is obviously of major importance to society. Table 24.1 lists some of the world's most notable earthquakes according to their impact on society. The 1906 San Francisco earthquake is notable because of its damage to the city and its inhabitants. But interestingly enough, in the winter of 1811–1812 a much greater earthquake on the New Madrid fault in Missouri changed the

*The death toll was largely due to great seismic sea waves, or *tsunami*. A tsunami is generated from the displacement of water as a result of an earthquake, submarine landslide, or an underwater volcanic eruption.

Figure 24.12 The relative movement of a strike-slip fault is horizontal.

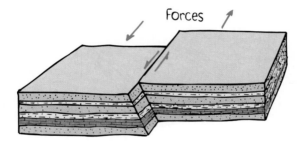

landscape beyond recognition. It also shifted the direction and course of the Mississippi River. Fortunately, due to the remote region and few settlers, human casualties were few. Going much further back, the presently inactive Appalachians and parts of the Rocky Mountains were once zones of intense earthquake activity, much like the places listed in Table 24.1 that have recently endured the awesome power of the quaking earth.

Figure 24.13 Offset orchard rows in an orange grove that straddles the right-lateral strike-slip San Andreas fault. The rows in the background have moved to the right relative to the rows in the foreground.

TABLE 24.1   Some of the World's Most Notable Earthquakes

Year	Location	Magnitude	Estimated Deaths	Comments
1556	Shensei, China	unknown	830,000	Possibly the greatest natural disaster in recorded history
1811	New Madrid, MI	unknown	few	
1906	San Francisco, CA	8.25	700	Fires caused extensive damage
1908	Messina, Italy	7.5	120,000	
1920	Kansu, China	8.5	180,000	
1923	Tokyo, Japan	8.2	150,000	Fire caused extensive destruction
1960	Southern Chile	8.7	5700	The largest earthquake ever recorded
1964	Anchorage, AK	8.5	131	More than $300 million in damage
1970	Peru	7.8	66,000	Great rockslide
1971	San Fernando, CA	6.5	65	More than $5 billion in damage
1975	Liaoning, China	7.5	few	First major earthquake to be predicted
1976	Tangshan, China	7.6	500,000	
1985	Mexico City, Mexico	8.5	7,000	
1989	San Francisco, CA	7.1	62	More than $6 billion in damage
1992	Ferndale, CA	6.9	0	More than $50 million in damage
1994	Los Angeles, CA	6.6	55	More than $30 billion in damage

Earthquake Measurements— Mercalli and Richter Scales

Every year hundreds of thousands of earthquakes occur. Although most of these earthquakes are small and go undetected, the danger of large earthquakes certainly exists. Earthquake prone regions, such as California, experience large earthquakes about every 50 to 100 years. Over the years, scientists have used several methods to measure the sizes of earthquakes. The Mercalli Scale measures the intensity of an earthquake in terms of the effects it produces on the local environment. The Mercalli Scale ranges from an intensity of I, which is barely detectable, to an intensity of XII, which results in total destruction.

Based purely on observation, the Mercalli Scale, though a valuable yardstick, does not provide a precise measurement of the size of an earthquake. Seismologists developed a more precise way to estimate the energy released in an earthquake by measuring the amplitudes of the seismic waves recorded on a seismogram. The Richter Scale, a magnitude scale, measures the severity of an earthquake in terms of the intensity of energy released. The magnitude scale is logarithmic, so each increase of 1 on the Richter scale has 10 times the intensity of the integer before it. For example, magnitude 7.4 means 10 times the intensity of magnitude 6.4.

The Mercalli Scale of Intensity

I   Not felt except by a very few under especially favorable circumstances.

II   Felt only by a few persons at rest, especially on upper floors of buildings.

III   Felt quite noticeably indoors, especially on upper floors of buildings, but many people do not recognize as an earthquake.

IV   Most people feel it indoors, a few outdoors. Dishes, windows, doors rattle.

V   Felt by nearly everyone. Disturbances of trees, poles, and other tall objects.

VI   Felt by all; many frightened and run outdoors. Some heavy furniture moved; few instances of fallen plaster or damaged chimneys. Damage slight.

VII   Everybody runs outdoors. Damage negligible in buildings of good design and construction; slight to moderate in well-built structures; considerable in poorly built structures.

VIII   Damage slight in specially designed structures; considerable in ordinary substantial buildings with partial collapse; great in poorly built structures (fall of chimneys, factory stacks, columns, monuments, walls).

IX   Damage considerable in specially designed structures. Buildings shifted off foundations. Ground conspicuously cracked.

X   Some structures destroyed. Most masonry and frame structures destroyed with foundations. Ground badly cracked.

XI   Few, if any, structures (masonry) remain standing. Bridges destroyed. Broad fissures in ground.

XII   Damage total. Waves seen on ground surfaces. Objects thrown up in air.

Source: U.S. Coast Guard and Geodetic Survey

Richter Magnitude

Magnitude	Number per Year	Mercalli	Characteristic Effects
< 3.4	800,000	I	Recorded only by seismographs
3.4–4.4	30,000	II and III	Felt by some people in the area.
4.4–4.8	4,800	IV	Felt by many people in the area.
4.8–5.4	1,400	V	Felt by everyone in the area.
5.4–6.0	500	VI and VII	Slight building damage.
6.0–7.0	100	VIII and IX	Much building damage.
7.0–7.4	15	X	Serious damage, bridges twisted, walls fractured.
7.4–8.0	4	XI	Great damage, buildings collapse.
> 8.0	one every 5–10 years	XII	Total damage, waves seen on ground, objects thrown in air.

Source: From B. Gutenberg, 1950

24.4 THE THEORY OF CONTINENTAL DRIFT

Although many new discoveries were made during the early twentieth century, scientists of that time still believed that oceans and continents were geographically fixed. They regarded the surface of the planet as a static skin spread over a molten, gradually cooling interior. They believed that the cooling of the planet resulted in its contraction, which caused the outer skin to contort and wrinkle into mountain and valley structures.

Many people had noticed that the shorelines between South America and Africa seemed to fit together like a jigsaw puzzle (Figure 24.14). One earth scientist who took this observation seriously was Alfred Wegener. He gathered evidence to support this concept. Wegener saw the earth as a dynamic planet with the continents in constant motion. He believed that all the continents had once been joined together in one great supercontinent he called **Pangaea,** meaning "all land."

Wegener proposed that the geological boundary of each continent lay not at its shoreline, but at the edge of its continental shelf (the gently sloping platform between the shoreline and the steep gradient that leads to the ocean floor). He supported his hypothesis with impressive geological, biological, and climatological evidence. When Wegener fit Africa and South America together along their continental shelves, he found that the rocks on the two continents were virtually identical. Many of the mountain systems in both Africa and South America show strong evidence of a previous connection. Similarly, fossils of identical land-dwelling animals are found in South America and Africa, but nowhere else. And fossils of identical trees such as the *glossopteris** are found in South America, India, Australia, and Antarctica.

*The term *glossopteris flora* is named after the dominant gymnosperm tree found in the prehistoric southern temperate forests of South America, India, Australia, and Antarctica. Because the seeds from these trees were too large to be distributed by air, the wide distribution of this flora supports Wegener's theory that the continents were once joined together.

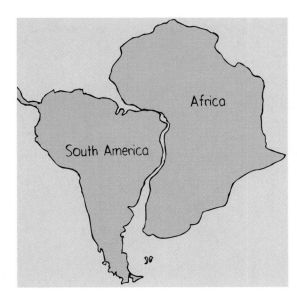

Figure 24.14 When you align the shorelines of South America and Africa, the continents fit together like pieces in a jigsaw puzzle.

Even stronger evidence for a supercontinent comes from paleoclimatic (ancient climate) data. Over 300 million years ago a huge continental ice sheet covered parts of South America, southern Africa, India, and southern Australia. The ice sheet left evidence of its existence in thousands of well preserved glacial striations that reveal the directions of ice flow. If these continents were in their present positions, the ice sheet would have had to cover the entire southern hemisphere, and in some places, cross the equator! If the ice sheet was that extensive the world climate would have been very cold. But, there is no evidence of glaciation in the northern hemisphere at that time. In fact, the time of glaciation in the southern hemisphere was a time of subtropical climate in the northern hemisphere. To account for the enigmatic paleoclimatic regions Wegener joined the landmasses as one supercontinent with South Africa centered over the South Pole. This reconstruction moves the extensive expanses of glacial ice over the South Pole and places the northern landmasses nearer the tropics.

Wegener published *The Origin of Continents and Oceans,* describing continental drift. Although he used evidence from different scientific disciplines, his well-founded hypothesis was ridiculed by the community of earth scientists. Antagonists complained that Wegener failed to provide a suitable driving force to account for the continental movements. Wegener proposed that the tidal influence of the moon could produce the drift of the continents. He also proposed that the continents broke through the crust like ice breakers cutting through ice. Without a convincing explanation for his theory, however, Wegener was laughed at. The scientific community of the early part of the century was not ready to believe that the continents had drifted to their present position. It is only recently, with new-found discoveries, that Wegener's concept has become accepted.

A Scientific Revolution

One of the first key discoveries in support of continental drift came about through studies of the earth's magnetic field. We know that the earth is a huge magnet with a magnetic north and south pole differing in location from the geographic poles, but in close proximity to them. Because certain minerals align themselves with the

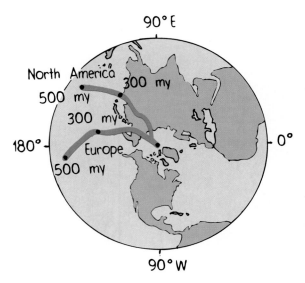

Figure 24.15 The plotted path of the magnetic north pole during the last 500 million years suggests that either the magnetic pole wanders or the continents have moved.

magnetic field, rocks have a preserved imprint of the earth's magnetism. This is known as **paleomagnetism**. Three essential bits of information are contained in the preserved magnetic record: (1) the polarity—whether the magnetic field was normal or reversed at the time of formation; (2) the direction of the magnetic pole at the time the rock formed; and (3) the latitude of origin—the angle of inclination of the magnetic field relative to the earth's surface (which steepens toward the poles) where the rock formed. Once the magnetic latitude of a rock and the direction of the magnetic poles are known, the position of the magnetic pole at the time of formation can be determined. During the 1950s a plot of the position of the magnetic north pole through time revealed that over the past 500 million years the position of the pole had gradually wandered all over the world (Figure 24.15)! It seemed that either the magnetic poles migrated through time or the continents had drifted. Since the apparent path of polar movement varied from continent to continent, it was more plausible that the continents, rather than the poles, had moved. Thus the hypothesis of continental drift was revived, but a mechanism to explain how the movement occurred was still lacking.

Figure 24.16 Detailed maps of the ocean floors reveal deep rift valleys and enormous mountain ranges in the middle of the oceans, with deep ocean trenches near the continental land masses.

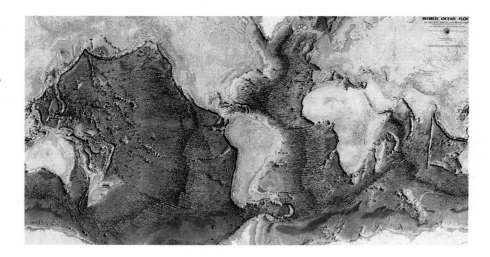

Figure 24.17 In conveyor-belt fashion, new crustal material is formed at the mid-ocean ridge as old crustal material is subducted into a deep ocean trench.

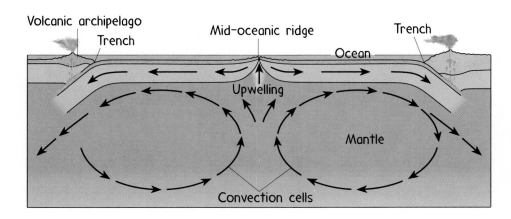

The 1950s were a time of extensive and detailed mapping of ocean floors. Topographic features revealed huge mountains and deep valleys in the middle of the oceans with deep ocean trenches near the continental land masses (Figure 24.16). Volcanism and high thermal energies were generated at ridge systems in the middle of the oceans. The oceanic crust was found to be thin and young near the central ridge region, and progressively thicker and older away from the ridge. With this new information, H. H. Hess, an American geologist, presented the hypothesis of **sea-floor spreading**. Hess proposed that the sea floor is not permanent, but is constantly being renewed. He theorized that the ocean ridges are located above upwelling convection cells in the mantle. As rising material from the mantle oozes upward, new crustal material is formed. The old crust is simultaneously destroyed in the deep ocean trenches near the edges of continents. Thus, in a conveyor belt fashion, new thin crust forms at the center, and older thicker crust is pushed from the ridge crest to be eventually subducted into a deep ocean trench (Figure 24.17).

Support for this theory of sea-floor spreading came from paleomagnetic analysis of the ocean floor. As new basalt is extruded at the oceanic ridge, it is magnetized according to the existing magnetic field. The magnetic surveys of the ocean's floor showed alternating strips of normal and reversed polarity, paralleling either side of the rift area (Figure 24.18). The magnetic history of the earth is thus recorded in

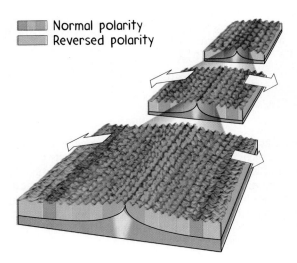

▨ Normal polarity
▨ Reversed polarity

Figure 24.18 As new material is extruded at the oceanic ridge, it is magnetized according to the existing magnetic field. Magnetic surveys show alternating strips of normal and reversed polarity paralleling both sides of the rift area. Like a very slow magnetic tape recording, the magnetic history of the earth is thus recorded in the spreading ocean floors.

Hot Spots and Lasers—A Measurement of Tectonic Plate Motion

Motion is relative. When we discuss the motion of something, we describe its motion relative to something else. We call the place from which motion is observed and measured a reference frame. With a world where everything is in motion, how can we measure rates of plate motion? What do we choose for our reference frame? The Canadian geophysicist, J. Tuzo Wilson, suggested that Pacific sea-floor movement is recorded by the age of the Hawaiian Volcanic islands. Wilson postulated that the islands are the tips of huge volcanoes that formed as the floor of the Pacific Ocean moved over a fixed "hot spot"—a magma source rising from the earth's interior. The concept of a fixed hot spot provides a stationary reference point on the surface of the earth against which plate motion can be determined. There are about a hundred such reference hot spots around the world used together to determine rates of plate movements.

A fascinating and more precise way to measure earth movement is by a laser beam reflecting off "mirrors" in outer space. Lasers are used to detect the broad movements of tectonic plates as well as movement on a local level. In order to measure broad movements, radio telescopes are keyed to a reference point in outer space—a quasar or a satellite—from which relative positions of points on earth can be plotted (Figure 24.19a). Laser pulses beamed to the reference point from a pair of ground stations located on opposite sides of a plate boundary (or on opposite sides of a fault) start timers that run until the reflected pulses are received back at the stations (Figure 24.19b). A computer combines this elapsed time with the known position of the reference point to determine the exact position of the ground station. Any movement of the ground station is registered as a change in the elapsed time. Neato!

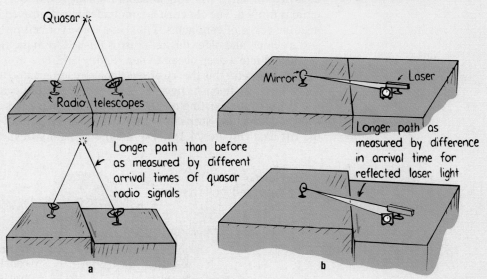

Figure 24.19 Earth movements are measured by radio telescopes and lasers that are keyed to a stationary reference point. (a) Broad movements in the range of one centimeter per year are detected by satellites and quasars as reference points. (b) Fault movements are measured by laser beams shot from opposite sides of a fault. The laser flashes a beam off a reflector; the bounce-back time is recorded. Because of light's constant speed, the time of bounce-back will change with earth movement.

spreading ocean floors as in a very slow magnetic tape recording, forming a continuous record of the movement of the sea floors. Since the dates of pole reversal can be calculated, the magnetic pattern of the spreading sea floor would document not only the sea floor's age, but also the rate at which it spreads.

With the theory of sea-floor spreading, a mechanism was provided for the concept of continental drift. The time was right for the revolutionary concepts to follow. The tide of scientific opinion had indeed switched in favor of a mobile earth.

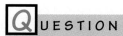UESTION

Why was Wegener's theory of continental drift not taken more seriously in the early part of this century?

24.5 THE THEORY OF PLATE TECTONICS

The theory of plate tectonics provides a framework for understanding how and why the various features of the earth constantly change. The theory describes the forces within the earth that give rise to continents, ocean basins, mountain ranges, earthquake belts, and other large-scale features of the earth's surface. The theory of plate tectonics holds that the earth's outer shell, the lithosphere, is divided into nine large plates and a number of smaller plates (Figure 24.20). These lithospheric plates ride atop the plasticlike asthenosphere below. Because each plate moves as a distinct unit in relation to other plates, all major interactions between plates are along plate boundaries. Thus most of the earth's seismic activity, volcanism, and mountain building occur along these dynamic margins.

Huge convection currents operating below the lithosphere generate slow but constant movements that greatly affect the surface features of our planet. These convection currents act as huge conveyor belts that create crust at one end of a plate and destroy it at another. So the earth's solid crust near one plate today may be plowed downward to become molten material tomorrow. And at the other end of the plate, molten material today rises to become solid crust tomorrow. The boundaries of these plates mark the edges of crustal formation or destruction. Three distinct types of plate boundaries, named for the movement they display, are shown in Figure 24.21.

Divergent Boundaries

Where two plates are moving apart, tensional forces act to stretch the crust and generate a spreading center. Hot molten rock from the earth's interior buoyantly

Ⓐ NSWER

Wegener failed to produce a suitable driving mechanism to support his theory. But even if he postulated the role of the convective interior, we can only speculate about how quickly the scientific community would have accepted continental drift as a viable hypothesis. Scientists, like all human beings, tend to identify with the ideas that characterize their time. Do advances in knowledge, scientific or otherwise, occur because they are accepted by the status quo, or because holders of the status quo eventually die off? Knowledge that is radical to the old guard is often easily accepted by newcomers who use it to push the knowledge frontier further. Hooray for the young (and the young-at-heart)!

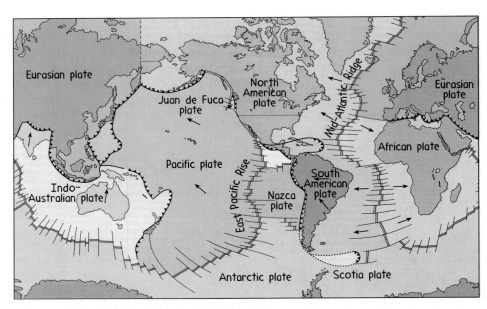

Figure 24.20 The lithosphere is divided into nine large plates and a number of smaller plates (more about the history of these plates and their motions in the next chapter).

rises up to form new crust. The crust near the spreading edge is thin, and has a relatively low density due to heat and expansion of the rising magma. As the new crust moves away from the spreading center it cools, contracts, and becomes more dense.

The Mid-Atlantic Ridge is a spreading center that has been producing the Atlantic Ocean floor for 160 million years. The spreading is accompanied by almost continuous earthquake activity that, except by scientists, goes largely unnoticed because of low magnitude and harmlessness to humans. With production of crustal material at the ridges, the ocean floor grows and the continents drift apart. The crust is younger near the central ridge region and progressively older away from the ridge.

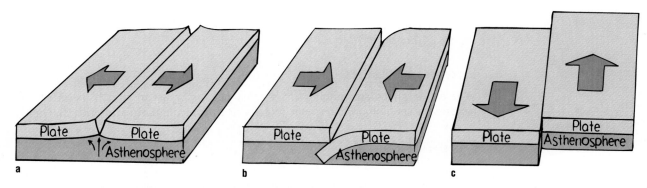

Figure 24.21 Plate boundaries mark the edges of crustal formation or crustal destruction. Named for the movement they accommodate, three types of plate boundaries are: (a) divergent boundaries, (b) convergent boundaries, and (c) transform fault boundaries.

The zone of active sea floor formation extends for tens of kilometers. The rate of sea floor spreading can be calculated if one knows the age of ocean floor and the distance of the dated ocean floor from the ridge crest. The spreading rate has been slow, about two centimeters each year (about how fast fingernails grow). This rate over 160 million years adds up to 3200 kilometers—the width of the Atlantic Ocean. So, it took only 160 million years for a mere fracture in an ancient continent to turn into the Atlantic Ocean!

Spreading centers are not restricted to the ocean floors, but develop also on land. Hot rising molten material in the earth's interior beneath continental land masses cause upwarping of the earth's crust. Gaps in the crust are produced and large slabs of rock sink and slide down into these gaps. The large down-faulted valleys generated by this process are **rifts,** or rift valleys. The Great Rift Valley of East Africa is a prime example of such a feature, and is the beginning of a new ocean basin (Figure 24.22).

Convergent Boundaries

Where two plates move toward each other compressional forces act to either push crust downward or shorten crust by folding and faulting. The regions of plate collisions, the *convergent boundaries,* are regions of great mountain building. Convergent zones, although similar, take on different manifestations, depending on

Figure 24.22 In a divergent boundary, sea-floor spreading leads to the formation of a new ocean basin. (a) Rising magma uplifts crustal surface. (b) Rift valley begins to form as crust is pulled apart. (c) The beginning of a narrow sea, which becomes an (d) ocean basin.

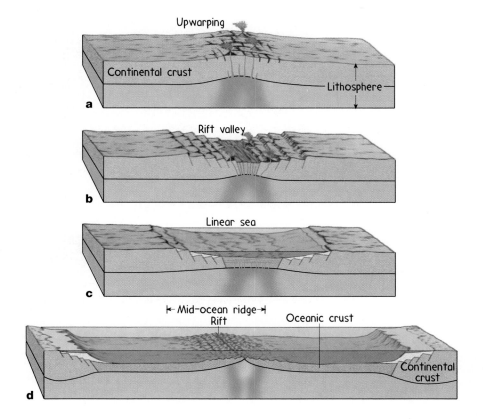

whether the leading edges of plates are continental or oceanic. There are three types of plate collisions: collisions between two plates of oceanic crust, between a plate of oceanic crust and a plate of continental crust, and between two plates of continental crust (Figure 24.23).

Oceanic-Oceanic Convergence Convergence (collisions) between two plates of oceanic crust results in the *subduction* of one plate beneath the other. During subduction one oceanic plate bends and descends beneath the other to produce a deep ocean trench. The deepest trench known is the Marianas Trench in the western Pacific Ocean, 11 kilometers (6.8 miles) below sea level. Partial melting of the subducted crust produces andesitic magma that buoyantly migrates upward to form a volcanic island arc system. The size and elevation of the arc increases over time with continued volcanic activity. Sediments scraped off the descending plate pile up to form an accretionary complex, which becomes folded and faulted into a thick deformed

Figure 24.23 The three types of convergent margins: (a) Oceanic to continental, (b) oceanic to oceanic, and (c) continental to continental.

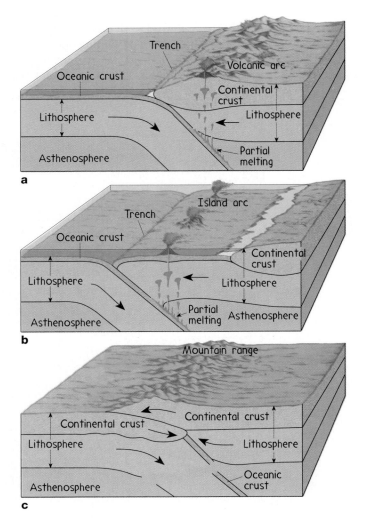

wedge. Such island arcs have formed the Aleutian, Mariana, and Tonga islands, and the island arc systems of the Alaskan Peninsula, the Philippines, and Japan. Deep ocean trenches mark the active subduction zones that border the island arcs.

Oceanic-Continental Convergence When a plate of oceanic crust and a plate of continental crust converge, the dense plate of oceanic crust is subducted beneath the lighter plate of continental crust, and a deep ocean trench is formed. The descending oceanic crust carries a mixture of oceanic basaltic rock and continental sediment. Similar to oceanic-oceanic convergence, andesitic magma rises upward, some to crystallize below the surface, and some to reach the surface and erupt as volcanic arcs. With the beginning of their existence as volcanic arcs, the Andes Mountains continue to grow. The subduction of the Nazca Plate beneath the continent of South America along the Peru-Chile Trench causes sediments from the Nazca Plate to be scraped off onto the granitic roots of the Andes. This adds thickness and buoyancy to the mountains so that they continually rise upward more rapidly than they are eroded by wind and rain. Remnants of the original volcanic arc are the exposed batholiths and metamorphic terrains that flank the Andes on the west coast of South America. In the western United States, examples of volcanic arc activity are found in the Sierra Nevada Mountains, an ancient volcanic arc, and the Cascade Range, which is currently active. The Sierra Nevada was produced by the subduction of the ancient Farallon Plate beneath the North American Plate. The Sierra Nevada batholith is a remnant of the original volcanic arc, while the Coast Ranges have remnants of the arc's accretionary complex. The Cascade Range, produced from the subduction of the Juan de Fuca Plate (a remnant of the Farallon Plate) beneath the North American plate, includes the volcanoes Mt. Rainier, Mt. Shasta, and Mt. St. Helens. The eruption of Mt. St. Helens gives testimony that the Cascade Range is still quite active.

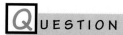

QUESTION

Erosion wears mountains down, yet the Andes Mountains grow taller each year. Why?

Continental-Continental Convergence The collision between two continental land masses is always preceded by oceanic-continental convergence. Continental crust, which is light and buoyant, does not undergo any appreciable amount of subduction. Instead, the convergence between two continental land masses is more like a head-on collision. Compressional forces cause the plates to buckle and fold upon each other, creating great thicknesses of the continental crust. Intensely compressed and metamorphosed rock defines the suture zone of plate collision. The collision between continental land masses has produced some of the most famous mountain ranges. A majestic example of such a collision is the snow-capped Himalayas, the highest mountain range in the world. This chain of towering peaks is still being

ANSWER

Subduction is still occurring at the Peru-Chile trench causing the uplift of the Andes Mountains. Uplift is not restricted to the Andes; some other mountain ranges continue to grow, such as the Sierra Nevada in North America, as uplift exceeds erosion.

Figure 24.24 The continent-to-continent collision of India with Asia produced the Himalayas.

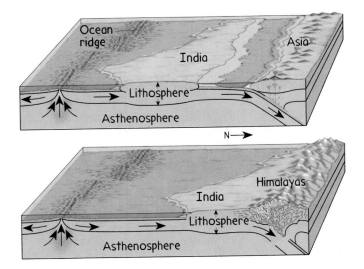

thrust upward, as India continues crunching against Asia (Figure 24.24). The European Alps were formed in a similar fashion, when fragments of the African Plate collided with the Eurasian Plate some 80 million years ago. Relentless pressure between the two plates continues, and is slowly closing up the Mediterranean Sea. In America, the Appalachian Mountains were produced from a continental-continental collision that ultimately resulted in the formation of the supercontinent of Pangaea.

Transform Fault Boundaries

Transform faults provide a link between plate boundaries that are not colliding head on, but slide past one another. They were first identified where they join segments of the oceanic ridge. The faults "transform" the motion from one ridge segment to another. Contact forces are neither tensional nor compressional, so there is no creation or destruction of the crust. It is a zone of strike-slip accommodation, with movement along the fault in the opposite direction of the apparent ridge offset. Transform faults provide the means by which crust created at the oceanic ridge is transported to the site of destruction at oceanic trenches (Figure 24.25).

The San Andreas Fault is one of the most famous transform faults, and stretches for 1500 kilometers (950 miles) from Cape Mendocino in northern California to the

Figure 24.25 Transform faults provide strike-slip accommodation between plate boundaries—the region where plates slide past one another in opposite directions.

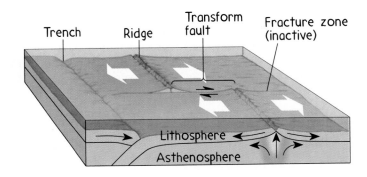

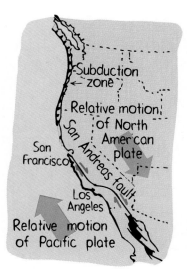

Figure 24.26 The San Andreas fault shows strike-slip movement as the Pacific Plate slides in a northward direction and the North American Plate slides in a southward direction.

East Pacific Rise in the Gulf of California (Figure 24.26). The Pacific Plate, on the west side of the fault, is moving northwest at a rate of about 5.0 centimeters per year relative to the North American Plate on the east side of the fault. The San Andreas fault accommodates about 70 percent of this motion, or about 3.5 centimeters per year. The remaining motion is taken up by other right-lateral strike-slip faults (Hayward and Calaveras faults, for example). Grinding and crushing take place as the two plates move past each other. When sections of the plates become locked, stress builds up until the friction is relieved in the form of a major earthquake. On April 18, 1906 the Pacific Plate lurched about 6 meters (20 feet) northward over a 434 kilometer (270 mile) stretch of the fault, releasing the built-up stress, and causing the catastrophic San Francisco earthquake.

So we see in summary that the tectonic interaction between the different lithospheric plate boundaries provides an explanation to the origin of mountain chains, the development and destruction of ocean basins, and the global distribution of earthquakes and volcanoes (Figure 24.27). The internal activities that change the earth's surface do so in a cyclical manner. In the next chapter we will see the effects of plate tectonic interaction through time. The study of geology uses processes that occur today to understand what may have occurred in the past. This concept is commonly stated as "the present is the key to the past." But also what has happened in the past provides clues as to what may happen in the future. The earth is indeed a dynamic planet.

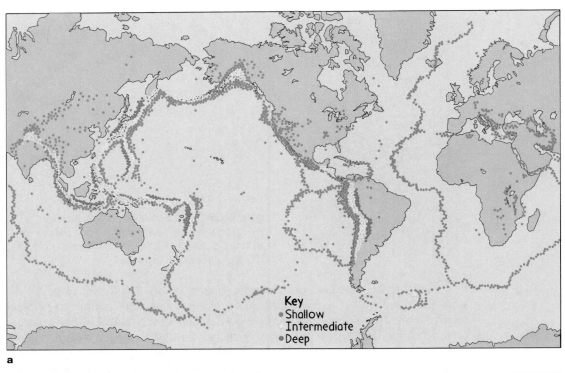

a

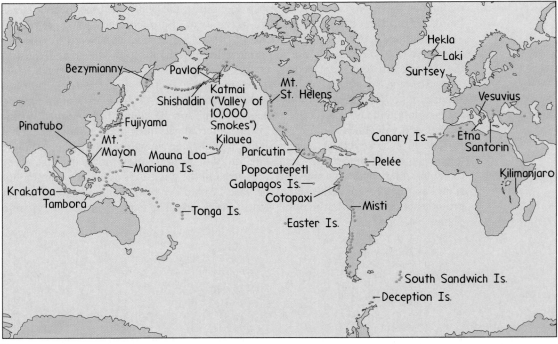

b

Figure 24.27 (a) The global distribution of earthquakes. (Source: NOAA) (b) The global distribution of recent volcanic centers. Note that earthquakes and volcanoes mark plate boundaries.

SUMMARY OF TERMS

Body wave A seismic wave that travels through the earth's interior.

Surface wave A seismic wave that travels along the surface of the earth.

Primary wave A body wave that involves compressional and expansional motion in the direction of propagation. P-waves travel through solids, liquids and gases and are the fastest of the seismic waves.

Secondary wave A body wave propagated by a shearing motion. Because S-waves do not travel through liquids they do not travel through the earth's outer core.

Rayleigh wave A surface wave with an elliptical motion.

Love wave A surface wave with a horizontal motion that is shear or transverse to the direction of propagation.

Mohorovičić discontinuity The boundary that separates the earth's crust from the mantle. This boundary marks the level at which P-wave velocities change abruptly from 6.7–7.2 km/sec (in the lower crust) to 7.6–8.6 km/sec (at the top of the upper mantle).

Core The most central layer in the earth's interior. The core itself is divided into an outer liquid core and an inner solid core.

Mantle The middle layer in the earth's interior. The mantle resides below the earth's crust and above the earth's core.

Asthenosphere A subdivision of the mantle situated below the lithosphere. This zone of weak material exists below a depth of about 100 kilometers and in some regions extends as deep as 700 kilometers. The rock within this zone is easily deformed.

Lithosphere A subdivision of the earth's interior which includes the entire crust and the uppermost portion of the mantle. The lithosphere is rigid and brittle, and is broken into many individual pieces called plates. These lithospheric plates are continually in motion as a result of the hot convection currents circulating in the asthenosphere and the mantle.

Crust The earth's outermost layer. The crustal surface represents less than 0.1 percent of the earth's total volume.

Isostasy The condition of equilibrium with all rock masses in balance.

Fold A series of ripples, large or small, that result from compressional deformation.

Syncline A fold in sedimentary strata that has stratigraphically younger rocks in the core of the fold. A syncline resembles a sag.

Anticline A fold in sedimentary strata that has stratigraphically older rocks in the core of the fold. Anticlines resemble an arch.

Fault A fracture or fracture zone along which visible displacement can be detected on both sides of the fracture or parallel to it.

Footwall The mass of rock beneath a fault.

Hanging wall The mass of rock above a fault.

Thrust fault Compressional fault in which the angle of the fault plane with the horizontal ground surface is less than 45°. Thrust faults push older, structurally deeper rocks on top of younger, structurally higher rocks.

Reverse fault Compressional fault in which the angle of the fault plane with the horizontal ground surface is greater than 45°. Reverse faults push older, structurally deeper rocks on top of younger, structurally higher rocks.

Normal fault Tensional fault in which rocks in the hanging wall drop down relative to the footwall. Normal faults place younger, structurally higher rocks on top of older, structurally deeper rocks.

Pangaea A great supercontinent that existed from about 300 to about 200 million years ago. Pangaea means "all land" and included most of the continental crust of the earth.

Paleomagnetism Rocks possess a preserved imprint of the earth's magnetism due to the ability of certain minerals to align themselves with the magnetic poles. Paleomagnetism is the study of natural magnetization in a rock in order to determine the intensity and direction of the earth's magnetic field at the time of the rock's formation.

Sea-floor spreading A hypothesis that the oceanic crust is constantly being renewed by convective upwelling of magma along the midoceanic ridges. At the same time new material is created old material is destroyed at the deep ocean trenches near the edges of continents.

Rift A long narrow trough that forms as a result of crustal divergence.

REVIEW QUESTIONS

Seismic Waves

1. Body waves, P-waves, and S-waves move through the earth's interior in two ways. What is the difference in their mode of propagation?

2. Can S-waves travel through liquids? Explain.

3. Surface waves form when body waves encounter the earth's surface. There are two types of surface waves. Name them and briefly discuss their motion.

4. List the different properties of seismic waves and how they contributed to the discovery of the earth's internal boundaries.

5. What technological activities spurred knowledge of the earth's interiors during the 1950s?

Earth's Internal Layers

6. What was Andrija Mohorovičić's major contribution?

7. What does the wave shadow that develops 105°–140° from the origin of an earthquake tell us about the earth's composition?

8. What is the evidence for the solidity of the earth's inner core?

9. Both the inner and outer core of the planet are composed predominantly of iron and nickel, yet the inner core is solid and the outer core is liquid. Why?

10. What is the evidence for the liquid phase of the earth's outer core?

11. Describe the asthenosphere and the lithosphere. In what way are they different from each other?

12. What "convectional movement" is responsible for plate movement on the crustal surface?

13. How does continental crust differ from oceanic crust?

14. Why do the continents "float" higher on the earth's surface than the crust of the ocean basins?

15. Define *isostasy* and relate it to continental and oceanic crust.

Structural Geology

16. When rocks are subjected to stress they begin to deform. Evidence of rock deformation is all around us. Describe three examples.

17. There are four major types of faults: thrust faults, reverse faults, normal faults, and strike-slip faults. What is the difference between these faults and how can geologists differentiate between them in the field?

18. Which kinds of faults result primarily from tension in the earth's crust? From compression in the earth's crust?

19. What are folds?

20. Are closely folded rocks the result of compressional or tensional forces?

21. Distinguish between anticlines and synclines.

22. What type of fault is associated with the 1964 earthquake in Alaska?

The Theory of Continental Drift

23. What key evidence did Alfred Wegener use to support his theory of continental drift?

24. How do the glacial striations found in parts of South America, southern Africa, India, and southern Australia support the concept of a supercontinent?

25. What was the stated reason for the scientific community rejecting Wegener's theory of continental drift?

A Scientific Revolution

26. What information can be learned from a rock's magnetic record?

27. What role does paleomagnetism play in the theory of continental drift?

28. What major discovery at the bottom of the sea was made by H. H. Hess?

29. How is the ocean floor similar to a giant, slow-moving tape recorder?

30. In what way does sea-floor spreading support the theory of continental drift?

The Theory of Plate Tectonics

31. What does the crust of the earth have in common with a conveyor belt?

32. According to the theory of plate tectonics, the lithosphere is broken into several rigid plates that float and slide over the partially molten asthenosphere. Because each plate moves in relation to the other plates all major interactions occur along the plate boundaries. Name and describe the three different types of plate boundaries.

Divergent Boundaries

33. How old is the Atlantic Ocean thought to be? For how many years has magma been extruding in the Mid-Atlantic?

34. What is a rift?

Convergent Boundaries

35. Where is the deepest part of the ocean?

36. What is the source of mountain building in coastal Western United States?

37. The Appalachian Mountains were produced at what type of plate boundary? What about the Andes Mountains?

38. Briefly relate the collision between two continental land masses as a continuing process of oceanic-continental convergence.

39. What clues could we use to recognize the boundaries between ancient plates no longer in existence?

Transform Fault Boundaries

40. What kind of plate boundary separates the North American Plate from the Pacific Plate? What kind separates the South American Plate from the African Plate?

HOME PROJECT

Look for a very old window, and note the lens effect of the bottom part of the glass. Glass has both solid and liquid properties; in fact it is often thought of as a very viscous liquid. Over many years, its downward flow due to gravity is evident by the increased thickness near the bottom of the pane.

EXERCISES

1. Are the present ocean basins a permanent feature on our planet? Are the present continents a permanent feature?

2. Why is it that the most ancient rocks are found on the continents, and not on the ocean's floor? Support your answer.

3. What is meant by pole reversals? What useful information can pole reversals tell us about the earth's history?

4. Using a photocopy of Figure 24.20, the map of the earth's tectonic plates, mark the different boundaries of plate interaction. Draw arrows showing direction of plate movement for convergent, divergent, and transform fault boundaries.

5. Earthquakes are the result of sudden motion or trembling in the earth caused by the abrupt release of slowly accumulated strain. This strain causes rock to fracture or fault. Relate faulting to strike-slip movement. Where does this type of movement occur?

6. Upon crystallization, certain minerals (the most important being magnetite) align themselves in the direction of the surrounding magnetic field. In what ways did this fossil imprint of the earth's magnetic record support the theory of continental drift?

7. How is the theory of sea-floor spreading supported by paleomagnetic data?

8. Does the fact that the mantle is beneath the crust necessarily mean that the mantle is more dense than the crust?

9. How does the age distribution of the Hawaiian Islands chain relate to the concept of plate movement?

10. What "convectional movement" is responsible for the earth's magnetic field?

11. Crustal material is continuously created and destroyed. Where does this creation and destruction take place? Why must they be in equilibrium?

12. Subduction is the process of one lithospheric plate descending beneath another. Why does the oceanic portion of the lithosphere undergo subduction while the continental portion does not?

13. In 1964, a large tsunami struck the Hawaiian Islands without warning. The coastal town of Hilo, Hawaii was devastated. Since that time a tsunami warning station has been established for the coastal areas of the Pacific. Why do you think these stations are located around the Pacific rim? Can you think of any other tsunamis that occurred in this area?

14. What type of volcano would you expect to find at continental subduction margins such as the Andes Mountains and the Cascade Range?

15. How did the Himalayan mountains originate? The San Andreas fault? The Andes Mountains?

16. Where is the world's longest mountain range located?

17. Briefly relate the collision between two continental land masses as a continuing process of oceanic-continental convergence.

PROBLEMS

1. The Richter Scale is logarithmic, meaning that each increase of 1 on the Richter scale corresponds to a wave amplitude that is ten times as great as the amplitude of the number before it. An earthquake of magnitude 8 is how many times the amplitude of a magnitude 6 earthquake?

2. The San Andreas Fault separates the northwest moving Pacific Plate, on which Los Angeles sits, from the southwest moving North American Plate, on which San Francisco sits. As the two plates slide past one another at a rate of 3.5 centimeters per year, how long will it take the two cities to form one large city? (The distance between Los Angeles and San Francisco is 600 km.)

3. The weight of ocean floor bearing down upon the lithosphere is increased by the weight of ocean water. Compared to the weight of the 10-km thick basaltic ocean crust (specific gravity 3), how much weight does the 3-km deep ocean (specific gravity 1) contribute? Express your answer as a percent of the crust's weight.

25

THE GEOLOGIC TIME SCALE

The earth is estimated to be four and a half billion years old. This vast span of time, called *geologic time,* is almost too long to comprehend. If we compress geologic time into a single year, so that our planet formed from matter surrounding the sun on January 1st, then the oldest rocks known would appear in mid-March. Life in the sea would first appear in May. Plants or animals would emerge in late November. Dinosaurs would rule the earth in mid-December, and disappear by December 26. Homo sapiens would appear at 11:00 PM on the evening of December 31, and all recorded human history would occur in the last few seconds of New Year's Eve!

Earth's history is recorded in the rocks of the crust. The rock record is like a long and detailed diary, containing the history of earth-shaping events. The book, however, is incomplete. Many pages, especially in the early part, are missing, and many other pages are tattered, torn, and difficult to read. But enough pages are preserved to give an account of the remarkable events of the earth's four and a half billion years of history.

25.1 RELATIVE DATING

The sequence of rock layers tells us their relative ages. In other words, since lower layers are established before the top layers, their relative positions indicate which are oldest. Also, fossils in rock layers indicate relative age. Fossils of primitive life forms indicate the surrounding rock is older than rock containing fossils of more advanced life forms.

Perhaps the world's most spectacular display of the rock record is the Grand Canyon of the Colorado River in Arizona (Figure 25.1). The pronounced horizontal layering is testimony to great geologic activity over millions of years. Like the layers of rings in the cross-section of a tree, the layers of sedimentary rock indicate the geologic conditions at the time of their deposition. The deposition of sediments has varied widely. The character of deposited sediment has changed from season to season and year to year. Yet some sedimentary layers reveal climatic cycles that span centuries. Some layers show periods of tremendous increase in rainfall accompanied by gradual uplift of the entire area. The resulting abrasive erosion has cut into the Grand Canyon river bed like a notched knife into a layer cake!

The nature and sequence of deposition is displayed before our eyes in the walls of the river's canyon and in the many small canyons cut by tributaries. The grandeur and complexity of the Grand Canyon has to be seen to be fully appreciated. In the Grand Canyon and elsewhere, earth scientists use five commonsense principles to discern the nature and sequence of geological events and the relative ages of rocks.

Original horizontality Layers of sediment are deposited in a rather even manner, with each new layer laid down nearly horizontally over older sediment. If layers are found inclined at an angle—from very slight to very steep—they must have been moved into that position by crustal disturbances after deposition.

Superposition In an undeformed sequence of sedimentary rocks, each layer is older than the one above and younger than the one below. Like the layers of a huge wedding cake, the record was formed from the bottom layer to the top. Upper layers are younger than lower layers.

Cross-cutting When an igneous intrusion or fault cuts through sedimentary rock, the intrusion or fault is younger than the rock it cuts. (Figure 25.2).

Figure 25.1 The lowermost layers of the Grand Canyon are older than the uppermost layers—the principle of superposition.

Figure 25.2 Dikes cutting into a rock body are younger than the rock they cut into. In the diagram, dike A cuts into dike B, and dike B cuts into dike C. From the principle of cross-cutting relationships, dike A is the youngest, followed by dike B, and dike C. The intruded rock body is older than dikes A, B, and C.

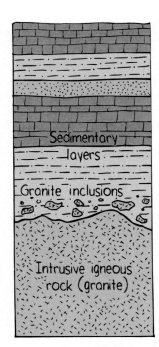

Figure 25.3 The rocks locked in the sedimentary layer existed before the sedimentary layer formed—the principle of inclusion.

Inclusion Inclusions are pieces of one rock type contained within another. The inclusion must be older than the rock containing it, just as pieces of rock that make up a slab of concrete were formed before the concrete was formed (Figure 25.3).

Faunal succession The evolution of life is recorded in the rock record in the form of fossils. Fossils provide a great tool for correlating rocks of similar age in different regions. Organisms succeed one another in a definite, irreversible, and determinable order, so any time period can be uniquely recognized by its fossil content.

It always comes as a surprise to see a fossil of a sea animal, known to be extinct for millions of years, encased in rock at an elevation of a kilometer or so above sea level. Such fossils are evidence that many of today's land surfaces were yesterday's sea bottoms. Finding fossils is a delight to casual and experienced fossil hunters alike (Figure 25.4).

Although most rock layers were deposited without interruption, nowhere is there a continuous sequence from earth's formation to the present time. The processes of weathering and erosion, crustal uplifts, and other geologic processes interrupt the normal sequence of deposition, creating breaks or gaps in the rock record (Figure 25.5). These gaps, or **unconformities,** are detected by observing the relationships of strata and fossils.

The most easily recognized of all unconformities is an **angular unconformity**—tilted or folded sedimentary rocks that are overlaid by younger, flatter rock layers. When overlying sedimentary rocks are found on an eroded surface of igneous or metamorphic rocks, the unconformity is defined as a **nonconformity.** Unfortunately, time gaps are difficult to identify in many cases.

Figure 25.4 Hunting for fossils can be a lot of fun. Finding one is a delightful experience, as the author indicates.

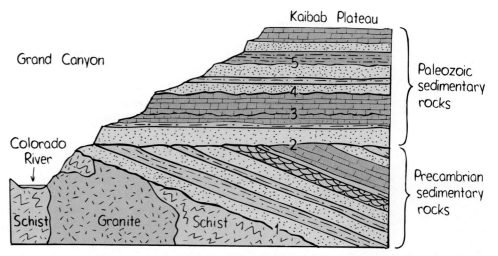

Figure 25.5 The age of the Grand Canyon can be deciphered by its sequence of rock layers. Like many places, the sequence is not continuous and there are time gaps. (1) A nonconformity—separating older metamorphic rocks from sedimentary layers. (2) An angular unconformity—separating older tilted layers from horizontally layered rocks above. (3, 4, and 5) Time gaps are represented between horizontal sedimentary layers. These unconformities are difficult to identify and often require a good eye and a knowledge of fossils.

QUESTION

If a granitic intrusion cuts into or across sedimentary layers, which is older: the granite or the sedimentary layers?

25.2 RADIOMETRIC DATING

Relative dating tells us what parts of the earth's crust are relatively old or new but it doesn't give us the ages of rock in absolute terms. The actual age of a rock can be estimated by **radiometric dating**. Radiometric dating of rock consists of measuring the proportions of radioactive isotopes and their decay products.

Recall from Chapter 15, that in a uranium-bearing rock, the radioactive isotope U-238 (half life 4.5 billion years) decays to Pb-206. The isotope Pb-206 found today was at one time U-238. So in an ore sample containing equal numbers of U-238 and Pb-206 atoms, the age of the ore is one radioactive half life, or 4.5 billion years. If uranium ore contains only a relatively small amount of Pb-206, it is relatively young. The age of a rock can be determined by the amount of radioactive material compared to the relative amount of its daughter products. Other radioactive isotopes frequently used to estimate geologic time are shown in Table 25.1.

ANSWER

The intrusion is actually new rock in the making. Therefore the sedimentary layers are older than the intrusions that cut into them.

TABLE 25.1   Isotopes Most Commonly Used for Radiometric Dating

Radioactive Parent	Stable Daughter Product	Currently Accepted Half-life Value
Uranium-238	Lead-206	4.5 billion years
Uranium-235	Lead-207	7.3 million years
Potassium-40	Argon-40	1.3 billion years
Carbon-14	Nitrogen-14	5730 years

a

b

Figure 25.6   (a) The oldest rock samples, dated at over 3.7 billion years old, occur in southwestern Greenland and in Minnesota. The age of these rocks does not represent the age of the earth for a great span of time passed before crustal formation occurred.
(b) Moon rocks have been dated at 4.2 billion years, which approaches the currently accepted 4.6 billion year age of the earth and solar system. Note the minicraters from meteorite impacts on the moon rock.

Radiometric dating with Uranium-238 is useful for very old rocks. The oldest rocks on earth are estimated to be 3.8 billion years old (Figure 25.6). Old, potassium-rich rocks (such as those containing micas and feldspars) are often dated by measuring the amounts of potassium-40 and argon-40. The radioactive isotope potassium-40 (half life 1.3 billion years) forms the inert gas argon-40 when it decays. Another isotope used in dating is carbon-14. But, because of its short half life (5730 years), it is useful only for geologically recent events. Carbon dating has other uses though. Climatology, studies of the movement of geologically young groundwater and ocean circulation, and the field of archeology utilize carbon-14 for dating.

Radiometric dating is based on the assumption that once a mineral has crystallized, any daughter product results only from the decay of the original unstable parent. If a mineral is reheated by metamorphism, for instance, the "time clock" of the mineral will be reset. This can complicate age estimation. Happily, cross checking by different radiometric methods increases accuracy. By radiometric dating we are able to know both the sequence of events and how long ago each occurred.

So we see that relative dating defines the geologic time scale by the relative order of sedimentary rocks and the fossils they contain, and radiometric dating, based on the natural radioactive decay of certain elements in rocks, assigns specific years to rock formations. Radiometric dating provides specific dates to refine the relative geologic time scale.

By convention the geologic time scale is divided into Eras, Periods, and Epochs. There are three eras, the **Paleozoic** (ancient life), the **Mesozoic** (middle life), and the **Cenozoic** (recent life). Table 25.2 shows each Era divided into Periods, which are further divided into Epochs. The largest span of time, the time period preceding the Paleozoic, is known as the **Precambrian,** the time of "hidden life."

Q UESTION

Could carbon-14 be used for dating rocks from Precambrian time?

A NSWER

No. Carbon-14 has a half life of 5730 years and can only be used to date relatively younger rocks. Any C-14 from Precambrian carbonaceous material would have long since been reduced to insignificant amounts.

TABLE 25.2

THE GEOLOGIC TIME SCALE

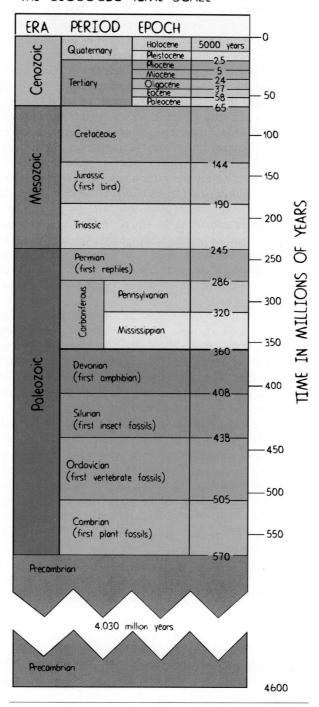

ERA	PERIOD	EPOCH	
Cenozoic	Quaternary	Holocene	5000 years — 0
		Pleistocene	
	Tertiary	Pliocene	2.5
		Miocene	5
		Oligocene	24
		Eocene	37
		Paleocene	58 — 50
			65
Mesozoic	Cretaceous		— 100
	Jurassic (first bird)		144 — 150
	Triassic		190 — 200
Paleozoic	Permian (first reptiles)		245 — 250
	Carboniferous	Pennsylvanian	286 — 300
		Mississippian	320 — 350
	Devonian (first amphibian)		360 — 400
	Silurian (first insect fossils)		408 — 450
	Ordovician (first vertebrate fossils)		438 — 500
	Cambrian (first plant fossils)		505 — 550
Precambrian			570

TIME IN MILLIONS OF YEARS

4,030 million years

Precambrian

4600

25.3 THE PRECAMBRIAN TIME

The Precambrian ranges from about four and one half billion years ago, when earth formed, to about 570 million years ago, when abundant macroscopic life appeared. The Precambrian makes up 85 percent of the earth's history—a time of which we know little.

The beginning of the earth's history is thought to have been a time of considerable volcanic activity and frequent meteorite bombardment. Imagine the earth as it was at that time: an oceanless planet covered with countless volcanoes belching forth gases and steam from its scorching interior. Huge holes and gashes left by meteorite bombardment scarred its surface. There was intense convection in the mantle and severe heat escaping from the interior left the surface of the earth's early crust in turmoil. The earliest crustal formations were short-lived ever-changing small lithospheric plates. About four billion years ago, heat dissipated, large meteorite impacts decreased, and crustal blocks began to survive. All were completely devoid of life however.

Gaseous transfer to the surface by volcanic processes created both a primitive atmosphere and an ocean. The first atmosphere was *anaerobic*—rich in water vapor but very poor in free oxygen. The first simple organisms were stromatolites, algal mats composed of wavy layers of algae that lived in shallow seas (Figure 25.7). Stromatolites played an important role in the gradual accumulation of atmospheric oxygen.

During the late Precambrian, organisms such as stromatolites and blue-green algae developed a simple version of photosynthesis. Photosynthetic organisms require CO_2 to utilize the sun's energy. They keep the carbon and expel the oxygen. With the release of free oxygen, a primitive ozone layer began to develop above the earth's surface. The ozone layer reduced the amount of harmful ultraviolet radiation reaching the earth. This protection and the accumulation of free oxygen in the earth's atmosphere permitted the emergence of new life.

The primitive blue-green algae and bacteria that lived during this time were composed of cells without a nucleus. Reproduction was by simple cell division. The first evidence of nucleated single-celled organisms in green algae occurs in rocks that are dated at 1.3–1.5 billion years ago. The discovery of multicellular plants and animals, dated at approximately 700 million years ago, shows evidence of major evolutionary change during the latter half of the Precambrian. Some rocks in Southern Australia contain diverse fossils of soft-bodied animals, ranging from jellyfish to wormlike forms. This area provides us with the first evidence of an animal community from shallow marine waters.

Figure 25.7 Primitive stromatolites found in western Australia are dated as old as 3.5 billion years. They are very similar in structure to the present day stromatolites pictured above. The first stromatolites were anaerobic but, with time, developed the ability to perform photosynthesis and contributed to making the earth's atmosphere rich in oxygen.

Precambrian Tectonics

Evidence from structural trends and radiometric ages indicates that the first continental crust movements took place about two billion years ago. During this period, continents formed from the accretion of smaller land masses. Speculation holds that about a billion and a half years ago Siberia fused to the western edge of North America while Europe was converging with the eastern region of North America. Other continents were converging from the south to form a large supercontinent. Large scale rifting in the central North American crust began about a billion years ago, resulting in extensive flood basalts and collision of the southeastern margin of North

America with southern continents. Thus culminated the creation of the first documented supercontinent (long before Pangaea).

So we see that the time of the Precambrian covered about four billion years or 85 percent of the earth's total history. During this long time period the earth cooled, slowing crustal movement just enough to allow the formation of continents. Because the interior of the planet remained hot, convection currents permitted movement of the continental plates. Toward the end of the Precambrian primitive photosynthesis developed, free oxygen was generated, and primordial life forms appeared.

25.4 THE PALEOZOIC ERA

The Paleozoic Era is better known than the Precambrian but was actually very short in comparison. It lasted about 375 million years. Sea level rose and fell several times worldwide. This allowed shallow seas to cover the continents and marine life to flourish. Changes in sea level greatly influenced the progression and diversification of life forms—from marine invertebrates to fishes, amphibians, and reptiles. The many different life forms of the Paleozoic mark it as one of the greatest times in the entire record of earth history. The Paleozoic began about 570 million years ago, and is divided into seven periods of geologic time. Each period is characterized by profound changes in life forms as well as by major tectonic changes. We will consider them in chronological order.

Cambrian Period

The Cambrian Period marks the base of the Paleozoic Era. The abundance of well preserved fossils helps to unravel the mysteries of the early earth's many historical changes. The most important event in the Cambrian period was the ability of organisms to secrete calcium carbonate and calcium phosphate for the formation of an outer skeleton. This ability helped organisms to become less vulnerable to predators and provided protection against ultraviolet rays, allowing them to move into shallower habitats.

The fossil record of the Cambrian is dominated by the skeletons of shallow marine organisms. A variety of these organisms flourished, including the *trilobite*, the armored "cockroaches" of the Cambrian sea (Figure 25.8).

Figure 25.8 The trilobite was the dominant fossil of the Cambrian Period.

Ordovician Period

Fossil records show the Ordovician Period was a time of great diversity and abundant marine life. Almost all major groups of marine organisms that could be preserved as fossils came into existence during this time. This includes coral, bryozoans, graptolites, and vertebrates. The most important feature of the Ordovician, as far as the history of evolution, is the appearance of the earliest vertebrate—the armored fish known as the *agnatha* (Figure 25.9). The end of the Ordovician brought many extinctions, likely a result of widespread glaciation. Tropical shallow water marine groups were the most affected, while high-latitude and deep-water organisms were relatively unaffected.

Figure 25.9 The hagfish is a descendant of the *agnatha*, a primitive armored fish that made its debut in the Ordovician Period.

Silurian Period

During the Silurian Period much of the North American continent was at or above sea level. Thick gypsum and other evaporites accumulated in the vanishing seas. The Silurian brought the emergence of terrestrial life. The earliest known terrestrial organisms were vascularized land plants (plants with a well developed circulatory system). These plants were closely tied to their water origins and inhabited only the low wetlands. As plants moved ashore, so did other terrestrial organisms. Air breathing scorpions and millipedes were common terrestrial animals.

Devonian Period

By the Devonian Period many dramatic changes had taken place. Vascular plants were spreading over the land surfaces while fishes proliferated in the seas. The Devonian Period, known as the age of the fishes, witnessed the diversification of the armored fish into many new groups. Some, such as the shark and bony fishes, are still present today. In the bony fish group, the lobe-finned fishes are of particular interest due to their development of internal nostrils. Internal nostrils enabled some species to breathe air. Today the lungfish and the "living fossil"* or *coelacanth* have such internal nostrils and breathe in a similar way. Another important characteristic of the lobe-finned fishes is that their fins were lobed and muscular with jointed appendages that enabled them to "walk." Animal life moved to land. Descended from the lobe-finned fishes, the first amphibians made their appearance during the late Devonian. This

Figure 25.10 Life in the Devonian sea.

---

*The coelacanth was thought to have become extinct after the Mesozoic. However, in 1938 the first living specimen was caught off the coast of East Africa. Since then other specimens have been discovered in the Madagascar area. The coelacanth is now considered to be a "living fossil."

event was of enormous importance on the evolutionary chain of all air breathing, vertebrate land animals. Amphibians, although able to live on land, needed to return to water to lay their eggs. Lowland forests of seed ferns, scale trees, and true ferns also flourished in the Devonian (Figure 25.10).

Carboniferous Period

The Carboniferous Period encompasses both the Mississippian and the Pennsylvanian Periods.* The warm and moist climatic conditions contributed to lush vegetation and dense swampy forests (Figure 25.11). These swamps were the source of the extensive coal beds that lie under North America, Europe, and northern China today. In the Carboniferous Period insects underwent rapid evolution that led to such diverse forms as giant cockroaches and dragonflies with wingspans of 80 centimeters. Some amphibians evolved the capability to live apart from their water environment. The evolution of the amniotic egg, a porous shell with membranes, provided a completely self-contained environment for an embryo. The shell protected the embryo from desiccation. This allowed animals to make a transition from aquatic environments to land and set the stage for the evolution of the first reptiles.

Permian Period

Evolution of the reptiles continued in the Permian Period. The reptiles must have been well suited to their environment for they ruled the earth for 200 million years. By comparison, humans have inhabited the earth for less than one half million years! Three groups of reptiles appeared during the Permian: the *anapsids,* the earliest reptiles; the *diapsids,* which include the lizards, snakes, and dinosaurs; and the *synapsids,* which include the mammal-like reptiles.

At the end of the Permian, one of the greatest extinctions of marine animals in the earth's history occurred. The cause of the extinction is not well understood. One hypothesis is that worldwide global cooling resulted in glaciation and an accompanying lowering of sea level. Climatic extremes ranging from glaciers to deserts are well recorded in the rocks of this time. The duration of low sea level, 20 to 25 million years, no doubt stressed the environments of marine organisms. Yet this alone cannot account for the magnitude of their extinction. Whatever happened took a less

Figure 25.11 Warm and moist climatic conditions contributed to the lush vegetation and swampy coal forests of the Carboniferous Period.

---

*The term *Carboniferous Period* originated in England but is now used around the world. In North America the terms *Mississippian Period,* named for the Mississippi River valley, and *Pennsylvanian Period,* named for the state of Pennsylvania, refer to localities where rocks of these periods are well exposed. Whatever the name, rocks from this time period are known for their distribution of coal beds.

drastic toll on terrestrial life. Terrestrial life, although affected, continued to evolve and expanded rapidly. As we shall see in the next section, one likely explanation for the Permian crisis is the tectonic activity and the formation of Pangaea.

Paleozoic Tectonics

Advances and retreats in sea level characterized the early Paleozoic Era. Recall that as plates move apart they generate a spreading center. When the spreading ridge and the adjacent sea floor thermally expand, the crustal surface stands higher. The breakup of the Precambrian supercontinent nearly 600 million years ago opened up new ocean basins, which, due to their high stand, resulted in a worldwide rise in sea level. During the Ordovician the eastern margin of North America was tectonically active. Eastward subduction of the seafloor beneath an offshore volcanic arc resulted in the collision of the arc with the continental margin. The European margin experienced similar activity. At the same time, on a much larger scale, the ocean floor was being consumed by subduction on both the North American margin and the European margin. As the ocean basin closed, and the continents moved closer to each other, water flooded the land while mountain-building activity occurred on both sides of the ocean (Figure 25.12). These two continents came together in the northern hemisphere to produce the supercontinent of *Laurasia*.

Figure 25.12 Sequence of events of the early to mid Paleozoic: (1) continued divergence of American and European Plates after the break-up of the Precambrian supercontinent; (2) Proto-Atlantic ocean begins to close as American and European Plates converge, causing the formation of a subduction zone and a volcanic arc off the North American margin; (3) continued plate convergence caused by the formation of a subduction zone and volcanic arc off the European margin; (4) accretion of volcanic arcs and plate collision completed the formation of the supercontinent.

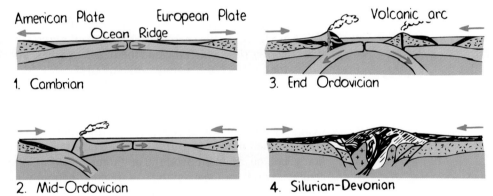

In the southern hemisphere, the *Gondwana* supercontinent was formed by the collision of several land masses—the early continents of Africa, Australia, Antarctica, South America, New Zealand, India, and southeast Asia. During the Ordovician Period, large ice sheets developed over much of Africa, which was then located at the South Pole (Figure 25.13). By the Silurian Period, western Africa and South America had shifted to the south pole.

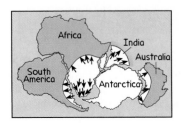

Figure 25.13 During the Ordovician Period, Gondwanaland was situated in the southern hemisphere with Africa over the South Pole. Glacial striations provide clues for the positioning of the continents.

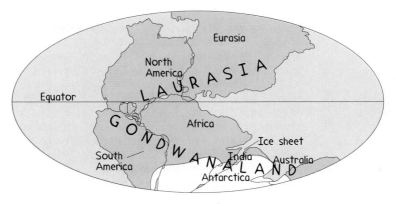

Figure 25.14 With the dramatic collision of Gondwanaland with Laurasia, the super-supercontinent Pangaea was formed.

In late Paleozoic time, the dramatic collision of Gondwanaland with Laurasia resulted in the super-supercontinent of *Pangaea* (Figure 25.14). Mountain-building activity was widespread throughout the Appalachian Mountains in America, the Hercynian and Caledonian Mountains in Europe, and the Ural Mountains in Russia. Disturbances affected not only continental margins but also inner regions. The ancestral Rocky Mountains were internal byproducts of the dramatic collision. By the end of the Paleozoic Era, all the continents were more or less interconnected to form the super-supercontinent of Pangaea.

The repositioning of land masses and a tropical seaway between them contributed greatly to the different climatic belts. North America, Europe, and North Africa were located close to the equator in the trade-winds belt. The climate ranged from humid uplands to drier lowlands. In the Appalachians the climate was monsoonal with seasonal rainfall. The Appalachians were the same height as the Himalayas are today; the high mountains blocked the moisture casting a rain shadow across western North America. The mid-African region was also very dry. The southern climate of Gondwana was dominated by widespread glaciation due to the close proximity to the South Pole. Paleomagnetic evidence suggests that Pangaea was drifting as a unit across the South Pole, accounting for the shift in centers of glaciation. Pangaea was very large, and greatly influenced the climate belts and the evolution of land life.

25.5 THE MESOZOIC ERA

The Mesozoic Era is known as "the age of the reptile." It's made up of three Periods: the *Triassic,* the *Jurassic,* and the *Cretaceous.* The Permian extinction at the end of the Paleozoic Era did not greatly affect the reptiles, so during the Mesozoic Era they became the rulers of the world. By far the most notable event of the Mesozoic is the rise of the dinosaurs.

Land plants greatly diversified during the Mesozoic Era. True pines and redwood forests appeared and rapidly spread throughout the land. Flowering plants arose and diversified so quickly that by the end of the period they were the dominant flora. The emergence of the flowering plants also accelerated the evolution and specialization of insects.

The Dinosaurs

Figure 25.15 Dinosaurs of the Mesozoic Era.

Figure 25.16 Archaeopteryx—the missing link between birds and dinosaurs.

During the late Triassic two orders of dinosaurs developed, the *Saurischia* and the *Ornithischia*. The earliest order, the Saurischia (which means "lizard hipped"), consisted of relatively small, bipedal animals that walked around on strong hind legs. These were the carnivorous dinosaurs. Animals of the order Ornithischia (which means "bird hipped"), derived from the saurischians, were smaller in size, with strong front and back legs. The ornithischian dinosaurs were exclusively herbivorous.

During the Jurassic the dinosaurs multiplied and diversified, and many became giants. These egg-laying vertebrates became the largest of all terrestrial animals. They ruled over the sea as well as the land and air (Figure 25.15). The *Seismosaurus,* the largest of all terrestrial dinosaurs, was over 37 meters long and weighed 100 tons. The *Plesiosaurus,* whose first debut was in the Triassic seas, multiplied and achieved worldwide distribution during the Jurassic. These "sea monsters" ranged from 3 to 18 meters in length. The *Pterodactyl,* one of the winged reptiles, ranged in size from that of a spar-row to that of a dragonlike creature with a wingspread of more than 6.1 meters. Interestingly enough, although unrelated to the *Pterodactyl,* the first birds were probably feathered dinosaurs.

Archaeopteryx, a Jurassic bird, was characterized by both reptilian and birdlike features. The reptilian features are the toothed beak, claws terminating the wing, and a long tail. The presence of feathers and the structure of the legs and wings are the more birdlike features (Figure 25.16). The *Archaeopteryx* provides paleontologists with the most conclusive evidence for the evolution of birds from reptiles.

During the Cretaceous Period the dinosaurs reached the final climax of their evolutionary development, at which time they were more numerous and more varied than at any other stage in their long history. *Tyrannosaurus rex,* the fiercest dinosaur of all, stood 6 meters tall, with a huge head armed with powerful jaws and daggerlike teeth. Some dinosaurs developed armor for protection, the most famous being the *Triceratops* (Figure 25.17).

Figure 25.17 Tyrannosaurus Rex, the fiercest of the dinosaurs, battles with the armored dinosaur Triceratops.

The end of the Cretaceous Period was a time of great extinction, and the dinosaurs were completely wiped out. The cause is unclear and controversy surrounds the many hypotheses.

Perhaps the best documented hypothesis was put forth by the father-son Alvarez team. Luis and Walter Alvarez have suggested that the extinction was caused by the impact of a large meteorite. Their hypothesis comes from their discovery of an abundance of iridium at the Cretaceous-Tertiary boundary. Iridium is not a common element of the earth's crust. Yet all over the world there is heavy deposition of it at the Cretaceous-Tertiary boundary. The Cretaceous-Tertiary boundary layer was deposited about 65 million years ago—the time of the great dinosaur extinction. The Alvarez team believed that the deposited iridium had an extraterrestrial origin. They postulated that an iridium-rich meteorite hit the earth with such force that a giant light-blocking cloud developed. "Nuclear winter" followed as iridium-rich dust filled the air. This stopped photosynthesis, terminated the food supply, and chilled the earth. Nuclear winter subsided when the iridium dust settled to earth. According to recent research, the site of the crater impact is located in the Mexican Yucatan peninsula.

An alternative to the Alvarez hypothesis suggests that the iridium layer may have been generated from massive volcanic eruptions.* The ash and debris from these eruptions also could have blocked out the sun. In any event, the cause for the Cretaceous extinction is unclear and controversial—one of the puzzles of our earth's history. The Cretaceous extinction marked the close of the Mesozoic Era.

Mesozoic Tectonics

The Mesozoic Era witnessed the breakup of Pangaea. The breakup began at the end of the Triassic Period with the eruption of extensive basalt flows associated with two major rift zones. The northern rift zone initiated the separation of North America from Gondwanaland, thus forming the central Atlantic ocean basin. The southern rift zone developed into a triple junction (a point where three lithospheric plates meet). During the Jurassic Period, India started on a northward journey from the triple junction, and simultaneously South America-Africa separated from Australia-Antarctica. The South Atlantic Ocean formed after the split of South America and Africa during the Cretaceous Period. A major fault developed between Africa and Europe which resulted in the fragmentation of the western continental shelf. As the continental shelf broke up, numerous microcontinents were created. The breakup of Pangaea was complex, and occurred during the entire Mesozoic Era. Of all the former continental unions begun in Paleozoic time, only that of Europe and Asia has survived to the present time (Figure 25.18).

For western North America, major compressional mountain building was occurring on the western margin. Subduction of the Farallon Plate and tectonic accretions to the continent began no later than the Triassic Period. This activity produced deformation and widespread volcanism in both the North American and the Andean mountain belts. Granitic batholiths of the Andes and the Sierra Nevadas are the remnants left behind from the numerous volcanic arcs that rimmed the eastern Pacific basin. Successive volcanic arc collisions with the continental margin triggered mountain-building activity across California and Nevada.

Probably due to the breakup of Pangaea, a worldwide rise in sea level occurred during the Cretaceous Period. Over one third of the present land area was submerged. With such an expanse of water covering the land, mild temperate to subtropical conditions dominated most continental areas. Mild ocean temperatures were spread worldwide; mid-latitude regions averaged 25°C, and north polar regions averaged 10°C. Paleomagnetic data reveal that the continents were approaching their present positions.

---

*The earth's mantle contains iridium. The iridium could have been extruded onto the earth's surface via a volcanic eruption.

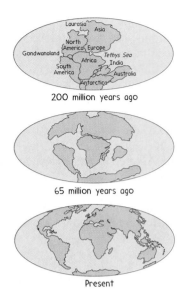

200 million years ago

65 million years ago

Present   **Figure 25.18**   Stages in the breakup of Pangaea.

25.6   THE CENOZOIC ERA

The Cenozoic Era is known as the "age of the mammals." It's made up of two Periods—the *Tertiary* and the *Quaternary*. From oldest to youngest, the Periods are broken up into the *Paleocene, Eocene, Oligocene, Miocene,* and *Pliocene* Epochs for the Tertiary; and the *Pleistocene* and *Holocene* Epochs for the Quaternary. We are currently in the Holocene Epoch. Early Cenozoic life forms, in many respects, are quite similar to those found today.

Another period of extinction occurred during the late Cenozoic. The extreme climatic variation could have affected some groups of mammals. However, the climate in many areas was relatively mild. In addition there is a strong correlation between the emergence of humans and the period of extinction of large mammals. In North America, many large mammal extinctions occurred after humans crossed the Bering land bridge. In Africa, mammalian extinctions can be related to the appearance of the Stone Age hunters. The cause of the Cenozoic extinction is a much debated issue. Was it due to climatic change? Was it due to human emergence? Like many ideas in science, ongoing research is necessary to pinpoint what the most important factors are. A single direct cause cannot be found for all events. In fact, the Cenozoic extinctions were probably the result of many factors. Only further research may prove the cause of the Cenozoic extinction.

Cenozoic Tectonics

Structural disturbances of great magnitude occurred rapidly throughout the world during the Tertiary Period of the Cenozoic Era. In the late Tertiary a collision between westward moving North America and the Pacific ridge system produced compressional forces within the continent leading to events that characterized this era.

Compressional mountain building in western North America culminated in the uplift of the Rocky Mountains and regional upwarping. This uplift rejuvenated

Life in the Cenozoic Era

Plant life of the Tertiary Period rapidly expanded and diversified. As the climate differentiated, specialized habitats for plants developed. Deciduous flowering plants dominated in the northern, colder regions. Evergreens prospered in the warmer tropic regions. When prairie grasses appeared, grazing animals benefitted greatly from their new habitat (Figure 25.19).

these giant beasts have been found at the La Brea tar pits in Los Angeles (Figure 25.20).

Descended from prosimian primates (tree shrews, lemurs, and tarsiers), true apes became well established in the rich tropical forests of the Nile Valley. They adapted to live in many habitats. The hominid primates lived in forested areas with stretches of open grassland. By the

Figure 25.19 Grasslands dominated the land during the Tertiary, providing new habitats for grazing animals.

Over time, mammals diversified. Their size also increased. The earliest mammals were quite small and rodent-like. By the Oligocene the mammals were more modern in appearance. Oligocene animals included pigs, deer, giraffes, camels, horses, and rhinoceroses. Some mammals grew to larger sizes than are typical of mammals today.

The diversification of the mammals continued into the Pleistocene. The mastodon and woolly mammoths were quite common in North America. Carnivorous mammals such as the saber tooth cat and the dire wolf achieved great physical size and strength. The remains of both of

end of the Miocene, molecular and morphological evidence shows a divergence between the hominid primates and the pongid (ape) primates.

Of all the Pleistocene events, the most important from our perspective was the emergence of early humans. Human evolution was complex. An incomplete pathway led from our primitive ancestor to modern humans. Along the path the guiding principle has been natural selection in favor of the development of intelligence. Human ancestors with large brains and reasoning capability had the advantage over their often ferocious competitors.

Figure 25.20 Ancient view of the La Brea tar pits in Los Angeles, California.

Early Humans

The earliest known hominid fossils are 3.2 million years old. Of particular interest is the nearly complete skeletal fossil of Lucy, a 20-year-old female of the *Australopithecus afarensis* species. This species was partially bipedal but perhaps not quite erect. She had a large overhung jaw with canine and incisor teeth, and an undetermined brain size. Later species of *Australopithecus* became more upright as bipedalism increased and a slightly larger brain (500 cm^3) developed. The first truly erect hominid, *Homo erectus,* is dated at 1.6 million years old. With a brain size of 1000 cubic centimeters, evidence of complex behavior, and the development of social patterns, it is quite likely that *Homo erectus* evolved into the modern human species, the *Homo sapiens*. Remnants of the first *Homo sapiens* were found in Africa and Europe in middle Pleistocene rocks. These early *Homo sapiens* are called **Neanderthal** people. They're characterized by heavy eyebrow ridges, a pronounced chin, and a stocky body. The brain size averaged 1200 cubic centimeters. The Neanderthal people were cave dwellers. Evidence of their well-developed society is indicated by their fabrication of stone tools, knowledge of fire, and burial of the dead.

Modern humans, *Homo sapiens sapiens*, first appeared some 90,000 years ago in South Africa, 50,000 years ago in the Middle East, and 35,000 years ago in Europe. These early humans, known as **Cro-Magnon,** are characterized by a high flattened brow, a shorter skull, and a rounded face. Cro-Magnon brain size averaged 1,300 cubic centimeters. Neanderthal and Cro-Magnon humans coexisted for about 10,000 years, but Cro-Magnon humans dominated and rapidly spread throughout the world.

As humans moved into new areas racial characteristics evolved to facilitate adaptation to particular environments. For example, different skin color evolved in response to the ultraviolet light intensity at different latitudes. Pigmented skin acts as an effective ultraviolet screen. As humans moved into higher latitudes the ultraviolet light grew less intense, dark skin pigmentation was not needed as a filter, and selective pressures operated toward lighter skin. No later than 10,000 years ago, all major races of modern humans had appeared and occupied their primary distribution areas.

Although modern humans appeared over 90,000 years ago, investigators today believe that it wasn't until about 25,000 years ago that humans reached North America. With the lowering of sea level due to glaciation, the Bering land bridge allowed the entrance of humans from Asia to the North American continent. The same route was likely used earlier by the woolly mammoth, reindeer, and other Late Cenozoic animals.

all streams and rivers by steepening the gradients. The reinforced streams and rivers brought about a long period of canyon cutting. The Grand Canyon testifies to the dramatic effects of this erosional period. Also, widespread flood basalts spilled out from deep fissures to make up the Columbia Plateau in Washington and the Craters of the Moon in Idaho. Crustal extension produced normal faulting in the Basin and Range province. The San Andreas fault, the main boundary between the Pacific and North American Plates, grew in length and extended northward as what is now northern Mexico encountered the Pacific ridge (Figure 25.21). In time, Baja California was torn away from the Mexican mainland and the Gulf of California was created. The process continues, and western California will eventually either become completely detached from the mainland or will find itself joined to western Canada.

The Hawaiian Island–Emperor Seamount chain (Figure 25.22) gives evidence of the change in movement of the Pacific Plate. The bend in the chain of islands

Figure 25.21 The San Andreas fault is the result of the North American Plate encountering the Pacific Ridge system. As the San Andreas grew in length Baja California was torn from the continental margin.

occurred about 40 million years ago when plate motion changed from nearly due north to a northwesterly direction. The change in direction corresponds with the collision of northern Mexico with the Pacific ridge.

In Eurasia there was also considerable tectonic activity. Recall the formation of microcontinents after the breakup of Pangaea. These microcontinents were transported northeastward to eventually collide with the basement of Southern Eurasia. The culmination of these collisions occurred in the mid-Cenozoic when Afro-Arabia collided with Europe to produce the Alps, and India collided with Asia to produce the Himalayas. The leading edge of the Indian Plate was forced partially under Asia, generating an unusually thick accumulation of continental lithosphere. Due to isostasy, the thick lithosphere gave additional uplift to the Himalayas (Figure 25.23).

During the Pleistocene, glaciation greatly impacted the surface features of the earth. With nearly one third of present land covered by ice, it is not difficult to comprehend that glaciation has influenced the fluctuation in sea level, depression of the land from ice load, and considerable changes in the drainage pattern of many streams and rivers. The Pleistocene glaciers left behind dramatic changes on many of our present day land surfaces. The fluctuating glacial-interglacial climates are responsible for carving out many of the land forms that we see today. Do you ever wonder what will happen next?

In summary we see that the history of our planet is long and complex. The rock record chronicles the evolution of many life forms. From the beginning of primitive life to the emergence of man, the history of life on earth is recorded in the rocks. The rocks were laid down in a sequence that reads like a book from beginning to end,

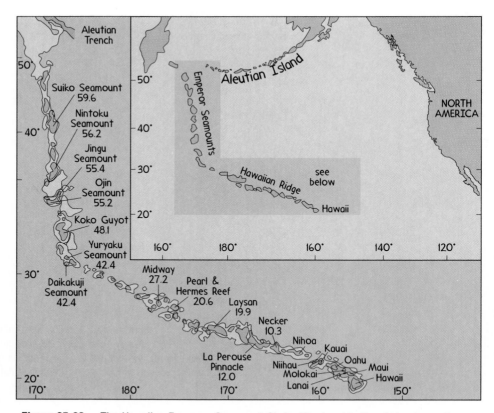

Figure 25.22 The Hawaiian Emperor–Seamount Chain. The bend in the chain shows the change in direction of the Pacific Plate as a result of the collision of northern Mexico with the Pacific Ridge.

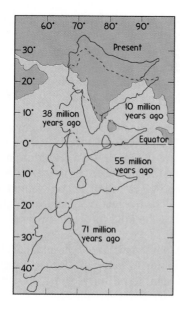

Figure 25.23 The formation of the Himalayas was a result of the collision of India with Asia. As this was a continent-to-continent collision, the Himalayas have an unusually thick accumulation of continental lithosphere. Like the bottom of an iceberg, the mountains run deeper than they are high.

with each interval of life and time unique. From the time that life first appeared on the earth, we see that birth, development, old age, and death follow in an unbroken succession. We as human beings are the only creatures in earth's history, for better or worse, to be a geologic force—we change our environment. Yet, as human beings we are still a part of the earth's unbroken succession.

Human Geologic Force

Perhaps the most important event in the history of earth science, at least from our perspective, was the emergence of humans. Although the "human age" amounts to only a brief 0.05 percent of geologic time, we are probably the most clever and adaptable organism that has evolved on the planet. We do more than adapt to our environment; we manipulate it to meet our needs. We have but to look at the irrigation systems of Mesopotamia, the cultivation of the Nile, the plowing of the prairies in the Great Plains, the invention of machines to further utilize the land, and the dams and locks on the Mississippi, Missouri, and Colorado rivers to pause and consider what the mind and hand of humanity has done! Our control of the environment has resulted in the destruction of the habitats of many other species. An average estimate of one species per year over the past century has become extinct due to human encroachment into new habitats. And now we are faced with the problems of PCBs and the ozone layer, hydrocarbon pollution, global warming, and poisoning of air, land, and sea. Will we awake some morning, terrified by the realization that human beings have, in a few short years, set our terrestrial home on a course that could return it to the condition described at the beginning of this chapter? Is it too late?

SUMMARY OF TERMS

Original horizontality Layers of sediment are deposited in a rather even manner, with each new layer laid down almost horizontally over the older sediment. If layers are found to be inclined at a steep angle, they must have been moved into that position by crustal disturbances at a time after their deposition.

Superposition In an undeformed sequence of sedimentary rocks, each bed or layer is older than the one above and younger than the one below.

Cross-cutting When an igneous intrusion or fault cuts through other rocks, the intrusion or fault must be younger than the rock it cuts.

Inclusions Pieces of one rock type contained within another. The inclusion must be older than the rock containing it in order to become included as part of the host rock.

Faunal succession Fossil organisms succeed one another in a definite, irreversible, and determinable order. Because of this any time period can be uniquely recognized by its fossil content. Fossils are a great tool for correlating rocks of similar age in different regions.

Unconformities A break or gap in the geologic record, such as by an interruption in the sequence of deposition, or by a break between eroded metamorphic or igneous rocks and a sedimentary rock formation.

Angular unconformity An unconformity in which the older strata dip at an angle different from that of the younger beds.

Nonconformity The condition of overlying sedimentary rocks on an eroded surface of igneous or metamorphic rocks.

Radiometric dating Calculating an age in years for geologic materials based on the nuclear decay of naturally occurring radioactive isotopes.

Neanderthal Early form of humans (*Homo sapiens*) who were cave dwellers and lived during the last glacial period, between 100,000 and 40,000 years ago. Neanderthals show evidence of a well developed society, stone tools, knowledge of fire, and burial of the dead.

Cro-Magnon Human beings (*Homo sapiens sapiens*) who lived about 35,000 years ago and assumed greater numbers than the Neanderthals.

REVIEW QUESTIONS

Relative Dating

1. Before the discovery of radioactivity, how did geologists determine the age of the earth?

2. The concept of relative dating states that rocks and their structures are formed in a definite sequence or order. What five principles are used in relative dating?

3. When we find a granitic dike in a massive bed of sandstone what can we say about the age of the dike and the age of the sandstone? What is this principle called?

4. In what type of rock do we find fossils?

5. What is a fossil? How are they used in the determination of geologic time?

6. In a sequence of sedimentary rock layers that have been folded we find the oldest layer to be on the bottom and the youngest layer at the top. What principle does this observation fit?

7. In a sequence of sedimentary rock layers we find the youngest rock layer at the bottom and the oldest rock layer at the top. What could this type of layering signify?

8. Why isn't it possible to find a rock formation with a continuous sequence from the beginning of time to the present?

9. How is it possible that fossils of fishes and other marine organisms are found at high elevations such as the Himalayas?

10. In an undeformed sequence of rocks fossil X is found in a limestone layer at the bottom of the formation and fossil Y is found in a shale layer at the top of the formation. What can we say about the ages of fossils X and Y?

Radiometric Dating

11. What is meant by half life? What are the half lives of uranium-238, potassium-40, and carbon-14?

12. What isotope is best for dating very old rocks?

13. What isotope is commonly used for dating rocks from the Pleistocene?

The Precambrian Time

14. Which of the geologic time units spans the greatest length of time?

15. How old is the earth?

The Paleozoic Era

16. The geologic time scale is divided into three eras. Each era signifies a major change. What is this change?

17. The Paleozoic Era experienced several fluctuations in sea level. What effect did this have on life forms?

18. There are seven periods of geologic time for the Paleozoic Era. What are these periods?

19. The Cambrian Period is often referred to as the "Cambrian Explosion." What is meant by this phrase?

20. What is the Silurian Period most known for?

21. The Devonian is known as "the age of fishes." Briefly discuss the different Devonian life forms.

22. During the Devonian the lobe-finned fishes developed internal nostrils. What is the significance of this development?

23. Why are the lobe-finned fishes considered to be an important group of fishes?

24. Coal beds are formed from the accumulation of plant material in an oxygen-poor environment. As trees and shrubs died and fell into swamps, much of their debris became submerged and buried, thus restricting oxygen and preventing decay. During what time period were these coal deposits laid down? Why was this period unique?

25. In what area of the United States do we find rich coal deposits?

26. Amphibians have a close link with their water environment since they require water for reproduction. The evolvement of an amniotic embryo had a great effect on the amphibians. What was this effect?

27. The end of the Permian Period was marked by one of the greatest extinctions of earth's history. What circumstances may have contributed to such a catastrophe?

28. What is Gondwanaland?

29. What is glossopteris? Where would you find it?

30. The collision of Gondwanaland and Laurasia resulted in the super-supercontinent of Pangaea. What effect did this collision have on the land features? On the different climatic regions? On sea level?

The Mesozoic Era

31. What is the Mesozoic Era known as?

32. Dinosaurs ruled the earth during what periods?

33. What is Archaeopteryx?

34. What are the possible reasons given for the Cretaceous extinction?

35. What does iridium have to do with the extinction of the dinosaurs?

36. The breakup of Pangaea was complex and occurred over the entire Mesozoic Era. What effect did this

breakup have on sea level? Do you think the breakup of Pangaea affected the Cretaceous extinction?

37. The breakup of Pangaea resulted in the formation of our present continents. What Pangaean land mass still survives to this day?

The Cenozoic Era

38. The Cenozoic Era is made up of two periods, the Tertiary and the Quaternary. What life forms characterize these two periods?

39. What epochs make up the Tertiary? The Quaternary?

40. Explain the formation of the Alps.

41. What route did humans use to enter the Western Hemisphere?

42. Why did the western margins of North and South America experience folding and episodes of volcanism during the Cenozoic Era?

43. The bend in the Hawaiian Island–Emperor Seamount chain was caused by what event?

44. Explain the formation of the Himalayan Plateau.

45. Relate the Grand Canyon to tectonic activity in the Cenozoic Era.

46. What role did tectonic activity play in the formation of the San Andreas fault?

47. How did the Pleistocene glaciation affect the land surface?

48. Some of the Cenozoic life forms grew to great proportions. What are some of the possible causes for their great size?

49. How was the Gulf of California formed?

50. What land bridge did the Ice Age mammals use to cross over into new continents?

• •

EXERCISES

1. In the field we encounter an outcrop of sedimentary rock that is overlaid by a basalt flow. A fault displaces the bedding of the sedimentary rock but does not intersect the basalt flow. Relate the fault to the ages of the rock.

2. If a sedimentary rock contains inclusions of metamorphic rock, which is older, the sedimentary rock or the metamorphic rock?

3. Refer to the figure below. Using the principles of relative dating, determine the relative ages of the rock bodies and other lettered features. Start with the question: What was there first?

Sequence of events

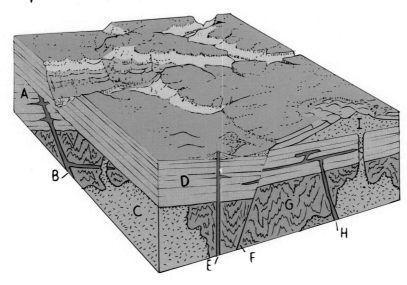

4. Absolute dating is the determination of the numerical age for a given geologic event using the relative abundances of radioactive parent and daughter elements. Which isotopes would be most appropriate for dating formations from the following ages?

   (a) early Precambrian;
   (b) Mesozoic;
   (c) late Pleistocene.

5. Has the amount of uranium in the earth increased in geologic time? Has the amount of lead changed?

6. In dating a mineral, how can its time clock be reset? What effect does this have on time determination?

7. Granitic pebbles within a conglomerate have a radiometric age determination of 300 million years. What can be said about the age of the conglomerate? Nearby, an outcrop of the same conglomerate is intruded by a dike. The dike has a radiometric age determination of 200 million years. With this information what can we say about the age of the conglomerate?

8. In an undeformed sequence of rocks we find a trilobite embedded in shale layers at the bottom of the formation and fossil leaves embedded in shale at the top of the formation. With only these two clues, what can we tell from our observations?

9. Just as sediments are deposited, the processes of erosion and crustal uplift act to create breaks or gaps in the rock record. These breaks or gaps are known as unconformities. Distinguish among the different types of unconformities.

10. If fine muds were laid down at a rate of 1 cm per 1000 years, how long would it take to accumulate a sequence 1 kilometer thick?

11. The Precambrian makes up over 85 percent of our earth's history. Yet we know few details about this time. What key developments occurred during this span of time?

12. The first atmosphere during the early Precambrian consisted of carbon dioxide, water vapor, and nitrogen, with very little free oxygen. What factors are believed to have contributed to the generation of free oxygen? In what way did the increase in oxygen affect our planet?

13. The earliest stromatolites were anaerobic. Are the modern day stromatolites anaerobic? Give reasons for your belief.

14. What evidence do we have of Precambrian life?

15. Why can we find Paleozoic sedimentary rocks, such as limestone and dolomite, widely distributed in the continental interiors?

16. Coal beds are formed from the accumulation of plant material that becomes trapped in swamp floors. Yet coal deposits are present in the continent of Antarctica, where no swamps or vegetation exist. How can this be?

17. How have recent humans affected geological processes?

18. Using flakes of mica for radiometric dating, what different events would be found in (a) a granite, (b) a schist, and (c) a sandstone?

19. During the earth's long history, life has emerged and life has perished. Briefly discuss the emergence of life and the extinction of life for each era.

20. Why does sea level go up (rise) when spreading rate increases?

21. What are two possible effects of rising sea level? If sea level were to rise today what areas would be most affected? Could this cause the extinction of any life forms?

22. In what ways could sea level be lowered? How might this affect existing life forms?

23. Other than an increase in spreading rate, what could cause a rise in sea level? Is this likely to happen in the future? Why or why not?

24. When humans spread out into new areas they changed as they adapted to their new environments. What were some of these changes? Are the effects of these changes still seen today?

25. What is the distinguishing difference between humans and other life forms from the past to the present? Is it possible that we too will one day become extinct?

26

METEOROLOGY

The earth is surrounded by a life-giving envelope of air—the **atmosphere**. This atmosphere protects life on the surface of the planet by blocking the lethal high frequency part of the sun's radiant energy. Without this envelope of air, life as we know it would never have evolved. If its composition were to change in the future, life may cease to exist. How did our atmosphere come to be, and how do its properties and functions affect us?

As we shall see in the next chapter, most scientists agree that the sun and planets of our solar system formed from an enormous cloud of cosmic gas and dust, the **solar nebula**. Initially, this gaseous cloud of hydrogen, helium, and a few heavier elements was extremely cold, less than 50 K. Molecular speeds were so slow and gas pressures so low that gravity exerted the decisive force and matter drifted toward the center of the solar nebula. As the cloud contracted due to gravity, escalating temperatures in the hot center led to the formation of the **protosun** (a pre-sun formation before the nuclear fusion stage). The central portion of the protosun eventually became our sun, with a surrounding disk of gas and dust that eventually formed the planets (more about this in Chapter 27).

This process has likely occurred countless times throughout the universe as stars like our sun were formed. During protostar formation young planets in the making collect most of their matter. When additional matter collides into the forming planets the kinetic energy is converted to heat. The heat of collision combined with the heat generated by gravitational contraction and radioactive decay produce a molten planet. Heavy elements such as iron and nickel sink to the center, and lighter elements such as silicon and aluminum rise to the surface. Different shells form, gases are belched, and if the planet is large enough, escaping gases will form an atmosphere. This is likely how the earth's earliest atmosphere came to be.

26.1 EVOLUTION OF EARTH'S ATMOSPHERE

Earth's atmosphere evolved in discrete stages. The earth probably had an atmosphere before the sun was fully formed. If so, it would have been composed principally of hydrogen and helium (the two most abundant gases in the universe), along with a few simple compounds such as ammonia and methane.

Meanwhile, as temperature and pressure in the contracting center of the proto-sun continued to rise, the temperature there got high enough to ignite thermonuclear reactions, turning the protosun into a star. Our sun was born!

The blast from this reaction must have produced solar winds strong enough to sweep the earth of its earliest atmosphere. The next stage in the formation of the atmosphere occurred when gases trapped in the earth's hot interior escaped through volcanoes and fissures at the earth's surface. The gases that were spewed out in these early eruptions were probably much like the gases found in the volcanic eruptions of today—about 85 percent water vapor, 10 percent carbon dioxide, and 5 percent nitrogen, by mass. As the earth cooled, the rich supply of water vapor condensed to form oceans. These oceans, essential to the evolution of life, and ultimately to the development of the present global environment, have remained for the entire history of the earth.

The early atmosphere had no free oxygen, and therefore was inhospitable to life as we know it. The production of free oxygen did not occur until green algae appeared. Green algae, like all higher forms of green plants, use photosynthesis to convert carbon dioxide and water to hydrocarbon and free oxygen.

$$CO_2 + H_2O + light \rightarrow CH_2O + O_2$$

With the production of free oxygen, an ozone (O_3) layer formed in the atmosphere. The ozone layer acted like a filter to reduce the amount of ultraviolet radiation reaching the earth's surface. Thus the surface became more hospitable to life. The evolution of this global envelope has been a vital step in the history of the earth and its life.

26.2 COMPOSITION OF EARTH'S ATMOSPHERE

If gas molecules in the atmosphere were not constantly moving, our atmosphere would lie dormant on the ground like popcorn at the bottom of a popcorn machine. But add heat to the popcorn, or the atmospheric gas, and both will bumble their way

TABLE 26.1   Composition of the Atmosphere

Permanent Gases			Variable Gases		
Gas	Symbol	% by Volume	Gas	Symbol	% by Volume
Nitrogen	N$_2$	78	Water vapor	H$_2$O	0 to 4
Oxygen	O$_2$	21	Carbon dioxide	CO$_2$	0.034
Argon	Ar	0.9	Ozone	O$_3$	0.000004*
Neon	Ne	0.0018	Carbon monoxide	CO	0.00002*
Helium	He	0.0005	Sulfur dioxide	SO$_2$	0.000001*
Methane	CH$_4$	0.0001	Nitrogen dioxide	NO$_2$	0.000001*
Hydrogen	H$_2$	0.00005	Particles (dust, pollen)		0.00001*

* Average value in polluted air.

up to higher altitudes. Popcorn attains speeds of a meter per second, and can rise a meter or two, but air molecules move at speeds of about 1600 kilometers per hour, and a few bumble up to more than 50 kilometers in altitude. If there were no gravity, both popcorn and atmosphere would fly into outer space. Fortunately there is sun, and gravity, so we have a friendly atmosphere. The atmosphere protects the earth and its inhabitants from harmful radiation and cosmic debris by absorbing and scattering radiation and causing solid matter to burn by heat generated from air friction.

Table 26.1 shows that the earth's present day atmosphere is a mixture of various gases—primarily nitrogen and oxygen with small percentages of argon and carbon dioxide, and minute traces of other elements and compounds.

Vertical Structure of the Atmosphere

If you have ever gone mountain climbing you probably noticed that the air grows cooler and thinner with increasing elevation. At lower elevations, such as sea level, the air is generally warmer and denser. The greater density near the earth's surface is due to gravity. Like a deep pile of feathers, the density is greatest at the bottom of the pile and least at the top. More than half the atmosphere's mass lies below an altitude of 5.6 kilometers (3.5 miles), and about 99 percent lies below an altitude of 30 kilometers (18 miles). Unlike a pile of feathers, the atmosphere doesn't have a distinct top. It gradually thins to the near vacuum of outer space.

We consider the atmosphere as divided into layers, each distinct in its characteristics (Figure 26.1). The lowest layer is the **troposphere,** which is the thinnest of the atmospheric layers, containing 90 percent of the atmospheric mass and essentially all of the atmosphere's water vapor and clouds. This is where weather occurs. Commercial jets generally fly at the top of the troposphere to minimize the buffeting and jostling caused by weather disturbances. The troposphere extends to a height of 16 kilometers (10 miles) over the equatorial region and 8 kilometers (5 miles) over the polar regions. Temperature of the troposphere decreases steadily (6°C per kilometer) with increasing altitude. At the top of the troposphere, temperature averages about −50°C.

Above the troposphere is the **stratosphere,** which reaches a height of 50 kilometers (31 miles). Ultraviolet radiation from the sun is absorbed by a very thin **ozone layer** in the stratosphere. As radiation is absorbed, temperature increases from about −50°C at the bottom to about 0°C at the top. The ozone layer acts as a sunscreen, protecting the surface below from harmful solar ultraviolet radiation (see the box on polar ozone depletion).

Figure 26.1 The lower atmospheric layers.

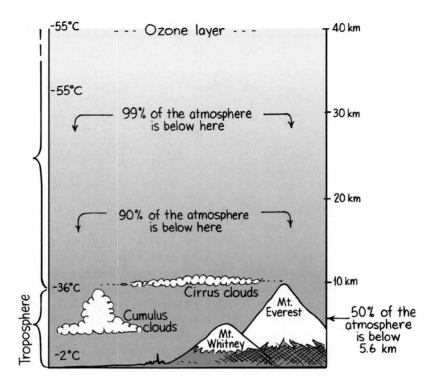

Above the stratosphere, the **mesosphere** extends upward to about 80 kilometers (50 miles). The gases that make up the mesosphere absorb very little of the sun's radiation. As a result, temperature decreases from about 0°C at the bottom of the layer to about −90°C at the top.

The situation is just the opposite in the layer above the mesosphere, the **thermosphere**. Extending upward to 500 kilometers (310 miles), it contains very little air. What air there is absorbs enough solar radiation to bring about a 2000°C temperature! This extreme temperature, however, has little significance because of low air density. Very little heat would be transferred to a slowly moving body in this region.

The **ionosphere** is an ion-rich region within the thermosphere and uppermost mesosphere. The ions in it are produced from the interaction of high frequency solar radiation and atmospheric atoms. The incoming solar rays strip electrons from nitrogen and oxygen atoms producing the large concentration of free electrons and positive ions in the ionosphere layer.

Radio waves from radio transmitters on the earth's surface reflect from the ionosphere (Figure 26.2). The number of ions increases with altitude. More ions correspond to greater reflection of radio waves. Thus higher regions have higher ion concentrations and are better reflectors of radio waves. In the lower region, the ions not only reflect radio waves but also weaken them by absorption. Layer settling occurs at night, and the reflection of radio waves, particularly lower-frequency AM signals, is accompanied with less energy loss. Thus, standard AM radio waves are able to reflect better at nighttime, and reception is extended thousands of kilometers.

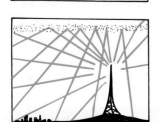

Figure 26.2 The effect of the ionosphere on radiowaves. During the daytime, solar radiation increases ionization, which hinders radio-wave reflection. At night the ions settle and allow the reflection of radio waves, particularly the lower-frequency AM signals. That's why AM reception over great distances is better at nighttime.

Figure 26.3 The aurora borealis over Alaska is created by solar gas particles that strike the upper atmosphere and light up the sky (just as similar particles on a smaller scale light up a fluorescent lamp).

Ions in the ionosphere cast a faint glow that prevents moonless nights from becoming stark black. Near the poles, fiery auroral displays occur as solar wind (high speed charged particles ejected by the sun) further agitates the ionosphere (Figure 26.3). These displays are particularly spectacular during times of solar disturbances such as solar flares. These disturbances also play havoc with radio reception, so when the aurora is brilliant, ham radio operators hear more static on their receivers (Figure 26.4).

Finally, above 500 kilometers (310 miles), in the **exosphere,** the thinning atmosphere gradually yields to the radiation belts and magnetic fields of interplanetary space.

Figure 26.4 Ham radio operators hear more static on their receivers during solar disturbances such as sunspots and solar flares.

26.3 SOLAR ENERGY

Why are some regions of the earth hot and others cold? Why are equatorial regions always warmer and polar regions always colder? The temperature of the earth's surface depends very much on the energy per surface area received from the sun each day. This depends on the angle between the sun's rays and the earth's surface. To see how, hold a flashlight vertically over a table, shining the light directly down on the flat surface. The light produces a bright circle. Now tip the light at various angles to the horizontal, and notice how the circle elongates into ellipses, spreading the same energy over more area and decreasing the intensity of illumination (Figure 26.5). Likewise for sunlight on the earth's surface. High noon in equatorial regions is akin to the vertically held flashlight; high noon at higher latitudes is akin to the flashlight held at an angle.

The sun is never directly overhead in the northern United States and Canada. These temperate regions have distinct summer and winter seasons because of the variation in the sun's rays. Figure 26.6 shows how the tilt of the earth and the corresponding different spreadings of solar radiation produce the yearly cycle of seasons. When the sun's rays are closest to perpendicular, the region experiences summer. Six months later the rays are incident upon the same region at a lower angle and we have winter. In between are the seasons fall and spring.

Figure 26.5 When the flashlight is held directly above at a right angle to the surface, the beam of light produces a bright circle. If the light is shone at an angle, the light beam is dispersed over a larger area and is less intense.

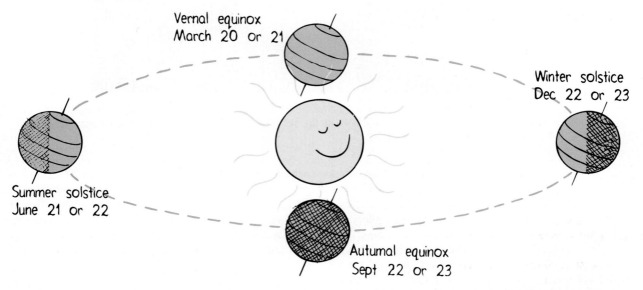

Vernal equinox
March 20 or 21

Winter solstice
Dec 22 or 23

Summer solstice
June 21 or 22

Autumal equinox
Sept 22 or 23

Figure 26.6 The tilt of the earth and the corresponding different spreading of solar radiation produce the yearly cycle of seasons.

Figure 26.7 Over each square meter that is perpendicular to the sun's rays at the top of the atmosphere, the sun pours 1400 J of radiant energy each second. Hence the solar constant is 1.4 kJ/s/m², or 1.4 kW/m².

Another effect of the tilting rays is the length of daylight each day. Can you see in the figure that a location in summer has more daylight per daily rotation of the earth than the same location when the earth is on the opposite side of the sun in winter? This is most pronounced at high latitudes. Consider the special latitude where daylight on it lasts 24 hours during peak summer, and nighttime lasts 24 hours at the peak of winter. This latitude in the northern hemisphere is the Arctic Circle. In the southern hemisphere it is the Antarctic Circle. Summer and winter are reversed, of course, in the two hemispheres. Daylight and nighttime hours are equal in mid-September and mid-March, between the peaks of winter and summer (not only at the Arctic and Antarctic Circles, but all over the world). Above the Arctic Circle (and below the Antarctic Circle) daylight occurs 24 hours per day in the summer months, and nighttime lasts 24 hours per day during the winter months. Day is never really bright, for the sun hangs low above the horizon. Likewise, nights are never really dark because the sun hangs low beneath the horizon. The polar regions are eerie places.

Solar Constant

So we see that the earth's surface is warmed by radiant energy from the sun, and we see how yearly variations in the spreading of sunlight produce the seasons. We feel the sun's energy when we step from the shade into the sunshine. The warmth we feel isn't so much because the sun is hot, for its surface temperature of 6000°C is no hotter than the flames of some welding torches; we are warmed mainly because the sun is so *big*. As a result, it emits enormous amounts of radiant energy, most of which of course misses the earth. Less than one part in a billion reaches the earth. The amount of energy received each second over each square meter at right angles to the sun's rays at the top of the atmosphere, the **solar constant,** is 1400 joules (Figure 26.7). This input, in power units, is 1.4 kilowatts per square meter (1.4 kW/m²). Solar intensity reaching the ground is much reduced by atmospheric absorption.

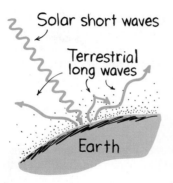

Figure 26.8 The hot sun emits short waves (high frequency). The cool earth re-emits long waves (low frequency). Radiation emitted from the earth is *terrestrial radiation*.

Terrestrial Radiation

Incoming solar radiation warms the earth's surface, which in turn warms the air around us. Recall from Chapter 7 that all objects both absorb and emit radiation. The earth absorbs solar radiation and emits **terrestrial radiation**. It is terrestrial radiation rather than solar radiation that directly warms the lower atmosphere (Figure 26.8). That's a major reason why air close to the ground is appreciably warmer than air at higher elevations. The temperature of the earth's surface depends on the amount of solar radiation coming in compared to the terrestrial radiation going out. In direct sunlight, the net effect is warming as the earth's surface absorbs more energy from the sun than it emits. At night, the net effect is cooling as the earth's surface emits more energy than it absorbs. Cloud cover acts to block either incoming solar radiation or outgoing terrestrial radiation. Can you see that cloudy days are cooler than sunny days, and cloudy nights are warmer than clear nights?

The earth's average temperature changes very little from one year to the next. In the equation of solar energy in and terrestrial energy out, starting with incoming solar radiation as 100 percent, about 19 percent is absorbed by the atmosphere and clouds, and 30 percent is reflected back to space by the earth's surface and atmosphere, particularly by clouds. This leaves 51 percent absorbed by the earth's surface (Figure 26.9).

Figure 26.9 Distribution of solar energy incident upon the earth.

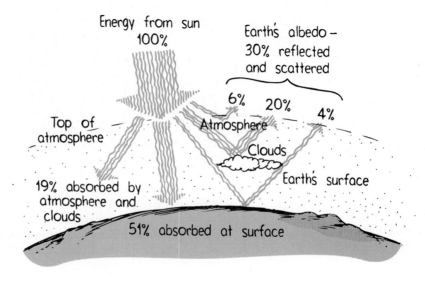

The atmosphere is more transparent to short-wavelength solar radiation than to longer-wavelength terrestrial radiation. Recall from Chapter 7 that it is these differences in atmospheric absorption of different wavelengths that produce the greenhouse effect, which plays a large role in global warming.

26.4 BEHAVIOR OF AIR

Air is a mixture of a few kinds of gas molecules moving haphazardly in all directions, colliding with one another like ricocheting marbles. Moving molecules exert a small push on whatever they hit. This force (push) spread over the area on which it is exerted is pressure. Because we are talking about air molecules we use the term *air pressure*. The higher the kinetic energy of the air molecules, the greater the air pressure. Or the greater the number of air molecules colliding, that is, the denser the air, the greater will be the air pressure. The behavior of air depends a lot on its temperature, pressure, and density.

Whatever the behavior of air, a key concept is energy. Recall the thermal version of the conservation of energy in Chapter 8—the *First Law of Thermodynamics:* Whenever thermal energy is added to a system, it transforms to an equal amount of some other form of energy. Meteorologists express the first law of thermodynamics in terms of energy, temperature, and pressure, in the following form:

Change in temperature ~ thermal energy added/subtracted + pressure change

The temperature of an air mass may be changed by adding or subtracting thermal energy, by changing the pressure, or by both. Thermal energy can be added or subtracted in several ways—added by solar radiation, by moisture condensation, or by contact with warm ground. Thermal energy can be subtracted by radiation to space, by evaporation of rain falling through dry air, or by contact with cold surfaces. When thermal energy is added, temperature rises; when thermal energy is subtracted, temperature drops.

There are many atmospheric processes, usually involving time scales of a day or less, in which the amount of thermal energy added or subtracted is very small—so small that the process is nearly **adiabatic** (no thermal energy enters or leaves the system). Then the change in temperature is due only to pressure changes. In this case the adiabatic form of the first law of thermodynamics is:

Temperature change ~ pressure change

Adiabatic processes in the atmosphere are characteristic of large air masses, or *parcels*. Since parcels are so large, mixing of different temperatures or pressures at their edges does not appreciably alter the overall composition of a parcel, which behaves as if it were enclosed in a giant very-thin-plastic garment bag. As an air parcel flows up the side of a mountain, its pressure decreases, allowing it to expand and cool. To convince yourself that expanding air cools, blow onto your hand through your puckered lips and note how they cool as your breath expands, as compared to blowing with your mouth open so your breath doesn't expand as much. Reduced pressure with accompanying expansion of air results in reduced temperature.

Measurements show that the temperature of a dry air parcel will decrease 10°C for a decrease in pressure corresponding to a 1-kilometer gain in altitude. So dry air cools 10°C for each kilometer it rises (Figure 26.10). Air flowing over tall mountains or rising in thunderstorms may change elevation by several kilometers. Thus,

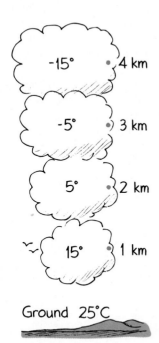

Figure 26.10 The temperature of a parcel of dry air that expands adiabatically changes by about 10°C for each km of elevation.

Ground 25°C

-15° 4 km
-5° 3 km
5° 2 km
15° 1 km

Figure 26.11 Chinooks, warm dry winds, occur when high altitude air descends and is adiabatically warmed.

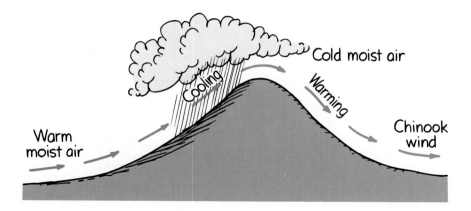

when a dry air parcel at ground level at a comfortable 25°C rises 6 kilometers, the temperature drops to a frigid −35°C. On the other hand, if air at a typical temperature of −20°C at 6 kilometers descends to the ground, its temperature rises to a whopping 40°C.

A dramatic example of this adiabatic warming is the **Chinook**—a dry wind that blows down from the Rocky Mountains across the Great Plains (Figure 26.11). Cold air moving down a mountain slope is compressed to a smaller volume, and is appreciably warmer. The effect of expansion or compression on gases is quite impressive.*

A rising parcel cools as it expands. But the surrounding air is also cooler at higher elevation. The parcel will continue to rise as long as it is warmer (less dense) than the surrounding air. If it gets cooler (denser) than its surroundings, it will sink. Under some conditions, large parcels of cold air sink, and remain at a low level with the result that the air above is warmer. When the upper regions of the atmosphere are warmer than the lower regions, we have a **temperature inversion**. Unless rising warm air is less dense than this upper layer of warm air, it will rise no farther. It is common to see evidence of this over a cold lake, when visible gas and particles, such as smoke, spread out in a flat layer above the lake rather than rising and dissipating higher in the atmosphere (Figure 26.12). Temperature inversions trap smog and

Figure 26.12 The layer of campfire smoke over the lake indicates a temperature inversion. The air above the smoke is warmer than the smoke, and the air below is cooler.

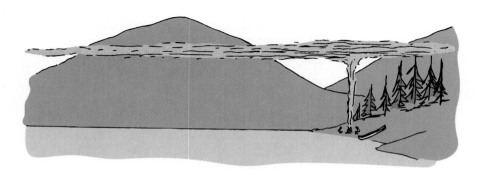

---

*Interestingly enough, when you're flying at high altitudes where outside air temperature is typically −35°C, you're quite comfortable in your warm cabin—but not because of heaters. The process of compressing outside air to a cabin pressure of nearly sea level would normally heat the air to a roasting 55°C (131°F). So air conditioners must be used to extract heat from the pressurized air.

Figure 26.13 Smog in Los Angeles is trapped by the mountains and a temperature inversion caused by warm air from the Mohave Desert overlying cool air from the Pacific Ocean.

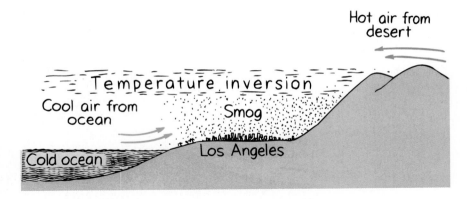

other thermal pollutants. The smog of Los Angeles is trapped by such an inversion, caused by low-level cold air from the ocean capped by a layer of hot air moving over the mountains from the hot Mojave Desert. The mountains help hold the trapped air (Figure 26.13). The mountains on the edge of Denver play a similar role in trapping smog beneath a temperature inversion.

Adiabatic parcels are not restricted to the atmosphere, and changes in these parcels do not necessarily happen quickly. Convection of some deep ocean currents takes thousands of years for circulation. The water masses are so huge, and conductivities are so low, that no appreciable quantities of thermal energy are transferred to or from these parcels during these long periods of time. They are warmed or cooled adiabatically by changes in internal pressure. Changes in adiabatic ocean convection, as evidenced by the recurring El Niño ocean current in the Pacific Ocean, have a great effect on the earth's climate. Ocean convection is influenced by the temperature of the ocean floor, which in turn is influenced by convection currents that operate beneath the earth's crust.

Questions

1. If a parcel of air initially at 0°C expands adiabatically while flowing upward alongside a mountain a vertical distance of 1 km, what will its temperature be? How about when it has risen 5 km?

2. What happens to the air temperature in a valley when dry cold air blowing across the mountain tops descends into the valley?

3. Imagine a giant dry-cleaner's garment bag full of air at a temperature of −10°C floating like a balloon with a string hanging from it 6 km above the ground. If you were able to yank it suddenly to the ground, what would its approximate temperature be?

Answers

1. At 1-km elevation, its temperature will be −10°C; at 5 km, −50°C.

2. The air is adiabatically compressed and the temperature in the valley is increased. In this way, residents of some valley towns in the Rocky Mountains, such as Salida, Colorado, experience "banana belt" weather in midwinter.

3. If it is pulled down so quickly that heat conduction is negligible, it would be adiabatically compressed by the atmosphere just as air is compressed in a bicycle pump, and its temperature would rise to a piping hot 50°C (122°F).

26.5 MOVEMENT OF AIR

Figure 26.14 The general cell-like circulation pattern is the result of unequal heating of the earth's surface. As heated air rises at the equator it moves toward the polar regions where it gradually cools in the upper air. The cooled air then sinks and is drawn back to the warmer regions of the equator.

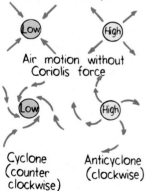

Figure 26.15 The Coriolis effect—winds deflected from straight line paths by the earth's rotation. In the northern hemisphere, rotation is counterclockwise around a low and clockwise around a high, as viewed from above. How would rotation differ in the southern hemisphere?

We know that as warm air rises, it expands and cools. As the air cools it sinks to occupy the region left vacant by the rising warm air. These motions result in the generation of a convection cycle and thermal circulation of the air. As convection currents stir the atmosphere, wind results. Wind is air in nearly horizontal motion, generated in response to pressure differences in the atmosphere, which in turn are the result of temperature differences. Pressure differences occur on a local level as well as on a larger, global level. Air moves toward a low pressure region, where it rises, or away from a high pressure region, where it sinks. The underlying cause of general air circulation comes from the unequal heating of the earth's surface.

Global Wind Circulation

We have seen that equatorial regions receive optimum radiant energy from the sun and as a result have higher average temperatures than other regions. As the heated air rises at the equator it moves out toward the polar regions, cooling gradually in the upper air. This cooled air then sinks and is drawn back to the warmer regions of the equator. If we assume the earth to be a nonrotating, water-covered sphere, the effect is a simple single cell-like circulation pattern (Figure 26.14).

The earth's rotation greatly affects the path of air circulation.* As the earth rotates towards the east, the air moves with it, and circulates in an eastward direction. Thus, the path of circulation in the northern hemisphere has an eastward component as well as a northward component in the middle latitudes, and its path is curved to the northeast. In the southern hemisphere the eastward component is the same but the air moves to the south, so its path turns to the southeast. This is the **Coriolis effect**. Winds are deflected from straight paths to curved paths, by virtue of rotational effects. Thus the air rushing in to a low pressure region does not move directly inward, but instead rotates counterclockwise in the northern hemisphere and clockwise in the southern hemisphere (Figure 26.15).†

Cell-like circulation patterns are responsible for the redistribution of heat and our global winds (Figure 26.16). Beginning at the equator, direct heat causes the air motion to flow upward with very little horizontal movement, resulting in a vast low-pressure zone. A narrow windless realm occurs with air that is still, hot, and stagnant. Seamen of long ago cursed the equatorial seas as their ships floated listlessly

---

*As the air circulates over the oceans, it causes the surface water to drift along with it. A severe storm in 1990 has given scientists an unusual tool for studying the currents of the Pacific Ocean. Five cargo containers of Nike shoes were washed overboard from freighters that ran into stormy seas en route from South Korea to the Pacific Northwest. Since then, in a confirmation of theories about currents in the Northwest Pacific, thousands of sneakers, hiking boots, children's sandals, and other shoes have been picked up along beaches from British Columbia to Oregon and as far into the Mid-Pacific as Hawaii. Despite months at sea, most shoes have been wearable after washing. The problem, though, is that the shoes were not tied together. Beachcombers have since formed "swap meets" to search for mates of found shoes!

†Whether winds are deflected clockwise or counterclockwise has to do with latitude, and whether the winds circle about high pressure regions or about low pressure regions. For brevity, we skip a detailed explanation of the Coriolis effect and its variations, and leave this to outside reading for those of you who will further pursue the study of meteorology.

Figure 26.16 Global winds are the result of several cell-like circulation patterns, brought about by unequal distribution of land masses, with consequent unequal heating of the earth's surface, and compounded by effects of the earth's rotation.

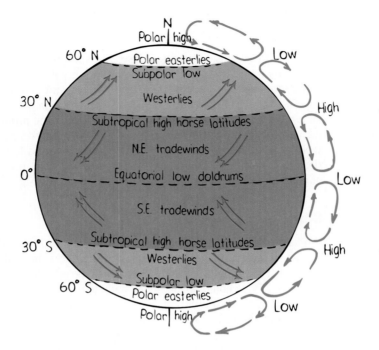

for lack of wind, and referred to the area as the **doldrums**. When the moist air from the doldrums rises, it cools and releases torrents of rain. Over the land areas, these frequent rains give rise to the tropical rain forests that characterize the equatorial region.

From the sweltering doldrums, the air rises to the boundary between the troposphere and stratosphere where it divides and spreads out to the north or south. Very little wind crosses the equator into the neighboring hemisphere. At about 30° N and 30° S latitudes, the air cools and descends toward the surface. The descending air is compressed and warms. A resulting high pressure zone girdles the earth, creating a belt of hot and dry surface air. On land, these high pressure zones account for the world's great deserts—the Sahara Desert in Africa, the Arabian Desert in the Middle East, the Mojave Desert in the United States, and the great Victoria Desert in Australia. At sea, the hot, descending air produces very weak winds. According to legend, early sailing ships were frequently stalled at this latitude, and as food and water supplies dwindled, horses on board were eaten for food or cast overboard to conserve fresh water. As a result, this region is now known as the **horse latitudes**. The thermal convection cycle that starts at the equator is completed when air flowing southward from the horse latitudes in the northern hemisphere and northward in the southern hemisphere is deflected westward to produce the **trade winds**. Air that flows northward from the horse latitudes in the northern hemisphere and southward in the southern hemisphere is deflected eastward to produce the prevailing **westerlies.***

From the polar regions frigid air continually sinks, pushing the surface air outward. The Coriolis effect is quite evident in the polar regions as the wind deflects to

---

*Meteorologists refer to wind direction as the direction from which the winds come. In the case of the westerlies the wind comes from the west but moves toward the east.

the west to create the **polar easterlies**. The cool polar air meets the warm air of the westerlies at latitudes 60° N and 60° S. This boundary, called the **polar front,** is a zone of low pressure where surface air converges and rises, often causing storms.

The mid-latitudes are noted for their unpredictable weather. Although the winds tend to be westerlies, they are often quite changeable as the temperature and pressure differences between the subtropical and polar air masses at the polar front produce powerful winds. As air moves from regions of high pressure, where air is denser, toward regions of low pressure, the result is a cyclone effect. Irregularities in the earth's surface also influence wind behavior. Mountains, valleys, deserts, forests, and great bodies of water all play a part in determining how the wind blows.

In the upper troposphere, "rivers" of rapidly moving air meander around the earth at altitudes of 9 to 14 kilometers (30,000 to 45,000 feet). These high-speed winds are the **jet streams**. With wind speeds averaging between 95 to 190 kilometers per hour (60 to 115 miles per hour) the jet streams play an essential role in the global transfer of thermal energy from the equator to the poles.

The two most important jet streams, the polar jet and the subtropical jet, form in both the northern and southern hemispheres in response to temperature and pressure contrasts. The formation of polar jet streams is a result of the contrast in temperature at the polar front where cool polar air meets warm tropical air. With a sudden change in temperature there is a sudden change in pressure, and the steep pressure gradient intensifies the wind speed. The polar jet stream shows seasonal variation. During the winter, the polar jet is strong and extensive as it migrates to lower latitudes bringing strong winter storms and blizzards to the United States. In summer, the jet stream is weaker and migrates to higher latitudes.

The subtropical jet streams are generated as warm air is carried from the equator to the poles, producing a sharp temperature contrast along the boundary (subtropical front). Once again the sharp contrast in temperature produces a corresponding contrast in pressure and generates strong winds. The subtropical jet stream above Southeast Asia, India, and Africa merits special mention. The formation of this jet stream is related to the warming of the air above the Tibetan highlands. During the summer the air above the continental highlands is warmer than the air above the ocean to the south—thus a temperature contrast and pressure gradient generates strong on-shore winds that contribute to the region's **monsoon** (rainy) climate. During winter, the winds change direction to produce a dry season.

This cycle of winds characterizes the climates of much of Southeast Asia (Figure 26.17). The predictable rain-bearing summer wind from the sea that moves over the heated land is called the *summer monsoon;* the prevailing wind from land to sea in winter is called the *winter monsoon.*

The jet stream significantly affects travel by airplane. Depending on whether tail winds or head winds are encountered, the progress of the flight is aided or impeded. The jet stream is sought by eastbound aircraft in order to gain speed and save fuel, and avoided by westbound aircraft. For this reason eastbound flights are usually faster.

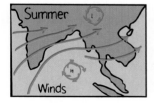

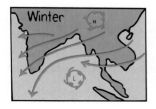

Figure 26.17 During winter months, air over continents is cooler than air over oceans. The winter monsoon has generally clear skies with winds that blow from land to sea. During the summer months the situation is reversed—air over the oceans is cooler than the air over land. The summer monsoon brings heavy rains as the winds blow from sea to land.

Polar Ozone Depletion

Within the stratosphere at an altitude of about 30 km is the ozone layer that protects life below from harmful solar ultraviolet radiation. As was discussed in Chapter 20, this ozone layer is threatened by the catalytic action of both natural and human-made pollutants. Evidence for this threat came in the early 1970s when ozone depletion over Antarctica was first observed. Interestingly enough, this depletion, which appears as an "ozone hole," occurs each southern spring during the months of September and October.

Why ozone depletion is so apparent over Antarctica at only a certain time of the year can be understood by considering the meteorology of this region. Like the equatorial and horse latitudes, the geographic poles are naturally regions of relatively light winds. During the Antarctic winter (June and July), however, the sun remains set beyond the horizon and darkness prevails. Without the heat from the sun, the temperatures drop and the air becomes unusually still. This condition allows the formation of rare ice-crystal-containing stratospheric clouds, which scientists believe facilitate the destruction of ozone by providing a surface upon which potential ozone-destroying reactants gather. When spring comes, the sun rises and solar energy greets the accumulated reactants thereby initiating an intense onslaught on the existing ozone. This continues until the stratospheric clouds are dissipated by solar heat and the ozone-destroying reactants are dispersed by the buildup of winds.

Since the early 1970s careful monitoring has shown that each year the Antarctic ozone hole has become more intense, due in part to increased concentrations of stratospheric chlorine, which is formed from chlorofluorocarbons (CFCs)—a human-made pollutant. Recently, initial signs of ozone depletion over the northern pole during the Arctic spring months (March and April) have also been observed. The extent of the northern Arctic ozone hole, however, has not been as great as in the Antarctic. A likely reason has to do with the distribution of land masses over the planet. In the northern hemisphere there is more land mass. Since land has a lower heat capacity (Chapter 6), it heats up and cools down quicker than the oceans. This gives rise to more turbulent weather patterns, which may serve to inhibit the formation of stratospheric clouds and the accumulation of ozone-destroying reactants.

QUESTION

What is the main underlying cause of the trade winds, the jet streams, the monsoons, and ultimately their bearing on the world's climates?

ANSWER

Simply enough, the main underlying cause of these air circulation patterns stems from the unequal heating of the earth's surface.

26.6 EVAPORATION, CONDENSATION, AND PRECIPITATION

The continual exchange of water between the planet's surface and the atmosphere is a major factor in determining weather conditions. Evaporation, as covered in Chapter 6, is the process whereby water molecules change from a liquid phase to a vapor phase. With over 70 percent of the earth covered by water, solar energy absorbed by the sea warms the water and enhances evaporation. The resulting water vapor then condenses to form clouds and eventually precipitates.

There is always some water vapor in the air. At any given temperature there is a limit to the amount of water vapor in the air. When this limit is reached, the air is **saturated**. A measure of the amount of water vapor in the air is called **humidity** (the mass of water per volume of air). Weather reports often use **relative humidity**—the ratio of water vapor in the air to the amount of water vapor the air could hold at that temperature. Relative humidity is a good indicator of comfort. Relative humidity of 50 percent, for example, means the water content in the air is half the amount it could be at that temperature.*

Saturation results when temperature falls so that water vapor molecules in the air condense. Recall from earlier chapters that water molecules are electric dipoles,

Fast-moving H$_2$O molecules rebound upon collision

Slow-moving H$_2$O molecules coalesce upon collision

Figure 26.18   Condensation of water molecules.

---

*For most people, conditions are ideal when the temperature is about 20°C and the relative humidity is about 50 to 60 percent. When too high, moist air feels "muggy" as condensation counteracts the evaporation of perspiration. Cold air with a high relative humidity feels colder than dry air of the same temperature because of increased conduction of heat from the body. When the relative humidity is high, hot weather feels hotter, and cold weather feels colder.

Figure 26.19 San Francisco is well known for its summer fog.

and tend to stick together. Because of their normally high average speeds in air, however, they rebound when they collide. Slow-moving water molecules are likely to stick upon collision (Figure 26.18 on page 675). In a similar way, a fly making slow-moving contact with flypaper sticks to the paper, whereas a fast-moving fly may rebound from the flypaper. The slower a water molecule moves, the more likely it will be to condense and form droplets. Although condensation in the air occurs more readily at low temperatures, it occurs to some extent at high temperatures too. There are always some molecules moving slowly enough to condense when they collide.

As air rises, expands, and cools, water vapor molecules make slower-moving collisions. If there are larger and slower-moving particles or ions present, water vapor condenses upon these particles, and we have a cloud. As the size of the cloud droplets grow, they fall as rain, sleet, or snow. This is **precipitation**. When condensation is at the earth's surface, we call it **dew, frost,** or **fog**. On cool clear nights objects near the ground cool more rapidly than the surrounding air. As the air cools below a certain temperature, the **dew point,** water condenses onto the nearest available surface. This may be a twig or blade of grass, or the windshield of a car, and we have early morning dew. When the dew point is at or below freezing, we have frost. When a large mass of air cools and the relative humidity approaches 100 percent, we have a cloud near the ground—fog (Figure 26.19).

26.7 CLOUDS

When condensation occurs above the earth's surface, we have a cloud.* Cloud formation takes place as rising moist air expands and cools. Clouds are formed by vertical air motion and are shaped and moved about by horizontal winds. Clouds are generally classified according to height and shape. There are ten principal cloud forms, each divided into four primary groups. Table 26.2 lists the four groups and their cloud types, which are shown in Figure 26.20.

TABLE 26.2   *The Four Major Cloud Groups and Their Cloud Types*

1. High clouds
  Cirrus
  Cirrostratus
  Cirrocumulus
2. Middle clouds
  Altostratus
  Altocumulus

3. Low clouds
  Stratus
  Stratocumulus
  Nimbostratus
4. Clouds (vertical development)
  Cumulus
  Cumulonimbus

*If the particles or ions on which water vapor normally condenses are not present, we can stimulate cloud formation by "seeding" the air with appropriate particles or ions.

Figure 26.20 The different cloud formations.
(a) The high clouds
(b) The low clouds
(c) The middle clouds

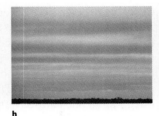

a b c

High Clouds High clouds generally form above 6000 meters. The air at this elevation is quite cold and dry so clouds this high are made up almost entirely of ice crystals. The most common high clouds are thin, wispy **cirrus** clouds, blown by high winds into the well known wispy mare's tail or "artist's brush." Cirrus clouds indicate fair-to-pleasant weather, but may also indicate approaching rain. **Cirrocumulus** clouds are the familiar rounded white puffs, in patches, seldom covering more than a small portion of the sky. Small ripples and a wavy appearance make the clouds resemble fish scale structures; hence, cirrocumulus clouds are often said to make up a *mackerel sky*. **Cirrostratus** clouds are thin and sheetlike, and often cover the whole sky. The ice crystals in these clouds act to refract light and produce a halo around the sun or the moon. When cirrostratus clouds thicken they give the sky a white glary appearance—an indication of coming rain or snow.

Middle Clouds Middle clouds form between 2000 and 6000 meters. These clouds are made up of water droplets, and when temperature allows, ice crystals. **Altostratus** clouds are gray to blue gray, often covering the entire sky for hundreds of square kilometers. Thin altostratus clouds are often confused with thick cirrostratus clouds. Altostratus clouds are often so thick that ground shadows are not produced. Since altostratus clouds often form ahead of storms, look on the ground and if you don't see your shadow, cancel that picnic. **Altocumulus** clouds appear as gray puffy masses in parallel waves or bands. The individual puffs are much larger than those found in cirrocumulus and the color is also much darker. The appearance of these clouds on a warm humid summer morning often indicate thunderstorms by late afternoon.

Low Clouds Low clouds form below 2000 meters. They are almost always made up of water droplets, but in cold weather may contain ice crystals and snow. **Stratus** clouds are uniformly gray and often cover the whole sky. They are very common in winter and as a consequence give rise to the sky's "hazy shade of winter." They resemble a high fog that doesn't touch the ground. Although stratus clouds are not directly associated with falling precipitation, they sometimes generate a light drizzle or mist. **Stratocumulus** clouds form a low lumpy layer that grows in horizontal rows or patches, or, with weak updrafts, appear as rounded masses. The color is generally light to dark gray. They are often confused with altocumulus clouds but can be distinguished by their size; hold your hand at arm's length and point toward the cloud. The altocumulus cloud will commonly appear as the size of a thumbnail, whereas the stratocumulus cloud will be about the size of your fist. Precipitation of rain or snow does not usually fall from stratocumulus clouds. **Nimbostratus** clouds are dark and foreboding. They are a wet looking cloud layer associated with light to moderate rain or snow.

Clouds with Vertical Development Vertical cloud development is caused by rising air currents. These are **cumulus** clouds—the most familiar of the many cloud forms. They resemble pieces of floating cotton with sharp outlines and a flat base, they are white to light gray in color, and are generally only about 1000 meters above the surface. The tops of the clouds are often in the form of rising towers, denoting the limit of the rising air. They are the clouds childhood dreams are made of. Using imagination one can see huge horses, dragons, and magic palaces in them. It is easy to imagine climbing on them or wandering through them. When the cumulus clouds turn dark and are accompanied by precipitation they are referred to as **cumulonimbus** clouds and indicate a coming storm. As we shall see, these clouds often become thunderheads.

26.8 WEATHER

Consider a world with a uniform smooth surface, with no thermal differences between land and water, with no cloud cover, and with no rotational spin. On such a world there would be no changes in air temperature, wind, or precipitation. There would be no changes in *weather*. Our world has a nonuniform surface, has great thermal differences between land and water, has a very variable cloud cover, and steadily rotates every 24 hours. All these factors contribute to the *weather*.

Weather is a description of the day-to-day changes in air temperature, wind, and precipitation at a specific locality. Weather is affected by a **front,** the contact zone between two air masses. If a cold air mass is moving into an area occupied by a warm air mass, the contact zone between them is called a *cold front,* and if warm air moves into an area occupied by cold air, the zone of contact is called a *warm front.* If neither of the air masses is moving, the contact zone is called a *stationary front.* The fronts between air masses are usually accompanied by wind, clouds, rain, and storms.

Air Masses, Fronts, and Storms

The characteristics of an air mass depend on where the air mass forms. An air mass formed over water in the tropics has different characteristics than an air mass formed, say, over land in the polar regions. Air masses are divided into four general categories according to the surface type and latitude of their source regions (Table 26.3 and

TABLE 26.3   Classification of Air Masses and Their Characteristics

Source Region	Classification	Symbol	Characteristics
Arctic regions	Maritime arctic	mA	Cool, moist, unstable
Greenland	Continental arctic	cA	Cold, dry, stable
N. Atlantic & Pacific oceans	Maritime polar	mP	Cool, moist, unstable
Alaska and Canada	Continental polar	cP	Cold, dry, stable
Caribbean Sea, Gulf of Mexico	Maritime tropical	mT	Warm, moist, usually unstable
Mexico, SW United States	Continental tropical	cT	Hot, dry; stable aloft, unstable at surface

Figure 26.21 Source regions of air masses for North America.

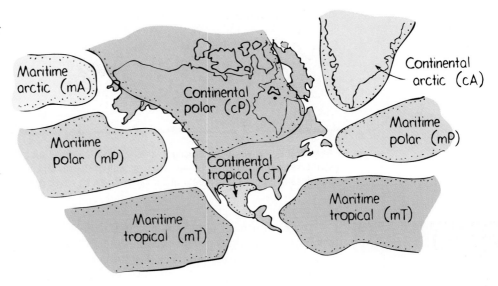

Figure 26.21). The surface type is designated by lower case letters (*m* for maritime, *c* for continental), and the source region is designated by upper case letters (*A* for Arctic, *P* for polar, *T* for tropical.)

Each air mass has its own characteristics. Continental polar and Arctic air masses generally produce very cold dry weather in winter, and cool, more pleasant weather in summer. Maritime polar and Arctic air masses, picking up moisture as they travel across the oceans, generally bring cool moist weather to a region. Continental tropical air masses are generally responsible for the hot dry weather of summer, while warm humid conditions are due to maritime tropical air masses. Where air masses with different properties meet, we find weather fronts.

When cold air moves into a warm air mass, forming a cold front, warm air is forced upward. As it rises, it cools, and water vapor condenses to clouds. Before the front, strong winds blow over cumulonimbus clouds that turn into cirrostratus and cirrus clouds. At the front, thunderstorms develop with heavy showers and gusty winds. Behind the front, the air cools and sinks, the pressure rises, and rain stops. Except for a few fair weather cumulus clouds (Figure 26.22), the skies clear and we have the calm after the storm.

Figure 26.22 A cold front occurs when a cold air mass moves into a warm air mass. The cold air forces the warm air upward where it condenses to form clouds. If the warmer air is moist and unstable, heavy rainfall and gusty winds develop.

Figure 26.23   A warm front occurs when a warm air mass moves into a cold air mass. The less dense warmer air rides up and over the colder denser air, resulting in widespread cloudiness and light to moderate precipitation that can cover great distances.

Figure 26.24   The mature stage of a thunderstorm cloud appears as a dark, towering cumulonimbus cloud that reaches into the tropopause—the boundary between the troposphere and the stratosphere. Strong horizontal winds and icy temperatures flatten and distend the cloud's crown into a characteristic anvil shape.

When warm air moves into a cold air mass, forming a warm front, the less dense warmer air rides up and over the colder denser air. High cirrus clouds indicate the coming of a warm front. The sky turns to an overcast gray. Light to moderate rain or snow develops, and winds become brisk. At the front, air gradually warms, and the rain or snow turns to drizzle. After the front, the air is warm and the clouds scatter (Figure 26.23).

Storms often develop as a result of vertical air movement within a single air mass, or as a result of frontal activity between two air masses.

Thunderstorms When warm humid air rises in an unstable environment, thunderstorms are produced. The cycle of a thunderstorm begins with humid air rising, cooling, and condensing into a single cumulus cloud. This cloud builds and grows upward as long as it is fed by rising air from below. Particles of precipitation grow larger and heavier, and eventually begin to fall. The falling rain creates a downdraft, chilling the air, making it both colder and denser than the air around it. Together, the updraft and the downdraft make a storm cell within the cloud. This is the mature stage where the thunderstorm cloud appears as a lonely giant, dark and brooding in the sky. It has a typical base several kilometers in diameter and towers over 12 kilometers into the tropopause,* where horizontal winds and lower temperatures flatten and distend its crown into a characteristic anvil shape (Figure 26.24). Thunder and lightning and heavy rain (and sometimes hail) fall from the cloud. After the thunderstorm dissipates, it leaves behind the cirrus anvil as a reminder of its once mighty presence.

At any time, there are about 1800 thunderstorms in progress in the earth's atmosphere. Wherever thunderstorms occur, there is lightning and its noisy companion, thunder. Lightning strikes the earth some 100 times every second, with some lightning bolts having an electric potential of as much as 100 million volts. Lightning claims more than 200 victims per year in the United States alone.

Tornadoes A revolving object, such as a ball on a string, speeds up when pulled toward its axis of revolution, thus conserving its angular momentum. Similarly, winds

---

*The boundary between the troposphere and the stratosphere.

Figure 26.25 Like a giant vacuum cleaner, the strong wind of a tornado can pick up and obliterate everything in its path.

slowly rotating over a large area will speed up when the radius of rotation decreases, producing a **tornado**. A tornado is a funnel-shaped cloud that extends from a large cumulonimbus cloud. The funnel cloud is called a tornado only after it touches the ground. The winds of a tornado travel at speeds of up to 800 kilometers per hour in a counterclockwise direction. As it moves across the land, advancing at forward speeds from 45 to 95 kilometers per hour, a tornado may bounce and skip, rising briefly from the ground and then touching down again. As it moves it acts like a giant vacuum cleaner picking up everything in its way. It hits the earth's surface in an explosion of flying dirt and debris (Figure 26.25).

Tornadoes occur in many parts of the world. In the flat central plains of the United States, a tornado zone extends from northern Texas through Oklahoma, Kansas, and Missouri, where over 300 tornadoes touch down each year—a zone known as Tornado Alley. Tornadoes are so frequent in this part of the country that some homes are built with underground shelters (do you recall the Wizard of Oz?). The power of a tornado is terrifying and devastating.

Hurricanes In the steamy tropics, as the sun warms the oceans, the transfer of thermal energy to the atmosphere by evaporation and conduction is so thorough that air and water temperatures are about equal. This high humidity favors the development of cumulus clouds and afternoon thunderstorms. Most of these individual storms are not severe. However, as moisture and thermal energy increase, and surface winds converge, a strong vertical wind shear can cause the updraft to tilt producing a more violent storm—a hurricane—with wind speeds up to 279 kilometers per hour (173 miles per hour). Gaining energy from its source area, a hurricane grows as more air rises and increasing winds rotate around a relatively calm low-pressure zone—the eye of the storm.

The forecast of hurricanes and other storms is a prime concern to meteorologists. Weather forecasting is, in part, a matter of determining air mass characteristics, predicting how and why they might change, and in what direction they might move. In the case of hurricanes and tornadoes, such predictions are lifesaving (Figure 26.26).

Figure 26.26 On August 24, 1992, Hurricane Andrew struck the southern peninsula of Florida with measured 164-mile-per-hour gusts. The storm was catastrophic, with 45 deaths and over 180,000 people left homeless. Without the timely hurricane watch issued by the National Weather Service, casualty losses could have been much more severe.

26.9 WEATHER FORECASTING

Meteorologists have a long and remarkable record of reducing property loss and saving many lives. All of our earth's weather occurs in the troposphere. Weather is always changing. To forecast how weather will change we must first know the present conditions of weather over a large area. The data needed include all the elements of the air conditions: the temperature, the air pressure, the humidity, the type of clouds, the level of precipitation, the visibility, and the direction and speed of the wind. This is a great deal of data, especially if measurements need to be taken throughout a large region. A network of observation stations has been set up with over 10,000 land-based stations, hundreds of ship-based stations, and several satellites that provide surface weather information to the World Meteorological Organization (WMO). The WMO is an organization of 130 nations that share and exchange international weather information. This information is then sent to the National Meteorological Center (NMC), located near Washington D.C., where the different data are analyzed, weather maps and charts are prepared, and the prediction of the weather on a global scale begins. The NMC then distributes the information to public and private agencies worldwide. In the United States, the information is used and adapted for regional and local weather forecasts. These forecast reports are then broadcast to the general public by radio or television.

Methods of Weather Forecasting

There are several methods of weather forecasting. Some forecasts are based on the continuity of a weather pattern, such as rain today likely means rain tomorrow. Or, because surface weather systems tend to move in the same direction and at the same speed, a forecast is based on the trend of the weather pattern. For example, if a cold front is moving eastward at an average speed of 20 kilometers per hour, it can be expected to affect the weather 80 kilometers away in four hours. Present weather features are also compared to past weather conditions, so if the features are similar enough, forecasters assume the present weather conditions will be the same as past weather conditions. Say, for example, the average weather condition for San Francisco during the summer months is fog. Then forecasters can predict

Weather Maps

The weather forecaster's primary tool is the surface weather map or chart. A weather map is essentially a representation of the frontal systems and the high- and low-pressure systems that overlie the areas outlined in the map. In order to communicate the data gathered from the various observation stations, meteorologists use weather codes.

This shorthand notation of weather codes compiles 18 distinct items of data into a very small area called a *station model* (Figure 26.27). When this entry is on a surface map, the symbols are replaced by actual measured entries, most of them in numerals. In Figure 26.27, the circle at the center describes the overall appearance of the sky; jutting from the circle is a wind arrow; its tail is in the direction from which the wind is coming and the feathers indicate the wind speed. The other fifteen weather elements are in

Figure 26.27
Surface-Station Model.

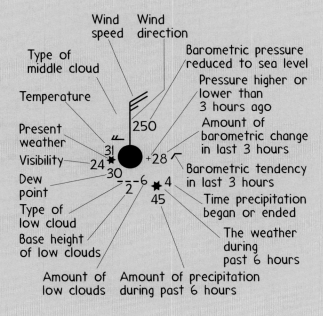

Wind speed · Wind direction · Barometric pressure reduced to sea level · Pressure higher or lower than 3 hours ago · Amount of barometric change in last 3 hours · Barometric tendency in last 3 hours · Time precipitation began or ended · The weather during past 6 hours · Amount of precipitation during past 6 hours · Amount of low clouds · Base height of low clouds · Type of low cloud · Dew point · Visibility · Present weather · Temperature · Type of middle cloud

Figure 26.28 Weather maps show atmospheric conditions. As warm air rises it expands and chills. As it chills the water vapor molecules condense to form clouds. Because air moves from high pressure to low pressure, low pressure zones are accompanied by cloud cover. In a high pressure zone air generally sinks. Because sinking air does not usually produce clouds, we find clear skies and fair weather.

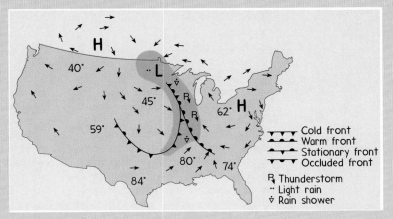

Cold front
Warm front
Stationary front
Occluded front

℞ Thunderstorm
·· Light rain
▿ Rain shower

TABLE 26.4 Weather Symbols

Total Sky Cover

- ◯ No clouds
- ◐ Less than one-tenth or one-tenth
- ◔ Two-tenths or three-tenths
- ◑ Four-tenths
- ◑ Five-tenths
- ◑ Six-tenths
- ◕ Seven-tenths or eight-tenths
- ◑ Nine-tenths or overcast with openings
- ● Completely overcast
- ⊗ Sky obscured

Pressure Tendency

- ⌃ Rising, then falling
- ⟋ Rising, then steady; or rising, then rising more slowly
- ⟋ Rising steadily, or unsteadily
- ⌄ Falling or steady, then rising; or rising, then rising more quickly

  Barometer no higher than 3 hours ago

- — Steady, same as 3 hours ago
- ⌄ Falling, then rising, same or lower than 3 hours ago
- ⟍ Falling, then steady; or falling, then falling more slowly
- ⟍ Falling steadily, or unsteadily
- ⌃ Steady or rising, then falling; or falling, then falling more quickly

  Barometer no lower than 3 hours ago

Wind Entries

	Miles (Statute) Per Hour	Knots	Kilometers Per Hour
◎	Calm	Calm	Calm
—	1–2	1–2	1–3
⌐	3–8	3–7	4–13
⌐	9–14	8–12	14–19
⌐	15–20	13–17	20–32
⌐	21–25	18–22	33–40
⌐	26–31	23–27	41–50
⌐	32–37	28–32	51–60
⌐	38–43	33–37	61–69
⌐	44–49	38–42	70–79
⌐	50–54	43–47	80–87
⌐	55–60	48–52	88–96
⌐	61–66	53–57	97–106
⌐	67–71	58–62	107–114
⌐	72–77	63–67	115–124
⌐	78–83	68–72	125–143
⌐	84–89	73–77	135–143
⌐	119–123	103–107	144–198

Common Weather Symbols

- • Light rain
- ∴ Moderate rain
- ∴ Heavy rain
- ✶ ✶ Light snow
- ✶ Moderate snow
- ✶ Heavy snow
- ,, Light drizzle
- △ Ice pellets (sleet)
- ∿ Freezing rain
- ∿ Freezing drizzle
- ▽ Rain shower
- ▽ Snow shower
- △ Showers of hail
- ⟶ Drifting or blowing snow
- ⟳ Dust storm
- = Fog
- ∞ Haze
- ∿ Smoke
- ⟨ Thunderstorm
- ↯ Hurricane

Front Symbols

- ▲▲▲▲ Cold front (surface)
- ●●●● Warm front (surface)
- ▲●▲● Occluded front (surface)
- ▲⌄▲⌄ Stationary front (surface)
- ⌒⌒⌒ Warm front (aloft)
- △△△△ Cold front (aloft)
- ——•• —— Squall line

---

standard position around the circle. The common weather elements are depicted in Table 26.4.

A weather map is covered with lines, **isobars,** connecting points of equal pressure. Air moves from high pressure to low, rises and cools, and the moisture in the air condenses into clouds. So, in the vicinity of the low (*L* on map) we see an extensive cloud cover. In the vicinity of the high (*H* on map) we see clear skies (Figure 26.28). In a high pressure region air sinks and warms adiabatically. Because sinking air does not produce clouds, we find clear skies and fair weather. The heavy lines depicted on a weather map correspond to frontal symbols. Because fronts generally depict a change in the weather, they are of great importance on weather maps. The movements and relationships among air masses, together with the different pressures within them, are basic factors in weather forecasting.

foggy weather conditions for San Francisco for any particular day in July with a high chance of being correct. We often hear about the probability of a weather condition, for example—the probability of rain is 70 percent. This is an expression of chance, meaning that there is a 70 percent chance that rain will fall somewhere in the forecast area. So you should probably carry an umbrella. Another forecast we often hear about is the extended forecast. This forecast is based on weather types that develop in certain areas. Recall the classification of air masses and their characteristics; if a continental polar air mass is approaching, we can expect cold dry weather, whereas if a maritime polar air mass is approaching, we can expect cold moist weather. All these methods of weather prediction are based on the statistical analysis of weather information.

Weather forecasting involves great quantities of data from all over the world. Before the 1960s most of this data was assembled, analyzed, and plotted on weather maps and charts by hand. This took thousands of calculations, a great work force, and long hours. Now, with modern day computers, the great quantities of data from around the world can be processed in a matter of minutes. Computers not only plot and analyze data, they also predict the weather. Meteorologists design atmospheric models using six to eight mathematical equations that describe how atmospheric temperature, pressure, and moisture will change with time. These models are then programmed into the computer, and measurements of temperature, air pressure, humidity, and wind direction are fed into the equations. Each equation is then solved for an increment of time, and for a large number of locations. The results of these equations are then fed back into the original equations of the atmospheric model and the computer analyzes and draws a chart of the projected weather conditions. The weather forecaster then uses these projections as a guide to predicting the weather. Even so, the many variables involved are not exactly predictable and it may unexpectedly rain on your parade!

In summary we see that the science of meteorology, a fairly new science, concerns the atmosphere and the important application of weather prediction. The atmosphere, like other parts of Planet Earth, is a dynamic system constantly changing. While major transformations of the earth's surface take millions and millions of years to occur, the state of the atmosphere can change in a matter of minutes. To see these changes all we need do is go outside and look upward.

SUMMARY OF TERMS

Atmosphere The gaseous envelope surrounding the earth.

Solar nebula The cloud of gas and dust from which the sun and solar system formed.

Protosun The youngest stage of the sun—the stage of formation.

Troposphere The atmospheric layer closest to the earth's surface. The troposphere is where our weather occurs; it contains 90 percent of the atmospheric mass and essentially all its water vapor and clouds. Temperature decreases with increasing altitude with an average temperature of −50°C at the top.

Stratosphere The second atmospheric layer above the earth's surface. Ultraviolet solar radiation is absorbed by a thin ozone layer within the stratosphere, which raises its temperature from about −50°C at the bottom of the layer to about 0°C at the top.

Ozone layer Layer formed in the stratosphere, which reduces the amount of ultraviolet radiation reaching the earth's surface.

Mesosphere The third atmospheric layer above the earth's surface. Very little of the sun's radiation is absorbed in this layer and as a result, there is a rapid decrease in temperature from about 0°C at the bottom of the layer to about −90°C at the top.

Thermosphere The fourth atmospheric layer above the earth's surface. The little bit of air in this layer absorbs enough incoming solar radiation to cause an increase in temperatures to about 2000°C.

Ionosphere An electrified region within the thermosphere and uppermost mesosphere where fairly large concentrations of ions and free electrons exist.

Exosphere The fifth atmospheric layer above the earth's surface, which extends into interplanetary space.

Solar constant The 1400 joules per square meter received from the sun each second at the top of the earth's atmosphere. Expressed in terms of power, it is 1.4 kilowatts per square meter.

Terrestrial radiation Radiant energy emitted from the earth after having been absorbed from the sun.

Adiabatic A process in which no heat enters or leaves a system.

Chinook A warm, dry wind that blows down from the eastern side of the Rocky Mountains across the Great Plains.

Temperature inversion When the upper regions of the atmosphere are warmer than the lower regions.

Coriolis effect The effect of the earth's rotation on the paths of air circulation.

Doldrums The region near the equator characterized by a low-pressure zone and very little horizontal air movement.

Horse latitudes The belt of latitudes at 30° N and 30° S where air movement is calm and weather is hot and dry.

Trade winds Tropical winds that blow from the subtropical highs to the equatorial low.

Westerlies The general wind pattern north of the horse latitudes in the northern hemisphere and south of the horse latitudes in the southern hemisphere. The winds blow from the west to the east. The westerlies are the general wind pattern in the United States.

Polar front Boundary where the cool polar air meets the warm air of the temperate zone.

Polar easterlies Winds that form behind the polar front, as cool air moves down into the warm latitudes. The winds blow from the east to the west.

Jet stream High speed winds in the upper troposphere. These winds play an essential role in the global transfer of heat energy from the equator to the poles.

Monsoon A wind system that changes direction with the seasons. In Southeast Asia the monsoons are associated with heavy rains in the summer and dry climates in the winter.

Saturated The limit at which the amount of water vapor in the air is the most it can be for a given temperature.

Humidity A measure of the amount of water vapor in the air.

Relative humidity The amount of water vapor in the air, compared to the amount at saturation at that temperature.

Precipitation Any form of water particles—rain, sleet, or snow—that falls from the atmosphere to the ground below.

Dew point The temperature to which air must be cooled for saturation to occur.

Front The contact zone between two different air masses.

Clouds The condensation of water droplets above the earth's surface. Clouds are generally classified by height. The high clouds include the cirrus, cirrocumulus, and cirrostratus clouds. The middle clouds include the altostratus and altocumulus clouds. The low clouds include the stratus, stratocumulus, and nimbostratus clouds. The clouds with vertical development are the cumulus and cumulonimbus clouds.

Weather The effect of the day-to-day changes in air temperature, wind, and precipitation for a given locality.

Front The contact zone between two air masses.

Thunderstorm A storm accompanied by thunder and lightning. Produced when warm humid air rises in an unstable environment.

Tornado A funnel-shaped cloud that extends from a large cumulonimbus cloud.

Hurricane A severe tropical storm with rotating wind speeds up to 279 kilometers per hour.

Isobars Lines on a weather map used to connect points of equal pressure.

• •

REVIEW QUESTIONS

Evolution of Earth's Atmosphere

1. If the temperature of the proto solar system was less than 50K, what caused the rise in temperature toward the intense temperature of today's sun?

2. The earth's atmosphere likely developed from gases that escaped from the interior of the earth during volcanic eruptions (outgassing). What, besides a primitive atmosphere, did this outgassing produce?

3. What is the connection between life on earth and the ozone layer? Explain.

4. Explain the importance of photosynthesis in the evolution of the atmosphere.

Composition of Earth's Atmosphere

5. Why doesn't gravity flatten the entire atmosphere against the earth's surface?

6. What elements made up the earth's earliest atmosphere? What elements make up today's atmosphere?

Vertical Structure of the Atmosphere

7. At what altitude and at what atmospheric layer does all our weather occur?

8. Why does the air temperature in the troposphere decrease with height, while in the stratosphere temperature increases with height?

9. As we climb higher into the thermosphere, the temperature climbs to about 2000°C. What produces this increase in temperature?

10. Within the thermosphere and uppermost mesosphere is the ionosphere. What is the significance of this region?

11. Why can you often pick up AM stations hundreds of kilometers away at night, but not at all during the day?

12. What causes the fiery displays of light called the auroras?

Solar Energy

13. What does the angle at which the sun strikes the earth have to do with the temperate and polar regions?

14. What does the tilt of the earth have to do with the change of seasons?

15. If it is winter and January in Chicago, what is the corresponding season and month in Sydney, Australia?

16. Why are the hours of daylight equal all around the world on the two equinoxes?

17. What is radiant energy?

18. About how much of the sun's radiant energy actually reaches the earth's surface?

19. Do all parts of the earth receive the same amount of solar energy? Explain.

20. What is meant by terrestrial radiation?

Behavior of Air

21. Explain why warm air rises and cools as it expands.

22. In what ways is thermal energy in air gained?

23. In what ways is thermal energy in air reduced?

24. What is an adiabatic process?

25. What is a temperature inversion? Give examples of where these inversions may occur.

Movement of Air

26. Explain how a convection cycle is generated.

27. When is a sea breeze more apt to occur? How about a land breeze?

28. How does the Coriolis effect determine the general path of air circulation?

29. What is the characteristic climate of the doldrums and why does it occur?

30. Why are most of the world's desert regions found in the area known as the horse latitudes?

31. What are the tradewinds?

32. What is the name given to the area where cold polar air meets the warm air of the temperate zone?

33. Why are eastbound aircraft flights usually faster?

34. Why do the temperate zones have unpredictable weather?

35. What are the jet streams and how do they form?

36. In summer, the regions of Southeast Asia, India, and Africa experience heavy flooding. Why?

Evaporation, Condensation, and Precipitation

37. What factors are responsible for the process of condensation?

38. Distinguish between humidity and relative humidity.

39. Does condensation occur more readily at high temperatures, or low temperatures? Why?

40. Distinguish between *dew* and *frost*.

Clouds

41. Name the cloud type associated with each of the following:

   (a) hazy shade of winter
   (b) mackerel sky
   (c) floating cotton
   (d) snowfall

42. Name the cloud family for each of the following cloud types:

   (a) altocumulus
   (b) cirrostratus
   (c) nimbostratus
   (d) cumulus

Weather

43. What is a front? Differentiate between a cold front and a warm front.

44. What are the characteristic properties of the air in the following air masses:

   (a) continental polar
   (b) continental tropical
   (c) maritime polar
   (d) maritime tropical

45. How do downdrafts form in thunderstorms?

46. Why are tornadoes usually associated with thunderstorms?

47. Where does the highest frequency of tornadoes occur in the United States?

48. When and where are hurricanes most likely to develop? Explain.

Weather Forecasting

49. In order to predict the weather what information must be known?

50. What are three different methods in weather forecasting?

. . . .

EXERCISES

1. It is said that a gas fills all the space available to it. Why, then, doesn't the atmosphere go off into space?

2. How would the density of air in a deep mine compare to the density of air at sea level? Explain.

3. Why do your ears pop when you ascend to higher altitudes? Explain.

4. In a still room, smoke from a cigarette will sometimes rise and then settle in the air before reaching the ceiling. Explain why.

5. If the composition of the upper atmosphere were changed so that it permitted a greater amount of terrestrial radiation to escape, what effect would this have on the earth's climate? How about if the atmosphere reduced the escape of terrestrial radiation?

6. As more energy is consumed on earth, the overall temperature of the earth tends to rise. Regardless of the increase in energy, however, the temperature does not rise indefinitely. By what process is an indefinite rise prevented? Explain your answer.

7. The earth is closest to the sun in January, but January is a winter month in the northern hemisphere. Why is January so cold?

8. How do the wavelengths of radiant energy vary with the temperature of the radiating source? How does this affect solar and terrestrial radiation?

9. How is global warming affected by the relative transparencies of the atmosphere to long and short wavelength electromagnetic radiation?

10. Why does a July day in the Gulf of Mexico generally feel appreciably hotter than a July day in Arizona?

11. Why is it important that mountain climbers wear sunglasses and use sunblock even when the temperature is below freezing?

12. If there were no water on the earth's surface, would weather occur? Defend your answer.

13. Is it possible for the temperature of an air mass to change if no thermal energy is added or subtracted? Defend your answer.

14. What distinguishing factor affects crops such as corn and wheat that are grown in northern parts of the United States and in Canada? Why are these crops not grown in equatorial regions, where higher temperatures and more abundant rainfall usually favor plant growth?

15. Ozone is a component of automobile exhaust. Does the pollution from automobiles help to alleviate the ozone hole problem above the south pole? Defend your answer.

16. Why does a drop in barometric pressure predict cloudy weather, and a rise in barometric pressure predict clear weather?

17. Which is more predictable, weather or climate? Why?

18. Antarctica is covered by glaciers and large ice sheets. Is the snowfall in Antarctica therefore heavy or light? Why?

19. What is the source of the enormous amount of energy released by a hurricane?

PART 8

ASTRONOMY

We began this book with the *physics* of the everyday world, then progressed to the microscopic realm, the *chemistry* of molecules. Then to the *geology* of the whole earth -- big by comparison. Now we conclude our study of *Conceptual Physical Science* by studying the very, very big -- the universe! We are of the stars, in matter and energy. Atoms form in stars, then disperse when the stars explode to become the material that becomes us. Most of our energy comes from the nearest star, the sun. So we are literally stardust -- the conscious part of nature looking at itself. Onward to *astronomy*!

27

THE SOLAR SYSTEM

Figure 27.1 Both the moon and the spherical fireworks display look two-dimensional because they are far from our eyes.

For thousands of years people have looked into the night sky and wondered about the stars. With only the unaided eye, they neither saw nor dreamed that the stars are greater in number than all the grains of sand on all the deserts and beaches of the world! Nor did they realize that the sun is a star—simply the nearest star of all in the universe. Most fascinating in the sky was the moon, which when full was perceived as a flat circular disk rather than as the 3-dimensional sphere we know it to be. This is because depth cues are lost at great distance—human eyes are not far enough apart compared to the moon's great distance to perceive its bulging face. Even fireworks that explode in three dimensions high in the sky look flat, and for the same reason. The fact that we only see one face of the moon further supported the idea that the moon was a disk. Why do we see only one side of the moon? Why does the moon go through phases of full to a thin crescent, while the sun remains round?

These questions about the moon have intrigued humans for many thousands of years. We begin our study of astronomy with a brief description of the moon and answers to these questions about our nearest celestial neighbor. Then we'll study our next closest neighbor, the sun, and how the solar system came to be.

We'll continue with a brief tour of the planets. We'll conclude the chapter with other bodies within the solar system—asteroids, meteoroids, and comets. Then in the next chapter we'll learn about stars, both ordinary stars like the sun and exotic black holes, and how they came to be. The final chapter deals with the bigger picture—the universe.

27.1 THE MOON

Figure 27.2 Edwin E. Aldrin, Jr., one of the three Apollo 11 astronauts, stands on the dusty lunar surface. Old Glory is rigged to appear flapping in the wind, for the moon is too small to have an atmosphere.

On July 20, 1969, Neil Armstrong was the first human to set foot on the moon. To date, 12 people have stood on the moon. We know more about the moon than any other celestial body. From nearly 400 kg of rock and soil samples brought back from lunar landings, we know the moon's age, its composition, and a lot about its history. From its low density (3.36 g/cm^3) we know the moon cannot have a substantial iron core. From its extremely low magnetic field (less than 0.0001 that of earth's) we know it cannot have a large molten core. But there is so much we don't know about the moon; how it formed, for example. Did it split off the earth while the earth was forming? Did the earth and moon condense from the same material during solar system formation? Did the moon form somewhere else and then fall into the earth's gravitational grip? Did it form from debris ejected into a disk around the earth by impact with some other body? Or perhaps the earth and moon are the result of a collision and merger of two very large planets in the making—a hypothesis presently gaining the favor of many astronomers. How the moon formed is a much-debated subject. We simply don't know. Most astronomers hold that the moon formed at about the same time the earth formed. The only thing all astronomers agree on in fact is that the moon exists.

The moon is small, its diameter about the distance from San Francisco to New York City. The moon began with a molten surface, cooled too rapidly for plate motion like the earth's, formed an igneous crust thicker than the earth's, and underwent intense meteoroid bombardment early in its evolution. Some three billion years ago basins that were formed by bombardment and volcanic activity filled with lava to produce a surface that has undergone very little change since. The moon is too small to have an atmosphere, so without weather, the only eroding agents have been meteoroid impacts.

Figure 27.3 The earth and moon as photographed in 1992 from the Galileo spacecraft. (NASA)

Figure 27.4 The moon rotates about its own polar axis just as often as it circles the earth; once every 28 days. So as the moon circles the earth, it rotates so that one side (shown yellow) always faces the earth. Note in the four successive positions shown, the moon has rotated $\frac{1}{4}$ turn.

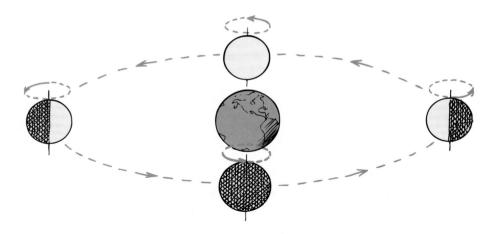

We know less about the surface of the far side of the moon, which is always directed away from the earth. Let's see why only one side of our satellite, the moon, faces earth.

One Side of the Moon

The first humans to see the back side of the moon were Russian cosmonauts who orbited the moon in 1968. From earth we see only a single lunar side. We can see this is true even with naked eye observations—the familiar facial features of the "man in the moon" are always turned toward us on earth. Since the same side of the moon is always facing us, it appears to be without axial spin, or rotation. But with respect to the stars, the moon clearly rotates—but quite slowly, about once per 28 days. This rotational rate matches the rate at which the moon revolves about the earth. This explains why the same side of the moon is always facing earth (Figure 27.4).* If the moon rotated faster or slower, all faces of the moon would reveal themselves to the earth. This matching of axial rotation and orbital revolution is not a coincidence. Let's see why.

We can better understand this phenomenon if we first understand why a magnetic compass aligns with a magnetic field. This involves a *torque*—a "turning force with leverage" (like that produced by a child at the end of a see-saw), that tends to produce rotation. The compass needle in Figure 27.5a rotates due to a pair of torques produced by off-axis forces. The needle rotates until the compass is aligned with the magnetic field (Figure 27.5b). Then the forces are no longer off axis and the compass is stable. Similarly the moon aligns with the earth's gravitational field.

We know that gravity weakens with distance (the inverse-square law, Chapter 4) so the side of the moon nearest the earth is pulled toward the earth with more force than the side farthest from the earth. This elongates the moon and makes it slightly football shaped. (Recall from our treatment of ocean tides in Chapter 4 that the earth is similarly elongated.) The greater pull on the near side means the half of the moon closest to the earth is "heavier," making the moon's center of gravity

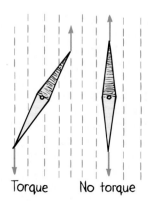

Torque   No torque

Figure 27.5 (a) When the compass is not aligned with the magnetic field (dashed lines) the forces at either end produce a pair of torques that rotate the compass. (b) When the compass is aligned with the magnetic field, the forces no longer produce torques.

---

*There is a difference between *rotation* and *revolution*—a body rotates about its own internal axis; a body revolves about some external axis. So the moon rotates (spins about its own axis), while it revolves (circles about the earth).

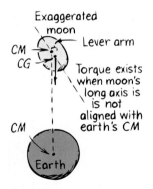

Figure 27.6 When the long axis of the moon is not aligned with the earth's gravitational field, a torque exists to rotate it into alignment.

(center of "weight") closer to us than its center of mass (geometrical center). The centers of gravity and mass lie along the oblong moon's long axis. If this axis does not point directly to the earth's center of gravity, that is, if this axis is not aligned with the earth's gravitational field, then a torque is set up that tends to produce alignment (Figure 27.6). Just as a compass lines up with a magnetic field, the oblong moon lines up with the earth's gravitational field. We say there is a *gravity lock* between the earth and moon. This gravity lock acts to some degree on all astronomical bodies close to each other. The degree to which it acts on the moon is enough to produce the result we see.

QUESTION

A friend says the moon does not spin about its polar axis, and cites as evidence the fact that only one side of the moon faces earth. What do you say?

Phases of the Moon

Rays of sunlight illuminate one half of the earth's surface, and because the earth rotates, we have day and night. Rays of sunlight also illuminate one half the moon's surface, and because the moon rotates and revolves about the earth, we have the moon's **phases**. Phases are the changes in the moon's visible shape that occur in monthly cycles. The first half of the moon cycle begins with the new moon and climaxes with the full moon. The **new moon** phase occurs when the sun, moon, and earth are lined up, with the moon in between. The new moon might have been better named "no moon" for it is not visible in this phase. The new moon is invisible because the side of the moon facing the earth is dark; the moon is in a position such that it cannot reflect any of the sun's rays toward the earth. The moon's illuminated side is the side we cannot see.

Figure 27.7 Positions of the moon in various phases. Since the same side of the moon always faces the earth, we see the same topography regardless of the moon's phase.

ANSWER

Help your friend distinguish between apparent spin and actual spin. With respect to the earth, the moon has no apparent spin for it appears fixed—nonspinning. But if we broaden our point of view and look at things from the stars, we see that the moon actually does spin—but just as slowly as it revolves about the earth. If your friend were correct and the moon did not spin, then different parts of the moon would face the earth throughout the month. Every 14 days opposite sides of the moon would face us. The fact that only one side of the moon faces the earth is evidence that the moon *does* rotate, not that it doesn't. The key concept is that it rotates as often as it revolves.

Figure 27.8   Sunlight illuminates one half of the moon while the other half is dark. As the moon orbits the earth we see varying amounts of the moon's sunlit side. The lunar phase cycle is $29\frac{1}{2}$ days.

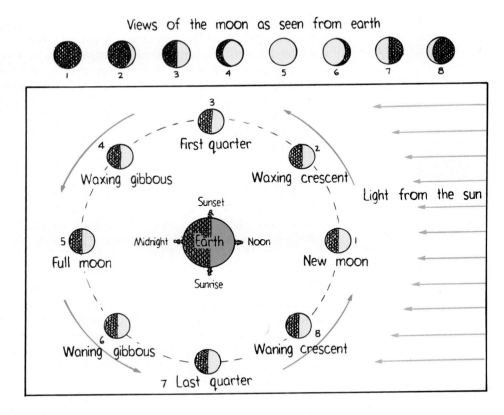

During the next seven days progressively more and more of the moon's side exposed to our view becomes illuminated. We see this in Figure 27.8. The moon is going though its *waxing crescent* phase. At the *first quarter* the angle between the sun, moon, and earth is 90 degrees. Then we see half the sunlit portion of the moon.

During the next week more and more of the sunlit part is exposed to us as the moon goes though its *waxing gibbous* phase. We see a **full moon** when the side of the moon that faces us is completely illuminated—when the sun, earth, and moon are lined up with the earth in between. Moonrise occurs at sunset during this phase.

The cycle reverses during the following two weeks as we see less and less of the sunlit side while the moon continues in its orbit. This movement produces the *waning gibbous, last quarter,* and *waning crescent* phases. The time for a complete cycle is about $29\frac{1}{2}$ days.*

So the phases of the moon depend on the relative positions of the sun, moon, and earth. The alignment affects whether the side of the moon facing the earth reflects the sun's rays toward us. Thus, contrary to a popular misconception, lunar phases are *not* caused by an earth shadow on the moon. When alignment is perfect, that is, when the sun, moon, and earth all lie along the same line of sight, we do have an event created by shadows—an eclipse.

---

*The moon actually orbits the earth once each 27.3 days with respect to the stars. The 29.5-day cycle is with respect to the sun, and is due to the motion of the earth-moon system as it revolves about the sun.

1. Can a full moon be seen at noon? Can a new moon be seen at midnight?
2. Astronomers prefer to view the stars when the moon is absent from the nighttime sky. When, and how often, is the nighttime sky moonless?

27.2 ECLIPSES

Although the sun is 400 times larger in diameter than the moon, it is also 400 times farther away. So from the earth both the sun and moon subtend the same angle (0.5°) and appear the same size in the sky. It is this coincidence that allows us to see solar eclipses.

Both the earth and the moon cast shadows when sunlight shines upon them. When the path of either of these bodies crosses into the shadow cast by the other, an eclipse occurs. A **solar eclipse** occurs when the moon's shadow falls on the earth. Because of the large size of the sun, the rays taper to provide an umbra and a surrounding penumbra (Figure 27.9). An observer in the umbra part of the shadow experiences darkness during the day—a total eclipse, *totality*. Totality begins when the sun disappears behind the moon, and ends when the sun reappears on the other edge of the moon. The average time of totality is about 2 or 3 minutes, with a maximum no longer than 7.5 minutes. The eclipse time in any location is brief because of the

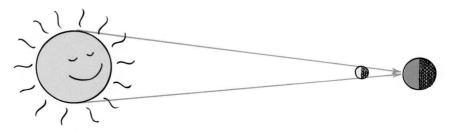

Figure 27.9 A solar eclipse occurs when the moon is directly between the sun and earth, and the moon's shadow is cast on the earth. Because of the small size of the moon and tapering of the umbra, the eclipse occurs only on a small part of the earth.

1. Inspection of Figure 27.8 will show that a noontime observer would be on the wrong side of the earth to see the full moon. So the full moon is out of sight to a noontime observer who would be on the daylight side of the earth. Likewise, an observer at midnight is out of sight of a new moon. The new moon is overhead in the daytime, not the nighttime.

2. At the time of the new moon, and during the week on either side of the new moon, the nighttime sky is without a moon. Unless an astronomer wishes to study the moon, these dark nights are the best time for viewing other objects. Astronomers usually view the night skies during two-week periods every two weeks.

Figure 27.10 Eclipsed view of the sun, showing the corona, a pearly white halo of solar gases that extends outward several million kilometers.

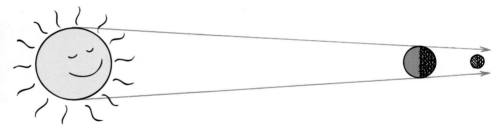

Figure 27.11 A lunar eclipse occurs when the earth is directly between the moon and the sun, and the earth's shadow is cast on the moon.

moon's motion. An observer in the penumbra experiences a partial eclipse, and can still see part of the sun.*

It is interesting to note that the darkness of totality is not complete because of the bright corona that surrounds the sun (Figure 27.10). The corona is not normally seen only because it is overwhelmed by the brightness of the sun's disk.

The lining up of earth, moon, and sun also produces a **lunar eclipse** when the moon passes into the shadow of the earth (Figure 27.11). Usually a lunar eclipse

Appearance of the Moon During a Lunar Eclipse

A fully eclipsed moon is not completely dark in the shadow of the earth, but is quite visible. This is because the earth's atmosphere acts as a lens and refracts light into the shadow region—enough to make the moon visible. More interestingly, the eclipsed moon often has a deep copper reddish color. To understand why it is reddish, recall the reason for red sunsets from Chapter 12: First, remember that the atmosphere scatters high-frequency light (which produces the daytime blue sky); second, remember that when high frequencies are scattered from a beam of white light, the low frequencies that aren't scattered make the beam reddish (which produces the redness of sunsets). Beams of sunlight through the air travel the longest distance when the sun is on the horizon—at sunset or sunrise. The

longer "filtering path" of sunlight at that time produces the redness of sunlight.

The next time you view a sunset (or sunrise), quickly move your head to one side so the light meeting your eye instead misses and continues to the horizon behind you (Figure 27.12). If someone else's head or nothing else is in the way, the light will continue through the atmosphere and refract into space. If the light was reddish when it passed you, it will be even redder by the time it travels the additional distance through the atmosphere before continuing into space. This is the light that shines on the eclipsed moon—hence its deep reddish color (Figure 27.13). So poetically enough, the redness of the eclipsed moon is the red light from all the sunsets and sunrises that completely circle the world.

---

*People are cautioned not to look at the sun at the time of a solar eclipse because the brightness and the ultraviolet light of direct sunlight are damaging to the eyes. This good advice is often misunderstood by those who then think that sunlight is more damaging at this special time. But staring at the sun when it is high in the sky is harmful whether or not an eclipse occurs. In fact, staring at the bare sun is more harmful than when part of the moon blocks it! The reason for special caution at the time of an eclipse is simply that more people are interested in looking at the sun during this time. And if they do so with binoculars or a telescope, they're really in trouble.

Figure 27.12 When the sun is low in the sky, the long path through the atmosphere to the observer filters high frequencies to make the sunlight reddish. Hence the red sunset. Light that continues past the observer travels through twice as much atmosphere and is even redder when it shines on the eclipsed moon.

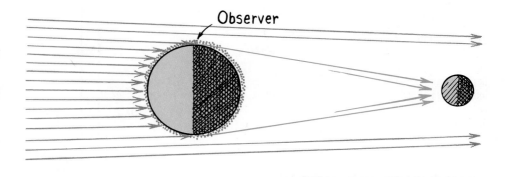

Figure 27.13 The fully eclipsed moon is often red because the earth's atmosphere acts like a lens and refracts light from sunsets and sunrises all around the world onto the otherwise dark moon.

precedes or follows a solar eclipse by two weeks. Just as all solar eclipses involve a new moon, all lunar eclipses involve a full moon. They may be partial or total. All observers on the dark side of the earth see a lunar eclipse at the same time. Interestingly enough, when the moon is fully eclipsed, it is still visible and is reddish in color (see box on lunar eclipses).

Why are eclipses a relatively rare event? This has to do with the different orbital planes of the earth and moon. The earth revolves around the sun in a flat planar orbit. The moon similarly revolves about the earth in a flat planar orbit. But the planes are slightly tipped to each other—a 5.2-degree tilt (Figure 27.14). If

Figure 27.14 The moon orbits the earth in a plane that is tipped 5.2° to the plane of the earth's orbit around the sun (not to scale). A solar or lunar eclipse occurs only when the moon intersects the earth-sun plane (points A or B) at the precise time of the 3-body alignment.

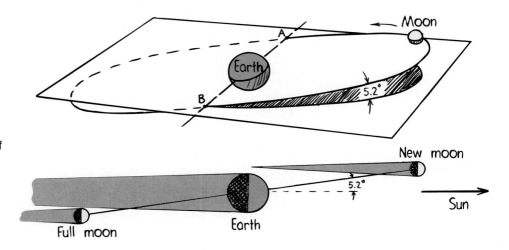

Figure 27.15 Eclipses can occur only when the earth and moon meet near a *node* (points A or B in Figure 27.14), where the tipped orbital planes intersect. (a) Partial solar eclipse; (b) partial lunar eclipse.

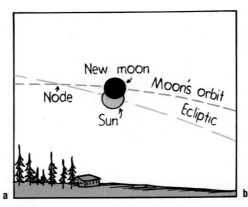
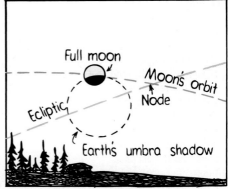

the planes weren't tipped, eclipses would occur monthly. Because of the tip, eclipses occur only when the moon intersects the earth-sun plane at the time of a three-body alignment (Figure 27.15). This occurs some two times per year, which is why there are at least two solar eclipses per year (visible only from certain locations on earth). Sometimes there are as many as seven solar and lunar eclipses in a year.

QUESTIONS

1. Does a solar eclipse occur at the time of a full moon or a new moon?

2. Does a lunar eclipse occur at the time of a full moon or a new moon?

3. Why are lunar eclipses more commonly seen than solar eclipses?

27.3 THE SUN

Beyond the moon is the sun—our nearest star. Ancients who worshipped the sun seem to have realized that the sun is the source of all earthly life. Our life energy originates in the sun; we are able to see, hear, touch, laugh, and love because every

ANSWERS

1. A solar eclipse occurs at the time of a new moon, when the moon is directly in front of the sun. Then the shadow of the moon falls on part of the earth. Because of the tilt in the moon's orbit around the earth, most of the time the shadow misses the earth.

2. A lunar eclipse occurs at the time of a full moon, when the moon and sun are aligned on opposite sides of the earth. Then the shadow of the earth falls on the full moon. Because of the tilt in the moon's orbit around the earth, most of the time the shadow misses the moon.

3. Relatively few people witness solar eclipses because the shadow of the small moon tapers to a very small part of the earth's surface. But during a lunar eclipse the similarly tapered shadow of the large earth completely covers the moon, so everybody on the dark side of the earth can see the shadow of the earth on the moon. That's why nearly all your friends have seen a total lunar eclipse, while relatively few of them have ever witnessed a total solar eclipse.

Figure 27.16 In every second, 4.5 million tons of mass are converted to radiant energy in the sun. The sun is so massive, however, that in a million years only one ten-millionth of its mass will have been converted to radiant energy.

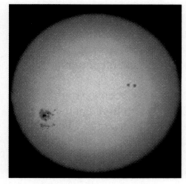

Figure 27.17 Sunspots on the solar surface are relatively cool regions. We say *relatively* cool, because they are hotter than 4000 K. They look dark only by contrast with the 5800-K surroundings.

second, $4\frac{1}{2}$ million tons of mass in the sun is being converted to radiant energy, a tiny fraction of which is intercepted by the earth.

This is the energy of thermonuclear fusion taking place in the interior of the sun, where hydrogen nuclei are being crushed together to form helium (Chapter 16). The resulting helium has mass 99.3 percent of the original hydrogen mass. The conversion of hydrogen to helium in the sun has been going on since it formed nearly 5 billion years ago and is expected to continue at this rate for another 5 billion years. If all the hydrogen in the sun's core were changed to helium, the core would still have 99.3 percent of its former mass.

The part of the sun visible to us is, of course, its surface and its atmosphere. The sun's tenuous surface is neither solid nor liquid nor gas but a glowing 5800 K plasma, probably no more than 500 kilometers thick. At nearly 6000 K the sun's surface is well above the temperature required to vaporize any known material. This transparent solar surface is the *photosphere* (sphere of light). On the surface are relatively cooler regions brought about by magnetic field concentrations that appear as **sunspots**. These can be seen by the unaided eye through protective filters or when the sun is low enough on the horizon not to hurt the eyes. Sunspots are typically twice the size of the earth, move around due to the sun's rotation, and last about a week or so. Often they cluster in groups (Figure 27.17).

The layer of the sun's atmosphere just above the photosphere is a transparent 10,000-kilometer-thick shell of plasma called the *chromosphere* (sphere of color), seen during an eclipse as a pinkish glow surrounding the eclipsed sun. Beyond the chromosphere are streamers and filaments of outward-moving, high temperature plasmas curved by the sun's magnetic field. This outermost region of the sun's atmosphere is the *corona* (Figure 27.10), extending out several million kilometers where it merges into a hurricane of high speed protons and electrons—the *solar wind*. It is the solar wind that powers the aurora borealis on earth and produces the tails of comets.

The sun spins on its axis, but slowly. Interestingly enough, different latitudes of the sun spin at different rates. Equatorial regions spin once in 25 days, whereas the higher latitudes take up to 36 days per rotation, which means the surface near the equator pulls ahead of the surface farther north or south.

Formation of the Solar System

How the sun formed is still a matter of some conjecture, but it is generally believed to have originated from the gravitational contraction of a huge amount of interstellar matter nearly five billion years ago. Although the universe began primarily as hydrogen, with some helium, by the time the solar system formed some 10 billion or so years later, it contained small amounts of all the elements known today. All elements beyond hydrogen and helium had to be formed in the cores of previous stars. When these stars underwent their death throes, their heavier elements were spewed into the interstellar mix, providing material for new stellar systems.

In the course of random movements of interstellar matter in space, pockets of gas are believed to condense and dissolve repeatedly, as wisps of fog similarly form in air. Sometimes these temporary condensations become permanent, because the atoms and molecules in the gas are held together by mutual gravity. The pocket adds mass to itself by attracting neighboring atoms and molecules. Because of gravity it falls in upon itself. Gravitational potential energy becomes thermal energy and the temperature at the center rises. Continued gravitational contraction and rising

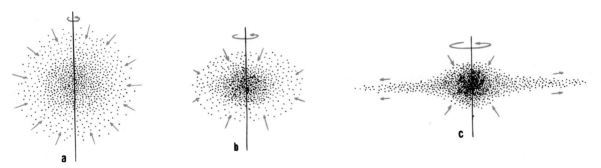

Figure 27.18 A slightly rotating ball of interstellar gas (a) contracts due to mutual gravitation and (b) conserves angular momentum by speeding up. The increased momentum of individual particles and clusters of particles causes them to (c) sweep in wider paths about the rotational axis, producing an overall disk shape.

temperatures contribute to a concentration of hot matter toward the center. Like the ice skater drawing arms inward when going into a spin, the contraction is accompanied by an increase in angular speed. As the collapsing matter spins faster, it flattens into a disk shape (Figure 27.18). A disk has more surface area per volume than its configuration before flattening and consequently radiates more of its energy into space. So this flattening results in cooling. In the formation of our solar system, this decreasing temperature probably was accompanied by the condensation of matter in swirling eddies—the birthplace of the planets.

Although temperatures near the center of the disk would have been too high for matter to solidify, farther out, earthy and less volatile material with high condensation temperatures likely solidified to become the four inner planets: Mercury, Venus, Earth, and Mars. At relatively cooler locations farther away from the hot center, condensations of larger amounts of more volatile matter, mostly hydrogen, are now the giant outer planets: Jupiter, Saturn, Uranus, and Neptune. Small and distant Pluto, as we shall see, is an exception.

So we have a solar system, our home in the universe. Of the countless generations who have wondered what and where we are in the universe, only those of our lifetime have begun to understand.

UESTION

What is the evidence that our sun is a second- or third-generation star?

ANSWER

Heavy elements are normally fused from lighter elements only in star cores. When stars explode, they contribute heavy elements to the primordial (original) hydrogen-helium mix. The fact there is an abundance of heavy elements in the solar system is evidence of heavy elements in the mixture from which the solar system formed, and therefore evidence of previous stars. So the sun is a newcomer. There are many stars twice as old as the sun in our galaxy. These older stars have lower abundances of heavy elements.

Kepler's Laws of Planetary Motion

Long before the invention of the telescope, the Danish astronomer Tycho Brahe spent his lifetime making accurate observations of the positions of the planets. Upon his death in 1601 his charts and record books were passed on to his gifted assistant, Johannes Kepler. At this time Kepler and other astronomers assumed the planets moved in perfect circles. Kepler performed the enormous task of transforming Brahe's earthbound observations into a path in space such as might be seen by a stationary observer outside the solar system. His conviction of circularly moving planets was shattered after years of effort. He found the paths to be ellipses.

Kepler also found that the planets do not go around the sun at a uniform speed but move faster when near the sun and more slowly when farther away. He found that they do this in such a way that an imaginary line or spoke joining the sun and the planet sweeps out equal areas of space in equal times. The triangular-shaped area swept out during a month when a planet is far from the sun (triangle ASB in Figure 27.19) is equal to the triangular area swept out during a month when the planet is closer to the sun (triangle CSD in Figure 27.19).

Ten years later Kepler discovered a third law. He had spent these years searching for a connection between the size of a planet's orbit and its period (the time for a complete revolution around the sun). From Brahe's data Kepler found that the square of a period is proportional to the cube of its average distance from the sun. He discovered this by noting that the fraction R^3/T^2 is the same for all planets, where R is the planet's average orbit radius and T is the planet's period measured in earth days. Thus, *Kepler's laws of planetary motion* are:

LAW 1: Each planet moves in an elliptical orbit with the sun at one focus.

LAW 2: The line from the sun to any planet sweeps out equal areas of space in equal time intervals.

LAW 3: The squares of the times of revolutions (periods) of the planets are proportional to the cubes of their average distances from the sun. ($T^2 \sim R^3$ for all planets.)

Kepler's laws apply not only to planets but to moons or any satellite in orbit around any body. Except for Pluto (which Kepler had no knowledge of), and Mercury, the elliptical orbits of the planets are very nearly circular. Only the precise measurements of Brahe showed the slight differences. Kepler preceded Newton and had no idea why the planets moved in elliptical paths, and no general explanation for the mathematical relationships he had discovered. He was familiar with Galileo's ideas about inertia and accelerated motion, but he failed to apply them to his own work. As a consequence, Kepler's thinking about forces was concerned with forces he supposed were directed in the same direction as the planets' paths to keep them moving. He never appreciated the concept of inertia, that the planets once moving would move without forces. And he didn't realize that the forces on the planets were directed not along their direction of motion, but to the sun around which they orbited. Kepler's work was soon to be an important stepping stone for Isaac Newton in his development of the law of universal gravitation.

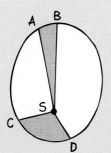

Figure 27.19 Equal areas are swept out in equal intervals of time.

27.4 THE PLANETS

The ancients distinguished between the planets and the stars because of their different apparent motions. The stars remain relatively fixed in their patterns in the sky, but the planets wandered. The planets were called the *wanderers*. Today we know that planets are relatively cool bodies orbiting about the sun. Planets emit no visible light of their own, and like the moon, simply reflect sunlight.

To informed minds a century and more ago, knowledge of the planets was very meager. Detailed knowledge of planets today is enormous. How human knowledge advanced from knowing planets as no more than starlike spots that crossed the nighttime skies is a fascinating detective story—made possible first by careful observation with the unaided eye, then with telescopes, and more recently with satellite probes launched from earth. For brevity, we leave this fascinating story to the recommended reading, and here offer only a brief description of the planets. We divide the planets into two groups; the inner planets and the outer planets.

Inner Planets

The group of planets closest to the sun are relatively close together. These are Mercury, Venus, Earth, and Mars, which are relatively small and dense worlds with sparse atmospheres. These inner planets have solid mineral crusts and earthlike compositions, which is why they are called the *terrestrial planets*. The only terrestrial planets to have natural satellites are Earth and Mars, which enabled knowledge of their masses before the era of space probes. It is interesting to note that the mass of a celestial body cannot be found unless it has a satellite (see box on Finding the Mass of a Planet). The masses of Mercury and Venus were only known after being circled by space probes in the early 1960s. Let's visit the inner planets in order.

Mercury Mercury, somewhat larger than the moon and similar in appearance, is the closest planet to the sun. Because of its closeness it is the fastest planet, taking

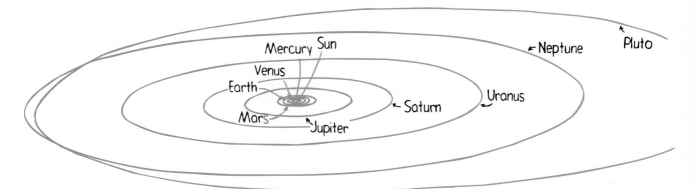

Figure 27.20 Scale drawing of the solar system that shows the four inner planets crowded around the sun, and the five outer planets orbiting at greater distances.

TABLE 27.1

	Mean Distance from Sun Earth Distances (Au)	Orbital Period (Years)	Diameter		Mass		Average Density (g/cm³)
			(km)	(Earth = 1)	(g)	(Earth = 1)	
Sun			1,392,000	109.1	1.99×10^{33}	3.3×10^{5}	1.41
Mercury	0.39	0.24	4,880	0.38	3.3×10^{26}	0.06	5.4
Venus	0.72	0.62	12,100	0.95	4.9×10^{27}	0.81	5.2
Earth	1.00	1.00	12,760	1.00	6.0×10^{27}	1.00	5.5
Mars	1.52	1.88	6,800	0.53	6.4×10^{26}	0.11	3.9
Jupiter	5.20	11.86	142,800	11.19	1.90×10^{30}	317.73	1.3
Saturn	9.54	29.46	120,700	9.44	5.7×10^{29}	95.15	0.7
Uranus	19.18	84.0	50,800	3.98	8.7×10^{28}	14.65	1.3
Neptune	30.06	164.79	49,600	3.81	1.0×10^{29}	17.23	1.7
Pluto	39.44	247.70	2,300	0.18	10^{25}	0.002	1.9

88 days to make a revolution. Its "years" are therefore very short. Mercury rotates only three times for each two revolutions about the sun, so "daytime" on Mercury is very long, and very hot, as much as 430°C. Because of its smallness and weak gravitational field, it holds very little atmosphere—it's about a trillionth as dense as the earth's atmosphere. So without a blanket of atmosphere, and because there are no winds to carry heat from one region to another, nighttime is very cold, about −170°C. Mercury is a fairly bright object in the nighttime sky, and is best seen as an evening star during March and April or a morning star during September and October. It is seen near the sun at sunrise or sunset.

Venus Venus is the next closest planet to the sun, appears brighter than Mercury, and is also seen near the sun at sunrise or sunset. Because Venus is often the first starlike object to appear after the sun goes down, it is often called the evening "star." Compared to the other planets, Venus most closely resembles Earth with respect to size, density, and distance from the sun. A major difference is its very dense atmosphere and opaque cloud cover, and a greenhouse effect that makes it an unbearably 460°C hot place—too hot for oceans. Another difference is that it rotates in a

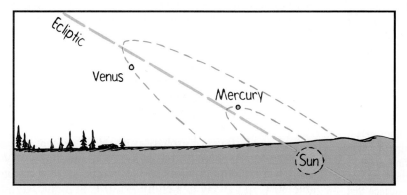

Figure 27.21 Because the orbits of Mercury and Venus lie inside the orbit of the earth, they always are near the sun. Near sunset (or sunrise) they are visible in the sky as evening "stars" (or morning "stars").

Figure 27.22 Earth, the blue planet.

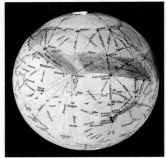

Figure 27.23 A globe of Mars showing some of the canals mapped by astronomer Percival Lowell in the late 1800s. The canals proved to be optical illusions produced by the brain's ability to assemble vague markings into a coherent image (the same ability that enables us to see TV images rather than swarms of incoherent dots). In 1965 Mariner 4, the first spacecraft to fly by Mars, and Mariner 9 in 1971, radioed back photos of the Martian surface that showed no canals where they were "seen to be."

direction opposite to Earth's, and very much slower with a period of 243 days, slightly longer than its 225 earth-day year.

Since Venus has been regarded as almost an earth twin, early speculations were that its surface is a steamy swamp inhabited by unfamiliar creatures. This was before the greenhouse effect was invoked to explain high "radio" temperatures in the 1960s. Speculations of life there have been severely dimmed in recent years by data from 17 probes that have landed on its surface and from surveillance of 18 flyby spacecraft (notably Pioneer Venus in 1978 and Magellan in 1993). Venus has been very active volcanically and is an extremely harsh place. Evidence gathered by satellite probes suggests that the surface temperature and atmosphere of Venus, however, were once very much like those of Earth. Whereas most of Earth's carbon dioxide is locked in limestone formations, and some in the oceans, with very little in the atmosphere, the greater sunlight on Venus produced the opposite result. More carbon dioxide was released into its atmosphere, which increased the greenhouse effect, which in turn released even more carbon dioxide, which now makes up about 95 percent of the Venus atmosphere. Venus serves as a model for Earth, for we wonder if a small temperature rise here on Earth could trigger a similar irreversible chain reaction.

Earth The earth is well described in Chapters 22–26, so our treatment here is brief. In answer to the question of what it must be like to live in outer space, we must not forget that we in fact *are* in outer space—gathered together on a hospitable planet in an inhospitable universe. Planet Earth is our haven, and deserves our greatest respect.

Ours is the blue planet, with more water surface than land. Not too close to the sun and not too far, with an average surface temperature delicately between that of freezing and boiling water, and a just-dense-enough atmosphere to keep the oceans in their liquid state. The insulating properties of our atmosphere and our relatively high rotational rate ensure only a brief and small lowering of temperature on the side of the earth away from the sun. So temperature extremes of day and night are conducive to life as we know it. Compared to the harshness of most of the universe, it's a nice place to live. Our activities ought to be consistent with keeping it that way.

Mars Our nearest neighbor, Venus, is too close to the sun and too hot for human habitation, and our next nearest neighbor, Mars, is farther from the sun than Earth, but not too cold for human habitation. Mars captures our fantasies as another world, perhaps a world with life. This is because of the similarities of Earth and Mars; Mars is a little more than half Earth's size, has about $\frac{1}{9}$ of the Earth's mass, and has a thin nearly cloudless atmosphere. Its axis tilts from the perpendicular to its orbital plane, which gives it polar ice caps and seasons that are nearly twice as long as Earth's because Mars takes nearly 2 earth years to orbit the sun. When Mars is closest to Earth, about every 15 to 17 years, its bright ruddy color far outshines the brightest stars.

The Martian atmosphere is about 95 percent carbon dioxide, with only about 0.15 percent oxygen. So bring your own air supply if you plan to visit there. Also bring warm clothing, for its surface temperature at the equator goes from a comfortable 30°C in the day to a chilly −130°C at night. Nighttime is only slightly longer than on earth, for the Martian day lasts 24 hours and 37.4 minutes. Never mind your raincoat, for there is far too little water vapor in the atmosphere for rain. Even the

Finding the Mass of a Planet

When the first artificial satellite Sputnik was launched by the Russians in 1958, U.S. President Dwight Eisenhower was surprised to learn from physicists that there was no way to determine Sputnik's mass. No amount of observation would reveal its mass. One would have to poke it with a probe of known mass and measure its rebound, and by momentum conservation calculate Sputnik's mass. Or, quite far fetched, one would have to send a smaller satellite in orbit about Sputnik and measure its period of revolution. Clearly Sputnik's mass was too small to employ this latter method under the most ideal conditions. But that is precisely how scientists measure the masses of larger celestial bodies. In fact, it is their only way of measuring such masses. Unless something orbits a planet or star, and unless the average distance and period of orbit is known, astronomers can only guess about the mass of that planet or star.

The mass of a planet, found from Newton's second law and the law of gravity, is directly proportional to the cube of the average distance between the planet and satellite, and inversely proportional to the square of the satellite's period. In symbols, $M \sim R^3/T^2$. Find this yourself by equating the centripetal force on the satellite, mv^2/R,* to the gravitational force, $G\,mM/R^2$, between the satellite (mass m) and the planet (mass M). The speed of the satellite v is actually the orbital distance $2\pi R$ per time of orbit, period T. After substituting $2\pi R/T$ for the satellite's speed v, and seeing that the mass m of the satellite cancels, you should get the planet's mass $M = \frac{4\pi^2}{G}\left(\frac{R^3}{T^2}\right)$. Try it and see.

Hooray, this also gives you Kepler's third law!

Centripetal force is the name given to any force that pulls an object into a circular path. In our present example, centripetal force is the force of gravity, but it may be the force supplied by a string on a whirling object, or the electrical force exerted on a circularly moving charge, etc. Any object moving at constant speed in a circle has an acceleration v^2/R, directed toward the center of the circle. From Newton's 2nd law, $F = ma$, centripetal force is therefore $F = mv^2/R$.

---

ice at the poles is primarily carbon dioxide. And never mind your rain boots, for the low atmospheric pressure won't allow any puddles or lakes.

Investigators speculate that water may have been more abundant in the Martian past, enough to carve some of the channels seen on the Martian surface. These channels intrigued early investigators, who saw them through their telescopes as canals, reinforcing speculation of a Martian civilization. Today we find Mars with no trace of life, past or present, and with no canals. It's a very dry place and windy too. Unequal heating produces winds normally about ten times the wind speeds on earth because of the small atmospheric density on Mars.

Mars has two small satellites—Phobos, the inner, and Deimos, the outer. Both are potato-shaped with cratered surfaces. Phobos orbits in the same easterly direction that Mars rotates (like our moon), at a distance of almost 6000 kilometers in a period of 7.5 hours. From Mars it appears about half the size of our moon. But because it revolves about Mars much faster than Mars rotates, it rises on the western horizon and sets on the eastern horizon 5.5 hours later. It appears to be "going backward."

Deimos is about half the size of Phobos, and orbits Mars' surface in a period of 30.3 hours at a distance of 20,000 kilometers. Its orbital period is somewhat longer than the rotational period of Mars, so it rises on the eastern horizon and sets on the western horizon nearly three days later. It appears small, subtending the same angle that a 25-cent piece subtends when it is 37 meters distant.

Beyond Mars is the first of the outer planets, Jupiter—but there's a large gap in between. This gap, as we shall see later in this chapter, is populated by a belt of many thousands of small rock and ice fragments called *asteroids*. If you're going to Jupiter, good luck in getting through the asteroid belt.

Outer Planets

The more widely spaced outer planets beyond the asteroid belt are considerably different than the inner planets—in size, composition, and the way they were formed. Jupiter, Saturn, Uranus, and Neptune are gigantic gaseous and low-density worlds. They are typified by Jupiter and are called *Jovian* planets. All have Saturnlike ring systems. Beyond these giants is outermost Pluto, the most different among the planets, being neither terrestrial nor Jovian. We consider the outer planets in turn.

Jupiter The largest of all the planets is Jupiter, whose yellow light seen in the night sky outshines the stars. In pre-spacecraft years it was thought of as a failed star, for its composition is closer to that of the sun than the terrestrial planets. Jupiter is more liquid than gaseous or solid. It rotates rapidly about its axis in about ten hours, which produces a flattening that makes the equatorial diameter about 6 percent greater than the polar diameter. Interestingly enough, like the sun, all parts do not rotate in unison. Equatorial regions complete a revolution several minutes before adjacent higher and lower latitude regions. Jupiter doesn't have a hard surface crust that an astronaut could walk on. And if there were a place to stand, atmospheric pressure would be a crushing millionfold the atmospheric pressure of Earth. Jupiter's atmosphere is about 82 percent hydrogen, 17 percent helium, and 1 percent methane, ammonia, and other molecules.

Figure 27.24 Relative sizes of the sun and planets. The sun is 13 times larger than Jupiter in diameter.

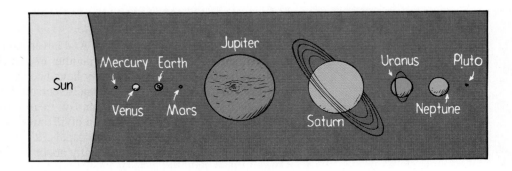

The mean diameter of Jupiter is about eleven times greater than the earth's, which means Jupiter is more than a thousand times larger in volume than the earth. Jupiter's mass exceeds the combined masses of all the other planets. But due to its low density of about one quarter of the earth's, its mass is barely more than three hundred times the earth's. Jupiter's core is a solid sphere about 20 times as massive as the earth, composed of iron, nickel, and other minerals.

More than half of Jupiter's volume is an ocean of liquid hydrogen, H_2. Beneath the hydrogen ocean lies an inner layer of hydrogen pressed into a liquid metallic state, permeated by abundant conduction electrons that flow to produce an enormous magnetic field. The strong magnetic field about Jupiter captures high energy particles

Figure 27.25 Jupiter, with two of its moons, Io and Europa, as seen from the Voyager I spacecraft in February 1979. The great red spot, lower left, is a cyclonic weather pattern of high winds and turbulence larger than Earth. (NASA)

and produces radiation belts 400 million times as energetic as Earth's (the Van Allen radiation belts). Radiation levels surrounding Jupiter are the highest ever recorded in space.

Surface temperatures are about the same in day and night. Jupiter radiates about twice as much heat as it receives from the sun. The excess heat likely comes from internal heat generated long ago by gravitational contraction at the time of Jupiter's formation. When forming planets contract, gravitational potential energy is converted to thermal energy.

If you're planning to visit Jupiter, choose one of its moons instead. At least 16 known satellites and a faint Saturn-type ring orbit Jupiter. Among the four largest satellites, discovered by Galileo in 1610, Io and Europa are about the size of our moon, while Ganymede and Callisto are somewhat larger than Mercury. The most interesting of Jupiter's moons seems to be Io, which has more volcanic activity than any body in the solar system.

Saturn Because its rings are clearly visible with binoculars, Saturn is one of the most remarkable objects in the sky. It is brighter than all but two stars, and is second among planets in mass and size. Saturn is twice as far from us as Jupiter. Its mean diameter, minus its ring system, is nearly ten times that of Earth and its mass nearly 100 times greater. It is composed primarily of hydrogen and helium and has the lowest density of any planet, 0.7 times that of water. Saturn would easily float in a bathtub if the bathtub were big enough. Because of Saturn's low density and its 10.2-hour rapid rotation, it has more polar flattening than any planet, about 11 percent. Like Jupiter, Saturn radiates about three times as much heat as it receives from the sun.

Saturn's rings, likely only a few kilometers thick, lie in a plane coincident with Saturn's equator. Four concentric rings have been known for many years, and spacecraft missions have detected others. The rings are composed of chunks of frozen

Figure 27.26 Saturn, surrounded by its famous rings believed to be composed of chunks of ice.

water and rocks, believed to be the remnants of a moon that never formed or a moon ripped apart by tidal forces. All the rocks and bits of matter that make up the rings pursue independent orbits about Saturn.* Inner parts of the ring travel faster than outer parts, just as any satellite near a planet travels faster than a more distant satellite.

Saturn has some nineteen satellites beyond its rings. The largest is Titan, 1.6 times larger than our moon, and even larger than the planet Mercury. It revolves once each 16 days, has a methane atmosphere with atmospheric pressure likely greater than the earth's, and its surface temperature is a cold −170°C. So bring a heavy coat and breathing gear if you plan to visit Titan. If that doesn't work out, try another of Saturn's large moons, Iapetus, that is partially snow covered. One side is very bright and the other very dark. Try the region between these two extremes.

Uranus Uranus, twice as far from Earth as Saturn, is barely perceptible to the naked eye and was unknown to the ancient astronomers. It has a diameter four times larger than that of the earth and a density slightly greater than water. So put Uranus in that bathtub and it would sink. The most unusual feature of Uranus is its tilt. Its axis is tilted 98 degrees to the perpendicular of its orbital plane. So it lies on its side. Unlike Jupiter and Saturn, it appears to have no internal source of heat. It is a cold place.

Uranus has at least 15 satellites, in addition to a complicated faint ring system. Many astronomers suspect the presence of a belt of cometary material between Saturn and Uranus.

All planets are held in the solar system by their gravitational interaction with the sun. But the planets interact with each other, and everything else, as well. When one planet is near another, the pull between the planets slightly disturbs the orbit. This disturbance is called a *perturbation*. Early in the nineteenth century, unexplained perturbations were observed for the planet Uranus. Either the law of gravitation was failing at this great distance from the sun or an unknown eighth planet was perturbing Uranus. An Englishman and a Frenchman, J. C. Adams and Urbain Leverrier, independently calculated where an eighth planet should be. At about the same time, both sent letters to their respective observatories with instructions to search a certain area of the sky. The request by Adams was delayed by misunderstandings at Greenwich, but Leverrier's request to the director of the Berlin observatory was heeded immediately. The planet Neptune was discovered one half hour later.†

Neptune Neptune's diameter is about 3.9 times that of Earth's, its mass is 17 times greater, and its mean density is about a third that of Earth's. Its atmosphere is mainly hydrogen and methane, with some helium and ammonia. Like Jupiter and Saturn, it emits about twice as much heat as it receives from the sun.

---

*The entire ring system lies within a critical distance called the *Roche limit,* equal to about 2.4 Saturn radii. The Roche limit, named after the nineteenth-century French mathematician Edouard Roche, is the distance where gravitational attraction by a planet on two adjacent orbiting particles is larger than the attraction of the two particles to each other. If our moon were within 2.4 earth radii, its expected fate would be a similar ring system about the earth.

†Interestingly enough, recent studies of Galileo's notebooks show that Galileo saw Neptune in December of 1612 and again in January 1613. Galileo was interested in Jupiter at the time, and plotted Neptune as a star in the background of Jupiter.

Neptune has at least eight satellites in addition to a ring system. The largest satellite is Triton, which orbits Neptune in 5.9 days in a direction opposite the planet's eastward rotation. Triton is 0.75 the moon's diameter, with a mass double the moon's. A smaller satellite, Nereid, takes nearly a year to orbit in a highly elongated elliptical path.

Pluto The relative positions of stars on photographs do not move. Star images taken one time will be in the same positions as when taken days later—but the images of planets or asteroids will be in different places. Careful examination of such photographic plates resulted in the discovery of Pluto in 1930 at the Lowell Observatory in Arizona.

Whereas most of the planetary orbits in the solar system are nearly circular, Pluto's is the most elliptical and most steeply inclined to the planetary plane. Pluto's orbit is so eccentric that it is presently closer to the sun than Neptune. Its most recent perihelion (near point) occurred in 1989, which found it more than 100 million kilometers closer than Neptune to the sun. After March 4, 1999, it will move farther than Neptune from the sun. Pluto takes 248 years to make a single revolution about the sun, so no one will see it in its discovered position again until the year 2178.

Pluto is smaller than our moon, and has a diameter about one-fifth Earth's diameter, and a mass of about 0.002 earth mass. Pluto has bright polar caps that are likely frozen methane. Pluto's rotational period is 6.4 days, and it has a satellite named Charon that is about 5 to 10 percent of the mass of Pluto. Charon also has a period of 6.4 days, so it appears motionless in Pluto's sky.

Not surprisingly, less is known about Pluto than the other planets. Pluto's uniqueness suggests to some astronomers that it may be a moon that escaped from Neptune. It is so distant that from its surface the sun would appear as an ordinary star among stars. Pluto must be a very cold place.

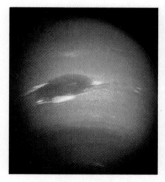

Figure 27.27 Cyclonic disturbances on Neptune produce a great dark spot, which is larger than the earth, and similar to Jupiter's great red spot. (NASA)

27.5 ASTEROIDS, METEORITES, AND COMETS

As mentioned earlier, there is a large gap between Mars and Jupiter. In this gap is the *asteroid belt*, populated by tens of thousands of small rock and ice fragments called **asteroids** that orbit the sun. The smaller ones are irregular in size, like boulders, and the larger ones are roughly spherical. Most asteroids are about 40 kilometers across although about 200 have diameters exceeding 100 kilometers, and about 2000 have diameters more then 10 kilometers. The largest is Ceres, often called a minor planet, which has a diameter of 750 kilometers. Asteroids are thought to be material that failed to become a planet during the formation of the solar system. If the planet had formed, it would have been small, for the combined masses of the asteroids are considerably less than the mass of the earth's moon.

Although many asteroids neatly circle the sun, others do not. Collisions among asteroids are common, sending some of them helter skelter. Some stray toward Earth. Asteroids smaller than a few hundred kilometers across are called **meteoroids**. A **meteor** is a high-speed meteoroid that strikes the earth's atmosphere, usually at an altitude of about 80 kilometers. A meteor is heated to incandescence by friction with the atmosphere and is seen as a flash of light—a "falling star." Most meteors we see are small meteoroids, about the size of a grain of sand. A chunk of rock that survives its fiery descent through the atmosphere and reaches the ground is called a **meteorite**.

Figure 27.28 Asteroids leave blurred trails on time-exposure photographs of the stars. The images of two asteroids are seen in this photograph.

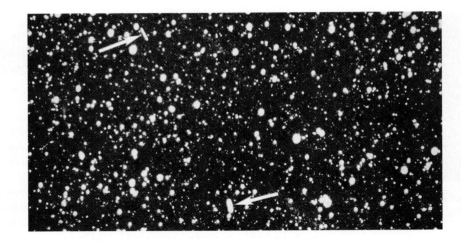

Most meteorites are small and strike the earth with no more energy than a falling hailstone. Some are big, and evidence of their impact is seen as craters. If the earth were without weather and other eroding elements, our surface would likely be as cratered as the moon's. Most impact craters on earth have long ago eroded or been covered by geologic processes. More recent impacts, however, leave telltale marks (Figure 27.30). Perhaps the most dramatic impact of all was one we have a record of. This would be the impact made near the Yucatan Peninsula in Mexico 65 million years ago, discussed earlier in Chapter 25. The effects of that impact are thought to be responsible for the extinction of dinosaurs and half the living species in the Cretaceous period.

Figure 27.29 A meteor is produced when a high-speed piece of interplanetary rock or dust encounters the earth's atmosphere, usually about 80 km high. These are "falling stars."

Q UESTION

A school project is to visit a science museum and view both a meteor and a meteoroid. Is this a reasonable project?

Meteoroids come not only from the asteroid belt, but also from comets. Most of the meteors we see are the small particles of comet debris. Unlike meteors that shoot briefly across the sky, a comet moves slowly and gracefully to display one of nature's most beautiful astronomical spectacles (Figure 27.32).

A **comet** is a dusty chunk of ice that becomes partly vaporized as it passes near the sun. Whereas asteroids travel between the planets in roughly circular orbits, the orbits of comets are highly elliptical, extending far beyond the planets. As a comet approaches the sun, solar heat begins to vaporize the ices. Escaping vapors glow to produce a fuzzy luminous ball called a *coma*, typically a million kilometers in

Figure 27.30 The Barringer Crater in Arizona, made 25,000 years ago by impact of a 50-m-across iron meteorite. The crater is 1.2 km in diameter, and 200 m deep.

A NSWER

Any extraterrestrial rocks you see in a museum are, by definition, meteorites. So unless the science museum is in outer space, you'll never view a meteoroid from there, for meteoroids exist only outside the earth's atmosphere. And the only meteor you can possibly see in a museum is one streaking across the sky through a window at night, for meteors are high-speed incandescent meteoroids.

Figure 27.31 When the earth crosses the orbit of a comet, we see a meteor shower. So meteors are either remnants of the asteroid belt or the debris of comets.

Figure 27.32 Halley's comet, which orbits the sun each 76 years, is composed mainly of water vapor, carbon dioxide, carbon monoxide, and carbon-nitrogen molecules. Halley's comet will return to the inner solar system in 2062.

diameter. Within the bright coma is the solid part of the comet, the *nucleus,* typically a chunk of ice and other materials measuring a few kilometers across.

Solar wind and radiation pressure blow luminous vapors from the coma outward, away from the sun, into a long flowing *tail.* A comet's tail can extend over 100 million kilometers. Most often the sun produces two tails on a comet: an ion tail and a dust tail (Figure 27.33). The ions are largely the remnants of water vapor too massive to be affected by the pressure of sunlight. The ions flow with the high-speed solar wind, directly away from the sun. The dust tail, on the other hand, is composed of micron-sized dust particles, large enough to be affected by radiation pressure. The lower-speed dust tail curves, much as a water stream curves from the nozzle of a moving hose. The density of material in these tails is quite low—less than the density achieved in typical industrial vacuums. So compared to the atmosphere, the tail of a comet is "nothing at all." When the tail of a comet crosses the earth directly, except for meteor showers high in the atmosphere, nothing at all changes at the earth's surface. The incidence of a comet nucleus, however, is a different story. "Meteor" craters are formed by the impact of comets as well as meteors. Only the impact debris indicates the difference.

Comets are plentiful. There is almost always a comet in the sky, but most are too faint to be seen without a good telescope. About half a dozen new comets are discovered each year, many by amateur astronomers. Most comets have no visible tails, for their supply of ice is eventually exhausted. After about 100 to 1000 passes around the sun, a comet is pretty well burned out.

Figure 27.33 The two tails of Comet West. A comet is always named after the person who first sees it. Guess who was the first person to see this comet?

SUMMARY OF TERMS

Moon phases The cycles of change of the face of the moon, changing from new to waxing, to full, to waning, and back to new.

New moon The phase of the moon when darkness covers the side facing Earth.

Full moon The phase of the moon when its sunlit side is the side facing Earth.

Solar eclipse The phenomenon whereby the shadow of the moon falls upon the earth producing a region of darkness in the daytime.

Lunar eclipse The phenomenon whereby the shadow of the earth falls upon the moon producing relative darkness of the full moon.

Sunspots Temporary, relatively cool and dark regions on the sun's surface.

Asteroid A small rocky planetlike fragment that orbits the sun. Tens of thousands of these objects make up an asteroid belt between the orbits of Mars and Jupiter.

Meteoroid A small rock in interplanetary space.

Meteor The streak of light produced by a meteoroid burning in the earth's atmosphere; a "shooting star."

Meteorite A meteoroid or part of a meteoroid that has survived passage through the earth's atmosphere to reach the ground.

Comet A body composed of ice and dust that orbits the sun, usually in a very eccentric orbit, and which casts a luminous tail produced by solar radiation pressure when it is close to the sun.

• •

REVIEW QUESTIONS

1. How numerous are stars compared to the grains of sand on all the deserts and beaches of the earth?

2. Why do the sun and moon appear disklike rather than spherelike?

The Moon

3. What is the evidence that the moon lacks an iron core?

4. What is the evidence that the moon lacks a large molten core?

5. What fact about the moon is noncontroversial among all astronomers?

6. How does the moon's rate of rotation about its own axis compare with its rate of revolution about the earth?

7. Why does only one side of the moon always face the earth?

8. What is the comparison between the action of a magnetic compass aligning with a magnetic field, and the moon's orientation to the earth?

9. What is meant by *gravity lock?*

10. What conditions account for the various phases of the moon?

Eclipses

11. In what alignment of sun, moon, and earth, does a solar eclipse occur?

12. Why is totality during a solar eclipse not altogether dark?

13. In what alignment of sun, moon, and earth, does a lunar eclipse occur?

14. Why is totality during a lunar eclipse not altogether dark?

15. What two conditions are necessary for either a solar or lunar eclipse?

The Sun

16. What happens to the amount of the sun's mass as it "burns"?

17. Distinguish between the *photosphere* and *chromosphere* of the sun.

18. What is the *solar wind?*

19. How does the rotation of the sun differ from the rotation of a solid body?

20. How old is the sun?

21. Where are elements heavier than hydrogen and helium formed?

The Planets

22. Why did the ancients call planets the "wanderers"?

23. Into what two major groups are the planets divided?

Inner Planets

24. List the inner planets. Then state why they are called the terrestrial planets.

25. Why are days on Mercury very hot, and nights very cold?

26. Why are Mercury and Venus seen as evening or morning "stars"?

27. What drastic change in atmosphere dramatically altered Planet Venus?

28. Why is Earth called the "blue planet"?

29. What predominant gas makes up the Martian atmosphere?

30. What is the cause of winds on Mars (and about every other planet)?

Outer Planets

31. What is meant by the term *Jovian* planets?

32. What is the major difference between the terrestrial and Jovian planets?

33. Why was Jupiter once thought to be a "failed star"?

34. What surface feature is common to both Jupiter and the sun?

35. Why does Jupiter bulge at the equator?

36. What is thought to produce the strong magnetic fields of Jupiter?

37. What distinguises the rings of Saturn from the rings of the other Jovian planets?

38. How do the inner parts of Saturn's rings travel compared to the outer parts?

39. How did Uranus lead to the discovery of the next planet, Neptune?

40. By what investigative method was Pluto discovered?

41. Why is Pluto sometimes closer to earth than Neptune?

Asteroids, Meteorites, and Comets

42. Distinguish between an *asteroid,* a *meteoroid,* and a *comet.*

43. Where, as far as we can tell, do most asteroids reside?

44. Distinguish between a *meteoroid,* a *meteor,* and a *meteorite.*

45. What is a "falling star"?

46. Why do the tails of comets point away from the sun?

47. Why do most comets actually have two tails?

48. What would be the consequences of a comet's tail sweeping across the earth?

49. A falling star is visible only once but a comet may be visible at regular intervals throughout its lifetime. Why?

50. Why does a comet eventually burn out?

• •

EXERCISES

1. Why does the moon lack an atmosphere? Defend your answer.

2. Is the fact we see only one side of the moon evidence that the moon rotates, or that it doesn't rotate? Defend your answer.

3. Distinguish between a *waning moon* and a *waxing moon,* and between *gibbous* and *crescent* phases.

4. Do star astronomers work during the full moon or new moon part of the month? Why?

5. Nearly everybody has seen a lunar eclipse, yet relatively few people have ever witnessed a solar eclipse. What are the principal reasons for this?

6. If the sun were only ten times the diameter of the moon, and only ten times farther away from earth than the moon, what differences would occur for *(a)* solar eclipses, and *(b)* lunar eclipses?

7. Because of the earth's shadow, a partially eclipsed moon looks like a cookie with a "bite" taken out of it. Explain with a sketch how the curvature of the bite indicates the size of the earth, compared to the size of the moon. What is the effect of the tapering of the sun's rays?

8. What are the energy processes that make the sun shine? In what sense can it be said that gravity is the prime source of solar energy?

9. When a contracting hot ball of gas spins into a disk shape, it cools. Why is this so? Defend your answer.

10. Why was the mass of Mercury not precisely known until 1974?

11. Peform the derivation described in the box on finding the mass of a planet and show by a series of algebraic steps that a planet's mass $M = \frac{4\pi^2}{G}\left(\frac{R^3}{T^2}\right)$.

12. Find the appropriate data for the variables in the preceding exercise and calculate the mass of Planet Earth.

13. Show that the results of Exercise 11 lead to Kepler's third law.

14. The greenhouse effect on Venus is very pronounced, but doesn't exist on Mercury. Why?

15. More than 70 percent of the earth's surface is ocean. How does Jupiter compare in this regard?

16. Jupiter has more than 100 times the mass of the earth, yet one's weight on the surface of Jupiter would be only about 3 times earth weight. Why is this so? (*Hint:* Go back to the equation for gravitational force in Chapter 4, and take into account Jupiter's size.)

17. It is said that Saturn would "float in a bathtub" (if the tub were big enough) because of Saturn's low density. Would it float in a water-filled bathtub in any gravitational field, or specifically the gravitational field near the surface of the earth? Defend your answer.

18. A friend suggests that the rings of Saturn are probably solid flat disks (obviously with a hole for Saturn in the center). Then your friend looks to you for confirmation or refutation—with your usual well-thought-out explation. What is your response?

19. Could the rings of Saturn be composed of a concentric series of thin rings and be consistent with your answer to the previous exercise?

20. Why are the seasons on Uranus different than the seasons on any other planet?

21. What were the similar historical circumstances that link the names of the planets Neptune and Pluto with the elements Neptunium and Plutonium?

22. On what continent on earth would you expect to find the largest number of stony meteorites with minimum weathering?

23. Chances are about 50-50 that in any nighttime sky there is at least one visible comet that has never been discovered. This keeps amateur astronomers busy, for the discoverer of a comet experiences a wide measure of approval and recognition, and the honor of having the comet named for him or her. With this high probability of nondiscovered comets, why aren't more comets found?

24. In terms of the conservation of energy, describe why comets eveually burn out.

28

· ·

THE STARS

The roots of astronomy reach back to prehistoric times when humans became familiar with star patterns in the night sky. They divided the night sky into groups of stars, such as the seven stars we now know as the Big Dipper. Star groups in large sections of the sky became the *constellations*. The names of the constellations today carry over mainly from the names assigned by early Greek, Babylonian, and Egyptian astronomers. The Greeks, for example, included the stars of

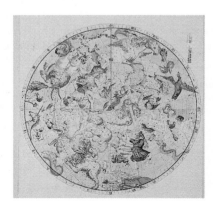

Figure 28.1 The constellations and Taurus represent figures from Greek mythology.

the Big Dipper into a larger group of stars that outlined a bear—the large constellation, Ursa Major (the Great Bear). The grouping of stars and the significance given to them varied from culture to culture. Humans in the southern hemisphere, of course, saw groups of stars not visible to their northern contemporaries. To some cultures, the constellations stimulated storytelling and the making of great myths; to others, the constellations honored great heroes like Hercules and Orion; to others, they served as navigational aids for travelers and sailors; and to others, they provided a guide for the planting and harvesting of crops—for the constellations were seen to move periodically in the sky, in concert with the seasons. Charts of this periodic movement became some of the first calendars.

Stars were thought to be points of light on a great revolving celestial sphere with the earth at the center. Positions of the sphere were believed to affect earthly events, so were carefully measured. Keen observations and logical reasoning gave birth to both astrology and, later, astronomy. So we find that astronomy and astrology—science and pseudoscience, share the same roots. Today we know that the earth orbits the sun, with its nightside always away from the sun. We see in Figure 28.2 why the background of stars varies in the nighttime sky throughout the year.*

Figure 28.2 The nightside of the earth always faces away from the sun. As the earth circles the sun, different parts of the universe are seen in the nighttime sky. Here the circle, representing one year, is divided into twelve parts—the monthly constellations. The stars seen in the nighttime sky move in a yearly cycle.

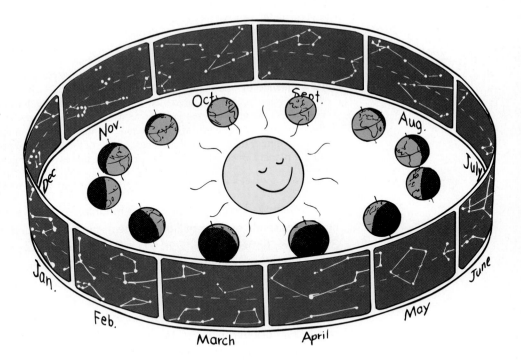

*Place a lamp on a table in the middle of your room, and move around the table, say from left to right, keeping your back to the lamp at all times. You'll see different parts of the room as you walk. Likewise, the night side of the earth is in view of different parts of the sky as the earth orbits about the sun. Looking at Figure 28.2, can you see why the background stars during a midday solar eclipse are of constellations normally seen six months earlier or later?

Astrology

There is more than one way to view the cosmos and its processes—astronomy is one and astrology is another. Astrology is a belief system that began more than 2000 years ago in Babylonia. Astrology has survived nearly unchanged since the second century AD, when some revisions were made by Egyptians and Greeks who believed that their gods moved heavenly bodies to influence the lives of people on earth. Astrology today holds that the position of the earth in its orbit around the sun at the time of birth, combined with the relative positions of the planets, has some influence over one's personal life. The stars and planets are said to affect such personal things as one's character, marriage, friendships, wealth, and death.

The question is raised as to whether the force of gravity exerted by these celestial bodies is a legitimate factor in human affairs. After all, the ocean tides are the result of the moon's and sun's positions, and the pulls between planets perturb one another's orbits. Since slight variations in gravity produce these effects, might not slight variations in the planetary positions at the time of birth affect a newborn? If the influence of stars and planets is in their gravity, then more credence must be given to the gravitational pull between the newborn and the earth, which is enormously greater than the combined pull of all the planets, even when lined in a row (as occasionally happens). The gravitational influence of the hospital building on the newborn would exceed that of the distant planets. So gravitation cannot be an underlying agent for astrology.

Astrology must look to another realm for its basis, for all attempts to find physical explanations to support it have failed. Astrology is not a science, for it doesn't change with new information as science does, nor are its predictions borne out by fact. So the realm of astrology may be spiritual, a religion of sorts. Or it may be a primitive psychology where the stars serve as a point of departure for musings about personality and personal decisions. Or astrology may be in the realm of numerology or phrenology—rigid and empty superstition that prevails because of its focus on what is very important to each of us—ourselves.

A common position is that astrology is a harmless belief—a little fun at minimum harm. But is astrology harmless when believers are led to think their personalities are fixed by the stars at birth, that weak people will remain weak, that sad people will remain sad, that one's fate is dictated by the stars? We must also question the harm dealt people whose astrological signs are deemed incompatible with the signs of others. The harmlessness of the belief that people are hostages to the stars is questionable.

When we look at the stars on a moonless night, we might guess we see many thousands or even millions of them, but the unaided eye sees at most about 3000 stars, horizon to horizon. We see many more stars with a telescope, of course, but disappointingly, stars appear as point sources with or without magnification. Telescopes will show the details of the moon or planets, but no details of stars. Stars are really far away. Many of the brightest are 10 to 1000 light-years distant.* Because of their great distance they appear equally remote, as on the celestial sphere imagined by the ancients. This chapter is about stars—how they are born, how they live, and how they die.†

---

*One light-year is the distance light travels in one year, about 9.5×10^{12} km. Another unit of distance popular with astronomers is the *parsec,* which is the same as 3.26 light-years.

†This chapter presents a brief "This is how it is" treatment of astronomy. For an expanded "This is how we know this is how it is" treatment, refer to the suggested readings at the end of the book.

The Big Dipper and the North Star

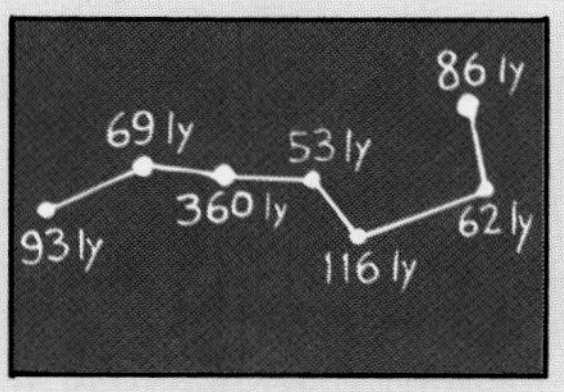

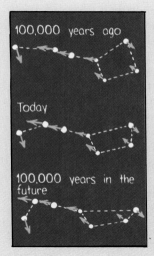

Figure 28.3 The familiar Big Dipper. The size of the dots represents the apparent brightness of the stars, which are not all the same distance from earth. Their distances are noted in light years.

Figure 28.4 The pattern of the Big Dipper is temporary. We see its pattern (a) 100,000 years ago; (b) as it appears at present; and (c) 100,000 years in the future.

Perhaps the most easily recognized star group in the northern hemisphere is the Big Dipper (Figure 28.3). Because of its great distance it seems to form a plane, but the seven stars actually lie at quite different distances from us. The Big Dipper and the larger groups that make up constellations, of course, would take on entirely different patterns if viewed from other locations in the universe. Because of the variety of speed and directions of stars, the familiar patterns of all groups are temporary. We see in Figure 28.4 how the Big Dipper looked from earth in the past and how it is projected to look in the future.

The Big Dipper is most useful for locating the North Star (Polaris), which happens to lie almost exactly on the earth's rotational axis. It is easily located by drawing a line through the two stars in the end of the bowl of the Dipper, and extending the line away from the bowl about five times the distance between these two stars (Figure 28.5). Because

the North Star lies very close to the projection of the earth's rotational axis, it appears stationary as the earth rotates. All the surrounding stars appear to move in circles around the North Star, as evidenced in long-exposure photographs (Figure 28.6).

One of the earliest tests of good eyesight for centuries has been to see if you can see which star in the Big Dipper is actually two closely spaced stars. That star is the next-to-last star in the Dipper's handle. Although they seem to be a double star from our point of view, they are actually quite far apart in space. They look close because they happen to lie approximately along the same line of sight from the solar system. Interestingly enough, the brighter of these two stars, Mizar, *is* actually a pair of stars—the first optical binary to be observed by telescope. No amount of good eyesight will show the double star Mizar without the aid of a good-sized telescope.

Figure 28.5 The pair of stars in the end of the Dipper's bowl point to the North Star. The earth rotates about its axis and therefore about the North Star, so over a 24-hour period the Big Dipper (and other surrounding star groups) make a complete revolution.

Figure 28.6 Time exposure of the northern night sky.

28.1 BIRTH OF STARS

Interstellar space is not empty, but contains faint amounts of elements, primarily hydrogen, and to a lesser degree, a wide variety of other molecules ranging in complexity from ammonia to ethyl alcohol. Among atoms and molecules are also specks of interstellar dust that are composed of carbon and silicates, sometimes coated with the frozen ices of water, carbon dioxide, methane, and ammonia. Dust particles may play the same condensing role in star formation that similar dust particles play in cloud formation. The density of all this interstellar material is a million times lower than the densities of the highest vacuums achieved in earthbound laboratories.

To make a star, begin with a giant cloud of low temperature interstellar material. The gas will not be perfectly uniform; there will be regions of gas density different from the overall average. Regions of slightly greater gas density will have slightly more mass and a slightly greater gravitational field; therefore, they will more strongly attract neighboring particles. This increases the mass and gravitational field of the region, which then attracts still more particles. In time we have an aggregation of matter many times the mass of the sun spread out over a volume many times larger than the solar system—a forming star, that is, a **protostar**.

Mutual gravitation between the gaseous particles in a protostar results in an overall contraction of this huge ball of gas, and the density at the center increases dramatically as matter is scrunched together with an accompanying rise in pressure and temperature. When the central temperature reaches about 10 million K, some of the hydrogen nuclei *fuse* to form helium nuclei. This **thermonuclear reaction,** converting hydrogen to helium, releases an enormous amount of radiant and thermal energy. This ignition of nuclear fuel marks the change from protostar to star. Outward moving radiant energy and the gas it pushes with it exert an outward pressure on the contracting matter, ultimately becoming strong enough to stop the contraction. Radiation and gas pressures balance gravitational pressure, resulting in a full-fledged star.

The material composing the star depends on the age of the universe during its formation. The very first protostars had only primordial hydrogen, with some helium, to work with. Stars run their life cycles and like living things, return their materials to the overall environment. Elements heavier than hydrogen and helium are manufactured in star cores, and when the stars run their life courses, they spew these heavier elements into the interstellar mix. So the protostars that follow are enriched with heavier elements. Heavy elements that make up the sun and its planets are testimony that many stars lived and died before the solar system came to be. All atoms on earth heavier than helium were once part of another star. So we are quite literally made of star dust.

Figure 28.7 The sun is about 5 billion years old, about half its expected life span of 10 billion years.

QUESTION

What do the processes of thermonuclear fusion and gravitational contraction have to do with the physical size of a star?

28.2   LIFE OF STARS

Stars have varying life spans which depend in good part on the rate at which they burn their fuel. We are most familiar with stars like our sun, a hydrogen burner. Our sun has an expected life span of some 10 billion years. Hydrogen fusion in more massive stars occurs at a more furious rate, and the stars are very bright and have relatively short lives. In low mass stars, hydrogen fusion occurs at a much slower rate, and the stars are dimmer and live longer.

Surprisingly, about half the stars seen in the sky do not live alone, but are actually two stars revolving about a common center just as the earth and the moon revolve about each other. These double stars are **binary stars**. By observing how the two stars in a binary revolve about their common center, we can calculate their masses. Recall from the box on Finding the Mass of a Planet in the previous chapter that the only way astronomers can find the masses of planets is by measuring the average distance and period of a satellite about a planet. This is also true of stars. The only way to determine the mass of a star is to find it in a binary system (sun excluded, for its planets provide this information). The sizes and periods of orbits of binary stars depend on the masses of the stars.* So binaries provide astronomers the basic means to determine how much matter is contained in stars.

There is speculation that our sun does not live a solitary life but is part of a binary system. If it is, its partner is very small and very distant. This star is thought to travel in a very large elliptical orbit, as far away as three light-years from the sun. At its closest approach, which would occur every 26 to 30 million years, it would pass near the fringes of the outermost planets. Its gravity would perturb billions of comets into the inner solar system, all within a few million years. This star has been named *Nemesis,* after the goddess of divine retribution, because some speculators credit it

ANSWER

The size of a star is the result of these two continually occurring processes. Thermonuclear fusion tends to blow the star up like an ongoing hydrogen bomb "explosion," and gravitation tends to compress its matter in an ongoing implosion. The outward thermonuclear expansion and inward gravitational contraction produce an equilibrium that accounts for the star's size.

---

*Recall from the box on page 691 that the mass of a planet is found from the ratio of (satellite distance)^3 to (satellite period)^2, R^3/T^2 (in accord with Kepler's third law). Since satellite mass was negligible compared to the mass of the planet, the center of rotation was at the center of the planet. In much the same way, the *sum of the masses* of two stars is proportional to R^3/T^2, where R is the average distance between the stars' centers, and T the period of revolution. Individual star masses are found by their relative distances from their center of revolution, somewhere between the stars. A star four times as massive as its companion, for example, will be four times closer to the center. Equal masses would be equidistant from the center.

with triggering the meteorite impact that may have led to the demise of the dinosaurs some 60 million years ago. This deadly companion has not been found and may not exist.

With or without a companion star, our sun certainly does not live alone. It has us, and at least eight other planets. The sun has 99 percent of all the mass in the solar system, but has only about 2 percent of the solar system's angular momentum.* Hence it has a slow spin with 98 percent of the solar system's angular momentum in the orbiting planets. Measurements of the spin rates of thousands of stars show that the hot massive stars are spinning at rates 100 times that of the sun; while the cooler, less massive stars spin slowly like our sun. Perhaps the massive stars spin rapidly because they alone posses angular momentum, while the less massive stars have formed planetary systems that absorb most of the system's angular momentum.

The H-R Diagram

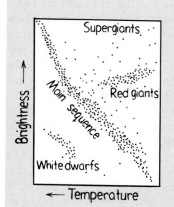

Figure 28.8
H-R diagram.

One of the most important tools of astronomers is the **Hertzsprung-Russell diagram,** or **H-R diagram,** developed early in this century by Danish astronomer Ejnar Hertzsprung and American astronomer Henry Norris Russell. The H-R diagram is a plot of stellar variables equivalent to brightness versus temperature. The temperature of a star is evident by its color. Cooler stars are red, while medium-temperature stars are white, and hotter stars glow bluish-white in color. Figure 28.8 is a typical H-R diagram. Each dot represents a star whose absolute magnitude of brightness has been determined. Bright stars are near the top of the diagram and dim stars toward the bottom. Hot stars are toward the left side of the diagram and cool stars are toward the right side.

The H-R diagram shows several distinct regions of stars. The band that stretches diagonally across the diagram represents a majority of stars seen in the night sky. This band is called the main sequence, and extends from hot bright bluish stars in the upper left to cool, dim, reddish stars in the lower right of the diagram. About 90 percent of all stars, including our sun, lie in the main sequence of stars.

Toward the upper right of the diagram are the group of stars called red giants. They are cooler stars and appear reddish in the night sky. Above these are a few rare stars, the supergiants, larger and brighter than the red giants. Toward the lower left are stars both hot and dim, the white dwarfs, which cannot be seen with the unaided eye.

When a spectroscope is attached to a telescope, stars are seen to have a variety of spectral patterns that indicate their elemental makeup and temperatures. Stars are arranged into various *spectral types,* according to their spectra. When brightness versus spectral type is plotted, the same H-R pattern results. The H-R diagram dramatically shows the fundamental fact that different types of stars exist. These stars represent different stages of stellar evolution.

---

*Whereas linear momentum, as studied in Chapter 3, is inertia × velocity, angular momentum is *rotational inertia × rotational velocity.* The conservation of angular momentum states that angular momentum is conserved during internal processes. So just as a spinning figure skater spins faster when arms are drawn in, a rotating ball of gas spins faster when it contracts (a decreased rotational inertia is compensated by an increased rotational velocity). And just as the spinning skater slows when arms are extended, the sun slows as its planets form.

If other stars with as little mass as our sun have planetary systems, the planets can't be seen with today's telescopes because of their nonluminosity and great distance. For many years it has seemed a fair speculation that some are located, like the earth, at a distance from their star that is not too hot and not too cold—at a location that would support life. This idea is rather appealing. However based upon recent "hard-science modeling," a growing number of astronomers contend that the range of distances from a star for life supporting conditions as we know them is *much* tinier than previously thought—perhaps as little as one suitable planet every ten galaxies or so. Nevertheless, a large NASA SETI (Search for Extraterrestrial Intelligence) program is presently in progress. Our own civilization is so young that there has hardly been enough time for it to have come to the attention of others. The most conspicuous evidence of life on earth—radio, TV, and radar broadcast—has by now reached some 65 light-years into space, a distance encompassing only a few hundred of the galaxy's 200 billion stars.*

In a similar way, most of the starlight from stars in the universe has not reached earth yet. That's how far away most of the universe is. And light from most far away stars that does reach us is Doppler shifted below the visible part of the spectrum and is invisible to us. Hence the night sky is black instead of ablaze with starlight!

28.3 DEATH OF STARS

All luminous stars "burn" nuclear fuel. A star's life begins when it ignites its nuclear fuel, and it ends when its nuclear fires go out. The first ignition in a star core is the fusion of hydrogen to helium. This hydrogen-fusing process may last from a few million to a few hundred billion years, depending on the star's mass. In the old age of an average mass star like our sun, the burned-out hydrogen core that has been converted to helium contracts due to gravity, raising its temperature. This ignites both the helium in the core and the unfused hydrogen outside the core, and the star expands to become a **red giant**. Our sun will eventually reach this stage about 5 billion years from now. On the way to reaching this stage the swelling and more luminous sun will cause earthly temperatures to escalate, first stripping the earth of its atmosphere and then boiling the oceans dry. Ouch!

The cores of solar-mass and lower-mass stars are not hot enough to fuse carbon, and lacking a source of nuclear energy, they shrink. In doing so, the outer stellar layers are sometimes ejected and form expanding shells that appear smoke-ring-like and eventually disperse and mix with the interstellar material. This expanding shell is a **planetary nebula** (Figure 28.9). The shrunken core that is left blazes white-hot and is a **white dwarf**. Here, matter is so compressed that a teaspoonful of it weighs tons.

The nuclear fires of a white dwarf have burned out, so it is not actually a star anymore. It's more accurate to call it a *compact object*. It may continue to radiate

Figure 28.9 The planetary ring nebula in Lyra, which can be seen in modest telescopes.

---

*On the positive side, one verified contact will prove that intelligent life exists elsewhere. On the negative side, we can never prove that extraterrestrial life *doesn't* exist. However intense our search, it could always be "just around the next corner." If after centuries of listening and looking we find no sign of extraterrestrial intelligence, we might then be justified in assuming that we are alone. If we are the sole heirs to the galaxy, then our present concern for tending Planet Earth should extend to being the guardians of the galaxy.

energy and change from white, to yellow, and then to red, until it slowly but ultimately fades to a cold, black lump of matter—a **black dwarf**. Its density is enormous. Into a volume no more than that of an average-size planet is concentrated a mass hundreds of thousands times greater than that of the earth. The star has a density comparable to that of a full fledged battleship squeezed into a pint jar!

There is another possible fate for a white dwarf, if it is part of a close binary system. Because of its strong gravity, the white dwarf may pull hydrogen from its companion star and deposit this material on its surface as a very dense hydrogen layer. Continued compacting increases the temperature of this layer which ignites to embroil the white dwarf's surface in a thermonuclear holocaust that we see as a **nova**. Then, a nova subsides until enough matter again accumulates to repeat the event. Novae flare up at irregular intervals spaced by decades or perhaps by as long as hundreds of thousands of years.

The Bigger They Are, the Harder They Fall

How a star evolves depends on its mass. The low- and medium-mass stars become white dwarfs. The fate of more massive stars is different. When a star with mass much greater than the sun's mass contracts, more heat is generated than in the contraction of a small star. Instead of shrinking to a white dwarf, carbon nuclei in the core fuse and liberate energy while synthesizing heavier elements such as neon and magnesium. Gas and radiation pressure halt further gravitational contraction until all the carbon is fused. Then the core of the star contracts again to produce even greater temperatures and a new fusion series that produces even heavier elements. The fusion cycles repeat until the element iron is formed. The fusion of elements beyond iron nuclei will require energy rather than liberating energy (recall from our treatment of nuclear fusion, Figure 16.11 in Chapter 16, that the mass of nucleons increase and absorb energy as elements beyond iron are fused). With no energy coming from the iron core, the center of the star collapses without rekindling. The entire star begins its final collapse.

The collapse is catastrophic. When the core density is so great that all the nuclei are compressed against one another, the collapse momentarily comes to a halt. The collapsed star, compressed like a spring, rebounds violently in a great explosion, hurling into space the elements manufactured over previous billions of years. The entire episode can last as briefly as a few minutes. It is during this brief time that the heavy elements beyond iron are synthesized, as protons and neutrons mesh with other nuclei to produce elements such as silver, gold, and uranium. Because the time available for making these heavy elements is so brief, they are not as abundant as iron and the lighter elements.

Such a stellar explosion is a **supernova,** one of nature's most spectacular cataclysms. Supernovae are fiery cauldrons that generate the elements essential to life, for all the elements beyond iron that make up our bodies originated in far-off, long-ago supernovae. A supernova flares up to millions of times its former brightness. In 1054, Chinese astronomers recorded their observation of a star so bright it could be seen by day as well as by night. This was a supernova, its glowing plasma remnants now making up the spectacular Crab nebula (Figure 28.10). The recent 1987 supernova (see box) was less spectacular, but afforded astronomers an exciting firsthand look at one of these seldom seen events.

The inner part of the supernova star implodes to form a core compressed to neutron density. Protons and electrons have been compressed together to form a core of neutrons just a few kilometers wide. This superdense, central remnant of a supernova

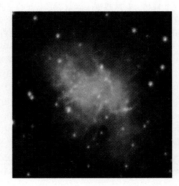

Figure 28.10   The Crab nebula, the remnant of a supernova explosion that was seen on earth in 1054 AD.

survives as a **neutron star**. In accord with conservation of angular momentum, these tiny bodies with densities hundreds of millions of times that of white dwarfs spin at fantastic speeds. Neutron stars are the explanation of **pulsars,** discovered in 1967 as rapidly varying sources of low-frequency radio emission. As they spin they sweep the beams around the sky, and if the beams sweep over the earth, we detect pulses. Of the approximately 300 known pulsars, only a few have been found emitting X-ray or visible light. One is in the center of the Crab nebula (Figure 28.11). It has one of the highest rotational speeds of any pulsar studied, rotating about 100 times in 3 seconds. This is a relatively young pulsar, and it is theorized that X-ray and optical radiation is emitted only during a pulsar's early history.

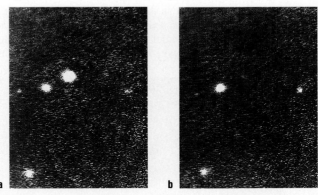

a   b

Figure 28.11   The pulsar in the Crab nebula rotates like a searchlight, beaming light and X rays toward earth about 30 times a second, blinking on and off; (a) shows the pulsar "on," (b) "off." The pulsar is a neutron-star core, the remnant of the supernova that created the Crab nebula.

Dying stars with cores greater than three to five solar masses collapse so violently that no physical forces are strong enough to inhibit continued contraction. The bigger they are, the harder they fall! The enormous gravitational field about the imploding concentration of mass makes explosion impossible. When the star has collapsed down to only 18 kilometers in diameter, collapse is unrestrained, and the star disappears from the observable universe. What is left is a *black hole*.

The 1987 Supernova

Johannes Kepler is credited in 1604 for spotting the last supernova before the invention of the telescope. Since then a dozen generations of astronomers have lived and died without ever witnessing such a stellar explosion. Astronomers have contented themselves studying the remnants of explosions that occurred before their time. This generation is more fortunate, for on February 24, 1987, Ian Shelton, graduate student dropout and resident observer at the University of Toronto telescope in the Andes Mountains of northern Chile, stumbled upon a curious blotch in a photograph he had just made of the Large Magellanic Cloud. He stepped outside and confirmed his finding firsthand. He saw a pinpoint of light less bright than others in the Magellanic Cloud that hadn't been there the night before. This was Supernova 1987A.

During the first few days after Shelton's sighting the supernova's brightness increased a thousandfold, much swifter than expected, and leveled off dimmer than expected. Although initially very hot and blue, it turned red very quickly, a change that occurred as the shell of debris raced outward at some 80,000,000 km/h and cooled. In March the supernova brightened again, powered by the decay of radioactive elements in the stellar remains. Enormous quantities of nickel and cobalt that were forged in the detonation were decaying into iron—enough to construct 20,000 earths. The burst reached its peak toward the end of May when it was as luminous as billions of suns. This colossal event ran its course during the summer, and a year later the visible fireworks were completely over. To the unaided eye Supernova 1987A had faded into oblivion in the southern sky.

Centuries from now astronomers will still be studying the wispy, incandescent filaments of Supernova 1987A, just as observers today meticulously examine the remains of the "new stars" observed centuries ago by Kepler and others. Theories rise and fall with each new measurement of the ever-expanding star remnants.

How awe-inspiring it is that men and women on a small planet in the outer reaches of a galaxy are able to investigate spots of light in the night sky and from their examinations arrive at a magnificent description of creation. This was well put in 1948 during the opening of the famous Hale telescope: "In the last analysis, the mind that encompasses the universe is more marvelous than the universe that encompasses the mind."

Figure 28.12   The 1987A Supernova, seen in the southern sky in the Large Magellanic Cloud.

28.4 BLACK HOLES

A **black hole** is what is left when a star has undergone gravitational collapse. The collapsed star is black because the enormous gravity at its surface does not allow light to escape. We can see why gravity is so great in the vicinity of a black hole by considering the change in the gravitational field at the surface of any star that collapses. In accord with Newton's law of gravity, any mass at the surface of a star, whether it be an object or simply a particle, has weight that depends on both its mass and the mass of the star. But more important, weight also depends on the distance between the object and the center of the star. So if a star collapses, the distance between the object and the star's center decreases. Weight increases, without a change in total mass. How much? That depends on the amount of collapse. If a star collapses to half size, then in accord with the inverse-square law, the weight of an object at its surface quadruples (Figure 28.13). Collapse to a tenth the size and the weight at the surface is 100 times as much. Along with the increase in gravitational field, the escape velocity from the surface of the collapsing star increases. If a star such as our sun collapsed to a radius of 3 kilometers, the escape velocity from its surface would exceed the speed of light, and nothing—not even light—could escape!* The sun would be invisible. It would be a black hole.

The sun, in fact, has too little mass to experience such a collapse, but when some stars with core masses greater than four suns reach the end of their nuclear resources, they undergo collapse; and unless rotation is high enough, the collapse continues until the stars reach infinite densities. Gravitation near the surfaces of these shrunken stars is so enormous that light cannot escape from them. They have crushed themselves out of visible existence. The results are completely invisible black holes.

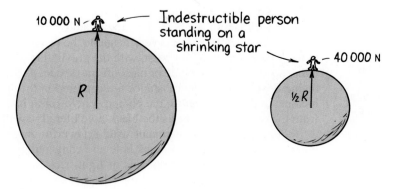

Figure 28.13   If a star collapses to half its radius and there is no mass change, gravitation at its surface would increase by four (inverse-square law). If the star collapses to one-tenth its radius, gravitation at its surface would be a hundredfold.

---

*In Chapter 29 we'll see that light, like massive things, is affected by gravity. Just as we fail to see the curvature of a high-speed bullet when viewed along short segments, we most often fail to see the curvature by gravity of light because of its very high speed. We'll see that light *does* curve in a gravitational field.

A black hole is no more massive than the star from which it collapsed, so the gravitational field in regions at and greater than the original star's radius is no different after the star's collapse than before. But closer distances near the vicinity of a black hole are nothing less than the collapse of space itself, with a surrounding warp into which anything that passes too close—light, dust, or a spaceship—is drawn. Astronauts could enter the fringes of this warp and with a powerful spaceship still escape. Below a certain distance, however, they could not, and they would disappear from the observable universe.

QUESTIONS

1. What determines whether a star becomes a white dwarf, a neutron star, or a black hole?

2. If the sun somehow suddenly collapsed to a black hole, what change would occur in the orbital speed of Planet Earth?

Black Hole Geometry

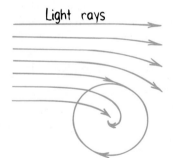

Light rays

Figure 28.14 Light rays deflected by the gravitational field around a black hole. Light passing far away is bent only slightly; light passing closer can be captured into circular orbit; and light passing closer still is sucked into the hole.

We will see in Chapter 29 when we study Einstein's general theory of relativity that light responds to gravity. We can understand the geometry of a black hole by considering the behavior of light in its vicinity.

If we shine a beam of light past a black hole, gravity is intense enough to noticeably deflect the beam. If the beam passes very far from the hole, where gravity is not as strong, the beam bends only slightly (Figure 28.14). The closer the light beam is to a black hole, the more it bends. If we shine a beam toward but slightly away from a black hole at precisely the right distance, we can put the light into *circular orbit* about the hole. This region above the black hole is called the *photon sphere*. The photon sphere is very unstable, however, because the slightest variation in the interaction of a light beam with the gravitational field will send the light beam either spiraling into the hole or back off into space. All beams of light that happen to be incident at this critical distance are captured in the sphere, while beams that are incident at distances within the photon sphere spiral into the black hole and are lost from the outside universe as the black hole literally swallows them up.

An indestructible astronaut with a powerful enough spaceship could venture into the photon sphere of a black hole and come out again. While inside the photon sphere, she could still send beams of light back into the outside universe (Figure 28.15). If she directed her flashlight in sideways directions and toward the black hole, the light would quickly spiral into the black hole; but light directed vertically

ANSWERS

1. The mass of a star is the principal factor that determines its fate. Stars having solarlike masses and less than solar masses evolve to become white dwarfs; more massive stars evolve to become neutron stars; and stars having more than from three to five solar masses in their cores ultimately become black holes.

2. None. This is best understood classically; nothing in Newton's law of gravitation, $F = G(mM/d^2)$, changes. The fact that the mass of the sun, M, is compressed doesn't change its mass, nor its distance from the earth, nor the force.

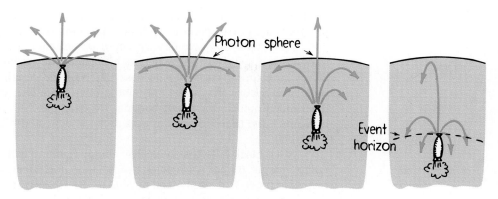

Figure 28.15 Just beneath the photon sphere, an astronaut could still shine light to the outside. But as she gets closer to the black hole, only light directed nearer the vertical gets out, until finally even vertically directed light is trapped. This distance is the event horizon.

and at angles close to the vertical would still escape. As she gets closer and closer to the black hole, however, she finds she must shine the light beams closer and closer to the vertical for escape. Moving closer still, our astronaut would find a particular distance where *no* light can escape. No matter what direction the flashlight points, all the beams are deflected into the black hole. Our unfortunate astronaut would have passed within the **event horizon**. Once inside the event horizon, she could no longer communicate with the outside universe; neither light waves, radio waves, nor any matter could escape from inside the event horizon. Our astronaut would have performed her last experiment in the universe as we conceive it.

The event horizon surrounding a black hole is often called the *surface* of the black hole, the diameter of which depends on the mass of the hole. For example, a black hole resulting from the collapse of a star ten times as massive as the sun has an event horizon diameter of about 30 kilometers. The radii of event horizons for black holes of various masses are shown in Table 28.1.

When a collapsing star contracts within its own event horizon, the star still has substantial size. There are no forces known that can stop the continued contraction, however, and the star quickly shrinks in size until finally it is crushed, presumably to the size of a pinhead, then to the size of a microbe, and finally to a realm of size

TABLE 28.1

Mass of Black Hole		Radius of Event Horizon	
1	Earth mass	0.8	centimeter
1	Jupiter mass	2.8	meters
1	Solar mass	3	kilometers
2	Solar masses	6	kilometers
3	Solar masses	9	kilometers
5	Solar masses	15	kilometers
10	Solar masses	30	kilometers
50	Solar masses	148	kilometers
100	Solar masses	296	kilometers
1000	Solar masses	2961	kilometers

smaller than ever measured by humans. At this point, according to theory, there is infinite density. This point is the **black hole singularity**.

Whereas a complete description of an ordinary star is very complicated, involving many physical properties such as chemical composition, densities, temperatures, and so on, the complete description of the simplest kind of black hole is quite straightforward. It has only one property—mass. And this can be precisely determined outside the event horizon, for example, by a physicist who measures how much the trajectory of a rocket probe is deflected when in the vicinity of the black hole.

If the black hole has an electric charge (either positive or negative) or a magnetic "charge" (either north pole or south pole), the effects of these charges, like the effects of mass, will also extend beyond the event horizon. A distant physicist could use a sensitive apparatus to detect these charges. So, in addition to mass, a charge on a black hole is not lost to the universe and is an additional physical property. It is unlikely that black holes with appreciable electric charges exist, however, for if a black hole did have a substantial charge, its electric field would soon tear apart atoms in nearby space and in a very short time become neutralized by particles of opposite charge.

More important is spin, for most stars are rotating and possess angular momentum. If a black hole is formed from a rotating star, the surrounding space and time will be dragged with it. (The connection between space, time, and gravity will be explored in Chapter 29 when we treat general relativity.) An observer sitting in a spaceship far from the hole would notice a gradual pull around the hole in the direction of black hole rotation. The closer to the rotating hole, the faster the ship is pulled around. We see that the angular momentum of a black hole is also information that extends beyond the event horizon and is a third property of black holes.*

Black holes, then, can have only three possible properties: mass, charge, and angular momentum. Whereas a complete description of a star involves all sorts of hairy things such as chemical composition, varying pressure, densities, temperatures at different depths, and so on, no such complications are involved in black holes. Physicists put it simply by stating, *"Black holes have no hair!"*

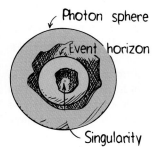

Figure 28.16 Structure of a simple, ideal black hole (uncharged and nonrotating).

28.5 GALAXIES

A **galaxy** is a large assemblage of stars, nebulae, and interstellar gas and dust. Galaxies are the breeding grounds of stars. Our own star, the sun, is an ordinary star among some 200 billion others in an ordinary galaxy known as the **Milky Way**. With unaided eyes we see the Milky Way as a faint band of light that stretches across the sky.

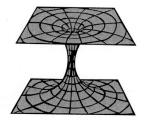

*Science types speculate that material drawn into a black hole may reappear elsewhere through a "white hole." What vanishes in one place might be spewed out at another. The downside to this speculation is that everything that falls into a black hole is reduced to elementary particles, and everything would emerge as particles. If black holes are a gate from one realm to another, and a temporary gate at that due to fluctuations, they are a gate through which nothing can pass intact. An upside presents itself, however, if the black hole is spinning fast enough to form a *ring,* whereupon astronauts approaching this ring along the rotational axis of the black hole might travel unharmed through the ring, and perhaps enter another universe! The merit of these intriguing speculations, however, is pretty much confined to science fiction stories. They're fun to think about!

Observations of Black Holes

We suspect that all dead stars with core masses greater than about three to five suns may be black holes. Finding these invisible star cores, however, is very difficult. One way is to look for a binary system in which a single luminous star appears to orbit about an invisible companion. If they are closely situated, matter ejected by the normal companion and accelerating into the neighboring black hole should emit X rays. The first convincing candidate for a black hole discovered by astronomers in 1972 was the X-ray star Cygnus X-1. Cygnus X-1 is about nine or ten solar masses and is a part of a binary system with a blue supergiant. Material streaming into the supposed black hole from the supergiant companion is found to be emitting X radiation. Similar radiation patterns have also been found in Circinus X-1 (3u 1516–56), another binary system, and more recently in V404 Cygni, both candidates for black holes. Observations by the NASA satellite *Copernicus* strongly suggest that a star in the constellation Scorpius is a black hole. This star, V861 Sco, is at the relatively close distance of 5000 light-years—slightly nearer than Cygnus X-1. The X ray source called LMC X-3 in the large Magellanic Cloud—a dwarf companion galaxy to our own—is very likely a black hole. Other massive black holes of 100 to 1000 solar masses are thought to exist at the centers of certain globular clusters (NGC 6624, NGC 1851, and NGC 6440).

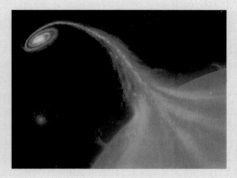

Figure 28.17 A black hole steals matter from a giant companion star, gathers it into a hot, X-ray emitting disk, and slowly swallows it.

The early Greeks called it the "milky circle" and the Romans called it the "milky road" or "milky way." The latter name has stuck.

Most astronomers believe that some 10 to 15 billion years ago galaxies formed from huge clouds of primordial gas pulled together by gravity, similar to our description in the previous chapter of the way the solar system formed. The steps of formation begin with gravitational attraction between distant particles, then contraction followed by an increased rotational rate, then in some cases, flattening to a disk due to this rotation. A most striking feature of our galaxy is the spiral arms that wind outward through the disk. These arms are made of swarms of hot, blue stars, clouds of dust and gas, and clusters of young stars.

Masses of galaxies range from about a millionth the mass of our galaxy to some 50 times more. Galaxies are calculated to have much more mass than has been detected. This nondetected mass is known as *dark matter*. The nature of this dark matter is still a question.

The millions of galaxies that are visible on long-exposure photographs can be distinguished into three main classes: *elliptical, spiral,* and *irregular.* In each class of galaxies there is variation in size and the relative amounts of gas and dust present.

Figure 28.18 This wide-angle photograph of the Milky Way, from Sagittarius on the left to Centaurus on the right. The dark lanes and blotches are caused by interstellar gas and dust that obscure the light of background stars.

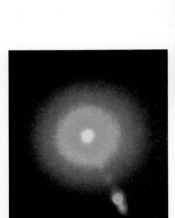

Figure 28.19 The giant elliptical galaxy M87, one of the most luminous galaxies in the sky, is located near the center of the Virgo cluster, some 50 million light years from earth. It is about 40 times more massive than our Milky Way galaxy.

Elliptical galaxies are the most common galaxies in the universe. Because most of them are relatively dim, they are difficult to see compared to other types of galaxies. An exception is the giant elliptical galaxy M87 (Figure 28.19). Elliptical galaxies contain little gas and dust and cannot make new stars.

Irregular galaxies are normally small and faint. They are difficult to detect. They are without obvious spiral arms or nuclei, and contain large clouds of gas and dust mixed with both young and old stars. The irregular galaxy first described by the navigator on Magellan's voyage around the world in 1521 is our nearest neighboring galaxy—the *Magellanic Clouds*. These consist of a large cloud dotted with hot young stars having a combined mass of some 20 billion solar masses, and a small cloud with about 2 billion solar masses (Figure 28.20). The combined mass is small for a galaxy. Irregular galaxies are probably as common as spiral galaxies. Spiral and irregular galaxies contain large amounts of gas and dust and are still forming stars.

Figure 28.20 (a) The Large Magellanic Cloud and (b) neighboring small Magellanic Cloud comprise an irregular galaxy that is only about 150 thousand light-years away. The Magellanic Clouds are our closest galactic neighbors and likely orbit our Milky Way.

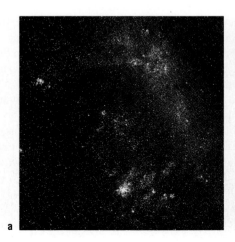

a

b

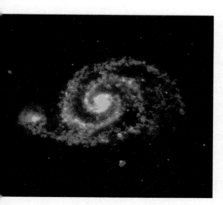

Figure 28.21 Spiral galaxy about 20 million light-years from earth.

Spiral galaxies seem the most beautiful arrangements of stars in the heavens. They are bright with the light of newly formed stars. The brightness of most spiral galaxies makes them easy to see at great distances. How their spiral arms form is still being investigated. Perhaps differential rotation of the galaxy stretches star-forming regions into elongated arches of stars and nebulae. Or maybe they are created by density waves that sweep around the galaxy. Investigators are still learning about spiral-arm formation. We know that about two-thirds of all known galaxies are spirals, although they make up probably 15 to 20 percent of all galaxies. We do not see the greater number of fainter elliptical galaxies thought to exist.

We know what it's like to live in a spiral galaxy, for our Milky Way is a typical one. When we look at the Milky Way that crosses the night sky we are looking through the disk of the galaxy. Interstellar dust obscures our view of most of the visible light that lies along the plane of this disk. Most of what we know of our galaxy is via infrared and radio telescopes. Infrared and radio observations reveal many details of the galactic nucleus, but astronomers are still puzzled by the processes occurring there. The nucleus seems to be crowded with stars and hot dust, and at the very center is thought to be a massive black hole with a mass of a million suns that generates energy by swallowing surrounding matter. Don't go too near the center of the Milky Way.

Galaxies collide. Stars are normally so far apart from one another that physical collisions of individual stars is a highly unlikely event. But interstellar gases and dust collide violently, with matter stripped from one galaxy and deposited in another. These collisions also prompt the formation of new stars. Low-speed collisions can result in the merger of a new galaxy. There is evidence that our own Milky Way galaxy may be presently consuming the Magellanic Clouds. At high velocities, collisions can distort each other through tidal forces and create tails and bridges. The collision of spiral galaxies are thought to form giant elliptical galaxies. Many giant elliptical galaxies are believed to contain the merged remains of several spiral galaxies. Galaxies are cannibals. Spiral galaxies are survivors that have experienced no collisions or very few collisions with large galaxies since their formation.

Galaxies are not the largest things in the universe. Galaxies come in **clusters**. And clusters of galaxies appear to be part of even larger clusters, the **superclusters**. It doesn't stop there; superclusters in turn seem to be part of a network of filaments surrounding empty voids. Comprehension of the universe becomes mind-boggling.

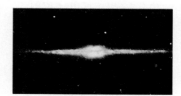

Figure 28.22 An edge-on view of our galaxy, the Milky Way. Our sun is about $\frac{5}{8}$ from the center.

Figure 28.23 The great nebula in Andromeda, a spiral galaxy about 2.3 million light years from earth.

Figure 28.24 An X-ray view of a cluster of galaxies.

28.6 QUASARS

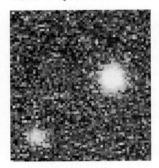

Figure 28.25 Colorized image of the quasar BR 1202-07, the brightest observed quasar to date.

Galaxies are not the brightest parts of the universe. Brighter still are **quasars**. The energy output of these objects is enormous—hundreds of times that of the entire Milky Way galaxy. Quasars were first thought to be relatively faint ordinary stars in our own galaxy, but in 1960 astronomers found that they emit radio waves. Radio waves are frequently observed coming from galaxies, but no star had been observed to emit strong radio signals. Further investigation revealed these "radio stars" had a pattern of spectral lines that could not be deciphered. These objects became known as "quasi-stellar sources," soon shortened to *quasars*.

The unusual spectra turned out to be a normal spectrum with an extremely large and unprecedented red shift, which indicated enormous recessional velocities—some more than 90 percent the speed of light. Clearly the objects couldn't be stars in our own galaxy, for we would have noticed a change in their positions against the background of the fixed stars, and quasars had been observed for years as faint stars with no noticeable change in position. At first some investigators thought that the red shift was not a Doppler shift, but a gravitational red shift characteristic of a small body with enormous mass and a correspondingly enormous gravitational field. But spectra from quasars revealed emission lines from normal atoms with normally orbiting electrons that wouldn't exist in a neutron star, black hole, or any body with gravitation intense enough to produce such a large red shift.

We do not know what quasars are. They appear to be as much as 15.5 billion light-years distant, putting them back to the beginnings of the universe. They may be gigantic black holes that pull enormous amounts of material toward them with resulting collisions that liberate immense energies. Current findings indicate that quasars are the brilliant cores of very distant spiral galaxies that we see as they were when they were young. Quasars are currently the most puzzling objects known to astronomers.

Figure 28.26 What do the atoms in Melissa's body have in common with the stars she contemplates?

SUMMARY OF TERMS

Protostar The aggregation of matter that goes into and precedes the formation of a star.

Thermonuclear reaction The fusion reaction brought about by high temperatures.

Binary star Pairs of stars that orbit about a common center of mass.

Nemesis The name of the hypothetical star companion to the sun.

H-R diagram (Hertzsprung-Russell diagram) A plot of intrinsic brightness versus surface temperature of stars. When so plotted, stars' positions take the form of a main sequence for average stars, with exotic stars above or below the main sequence.

Red giant Cool giant stars above main sequence stars on the H-R diagram.

Planetary nebula An expanding shell of gas ejected from a low-mass star during the latter stages of its evolution.

White dwarf Dying star that has collapsed to the size of the earth and is slowly cooling off; located at the lower left of the H-R diagram.

Black dwarf The presumed end state of a white dwarf that has cooled off.

Nova A star that suddenly brightens, appearing as a "new" star; believed to be associated with the eruptions from the surfaces of white dwarfs in binary systems.

Supernova An exploding star, caused either by transfer of matter to a white dwarf, or by gravitational collapse of a massive star, where enormous quantities of matter are emitted.

Neutron star A small, highly dense star composed of tightly packed neutrons formed by the welding of protons and electrons.

Pulsar Likely a neutron star that rapidly spins, sending short precisely timed bursts of electromagnetic radiation.

Black hole The remains of a giant star that has collapsed upon itself, so dense and gravitational field so intense, that light itself cannot escape.

Event horizon The boundary region of a black hole from which no radiation may escape. Any events within the event horizon are invisible to distant observers.

Black hole singularity The object of zero radius into which the matter of a black hole is comprised.

Milky Way The name of the galaxy to which we belong. Our cosmic home.

Elliptical galaxy A galaxy that is round or elliptical in outline. It has little gas and dust, no disk or spiral arms, and few hot and bright stars.

Irregular galaxy A galaxy with a chaotic appearance, large clouds of gas and dust, without spiral arms.

Spiral galaxy A disk-shaped galaxy with hot bright stars, and spiral arms. Our Milky Way is a spiral galaxy.

Galaxy cluster Pertains to a group of more than one galaxy.

Galaxy supercluster A group of an enormous number of galaxies.

Quasar (Quasi-stellar object) A small powerful source of energy believed to be the active core of very distant galaxies.

• •

REVIEW QUESTIONS

1. What are constellations?

2. Why are winter constellations different from summer constellations?

Birth of Stars

3. What is a protostar?

4. What process changes a protostar to a full fledged star?

5. What are the outward forces that act on a star?

6. What are the inward forces that act on a star?

7. What do outward and inward forces on a star have to do with its size?

8. Where did atoms heavier than helium on earth originate?

Life of Stars

9. Compare the lifetimes of high-mass stars with those of low-mass stars.

10. How common are binary stars in the universe?

11. What measurements of binary stars provide data for finding their masses?

12. What is the companion star to the hypothetical star, *Nemesis?*

13. Where is most of the solar system's angular momentum?

14. What is the goal of SETI programs?

Death of Stars

15. What event marks the birth of a star, and what event marks its death?

16. When will our sun reach the stage of *red giant?*

17. What is the relationship between a *planetary nebula* and a *white dwarf?*

18. What is the relationship between a *compact object* and a *black dwarf?*

19. What is the relationship between a *white dwarf* and a *nova?*

20. Is a white dwarf a former low-mass star, or a former high-mass star?

21. What is the relationship between the heavy elements we find on earth and supernovae?

22. When was the last supernova seen?

23. What is the relationship between a *neutron star* and a *pulsar?*

Black Holes

24. What is the relationship between an ordinary star and a *black hole?*

25. How far would the sun have to collapse so that its light couldn't escape?

26. How does the mass of a star before collapse compare to the mass of the black hole it becomes?

27. Make a comparison of the circular orbit of light around a black hole with the circular orbit of a satellite about Planet Earth.

28. What is the relationship between the *photon sphere* and *event horizon* of a black hole?

29. What is the relationship between the *event horizon* and the *surface* of a black hole?

30. What is a black hole singularity?

31. What are the three possible properties of a black hole?

32. Are white holes facts or speculations?

33. Since black holes are invisible, what is the evidence for their existence?

Galaxies

34. What type of galaxy is the Milky Way?

35. What are the three types of galaxies?

36. What are the consequences of galaxies colliding?

Quasars

38. Which are brighter, galaxies or quasars?

39. Are quasars thought to be relatively close or relatively distant objects? Why?

40. We don't really know what quasars are. What do we think they are?

• •

EXERCISES

1. Why do we not see stars in the daytime?

2. Make a sketch similar to Figure 28.2 to show that the constellation seen in the background of a solar eclipse is a constellation normally seen six months earlier or later in the nighttime sky.

3. Thomas Carlyle wrote, "Why did not somebody teach me the constellations and make me at home in the starry heavens, which are always overhead and which I don't half know to this day?" What, beside the names of the constellations, did Thomas Carlyle not know?

4. We see the constellations as distinct groups of stars. Discuss why the constellations would have entirely different appearances if viewed from other distant locations in the universe.

5. The Big Dipper is sometimes right-side-up (where it could "hold water"), and at other times upside-down, (where like an upside-down bowl, it could not "hold water"). What length of time is required for the dipper to change from a position of "right-side-up" to "upside-down?"

6. In what sense are we all made of "star dust"?

7. How is the gold in your sweetheart's ring evidence of ancient stars that ran their life cycles long before the solar system came into being?

8. Would you expect metals to be more abundant in old stars or new stars? Defend your answer.

9. Why is there a lower limit on the mass of a star?

10. What ordinarily keeps a star from collapsing?

11. How does the energy of a protostar differ from the energy that powers a star?

12. Why do nuclear fusion reactions not occur on the outer layers of stars?

13. Why are massive stars generally shorter lived than low-mass stars?

14. What is meant by the statement, "The bigger they are, the harder they fall," with respect to stellar evolution?

15. Why will the sun not be able to fuse carbon nuclei in its core?

16. The conservation of angular momentum states that the angular momentum of a star (rotational inertia × rotational speed) remains the same before and after any event that does not involve external influences. How does this conservation principle explain why neutron stars spin so rapidly?

17. What does the spin rate of a star have to do with whether or not it has a system of planets?

18. In what way is a black hole blacker than black ink?

19. If you were to fall into a black hole, you'd likely die as a result of tidal forces. Explain.

20. A black hole is no more massive than the star from which it collapsed. Why, then, is gravitation so intense near a black hole?

21. What happens to the radial distance of a photon sphere as more and more mass falls into the black hole?

22. Make a comparison of the circular orbit of light around a black hole with the circular orbit of a satellite about Planet Earth.

23. What is the relationship between the *photon sphere* and *event horizon* of a black hole?

24. What is the evidence for the existence of black holes?

25. The Milky Way is far more prominent in July than in December. Suggest a reason.

26. Are there galaxies other than the Milky Way that can be seen with the unaided eye? Discuss.

27. How can collisions of galaxies affect their shapes?

28. What does it mean to say that galaxies are cannibals? Explain.

29. Quasars are the most distinct objects we know of in the universe. Why do we therefore say their existence goes back to the earliest times in the universe?

30. In any direction one looks in the night sky, there is a star. Why, then, isn't the night sky ablaze with light?

29

RELATIVITY AND THE UNIVERSE

We began this book with the physics concepts of motion, force, and energy, in the everyday world. In the chemistry chapters we studied these concepts in the micro world, and in the geology chapters they were applied to the processes of change on Planet Earth. Then we expanded discussion of these concepts to our solar system and the stars beyond. We conclude with the widest view of all—the universe—the whole show.

Except for an occasional exploding star here and there, earlier astronomers thought of the universe as an unchanging place. This thinking is carried over today when we speak of the "fixed stars." But today's scientific understanding describes a violently changing universe—one that had its beginnings in the violent fireball we call the Big Bang, and that continues to change rapidly. In modern times we have discovered objects that stretch the imagination such as pulsars, quasars, and black holes, which were briefly treated in the previous chapter. In this chapter we consider the evolving universe and its fabric—space-time. We'll learn about the relationship between space and time developed early in this century by Albert Einstein in both his *special theory of relativity* and his *general theory of relativity*. Our treatment of the universe starts with its beginning.

29.1 THE BIG BANG

No one knows how the universe began. Evidence suggests that about 15 to 20 billion years ago, the universe existed in an extremely high-temperature, high-density state. A primordial explosion occurred, called the **Big Bang**. Theorists generally believe that within the first three minutes after the Big Bang, great quantities of hydrogen and helium were created, spewing apart at great speeds. About 3 million years later, huge clouds of this matter, some 500 million light-years across, began to gradually condense. After about 200 million years these condensations formed the first galaxies—the birthplace of the stars—and heavier elements. The universe today is the remnant of the Big Bang.

The concept of the Big Bang came into focus in the 1930s, after the American astronomer Edwin P. Hubble, for whom the Hubble Telescope is named, showed that the universe is expanding. Further findings implied the cosmos was once concentrated in a very small hot place at a definite time. The concept of the Big Bang holds that this special time was the *beginning of time.*

The space formed by the Big Bang was filled by intense extremely energetic high-frequency radiation called the **primeval fireball**. Radiation from the dying embers of the primeval fireball now permeates all space in the form of long-wavelength microwaves, lengthened by the expansion of the universe. These microwaves were inadvertently discovered in 1964–65 by Arno A. Penzias and Robert W. Wilson of Bell Laboratories while they were trying to rid their radio antenna of microwave noise. The microwave background radiation they discovered was predicted by Big Bang theory. Then in 1992, measurements of minuscule variations in the background radiation vindicated another prediction of the Big Bang theory—that only such variations could account for the accumulation of matter to form galaxies.

The present expansion of the universe is evident in a Doppler red shift in the light from its galaxies. Recall that sound and light waves received by an observer are stretched out when a source recedes, and compressed when a source approaches. Stretched-out visible light waves are red-shifted, and indicate a receding light source. So the red shift shows that the distance between us and each galaxy, and the distance between galaxies, is increasing. This does not, however, place our own galaxy in a central position. Consider a balloon with ants on it: As the balloon is inflated, every ant will see every neighboring ant getting farther away, which certainly doesn't suggest a central position for each ant. In an expanding universe, any observer sees all other galaxies receding.

Figure 29.1 Every ant on the expanding balloon sees every other ant getting farther away.

Red-shifted galaxy light shows us not only that the universe is expanding but that it is expanding ever more slowly. Toss a rock skyward and it slows due to its gravitational interaction with the earth. Similarly, every bit of matter blown apart in the primordial explosion is attracted by gravity toward every other bit of matter, resulting in a continual slowing down of the expansion. Evidence for this slowdown is the greater red shift for the most distant galaxies. When we look at the stars and faraway galaxies, we are actually looking backward in time since it takes considerable time for the light from those bodies to reach all the way to earth. The stars farthest away are the stars we are seeing as they were longest ago. Since the degree of red shift indicates how quickly a light source is receding, the greater red shift of the farther galaxies shows that the universe was expanding faster in the beginning. The expansion of the universe is slowing with time.

Our analogies of the expanding balloon and the tossed rock give a visual picture of expansion and its slowing. An important point these analogies don't make is that the expansion of the universe is an expansion of space itself. The acts of blowing up a balloon or tossing a rock skyward are done in a surrounding space in some specified time—the balloon is blown up against space, and the rock tossed into space. There is a space "waiting" for each event. Similarly with time. There was a time before, during, and after the balloon was blown up and the rock tossed. But not so with the universe. The fundamental difference between the Big Bang explosion and ordinary explosions is that there was no space for the explosion to go into—space itself was exploding. The universe does not "exist" in space nor does the universe "exist" in time; rather both space and time "exist" within the universe. Without the universe, there would be no space and time. Space and time are in the universe, and not the other way around. The Big Bang was the expansion of space itself at the beginning of time.

To gain a perspective of the universe, we first examine the relationship between space and time.

29.2 SPACE-TIME

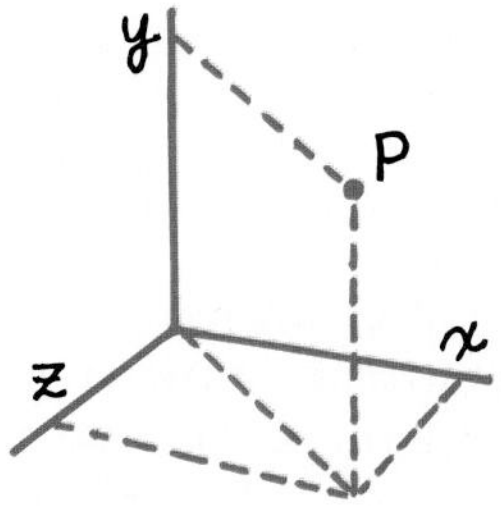

Figure 29.2 Point *P* can be specified with three numbers: the distances along the *x* axis, *y* axis, and *z* axis.

When we look up at the stars, we realize that we are looking backward in time. Some of the stars we see may have died long ago. We measure their distances in light-years,* which indicates space and time may be bound together. Einstein showed that space and time are indeed very intimately bound together.

The space we live in is three-dimensional; that is, we can specify any place in space by three dimensions. Loosely speaking, these dimensions are how far over, how far across, and how far up or down. For example, if we are at the corner of a rectangular room and wish to specify the position of any point in the room, we can do so with three numbers. The first is the number of meters to the point along a line joining the adjacent left wall and the floor; the second is the number of meters the point is along a line joining the adjacent right wall and the floor; and the third is the number of meters the point lies above the floor or along the vertical line joining the walls at the corner. Physicists speak of these three lines as the *coordinate axes* of a reference frame (Figure 29.2). Three numbers—the distances along the *x* axis, the *y* axis, and the *z* axis—will specify the position of a point in space.

We specify the size of objects with three dimensions. A box, for example, is described by its length, width, and height. But the three dimensions do not give a complete picture. There is a fourth dimension—time. The box was not always a box of given length, width, and height. It began as a box only at a certain point in time, on the day it was made. Nor will it always be a box. At any moment it may be crushed, burned, or destroyed. So the three dimensions of space are a valid description of the box only during a certain period of time. We cannot speak meaningfully about space without implying time. Things exist in **space-time**. Each atom, each object, each person, each planet, each star, each galaxy exists in "the space-time continuum."†

---

*Recall that a light-year is the distance light travels in 1 year, about 10^{16} m.

†Points in space and time may be quantized points in a four-dimensional space-time lattice. From the sizes of elementary particles and minimum separation between colliding particles, there seems to be an elemental unit of distance (9.05×10^{-35} m), and the lifetimes of all known elementary particles are consistent with being an integral number of "chronons" (1.35×10^{-43} s).

29.3 SPECIAL RELATIVITY

Einstein's concepts about space and time are part of a larger picture, a revolutionary one that predicts that motion through space causes a "slowing of time," that objects in motion are shorter and more massive than the same objects at rest, and that mass is actually congealed energy. These are the ideas of the **special theory of relativity,** which Einstein developed in 1905. The special theory of relativity is based upon two postulates. The first can be stated:

> **Observers can never detect their *uniform* motion except relative to other objects.**

We've all noticed that from inside a car at a traffic light, we see a nearby car move, only to find it is at rest and we are moving. We can tell which is moving by the background of trees or other objects. Imagine, however, we are in a spaceship in interstellar space and another spaceship coasts by at constant velocity. Which spaceship is moving? Without a background, all we can say is the ships are moving relative to each other. And even with a background, how could we say the background wasn't moving? These are Newtonian ideas. Einstein thought about them and added his conclusion—that there is no experiment you can perform to decide which ship is moving and which is not. This means there is no such thing as absolute rest—all motion is relative.

If there is no experiment that can be performed to detect absolute motion through space, the laws of physics must be the same on both spaceships. The more general form of the first postulate is:

> **All laws of nature are the same in all uniformly moving reference frames.**

On a jet airplane going 700 km/h, for example, coffee pours as it does when the plane is at rest; we swing a pendulum and it swings as it would if the plane were on the runway. There is no physical experiment we can perform to determine our state of uniform motion. By uniform motion, we mean nonaccelerated motion. The laws of physics within the uniformly moving cabin (constant velocity; zero acceleration) are the same as those in a stationary laboratory.

According to the first postulate, any measurements of the speed of light would show the same value in uniformly moving reference frames. This constancy of the speed of light is the second postulate of special relativity:

> **The speed of light in free space will have the same value to all observers, regardless of the motion of the source or the motion of the observer. The speed of light is a constant.**

Every measurement of the speed of light, and there have been many, has confirmed the second postulate. Move away from a tossed baseball when you catch it and you'll catch it at a slower speed. Do this for light and you have a different story. Pretend we're in a high-speed rocket moving away from a light source at nearly the speed of light. Good old common sense will tell us that the light that catches up to us and passes us is slower than if we weren't moving. But not so. While we move we in effect stretch out the space between us and the light source. But Einstein says we can't do that without also stretching out time. How much stretch? Just enough so that when we divide the space traveled by the time taken, the value for the speed of light is the same as if we weren't moving at all.

Clockwatching on a Trolleycar Ride

Figure 29.3 Space and time are intertwined. All space and time measurements of light are unified by *c*.

Pretend you are Einstein at the turn of the century in a trolleycar that provided the high-speed travel back then. Suppose the trolleycar is moving in a direction away from a huge clock in a village square. The clock reads 12 noon. To say it reads 12 noon is to say that light that carries the information "12 noon" is reflected by the clock and travels toward you in the direction of your line of sight. If you suddenly move your head to the side, instead of meeting your eye, the light carrying the information continues past, presumably out into space. Out there an observer who *later* receives the light says, "Oh, it's 12 noon on earth now." But from your point of view it isn't. You and the distant observer will see 12 noon at different times. You wonder more about this idea. If the trolleycar traveled as fast as the light, then it would keep up with its information that says "12 noon." Traveling at the speed of light, then, tells you it's always 12 noon at the village square. Time at the village square is frozen! So if the trolleycar is not moving, you see the village square clock move into the future at the rate of 60 seconds per minute; if you move at the speed of light, you see seconds on the clock taking infinite time. These are two extremes. What's in between? How about at speeds less than the speed of light? A little thought will show that the clock will be seen to run somewhere between the rate of 60 seconds per minute and 60 seconds per an infinity of time if your speed is between zero and the speed of light. From your high-speed (but less than *c*) moving frame of reference, the clock and all events in the reference frame of the clock will be seen in slow motion. Time will be stretched; how much it is stretched depends on speed. This is time dilation.

29.4 TIME DILATION

Let's examine this notion that time can be "stretched." Imagine that we are somehow able to observe a flash of light bouncing to and fro between a pair of parallel mirrors, like a ball bouncing to and fro between a floor and a ceiling. If the distance between the mirrors is fixed, then the arrangement constitutes a sort of "light clock," because the back-and-forth trips of the light flash take equal time intervals (Figure 29.4). Suppose our light clock is inside a transparent high-speed spaceship. If we travel

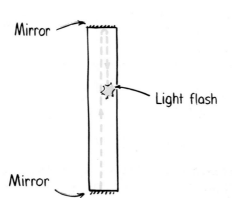

Figure 29.4 A light clock. Light will bounce up and down between parallel mirrors and "tick off" equal intervals of time.

Figure 29.5 (a) An observer moving with the spaceship observes the light flash moving vertically between the mirrors of the light clock. (b) An observer who sees the moving ship pass by observes the flash moving along a diagonal path.

along with the ship and watch the light clock (Figure 29.5a), we will see the flash of light reflecting straight up and down between the two mirrors, just as it would if the spaceship were at rest. Our observations will show no unusual effects. Note that there is no relative motion between us and our light clock; we say that we share the same reference frame in space-time.

If we instead make our observations from some relative rest position as the spaceship whizzes by us at high speed—say, half the speed of light—things are quite different. We will not see the path of light in simple up-and-down motion as before. Because the light flash keeps up with the horizontally moving light clock, we will see the flash follow a diagonal path (Figure 29.5b). Notice that from our frame of reference the flash travels a *longer distance* as it moves between the mirrors, considerably longer than it does in the reference frame of an observer riding with the ship. Since the speed of light is the same in all reference frames (Einstein's second postulate), the flash must travel for a correspondingly longer time between the mirrors in our frame than in the reference frame of the on-board observer. This follows from the definition of speed—distance divided by time. *The longer diagonal distance must be divided by a correspondingly longer time interval to yield an unvarying value for the speed of light.* This stretching out of time is called **time dilation**. We have considered a light clock in our example, but the same is true for any kind of clock. All moving clocks run slow. Time dilation has nothing to do with the mechanics of clocks, but with the nature of time itself.

The exact relationship of time dilation for different frames of reference in space-time can be derived from Figure 29.6 with simple geometry and alge-

Figure 29.6 The longer distance taken by the light flash in following the diagonal path must be divided by a correspondingly longer time interval to yield an unvarying value for the speed of light.

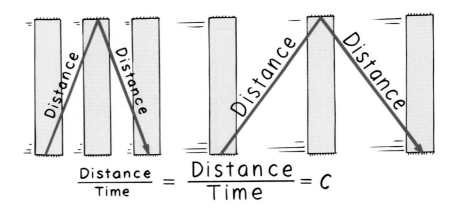

$$\frac{Distance}{Time} = \frac{Distance}{Time} = c$$

bra.* The relationship between the time t_0 (proper time) in the observer's own frame of reference and the relative time t measured in another frame of reference is

$$t = \frac{t_0}{\sqrt{1 - v^2/c^2}}$$

where v represents the relative velocity between the observer and the observed and c is the speed of light. Because no material object can travel at or beyond the speed of light, the ratio v/c is always less than 1; likewise for v^2/c^2. For $v = 0$, this ratio is zero, and for everyday speeds where v is negligibly small compared to c, it's practically zero. Then $1 - (v^2/c^2)$ has a value of 1, as has $\sqrt{1 - (v^2/c^2)}$, and we find $t = t_0$ and time intervals appear the same in both systems. For higher speeds, v/c is between zero and 1, and $1 - (v^2/c^2)$ is less than 1; likewise, $\sqrt{1 - (v^2/c^2)}$. So t_0 divided by a value less than 1 produces a value greater than t_0, an elongation, a dilation of time.

Figure 29.7 When we see the rocket at rest, we see it traveling at the maximum rate in time: 24 hours per day. If we see the rocket traveling at the maximum rate through space (the speed of light), we see its time standing still.

---

*The light clock is shown in three successive positions in the figure below. The diagonal lines represent the path of the light flash as it starts from the lower mirror at position 1, moves to the upper mirror at position 2, and then moves back to the lower mirror at position 3. Distances on the diagram are marked ct, vt, and ct_0, which follows from the fact that distance traveled equals speed multiplied by time.

The symbol t_0 represents the time it takes the flash to move between mirrors as measured from a frame of reference fixed to the light clock. This is the time for straight up or down motion. The speed of light is c, and the path of light is seen to move a vertical distance ct_0. This distance between mirrors is at right angles to the motion of the light clock and is the same in both reference frames.

The symbol t represents the time it takes the flash to move from one mirror to the other as measured from a frame of reference in which the light clock moves with speed v. Since the speed of the flash is c and the time to go from position 1 to position 2 is t, the diagonal distance traveled is ct. During this time t, the clock (which travels horizontally at speed v) moves a horizontal distance vt from position 1 to position 2.

These three distances make up a right triangle in the figure, in which ct is the hypotenuse and ct_0 and vt are legs. A well-known theorem of geometry (the Pythagorean theorem) states that the square of the hypotenuse is equal to the sum of the squares of the two sides. If we apply this to the figure, we obtain:

$$c^2t^2 = c^2t_0^2 + v^2t^2$$
$$c^2t^2 - v^2t^2 = c^2t_0^2$$
$$t^2[1 - (v^2/c^2)] = t_0^2$$
$$t^2 = \frac{t_0^2}{1 - (v^2/c^2)}$$
$$t = \frac{t_0}{\sqrt{1 - (v^2/c^2)}}$$

Path of light as seen from a position of rest

ct ct_0 vt

Mirrors at position 1   Mirrors at position 2   Mirrors at position 3

To consider some numerical values, assume that v is 50 percent the speed of light. Then we substitute $0.5c$ for v in the time dilation equation and after some arithmetic find that $t = 1.15t_0$. This means that if we viewed a clock on a spaceship traveling at half the speed of light, we would see the second hand take 1.15 minutes to make a revolution, whereas if it were at rest, we would see it take 1 minute. If the spaceship passes us at 87 percent the speed of light, we find $t = 2t_0$, and we measure events on the spaceship taking twice the usual time intervals—hands on the ship clock turn only half as fast as those on our own clock. Events on the ship are seen in slow motion. At 99.5 percent the speed of light, $t = 10t_0$, we see the second hand of the spaceship's clock take 10 minutes to sweep through a revolution requiring 1 minute on our clock.

To put these figures another way, at 99.5 percent c, the moving clock would run at a tenth of our rate; its hands would tick only 6 seconds while our clock's second hand ticks 60 seconds. At 87 percent c, the moving clock ticks at half our rate and shows 30 seconds to our 60 seconds; at 50 percent c, the moving clock ticks 1/1.15 as fast and ticks 52 seconds to our 60 seconds. We see that moving clocks run slow.

Nothing is unusual about a moving clock itself; it is simply ticking to the rhythm of a different time. The faster a clock moves, the slower it runs as viewed by an observer not moving with the clock. If it were possible for an observer to watch a clock pass by at the speed of light, the clock would not appear to be running at all. This observer would measure the interval between ticks to be infinite. Time would be frozen and the clock would be ageless! If our observer were moving with the clock, however, the clock would not show any slowing of time at all. To the observer the clock would be operating normally. This is because there would be no motion between the observer and the observed. The v in the time dilation equation would then be zero, and $t = t_0$; they share the same reference frame in space-time.

If the person who whizzes past us checked a clock in our reference frame, however, she would find our clock to be running as slowly as we find hers to be. We each see each other's clock running slow. There is really no contradiction here, for it is physically impossible for two observers moving at different velocities to refer to one and the same realm of space-time. All measurements made in one realm of space-time need not agree with all measurements made in another realm of space-time. The measurement they will always agree on, however, is the speed of light.

Time dilation has been confirmed in the laboratory innumerable times with atomic particle accelerators. The lifetimes of fast-moving radioactive particles increase as the speed goes up, and the amount of increase is just what Einstein's equation predicts.

Time dilation has been confirmed also for not-so-fast motion. In 1971, to test Einstein's theory, four cesium-beam atomic clocks were twice flown on regularly scheduled commercial jet flights around the world, once eastward and once westward. The clocks indicated different times after their round trips. Relative to the atomic time scale of the U.S. Naval Observatory, the observed time differences, in billionths of a second, were in accord with relativistic prediction.

This all seems very strange to us only because it is not our common experience to deal with measurements made at relativistic speeds or atomic-clock-type measurements at ordinary speeds. Due to this inexperience, the theory of relativity does not make common sense. But common sense, according to Einstein, is that layer of prejudices laid down in the mind prior to the age of 18. If we spent our youth zapping through the universe in high-speed spaceships, we would probably be quite comfortable with the results of relativity.

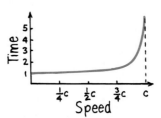

Figure 29.8 The graph shows how 1 second on a stationary clock is stretched out, as measured on a moving clock. Note the stretching becomes significant only at speeds near the speed of light.

Figure 29.9 From the earth frame of reference, light takes 25,000 years to travel from the center of the galaxy to our solar system. From the frame of reference of a high-speed spaceship, the trip takes less time. From the frame of reference of light itself, the trip takes no time. There is no time in a speed-of-light frame of reference.

QUESTIONS

1. If you are moving in a spaceship at a high speed relative to the earth, would you notice a difference in your pulse rate? In the pulse rate of the people back on earth?

2. Will observers A and B agree on measurements of time if A moves at half the speed of light relative to B? If both A and B move together at half the speed of light relative to the earth?

3. Does time dilation mean that time really passes more slowly in moving systems or that it only seems to pass more slowly?

Space Travel

One of the old arguments advanced against the possibility of human interstellar travel was that our life span is too short. It was argued, for example, that the nearest star (after the sun), Alpha Centauri, is 4 light-years away, and a round trip even at the speed of light would require 8 years. And even a speed-of-light voyage to the center of our galaxy, 25,000 light-years distant, would require a 25,000-year lifetime. But these arguments fail to take into account time dilation. Time for a person on earth and time for a person in a high-speed rocket ship are not the same.

ANSWERS

1. There would be no relative speed between you and your own pulse, which share the same frame of reference, so you would notice no relativistic effects in your own pulse. There would, however, be a relativistic effect between you and people back on earth. You would find their pulse rate slower than normal (and they would find your pulse rate slower than normal). Relativity effects are always attributed to the other guy.

2. When A and B move relative to each other, each observes a slowing of time in the frame of reference of the other. So they will not agree on measurements of time. When they are moving in unison, however, they share the same frame of reference and will agree on measurements of time. They will see each other's time as passing normally, and they will each see events on earth in the same slow motion.

3. The slowing of time in moving systems is not merely an illusion resulting from motion. Time really does pass more slowly in a moving system compared to one at relative rest. (This is dramatically shown in "The Twin Trip," in the *Conceptual Physical Science Practice Book*.)

A person's heart beats to the rhythm of the realm of space-time it is in; and one realm of space-time seems the same as any other to the person, but not to an observer who stands outside the person's frame of reference. For example, astronauts traveling at 99 percent *c* could go to the star Procyon (10.4 light-years distant) and back in 21 years. It would take light itself 20.8 years to make the same round trip. Because of time dilation, it would seem that only 3 years had gone by to the astronauts. This is what all their clocks would tell them—and biologically they would be only 3 years older. It would be the space officials greeting them on their return who would be 21 years older!

At higher speeds the results are even more impressive. At a rocket speed of 99.99 percent *c*, travelers could travel slightly more than 70 light-years in a single year of their own time; at 99.999 percent *c*, this distance would be pushed appreciably farther than 220 light-years. A 5-year trip for them would take them farther than light travels in 1100 earth-time years!

Present technology does not permit such journeys. The main problems are radiation and energy. Spaceships traveling at relativistic speeds would encounter hails of interstellar particles just as if the ship were at rest on the launching pad and bombarded by a steady stream of particles shot by a particle accelerator. No way of shielding such intense particle bombardment for prolonged periods of time is presently known. And if somehow a way were devised to solve this problem, there would be the problem of energy and fuel. Spaceships traveling at relativistic speeds would require billions of times the energy used to put a space shuttle into orbit. To send a shuttle-type craft at 70 percent *c* to and from Alpha Centauri, for example, would require some 500,000 times the amount of energy presently consumed in the United States in a year. Even if the spaceship scooped and fused interstellar hydrogen by some kind of interstellar ramjet, it would have to overcome the enormous retarding effect of scooping up the hydrogen at high speeds. The practicalities of such space journeys are enormously prohibitive. So for the time being, interstellar space travel must be relegated to science fiction.

Century Hopping

We can speculate about human space-faring possibilities if the prohibitive problems of radiation and energy are overcome and space travel one day becomes a routine experience. People will have the option of taking a trip and returning to any future century of their choosing. For example, one might depart from earth in a high-speed ship in the year 2100, travel for 5 years or so, and return in the year 2500. One could live among the earthlings of that period for a while and depart again to try out the year 3000 for style. People could keep jumping into the future with some expense of their own time—but they could not trip into the past. They could never return to the same era on earth that they bid farewell to. Time, as we know it, travels one way—forward. Here on earth we move constantly into the future at the steady rate of 24 hours per day. A deep-space astronaut leaving on a deep-space voyage must live with the fact that, upon return, much more time will have elapsed on earth than the astronaut has subjectively and physically experienced during voyage. The credo of all star travelers, whatever their physiological condition, will be permanent farewell.

29.5 LENGTH CONTRACTION

As objects move through space-time, space as well as time undergoes changes in measurement. The lengths of objects contract when they move by us at relativistic speeds. This **length contraction** was first proposed by the physicist George F. FitzGerald and mathematically expressed by another physicist, Hendrik A. Lorentz. It is referred to as the *Lorentz-FitzGerald contraction.* We express it mathematically as

$$L = L_0 \sqrt{1 - v^2/c^2}$$

where v is the relative velocity between the observed object and the observer, c is the speed of light, L is the measured length of the moving object, and L_0 (proper length) is the measured length of the object at rest.

Suppose that an object is at rest so $v = 0$. Upon substitution of $v = 0$ in the equation, we find $L = L_0$, as we would expect. At 87 percent c, an object would be contracted to half its original length. At 99.5 percent c, it would contract to one-tenth its original length. If the object moved at c, its length would be zero. This is one of the reasons we say that the speed of light is the upper limit for the speed of any moving object. A ditty popular with the science types is:

> There was a young fencer named Fisk,
> Whose thrust was exceedingly brisk.
> So fast was his action
> The Lorentz-FitzGerald contraction
> Reduced his rapier to a disk.

As Figure 29.10 indicates, contraction takes place only in the direction of motion. If an object is moving horizontally, no contraction takes place vertically.

Figure 29.10 The Lorentz-FitzGerald contraction. As speed increases, length in the direction of motion decreases. Lengths in the perpendicular direction do not change.

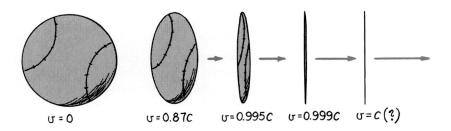

$\upsilon = 0$ $\upsilon = 0.87c$ $\upsilon = 0.995c$ $\upsilon = 0.999c$ $\upsilon = c\,(?)$

Length contraction should be of considerable interest to space voyagers. The center of our Milky Way is some 25,000 light-years distant. Does this mean that if we traveled in that direction at the speed of light it would take 25,000 years to get there? From an earth frame of reference, yes; but to the space voyagers themselves, decidedly not! At the speed of light the 25,000 light-year distance would be contracted to no distance at all. Space voyagers would arrive there instantly!*

---

*Did songwriter Leon Russell have this in mind when he sang, "I'll love you in a place where there's no space and time; I'll love you forever, you're a friend of mine."?

Figure 29.11 The meter stick is measured to be half as long when traveling at 87 percent the speed of light relative to the observer.

Space voyagers at best will travel at speeds less than the speed of light. Nevertheless, we see that at very high speeds, distances such as those from one galaxy to another will be shortened. If space voyagers are ever able to boost themselves to relativistic speeds, perhaps no part of the universe will be inaccessible to them.

QUESTION

A rectangular billboard in space has the dimensions 10 m × 20 m. How fast and in what direction with respect to the billboard would a space traveler have to pass for the billboard to appear square?

29.6 RELATIVISTIC MASS

If we push on an object, it accelerates. If we maintain a steady push it will accelerate to higher and higher speeds. But there is a speed limit in the universe—c. In fact, we cannot accelerate any material object enough to reach the speed of light, let alone surpass it.

Why this is so can be understood from Newton's second law. Recall that the acceleration of an object depends not only on the impressed force, but on the mass as well: $a = F/m$. Einstein reasoned that when work is done to increase the velocity of an object, its mass increases also. So an impressed force produces less and less acceleration as speed increases. The relationship between mass and speed is given by

$$m = \frac{m_0}{\sqrt{1 - v^2/c^2}}$$

Here m represents the measured mass (relativistic mass) of an object pushed to any speed v. The symbol m_0 is the *rest mass*, the mass the object would have at rest. Again, v represents the relative velocity between the observer and the observed.

Investigation of the relativistic mass equation shows that as v approaches c, m approaches infinity! A particle pushed to the speed of light would have infinite mass and would require an infinite force, which is clearly impossible. Therefore, we say that no material particle can be accelerated to the speed of light.

Atomic particles have been accelerated to speeds in excess of 99 percent the speed of light. Their masses increase thousands fold, as evidenced when a beam of particles, usually electrons or protons, is directed into a deflecting magnetic field. The particles do not bend as much as they would if they did not undergo the relativistic mass increase (Figure 29.12). They strike their targets at positions predicted

ANSWER

The space traveler would have to travel at 0.87c in a direction parallel to the longer side of the board.

Figure 29.12 If the mass of the electrons did not increase with speed, the beam would follow the dashed line. But because of the increased inertia, the high-speed electrons in the beam are not deflected as much.

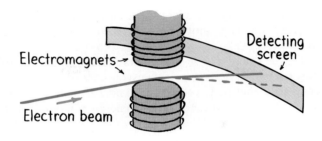

by the relativistic equation. Mass increase with velocity is an everyday fact of life to physicists working with high-energy particles.*

The reason that energy pumped into nuclear particles increases the mass of the particles is simply that energy and mass are equivalent to each other. This is one of the most remarkable results of special relativity. The amount of energy E is related to the amount of mass m by the most celebrated equation of the twentieth century:†

$$E = mc^2$$

The c^2 is the conversion factor between energy units and mass units.†† Because of the large magnitude of c, the speed of light, we can see that a small mass corresponds to an enormous quantity of energy.

So when gravitation crunches mass in the sun and stars and ignites thermonuclear fusion, the energy that emerges is accompanied by a corresponding lowering of mass—but only a tiny bit, about 1 part in 10^9. The helium nucleus produced by the fusion of a pair of deuterium nuclei is only a billionth less massive. Sunlight and starlight, then, is this tiny bit of mass transformed by thermonuclear fusion into radiant energy.

The transformation can go the other way. Radiant energy can be transformed into mass. The first direct proof of this was found in 1932 in a photograph emulsion used in high-altitude balloons to catch cosmic rays. C. D. Anderson, an American physicist, found that a photon of gamma radiation that had entered the emulsion had transformed into a pair of particles. One of the particles was an electron. The other particle, which had the same mass as an electron but had a *positive* charge rather than a negative charge, was given the name *positron*. A positron is the *antiparticle* of an electron. So when a gamma ray transforms from energy to mass, a pair of particles is produced. One of the particles is the antiparticle of the other, with the same mass and

---

*Physicists measure changes in momentum of particles in a beam, rather than mass directly. Advanced textbooks on this subject speak of relativistic momentum increase rather than relativistic mass increase.

†If you take a follow-up physics or astronomy course, you will likely encounter this equation in your next textbook as $E_o = mc^2$ or $E_o = m_o c^2$. These versions emphasize that it is an object's rest energy E_o, not its total energy, which is proportional to its mass. Rest energy, the energy an object has just by existing, is nonzero even if the object's kinetic and potential energy *are* zero. That nonzero energy is proportional to the object's mass. We'll stay with the familiar $E = mc^2$, to simply communicate the concept that mass and energy are related, whether at rest or not.

††When c is in meters per second and m is in kilograms, then E will be in joules. If the equivalence of mass and energy had been understood long ago when physics concepts were first being formulated, there would probably be no separate units for mass and energy. Furthermore, with a redefinition of space and time units, c could equal 1, and $E = mc^2$ would simply be $E = m$.

spin but with opposite charge. The antiparticle doesn't last long. It soon encounters a particle that is its antiparticle, and the pair is annihilated, sending out a gamma ray in the process.

The equation $E = mc^2$ is more than a formula for the conversion of mass into pure energy, or vice versa. It states that energy and mass are the same. Mass is simply congealed energy. If you want to know how much energy is in a system, measure its mass. For an object at rest, its energy is its mass. It is energy itself that is hard to shake.

UESTION

Can we look at the equation $E = mc^2$ another way and say that matter transforms into pure energy when it is traveling at the speed of light squared?

29.7 GENERAL RELATIVITY

The special theory of relativity is about uniform motion, which is why it is called special. Einstein's conviction that the laws of nature should be expressed in the same form in every frame of reference, accelerated as well as nonaccelerated, was the primary motivation that led him to the **general theory of relativity**—a new theory of gravitation, where gravity causes space to become curved and time to slow down.

Einstein was led to this new theory of gravity by thinking about observers in accelerated motion. He imagined himself in a spaceship far away from gravitational influences. In such a spaceship at rest or in uniform motion relative to the distant stars, he and everything within the ship would float freely; there would be no "up" and no "down." But if rocket motors were activated to accelerate the ship, things would be different; phenomena similar to gravity would be observed. The wall adjacent to the rocket motors would push up against any occupants and become the floor, while the opposite wall would become the ceiling. Occupants in the ship would be able to stand on the floor and even jump up and down. If the acceleration of the spaceship were equal to g, the occupants could well be convinced the ship was not accelerating, but was at rest on the surface of the earth.

Einstein concluded that gravity and motion through space-time are related, a conclusion now called the principle of equivalence:

Local observations made in an accelerated frame of reference cannot be distinguished from observations made in a Newtonian gravitational field.

Ⓐ NSWER

No, no, no! This is all wrong! As matter is propelled faster, its mass increases rather than decreases. As its speed approaches c, its mass, in fact, approaches infinity. At the same time, its energy approaches infinity. It has more mass and more energy. Matter cannot be made to move at the speed of light, let alone the speed of light squared (which is not a speed!). $E = mc^2$ simply means that energy and mass are two sides of the same coin.

To examine this new "gravity" in the accelerating spaceship, Einstein considered the consequence of dropping two balls, say one of wood and the other of lead. When released, the balls would continue to move upward side by side with the velocity of the ship at the moment of release. If the ship were moving at *constant velocity* (zero acceleration), the balls would remain suspended in the same place since both the ship and the balls move the same amount. But since the ship was accelerating, the floor moves upward faster than the balls, which are soon intercepted by the floor (Figure 29.14). Both balls, regardless of their masses, would meet the floor at the same time. Remembering Galileo's demonstration at the Leaning Tower of Pisa, occupants of the ship might be prone to attribute their observations to the force of gravity.

This equivalence would be relatively unimportant if it applied only to mechanical phenomena, but Einstein went further and stated that the principle holds for all natural phenomena; it holds for optical and all electromagnetic phenomena as well.

Just as a tossed ball curves in a gravitational field, so does a light beam. Consider the ball thrown sideways in the stationary spaceship in a gravity-free region in Figure 29.15. The ball will follow a straight-line path relative both to an observer inside the ship and to a stationary observer outside the spaceship. But if the ship is accelerating, the floor overtakes the ball and it hits the wall below a point opposite to the window (Figure 29.16). An observer outside the ship still sees a straight-line path, but to an observer in the accelerating ship, the path is curved; it is a parabola. The same holds true for a beam of light (Figure 29.17). The only difference is the curvatures of both. If the ball were somehow thrown at the speed of light, both curvatures would be the same.

According to Newton, tossed balls curve because of a force of gravity. According to Einstein, tossed balls and light curve not because of any force, but because the space-time in which they travel is curved.

Figure 29.13 (a) Everything is weightless on the inside of a nonaccelerating spaceship far away from gravitational influences. (b) When the spaceship accelerates, an occupant inside feels "gravity."

Figure 29.14 To an observer inside the accelerating ship, a lead ball and a wood ball appear to fall together when released.

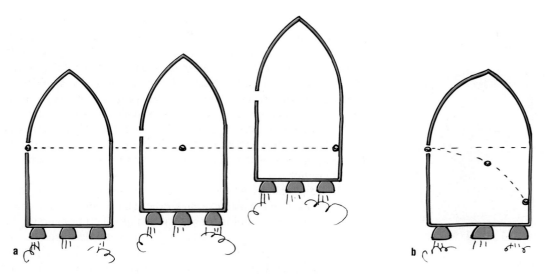

Figure 29.15   (a) An outside observer sees a horizontally thrown ball travel in a straight line, and since the ship is moving upward while the ball travels horizontally, the ball strikes the wall below a point opposite the window. (b) To an inside observer, the ball bends as if in a gravitational field.

Gravity, Space, and a New Geometry

Einstein perceived a gravitational field as a geometrical warping of four-dimensional space-time. Four-dimensional geometry is altogether different than three-dimensional Euclidean geometry. The laws of Euclidean geometry taught in high school are no longer valid when applied to objects in the presence of strong gravitational fields.

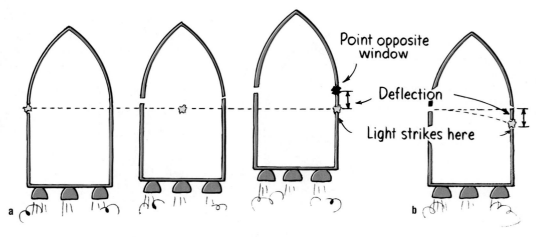

Figure 29.16   (a) An outside observer sees light travel horizontally in a straight line, but like the ball in the previous figure, it strikes the wall slightly below a point opposite the window. (b) To an inside observer, the light bends as if responding to a gravitational field.

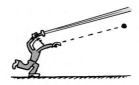

Figure 29.17 The trajectory of a flashlight beam is identical to the trajectory of a baseball "thrown" at the speed of light. Both paths curve equally in a uniform gravitational field.

Figure 29.18 The sum of the angles for a triangle drawn (a) on a plane surface = 180°, (b) on a spherical surface is greater than 180°, and (c) on a saddle-shaped surface is less than 180°.

The familiar rules of Euclidean geometry pertain to various figures you can draw on a flat surface. The ratio of the circumference of a circle to its diameter is equal to π; all the angles in a triangle add up to 180°; the shortest distance between two points is a straight line. The rules of Euclidean geometry are valid in flat space, but if you draw these figures on a curved surface like a sphere or a saddle-shaped object, the Euclidean rules no longer hold (Figure 29.18). If you measure the sum of the angles for a triangle in space, you call the space flat if the sum is equal to 180°, spherelike or positively curved if the sum is larger than 180°, and saddlelike or negatively curved if it is less than 180°.

Of course the lines forming the triangles in Figure 29.18 are not "straight" from the three-dimensional view, but they are the "straightest" or *shortest* distances between two points if we are confined to the curved surface. These lines of shortest distance are called *geodesic lines* or simply **geodesics**.

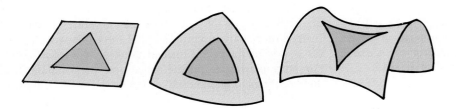

The path of a light beam follows a geodesic. Suppose three experimenters on Earth, Venus, and Mars measure the angles of a triangle formed by light beams traveling between these three planets. The light beams bend when passing the sun, resulting in the sum of the three angles being larger than 180° (Figure 29.19). So the space around the sun is positively curved. The planets that orbit the sun travel along four-dimensional geodesics in this positively curved space-time. Freely falling objects, satellites, and light rays all travel along geodesics in four-dimensional space-time.

The whole universe may have an overall curvature. If it is negatively curved, it is open-ended and extends without limit; if it is positively curved, it closes in on itself. The surface of the earth, for example, forms a closed curvature, so if you travel along a geodesic, you come back to your starting point. Similarly, if the universe were positively curved, it would be closed, so if you could look infinitely into space through an ideal telescope, you would see the back of your own head! (This is assuming that you waited a long enough time or that light traveled infinitely fast.)

Figure 29.19 The light rays joining the three planets form a triangle. Since light passing near the sun bends, the sum of the angles of the resulting triangle is greater than 180°.

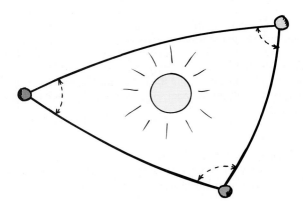

General relativity, then, calls for a new geometry: a geometry not only of curved space but of curved time as well—a geometry of curved four-dimensional space-time.* The mathematics of this geometry is too formidable to present here.

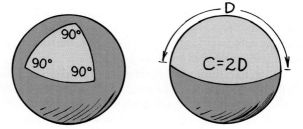

Figure 29.20 The geometry of the curved surface of the earth differs from the Euclidean geometry of flat space. Note that the sum of the angles for an equilateral triangle where the sides equal $\frac{1}{4}$ the earth's circumference is clearly greater than 180°, and the circumference is only twice its diameter instead of 3.14 times its diameter. Euclidean geometry is also invalid in curved space.

The essence, however, is that the presence of mass produces the curvature or warping of space-time; by the same token, a curvature of space-time reveals itself as mass. Instead of visualizing gravitational forces between masses, we abandon altogether the notion of force and instead think of masses responding in their motion to the curvature or warping of the space-time they inhabit. It is the bumps, depressions, and warpings of geometrical space-time that *are* the phenomena of gravity.

Gravitational Waves

Every object has mass and therefore makes a bump or depression in the surrounding space-time. When an object moves, the surrounding warp of space and time moves to readjust to the new position. These readjustments produce ripples in the overall geometry of space-time. This is similar to moving a ball that rests on the surface of the waterbed. A disturbance ripples across the waterbed surface in waves; if we move a more massive ball, then we get a greater disturbance and the production of even stronger waves. The ripples travel outward from the gravitational sources at the speed of light and are **gravitational waves**.

Any moving object produces a gravitational wave. In general, the more massive the moving object and the more violent its motion, the stronger the resulting gravitational wave. But even the strongest waves produced by ordinary astronomical events are extremely weak—the weakest known in nature. For example, the gravitational waves emitted by a vibrating electric charge are a trillion-trillion-trillion times weaker than the electromagnetic waves emitted by the same charge. Detecting gravitational waves is enormously difficult, and no confirmed detection has occurred to date. A new generation of wave detectors, soon to be built, is expected to detect gravitational waves from supernovae, where as much as 0.1 percent of their mass may be radiated away as gravitational waves.

As weak as they are, they are everywhere. Shake your hand back and forth: you have just produced a gravitational wave. It is not very strong, but it exists.

---

*Don't be discouraged if you cannot visualize four-dimensional space-time. Einstein himself often told his friends, "Don't try. I can't do it either." Perhaps we are not too different from the great thinkers around Galileo who couldn't think of a moving earth!

We cannot visualize the four-dimensional bumps and depressions in space-time because we are three-dimensional beings. We can get a glimpse of this warping by considering a simplified analogy in two dimensions: a heavy ball resting on the middle of a waterbed. The more massive the ball, the greater it dents or warps the two-dimensional surface. A marble rolled across a small-dented surface may trace an elliptical curve and orbit the ball. The planets that orbit the sun similarly travel along four-dimensional geodesics in the warped space-time about the sun.

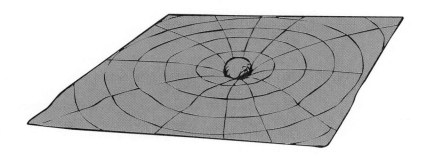

Figure 29.21 A two-dimensional analogy of four-dimensional warped space-time. Space-time near a star is curved in a way similar to the surface of a waterbed when a heavy ball rests on it.

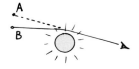

Figure 29.22 A precessing elliptical orbit.

Tests of General Relativity

Using four-dimensional field equations, Einstein recalculated the orbits of the planets about the sun. Beyond the planets, space is almost flat and objects travel along nearly straight-line paths. Near the sun, planets and comets travel along curved paths because of the curvature of space. With only one minor exception, his theory gave almost exactly the same results as Newton's law of gravity. The exception was that Einstein's theory predicted that the elliptical orbits of the planets should *precess* (Figure 29.22)—independently of the Newtonian influence of other planets. This precession would be very slight for distant planets, and more pronounced close to the sun. Mercury is the only planet close enough to the sun for the curvature of space to produce an effect on it not predicted by Newton's law.

Precession in the orbits of planets caused by perturbations of other planets was well known. Since the early 1800s astronomers have measured a precession of Mercury's orbit—about 574 seconds of arc per century. Perturbations by the other planets were found to account for the precession—except for 43 seconds of arc per century more than the calculated value. Even after all known corrections due to possible perturbations by other planets had been applied, the calculations of physicists and astronomers failed to account for the extra 43 seconds of arc. Either Venus was extra massive or a never-discovered other planet (called Vulcan) was pulling on Mercury. And then came the explanation of Einstein, whose general relativity field equations applied to Mercury's orbit predict the extra 43 seconds of arc per century!

As a second test of his theory, Einstein predicted that measurements of starlight passing close to the sun would be deflected by an angle of 1.75 seconds of arc—large enough to be measured. This deflection of starlight can be observed during an eclipse of the sun. (Measuring this deflection has become a standard practice at every total eclipse since the first measurements were made during the total eclipse of 1919.) A photograph taken of the darkened sky around the eclipsed sun reveals the presence of the nearby bright stars. The positions of the stars are compared with those in other photographs of the same area taken at other times in the night with the same telescope. In every instance, the deflection of starlight has supported Einstein's prediction (Figure 29.23).

Figure 29.23 Starlight bends as it grazes the sun. Point A shows the apparent position; point B shows the true position.

Figure 29.24 The stronger the gravitational field, the slower a clock runs. A clock at the surface of the earth runs slower than a clock farther away.

Why do we not notice the bending of light by gravity in our everyday environment?

Einstein made a third prediction—that gravity causes clocks to run slow. He predicted that clocks on the first floor of a building should tick slightly slower than clocks at the top floor, which are farther from the earth and therefore in a weaker gravitational field. For the tallest skyscraper the difference is very small, only a few millionths of a second per decade, because the difference in the earth's gravitational field at the bottom and top of the skyscraper is very small. For larger differences, like those at the surface of the sun compared to the surface of the earth, the differences in time should be more pronounced. A clock at the surface of the sun should run measurably slower than a clock at the surface of the earth. Einstein suggested a way to measure this.

All atoms emit light at specific frequencies characteristic of the vibrational rate of electrons within the atom. Every atom is therefore a "clock," and a slowing down of atomic vibration indicates the slowing down of such clocks. An atom on the sun, where gravitation is strong, should emit light of a lower frequency (slower vibration) than light emitted by the same kind of atom on the earth. Since red light is at the low-frequency end of the visible spectrum, a lowering of frequency shifts the color toward the red. This effect is the **gravitational red shift**. Although it is weak in the sun, it is stronger in more compact stars with greater surface gravities. The culminating experiment was performed in 1960, as Einstein first suggested, between the top and bottom floors of a building. The gravitational red shift of this very small gravitational field difference was detected using high-frequency gamma rays sent between the top and bottom floors of a laboratory building at Harvard University.* Incredibly precise measurements confirmed the gravitational slowing of time.

So measurements of time depend not only on relative motion, as we learned in special relativity, but also on the relative gravitational field strengths of the regions in which the events are taking place. It is important to note the relativistic nature of time in both special relativity and general relativity. In both theories, there is no way that you can extend the duration of your own experience. Others moving at different

ANSWER

The reason is not because the curvature of earth's space-time is relatively slight, for it is enough to bend a tossed baseball 5 meters below a Euclidean straight line in its first second. It would bend a beam of light 5 meters also, if the light remained in the same gravitational field for 1 second (like bouncing to and fro between parallel mirrors). We easily notice the curve of the tossed ball, and we barely notice the curvature of a high-speed bullet, but we don't notice the bending of light only because it travels so fast.

---

*In the late 1950s, shortly after Einstein's death, the German physicist Rudolph Mössbauer discovered a way to use atomic nuclei as atomic clocks, called the *Mössbauer effect,* for which he was awarded the Nobel Prize. In 1960 Professors Pound and Rebka at Harvard University used the Mössbauer effect to perform the confirming experiment.

Newton's and Einstein's Gravity Compared

When Einstein formulated his new theory of gravitation, he realized that if his theory was valid, his field equations must reduce to Newtonian equations for gravitation in the weak-field limit. He showed that Newton's law of gravitation is a special case of the broader theory of relativity. Newton's law of gravitation is still an accurate description of most of the interactions between bodies in the solar system and beyond. From Newton's law, one can calculate the orbits of comets and asteroids and even predict the existence of undiscovered planets. Even today, when computing the trajectories of space probes throughout the solar system and beyond, only ordinary Newtonian theory is used.

This is because the gravitational field of these bodies is very weak, and from the viewpoint of general relativity, the surrounding space-time is essentially flat. But for regions of more intense gravitation, where space-time is more appreciably curved, Newtonian theory cannot adequately account for various phenomena—like the precession of Mercury's orbit close to the sun and, in the case of stronger fields, the gravitational red shift and other apparent distortions of space and time. These distortions reach their limit in the case of a star that collapses to a black hole, where space-time completely folds over on itself. Only Einsteinian gravitation reaches into this domain.

speeds or in different gravitational fields may attribute a great longevity to you, but your longevity is seen from *their* frame of reference—never your own. Changes in time and other relativistic effects are always attributed to "the other guy."

We saw in Chapter 13 that Newtonian physics is linked at one end with quantum theory, whose domain is the very light and very small—tiny particles and atoms. And now we have seen that Newtonian physics is linked at the other end with relativity theory, whose domain is the very massive and very large.

29.8 OUR EXPANDING UNIVERSE

We live in an expanding universe. And what is the end result of this expansion? Does the universe expand forever, or does it finally slow down, stop, and then fall in on itself again? And if it does fall in on itself, does it re-explode, start a new cycle, and keep repeating this cycle ad infinitum?

Before the 1960s one school of thought was that the universe was in a *steady-state* condition, where the density of the expanding universe remained unchanged instead of thinning out. This was thought to be accomplished by the spontaneous creation of hydrogen out of nothing. The steady-state idea has been pretty well discredited, and most cosmologists now support the *standard model*—the theory of an expanding universe of constant mass-energy emanating from the Big Bang nearly 20 billion years ago. The question is whether or not this expansion continues forever.

A popular school of thought today is the *oscillating theory of the universe.* You know that if you throw a rock skyward, two things can happen: the rock can simply go up, come to a stop, and come back down; or, if you throw it faster than 11.2 kilometers per second (escape velocity for earth), it will continue its upward motion and never come down again. The same thing may happen on a universal scale. If the expanding rate of the universe were less than the escape velocity for the universe, the expansion would finally run its course, momentarily come to a stop, and then fall back in on itself. The time for this cycle has been estimated to be somewhat less than 100 billion years. Being now close to 20 billion on the way out, we'd continue our outward expansion for some 30 billion years and momentarily come to rest. The universe would then be at its maximum extent. Galaxies would be farthest apart and show no Doppler shift in their light. Then the contraction would begin. After some 30 billion years of contracting, the universe would again be the size it is now, but galaxies would all show blue shifts instead of red shifts Then after some 20 billion more years would come the Big Crunch, as the universe collapsed into its own black hole—perhaps to gush out into a new universe for another 100-or-so billion years. Very enchanting.

Whether the universe oscillates or whether it expands indefinitely depends on the mass density of the universe. If the mass of the universe is less than a critical value, the present expansion exceeds escape velocity and the universe expands indefinitely. If the mass of the universe is greater than this critical value, then the present expansion is less than the escape velocity and finally comes to a halt, and then contracts to complete a cycle of its possible oscillation. Present indications are that the mass density of the universe is too small for oscillation. A search for dark matter or other mass to bring it up to critical have thus far been unsuccessful. We need to know more before we can say whether the mass density of the universe will turn out to be below or above this critical value. We have so much to find out.

We do not see the world as did the ancient Egyptians, Greeks, and Chinese. It is unlikely that people in the future will see the universe as we do. Our view of the universe may be wrong, but it is most likely less wrong than the views of others before us. Our view today stems from the findings of Copernicus and Galileo—findings that were opposed on the grounds that they diminished the importance of humans in the universe. The idea of importance then was in being able to rise above nature—in being apart from nature. We have expanded our vision since then by enormous effort, painstaking observation, and an unrelenting desire to comprehend our surroundings. Seen from today's understanding of the universe, we find our importance in being very much a part of nature, not apart from it. We are the part of nature that is becoming more and more conscious of itself.

SUMMARY OF TERMS

Big Bang The primordial explosion of space at the beginning of time.

Primeval fireball The burst of energy during the Big Bang.

Space-time The four-dimensional continuum in which all things exist; three dimensions of space and one of time.

Special theory of relativity The first of Einstein's theories of relativity, which discusses the effects of uniform motion on space, time, energy, and mass.

Time dilation The slowing of time as a result of speed.

Length contraction The contraction of objects in their direction of motion as a result of speed.

General theory of relativity The second of Einstein's theories of relativity, which discusses the effects of gravity on space and time.

Geodesic The shortest distance between two points in various models of space.

Gravitational wave The transport of energy by the motion of waves in a gravitational field.

Gravitational red shift The lengthening of the waves of electromagnetic radiation due to escape from a gravitational field.

• •
REVIEW QUESTIONS

The Big Bang

1. How does the Big Bang explosion differ from ordinary explosions?

2. Why is the present expansion of the universe slowing down?

Space-Time

3. What are the four dimensions of space-time?

4. Under what conditions will you and a friend share the same realm of space-time?

5. What is special about the ratio of space traveled and time taken for light?

Special Relativity

6. Cite at least three examples of Einstein's first postulate.

7. Cite at least three examples of Einstein's second postulate.

Time Dilation

8. What do we call the "stretching out of time"?

9. Suppose the parallel mirrors of a light clock were 150,000 km apart. In the frame of reference of the light clock, how much time would be required for a pulse of light to make a round trip between the mirrors?

10. Suppose the parallel mirrors of a light clock were 150 km apart. In the frame of reference of the light clock, how much time would be required for a pulse of light to make a round trip between the mirrors?

11. Would your answers to the previous two questions be different if your measurements were made from a frame of reference that is moving relative to the light clock? Explain.

12. Time is required for light to travel along a path from one point to another. If this path is seen to be longer because of motion, what happens to the time it takes for light to travel this longer path?

13. How do measurements of time differ for events in a frame of reference that moves at 50 percent the speed of light relative to us?

14. How do measurements of time differ for events in a frame of reference that moves at 99.5 percent the speed of light relative to us?

15. Suppose a particular clock accurately shows time passing half as fast in a particular frame of reference as in our own. What is the velocity of this frame of reference relative to us?

16. What is the evidence for time dilation?

17. What does Einstein say about common sense?

18. What are the present-day obstacles to interstellar space travel?

Length Contraction

19. How long would a meter stick appear if it were traveling like a properly thrown spear at 99.5 percent the speed of light?

20. How long would the meter stick in the previous question appear if it were traveling with its length perpendicular to its direction of motion? (Why is your answer different from the previous answer?)

21. If you were traveling in a high-speed rocket ship, would meter sticks on board appear to you to be contracted? Defend your answer.

Relativistic Mass

22. What happens to the mass of an object pushed to higher speeds?

23. Will an object at rest, with a mass of one kilogram, have a greater mass if it is accelerated to a higher speed?

24. What would be the mass of an object pushed to the speed of light?

25. What does $E = mc^2$ mean?

26. What is the evidence for $E = mc^2$ in cosmic ray investigations?

27. What is an *antiparticle?*

General Relativity

28. What is the principal difference between *special relativity* and *general relativity?*

29. How would the number of pushups one could perform at the earth's surface compare to the number of pushups one could perform in an elevator accelerating at *g* far from the earth's gravitational field?

30. Compare the bending of the paths of baseballs and photons by a gravitational field.

31. What does it mean to say that space is curved?

32. What is a *geodesic?*

33. Of all the planets, why is Mercury the best candidate for finding evidence of the relationship of gravitation to space?

34. What is the evidence for light bending near the sun?

35. Which runs faster, a clock at the top of the Sears Tower in Chicago or a clock on the shore of Lake Michigan?

36. What is the effect of a strong gravitational field on the frequency of light? On its wavelength?

37. What is the effect of a strong gravitational field on measurements of time?

38. Does Einstein's theory of gravitation invalidate Newton's theory of gravitation? Explain.

Our Expanding Universe

39. What is the estimated period for the oscillating universe?

40. What is a necessary condition for the expansion of the universe to turn around?

• •

EXERCISES

1. Why are the long-wavelength microwaves that permeate the universe considered evidence for the Big Bang?

2. What is meant by saying that the Big Bang is the explosion of space at the beginning of time? Why do we say explosion *of* space, rather than explosion *in* space?

3. Faraway galaxies show a red shift in their spectra. If the universe oscillates, and the galaxies one day come to a halt, what shift (if any) will their spectra show then? When they approach?

4. In Chapter 12 we learned that light travels more slowly in glass than in air. Does this contradict the special theory of relativity?

5. If we see somebody's clock running slow due to relative motion, will they see our clocks running slow also? Or will they see our clocks running fast? Explain.

6. Since there is an upper limit on the speed of a particle, does it follow that there is also an upper limit on its momentum? On its kinetic energy? Explain.

7. Light travels a certain distance in, say, 20,000 years. How is it possible that an astronaut could travel slower than the speed of light and travel 20,000 light-years in a 20-year trip?

8. Could a human being who has a life expectancy of 70 years possibly make a round-trip journey to a part of the universe thousands of light-years distant? Explain.

9. A twin who makes a long trip at relativistic speeds returns younger than his stay-at-home twin sister. Could he return before his twin sister was born? Defend your answer.

10. Is it possible for a son or daughter to be biologically older than his or her parents? Explain.

11. If you were in a rocket ship traveling away from the earth at a speed close to the speed of light, what changes would you note in your pulse? In your mass? In your volume? Explain.

12. If you were on earth monitoring a person in a rocket ship traveling away from the earth at a speed close to the speed of light, what changes would you note in his pulse? In his mass? In his volume? Explain.

13. How does the measured density of a body compare when at rest to when it is moving?

14. Is this ditty, popular with relativity types, consistent or inconsistent with relativity theory? Defend your answer.

> There was a young lady named Bright
> Who traveled much faster than light.
> She departed one day
> In an Einsteinian way
> And returned on the previous night.

15. As a meter stick that has a rest mass of 1 kg moves past you, your measurements show it to have a mass of 2 kg. If your measurements show it to have a length of 1 m, in what direction is the stick pointing?

16. In the preceding exercise, if the stick is moving in a direction along its length (like a properly thrown spear), how long will it appear to you?

17. If a high-speed spaceship appears shrunken to half size, by how much will measurements of its mass differ?

18. The "2-mile" linear accelerator at Stanford University in California "appears" to be less than a meter long to the electrons that travel in it. Explain.

19. Electrons that are accelerated in the Stanford accelerator gain thousands of times more mass by the time they reach the end of their trip than they had when they started. In theory, if you could travel with them, would you notice an increase in their mass? In the mass of the target they are about to hit? Explain.

20. The electrons that illuminate the screen in a typical television picture tube travel at nearly one-fourth the speed of light and have an increased mass of nearly three percent. Does this relativistic effect tend to increase or decrease your electric bill?

21. Muons are elementary particles that are formed high in the atmosphere by the interactions of cosmic rays with gases in the upper atmosphere. Muons are radioactive and have average lifetimes of about two-millionths of a second. Even though they travel at almost the speed of light, they are so high that very few should be detected at sea level—at least according to classical physics. Laboratory measurements, however, show that muons in great proportions do reach the earth's surface. What is the explanation?

22. When we look out into the universe, we see into the past. John Dobson, founder of the San Francisco Sidewalk Astronomers, says that we cannot even see the backs of our own hands *now*—in fact, we can't see anything *now*. Do you agree? Explain.

23. An astronaut is provided a "gravity" when the ship's engines are activated to accelerate the ship. This requires the use of fuel. Is there a way to accelerate and provide "gravity" without the sustained use of fuel? Explain.

24. In his famous novel *Journey to the Moon,* Jules Verne stated that occupants in a spaceship would shift their orientation from up to down when the ship crossed the point where the moon's gravitation became greater than the earth's. Is this correct? Defend your answer.

25. What happens to the separation distance between two people if they both walk north at the same rate from two different places on the earth's equator? And just for fun, where in the world is a step in every direction a step south?

26. We readily note the bending of light by reflection and refraction, but why is it we do not ordinarily notice the bending of light by gravity?

27. Why do we say that light travels in straight lines? Is it strictly accurate to say that a laser beam provides a perfectly straight line for purposes of surveying? Explain.

28. Light changes its energy when it "falls" in a gravitational field. This change in energy is not evidenced by a change in speed, however. What is the evidence for this change in energy?

29. Splitting hairs, should a person who worries about growing old live at the top or at the bottom of a tall apartment building?

30. Why does the gravitational attraction between the sun and Mercury vary? Would it vary if the orbit of Mercury were perfectly circular?

PROBLEMS

1. You observe a spaceship moving away from you at speed v_1, half the speed of light. A rocket is fired from the spaceship, straight ahead so that it also moves away from you. Suppose that from the spaceship it is fired at half the speed of light, v_2, relative to the spaceship. It so happens the speed of the rocket relative to you is not the speed of light. The relativistic addition of velocities is given by

$$V = \frac{v_1 + v_2}{1 + \dfrac{v_1 v_2}{c^2}}$$

Substitute $0.5c$ for both v_1 and v_2 and show that the velocity V of the rocket relative to you is $0.8c$.

2. Pretend that the spaceship of the previous question is somehow traveling at c with respect to you, and it fires a rocket at speed c with respect to itself. Use the equation to show that the speed of the rocket with respect to you is still c!

3. Substitute small values of v_1 and v_2 in the preceding equation and show that for everyday velocities V is practically equal to $v_1 + v_2$.

4. At the end of 1 s, a horizontally fired bullet has dropped a vertical distance of 4.9 m from its otherwise straight-line path in a gravitational field of 1 g. By what distance would a beam of light drop from its otherwise straight-line path if it traveled in a uniform field of 1 g for 1 s? For 2 s?

Appendix I
SYSTEMS OF MEASUREMENT

Two major systems of measurement prevail in the world today: the *United States Customary System* (USCS, formerly called the British system of units), used in the United States of America and in Burma, and the *Système International* (SI) (known also as the international system and as the metric system), used everywhere else. Each system has its own standards of length, mass, and time. The units of length, mass, and time are sometimes called the *fundamental units* because, once they are selected, other quantities can be measured in terms of them.

UNITED STATES CUSTOMARY SYSTEM

Based on the British Imperial System, the USCS is familiar to everyone in the United States. It uses the foot as the unit of length, the pound as the unit of weight or force, and the second as the unit of time. The USCS is presently being replaced by the international system—rapidly in science and technology (and in Department of Defense contracts) and some sports (track and swimming), but so slowly in other areas and in some specialties it seems the change may never come. For example, we will continue to buy seats on the 50-yard line. Camera film is in millimeters but computer disks are in inches.

For measure time there is no difference between the two systems except that in pure SI the only unit is the *second* (s, not sec) with prefixes; but, in general, minute, hour, day, year, and so on, with two or more lettered abbreviations (h, not hr), are accepted in the USCS.

SYSTÈME INTERNATIONAL

During the 1960 International Conference on Weights and Measures held in Paris, the SI units were defined and given status. Table I.1 shows SI units and their

TABLE I.1   SI Units

Quantity	Unit	Symbol
Length	meter	m
Mass	kilogram	kg
Time	second	s
Force	newton	N
Energy	joule	J
Current	ampere	A
Temperature	kelvin	K

TABLE I.2   Table Conversions between Different Units of Length

Unit of Length	Kilometer	Meter	Centimeter	Inch	Foot	Mile
1 kilometer	= 1	1000	100 000	39 370	3280.84	0.62140
1 meter	= 0.00100	1	100	39.370	3.28084	6.21×10^{-4}
1 centimeter	= 1.0×10^{-5}	0.0100	1	0.39370	0.032808	6.21×10^{-6}
1 inch	= 2.54×10^{-5}	0.02540	2.5400	1	0.08333	1.58×10^{-5}
1 foot	= 3.05×10^{-4}	0.30480	30.480	12	1	1.89×10^{-4}
1 mile	= 1.60934	1609.34	160 934	63 360	5280	1

symbols. SI is based on the *metric system*, originated by French scientists after the French revolution in 1791. The orderliness of this system makes it useful for scientific work, and it is used by scientists all over the world. The metric system branches into two systems of units. In one of these the unit of length is the meter, the unit of mass is the kilogram, and the unit of time is the second. This is called the *meter-kilogram-second* (mks) system and is preferred in physics. The other branch is the *centimeter-gram-second* (cgs) system, which because of its smaller values is favored in chemistry. The cgs and mks units are related to each other as follows: 100 centimeters equal 1 meter, 1000 grams equal 1 kilogram. Table I.2 shows several units of length related to each other.

One major advantage of a metric system is that it uses the decimal system, where all units are related to smaller or larger units by dividing or multiplying by 10. The prefixes shown in Table I.3 are commonly used to show the relationships among units.

TABLE I.3   Some Prefixes

Prefix	Definition
micro-	One-millionth: a microsecond is one-millionth of a second
milli-	One-thousandth: a milligram is one-thousandth of a gram
centi-	One-hundredth: a centimeter is one-hundredth of a meter
kilo-	One thousand: a kilogram is 1000 grams
mega-	One million: a megahertz is 1 million hertz

Meter

The standard of length for the metric system originally was defined in terms of the distance from the north pole to the equator. This distance was thought at the time to be close to 10 000 kilometers. One ten-millionth of this, the meter, was carefully determined and marked off by means of scratches on a bar of platinum-iridium alloy. This bar is kept at the International Bureau of Weights and Measures in France. The standard meter in France has since been calibrated in terms of the wavelength of light—it is 1 650 763.73 times the wavelength of orange light emitted by the atoms of the gas krypton-86. The meter is now defined as being the length of the path traveled by light in a vacuum during a time interval of 1/299 792 458 of a second.

Kilogram

Figure I.1 The standard kilogram.

The standard unit of mass, the kilogram, is a block of platinum, also preserved at the International Bureau of Weights and Measures located in France (Figure I.1).

The kilogram equals 1000 grams. A gram is the mass of 1 cubic centimeter (cc) of water at a temperature of 4° Celsius. (The standard pound is defined in terms of the standard kilogram; the mass of an object that weighs 1 pound is equal to 0.4536 kilogram.)

Second

The official unit of time for both the USCS and the SI is the second. Until 1956 it was defined in terms of the mean solar day, which was divided into 24 hours. Each hour was divided into 60 minutes and each minute into 60 seconds. Thus, there were 86 400 seconds per day, and the second was defined as 1/86 400 of the mean solar day. This proved unsatisfactory because the rate of rotation of the earth is gradually becoming slower. In 1956 the mean solar day of the year 1900 was chosen as the standard on which to base the second. In 1964, the second was officially defined as the time taken by a cesium-133 atom to make 9 192 631 770 vibrations.

Newton

One newton is the force required to accelerate 1 kilogram at 1 meter per second per second. This unit is named after Sir Isaac Newton.

Joule

One joule is equal to the amount of work done by a force of 1 newton acting over a distance of 1 meter. In 1948 the joule was adopted as the unit of energy by the International Conference on Weights and Measures. Therefore, the specific heat of water at 15°C is now given as 4185.5 joules per kilogram Celsius degree. This figure is always associated with the mechanical equivalent of heat—4.1855 joules per calorie.

Ampere

The ampere is defined as the intensity of the constant electric current that, when maintained in two parallel conductors of infinite length and negligible cross section and placed 1 meter apart in a vacuum, would produce between them a force equal to 2×10^{-7} newton per meter length. In our treatment of electric current in this text, we have used the not-so-official but easier-to-comprehend definition of the ampere as being the rate of flow of 1 coulomb of charge per second, where 1 coulomb is the charge of 6.25×10^{18} electrons.

Kelvin

The fundamental unit of temperature is named after the scientist William Thomson, Lord Kelvin. The kelvin is defined to be 1/273.15 the thermodynamic temperature of the triple point of water (the fixed point at which ice, liquid water, and water vapor coexist in equilibrium). This definition was adopted in 1968 when it was decided to change the name *degree Kelvin* (°K) to *kelvin* (K). The temperature of melting ice at atmospheric pressure is 273.15 K. The temperature at which the vapor pressure of pure water is equal to standard atmospheric pressure is 373.15 K (the temperature of boiling water at standard atmospheric pressure).

Area

The unit of area is a square that has a standard unit of length as a side. In the USCS it is a square whose sides are each 1 foot in length, called 1 square foot and written 1 ft^2. In the international system it is a square whose sides are 1 meter in length, which makes a unit of area of 1 m^2. in the cgs system it is 1 cm^2. The area of a given surface is specified by the number of square feet, square meters, or square centimeters that would fit into it. The area of a rectangle equals the base times the height. The area of a circle is equal to πr^2, where $\pi = 3.14$ and r is the radius of the circle. Formulas for the surface areas of other objects can be found in geometry textbooks.

Figure I.2   Unit square.

Volume

The volume of an object refers to the space it occupies. The unit of volume is the space taken up by a cube that has a standard unit of length for its edge. In the USCS one unit of volume is the space occupied by a cube 1 foot on an edge and is called 1 cubic foot, written 1 ft^3. In the metric system it is the space occupied by a cube whose sides are 1 meter (SI) or 1 centimeter (cgs). It is written 1 m^3 or 1 cm^3 (or cc). The volume of a given space is specified by the number of cubic feet, cubic meters, or cubic centimeters that will fill it.

In the USCS volumes can also be measured in quarts, gallons, and cubic inches as well as in cubic feet. There are 1728 ($12 \times 12 \times 12$) cubic inches in 1 ft^3. A U.S. gallon is a volume of 231 in^3. Four quarts equal 1 gallon. In the SI volumes are also measured in liters. A liter is equal to 1000 cm^3.

Figure I.3   Unit volume.

SCIENTIFIC NOTATION

It is convenient to use a mathematical abbreviation for large and small numbers. The number 50 000 000 can be obtained by multiplying 5 and 10, and again by 10, and again by 10, and so on until 10 has been used as a multiplier seven times. The short way of showing this is to write the number 5×10^7. The number 0.0005 can be obtained from 5 by using 10 as a divisor four times. The short way of showing this is to write 5×10^{-4} for 0.0005. Thus, 3×10^5 means $3 \times 10 \times 10 \times 10 \times 10 \times 10$, or 300 000; and 6×10^{-3} means $6/(10 \times 10 \times 10)$, or 0.006. Numbers expressed in this shorthand manner are said to be in scientific notation.

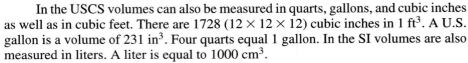

$$
\begin{aligned}
1\,000\,000 &= 10 \times 10 \times 10 \times 10 \times 10 \times 10 &&= 10^6 \\
100\,000 &= 10 \times 10 \times 10 \times 10 \times 10 &&= 10^5 \\
10\,000 &= 10 \times 10 \times 10 \times 10 &&= 10^4 \\
1000 &= 10 \times 10 \times 10 &&= 10^3 \\
100 &= 10 \times 10 &&= 10^2 \\
10 &= 10 &&= 10^1 \\
1 &= 1 &&= 10^0 \\
0.1 &= 1/10 &&= 10^{-1} \\
0.01 &= 1/100 = 1/10^2 &&= 10^{-2} \\
0.001 &= 1/1000 = 1/10^3 &&= 10^{-3} \\
0.0001 &= 1/10\,000 = 1/10^4 &&= 10^{-4} \\
0.00001 &= 1/100\,000 = 1/10^5 &&= 10^{-5} \\
0.000001 &= 1/1\,000\,000 = 1/10^6 &&= 10^{-6}
\end{aligned}
$$

We can use scientific notation to express some of the physical data often used in physics.

$$\text{Speed of light in a vacuum} = 2.9979 \times 10^8 \text{ m/s}$$
$$\text{1 astronomical unit (A.U.),}$$
$$\text{(average earth-sun distance)} = 1.50 \times 10^{11} \text{ m}$$
$$\text{Average earth-moon distance} = 3.84 \times 10^8 \text{ m}$$
$$\text{Equatorial radius of the sun} = 6.96 \times 10^8 \text{ m}$$
$$\text{Equatorial radius of Jupiter} = 7.14 \times 10^7 \text{ m}$$
$$\text{Equatorial radius of the earth} = 6.37 \times 10^6 \text{ m}$$
$$\text{Equatorial radius of the moon} = 1.74 \times 10^6 \text{ m}$$
$$\text{Average radius of hydrogen atom} = 5 \times 10^{-11} \text{ m}$$
$$\text{Mass of the sun} = 1.99 \times 10^{30} \text{ kg}$$
$$\text{Mass of Jupiter} = 1.90 \times 10^{27} \text{ kg}$$
$$\text{Mass of the earth} = 5.98 \times 10^{24} \text{ kg}$$
$$\text{Mass of the moon} = 7.36 \times 10^{22} \text{ kg}$$
$$\text{Proton mass} = 1.6726 \times 10^{-27} \text{ kg}$$
$$\text{Neutron mass} = 1.6749 \times 10^{-27} \text{ kg}$$
$$\text{Electron mass} = 9.1 \times 10^{-31} \text{ kg}$$
$$\text{Electron charge} = 1.602 \times 10^{-19} \text{ C}$$

Appendix II
EXPONENTIAL GROWTH AND DOUBLING TIME*

One of the most important things we seem not to perceive is the process of exponential growth. We think we understand how compound interest works, but we can't get it through our heads that a fine piece of tissue paper folded upon itself fifty times (if that were possible) would be more than 20 million kilometers thick. If we could, we could "see" why our income buys only half what it did 4 years ago, why the price of everything has doubled in the same time, why populations and pollution proliferate out of control.†

When a quantity such as money in the bank, population, or the rate of consumption of a resource steadily grows at a fixed percent per year, we say the growth is exponential. Money in the bank may grow at 8 percent per year; electric power generating capacity in the United States grew at about 7 percent per year for the first three-quarters of the century. The important thing about exponential growth is that the time required for the growing quantity to increase its size by a fixed fraction is constant. So the time required for the quantity to double in size (increase by 100 percent) is also constant. For example, if the population of a growing city takes 12 years to double from 10 000 to 20 000 inhabitants and its growth remains steady, in the next 12 years the population will double to 40 000, in the next 12 years to 80 000, and so on.

There is an important relationship between the percent growth rate and its *doubling time,* the time it takes to double a quantity.††

$$\text{Doubling time} = \frac{69.3}{\text{percent growth per unit time}} \approx \frac{70}{\%}$$

So to estimate the doubling time for a steadily growing quantity, we simply divide the number 70 by the percentage growth rate. For example, the 7 percent growth rate of electric power generating capacity in the United States means that in the past the capacity has doubled every 10 years (70%/[7%/year] = 10 years). A 2 percent growth rate for world population means the population of the world doubles every 35

---

*This appendix is drawn from material by University of Colorado physics professor Albert A. Bartlett, who strongly asserts, "The greatest shortcoming of the human race is man's inability to understand the exponential function." See Professor Bartlett's still-timely article, "Forgotten Fundamentals in the Energy Crisis" (*American Journal of Physics,* September 1978) or his revised version (*Journal of Geological Education,* January 1980).

†K. C. Cole, *Sympathetic Vibrations* (New York: Morrow, 1984).

††For exponential decay we speak about half-life, the time required for a quantity to reduce to half its value. This case is treated in Chapter 15.

Figure II.1 An exponential curve. Notice that each of the successive equal time intervals noted on the horizontal scale corresponds to a doubling of the quantity indicated on the vertical scale. Such an interval is called the doubling time.

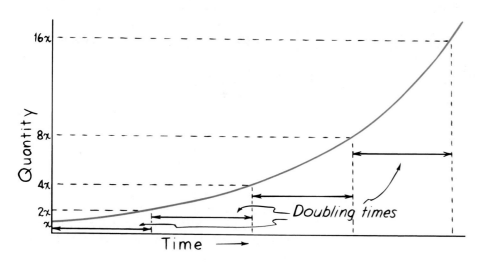

years (70%/[2%/year] = 35 years). A city planning commission that accepts what seems like a modest 3.5 percent growth rate may not realize that this means that doubling will occur in 70/3.5 or 20 years; that's double capacity for such things as water supply, sewage treatment plants, and other municipal services every 20 years.

What happens when you put steady growth in a finite environment? Consider the growth of bacteria that grow by division, so that one bacterium becomes two, the two divide to become four, the four divide to become eight, and so on. Suppose the division time for a certain strain of bacteria is 1 minute. This is then steady growth—the number of bacteria grows exponentially with a doubling time of 1 minute. Further, suppose that one bacterium is put in a bottle at 11:00 AM and that growth continues steadily until the bottle becomes full of bacteria at 12:00 noon. Consider seriously the following question.

Q UESTION

When was the bottle half-full?

It is startling to note that at 2 minutes before noon the bottle was only $\frac{1}{4}$ full. Table II.1 summarizes the amount of space left in the bottle in the last few minutes before noon. If you were an average bacterium in the bottle, at which time would you first realize that you were running out of space? For example, would you sense there was a serious problem at 11:55 AM when the bottle was only 3 percent filled ($\frac{1}{32}$) and had 97 percent of open space (just yearing for development)? The point here is that there isn't much time between the moment that the effects of growth become noticeable and the time when they become overwhelming.

Suppose that at 11:58 AM some farsighted bacteria see that they are running out of space and launch a full-scale search for new bottles. Luckily, at 11:59 AM they

Figure II.2

A NSWER ·

11:59 AM; the bacteria will double in number every minute!

TABLE II.1   The Last Minutes in the Bottle

Time	Part Full (%)	Part Empty
11:54 AM	$\frac{1}{64}$ (1.5%)	$\frac{63}{64}$
11:55 AM	$\frac{1}{32}$ (3%)	$\frac{31}{32}$
11:56 AM	$\frac{1}{16}$ (6%)	$\frac{15}{16}$
11:57 AM	$\frac{1}{8}$ (12%)	$\frac{7}{8}$
11:58 AM	$\frac{1}{4}$ (25%)	$\frac{3}{4}$
11:59 AM	$\frac{1}{2}$ (50%)	$\frac{1}{2}$
12:00 noon	Full (100%)	None

discover three new empty bottles, three times as much space as they had ever known. This quadruples the total resource space ever known to the bacteria, for they now have a total of four bottles, whereas before the discovery they had only one. Further suppose that thanks to their technological proficiency, they are able to migrate to their new habitats without difficulty. Surely, it seems to most of the bacteria that their problem is solved—and just in time.

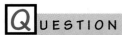

UESTION

If the bacteria growth continues at the unchanged rate, what time will it be when the three new bottles are filled to capacity?

We see from Table II.2 that quadrupling the resource extends the life of the resource by only two doubling times. In our example the resource is space—but it could as well be coal, oil, uranium, or any nonrenewable resource.

TABLE II.2   Effects of the Discovery of Three New Bottles

Time	Effect
11:58 AM	Bottle 1 is $\frac{1}{4}$ full
11:59 AM	Bottle 1 is $\frac{1}{2}$ full
12:00 noon	Bottle 1 is full
12:01 AM	Bottles 1 and 2 are both full
12:02 AM	Bottles 1, 2, 3, and 4 are all full

Continued growth and continued doubling lead to enormous numbers. In two doubling times, a quantity will double twice ($2^2 = 4$; quadruple) in size; in three doubling times, its size will increase eightfold ($2^3 = 8$); in four doubling times, it will increase sixteenfold ($2^4 = 16$); and so on.

Answer

12:02 PM!

Figure II.3 A single grain of wheat placed on the first square of the chessboard is doubled on the second square, this number is doubled on the third, and so on, presumably for all 64 squares. Note that each square contains more grain than all the preceding squares combined. Does enough wheat exist in the world to fill all 64 squares in this manner?

This is best illustrated by the story of the court mathematician in India who years ago invented the game of chess for his king. The king was so pleased with the game that he offered to repay the mathematician, whose request seemed modest enough. The mathematician requested a single grain of wheat on the first square of the chessboard, two grains on the second square, four on the third square, and so on, doubling the number of grains on each succeeding square until all squares had been used. At this rate there would be 2^{63} grains of wheat on the sixty-fourth square. The king soon saw that he could not fill this "modest" request, which amounted to more wheat than had been harvested in the entire history of the earth!

It is interesting and important to note that the number of grains on any square is one grain more than the total of all grains on the preceding squares. This is true anywhere on the board. Note from Table II.3 that when eight grains are placed on the fourth square, the eight is one more than the total of seven grains that were already on the board. Or the thirty-two grains placed on the sixth square is one more than the total of thirty-one grains that were already on the board. We see that in one doubling time we use more than all that had been used in all the preceding growth!

So if we speak of a doubling energy consumption in the next however many years, bear in mind that this means in these years we will consume more energy than has heretofore been consumed during the entire preceding period of steady growth. And if power generation continues to use predominantly fossil fuels, then except for some improvements in efficiency, we would burn up in the next doubling time

TABLE II.3 Filling the Squares on the Chessboard

Square Number	Grains on Square	Total Grains Thus Far
1	1	1
2	2	3
3	4	7
4	8	15
5	16	31
6	32	63
7	64	127
⋮	⋮	⋮
64	2^{63}	$2^{64} - 1$

a greater amount of coal, oil, and natural gas than has already been consumed by previous power generation; and except for improvements in pollution control, we can expect to discharge even more toxic wastes into the environment than the millions upon millions of tons already discharged over all the previous years of industrial civilization; and we would expect more human-made calories of heat to be absorbed by the earth's ecosystem than have been absorbed in the entire past! At the previous 7 percent annual growth rate in energy production, all this would occur in one doubling time of a single decade. If over the coming years the annual growth rate remains at half this value, 3.5 percent, then all this would take place in a doubling time of 2 decades. Clearly this cannot continue!

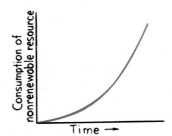 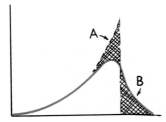 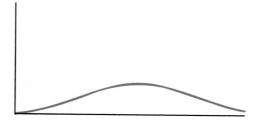

Figure II.4 (a) If the exponential rate of consumption for a non-renewable resource continues until it is depleted, consumption falls abruptly to zero. The shaded area under this curve represents the total supply of the resource. (b) In practice, the rate of consumption levels off and then falls less abruptly to zero. Note that the crosshatched area A is equal to the crosshatched area B. Why? (c) At lower consumption rates, the same resource lasts a longer time.

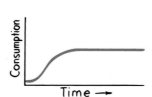

Figure II.5 A curve showing the rate of consumption of a renewable resource such as agricultural or forest products, where a steady rate of production and consumption can be maintained for a long period, provided this production is not dependent upon the use of a non-renewable resource that is waning in supply.

The consumption of a nonrenewable resource cannot grow exponentially for an indefinite period, because the resource is finite and its supply finally expires. The most drastic way this could happen is shown in the graph in Figure II.4(a), where the rate of consumption, such as barrels of oil per year, is plotted against time, say in years. In such a graph the colored area under the curve represents the supply of the resource. We see that when the supply is exhausted, the consumption ceases altogether. This sudden change is rarely the case, for the rate of extracting the supply falls as it becomes more scarce. This is shown in Figure II.4(b). Note that the area under the curve is equal to the area under the curve in (a). Why? Because the total supply is the same in both cases. The principal difference is the time taken to finally extinguish the supply. History shows that the rate of production of a nonrenewable resource rises and falls in a nearly symmetric manner, as shown by the curve in c. The time during which production rates rise is approximately equal to the time during which these rates fall to zero or near zero. If we fit the data for U.S. oil production in the lower forty-eight states to such a curve, we find that we are just to the right of the peak. This suggests that one-half of the recoverable petroleum that was ever in the ground in the U.S. has already been used and that in the future the domestic petroleum rate of production can only decrease. The U.S. production curve peaked in 1970, and by 1979 nearly half the U.S. consumption was imported. Each year we consume more oil than during the previous year.

Production rates for all nonrenewable resources decrease sooner or later. Only production rates for renewable resources, such as agriculture or forest products, can be maintained at steady levels for long periods of time (Figure II.5), provided such production does not depend on waning nonrenewable resources such as petroleum.

Much of today's agriculture is so petroleum-dependent that it can be said that modern agriculture is simply the process whereby land is used to convert petroleum into food. The implications of petroleum scarcity go far beyond rationing of gasoline for cars or fuel oil for home heating.

Power production from whatever sources will not meet the present increasing demands in this growing world. Fusion power may characterize the next century, but only the most optimistic forecasters see it as providing even a tiny fraction of our power needs during the first half of the twenty-first century. Even though the production of nuclear fission power had spectacular growth after its harnessing in 1942, it took 30 years of development to equal the annual energy consumption of firewood in the United States. Nuclear fusion will be vastly more complicated than fission, so even if nuclear fusion were successfully harnessed today, how long would it take before it could play a significant role in major power production?

However, the important question are more basic: Is growth really good? Is bigger really better? Is it true that if we don't grow, we will stagnate? In answering these questions, bear in mind that human growth is an early phase of life that continues normally through adolescence. Physical growth stops when physical maturity is reached. What do we say of growth that continues in the period of physical maturity? We say that such growth is obesity—or, worse, cancer.

● ●

QUESTIONS TO PONDER

1. According to a French riddle, a lily pond starts with a single leaf. Each day the number of leaves doubles, until the pond is completely covered by leaves on the 30th day. On what day was the pond half covered? One-quarter covered?

2. In an economy that has a steady inflation rate of 7 percent per year, in how many years does a dollar lose half its value?

3 At a steady inflation rate of 7 percent, what will be the price every 10 years for the next 50 years for a theater ticket that now costs $10? For a suit of clothes that now costs $100? For a car that now costs $10 000? For a home that now costs $100 000?

4. If the sewage treatment plant of a city is just adequate for the city's current population, how many sewage treatment plants will be necessary 42 years later if the city grows steadily at 5 percent annually?

5. In 1986 the population growth for the United States was 0.6 percent, for Mexico 2.6 percent, and for Kenya (the highest growth rate in the world) 4.1 percent (taking into account births, deaths, and immigration). At these rates, how long would it take for the population in each of these countries to double?

6. If world population doubles in 40 years and world food production also doubles in 40 years, how many people then will be starving each year compared to now?

7. A continued world population growth rate of 1.9 percent per year would produce a density of one person per square meter in 550 years. True or false: World population growth rate will sooner or later be zero.

8. Suppose you get a prospective employer to agree to hire your services for wages of a single penny for the first day, 2 pennies the second day, and double each day thereafter providing the employer keeps to the agreement for a month. What will be your total wages for the month?

9. In the preceding exercise, how will your wages for only the 30th day compare to your total wages for the previous 29 days?

10. We hear often that reserves of nonrenewable resources such as coal, oil and natural gas are "scarce," "abundant," or "superabundant." Why are these terms meaningless without referring also to their consumption rates?

11. Oil has been produced in the United States for about 125 years. If there remains undiscovered in the country as much oil as all that has been used, what is wrong with the argument that the remaining oil would be sufficient for another 125 years?

12. Present estimates are that one-eighth of the oil in the world has already been consumed. With respect to the example of the multiplying bacteria discussed in this appendix, how many "minutes are there until noon"?

13. How would your answer to the preceding exercise be different if new oil deposits were discovered that were equal in size to all those ever known?

14. Coal is relatively "abundant" in the United States today only because the growth in annual production of coal was zero from 1910 to the mid-1970s. In the previous half-century prior to 1910, coal production grew at a steady rate of almost 7 percent per year. Had this rate continued, the expiration of United States coal reserves would have occurred between the years 1965 and 1990, depending on low and high estimates of reserve sizes. Why does it violate good sense to talk of "abundant reserves" and continued growth at the same time?

15. When dealing with steady growth, is it necessary to have an accurate estimate of the size of a resource in order to make a reliable estimate of how long the resource will last? (Use Figure II.4 to explain your answer.)

16. If fusion power were harnessed today, the abundant energy resulting would probably sustain and even further encourage our present appetite for continued growth and in a relatively few doubling times produce an appreciable fraction of the solar power input to the earth. Make an argument that the current delay in harnessing fusion is a blessing for the human race.

Appendix III
VECTORS

· ·

VECTORS AND SCALARS

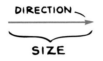

Figure III.1

A *vector* quantity is a directed quantity—one that must be specified not only by magnitude (size) but by direction as well. Recall from Chapter 3 that velocity is a vector quantity. Other examples are force, acceleration, and momentum. In contrast, a *scalar* quantity can be specified by magnitude alone. Some examples of scalar quantities are speed, time, temperature, and energy.

Vector quantities may be represented by arrows. The length of the arrow tells you the magnitude of the vector quantity, and the arrowhead tells you the direction of the vector quantity. Such an arrow drawn to scale and pointing appropriately is called a *vector*.

ADDING VECTORS

Vectors that add together are called *component vectors. The sum of component vectors is called a resultant.*

To add two vectors, make a parallelogram with two component vectors acting as two of the adjacent sides (Figure III.2). (Here our parallelogram is a rectangle.) Then draw a diagonal from the origin of the vector pair; this is the resultant (Figure III.3).

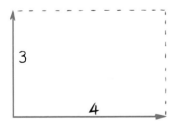

Figure III.2

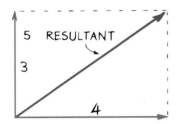

Figure III.3

Caution: Do not try to mix vectors! We cannot add apples and oranges, so velocity vector combines only with velocity vector, force vector combines only with force vector, and acceleration vector combines only with acceleration vector—each on its own vector diagram. If you ever show different kinds of vectors on the same diagram, use different colors or some other method of distinguishing the different kinds of vectors.

FINDING COMPONENTS OF VECTORS

Recall from Chapter 3 that to find a pair of perpendicular components for a vector, first draw a dotted line through the tail of the vector (in the direction of one of the desired components). Second, draw another dotted line through the tail end of the vector at right angles to the first dotted line. Third, make a rectangle whose diagonal is the given vector. Draw in the two components. Here we let **F** stand for "total force," **U** stand for "upward force," and **S** stand for "sideways force."

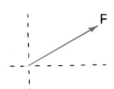

Figure III.4

Figure III.5

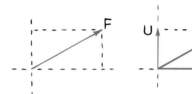

Figure III.6

Examples

1. A man pushing a lawnmower applies a force that pushes the machine forward and also against the ground. In Figure III.7, **F** represents the force applied by the man. We can separate this force into two components. The vector **D** represents the downward component, and **S** is the sideways component, the force that moves the lawnmower forward. If we know the magnitude and direction of the vector **F**, we can estimate the magnitude of the components from the vector diagram.

Figure III.7

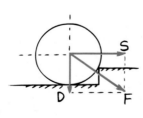

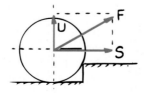

Figure III.8

2. Would it be easier to push or pull a wheelbarrow over a step? Figure III.8 shows a vector diagram for each case. When you push a wheelbarrow, part of the force is directed downward, which makes it harder to get over the step. When you pull, however, part of the pulling force is directed upward, which helps to lift the wheel over the step. Note that the vector diagram suggests that pushing the wheelbarrow may not get it over the step at all. Do you see that the height of the step, the radius of the wheel, and the angle of the

applied force determine whether the wheelbarrow can be pushed over the step? We see how vectors help us analyze a situation so that we can see just what the problem is!

3. If we consider the components of the weight of an object rolling down an incline, we can see why its speed depends on the angle. Note that the steeper the incline, the greater the component **S** becomes and the faster the object rolls. When the incline is vertical, **S** becomes equal to the weight, and the object attains maximum acceleration, 9.8 meters per second squared.

   There are two more force vectors that are not shown: the normal force **N**, which is equal and oppositely directed to **D**, and the friction force **f**, acting at the barrel-plane contact.

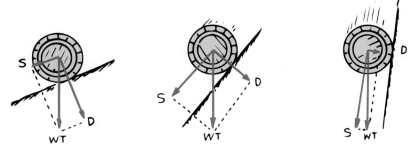

Figure III.9

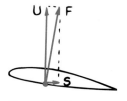

Figure III.10

4. When moving air strikes the underside of an airplane wing, the force of air impact against the wing may be represented by a single vector perpendicular to the plane of the wing (Figure III.10). We represent the force vector as acting midway along the lower wing surface, where the dot is, and pointing above the wing to show the direction of the resulting wind impact force. This force can be broken up into two components, one sideways and the other up. The upward component, **U**, is called *lift*. The sideways component, **S**, is called *drag*. If the aircraft is to fly at constant velocity at constant altitude, then lift must equal the weight of the aircraft and the thrust of the plane's engines must equal drag. The magnitude of lift (and drag) can be altered by changing the speed of the airplane or by changing the angle (called *angle of attack*) between the wing and the horizontal.

5. Consider the satellite moving clockwise in Figure III.11. Everywhere in its orbital path, gravitational force **F** pulls it toward the center of the host planet. At position A we see **F** separated into two components: *f*, which is tangent to the path of the projectile, and *f'*, which is perpendicular to the path. The relative magnitudes of these components in comparison to the magnitude of **F** can be seen in the imaginary rectangle they compose: *f* and *f'* are the sides, and **F** is the diagonal. We see that component *f* is along the orbital path but against the direction of motion of the satellite. This force component reduces the speed of the satellite. The other component, *f'*, changes the direction of the satellite's motion and pulls it away from its tendency to go in a straight line. So the path of the satellite curves. The satellite loses speed until it reaches position B. At this farthest point from the planet (apogee), the gravitational force is somewhat weaker but perpendicular to the satellite's motion, and component *f* has reduced to zero. Component *f'*, on the other hand, has increased and is now fully merged to become **F**. Speed at this

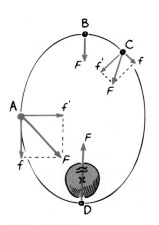

Figure III.11

point is not enough for circular orbit, and the satellite begins to fall toward the planet. It picks up speed because the component f reappears and is in the direction of motion as shown in position C. The satellite picks up speed until it whips around to position D (perigee), where once again the direction of motion is perpendicular to the gravitational force, f' blends to full **F**, and f is nonexistent. The speed is in excess of that needed for circular orbit at this distance, and it overshoots to repeat the cycle. Its loss in speed in going from D to B equals its gain in speed from B to D. Kepler discovered that planetary paths are elliptical, but never knew why. Do you?

6. Refer to the Polaroids held by Erika back in Chapter 12, in Figure 12.40. In the first picture (a), we see that light is transmitted through the pair of Polaroids because their axes are aligned. The emerging light can be represented as a vector aligned with the polarization axes of the Polaroids. When the Polaroids are crossed (b), no light emerges because light passing through the first Polaroid is perpendicular to the polarization axes of the second Polaroid, with no components along its axis. In the third picture (c), we see that light is transmitted when a third Polaroid is sandwiched at an angle between the crossed Polaroids. The explanation for this is shown in Figure III.12.

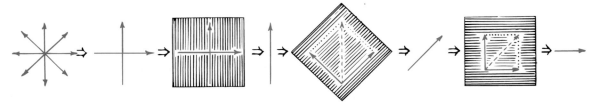

Figure III.12

SAILBOATS

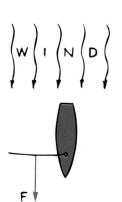

Figure III.13

Sailors have always known that a sailboat can sail downwind, in the direction of the wind. Sailors have not always known, however, that a sailboat can sail upwind, against the wind. One reason for this has to do with a feature that is common only to recent sailboats—a finlike keel that extends deep beneath the bottom of the boat to ensure that the boat will knife through the water only in a forward (or backward) direction. Without a keel, a sailboat could be blown sideways.

Figure III.13 shows a sailboat sailing directly downwind. The force of wind impact against the sail accelerates the boat. Even if the drag of the water and all other resistance forces are negligible, the maximum speed of the boat is the wind speed. This is because the wind will not make impact against the sail if the boat is moving as fast as the wind. The sail would simply sag. If there is no force, then there is no acceleration. The force vector in Figure III.13 *decreases* as the boat travels faster. The force vector is maximum when the boat is at rest and the full impact of the wind fills the sail, and is minimum when the boat travels as fast as the wind. If the boat is somehow propelled to a speed faster than the wind (by way of a motor, for example), then air resistance against the front side of the sail will produce an oppositely directed force vector. This will slow the boat down. Hence, the boat when driven only by the wind cannot exceed wind speed.

Figure III.14

Figure III.15

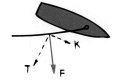

Figure III.16

If the sail is oriented at an angle, as shown in Figure III.14, the boat will move forward, but with less acceleration. There are two reasons for this:

1. The force on the sail is less because the sail does not intercept as much wind in this angular position.
2. The direction of the wind impact force on the sail is not in the direction of the boat's motion, but is perpendicular to the surface of the sail. Generally speaking, whenever any fluid (liquid or gas) interacts with a smooth surface, the force of interaction is perpendicular to the smooth surface.* The boat does not move in the same direction as the perpendicular force on the sail, but is constrained to move in a forward (or backward) direction by its keel.

We can better understand the motion of the boat by resolving the force of wind impact, *F*, into perpendicular components. The important component is that which is parallel to the keel, which we label *K*, and the other component is perpendicular to the keel, which we label *T*. It is the component *K*, as shown in Figure III.15, that is responsible for the forward motion of the boat. Component *T* is a useless force that tends to tip the boat over and move it sideways. This component force is offset by the deep keel. Again, maximum speed of the boat can be no greater than wind speed.

Many sailboats sailing in directions other than exactly downwind Figure (III.16) with their sails properly oriented can exceed wind speed. In the case of a sailboat cutting across the wind, the wind may continue to make impact with the sail even after the boat exceeds wind speed. A surfer, in a similar way, exceeds the velocity of the propelling wave by angling his surfboard across the wave. Greater angles to the propelling medium (wind for the boat, wave for the surfboard) result in greater speeds. A sailcraft can sail faster cutting across the wind than it can sailing downwind.

As strange as it may seem, maximum speed for most sailcraft is attained by cutting into (against) the wind, that is, by angling the sailcraft in a direction upwind! Although a sailboat cannot sail directly upwind, it can reach a destination upwind by angling back and forth in a zigzag fashion. This is called *tacking*. Suppose the boat and sail are as shown in Figure III.17. Component *K* will push the boat along in a forward direction, angling into the wind. In the position shown, the boat can sail faster than the speed of the wind. This is because as the boat travels faster, the impact of wind is increased. This is similar to running in a rain that comes down at an angle. When you run into the direction of the downpour, the drops strike you harder and more frequently; but when you run away from the direction of the downpour, the

---

*You can do a simple exercise to see that this is so. Try bouncing a coin off another on a smooth surface, as shown. Note that the struck coin moves at right angles (perpendicular) to the contact edge. Note also that it makes no difference whether the projected coin moves along path A or path B. See your instructor for a more rigorous explanation, which involves momentum conservation.

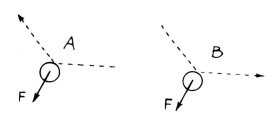

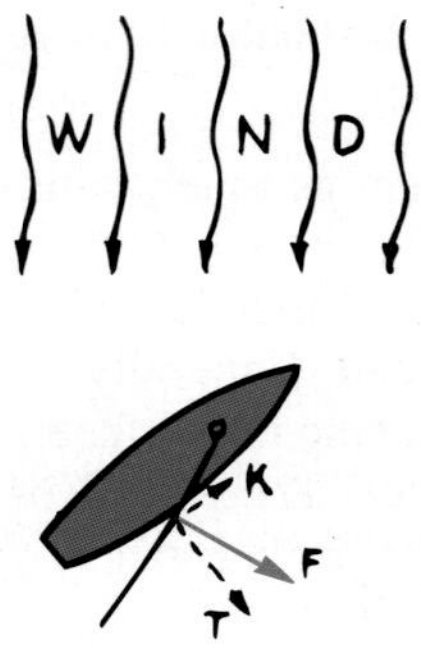

Figure III.17

drops don't strike you as hard or as frequently. In the same way, a boat sailing upwind experiences greater wind impact force, while a boat sailing downwind experiences a decreased wind impact force. In any case the boat reached its terminal speed when opposing forces cancel the force of wind impact. The opposing forces consist mainly of water resistance against the hull of the boat. The hulls of racing boats are shaped to minimize this resistive force, which is the principal deterrent to high speeds.

Iceboats (sailcraft equipped with runners for traveling on ice) encounter no water resistance and can travel at several times the speed of the wind when they tack upwind. Although ice friction is nearly absent, an iceboat does not accelerate without limits. The terminal velocity of a sailcraft is determined not only by opposing friction forces but also by the change in relative wind direction. When the boat's orientation and speed are such that the wind seems to shift in direction, so the wind moves parallel to the sail rather than into it, forward acceleration ceases—at least in the case of a flat sail. In practice, sails are curved and produce an airfoil that is as important to sailcraft as it is to aircraft. The effects are discussed in Chapter 5.

GLOSSARY

Aa A basaltic lava flow characterized by a rough jagged surface with dangerously sharp edges and spiny projections.

Aberration The distortion in an image produced by a lens or system of lenses.

Absolute temperature scale A temperature scale that has its zero point at 273.16°C. Temperatures in the absolute scale are designated in kelvins (K).

Absolute zero The lowest possible temperature that any substance may have; the temperature at which the molecules of a substance have their minimum kinetic energy. Absolute zero is 0 K; 0°C is 273.16 K.

Absorption spectrum A continuous spectrum, like that of white light, interrupted by dark lines or bands that result from the absorption of certain frequencies of light by a substance through which the light passes.

Acceleration due to gravity (g) The acceleration of a freely falling object. Its value near the earth's surface is about 9.8 meters per second each second.

Acceleration The rate at which an object's velocity changes with time; the change in velocity may be in magnitude (speed) or direction or both.

Acid A substance that produces or donates hydrogen ions in solution.

Acidic A solution in which the hydronium ion concentration is greater than the hydroxide ion concentration.

Actinides The inner transition metals within the seventh period.

Additive primary colors The three colors red, blue, and green that when added together in various proportions will produce any color in the spectrum.

Adhesion Molecular attraction between two surfaces making contact.

Adhesive forces Chemical interactions that arise between two different substances.

Adiabatic A process in which no heat enters or leaves a system.

Adiabatic process A change in gas volume with no heat entering or leaving the system.

Air resistance The friction that acts on something moving through air.

Alcohols A class of organic molecules that contain the hydroxyl group.

Aldehyde A class of organic molecules containing a carbonyl group in which the carbon of the carbonyl is bonded to one carbon atom and one hydrogen atom.

Alkali metals Group 1 elements.

Alkali-earth metals Group 2 elements.

Alkaloids Molecules found in nature that are alkaline because of the nitrogen atoms they contain. Many alkaloids have biological effects on humans and other organisms.

Alloy A substance composed of two or more metals or of a metal and a nonmetal.

Alpha particle Nucleus of a helium atom, composed of two protons and two neutrons, ejected by radioactivity.

Alpha ray A stream of alpha particles ejected by certain radioactive nuclei.

Alternating current (ac) Electric current that rapidly reverses in direction. The electric charges vibrate about relatively fixed positions, usually at the rate of 60 hertz.

Amide A class of organic molecules containing a carbonyl group in which the carbon of the carbonyl is bonded to one carbon atom and one nitrogen atom.

Amines A class of organic molecules that contain the element nitrogen.

Amino acid The fundamental building block of proteins. In an amino acid, amine and carboxylic acid groups are bonded to a central carbon atom. About 20 different types of amino acids are primary to life.

Amorphous A term used to describe a solid in which atomic or molecular particles are randomly oriented.

Ampere The unit of electric current, the flow of 1 coulomb of charge per second.

Amphoteric A substance that can behave as either an acid or a base.

Amplitude For a wave or vibration, the maximum displacement on either side of the equilibrium (midpoint) position.

Amplitude modulation (AM) A type of modulation in which the amplitude of the carrier wave is varied above and below its normal value by an amount proportional to the amplitude of the impressed signal.

Aneroid barometer An instrument used to measure atmospheric pressure; based on the movement of the lid of a metal box, rather than on the movement of a liquid.

Angstrom An outdated unit of length equal to 10^{-10} meter. Atoms have a radius of 1 to 2 angstroms.

Angular momentum The product of an object's rotational inertia and angular velocity about a particular axis. For an object that is small compared to the radial distance, it is the product of its mass, linear speed, and radius of curvature.

Angular unconformity An unconformity in which older strata dip at an angle different from that of the younger beds.

Anticline A fold in sedimentary strata that has stratigraphically older rocks in the core of the fold. An anticline resembles an arch.

Antimatter Matter composed of atoms with negative nuclei and positive electrons.

Antiparticle Particle having the same mass as a normal particle but a charge of the opposite sign. The antiparticle of the electron is the positron.

Apogee The point in an elliptical orbit farthest from the focus around which orbiting takes place.

Aquifer A rock body (or sediment body) through which groundwater easily moves.

Archimedes' principle An immersed object, submerged or floating, is buoyed up by a force equal to the weight of fluid displaced.

Archimedes' principle for air An object surrounded by the atmosphere is buoyed upward by a force equal to the weight of displaced air.

Aromatic compound Any organic molecule containing a benzene ring.

Artesian system A system in which groundwater is under great enough pressure to rise above the level of an aquifer. The confined groundwater will flow out of the ground at any opening that taps the aquifer.

Asteroid A small rocky planetlike fragment that orbits the sun. Tens of thousands of these object make up an asteroid belt between the orbits of Mars and Jupiter.

Asthenosphere A subdivision of the mantle situated below the lithosphere. This zone of weak material exists below a depth of about 100 kilometers and in some regions extends as deep as 700 kilometers. The rock within this zone is easily deformed.

Astigmatism A defect of the eye in which the cornea is curved more in one direction than another.

Atmosphere The gaseous envelope surrounding the earth.

Atmospheric pressure At the surface of the earth, the weight of air above that presses on surfaces; 1.01×10^5 newtons per square meter, or 101 kPa (14.7 pounds per square inch); used as a gas pressure unit.

Atom The smallest particle of an element that has all the element's chemical properties, composed of a nucleus and a number of surrounding electrons.

Atomic group A vertical column in the periodic table.

Atomic mass The mass of an element given in amu.

Atomic mass unit (amu) A very small unit of mass used for atoms and molecules. One atomic mass unit is equal to 1/12 the mass of the carbon-12 atom, or 1.661×10^{-24} grams.

Atomic mass number The number associated with an atom that is the same as the number of nucleons in its nucleus.

Atomic nucleus The core of an atom, consisting of two basic nucleons—protons and neutrons. The protons have positive electric charge, giving the nucleus a positive electric charge; the neutrons have no electric charge.

Atomic number A number designating an atom, which is the same as the number of protons in the nucleus or the same as the number of extranuclear electrons in a neutral atom.

Atomic period A horizontal row in the periodic table.

Atomic radius A measure of the size of an atom.

Atomic shell A spherical region of space about the atomic nucleus in which electrons may reside.

Atomic symbol The first letter or letters of the name of an element used to denote that element.

Avogadro's Number A very large number: 6.02×10^{23}. This is the number of atoms in exactly 12 grams of carbon-12.

Avogadro's principle Equal volumes of all gases at the same temperature and pressure contain the same number of molecules, 6.02×10^{23} in one mole (a mass in grams equal to the molecular mass of the substance in atomic mass units).

Axis The straight line about which rotation takes place.

Barometer A device that measures atmospheric pressure.

Base A substance that produces hydroxide ions in solution or accepts hydrogen ions.

Basic A solution in which the hydroxide ion concentration is greater than the hydronium ion concentration.

Basin The total area that contributes water to a stream.

Battery A device in which the chemical energy from oxidation and reduction reactions is transformed into electrical energy.

Beats A sequence of alternating reinforcement and cancellation of two sets of superimposed waves differing in frequency, heard as a throbbing sound.

Bernoulli's principle Pressure of a fluid on a surface decreases as the fluid's velocity relative to the surface increases.

Beta particle An electron (or a positron) emitted during the radioactive decay of an atomic nucleus.

Beta ray   A stream of beta particles.

Big Bang   The primordial explosion thought to have resulted in an expanding universe.

Bimetallic strip   Two strips of different metals welded or riveted together, used in thermostats.

Binary star   Pair of stars that orbit about a common center of mass.

Biomolecule   A chemical substance important for life.

Black body radiation   Radiation emitted by a perfect emitter of radiation (a black body); an ideal black body absorbs all the radiation it receives, and hence appears black at low temperatures.

Black dwarf   The end state of a white dwarf that has cooled off.

Black hole   The configuration of a massive star that has undergone gravitational collapse, in which gravitation is so strong that even the star's own light cannot escape.

Black hole singularity   The object of zero radius into which the matter of a black hole is comprised.

Body wave   A seismic wave that travels through the earth's interior.

Boiling   Change from liquid to gas occurring beneath the surface of a liquid; rapid vaporization. The liquid loses energy, the gas gains it.

Bow wave   The V-shaped wave made by an object moving across a liquid surface at a speed greater than the wave speed.

Boyle's law   The product of pressure and volume is a constant for a given mass of confined gas regardless of changes in either pressure or volume individually, so long as temperature remains unchanged: $P_1V_1 = P_2V_2$

Breeder reactor   Nuclear fission reactor producing power and more fuel than it consumes by transmuting nonfissionable isotopes into fissionable isotopes.

Brownian motion   Random movement of very small particles suspended in a fluid that results from collisions with molecules.

Buoyancy   The apparent loss of weight of an object due to buoyant force when it is immersed in a fluid.

Buoyant force   The net upward force that a fluid exerts on an immersed object, floating or submerged, due to the weight of displaced fluid, independent of the object's weight.

Calorie   Energy required to raise the temperature of one gram of water one degree Celsius; 4.186 joules. 1000 calories, c, equal one Calorie, C.

Capacitor   An electrical device; in its simplest form a pair of parallel plates separated by a small distance, that stores electrical charge, and that is used to smooth out irregular pulses in electric current.

Capillary action   The rising of liquid into a small vertical space due to adhesion between the liquid and the sides of the container and to cohesive forces within the liquid.

Carbohydrates   Biomolecules made of hydrated carbon. Carbohydrates are produced in plants by photosynthesis and used primarily as a source of energy in humans.

Carbon dating   The process of determining the time that has elapsed since death by measuring the radioactivity of remaining carbon-14 isotopes.

Carbonyl   A carbon atom double bonded to an oxygen atom, C=O. The carbonyl group is found in ketones, aldehydes, amides, and carboxylic acids.

Carboxylic acid   A class of organic molecules containing a carbonyl group in which the carbon of the carbonyl is bonded to one carbon atom and one hydroxyl group.

Carrier wave   A high-frequency radio wave modified by a lower-frequency wave.

Catalytic   A process whereby larger hydrocarbons are broken down into smaller less viscous hydrocarbons.

Cellulose   The most abundant polysaccharide on earth. Its purest form is cotton. Cellulose is found as the woody part of trees and the supporting material in plants and leaves.

Celsius temperature   Temperature on a scale that assigns the value 0°C to the freezing point of water and the value of 100°C to the boiling point of water both at standard pressure.

Center of gravity   The average position of weight or the point associated with an object where gravity force is considered to act.

Center of mass   The average position of mass or the point associated with an object where its mass can be considered to be concentrated.

Centrifugal force   An outward force caused by rotation.

Centripetal force   A center-directed force that causes an object to follow a curved or circular path.

Ceramic   Water- and heat-resistant materials usually made of elements from groups 13 or 14.

Chain reaction   A self-sustaining reaction that, once started, steadily provides the energy and matter necessary to continue the reaction.

Charge polarization   The spatial separation of positive and negative charges by the electrical alignment of molecules.

Charging by contact   Transfer of charge between objects by rubbing or by simple touching.

Charging by induction   Redistribution of charges in and on objects caused by the electrical influence of a charged object close by but not in contact.

Chemical bond   The electrical force that holds atoms together. Energy is required to break a chemical bond, and it is released when a chemical bond is formed.

Chemical change   The change of a substance into other substances through a reorganization of atoms.

Chemical compound   A material in which atoms of different elements are chemically bonded to one another. See elemental compound.

Chemical equation   A representation of a chemical reaction showing the relative numbers of reactants and products.

Chemical equilibrium   A dynamic state in which the rate of the forward chemical reaction is equal to the rate of the reverse chemical reaction. At chemical equilibrium the concentrations of reactants and products remain constant.

Chemical formula   A notation used to denote the composition of a chemical compound. In a chemical formula the atomic symbols for the different elements of the compound are written together along with numerical subscripts that indicate their proportions.

Chemical interaction   The electrical force of attraction or repulsion between two or more molecules or other chemical units.

Chemical mixture   The combination of different elements, elemental or chemical compounds.

Chemical property   The chemical characteristics of a substance that tend to change chemical identity. For example, it is a chemical property of iron to change into rust.

Chemical reaction   The energetic process whereby atoms are rearranged to give rise to new substances; a chemical change.

Chemical weathering   The chemical alteration of rock into substances more in balance with the surrounding environment.

Chinook   A warm, dry wind that blows down from the eastern side of the Rocky Mountains across the Great Plains.

Clouds   The condensation of water droplets above the earth's surface. Clouds are generally classified by height—high clouds, middle clouds, low clouds, and clouds with vertical development.

Cluster   Pertains to an astronomical group of more than one galaxy.

Coefficients   Numbers used in a chemical equation to show the ratio in which reactants combine and products form. In balancing a chemical equation, coefficients are used to make the number of times an atom appears before and after the arrow the same.

Cohesive forces   The attractive forces within a substance.

Combining volumes   A principle that states that volumes of gases, at the same temperature and pressure, combine chemically with one another in a ratio of small whole numbers.

Comet   A body composed of ice and dust that orbits the sun, usually in a very eccentric orbit, and which casts a luminous tail produced by solar radiation pressure when it is close to the sun.

Complementarity   The principle that the wave and particle models of matter and radiation complement each other.

Complementary colors   Any two colors that when combined produce white light.

Components   The parts into which a vector can be separated and that act in different directions from the vector.

Compression   The region of increased pressure in a longitudinal wave.

Concentrated   A solution containing a relatively large amount of solute.

Concentration   A quantitative measure of the amount of solute in a solution.

Condensation   The change of phase from vapor to liquid; the opposite of evaporation. Warming of the liquid results.

Conduction, electrical   The easy flow of electric charge through a material subjected to an impressed electrical force.

Conduction, heat   The transfer and distribution of energy from molecule to molecule in an object by means of electron and molecular collisions.

Conductor   Material that conducts heat or electricity easily.

Conservation   Principle that the amount of a quantity such as energy, mass, momentum, or electrical charge within a system remains constant when the system otherwise undergoes changes.

Conservation of mass   A principle stating that matter is neither created nor destroyed in a chemical reaction, as far as we are able to detect.

Continental divide   The continuous line running north to south down the length of North America separating the Pacific basin on the west from the Atlantic basin on the east. Water west of the line flows to the Pacific Ocean, and water east of the line flows to the Atlantic Ocean.

Convection   Transfer of energy in a fluid by means of currents in the heated fluid.

Converging lens   A lens that is thicker in the middle than at the edges and that refracts parallel rays passing through it to a focus.

Core   The most central layer in the earth's interior. The core itself is divided into an outer liquid core and an inner solid core.

Coriolis effect   The effect of the earth's rotation on the paths of air circulation.

Cornea   The transparent covering over the eye.

Correspondence principle   A new theory is valid provided that when it overlaps with the old theory, it conforms to the verified results of the old theory.

Cosmic ray One of various high-speed particles that travel throughout the universe and originate in violent events in stars.

Cosmology Study of the origin and development of the entire universe.

Coulomb The SI unit of electrical charge; the charge of 6.25×10^{18} electrons.

Coulomb's law The relationship among electric force, charge, and distance: $F = k\frac{q_1 q_2}{d^2}$. If the charges are alike in sign, the force is repulsive; if the charges are unlike, the force is attractive.

Covalent bond A chemical bond in which atoms are held together by their mutual attraction for two electrons that they share.

Covalent compound An elemental or chemical compound in which atoms are held together by the covalent bond.

Covalent crystal A group of molecules arranged in an orderly and periodic fashion.

Critical angle The angle at which light no longer is refracted when it meets a surface, but is instead totally internally reflected.

Critical mass Minimum mass of fissionable material in a reactor or nuclear bomb that will sustain a chain reaction.

Cro-Magnon Human being (*Homo sapiens*) that lived about 35,000 years ago and assumed greater numbers than the Neanderthals.

Cross-cutting When an igneous intrusion or fault cuts through sedimentary rock, the intrusion or fault is younger than the rock it cuts.

Crust The earth's outermost layer. The crustal surface represents less than 0.1% of the earth's total volume.

Crystal A material, usually a solid, in which atomic or molecular particles are arranged in an ordered periodic fashion.

Current Fluid flow (gas or liquid); also electron flow (electric). Alternating current (ac) rapidly and repeatedly reverses direction; direct current (dc) flows in one direction.

de Broglie matter waves All particles have wave properties; in de Broglie's equation, the product of momentum and wavelength equals Planck's constant.

Density Mass density is mass per volume; weight density is weight per volume; in general, any item per space element (e.g., number of dots per area).

Deuterium An isotope of hydrogen in which the nucleus consists of a single neutron in addition to the single proton.

Dew point The temperature to which air must be cooled for saturation to occur.

Diffraction grating A series of closely spaced parallel slits used to separate colors of light by interference.

Diffraction The bending of waves around a barrier, such as an obstacle or the edges of an opening.

Dilute A solution containing a relatively small amount of solute.

Diode An electrical device that allows current to flow in it in one direction.

Dipole A chemical unit wherein electric charge is separated. For example, a molecule negatively charged on one side and positively charged on the other is a dipole.

Dipole-dipole The chemical interaction involving dipoles.

Dipole-induced dipole The chemical interaction involving a dipole and an induced dipole.

Disaccharide A carbohydrate consisting of two monosaccharides. An example is sucrose, commonly known as table sugar.

Dispersion The separation of light into colors arranged according to their frequency—for example, by interaction with a prism or diffraction grating.

Dissolving The process of mixing a solute in a solvent.

Distillation The process of recollecting a vaporized substance.

Diverging lens A lens that is thinner in the middle than at the edges, causing parallel rays passing through it to diverge.

Divide The line that separates adjacent basins.

Doldrums The region near the equator characterized by a low-pressure zone and very little horizontal air movement.

Doppler effect Change in frequency of sound or light due to relative motion of source and receiver.

Double bond Two covalent bonds between the same two atoms.

Efficiency Ratio of result to effort, output to input, involving energy.

Effective nuclear charge The nuclear charge sensed when one or more electrons partially shield the nucleus.

Elastic collision A collision in which no energy is transformed to heat.

Elastic limit The distance of stretching or compressing beyond which an elastic material will not return to its original state.

Elasticity Property of a solid of returning to the original size and shape after deformation.

Electric current The flow of electric charge that transports energy from one place to another. Measured in amperes, where 1 ampere is the flow of 6.25×10^{18} electrons (or protons) per second.

Electric field The energetic region of space surrounding a charged object. About a charged point, the field decreases with distance according to the inverse-square law. Between oppositely charged parallel plates the electric field is uniform. A charged object placed in the region of an electric field experiences a force.

Electric polarization The separation of charge in an object so that one part bears a positive charge and another part bears an equal negative charge.

Electric potential The electric potential energy per amount of charge, measured in volts, and often called voltage: Voltage = F(electrical energy,charge).

Electric power The rate of electrical energy transfer or the rate of doing work, which can be measured by the product of current and voltage: Power = current × voltage. Measured in watts (or kilowatts), where 1 ampere × 1 volt = 1 watt.

Electric resistance The property of a material that causes it to resist the flow of an electric current through it. Measured in ohms.

Electrochemistry A branch of chemistry concerned with the relationship between electrical energy and chemical change.

Electrodynamics The study of moving electric charge, as opposed to electrostatics.

Electrolysis The use of electrical energy to produce chemical change.

Electromagnetic induction The induction of voltage when a magnetic field changes with time. If the magnetic field within a closed loop changes in any way, a voltage is induced in the loop: Voltage induced $\sim \frac{\text{no. of loops} \times \text{mag. field change}}{\text{change in time}}$. For the more general case of field induction, see Faraday's law.

Electromagnetic radiation The transfer of energy by the rapid oscillations of electromagnetic fields, which travel in the form of waves called electromagnetic waves.

Electromagnetic spectrum The range of frequencies over which electromagnetic radiation can be propogated. The lowest frequencies are associated with radio waves; microwaves have a higher frequency, and then infrared waves, visible light, ultraviolet radiation, X rays, and gamma rays in sequence.

Electromotive force (emf) Any force that gives rise to an electric current. A battery or a generator is a source of emf.

Electron affinity The ability of an atom to attract one or more additional electrons.

Electron The negative particles in the shell of an atom.

Electronegativity The ability of an atom to attract electrons to itself when bonded to another atom.

Electrostatics The study of electric charge at relative rest, as opposed to electrodynamics.

Element A substance composed of atoms that all have the same atomic number and therefore the same chemical properties.

Elemental compound A material in which only atoms of the same element are chemically bonded to one another.

Elemental formula A notation that uses the atomic symbol and a numerical subscript to denote the composition of an elemental compound.

Ellipse A closed curve of oval shape wherein the sum of the distances from any point on the curve to two internal focal points is a constant.

Elliptical galaxy A galaxy that is round or elliptical in outline. It has little gas and dust, no disk or spiral arms, and few hot and bright stars.

Emission spectrum A continuous or partial spectrum of wavelengths resulting from the characteristic frequencies of light from a luminous source.

Endothermic reaction A chemical reaction that has the net effect of absorbing heat.

Energy Anything that can change the condition of matter. Commonly defined circularly as the ability to do work; actually only describable (like pornography) by examples.

Entropy A measure of the degree of disorder in a substance or system.

Equilibrium The state of an object when not acted upon by a net force or net torque. An object in equilibrium may be at rest or moving at uniform velocity—that is, not accelerating.

Erosion The process by which rock particles are transported away by water, wind, or ice.

Escape velocity The velocity that a projectile, space probe, etc., must reach to escape the gravitational influence of the earth or celestial body to which it is attracted.

Ether A hypothetical medium that was formerly thought to be required for the propagation of electromagnetic waves.

Ethers A class of organic molecules in which two carbon atoms are bonded to a single oxygen atom.

Evaporation The change of phase at the surface of a liquid as it becomes vapor, resulting from the random motion of molecules that occasionally escape from the liquid surface; the opposite of condensation. Cooling of the liquid results.

Event horizon The boundary region of a black hole from which no radiation may escape. Any events within the event horizon are invisible to distant observers.

Excitation The process of boosting one or more electrons in an atom or a molecule from a lower to a higher energy level. An atom in an excited state will usually decay (deexcite) rapidly into a lower state by the emission of radiation. The frequency and energy of emitted radiation are related by $E = hf$ where h is Planck's constant.

Exosphere The fifth atmospheric layer above the earth's surface, which extends into interplanetary space.

Exothermic reaction A chemical reaction that has the net effect of producing heat.

Extrusive rocks Igneous rocks that form at the earth's surface.

Fact A close agreement by competent observers of a series of observations of the same phenomenon.

Fahrenheit temperature scale A temperature scale with the freezing point of water assigned the value 32°F and the boiling point of water 212°F.

Faraday's law An electric field is induced in any region of space in which a magnetic field is changing with time. The magnitude of the induced electric field is proportional to the rate at which the magnetic field changes; the direction of the induced field is at right angles to the changing magnetic field.

Fat A biomolecule that packs a lot of energy per gram. A typical fat molecule consists of a glycerol molecule with three fatty acid molecules attached to it (see Figure 21.34).

Fault A fracture or fracture zone along which visible displacement can be detected on both sides of the fracture or parallel to it.

Faunal succession Fossil organisms succeed one another in a definite, irreversible, and determinable order. Because of this any time period can be uniquely recognized by its fossil content.

Fluid Anything that flows; in particular, a liquid or a gas.

Fluorescence The property of absorbing radiation at one frequency followed by its re-emission at a lower frequency.

Focal length Distance between the center of a mirror or lens and its focal point.

Focal point Point at which light rays parallel to the axis of a mirror or lens come to a focus.

Focus Point at which straight lines intersect.

Fold A series of ripples, large or small, that result from compressional deformation.

Footwall The mass of rock beneath a fault.

Force Any influence that can cause an object to be accelerated, measured in newtons.

Forced vibration The setting up of vibrations in an object by a vibrating force.

Formula mass The mass of a chemical compound given in amu.

Fourier analysis A mathematical method that will resolve any periodic wave form into a sum of simple sine waves.

Fractional crystallization The process and sequence by which minerals crystallize. Minerals with the highest melting points crystallize first, followed by minerals with lower melting points. The sequence of crystallization allows a single magma to generate several different igneous rocks.

Fractional distillation A method whereby the components of crude petroleum are separated into fractions according to their boiling temperatures.

Frame of reference A vantage point (usually a set of coordinate axes) with respect to which the position and motion of an object may be described.

Free fall Motion of an object under the influence of gravitational pull only.

Freezing The change of phase from the liquid to the solid form; the opposite of melting. Energy is released by the substance undergoing freezing.

Frequency For a vibrating object, the number of vibrations it makes per unit time. For a wave, the number of waves that pass a particular point per unit time.

Frequency modulation (FM) A type of modulation in which the frequency of the carrier wave is varied above and below its normal frequency by an amount proportional to the amplitude of the impressed signal and in which the amplitude of the modulated carrier wave remains constant.

Friction Forces resisting motion between one set of molecules and another due to electrical attraction and repulsion, usually between two solid surfaces; static before motion starts and kinetic during motion.

Front The contact zone between two air masses.

Fulcrum The pivot point of a lever.

Full moon The phase of the moon when its sunlit side squarely faces earth.

Galvanometer An instrument used to measure electric current. With the proper combination of resistors, it can be converted to an ammeter or a voltmeter.

Gamma ray High-frequency electromagnetic radiation emitted by the nuclei of radioactive atoms.

Gas A phase of matter characterized by indefinite volume and shape.

General theory of relativity Einstein's generalization of special relativity, which features a geometric theory of gravitation.

Generator A device that produces electric current by rotating a coil within a stationary magnetic field.

Geodesic Shortest path between points on a surface.

Glacier A large mass of ice formed by the compaction and recrystallization of snow. The ice does not become a glacier until it is able to move downslope under its own weight.

Gradient The ratio of vertical drop compared to horizontal distance.

Gravitation Attraction between objects due to mass.

Gravitational field The space surrounding a massive body in which another massive body experiences a force of attraction.

Gravitational potential energy The stored energy that an object possesses by virtue of its elevated position in a gravitational field.

Gravitational red shift The shift toward longer wavelength experienced by light leaving the surface of a massive object, as predicted by the general theory of relativity.

Gravitational wave A gravitational disturbance that propagates through space-time, produced by a moving mass.

Greenhouse effect The result of sunlight (high frequency) energy that passes through the atmosphere, is absorbed, and then is radiated at a lower frequency that cannot escape through the atmosphere, thus increasing the temperature of the atmosphere.

Groundwater Subsurface water that is in the zone of saturation.

H-R diagram (Hertzsprung-Russell diagram) A plot of intrinsic brightness versus surface temperature of stars. When so plotted, stars' positions take the form of a main sequence for average stars, with exotic stars above or below the main sequence.

Half-life Time required for half the atoms in a radioactive element to decay; can be applied to any exponentially decreasing process.

Halogens Group 17 elements.

Hang time The time that a jumper's feet are off the ground during a jump.

Hanging wall The mass of rock above a fault.

Harmonics Modes of vibrations that begin with a lowest or fundamental vibrating frequency and continue as a sequence of tones that are integral multiples of the fundamental frequency.

Heat The thermal energy that flows from a substance of higher temperature to a substance of lower temperature, commonly measured in calories or joules.

Heat capacity The ability of a substance to retain thermal energy. Water has a large heat capacity because much of the heat added to it is stored in the breaking of hydrogen bonds.

Heat of fusion The heat released as a substance freezes or the heat absorbed as it melts.

Hertz Unit of frequency; one cycle (complete vibration) per second.

Heteroatom Any non-carbon or non-hydrogen atom appearing in an organic molecule.

Heterogeneous A chemical mixture in which the different components can be seen as individual substances.

Holography The process in which three-dimensional optical images are produced by illuminating a microscopic interference pattern (hologram) with monochromatic light.

Homogeneous A chemical mixture composed of components so finely mixed that the composition throughout is the same.

Hooke's law The extension x of an elastic object is directly proportional to the stretching force F that is applied: $F \sim x$, or $F = kx$.

Horse latitudes The belt of latitudes at 30°N and 30°S where air movement is calm and weather is hot and dry.

Humidity Absolute humidity is the mass of water per volume of air. Relative humidity is absolute humidity at that temperature divided by the maximum possible, usually given as a percent.

Hurricane A severe tropical storm with rotating wind speeds up to 279 kilometers per hour.

Huygens' principle Light waves spreading out from a light source can be regarded as a superposition of tiny secondary wavelets.

Hydrocarbon A chemical compound containing only carbon and hydrogen atoms.

Hydrogen bond A dipole-dipole interaction that involves a hydrogen atom chemically bonded to a strongly electronegative element, such as nitrogen, oxygen, or fluorine.

Hydrologic cycle The natural circulation of water from the oceans to the air, to the ground, then to the oceans and then back to the atmosphere.

Hydronium ion A water molecule after accepting a hydrogen ion, H_3O^+.

Hydroxide ion A water molecule after donating a hydrogen ion, HO^-.

Hydroxyl group An oxygen atom bonded to a hydrogen atom, $-OH$. The hydroxyl group is found in alcohols.

Hypothesis An educated guess; a reasonable explanation of an observation or experimental result that is not fully accepted as factual.

Igneous rocks Rocks formed by the cooling and consolidation of hot, molten rock material called magma. Igneous rocks make up about 95 percent of the earth's crust. Igneous means "formed by fire."

Impulse The product of the force acting on an object and the time during which it acts, equal to the change in momentum that results.

Impure The state of a material that consists of more than one element, elemental or chemical compound. A chemical mixture is impure.

Incandescence The state of an object glowing while at a high temperature, caused by electrons in vibrating atoms and molecules that are shaken in and out of their stable energy levels, emitting radiation in the process. The peak frequency of radiation is proportional to the absolute temperature of a heated substance: $\bar{f} \sim T$.

Inclusions   Pieces of one rock type contained within another. The inclusion must be older than the rock containing it in order to become included as part of the host rock.

Induced dipole   A dipole temporarily created in an otherwise nonpolar molecule. It is induced by a neighboring charge or dipole.

Induced dipole-induced dipole   The chemical interaction involving only induced dipoles. This is a relatively weak chemical interaction.

Inertia   The fundamental property of inert material tending to resist changes in its state of motion.

Inertial frame of reference   An unaccelerated vantage point in which Newton's laws hold exactly.

Inner shell shielding   Inner shell electrons are able to partially shield outer shell electrons from the nuclear charge. This is called inner shell shielding.

Inner transition metals   Two subgroups of metals within the transition metals.

Insoluble   Not capable of dissolving to any appreciable extent in a solvent.

Insulator   Any material through which charge strongly resists flow when subject to an impressed electrical force; a nonconductor.

Interference   Superposition of waves, producing regions of reinforcement (constructive) and regions of cancellation (destructive).

Internal energy   The total of all molecular energies, both kinetic energy and potential energy, internal to a substance. Changes in internal energy are of principal concern in thermodynamics.

Intrusive rocks   Rocks that crystallize below the earth's surface.

Inverse-square law   The intensity of an effect is related to the inverse square of the distance from the cause: Intensity $\sim \frac{1}{\text{distance}^2}$ Gravity follows an inverse-square law, as do electric, magnetic, light, sound, and radiation phenomena.

Ion   An atom that has an excess or deficiency of electrons, compared to protons. An ion is electrically charged positive if it lacks electrons, and negative if it has excess electrons.

Ion-dipole   The chemical interaction involving an ion and a dipole.

Ionic bond   The electrical force of attraction that holds ions of opposite charge together.

Ionic compound   Any chemical compound containing ions.

Ionic crystal   A group of many ions held together in an orderly and periodic 3-dimensional array.

Ionization   The process of removing or adding electrons to or from the atomic nucleus.

Ionization energy   The amount of energy required to pull an electron away from an atom.

Ionosphere   An electrified region within the thermosphere and uppermost mesosphere where fairly large concentrations of ions and free electrons exist.

Iris   The colored part of the eye that surrounds the pupil.

Isotopes   Atoms whose nuclei have the same number of protons but various numbers of neutrons.

Irregular galaxy   A galaxy with a chaotic appearance, large clouds of gas and dust, without spiral arms.

Isobars   Lines on a weather map used to connect points of equal pressure.

Isostasy   The condition of equilibrium with all rock masses in balance.

Jet stream   High-speed winds in the upper troposphere. These winds play an essential role in the global transfer of heat energy from the equator to the poles.

Joule   The SI unit of work or energy in any form.

Karst   Topography formed by the dissolution of carbonate rock. Karst topography is characterized by sinkholes, caves, and underground drainage.

Kelvin   The SI unit of temperature.

Kepler's laws of planetary motion   Law 1: Each planet moves in an elliptical orbit with the sun at one focus. Law 2: The line from the sun to any planet sweeps out equal areas of space in equal time intervals. Law 3: The squares of the times of revolution (or years) (T) of the planets are proportional to the cubes of their average distances (R) from the sun. ($T^2 \sim R^3$ for all planets.)

Ketone   A class of organic molecules containing a carbonyl group in which the carbon of the carbonyl is bonded to two carbon atoms.

Kilogram   The standard unit of mass; 1 kilogram is the amount of mass that a force of 1 newton will accelerate 1 meter per second squared.

Kinetic energy   Energy of motion, described by the relationship Kinetic energy $= \frac{1}{2}mv^2$.

Laminar   Steady flow of water in a straight line path with no mixing of sediment.

Lanthanides   The inner transition metals within the sixth period.

Laser (light amplification by stimulated emission of radiation)   An optical instrument that produces a beam of coherent light.

Lava   The fluid rock that flows onto the earth's surface. Magma at the earth's surface.

Law   A general hypothesis or statement about the relationship of natural quantities that has been tested over and over again and has not been contradicted. Also known as a principle.

Law of reflection   The angle of incident radiation equals the angle of reflected radiation.

Law of universal gravitation   Every object in the universe attracts every other object with a force that for two objects is proportional to the masses of the objects and inversely proportional to the square of the distance separating them: $F \sim \frac{m_1 m_2}{d^2}$, or $F = G \frac{m_1 m_2}{d^2}$.

Leachate   A solution formed by water that has percolated through soil containing soluble substances.

Length contraction   The apparent shrinking of an object moving at a speed that is a significant fraction of the speed of light.

Lens   A piece of glass or other transparent material that can bring light to a focus.

Lever   A simple machine, made of a bar that pivots around a fixed point.

Light   The visible part of the electromagnetic spectrum.

Linear motion   Motion along a straight-line path.

Liquid   The phase of matter between the solid and gaseous phases in which the matter possesses a definite volume but no definite shape: it takes on the shape of its container.

Liquid crystal   A liquid in which atomic or molecular particles have a resemblance of order and periodicity.

Liquid pressure   = weight density × depth

Lithification   The process by which sediment turns into sedimentary rock.

Lithosphere   A subdivision of the earth's interior which includes the entire crust and the uppermost portion of the mantle. The lithosphere is rigid and brittle, and is broken into many individual pieces called plates. These lithospheric plates are continually in motion as a result of the hot convection currents circulating in the asthenosphere and the mantle.

Longitudinal wave   A wave in which the individual particles of a medium vibrate back and forth in the direction in which the wave travels—for example, sound.

Lorentz contraction   Contraction of an object in its direction of motion due to that motion; part of Einstein's special theory of relativity (see also Length contraction).

Loudness   The physiological sensation directly related to sound intensity or volume. Relative loudness, or noise level, is measured in decibels.

Love wave   A surface wave with a horizontal motion that is shear or transverse to the direction of propagation.

Lunar eclipse   The phenomenon whereby the shadow of the earth falls upon the moon, producing relative darkness of the full moon.

Mach number   The ratio of the speed of an object to the speed of sound. For example, an aircraft traveling at the speed of sound is rated Mach 1.0; traveling at twice the speed of sound, Mach 2.0.

Magma   Rock heated to its melting point, molten rock.

Magnetic field   The region of "altered space" that will interact with the magnetic properties of a magnet. It is located mainly between the opposite poles of a magnet or in the energetic space about an electric charge in motion.

Magnetic domain   A clustered region of aligned magnetic atoms. When numerous regions themselves are aligned with each other, the substance containing them is a magnet.

Magnetic force   Between magnets, it is the attraction of unlike magnetic poles for each other and the repulsion of like magnetic poles. Between a magnetic field and a moving charge, the moving charge is deflected from its path in the region of a magnetic field; the deflecting force is perpendicular to the magnetic field lines. This force is maximum when the charge moves perpendicularly to the field lines and is minimum (zero) when the charge moves parallel to the field lines.

Magnetic monopole   A hypothetical particle having a single north or south magnetic pole, analogous to a positive or negative electric charge.

Magnetic pole   One of the regions on a magnet that produce magnetic force.

Mantle   The middle layer in the earth's interior. The mantle resides below the earth's crust and above the earth's core.

Mass   The quantity of matter in an object; the measurement of the inertia or sluggishness that an object exhibits in response to any effort made to start it, stop it, or change in any way its state of motion; a form of energy.

Mass-energy equivalence   The relationship between mass and energy as given by the equation $E = mc^2$.

Maxwell's counterpart to Faraday's law   A magnetic field is induced in any region of space in which an electric field is changing with time. The magnitude of the induced magnetic field is proportional to the rate at which the electric field changes. The direction of the induced magnetic field is at right angles to the changing electric field.

Mechanical deformation   The process of metamorphism that results as a rock is subjected to stress, such as increased pressure.

Mechanical weathering   The physical disintegration and fragmention of rocks into smaller and smaller pieces.

Melting   The change of phase from the solid to the liquid form; the opposite of freezing. Energy is absorbed by the substance that is melting.

Meniscus   The curving of a liquid at the interface of its container.

Mesosphere   The third atmospheric layer above the earth's surface. Very little of the sun's radiation is absorbed in this layer and as a result, there is a rapid decrease in temperature from about 0°C at the bottom of the layer to about −90°C at the top.

Metallic bond A non-specific chemical bond in which metal atoms are held together by their attraction to a common pool of electrons.

Metallic compound An elemental or chemical compound containing metal atoms that are held together by the metallic bond.

Metalloid Elements such as silicon and germanium that exhibit both the properties of a metal and a non-metal.

Metamorphic rocks Rocks formed from pre-existing rocks that have been changed or transformed by high temperature, high pressure, or both. The word metamorphic means "changed in form."

Meteor The streak of light produced by a meteoroid burning in the earth's atmosphere; a "shooting star."

Meteorite A meteoroid or part of a meteoroid that has survived passage through the earth's atmosphere to reach the ground.

Meteoroid A small rock in interplanetary space.

MHD (magnetohydrodynamic) power The generation of electric power by interaction of a plasma and a magnetic field.

Milky Way The name of the galaxy to which we belong. Our cosmic home.

Mineral A naturally formed, inorganic solid composed of an ordered array of atoms chemically bonded to form a particular crystalline structure.

Mirage A floating image that appears in the distance and is caused by refraction of light in the atmosphere.

Model A representation of an idea created to make the idea more understandable.

Modulation Impressing a signal wave system on a higher frequency carrier wave, AM for amplitude signals and FM for frequency signals.

Mohorovičíc discontinuity The boundary that separates the earth's crust from the mantle. This boundary marks the level at which P-wave velocities change abruptly from 6.7–7.2 km/sec (in the lower crust) to 7.6–8.6 km/sec (at the top of the upper mantle).

Molarity A common unit of concentration measured by the number of moles in one liter of solution.

Mole A very large number equal to 6.02×10^{23}. This number is a unit commonly used when describing a number of molecules.

Molecule Any group of atoms held together by covalent bonds.

Momentum The product of the mass of an object and its velocity.

Monosaccharide The simplest carbohydrate consisting of a single saccharide unit, which can be identified by a single ring of 5 or 6 atoms. Examples are α-glucose and β-fructose.

Monsoon A wind system that changes direction with the seasons. In Southeast Asia the monsoons are associated with heavy rains in the summer and dry climates in the winter.

Moon phases The cycles of change of the face of the moon, changing from new to waxing, to full, to waning, and back to new.

Neanderthal Early form of humans (*Homo sapiens*) who were cave dwellers and lived during the last glacial period, between 100,000 to 40,000 years ago. Neanderthals show evidence of a well-developed society, stone tools, knowledge of fire, and burial of the dead.

Nemesis The name of the hypothetical star companion to the sun.

Net force The resultant of all the forces that act on an object.

Neutral A solution in which the hydronium ion concentration is equal to the hydroxide ion concentration.

Neutrino Near-massless, uncharged particle emitted with an electron during beta decay.

Neutron star A small, highly dense star composed of tightly packed neutrons formed by the welding of protons and electrons.

New moon The phase of the moon when its dark side squarely faces earth.

Newton SI unit of force; a force of 1 newton accelerates a mass of 1 kilogram 1 meter per second each second.

Newton's law of cooling The rate of heat loss from an object is proportional to the excess temperature of the object over the temperature of its surroundings.

Newton's laws of motion Law 1: Every object continues in its state of rest or of uniform motion in a straight line unless it is compelled to change that state by forces impressed upon it. Law 2: The acceleration of an object is directly proportional to the net force acting on the object and inversely proportional to the mass of the object. Law 3: To every action force, there is an equal and opposite reaction force.

Nitrogen fixation A biological process whereby ammonia, NH_3, is produced from atmospheric nitrogen, N_2.

Noble gases Group 18 elements.

Node Point of zero amplitude in a standing wave.

Nonconformity The condition of overlying sedimentary rocks on an eroded surface of igneous or metamorphic rocks.

Nonlinear motion Any motion not along a straight-line path.

Nonpolar   The state of having no dipole. A molecule may be described as *nonpolar* when it has no dipole.

Normal   A line that is perpendicular to a surface.

Normal fault   Tensional fault in which rocks in the hanging wall drop down relative to the footwall. Normal faults place younger, structurally higher rocks on top of older, structurally deeper rocks.

Normal force   The component of support force perpendicular to a supporting surface. For an object resting on a horizontal surface, it is the upward force that balances the weight of the object.

Nova   A star that suddenly brightens, appearing as a "new" star; believed to be associated with the eruptions of white dwarfs in binary systems.

Nuclear fission   Splitting of the nucleus of a heavy atom, such as uranium-235, into parts, releasing much energy.

Nuclear fusion   The combining of the nuclei of light atoms to form heavier nuclei, releasing much energy.

Nucleon   A nuclear proton or neutron; the collective name for either or both.

Nucleus   The positively charged core of an atom.

Ohm   The SI unit of electrical resistance.

Ohm's law   The current in a circuit varies in direct proportion to the potential difference or emf and in inverse proportion to resistance: Current $= \frac{\text{voltage}}{\text{resistance}}$. A potential difference of 1 volt across a resistance of 1 ohm produces a current of 1 ampere.

Organic chemistry   The chemistry of carbon compounds.

Original horizontality   Layers of sediment are deposited in a rather even manner, with each new layer laid down almost horizontally over the older sediment.

Overtones   Tones produced by vibrations that usually are multiples of the lowest, or fundamental, vibrating frequency.

Oxidation   The process whereby a reactant loses one or more electrons.

Ozone layer   Layer formed in the stratosphere, which reduces the amount of ultraviolet radiation reaching the earth's surface.

Pahoehoe   A basaltic lava flow characterized by a smooth wrinkly skin that resembles the twisting braids in a rope.

Paleomagnetism   The study of natural magnetization in a rock due to the ability of certain minerals to align themselves with the magnetic poles. This preserved imprint allows scientists to determine the intensity and direction of the earth's magnetic field at the time of the rock's formation.

Pangaea   A great supercontinent that existed from about 300 to about 200 million years ago. Pangaea means "all land" and included most of the continental crust of the earth.

Parabola   The curved path followed by a projectile acting only under the influence of gravity.

Parallel circuit   An electric circuit with two or more resistances arranged in branches in such a way that any single one completes the circuit independently of all others providing two or more paths.

Partial melting   The incomplete melting of rocks resulting in magmas of different compositions.

Pascal   The SI units of pressure.

Pascal's principle   The pressure applied to a fluid confined in a container is transmitted undiminished throughout the fluid and acts in all directions.

Perigee   The point in an elliptical orbit closest to the focus about which orbiting takes place.

Period   The time required for a vibration or a wave to make a complete cycle; equal to 1/frequency.

Periodic table   A scheme of ordering the elements to show the periodicity of similar chemical properties, as shown on the inside front cover of this book.

Permeability   The ability of a material to transmit fluid.

Perturbation   The deviation of an orbiting object from its normal path, caused by an additional gravitational interaction.

pH   A measure of the acidity of a solution. The pH is equal to the negative logarithm of the hydronium ion concentration. The lower the pH, the greater the acidity.

Phlogiston   A long-since disproven hypothetical substance that is released by a material as it burns.

Phosphorescence   A type of light emission that is the same as fluorescence except for a delay between excitation and de-excitation. The delay is caused by atoms being excited to energy levels that do not decay rapidly. The afterglow thus provided may last from fractions of a second to hours or even days, depending on the type of material, temperature, and other factors.

Photoelectric effect   The emission of electrons from a metal surface when light shines on it.

Photon   A light corpuscle, or the basic packet of electromagnetic radiation. Just as matter is composed of atoms, light is composed of photons (quanta).

Physical change   A change in which a substance changes its physical properties without changing its chemical identity. The vaporization of water is an example of a physical change.

Physical property   The physical characteristics of a substance such as color, density, and hardness.

Pigment   A material that selectively absorbs colored light.

Pitch   The "highness" or "lowness" of a tone, as on a musical scale, which is governed by frequency. A high-frequency vibrating source produces a sound of high pitch; a low-frequency vibrating source produces a sound of low pitch.

Planck's constant A fundamental constant, h, that relates the energy and frequency of light quanta: $h = 6.6 \times 10^{-34}$ J·s.

Planetary nebula An expanding shell of gas ejected from a star during the latter stages of its evolution.

Plasma Hot matter (beyond the gaseous phase) that is composed of electrically charged particles. Most of the matter in the universe is in the plasma phase.

Pluton A large intrusive body formed below the earth's surface.

Polar easterlies Winds that form behind the polar front, as cool air moves down into the warm latitudes. The winds blow from the east to the west.

Polar front Boundary where the cool polar air meets the warm air of the temperate zone.

Polar The state of having a dipole. A molecule may be described as *polar* when it has a dipole.

Polarization The alignment of the electric vectors that make up electromagnetic radiation. Such waves of aligned vibrations are said to be polarized.

Polyatomic ion An ionized molecule.

Polymorph Minerals that have the same chemical composition but different crystal structures.

Polysaccharide A complex carbohydrate consisting of many monosaccharides.

Porosity The amount of open space in a rock or in sediment.

Positron The antiparticle of an electron; a positively charged electron.

Postulate A fundamental assumption.

Postulates of the special theory of relativity (1) All laws of nature are the same in all uniformly moving frames of reference. (2) The velocity of light in free space will be found to have the same value regardless of the motion of the source or the motion of the observer; that is, the velocity of light is invariant.

Potential difference The difference in electric potential, or voltage, between two points, which can be compared to a difference in water level between two containers. Measured in volts.

Potential energy The stored energy that an object possesses because of its position with respect to other objects.

Power The time rate at which work is performed: Power = $\frac{\text{work}}{\text{time}}$.

Precipitate A solute that has come out of solution.

Precipitation Any form of water particles—rain, sleet, or snow—that falls from the atmosphere to the ground below.

Pressure The ratio of the amount of force to the area over which the force is distributed: Pressure = $\frac{\text{force}}{\text{area}}$.

Primary colors The three colors red, blue, and green that, when added in certain proportions, will produce any color in the spectrum. (The effect is due to three types of retinal cells in the eye.)

Primary wave A body wave that involves compressional and expansional motion in the direction of propagation. P-waves travel through solids, liquids, and gases and are the fastest of the seismic waves.

Primeval fireball The burst of energy during the Big Bang.

Principle of equivalence Observations in a gravitational reference frame are equivalent to those made in an accelerating reference frame.

Prism A triangular piece of material such as glass that separates incident light by refraction into its component colors.

Product The new material formed by a chemical reaction. It appears after the arrow in the chemical equation.

Projectile Any object that is projected by some force and continues in motion by virtue of its own inertia.

Protein A structural biomolecule consisting of many amino acid units linked together.

Protostar The aggregation of matter that goes into and precedes the formation of a star.

Protosun The youngest stage of the sun—the stage of formation.

Pulsar Likely a neutron star that rapidly spins, sending short, precisely timed bursts of electromagnetic radiation.

Pupil The part of the eye that admits the passage of light.

Pure The state of a material that consists of only a single element or chemical compound.

Quality The characteristic timbre of a musical sound, governed by the number and relative intensities of the overtones.

Quantum An elemental unit of a quantity.

Quantum mechanics The branch of quantum physics that deals with finding the probability amplitudes of matter waves; basic departure from classical mechanics.

Quantum theory The physical theory based on the idea that energy is radiated in definite units called quanta or photons. Just as matter is composed of atoms, radiant energy is composed of quanta. Further, all material particles have wave properties.

Quarks The elementary constituent particles or building blocks of nuclear matter.

Quasar (Quasi-stellar object) A small powerful source of energy believed to be the active core of very distant galaxies.

Radiation The transfer of energy by means of electromagnetic waves or high-speed particles.

Radiometric dating Calculating an age in years for geologic materials based on the nuclear decay of naturally occurring radioactive isotopes.

Rarefaction The region of reduced pressure in a longitudinal wave.

Rate How fast something happens or how much something changes per unit of time.

Rayleigh wave A surface wave with an elliptical motion.

Reflection The bouncing of light rays from a surface in such a way that the angle at which a given ray is returned is equal to the angle at which it strikes the surface. When the reflecting surface is irregular, light is returned in irregular directions and is called diffuse reflection.

Reactant The starting material for a chemical reaction. It appears before the arrow in the chemical equation.

Real image An image formed by the actual convergence of light rays, which can be displayed on a screen.

Recrystallization The process of metamorphism that results when a rock is subjected to such high temperatures that the minerals go through a change in mineralogy, usually by the loss of H_2O or CO_2. Recrystallization generally results in larger grain size.

Rectifier A diode that converts ac to dc, which operates by either suppressing or alternating half-cycles of the current wave form or by reversing them.

Red giant Cool giant stars above main sequence stars on the H-R Diagram.

Reduction The process whereby a reactant gains one or more electrons.

Refraction The bending of an oblique ray of light when it passes from one transparent medium to another, caused by a difference in the speed of light in the transparent media. When the change in medium is abrupt (say, from air to water), the bending is abrupt; when the change in medium is gradual (say, from cool air to warm air), the bending is gradual, accounting for mirages.

Regelation The process of melting under pressure and the subsequent refreezing when the pressure is removed.

Relative humidity The ratio of the amount of water vapor in a sample of air to the amount of water vapor the sample of air is capable of supporting at a given temperature.

Relative Regarded in relation to something else.

Residence time The average time that a water molecule spends in a region.

Resolution A method of separating a vector into its component parts.

Resonance The setting up of vibrations in an object at its natural vibration frequency by a vibrating force or wave having the same (or submultiple) frequency.

Resultant The geometric sum of two or more vectors.

Retina The layer of light-sensitive tissue at the back of the eye.

Reverberation Re-echoed sound.

Reverse fault Compressional fault in which the angle of the fault plane with the horizontal ground surface is greater than 45°. Reverse faults push older, structurally deeper rocks on top of younger, structurally higher rocks.

Rifts A long narrow trough that forms as a result of divergence.

Ritz combination principle The spectral lines of elements have frequencies that are either the sums or the differences of frequencies of two other lines.

Rock cycle A sequence of events involving the formation, destruction, alteration, and reformation of rocks as a result of the generation and movement of magma, the weathering, erosion, transportation, and deposition of sediment, and the metamorphism of pre-existing rocks.

Rotational inertia The property of an object that resists any change in its state of rotation. If at rest, it tends to remain at rest; if rotating, it tends to remain rotating and will continue to do so unless interrupted.

Satellite A projectile or small celestial body that orbits a larger celestial body.

Saturated A solution containing as much solute as will dissolve.

Saturated hydrocarbon A hydrocarbon containing no multiple covalent bonds. In such a hydrocarbon, carbon atoms are "saturated" with hydrogen atoms.

Saturated The limit at which the amount of water vapor in the air is the most it can be for a given temperature.

Scalar quantity A quantity that may be specified by magnitude, without regard to direction. Examples are mass, volume, speed, and temperature.

Scattering The emission in random directions of light that encounters particles that are small compared to the wavelength of light; more often at short wavelengths (blue) than at long wavelengths (red).

Schrödinger wave equation The fundamental equation of quantum mechanics, which interprets the wave nature of material particles in terms of probability wave amplitudes. It is as basic to quantum mechanics as Newton's laws of motion are to classical mechanics.

Scientific method An orderly method for gaining, organizing, and applying new knowledge.

Sea floor spreading A hypothesis that the oceanic crust is constantly being renewed by convective upwelling of magma along the midoceanic ridges. At the same time new material is created old material is destroyed at the deep ocean trenches near the edges of continents.

Secondary wave   A body wave propagated by a shearing motion. Because S-waves do not travel through liquids they do not travel through the earth's outer core.

Sedimentary rocks   Rocks formed from weathered material (sediments) carried by water, wind, or ice. Sedimentary rocks are the most common rocks in the uppermost part of the earth's crust and cover over two thirds of the earth's surface.

Semiconductor   A device made of material not only with properties that fall between a conductor and an insulator but with resistance that changes abruptly when other conditions change, such as temperature, voltage, and electric or magnetic field. Semiconductors are usually fabricated from metalloid elements.

Series circuit   An electric circuit with devices that have resistance arranged so that the same electric current flows through all of them.

Shock wave   The cone-shaped wave made by an object moving at supersonic speed through a fluid.

SI (Système International)   Modern system of definitions and metric notation, now spreading throughout the academic, industrial, and commercial community—of which the United States of America is last.

Silicates   The most common mineral group. Silicates are constructed of four oxygen atoms (O^{2-}) with a single silicon atom (Si^{4+}) at the center in a structure called a tetrahedron.

Simple harmonic motion   A vibratory or periodic motion, like that of a pendulum, in which the force acting on the vibrating body is proportional to its displacement from its central equilibrium position and acts toward that position.

Sine curve   A wave form traced by simple harmonic motion that is uniformly moving in a direction perpendicular to the vibration direction, like the wavelike path traced on a moving conveyor belt by a pendulum swinging at right angles above the moving belt.

Solar constant   The 1400 joules per square meter received from the sun each second at the top of the earth's atmosphere. Expressed in terms of power, it is 1.4 kilowatts per square meter.

Solar eclipse   The phenomenon whereby the shadow of the moon falls upon the earth producing a region of darkness in the daytime.

Solar nebula   The cloud of gas and dust from which the sun and solar system formed.

Solar power   Energy per unit time derived from the sun.

Solid   A phase of matter characterized by definite volume and shape.

Solubility   The ability of a solute to dissolve, which depends not only on chemical interactions between the solute and the solvent, but upon the interactions among both solute molecules and among solvent molecules.

Soluble   Capable of dissolving in a solvent.

Solute   Any component in a solution that is not the solvent.

Solution   A homogeneous mixture in which all components are of the same phase. Salt water is an example of a solution.

Solvent   The component in a solution present in the largest amount.

Sonic boom   The loud sound resulting from the incidence of a shock wave.

Sound   A longitudinal wave phenomenon that consists of successive compressions and rarefactions of the medium through which the wave travels.

Space-time   The four-dimensional continuum in which all things exist; three dimensions are the coordinates of space, and the fourth dimension is time.

Specific heat   The quantity of heat per unit mass required to raise the temperature of a substance by 1K or, equivalently, 1°C. Measured in joules per kilogram kelvin or calories per gram Celsius degree.

Special theory of relativity   A formulation of the consequences of the absence of a universal frame of reference. It has two postulates: (1) All laws of nature are the same in all uniformly moving frames of reference. (2) The velocity of light in free space will be found to have the same value regardless of the motion of the observer; that is, the velocity of light is invariant.

Spectroscope   An optical instrument that separates light into its frequencies in the form of spectral lines; a spectrometer is an instrument that can also measure the frequencies.

Speed   The time rate at which distance is covered by a moving object.

Spiral galaxy   A disk-shaped galaxy with hot bright stars, and spiral arms. Our Milky Way is a spiral galaxy.

Standing wave   A stationary wave pattern formed in a medium when two sets of identical waves pass through the medium in opposite directions.

Starch   A digestible polysaccharide for humans and most species.

Stick structure   A short-hand notation of the structure of an organic molecule where the carbon framework is represented as a series of connected sticks.

Stratosphere   The second atmospheric layer above the earth's surface. Ultraviolet solar radiation is absorbed by a thin ozone layer within the stratosphere, which raises its temperature from about −50°C at the bottom of the layer to about 0°C at the top.

Streamline   The smooth path of a small region of fluid in steady flow.

Striations   Long parallel scratches produced in a rock by geological forces such as glaciers, stream flow, or faults. In the case of glaciers, striations are aligned in the direction of ice flow.

Structural isomer Molecules that have the same molecular formula yet differ by their chemical structures.

Subduction The process of one lithospheric plate descending beneath another.

Sublimation The direct conversion of a substance from the solid to the vapor phase, or vice versa, without passing through the liquid phase.

Subsidence Sinking or downward settling of the earth's surface due to compaction, the natural withdrawal of fluid, or human withdrawal of subsurface material by pumping of groundwater or oil.

Subtractive primary colors The three colors of absorbing pigments magenta, yellow, and cyan that, when mixed in certain proportions, will reflect any color in the spectrum.

Sunspots Temporary and relatively cool and dark regions on the sun's surface.

Supercluster A group of an enormous number of galaxies.

Superconductor A material that is a perfect conductor with zero resistance to the flow of electric charge.

Supernova An exploding star, caused either by transfer of matter to a white dwarf, or by gravitational collapse of a massive star, where enormous quantities of matter are emitted.

Superposition In an undeformed sequence of sedimentary rocks, each bed or layer is older than the one above and younger than the one below.

Supersaturated A solution that contains more solute than it normally contains.

Surface tension The tendency of the surface of a liquid to contract in area and thus behave like a stretched rubber membrane.

Surface waves A seismic wave that travels along the surface of the earth.

Suspension A homogeneous mixture consisting of more than a single phase. Examples include milk, blood, and clouds.

Syncline A fold in sedimentary strata that has stratigraphically younger rocks in the core of the fold. A syncline resembles a sag.

Technology A method and means of solving practical problems by implementing the findings of science.

Temperature A measure of the average kinetic energy per molecule in an object given in degrees Celsius or Fahrenheit or in kelvins.

Temperature inversion The condition wherein the upper regions of the atmosphere are warmer than the lower regions.

Terminal velocity The velocity attained by an object wherein the resistive forces counterbalance the driving forces, so motion is without acceleration.

Terrestrial radiation Radiant energy emitted from the earth after having been absorbed from the sun.

Theory A synthesis of a large body of information that encompasses well-tested and verified hypotheses about certain aspects of nature.

Thermal energy The total of all molecular energies, kinetic plus potential energy, internal to a substance.

Thermodynamics The physics of the interrelationships between heat and other forms of energy, characterized by two principal laws: First law: A restatement of the law of conservation of energy as it applies to systems involving changes in temperature. Whenever heat is added to a system, it transforms to an equal amount of some other form of energy. Second law: Heat cannot be transferred from a colder body to a hotter body without work being done by an outside agent.

Thermonuclear Pertaining to nuclear fusion caused by high temperatures.

Thermonuclear reaction The fusion reaction brought about by high temperatures.

Thermosphere The fourth atmospheric layer above the earth's surface. The little bit of air in this layer absorbs enough incoming solar radiation to cause an increase in temperatures to about 2000°C.

Thrust fault Compressional fault in which the angle of the fault plane with the horizontal ground surface is less than 45°. Thrust faults push older, structurally deeper rocks on top of younger, structurally higher rocks.

Thunderstorm A storm accompanied by thunder and lightning. Produced when warm humid air rises in an unstable environment.

Time dilation The apparent slowing down of time for an object moving at relativistic speeds.

Tornado A funnel-shaped cloud that extends from a large cumulonimbus cloud.

Torque The product of force and lever-arm distance, which tends to produce rotation.

Total internal reflection The total reflection of light traveling in a medium when it is incident on the surface of a less dense medium at an angle greater than the critical angle.

Trade winds Tropical winds that blow from the subtropical highs to the equatorial low.

Transformer A device for transforming electrical power from one coil of wire to another by means of electromagnetic induction.

Transition metals The elements of groups 3–12.

Transmutation Changing an element into another by changing the number of protons in the atom nucleus.

Transverse wave   A wave in which the individual particles of a medium vibrate from side to side perpendicularly (transversely) to the direction in which the wave travels. The vibrations along a stretched string are transverse waves. The term applies also to nonmaterial waves where the periodically changing quantity (electric field) has a direction at right angles to the direction of the wave propagation.

Triple bond   Three covalent bonds between the same two atoms.

Troposphere   The atmospheric layer closest to the earth's surface. The troposphere is where our weather occurs, it contains 90 percent of the atmospheric mass and essentially all its water vapor and clouds. Temperature decreases with increasing altitude with an average temperature of $-50°C$ at the top.

Turbulent   Erratic flow of water that moves in a jumbled manner stirring up everything with which it comes in contact.

Uncertainty principle   The ultimate accuracy of measurement is given by the magnitude of Planck's constant, h. Further, it is not possible to measure exactly both the position and the momentum of a particle at the same time, nor the energy and the time associated with a particle simultaneously.

Unconformity   A break or gap in the geologic record, such as by an interruption in the sequence of deposition, or by a break between eroded metamorphic or igneous rocks and a sedimentary rock formation.

Unsaturated   A solution that will dissolve additional solute if added.

Unsaturated hydrocarbon   A hydrocarbon containing at least one multiple covalent bond.

Valence electron   Any electron in the outermost shell of an atom.

Vector   An arrow drawn to scale, used to represent a vector quantity.

Vector quantity   A quantity that has both magnitude and direction. Examples are force, velocity, acceleration, torque, and electric and magnetic fields.

Velocity   The specification of the speed of an object and its direction of motion, a vector quantity.

Vibration   A "wiggle in time"; an oscillation.

Virtual image   An illusionary image that is seen by an observer through a lens but cannot be projected on a screen.

Volcano   The central vent through which lava, gases, and ash erupt and flow.

Volt   The SI unit of electric potential: $1 V = 1 J/C$.

Voltage   Electrical "pressure" or a measure of electrical potential difference.

Volume   The quantity of space an object occupies.

Water table   The upper boundary of the zone of saturation where every pore space is completely filled with water.

Watt   The SI unit of power: $1 W = 1 J/s$.

Wave   A "wiggle in space and time"; a disturbance propagated from one place to another with no actual transport of matter.

Wave velocity   The speed with which waves pass by a particular point: Wave velocity = frequency × wavelength.

Wavelength   The distance between successive crests, troughs, or identical parts of a wave.

Weather   The effect of the day-to-day changes in air temperature, wind, and precipitation for a given locality.

Weathering   The breakdown and decomposition of the earth's surface.

Weight   The force of the earth's gravitational attraction for any object on, below, or above the earth's surface.

Weightlessness   A condition wherein apparent gravitational pull is lacking.

Westerlies   The general wind pattern north of the horse latitudes in the northern hemisphere and south of the horse latitudes in the southern hemisphere. The winds blow from the west to the east. The westerlies are the general wind pattern in the United States.

White dwarf   Dying star that has collapsed to the size of the earth and is slowly cooling off; located at the lower left of the H-R diagram.

Work   The product of the force exerted and the distance through which the force moves: $W = Fd$.

X ray   Electromagnetic radiation emitted by atoms when the innermost orbital electrons undergo excitation.

PHOTO CREDITS

Unless otherwise acknowledged, all photographs are the property of Scott, Foresman and Company. Page abbreviations are as follows: (T) top, (C) center, (B) bottom, (L) left, (R) right, (INS) inset.

Cover   Robert Ewell

Title Page Robert Ewell

Prologue Opener Charles A. Spiegel

7L	Jeu de Paume, Paris/Art Resource, NY
15	Meidor Hu
16	Grafton M. Smith/The Image Bank
17T	The Granger Collection, New York
17B	Terry Murphy/ANIMALS ANIMALS
19	Erich Lessing/Art Resource, NY
28	David L. Johnson/Sports Photo Masters, Inc.
34	Larry Sergean/Comstock Inc.
38	National Portrait Gallery, London
43	R. Mackson/FPG
46	Fundamental Photographs
52	John Suchocki
57	SuperStock, Inc.
60	Terje Rakke/The Image Bank
61B	Meidor Hu
64	Paul G. Hewitt
65	Hubrich/The Image Bank
66	Paul G. Hewitt
71	Meidor Hu
73T	NASA
81	P. Wallick/H. Armstrong Roberts
86T	Richard Megna/Fundamental Photographs
90T	NASA
91	Richard Megna/Fundamental Photographs
96	Noren Trotman/Sports Photo Masters, Inc.
98	NASA
102	Diane Schiumo/Fundamental Photographs
106	NASA
111	Chuck O'Rear/West Light
114All	Jack Hancock
120All	Milo Patterson
123	The Granger Collection, New York
126	David Hewitt
131	William Waterfall/The Stock Market
139	Olan Mills
140	B. Brad Lewis
147B	Meidor Hu
149	Meidor Hu
150	Lynn M. Stone/The Image Bank
161	Jim Wallace
162	Tracy Suchocki
163	Milt & Joan Mann/Cameramann International, Ltd.
164	Nancy Rogers
169All	Robert D. Carey
170	David Cavagnaro
174	Sylvia Means
175	Reprinted by permission of the Boeing Company
181	Vanscan (R) Thermogram By Daedalus Enterprises, Inc.
184	Meidor Hu
194	Paul G. Hewitt
199	L. J. Hafencher/H. Armstrong Roberts
200	Steven Hunt/The Image Bank

206All	Palmer Physical Laboratory
207	Paul G. Hewitt
210	Zig Leszczynski/ANIMALS ANIMALS
213	Dan McCoy/Rainbow
227	John Suchocki
229	Richard Megna/Fundamental Photographs
230L	Richard Megna/Fundamental Photographs
230R	Richard Megna/Fundamental Photographs
231	Meidor Hu
233TL	Richard Megna/Fundamental Photographs
233TC	Richard Megna/Fundamental Photographs
233TR	Richard Megna/Fundamental Photographs
233B	Japanese National Railways
234	Argonne National Laboratory
243	Paul G. Hewitt
247	David Hewitt
248	Hermann Schlenker/Photo Researchers
250	Gabe Palmer/Mug Shots/The Stock Market
253	Meidor Hu
254	Terrence McCarthy/San Francisco Symphony
255	John Suchocki
256	Eric Meola/The Image Bank
258	AP/Wide World
259All	Education Development Center
261All	Paul G. Hewitt
265	Official U.S. Navy Photograph
267	Courtesy Palm Press, Inc./Courtesy estate of Dr. Harold Edgerton
271	Meidor Hu
276	Dr. Jeremy Burgess/SPL/Photo Researchers
281	Meidor Hu
285	Dave Vasquez
289	Meidor Hu
290	Camerique/H. Armstrong Roberts
291	Don King/The Image Bank
292All	Education Development Center
293	Ken Kay/Fundamental Photographs
301T	Diane Schiumo/Fundamental Photographs
306	Nina Leen/*Life*, Time Warner Inc.
309	Roger Ressmeyer/Starlight
310	Courtesy Institute of Paper
312T	Ted Mahieu
312B	Robert Greenler
319T	Will & Deni McIntyre/Photo Researchers
321	Ronald B. Fitzgerald
324	Paul G. Hewitt
330	Courtesy Albert Rose
333	Camerique/H. Armstrong Roberts
334	Milo Patterson
337	Leslie Hewitt
338	IBM Research/Peter Arnold, Inc.
339T	Paul G. Hewitt
339B	Courtesy The Lawrence Berkeley Laboratory
341	Courtesy of IBM
347	Courtesy Sargent-Welch Scientific Company
349	Courtesy of AIP Niels Bohr Library
350T	H. Rather, "Elektroninterferenzen HANDBUCH DER PHYSIK, Vol. 32, 1957, Springer-Verlag, Berlin, Heidelberg, New York.
350BL	H. Armstrong Roberts

350BR	Dr. Tony Brain/SPL/Photo Researchers
358BL	Breck P. Kent/Earth Scenes
361	Bettman Archive
366All	Courtesy of Bicran, Newberry, Ohio
370All	Courtesy The Lawrence Berkeley Laboratory
379	Photo by Walter Dickenman/Courtesy Sandia Laboratories
383	Roger Ressmeyer/Starlight
390	Courtesy General Atomics
391	Lawrence Livermore National Laboratory
397	John Suchocki
398	L. J. Hafencher/H. Armstrong Roberts
400	The Louvre/Art Resource, NY
401	The Granger Collection, New York
402	The Granger Collection, New York
404	The Granger Collection, New York
405	The Granger Collection, New York
406L	E. R. Degginger
406R	John Suchocki
407L	Paul G. Hewitt
407R	Joan Lucas
417B	E. R. Degginger
419	E. R. Degginger
424	John Suchocki
428	John Suchocki
429L	JPL/NASA
430T	E. R. Degginger
430B	SuperStock, Inc.
431	SuperStock, Inc.
432	John Suchocki
433B	Phil Degginger
434T	E. R. Degginger
434B	E. R. Degginger
435	National Museum of Natural History/Smithsonian Institution
436	Dan McCoy/Rainbow
450	Diane Schiumo/Fundamental Photographs
468T	Kip Peticolas/Fundamental Photographs
468B	John Suchocki
469	Leland C. Clark, Jr., Ph.D./Professor of Biological Sciences
471T	John Suchocki
471B	Four by Five Inc./SuperStock, Inc.
474	John Suchocki
476	John Suchocki
477	John Suchocki
480T	John Suchocki
480B	NASA
481B	E. R. Degginger
482	Four by Five Inc./SuperStock, Inc.
484	John Suchocki
485	John Suchocki
486	E. R. Degginger
487T	Ken Biggs/Tony Stone Images
490All	John Suchocki
495	E. R. Degginger
497	John Suchocki
507	John Suchocki
513T	Larry Voigt/Photo Researchers
513B	SuperStock, Inc.
531	Robert Ewell
531INS	Leslie Hewitt

532T Susan Middleton, 1986/California Academy of Sciences
532B Harold W. Hoffman/Photo Researchers
533TL Breck P. Kent/ANIMALS ANIMALS
533TR Breck P. Kent/ANIMALS ANIMALS
533CL E. R. Degginger
533CR Lee Boltin
533BL E. R. Degginger
533BC E. R. Degginger
533BR E. R. Degginger
535(A) Leslie Hewitt
535(B) Breck P. Kent/Earth Scenes
535(C) Roy Eisenhardt/California Academy of Sciences
535(D) Roy Eisenhardt/California Academy of Sciences
535(E) Roy Eisenhardt/California Academy of Sciences
535BL Lee Boltin
535BR Lee Boltin
535(F) Susan Middleton, 1986/California Academy of Sciences
536T E. R. Degginger
536B Breck P. Kent/Earth Scenes
538TL E. R. Degginger
538TR E. R. Degginger
538BL Breck P. Kent/ANIMALS ANIMALS
539T Breck P. Kent/Earth Scenes
539B E. R. Degginger
545T Charles R. Belinky/Photo Researchers
545TC E. R. Degginger
545C E. R. Degginger
545BC E. R. Degginger
545B E. R. Degginger
546T Jim Wark/Peter Arnold, Inc.
546B Fred Grassle/Woods Hole Oceanographic Institution
548L E. R. Degginger/Earth Scenes
548R E. R. Degginger
549B W. H. Hodge/Peter Arnold, Inc.
550 Breck P. Kent/Earth Scenes
552T Lee Boltin
552TC Breck P. Kent/Earth Scenes
552BC Breck P. Kent/Earth Scenes
552B A. T. Copley/Visuals Unlimited
553T E. R. Degginger
554T Adrienne T. Gibson/Earth Scenes
554B Breck P. Kent/Earth Scenes
555T Steve Lissau/Rainbow
556B Courtesy NASA/E. R. Degginger
556T Richard Weiss/Peter Arnold, Inc.
557TL Lee Boltin
557CL Paul Silverman/Fundamental Photographs
557CR E. R. Degginger/Earth Scenes
557TR E. R. Degginger
557B E. R. Degginger
559L Bob Abrams
559R Meidor Hu
560L A. J. Cunningham/Visuals Unlimited
560CL Alex Kerstitch/Visuals Unlimited
560CR Paul Silverman/Fundamental Photographs
560R Cabisco/Visuals Unlimited
561T Leslie Hewitt
562T John D. Cunningham/Visuals Unlimited
562TC Breck P. Kent/Earth Scenes
562BC Paul Silverman/Fundamental Photographs
562B Paul Silverman/Fundamental Photographs
563T E. R. Degginger

563C E. R. Degginger/Earth Scenes
569 E. R. Degginger
577 Joseph Burke/Rainbow
578B C. Allan Morgan/Peter Arnold, Inc.
579T Phil Degginger
579C Mammoth Cave National Park/U.S. Dept. of the Interior, National Park Service
579B Paul G. Hewitt
580T Bob Abrams
584 John Lemker/Earth Scenes
585T E. R. Degginger
587B Leslie Hewitt
589T USGS Anonymous
589C E. R. Degginger
589B W. H. Hodge/Peter Arnold, Inc.
590T E. R. Degginger
590B E. R. Degginger
591 E. R. Degginger
596 Juan Guzman/*Life Magazine*/Time Warner Inc.
604 John S. Shelton
606 John S. Shelton
610 ©Marie Tharp
624 Charles R. Knight/Field Museum of Natural History, Chicago
625 E. R. Degginger
626 Bob Abrams
628T Cold Spring Granite, Morton MN
628B National Air and Space Museum/Smithsonian Institution
630 Francois Gohier/Photo Researchers
631 Alex Kerstitch/Visuals Unlimited
632T Tom McHugh/Photo Researchers
632B Courtesy Department of Library Services/American Museum of Natural History
633 Courtesy Department of Library Services/American Museum of Natural History
636T Courtesy Department of Library Services/American Museum of Natural History
636C The Cleveland Museum of Natural History
636B Field Museum of Natural History, Chicago
639T American Museum of Natural History
639B Natural History Museum of Los Angeles County
647 William S. Bickel, University of Arizona
651 E. R. Degginger
662 Mike J. Howell/Stock Boston
663L Joyce Photographics/Photo Researchers
663C E. R. Degginger
663R E. R. Degginger
666 Thomas Nes/The Stock Market
667 E. R. Degginger
668 AP/Wide World
669B Howard Bluestein/SS/Photo Researchers
675 Richard Crowe
676 NASA
677All NASA
679All Lick Observatory
682 Roger Ressmeyer/Starlight
683 Dennis di Cicco
685 Mark Martin/NASA/Photo Researchers
690T NASA
690B Lowell Observatory
692 International Bureau of Weights and Measures
693T JPL/NASA
693B NASA
695 JPL/NASA

696T Yerkes Observatory
696C Dennis Milon/SPL/Photo Researchers
696B John Sanford/SPL/Photo Researchers
697T SPL/Photo Researchers
697B Rev. Ronald Royer/SPL/Photo Researchers
700T Paul G. Hewitt
700B San Diego State University
703 Roger Ressmeyer/Starlight
707 Hale Observatory/Photo Researchers
709T Ronald Royer/Photo Researchers
709B Lick Observatory
710 Royal Observatory, Edinburgh
715 Julian Baum/New Scientist/SPL/Photo Researcher
716TR Ronald Royer/SPL/Photo Researchers
716CL Dr. Steve Gull/SPL/Photo Researchers
716BL National Optical Astronomy Observatories
716BR Royal Observatory, Edinburgh/SPL/Photo Researchers
717T NRAO/SPL/Photo Researchers
717C NASA/SS/Photo Researchers
717BL Hale Observatory/Photo Researchers
717BR Dennis Milon/SPL/Photo Researchers
718T Royal Greenwich Observatory/SPL/Photo Researchers
718B Photo by Dennis McNelis
721 NASA
729 NOAO/Phil Degginger
Chapter 18 insert: Adaptation from diagram by Allen Carrol and Dale D. Glasgow; ©1989 National Geographic Society, redrawn by John and Tracy Suchocki.

LITERARY PERMISSIONS
69 Adapted from *Physics in Your World,* Second Edition by Karl F. Kuhn and Jerry S. Faughn. Copyright ©1980 by Saunders College Publishing. Reprinted by permission of the publisher.
343 From "The Prospects of Fusion Power" by Wm. C. Gough & B. J. Eastlund, *Scientific American,* Feb. 1971. Copyright ©1971 by Scientific American, Inc., All rights reserved.
551, 558, 604 Adapted from *Earth* by Frank Press and Raymond Siever. Copyright ©1986 by W. H. Freeman and Company. Reprinted with permission.
554, 555, 573, 575, 603, 627, 635, 641, 642 Adapted from *Physical Geology* by Brian J. Skinner and Stephen C. Porter. Copyright ©1987 by John Wiley & Sons, Inc. Reprinted by permission of John Wiley & Sons, Inc.
541, 545, 561, 564, 574, 583–605, 610–618, 634 From *The Earth: An Introduction to Physical Geology*, 2/e by Edward J. Tarbuck and Frederick K. Lutgens. Copyright ©1987 by Merrill Publishing Company. Reprinted with the permission of Merrill, an imprint of Macmillan Publishing Company.
582 From *Applied Hydrogeology*, 2/e by C. W. Fetter. Copyright ©1988 by Merrill Publishing Company. Reprinted with the permission of Merrill, an imprint of Macmillan Publishing Company.
731 From *A Song for You* by Leon Russel. Lyrics and Music by Leon Russell. Copyright ©1970 Irving Music, Inc. (BMI) Reprinted by permission. All Rights Reserved. International Copyright Secured.

INDEX

PHYSICAL CONSTANTS

Name	Symbol	Value
Speed of light	c	$2.997\ 924\ 58 \times 10^8$ m/s (exact)
Planck constant	h	$6.626\ 075\ 5 \times 10^{-34}$ J $\cdot$ s
		$4.135\ 669\ 2 \times 10^{-15}$ eV $\cdot$ s
Gravitational constant	G	$6.672\ 59 \times 10^{-11}$ N $\cdot$ m^2/kg^2
Charge of electron	e	$1.602\ 177\ 33 \times 10^{-19}$ C
Mass of electron	m_c	$9.109\ 389\ 7 \times 10^{-31}$ kg
		$0.510\ 999\ 06$ MeV
Mass of proton	m_p	$1.672\ 623\ 1 \times 10^{-27}$ kg
		$938.272\ 31$ MeV
Mass of neutron	m_n	$1.674\ 928\ 6 \times 10^{-27}$ kg
		$939.565\ 63$ MeV
Avogadro's number	N_A	$6.022\ 136\ 7 \times 10^{23}$ mol^{-1}
		1 mole $= 6.022 \times 10^{23}$ particles
Unified atomic mass unit	u	$1.660\ 540\ 2 \times 10^{-27}$ kg
		$931.494\ 32$ MeV

PHYSICAL PROPERTIES

Name	Value
Acceleration of gravity at earth's surface, g	9.81 m/s^2
Mass of sun	1.99×10^{30} kg
Radius of sun	6.96×10^8 m
Mass of earth	5.98×10^{24} kg
Radius of earth (ave.)	6.38×10^6 m
Radius of earth's orbit	1.50×10^{11} m $= 1$ AU
Mass of moon	7.36×10^{22} kg
Radius of moon	1.74×10^6 m
Radius of moon's orbit	3.84×10^8 m

CONVERSION FACTORS

Length, Area, Volume
1 inch = 2.54 cm (exact)
1 ft = 30.48 cm (exact)
1 m = 39.37 in.
1 mi = 1.609 344 0 km
1 liter = 10^3 cm^3 = 10^{-3} m^3

Time
1 year = $365\frac{1}{4}$ day = 3.1558×10^7 s
1 d = 86,400 s
1 h = 3600 s

Mass
1 kg = 1000 g
1 kg weighs about 2.205 lb

Pressure
1 Pa = 1 N/m^2
1 atm = $1.013\ 25 \times 10^5$ Pa
1 lb/in.^{2} = 6895 Pa

Energy and Power
1 cal = 4.187 J
1 kWh = 3.60×10^6 J
1 eV = 1.602×10^{-19} J
1 u = 931.494 32 MeV
1 hp = 746 W

Speed
1 m/s = 3.60 km/h = 2.24 mi/h
1 km/h = 0.621 mi/h

Force
1 lb = 4.448 N